Predictability, Stability, and Chaos in N-Body Dynamical Systems

NATO ASI Series

Advanced Science Institutes Series

A series presenting the results of activities sponsored by the NATO Science Committee, which aims at the dissemination of advanced scientific and technological knowledge, with a view to strengthening links between scientific communities.

The series is published by an international board of publishers in conjunction with the NATO Scientific Affairs Division

A	**Life Sciences**	Plenum Publishing Corporation
B	**Physics**	New York and London
C	**Mathematical and Physical Sciences**	Kluwer Academic Publishers
D	**Behavioral and Social Sciences**	Dordrecht, Boston, and London
E	**Applied Sciences**	
F	**Computer and Systems Sciences**	Springer-Verlag
G	**Ecological Sciences**	Berlin, Heidelberg, New York, London,
H	**Cell Biology**	Paris, Tokyo, Hong Kong, and Barcelona
I	**Global Environmental Change**	

Recent Volumes in this Series

Volume 268—The Global Geometry of Turbulence: Impact of Nonlinear Dynamics
edited by Javier Jiménez

Volume 269—Methods and Mechanisms for Producing Ions from Large Molecules
edited by K. G. Standing and Werner Ens

Volume 270—Complexity, Chaos, and Biological Evolution
edited by Erik Mosekilde and Lis Mosekilde

Volume 271—Interaction of Charged Particles with Solids and Surfaces
edited by Alberto Gras-Martí, Herbert M. Urbassek,
Néstor R. Arista, and Fernando Flores

Volume 272—Predictability, Stability, and Chaos in N-Body Dynamical Systems
edited by Archie E. Roy

Volume 273—Light Scattering in Semiconductor Structures and Superlattices
edited by David J. Lockwood and Jeff F. Young

Volume 274—Direct Methods of Solving Crystal Structures
edited by Henk Schenk

Volume 275—Techniques and Concepts of High-Energy Physics VI
edited by Thomas Ferbel

Series B: Physics

Predictability, Stability, and Chaos in N-Body Dynamical Systems

Edited by

Archie E. Roy

University of Glasgow
Glasgow, United Kingdom

Plenum Press
New York and London
Published in cooperation with NATO Scientific Affairs Division

Proceedings of a NATO Advanced Study Institute on
Predictability, Stability, and Chaos in N-Body Dynamical Systems,
held August 6–17, 1990,
in Cortina d'Ampezzo, Italy

Library of Congress Cataloging-in-Publication Data

NATO Advanced Study Institute on Predictability, Stability, and Chaos
  in N-Body Dynamical Systems (1990 : Cortina d'Ampezzo, Italy)
    Predictability, stability, and chaos in N-body dynamical systems /
  edited by Archie E. Roy.
       p.   cm. -- (NATO ASI series. Series B, Physics ; vol. 272)
    "Proceedings of a NATO Advanced Study Institute on Predictability,
  Stability, and Chaos in N-Body Dynamical Systems, held August 6-17,
  1990, in Cortina d'Ampezzo, Italy"--T.p. verso.
    "Published in cooperation with NATO Scientific Affairs Division."
    Includes bibliographical references and indexes.
    ISBN-13: 978-1-4684-5999-9   e-ISBN-13: 978-1-4684-5997-5
    DOI: 10.1007/978-1-4684-5997-5
    1. Many-body problem--Congresses. 2. Astronomy--Congresses.
  I. Roy, A. E. (Archie E.), 1924-   . II. North Atlantic Treaty
  Organization. Scientific Affairs Division. III. Title.
  IV. Series: NATO ASI series. Series B, Physics ; v. 272.
  QB362.M3N38 1990
  521--dc20   91-28278
                                                      CIP

ISBN-13: 978-1-4684-5999-9

© 1991 Plenum Press, New York
Softcover reprint of the hardcover 1st edition 1991
A Division of Plenum Publishing Corporation
233 Spring Street, New York, N.Y. 10013

SPECIAL PROGRAM ON CHAOS, ORDER, AND PATTERNS

This book contains the proceedings of a NATO Advanced Research Workshop held within the program of activities of the NATO Special Program on Chaos, Order, and Patterns.

Volume 208—MEASURES OF COMPLEXITY AND CHAOS
edited by Neal B. Abraham, Alfonso M. Albano, Anthony Passamante, and Paul E. Rapp

Volume 225—NONLINEAR EVOLUTION OF SPATIO-TEMPORAL STRUCTURES IN DISSIPATIVE CONTINUOUS SYSTEMS
edited by F. H. Busse and L. Kramer

Volume 235—DISORDER AND FRACTURE
edited by J. C. Charmet, S. Roux, and E. Guyon

Volume 236—MICROSCOPIC SIMULATIONS OF COMPLEX FLOWS
edited by Michel Mareschal

Volume 240—GLOBAL CLIMATE AND ECOSYSTEM CHANGE
edited by Gordon J. MacDonald and Luigi Sertorio

Volume 243—DAVYDOV'S SOLITON REVISITED: Self-Trapping of Vibrational Energy in Protein
edited by Peter L. Christiansen and Alwyn C. Scott

Volume 244—NONLINEAR WAVE PROCESSES IN EXCITABLE MEDIA
edited by Arun V. Holden, Mario Markus, and Hans G. Othmer

Volume 245—DIFFERENTIAL GEOMETRIC METHODS IN THEORETICAL PHYSICS: Physics and Geometry
edited by Ling-Lie Chau and Werner Nahm

Volume 256—INFORMATION DYNAMICS
edited by Harald Atmanspacher and Herbert Scheingraber

Volume 260—SELF-ORGANIZATION, EMERGING PROPERTIES, AND LEARNING
edited by Agnessa Babloyantz

Volume 263—BIOLOGICALLY INSPIRED PHYSICS
edited by L. Peliti

Volume 264—MICROSCOPIC ASPECTS OF NONLINEARITY IN CONDENSED MATTER
edited by A. R. Bishop, V. L. Pokrovsky, and V. Tognetti

Volume 268—THE GLOBAL GEOMETRY OF TURBULENCE: Impact of Nonlinear Dynamics
edited by Javier Jiménez

Volume 270—COMPLEXITY, CHAOS, AND BIOLOGICAL EVOLUTION
edited by Erik Mosekilde and Lis Mosekilde

Volume 272—PREDICTABILITY, STABILITY, AND CHAOS IN N-BODY DYNAMICAL SYSTEMS
edited by Archie E. Roy

PREFACE

The reader will find in this volume the Proceedings of the NATO Advanced Study Institute held in Cortina d'Ampezzo, Italy between August 6 and August 17, 1990 under the title "Predictability, Stability, and Chaos in N-Body Dynamical Systems".

The Institute was the latest in a series held at three-yearly intervals from 1972 to 1987 in dynamical astronomy, theoretical mechanics and celestial mechanics. These previous institutes, held in high esteem by the international community of research workers, have resulted in a series of well-received Proceedings. The 1990 Institute attracted 74 participants from 16 countries, six outside the NATO group. Fifteen series of lectures were given by invited speakers; additionally some 40 valuable presentations were made by the younger participants, most of which are included in these Proceedings.

The last twenty years in particular has been a time of increasingly rapid progress in tackling long-standing and also newly-arising problems in dynamics of N-body systems, point-mass and non-point-mass, a rate of progress achieved because of correspondingly rapid developments of new computer hardware and software together with the advent of new analytical techniques. It was a time of exciting progress culminating in the ability to carry out research programmes into the evolution of the outer Solar System over periods of more than 10^8 years and to study star cluster and galactic models in unprecedented detail. Notwithstanding these successes, however, in recent years it has become increasingly acknowledged that the influence of chaos in dynamical systems is of supreme importance in understanding their behaviour.

The present Institute therefore had as its main goal the examination of the relationships between predictability, stability and chaos in N-body dynamical systems. Among the systems studied were the Solar System's planets and satellites, comets, asteroids, and meteors, and the dynamics of star clusters and galaxies.

In real dynamical systems, imprecision of starting conditions and the inevitable accumulation of round-off error conspire with the non-linearity of the differential equations to produce limits on the predictability of the future behaviour of such systems. The 'predictability horizon' depends in a hypersensitive manner on the proximity of the system to chaotic regions.

In the few-body problem, as in sub-systems within the Solar System, it was recognised that resonance also played a major part in limiting predictability in, for example, the evolution of the orbits of asteroids disturbed by Jupiter and Saturn, where such asteroids evolved to become Mars- or Earth-crossers, becoming essentially chaotic systems.

In models of star clusters, it was evident that in spite of the powerful computer power now available, the predictability horizon for individual stars was surprisingly limited though a number of properties of the cluster's evolution could be explored for longer periods of time that were surprisingly independent of the precision aimed at for the individual stars.

The Institute concluded that for periods of time less than clearly identified predictability horizons, modern analytical methods, together with computer hardware and software, could produce results accurate enough for all practical purposes. For many of the most interesting and important problems, however, such as the long-term evolution of the Solar System and of star clusters, our modern techniques have demonstrated their inadequacy in the presence of chaos, necessitating a search for new approaches. Dynamics, in short, has recognised its arrival at, hopefully, the beginning of a new era in its history, one perhaps leading to new and fruitful analytical techniques. Ironically, the modern concept of chaos was found by Poincaré in his study of the three-body problem almost 100 years ago; his brilliant farsightedness and insight were highlighted by the number of times participants had occasion to refer to his work in this Institute. He himself predicted that a long time would pass before this kind of concept would be relevant for actual computations to be used in dynamical astronomy. Quoting Andrea Milani, in this volume: "Whether we like it or not, this time is now, and therefore the basic tools needed to build a theory of chaotic dynamics ... are now becoming essential tools of our trade."

Finally, let me record my thanks to my colleagues.

The high quality of preparation and presentation of invited lectures and other contributions was particularly pleasing. As pleasing were the frequent and sustained discussions and the warm international friendship enjoyed by the participants.

I am also glad to have this opportunity to pay grateful tribute for the help and support I received from the Organising Committee (Professor V. Szebehely, Dr. P.J. Message, Professor A. Milani). I also want to thank Dr. G. Volpi and Dr. B.A. Steves for their unstinting efforts before and during the meeting. The staff of the Antonelli Institute, where the ASI was held, were always more than helpful: they made our stay a very pleasant one and we extend our thanks to them.

The Committee are also very grateful to the Scientific Affairs Division of NATO for their guidance, counsel and support.

Lastly, I would like to take this opportunity to thank all those who came to Cortina and helped to make this ASI so enjoyable and productive. Their support, cooperation and friendship were very much appreciated.

Cortina D'Ampezzo, Italy
and Glasgow, United Kingdom

Archie E. Roy
Director, NATO Advanced Study
Institute and Editor of the
Proceedings

CONTENTS

PART I: ASPECTS OF CHAOS

Chaos in a Restricted, Charged Four-Body Problem 3
   J. Casasayas and A. Nunes

Chaos in the Three-Body Problem 11
   A. Milani

A New Route to Chaos: Generation of Spiral Characteristics 35
   G. Contopoulos

Chaos in the N-Body Problem of Stellar Dynamics 47
   D.C. Heggie

Chaos, Stability, and Predictability in Newtonian Dynamics 63
   V. Szebehely

Predictability, Stability, and Chaos in Dynamical Systems 73
   C. Marchal

Analytical Framework in Poincaré Variables for the Motion
      of the Solar System 93
   J. Laskar

Origin of Chaos and Orbital Behaviour in Slowly Rotating
      Triaxial Models 115
   S. Udry and L. Martinet

PART II: DYNAMICS OF ASTEROIDS, COMETS, AND METEORS

Modelling: An Aim and a Tool for the Study of the Chaotic
      Behaviour of Asteroidal and Cometary Orbits 125
   C. Froeschlé

Mapping Models for Hamiltonian Systems with Application
      to Resonant Asteroid Motion 157
   J.D. Hadjidemetriou

A Model for the Study of Very-High-Eccentricity Asteroidal Motion:
      The 3:1 Resonance 177
   S. Ferraz-Mello and J.C. Klafke

The Location of Secular Resonances 185
   A. Morbidelli

Temporary Capture into Resonance 193
    J. Henrard

Applications of the Restricted Many-Body Problem to
    Binary Asteroids 197
    V. Szebehely and J.R. Pojman

The Wavelet Transform as Clustering Tool for the Determination
    of Asteroid Families 205
    Ph. Bendjoya, E. Slezak, and Cl. Froeschlé

Delivery of Meteorites from the ν_6 Secular Resonance
    Region Near 2 AU 215
    Ch. Froeschlé and H. Scholl

The Dynamics of Meteoroid Streams 225
    I.P. Williams and Z. Wu

Perturbation Theory, Resonance, Librations, Chaos,
    and Halley's Comet 239
    P.J. Message

Rotational Behaviour of Comet Nuclei 249
    P. Oberti, E. Bois, and C. Froeschlé

PART III: DYNAMICS OF NATURAL AND ARTIFICIAL SATELLITES

The Moon's Physical Librations – Part 1: Direct Gravitational
    Perturbations 257
    I. Wytrzyszczak and E. Bois

The Moon's Physical Librations – Part II: Non-Rigid Moon and
    Direct Non-Gravitational Perturbations 265
    E. Bois and I. Wytrzyszczak

Significant High Number Commensurabilities in the Main Lunar
    Problem: A Postscript to a Discovery of the Ancient
    Chaldeans 273
    A.E. Roy, B.A. Steves, G.B. Valsecchi, and E. Perozzi

Moon's Influence on the Transfer from the Earth to a Halo
    Orbit Around L_1 283
    G. Gomez, À. Jorba, J. Masdemont, and C. Simó

First Order Theory of Perturbed Circular Motion: An Application
    to Artificial Satellites 291
    E. Bois and I. Wytrzyszczak

Poincaré-Similar Variables Including J_2-Secular Effects 297
    L. Floría and J.M. Ferrándiz

Measuring the Lack of Integrability of the J_2 Problem
    for Earth's Satellites 305
    C. Simó

The Effects of the J_3-Harmonic (Pear Shape) on the Orbits of
    a Satellite 311
    R.A. Broucke

Stability of Satellites in Spin-Orbit Resonances and
    Capture Probabilities 337
  A. Celletti

Statistical Analysis of the Effects of Close Encounters of
    Particles in Planetary Rings 345
  F.P. Gama and J.-M. Petit

The Three-Dipole Problem 355
  C.L. Goudas and E.G. Petsagourakis

The N-Dipole Problem and the Rings of Saturn 371
  C.L. Goudas

Long-Time Predictions of Satellite Orbits by Numerical Integration . . 387
  J.M. Ferrándiz, M.E. Sansaturio, and J. Vigo

Chaos in Coorbital Motion 395
  F. Spirig and J. Waldvogel

PART IV: THE THREE-BODY PROBLEM

Remarkable Termination Orbits of the Restricted Problem 413
  V.V. Markellos

Periodic Orbits in the Isosceles Three-Body Problem 425
  C.C. Monleon and J.M. Alfaro

Quasi-Periodic Orbits as a Substitute of Libration Points
    in the Solar System 433
  G. Gomez, À. Jorba, J. Masdemont, and C. Simó

Stability Zones Around the Triangular Lagrangian Points 439
  R. Dvorak and E. Lohinger

Chaotic Trajectories in the Restricted Problem of Three Bodies 447
  R.H. Smith and V. Szebehely

New Formulations of the Sitnikov Problem 457
  K. Wodnar

Periodic Solutions for the Elliptic Planar Restricted Three-Body
    Problem: A Variational Approach 467
  M.L. Bertotti

Hill-Type Stability and Hierarchical Stability of the
    General Three-Body Problem 475
  Y.-C. Ge

Equilibrium Connections on the Triple Collision Manifold 481
  A. Susín and C. Simó

Orbits Asymptotic to the Outermost KAM in the Restricted
    Three-Body Problem 493
M. Sekiguchi and K. Tanikawa

PART V: SELECTED TOPICS IN DYNAMICS

A New Interpretation of Collisions in the N-Body Problem 501
  J.G. Bryant

An Impulsional Method to Estimate the Long-Term Behaviour of a
    Perturbed System: Application to a Case of
    Planetary Dynamics 509
  B. Chauvineau

Improved Bettis Methods for Long-Term Prediction 515
  J.M. Ferrándiz and S. Novo

Application of Spherically Exact Algorithms to Numerical
    Predictability in Two-Body Problems 523
  J.M. Ferrándiz and M.T. Pérez

Are There Irregular Families of Characteristic Curves? 531
  J. Font and C. Simó

Non-Linearity in the Angles-Only Initial Orbit
    Determination Problem 541
  D. Kaya and D. Snow

A Perturbation of the Relativistic Kepler Problem 547
  A. Nunes, J. Casasayas, and L. Llibre

Integrable Three-Dimensional Dynamical Systems and the
    Painlevé Property 555
  C. Polymilis

Generic and Nongeneric Hopf Bifurcation 565
  F. Spirig

The Chaotic Motion of a Rigid Body Rotating About a Fixed Point . . . 573
  F. El-Sabaa and M. El-Tarazi

Participants and Speakers 583

Author Index 589

Subject Index 597

PART I
ASPECTS OF CHAOS

CHAOS IN A RESTRICTED CHARGED FOUR-BODY PROBLEM

J.Casasayas[1]* and A.Nunes[2]**

(1) Departament de Matemàtica Aplicada i Anàlisi, Facultat de Matemàtiques
Universitat de Barcelona, Barcelona, Spain

(2) Departamento de Física, Faculdade de Ciencias, Universidade de Lisboa
Lisboa, Portugal

Abstract. A restricted charged four body problem is considered which reduces to a two degrees of freedom Hamiltonian system. It is shown that an appropiate restriction of a Poincaré map of the system is conjugate to the shift homeomorphism on a certain symbolic alphabet.

1. Equations of motion

Let us consider two pairs of identical particles of masses $m_1 = m_2 = m$, $m_3 = m_4 = M$, and charges $e_1 = e_2 = E$, $e_3 = e_4 = e$. The four particles lie in the plane and their relative positions are such that each particle forms an isosceles triangle with the pair of particles which differ from it. Moreover, particles 1 and 2 (resp. 3 and 4) are given initial velocities along the straight line through them and symmetrical with respect to the straight line through particles 3 and 4 (resp. 1 and 2). See Figure 1.

Taking into account electrical as well as gravitational interactions, the equations of motion preserve the symmetry of the initial conditions described before. Therefore this physical problem has two degrees of freedom, and we may take as independent coordinates the oriented half distances between the particles: $x = (q_2 - q_1)/2$, $y = (q_3 - q_4)/2$. Finally, assuming the following two aditional conditions:

(a) $GM^2 - e^2 = 0$, i.e., the gravitational and the electrostatic interactions between particles 3 and 4 cancel each other,

(b) e and E have opposite signs, so that $GMm - eE$ is positive,

and after an appropiate reescaling of time, the equations of motion become,

---

* Partially supported by a grant of the CGICT no.PB86-0351 and DGICYT no.BE90-135.

** Partially supported by Instituto Nacional de Investigaçao Cientifica

$$\ddot{x} = -\frac{x}{(x^2 + y^2)^{3/2}} - \frac{Ax}{|x|^3},$$

(1)

$$\ddot{y} = -\frac{By}{(x^2 + y^2)^{3/2}},$$

where

$$A = \frac{GM^2 - E^2}{8(GMm - eE)}, \quad B = \frac{M}{m}.$$

Now we note that equations (1) have the same form that equations of well-known n-body problems. In fact, the isosceles three body problem[1] (I3BP) corresponds to $A > 0$, the charged isosceles three body problem[2] (CI3BP) to $A < 0$, the anisotropic Kepler problem[3] (AKP) to $A = 0$ and $B \neq 1$, and the Kepler problem (KP) to $A = 0$ and $B = 1$. See Figure 2. Thus, the four body symmetrical setting proposed here gives a new physical interpretation of the I3BP, CI3BP, AKP and the KP, as well as a model system for the strip in parametre space corresponding to $0 < B < 1$.

We shall study system (1) for $B \in (0,1)$ and $A \in (-1,-1/4)$, see the shadowed region of Figure 2. It corresponds to a Hamiltonian system with Hamiltonian $H : \Re^+ \times \Re^3 \longrightarrow \Re$ given by:

(2)

$$H(x, y, p_x, p_y) = \frac{p_x^2}{2} + \frac{bp_y^2}{2} + \frac{1}{x} - \frac{a}{(x^2 + y^2)^{1/2}},$$

where $b = B \in (0,1)$ and $a = -A^{-1} \in (1,4)$.

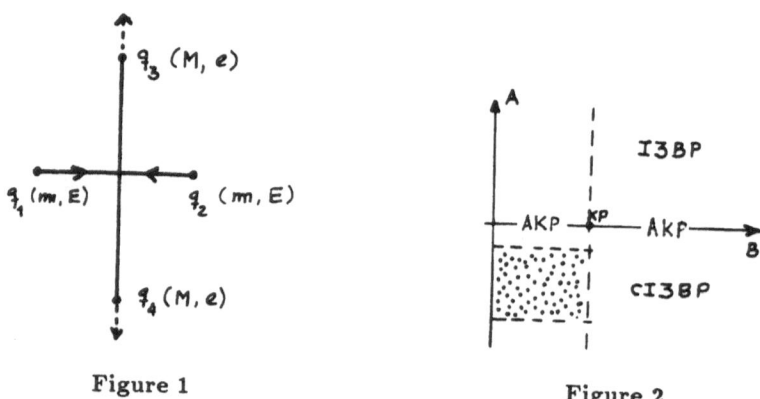

Figure 1

Figure 2

2. Some general tricks

T1. REGULARIZATION OF THE SINGULARITIES: Notice that the only singularity of system (1) corresponds to total collision configuration ($x = y = 0$). In fact, triple collisions are forbidden by the symmetry of the particle configuration (see Figure 1). As to the double collisions allowed by the symmetry restrictions, we have that particle 1 and particle 2 cannot collide since they repel each other ($x = 0$ is not accessible for finite energies), while the collision between particle 3 and particle 4 does not correspond to a singularity because these two particles do not interact (the system is regular for $y = 0$, $x \neq 0$).

In order to regularize the singularity corresponding to total (quadruple) collision we use McGehee's coordinates[4] (r, θ, v, u, τ) given by:

$$r = (x^2 + b^{-1}y^2)^{1/2},$$

$$\theta = arctan\left(\frac{b^{-1/2}y}{x}\right),$$

$$v = r^{-1/2}(xp_x + yp_y),$$

$$u = r^{-1/2}(b^{1/2}xp_y - b^{-1/2}yp_x),$$

$$\frac{dt}{d\tau} = r^{3/2}.$$

Hence, (r, θ) are polar type coordinates on the (x, y) plane (positions plane), v is a rescaled radial velocity, u is a reescaled angular velocity and τ is a reescaled time. Moreover, $r = 0$ corresponds to total quadruple collision.

Equations (1) in McGehee's coordinates are:

(3)
$$\bar{r} = rv,$$

$$\bar{\theta} = u.$$

$$\bar{v} = \frac{v^2}{2} + u^2 + V(\theta),$$

$$\bar{u} = -\frac{uv}{2} - V'(\theta),$$

where

$$V(\theta) = \frac{-a}{(\cos^2(\theta) + b\sin^2(\theta))^{1/2}} + \frac{1}{\cos(\theta)}, \quad V'(\theta) = \frac{dV}{d\theta},$$

and the energy relation (2) reads:

(4)
$$rh = \frac{u^2 + v^2}{2} + V(\theta).$$

Now $r = 0$ (total quadruple collision) is not a singularity of ystem (3). The total collision manifold Λ is defined by (4) when $r = 0$,

$$\Lambda = \{(\theta, v, u) \in S^1 \times \Re^2 : u^2 + v^2 = -2V(\theta)\},$$

the flow (3) is analytical on Λ and leaves Λ invariant (so that it is posible to study the flow on Λ obtained by letting $r = 0$ in (3)). It is usual to say that the singularity $r = 0$ has been blown up to the collision manifold.

It is not difficult to prove that if $a(1 - b) > 1$ then Λ and the flow on Λ has the following properties (see Figure 3):

(a) The collision manifold Λ is topologically a sphere.

(b) The only equilibrium points on Λ are

$$P^{\pm}(\theta_0) = (\theta = \theta_0, v = \pm\sqrt{-2V(\theta_0)}, u = 0),$$

where $\theta_0 \in \{0, arccos(d), -arccos(d)\}$ and $d = b^{1/2}\left(b - 1 + (a(1 - b))^{2/3}\right)^{-1/2}$. Points $P^{\pm}(0)$ are hyperbolic saddles, $P^+(\pm arccos(d))$ sinks and $P^-(\pm arccos(d))$ sources.

(c) The flow on Λ is gradient–like with respect to the v–coordinate.

(d) For every value of the parametre a there exists a countable set B such that if $b \in (0,1) \setminus B$ then the unstable manifold of $P^-(0)$ misses the stable manifold of $P^+(0)$.

In Figure 3 it is represented the global flow on Λ when $a(1-b) > 1$ and $b \in (0,1) \setminus B$. So trick 1, besides regularizing the singularity, gives us an extended system ((3)) which has hyperbolic equilibrium points, unstable and stable invariant manifolds associated to them,... .

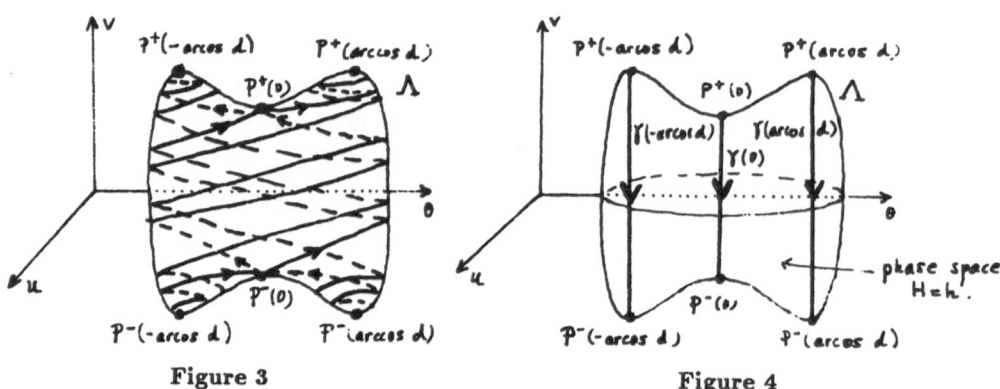

Figure 3   Figure 4

T2. SPECIAL SINGULAR ORBITS : Recall that a solution $(r(\tau), \theta(\tau), v(\tau), u(\tau))$ of system (2) is called homothetic solution if $\theta(\tau) = \theta_0$ for every $\tau \in (-\infty, +\infty)$.

After straightforward calculations one concludes that if $a(1-b) > 1$ then there exists a homothetic solution $\gamma(\theta_0)$ at $\theta = \theta_0$ if and only if $\theta_0 \in \{0, \arccos(d), -\arccos(d)\}$, see Figure 4. Moreover,

$$\gamma(0) \subset E(0) \cap C(0), \quad \gamma(\arccos(d)) = E(+) = C(+), \gamma(-\arccos(d)) = E(-) = C(-),$$

where $E(0)$, $E(+)$, $E(-)$ (resp. $C(0)$, $C(+)$, $C(-)$) denotes the ejection (resp. collision) orbits at $\theta = 0$, $\theta = \arccos(d)$, $\theta = -\arccos(d)$.

3. Chaotic behaviour

The main result of this paper is the following.

Theorem. It is possible to construct a Poincaré map F defined on a local surface of section S such that S contains an invariant set I on which F is conjugated to the shift homeomorphism σ over the space of the doubly infinite sequences of symbols of a countably infinite alphabet A.

This result is the strongest definition of chaos, see for instance Moser[5]. It also has relevant consequences as the existence of infinitely many hyperbolic periodic orbits and the non–integrability of the system (in the sense that there is no analytic integral of motion independent from the Hamiltonian).

The general idea is to verify the hypothesis of Smale's horseshoe theorem[6]. More precisely the proof is divided in three steps.

FIRST STEP: We start by proving that the two–dimensional invariant manifolds $W^u(P^+(0))$ and $W^s(P^-(0))$ intersect transversally along the homothetic orbit $\gamma(0) \subset W^u(P^+(0)) \cap W^s(P^-(0))$. Then, we use this fact in order to consider local coordinates (x, y) in D_0 (small disc contained in $\{v = 0\}$ and centered in $\gamma(0) \cap \{v = 0\}$) such that:

$$D_0 = \{(x, y) \in [-1, 1] \times [-1, 1]\},$$
$$A = W^u(P^+(0)) \cap D_0 = \{x = 0\},$$
$$B = W^s(P^-(0)) \cap D_0 = \{y = 0\},$$

see Figure 5.

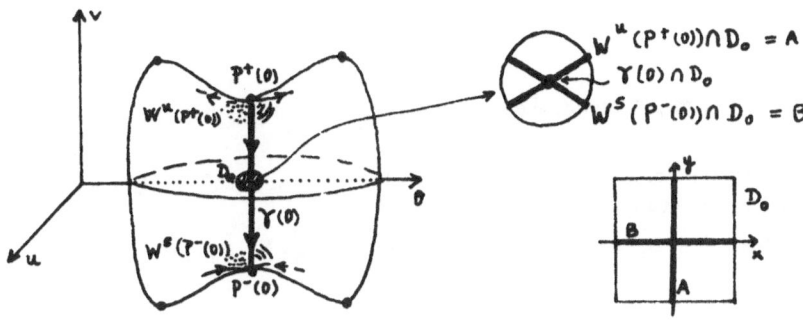

Figure 5

SECOND STEP: Let f be the map that takes in forward time $p \in D_0$ to $p' \in D_1 \cup D_2$ and g be the map that takes in backward time $p \in D_0$ to $p'' \in D_1 \cup D_2$. Note that f is defined in $D_0 \setminus B$ and g is defined in $D_0 \setminus A$. See Figure 6. Let A_μ, $\mu \in (-1, 1)$, be a foliation of D_0 by curves parallel to A such that $A_0 = A$, and let B_μ, $\mu \in (-1, 1)$, be a foliation of D_0 by curves parallel to B such that $B_0 = B$.

We prove that $f(A_\mu)$ is formed by two spirals, one in D_1 and other in D_2, accumulating on x_1 and x_2 (see Figure 7) and that, moreover

$$\left| \frac{d}{dy} f(A_\mu(y)) \right| \to +\infty \quad when \quad y \to 0.$$

Using the symmetries, we get an analogous result for $g(B_\mu)$.

THIRD STEP: We consider a sector C_i in D_i, $i = 1, 2$, containing the zero velocity curve Z.

Using the two previous results, we show that $f(D_0 \setminus B) \cap C_i$ and $g(D_0 \setminus A) \cap C_i$ are formed by an infinite number of bands which intersect transversally on Z, accumulating on x_i. See Figure 8.

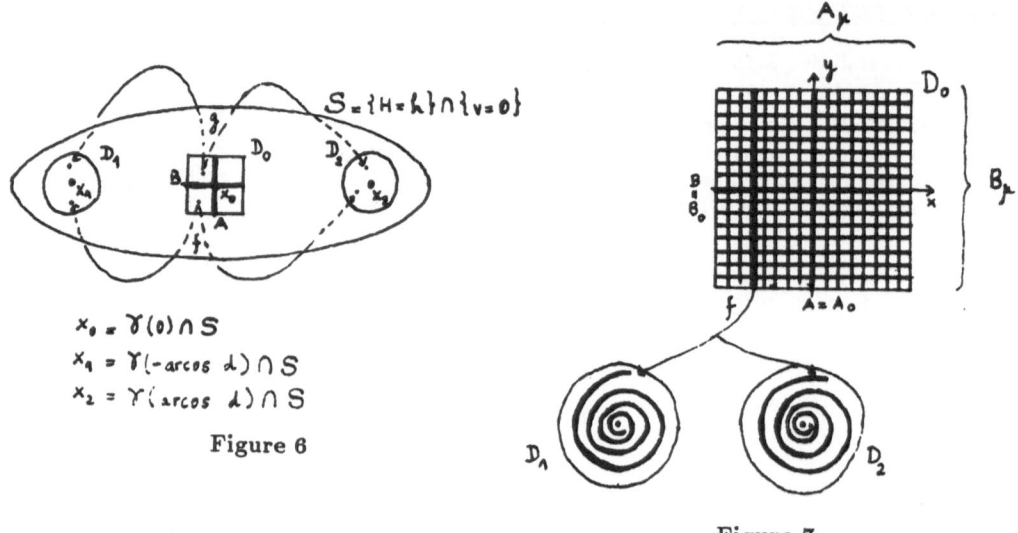

$$S = \{H = h\} \cap \{v = 0\}$$

$x_0 = \gamma(0) \cap S$

$x_1 = \gamma(-\arccos \lambda) \cap S$

$x_2 = \gamma(\arccos \lambda) \cap S$

Figure 6

Figure 7

Finally, using steps 1, 2 and 3, the hypothesis of Smale's horseshoe theorem are easily proved because:

(i) The image by the Poincaré map $h = g^{-1} \circ f$ of a "vertical band" in D_0 is an infinite number of vertical bands accumulating to A.

(ii) The "stretching" of the vertical bands is arbitrarily large for D_0 small enough. ■.

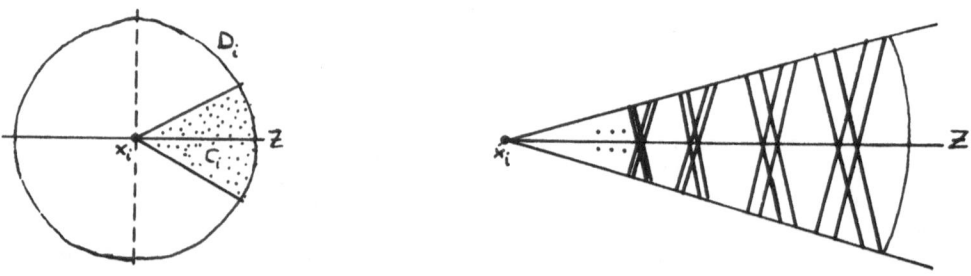

Figure 8

References

1. Simó, C. and Martinez, R.: 1988, 'Qualitative study of the planar isosceles three-body Problem', Celestial Mechanics, 41, 179-251.

2. Atela, P.: 'The Charged Isosceles 3-Body Problem', preprint.

3. Casasayas, J. and Llibre, J.: 1984, 'Qualitative analysis of the anisotropic Kepler problem', Memoirs of the AMS, 52 (312), November.

4. McGehee, R.: 1974, 'Triple collision in the collinear three-body problem', Invent. Math. 27, 191-227.

5. Moser, J.: 1973, Stable and Random Motions in Dynamical Systems, Princeton Univ. Press (Study 77), N.J. University Press.

6. Devaney, R.L.: 1978, 'Collision Orbits in the Anisotropic Kepler problem', Invent. Math. 45, 221-251.

7. Casasayas, J. and Nunes, A.: 1990, 'A restricted charged four-body problem', Celestial Mechanics and Dynamical Astronomy 47, 245-266.

CHAOS IN THE THREE BODY PROBLEM

Andrea Milani

Space Mechanics group
Department of Mathematics
University of Pisa
Via Buonarroti 2
56127 Pisa, Italy

INTRODUCTION

The purpose of this series of lecture notes is to give an outline of the basic tools required to show the occurrence of chaotic motions in the simplest non--integrable problems in Celestial Mechanics, such as the circular restricted planar 3--body problem. No formal proofs will be given here; they can be found in the references given for each section. Section 1 describes the linear and local theory of ordinary differential equations in the neighbourhood of a fixed point; the problems arising in the embedding of invariant stable and unstable manifolds are also discussed. Section 2 is about periodic orbits; the subjects discussed include variational equations, surfaces of section, the continuation of periodic orbits in the restricted 3--body problem, and bifurcation of hyperbolic periodic orbits from resonant periodic orbits. Section 3 covers fundamental models of resonance, the global behaviour of separatrices, and their intersections; all this allows to give at least an outline of the proof of the fundamental result --presented by Poincaré in his book *Les méthodes nouvelles de la mécanique céleste*-- by which homoclinic points must necessarily occur in the restricted problem. A short conclusion underscores the fact --shown in a rigorous way much later-- that this in turn implies that chaos in the strongest possible sense occurs in the restricted problem, and is an essential feature of every non--integrable system, even very simple ones with only two degrees of freedom. I apologize for reporting here my lectures in a very short format, almost without comments in between the formulas and statements of the main results; my understanding of the purpose of this notes is that they should serve as a reminder of the existence of many subjects to be studied, rather than a complete presentation which could not be contained in this format.

1 LINEAR AND LOCAL THEORY

1.1 LINEAR EQUATIONS

Our first problem is the study of the local behaviour of the solutions of an ordinary differential equation in the linear approximation.

Predictability, Stability, and Chaos in N-Body Dynamical Systems
Edited by A.E. Roy, Plenum Press, New York, 1991

If the equation is:

$$\frac{dz}{dt} = f(z) \quad ; \quad z \in \mathcal{R}^n, \; manifold$$

$$f(\underline{0}) = \underline{0} \quad ; \quad f(z) = Az + \mathcal{O}(|z|^2)$$

(1.1)

Linearisation at the equlibrium point $\underline{0}$ results from neglecting the order two part:

$$\frac{dz}{dt} = Az \quad ; \quad z(0) = z_0$$

(1.2)

The solution to the initial value problem (1.2) is obtained by the sequence of successive approximations:

$$z_1 = z_0 + \int_0^t Az_0 \, dt = (I + At)z_0$$

$$z_2 = z_0 + \int_0^t Az_1 \, dt = (I + At + \frac{A^2 t^2}{2})z_0$$

$$\cdots$$

$$z_m = z_0 + \int_0^t Az_{m-1} \, dt = \sum_{j=0}^{m} \frac{A^j t^j}{j!} z_0$$

that is, the solution is given as the sum of a series:

$$\exp(At) = \sum_{j=0}^{\infty} \frac{A^j t^j}{j!} \quad ; \quad z(t) = \exp(At)z_0$$

(1.3)

The series (1.3) defines the *matrix exponential* and it is always convergent: given the matrix norm $||A|| = Sup_{|z|=1}|Az|$

$$|| \exp(At)|| \leq \sum_{j=0}^{\infty} \frac{||A||^j \, |t|^j}{j!} = e^{||A|| \, |t|}$$

Matrix exponentials need to be computed explicitly in three cases only:

Example 1: diagonal matrix, real eigenvalues:

$$A = diag[\lambda_1, \ldots, \lambda_n] \quad ; \quad \exp(At) = diag[e^{\lambda_1 t}, \ldots, e^{\lambda_n t}]$$

and the fundamental solution $\exp(At)$ is also diagonal.

Example 2: nilpotent matrix:

$$\exp \begin{pmatrix} 0 & 1 \\ 0 & 0 \end{pmatrix} t = \begin{pmatrix} 1 & t \\ 0 & 1 \end{pmatrix}$$

In general, if $N^k = \underline{0}$ then $\exp(Nt) =$ Polynomial in t.

Example 3: imaginary eigenvalues:

$$\exp \omega J t = \exp \begin{pmatrix} 0 & \omega \\ -\omega & 0 \end{pmatrix} t = \begin{pmatrix} \cos \omega t & \sin \omega t \\ -\sin \omega t & \cos \omega t \end{pmatrix} = R_{\omega t}$$

and the fundamental solution is a rotation with constant angular velocity ω

All the other computations can be performed by exploiting the following formal properties of the exponential:

Property 1: if $AB = BA$ then $\exp[(A+B)t] = \exp(At) \cdot \exp(Bt)$.

Example 1+2 : multiple real eigenvalues, with nilpotent

$$\exp \begin{pmatrix} \lambda & \epsilon \\ 0 & \lambda \end{pmatrix} t = \begin{pmatrix} e^{\lambda t} & \epsilon t e^{\lambda t} \\ 0 & e^{\lambda t} \end{pmatrix}$$

Example 1+3 : complex eigenvalues

$$\exp \begin{pmatrix} \alpha & \omega \\ -\omega & \alpha \end{pmatrix} t = e^{\alpha t} R_{\omega t}$$

Property 2: $\exp(BAB^{-1}t) = B \exp(At) B^{-1}$

This allows to exploit the following:

Jordan real canonical form: If $\lambda_1, \ldots, \lambda_s$ are the real eigenvalues of A, and $\alpha_1 \pm i\omega_1, \ldots, \alpha_r \pm i\omega_r$ are the complex conjugate eigenvalues, then there is a matrix B such that:

$$BAB^{-1} = \begin{pmatrix} C & \underline{0} \\ \underline{0} & D \end{pmatrix} \quad ; \quad C = diag[\lambda_1, \ldots, \lambda_s] + N$$

$$N \text{ nilpotent} \quad ; \quad N_{jk} \neq 0 \; : k > j \text{ and } \lambda_k = \lambda_j$$

$$D = diag[\begin{pmatrix} \alpha_1 & \omega_1 \\ -\omega_1 & \alpha_1 \end{pmatrix}, \ldots, \begin{pmatrix} \alpha_r & \omega_r \\ -\omega_r & \alpha_r \end{pmatrix}] + N'$$

$$N' \text{ nilpotent}; \quad N'_{2j-1,2j+1}, \; N'_{2j,2j+2} \neq 0 : \alpha_j \pm i\omega_j = \alpha_{j+1} \pm i\omega_{j+1}$$

Conclusion: all the matrix exponentials can be computed by means of the exponentials of the Jordan real canonical form: $\exp(At) = B F(t) B^{-1}$ with $F(t)$ direct sum of examples 1, 2, 3, 1+2, 1+3, 2+3, 1+2+3.

The Hamiltonian case can be described in a simple notation:

$$z = \begin{pmatrix} x \\ y \end{pmatrix} \quad ; \quad H = H(x,y) \quad ; \quad \frac{dz}{dt} = J\nabla H(z) \quad ; \quad J = \begin{pmatrix} \underline{0} & I \\ -I & \underline{0} \end{pmatrix} \quad (1.4)$$

The linearisation at an equilibrium point is performed in the same way as in (1.2):

$$\nabla H(\underline{0}) = \underline{0} \quad ; \quad C = \nabla\nabla H(\underline{0}) \quad ; \quad \frac{dz}{dt} = JCz + \mathcal{O}(|z|^2)$$

that is the linearized equation is $dz/dt = JCz$.

In this case, the characteristic equation of the matrix JC of the linear system has special properties, resulting from $J^T = J^{-1} = -J$, $C = C^T$:

$$det[JC - \lambda I] = det[JC - \lambda I]^T = det[-CJ - \lambda I] =$$
$$= det[J(-CJ - \lambda I)J^{-1}] = det[J(-CJ)(-J) - \lambda I] = det[-JC - \lambda I] =$$
$$= (-1)^n det[JC + \lambda I]$$

that is, if λ is an eigenvalue, so is $-\lambda$ and with the same multiplicity.

As a result of the possibility of reduction to the Jordan canonical form, the qualitative theory of the linearized equations around a fixed point is essentially reduced to a very simple case:

$$\frac{dz}{dt} = (\alpha I + \omega J)z \quad ; \quad z(t) = e^{\alpha t}R_{\omega t}z_0 \quad (1.5)$$

and the behaviour of the solutions is as follows: $\alpha > 0$ implies $z \to \infty$ $\alpha < 0$ implies $z \to \underline{0}$ and $\alpha = 0$ results in $|z| = |z_0|$. This can be summarized in a definition of Lyapounov characteristic exponent:

$$\chi(\underline{0}, z_0) = \lim_{t \to +\infty} \frac{1}{t} \log \frac{|z(t, z_0)|}{z_0|}$$

As an example, for the system (1.5) $\chi(\underline{0}, z_0) = \alpha$. Lyapounov exponents are easily computed by means of the Jordan canonical form, essentially because nilpotents do not matter: $\lim_{t \to +\infty} \frac{1}{t} \log t^n = 0$.

If $V_\lambda = \mathcal{K}er(A - \lambda I)^n$ is the generalised eigenspace for the real eigenvalue λ, then $z_0 \in V_\lambda$ implies $\chi(\underline{0}, z_0) = \lambda$.

If $V_{\alpha \pm i\omega} = $ Real part of $\mathcal{K}er(A - (\alpha + i\omega)I)^n \oplus \mathcal{K}er(A - (\alpha - i\omega)I)^n$ is the generalised eigenspace for the complex conjugate eigenvalues $\alpha \pm i\omega$, then $z_0 \in V_{\alpha \pm i\omega}$ implies $\chi(\underline{0}, z_0) = \alpha$.

Definition: stable, unstable and center manifold for the linearised equations at an equilibrium point

$$W^+ = \{z_0 | \chi(\underline{0}, z_0) < 0\} \cup \{\underline{0}\}$$

$$W^- = \{z_0 | \chi(\underline{0}, z_0) > 0\} \cup \{\underline{0}\}$$

$$W^0 = \{z_0 | \chi(\underline{0}, z_0) = 0\} \cup \{\underline{0}\}$$

Properties: $W^+ =$ is the direct sum of all the generalised eigenspaces with eigenvalues μ with $Real(\mu) < 0$; $W^- =$ idem with $Real(\mu) > 0$. W^- is the W^+ of the system obtained by reversing time.

In the Hamiltonian case: $dimW^+ = dimW^-$

References: the proofs can be found in many textbooks; we recommend Hirsch and Smale (1974).

1.2 LOCAL THEORY

If the equation (1.1) is now considered with the nonlinear part, but only in a neigbourhood of the fixed point:

$$\frac{dz}{dt} = Az + g(z) \quad ; \quad g(\underline{0}) = \underline{0} \quad ; \quad \frac{\partial g}{\partial z}(\underline{0}) = \underline{\underline{0}}$$

the problem is: given the eigenvalues of A, can we describe the behaviour near $\underline{0}$? This question has a positive answer, at least in some cases:

Case 1: sink

$Real(\mu_j) < 0$ for every eigenvalue μ_j: A a contraction (in suitable norm).

$$w = Bz \quad ; \quad D = BAB^{-1} \; Jordan \; can. \; f.$$

$$\frac{dw}{dt} = Dw + g_1(w) \quad ; \quad w \cdot Dw < k|w|^2 \quad k < 0$$

$$r^2 = w \cdot w \quad ; \quad \frac{dr}{dt} \le \frac{k}{2}r + \mathcal{O}(r^2) \le k'r$$

Lyapounov exponents (as defined in the linear case) are $Real(\mu_j)$

Case 2: hyperbolic point

$$Real(\mu_1) \le \ldots Real(\mu_s) < 0 < Real(\mu_{s+1}) \le \ldots \le Real(\mu_n)$$

with $s \neq 0, n$. There is a coordinate change such that $z = (x, y)^T$:

$$\frac{dx}{dt} = \Lambda x + X(x, y) \quad ; \quad \Lambda = diag[\mu_1, \ldots, \mu_s] + nilpotent$$

$$\frac{dy}{dt} = \Gamma y + Y(x, y) \quad ; \quad \Gamma = diag[\mu_{s+1}, \ldots, \mu_n] + nilpotent$$

The flow of $\frac{dx}{dt} = \Lambda x$ is a contraction, the flow of $\frac{dy}{dt} = \Gamma y$ is an expansion (=contraction for negative time).

Theorem: (Poincaré–Hadamard–Perron)

For $|x|$ small there is a (locally unique) graph $y = h(x)$ with $h(\underline{0}) = \underline{0}$, $\partial h / \partial x(\underline{0}) = \underline{0}$ invariant for the flow.

Idea of the proof: (Hadamard) Take any initial graph (e.g. $y = h_0(x) = \underline{0}$); wait time $\delta t < 0$ along the flow F^t, the graph becomes $y = h_1(x)$; wait for δt again, get $y = h_2(x)$, ...; if $h_n(x) \to h(x)$ uniformly for $n \to +\infty$, then $y = h(x)$ is the invariant manifold. Convergence proven by the contraction lemma in the space of Lipschitz graphs. To prove that $h(x)$ is also differentiable some functional analysis required (e.g. Ascoli--Arzelà).

Alternative proof in the analytic case (Poincaré–Perron): write recursive equations for Taylor series of h, then prove convergence. Both proofs are constructive, lead to effective algorithms.

Corollary: stable manifold

The graph $y = h(x)$ is contained in $W^+(\underline{0})$, the stable manifold of $\underline{0}$ as defined in the linear case. The points of the graph have Lyapounov exponents equal to one of the $Real(\mu_j)$, $j = 1, \ldots, s$. .

$$W^+(\underline{0}) = \{z | \lim_{t \to +\infty} \frac{1}{t} \log \frac{|F^t(z) - \underline{0}|}{|z - \underline{0}|} < 0\}$$

There is also a (locally unique) unstable manifold defined as:

$$W^-(\underline{0}) = \{z | \lim_{t \to -\infty} \frac{1}{t} \log \frac{|F^t(z) - \underline{0}|}{|z - \underline{0}|} < 0\}$$

(proof by reversing time). However, the Lyapounov exponents as defined in the linear case are not necessarily positive for $z \in W^-(\underline{0})$. If z is near $\underline{0}$ it first goes away exponentially, but then the linear part is no more dominant and the distance can stop increasing (saturation distance). Lyapounov exponents need to be defined by the variational equations.

Generalisation:

$$Real(\mu_1) \leq \ldots Real(\mu_s) < c < Real(\mu_{s+1}) \leq \ldots \leq Real(\mu_n)$$

with c any number; $s \neq 0, n$. Then there is a (locally unique) invariant manifold tangent in $\underline{0}$ to the sum of the generalised eigenspaces of $\mu_1, \ldots, \mu_s$

Proof: the contraction lemma does not apply, convergence has to be shown by explicit majoration.

Corollary: Lyapounov exponents

If $Real(\mu_j) < 0$ there is a z (actually, an invariant manifold of z) such that the Lyapounov exponent $\chi(\underline{0}, z) = Real(\mu_j)$.

Corollary: center manifold

There is a (locally unique) invariant manifold tangent in $\underline{0}$ to the sum of the generalized eigenspaces of the eigenvalues with real part zero.

Proof: There are a center stable and a center unstable manifold tangent to the eigenspaces with real part ≤ 0 and ≥ 0 respectively; they are transversal and their intersection is locally a manifold.

Linearisation: Hartmann–Grobman theorem

If $\underline{0}$ is an hyperbolic fixed point (no eigenvalue with real part zero for the matrix A of the linearized system in $\underline{0}$) then there is a bicontinous coordinate change in a neigbourhood of $\underline{0}$ which changes the flow of the nonlinear system in the flow of the linear system defined by A.

All the proofs of the statements in this section are given in Hartmann (1964).

1.3 INVARIANT MANIFOLDS

Problem: global behaviour of an invariant manifold

$W^+(\underline{0})$, $W^-(\underline{0})$ are constructed by the proofs of the Hadamard--Perron theorem in a neighbourhood of the fixed point $\underline{0}$. Apply the flow to these open manifolds; the invariant sets obtained are global $W^+(\underline{0})$, $W^-(\underline{0})$ with the same definition. They are differentiable manifolds (the flow is a diffeomorphism). Globally they are not submanifolds of the phase space Z : $V \mapsto W^+(\underline{0}) \subset Z$ has maximal rank, but the neighbourhood of a point in $W^+(\underline{0})$ in the topology of Z is not necessarily a ball. To make this discussion clear, we need to remind the classical definition of a submanifold:

Definition: Manifold, submanifold

A *submanifold* V is a subset of $\mathcal{R}^n$ (another manifold) such that every point in V has a neighbourhood in V image of a *local map*: $z \in V \longrightarrow B \subset \mathcal{R}^r$; coordinate changes between intersecting maps are regular; V is closed and the induced topology in the ambient space is the topology of the local maps.

Thus a subset V of $\mathcal{R}^n$ can be a manifold in itself, without being a submanifold; this occurs whenever one of the following ''pathologies'' occur:

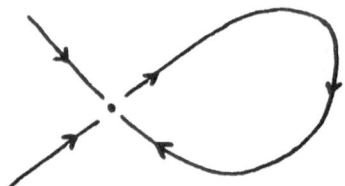

1: Self--intersections; they do not occur in the case of the stable/unstable manifolds, unless at the equilibrium point.

2: loose ends; they can occur when an invariant manifold is simultaneously stable and unstable manifold, e.g. of two different equilibrium points (heteroclinic connection).

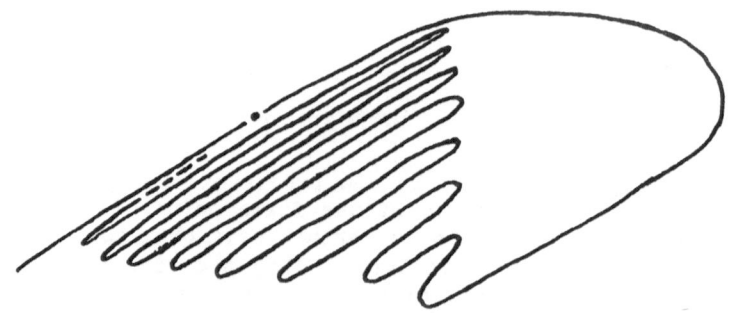

3: accumulation points of different coordinate patches; this does occur to the stable/unstable manifolds of dynamical systems.

2 PERIODIC ORBITS

2.1 VARIATIONAL EQUATIONS

Given a differential equation such as (1.1), the first order changes with respect to a given solution are described by the linearisation of the integral flow:

$$\frac{dz}{dt} = f(z) \quad ; \quad F^t(z_0) = z(t) \quad ; \quad A^t = \frac{\partial F^t}{\partial z_0}(z_0)$$

If the equation (1.1) is regular, the linearized flow L^t can be shown to fulfill the equations:

$$\frac{\partial}{\partial z_0}\frac{d}{dt}z = \frac{d}{dt}L = \frac{\partial}{\partial z_0}f(z) = \frac{\partial f}{\partial z}(z(t)) \cdot L$$

Thus the linearised flow is the unique matrix solution of the initial conditions problem (variational equation):

$$\frac{dL}{dt} = \frac{\partial f}{\partial z}(z(t)) \cdot L \quad ; \quad L(0) = I$$

Since the variational equations are linear, the saturation distance argument does not apply and it is possible to give a consistent definition of Lyapounov exponent which works equally for positive and negative exponents:

Definition: Lyapounov characteristic exponent

$$\chi(z_0, v) = \lim_{t \to +\infty} \frac{1}{t} \log \frac{|v(t)|}{|v(0)|}$$

where v is solution of the variational equation:

$$\frac{dv}{dt} = \frac{\partial f}{\partial z}(F^t(z_0)) \; v$$

Then the linear stable and unstable manifolds:

$$W^+(z_0) = \{v | \chi(z_0, v) < 0\} \quad ; \quad W^-(z_0) = \{v | \chi(z_0, v) > 0\}$$

are linear subspaces of the tangent space at z_0. For an equilibrium point and for a periodic orbit Lyapounov exponents are always well defined (see below). For general z_0 this is not always a good definition (see Froeschlé, 1984).

We shall now discuss the case of a Periodic orbit: $F^T(z_0) = z_0$ for some $T \neq 0$. The monodromy matrix is the matrix of partials of the one--period flow:

$$\frac{\partial F^T}{\partial z_0}(z_0) = L^T = M$$

and is the solution for time=period of the variational equation, a linear time dependent equation:

$$\frac{dL}{dt} = C(t)L \quad ; \quad L(0) = I \quad ; \quad C(t) \doteq \frac{\partial f}{\partial z}(F^t(z_0))$$

with coefficients $C(t)$ also periodic of period T

The eigenvalues of the monodromy matrix, $\gamma : det(M - \gamma I) = 0$ are the Floquet multipliers of the periodic orbit; one Floquet multiplier is necessarily $= 1$, because the velocity vector is an eigenvector with eigenvalue 1 of the monodromy matrix:

$$v = \frac{dz}{dt} \quad : \quad v(0) = v(T) = Mv(0)$$

On the contrary, 0 is not a multiplier (M^{-1} exists).

In this context, it is possible to compute a kind of *Matrix logarithm*, that is there is a matrix A such that $M = \exp(AT)$. Then by a periodic change of coordinates:

$$Z(t) = \exp(At)L^{-t} \quad ; \quad Z(0) = Z(T) = I$$

$$L_*^t = Z(t)L^t \quad ; \quad \frac{dL}{dt}_* = AL_*$$

the variational equation is reduced to constant coefficients. Then the Lyapounov exponents are the real parts of the eigenvalues of A, that is the real parts of the logarithms of the Floquet multipliers divided by the period.

Hamiltonian case: the equations are of the form: $\frac{dz}{dt} = J\nabla H(z)$ and the variational equations are also Hamiltonian:

$$\frac{d}{dt}L = J[\nabla\nabla H(z(t))]L$$

and the linearized flow L^t is $L^t = \exp(JAt)$ with A a symmetric matrix; then it is *symplectic* $(L^t)'JL^t = J$:

$$(L^t)'JL^t = \exp(JAt)'J\exp(JAt) = \exp(AJ't)J\exp(JAt)J^{-1}J =$$

$$= \exp(-AJt)\exp(JJAtJ')J = \exp(-AJt)\exp(AJt)J = J$$

This aplies also to the monodromy matrix M: $M' = JM^{-1}J^{-1}$ thus M and M^{-1} have the same eigenvalues, in couples $\gamma, 1/\gamma$; 1 is a Floquet multiplier with even multiplicity.

For the proof of the statements in this section see e.g. Lefschetz (1963).

2.2 SURFACES OF SECTION

The method of the surfaces of section has been introduced by Poincaré and is now a standard tool; its rigorous definition requires only the use of the implicit function theorem.

Let Σ be an hypersurface (e.g. defined by $g(z) = 0$; $\nabla g \neq 0$) such that the orbit through z_0 crosses transversally ($g(z_0) = 0$; $\nabla g(z_0) \cdot \nabla f(z_0) \neq 0$). Let $\Theta(z_0) = z_1 \in \Sigma$ be the next crossing of Σ by the same orbit; the map Θ is the Poincarè map. If the crossing of the hypersurface Σ by the flow is transversal (that is $\nabla g(z_1) \cdot \nabla f(z_1) \neq 0$) then the map Θ is smooth at z_0.

Linearisation of the Poincaré map: to compute the derivatives of the Poincaré map we need to take into account not only the derivatives of the flow with respect to the initial conditions, but also that the ''time for return to Σ'' depends upon the initial point on Σ:

$$\Theta(z) = F^{t(z)}(z) \quad ; \quad g[F^{t(z)}(z)] = 0$$

and by taking the differential of the last equation:

$$0 = \nabla g \cdot \left[\frac{\partial F^{t(z)}(z)}{\partial z} + \frac{\partial F^{t(z)}}{\partial t} \nabla t \right] = \nabla g M + [\nabla g \cdot f] \nabla t$$

$$\nabla t = -\frac{1}{\nabla g \cdot f} \nabla g M$$

$$\frac{\partial \Theta}{\partial z} = \left[\frac{\partial F^{t(z)}(z)}{\partial z} + \frac{\partial F^{t(z)}(z)}{\partial t} \nabla t \right]\big|_{\Sigma} = [M + f \nabla t]\big|_{\Sigma}$$

If $f(z_0)$ is parallel to $\nabla g(z_0)$ then $D\Theta = M|_{\Sigma}$ and the eigenvalues are the Floquet multipliers with one 1 removed. The Lyapounov exponents for a periodic orbit are redefined as Lyapounov exponents of the Poincaré map: in this way a zero exponent is removed (variations along the velocity are not allowed).

Hamiltonian case: Since the Hamiltonian is an integral:

$$\frac{dH}{dt} = \nabla H \cdot J \nabla H = 0$$

21

then it is constant along the flow:

$$H(F^{t(z)}(z)) = H(z) \qquad ; \qquad \nabla H[f\nabla t + M] = \nabla H$$

and the projection of ∇H on Σ is an eigenvector with eigenvalue 1 of $D\Theta$. To remove it we need eliminate one variable with the equation $H = h$.

Let Σ be a coordinate hyperplane $x_1 = 0$, and let us eliminate y_1 from $H(x_1, y_1, X, Y) - h = 0$ (possible since $\partial H/\partial y_1 = dx_1/dt \neq 0$).

Theory of consequents: the jacobian of the reduced Poincaré map $D\Theta_h$ is symplectic.

Proof: suppose $dx_1/dt > 0$ along the periodic orbit, and that the section is $x_1 = 0$ at the first intersection, $x_1 = \Delta$ at the second (e.g. angle variable). Solve for y_1 from $H(x_1, y_1, X, Y) - h = 0$ and get $y_1 = c(h, x, X, Y)$ from the implicit function theorem:

$$\frac{\partial c}{\partial X} = -\left(\frac{\partial H}{\partial y_1}\right)^{-1} \frac{\partial H}{\partial X} \qquad ; \qquad \frac{\partial c}{\partial Y} = -\left(\frac{\partial H}{\partial y_1}\right)^{-1} \frac{\partial H}{\partial Y}$$

Use as Hamiltonian with *time s* and parameter h:

$$K(X, Y, s, h) = -c(h, s, X, Y)$$

$$\frac{dX}{ds} = \frac{\partial K}{\partial Y} = -\frac{\partial c}{\partial Y} = \left(\frac{\partial H}{\partial y_1}\right)^{-1} \frac{\partial H}{\partial Y} = \frac{dX}{dt}\frac{dt}{ds}$$

$$\frac{dY}{ds} = -\frac{\partial K}{\partial X} = \frac{\partial c}{\partial X} = -\left(\frac{\partial H}{\partial y_1}\right)^{-1} \frac{\partial H}{\partial X} = \frac{dY}{dt}\frac{dt}{ds}$$

then the new time s is:

$$\frac{ds}{dt} = \left(\frac{dt}{ds}\right)^{-1} = \frac{\partial H}{\partial y_1}$$

and $s = x_1 + const$. Let G be the integral flow of the new equations (time dependent case):

$$G^{s,s'}(X_0, Y_0, h) = (X(s), Y(s)) \qquad ; \qquad G^{s,s}(X_0, Y_0, h) = (X_0, Y_0)$$

The reduced Poincaré map is then an integral flow:

$$\Theta_H(X, Y) = G^{\Delta,0}(X, Y, h)$$

and is canonical, as the flow of an Hamiltonian system is always ($D\Theta_H$ is solution of the variational equation).

In this section we have been following Poincaré, volume II of the *Méthodes nouvelles*; there are of course more modern presentations, such as Abraham and Marsden (1967).

2.3 CONTINUATION OF PERIODIC ORBITS

From now on we shall focus on one of the simplest non--integrable problems of nonlinear dynamics, namely the circular restricted planar 3-body problem: in heliocentric coordinates the problem is defined by the hamiltonian:

$$H = H_0 + \mu R = H(x, y, p_x, p_y)$$

$$H_0 = \frac{1}{2}(p_x^2 + p_y)^2 + (yp_x - xp_y) - \frac{k^2}{r_1} \quad ; \quad r_1 = \sqrt{x^2 + y^2}$$

$$R = x - \frac{1}{r_2} \quad ; \quad r_2 = \sqrt{(x-1)^2 + y^2}$$

where μ is the mass of the second body divided by the mass of the first one and the units have been chosen in such a way that:

$$k = \sqrt{1-\mu} \quad ; \quad G = 1 \quad ; \quad a' = 1 \quad ; \quad n' = 1$$

The problem is best studied by transforming to Delaunay variables in the rotating frame:

$$\lambda = \ell + \varpi - t \quad ; \quad \zeta = -\varpi + t \quad ; \quad \Lambda = k\sqrt{a} \quad ; \quad Z = k\sqrt{a}(1 - \sqrt{1-e^2})$$

by which the unperturbed Hamiltonian takes the simple form:

$$H_0 = -\frac{k^4}{2\Lambda^2} - \Lambda + Z$$

For near--zero eccentricities it is expedient to use the Poincaré variables $(\lambda, \eta, \Lambda, \xi)$, obtained from the Delaunay ones by the canonical analogue of a polar--to--cartesian transformation:

$$\eta = \sqrt{2Z}\cos\zeta \quad ; \quad \xi = \sqrt{2Z}\sin\zeta$$

$$H = -\frac{k^4}{2\Lambda^2} - \Lambda + \frac{1}{2}(\eta^2 + \xi^2) + \mu R(\lambda, \eta, \Lambda, \xi)$$

The technique of the Poincaré map can be applied in a straightforward way: let us chose as cross section: $\lambda = 2\pi j$, with j integer:

$$\frac{d\lambda}{dt} = n - 1 + \mu\frac{\partial R}{\partial \Lambda} \neq 0 \quad ; \quad n = \frac{k^4}{\Lambda^3}$$

23

As prescribed by the theory of consequents of the previous subsection, let us solve for Λ in $H = h$:

$$\Lambda = c(h, \lambda, \eta, \xi)$$

$$\frac{\partial c}{\partial \eta} = -\Big[\frac{\partial H}{\partial \Lambda}\Big]^{-1}\frac{\partial H}{\partial \eta} \quad ; \quad \frac{\partial c}{\partial \xi} = -\Big[\frac{\partial H}{\partial \Lambda}\Big]^{-1}\frac{\partial H}{\partial \xi}$$

Use as new Hamiltonian K with time s and parameter h:

$$K(\eta, \xi, s, h) = -c(h, s, \eta, \xi)$$

and the reduced equations are:

$$\frac{d\eta}{ds} = \frac{\xi}{n-1} + \mathcal{O}(\mu) \quad ; \quad \frac{d\xi}{ds} = \frac{-\eta}{n-1} + \mathcal{O}(\mu)$$

The unperturbed case for $\mu = 0$ can be solved explicitly:

$$\frac{d}{ds}\begin{pmatrix}\eta \\ \xi\end{pmatrix} = \frac{1}{n-1}J\begin{pmatrix}\eta \\ \xi\end{pmatrix}$$

$$\begin{pmatrix}\eta(s) \\ \xi(s)\end{pmatrix} = \exp\Big(\frac{s}{n-1}J\Big)\begin{pmatrix}\eta(0) \\ \xi(0)\end{pmatrix} = R_{-s/(n-1)}\begin{pmatrix}\eta(0) \\ \xi(0)\end{pmatrix}$$

The above equation is a typical example of a twist map: n depends upon $Z = \frac{1}{2}(\eta^2 + \xi^2)$:

$$\frac{\partial n}{\partial Z} = \frac{\partial n}{\partial \Lambda}\frac{\partial c}{\partial Z} = \frac{-3n - \frac{\partial H_0}{\partial Z}}{c} \frac{\partial H_0}{\partial \Lambda} = \frac{-3n}{\Lambda(n-1)}$$

We can express the reduced Poincaré map in polar coordinates: $(\zeta, Z) \mapsto (q, p)$

$$p = Z + \mathcal{O}(\mu) \quad ; \quad q = \zeta + \theta(Z) + \mathcal{O}(\mu)$$

$$\frac{\partial \theta}{\partial Z} = \frac{\partial \theta}{\partial n}\frac{\partial n}{\partial Z} = \frac{-6\pi j n}{(n-1)^3}$$

Continuation of periodic orbit (first kind): There is a fixed point $\eta = \xi = 0$ for $\mu = 0$: if the jacobian matrix of the fixed point equation is non singular:

$$det(D\Theta_H - I) \neq 0 \quad that\ is:$$

$$\theta(0) \neq 2\pi m \quad ; \quad \frac{-2\pi j}{n-1} \neq 2\pi m \quad ; \quad n \neq 1 - \frac{j}{m} \quad ; \quad j, m \in \mathcal{Z}$$

then the periodic orbit is preserved for small enough μ.

However these are all elliptic periodic orbits (Floquet multipliers are of modulus 1), since the linearised Poincaré map is a rotation (by an angle not multiple of 2π). Our next problem is then: are there hyperbolic periodic orbits in the perturbed problem?

This section follows Poincaré, vol I of *Méthodes nouvelles*; for a presentation with more similar notations, see e.g. Hadjidemetriou (1982).

2.4 BIFURCATION OF PERIODIC ORBITS

Continuation of periodic orbits is the rule, bifurcation an exception, since the condition to be fulfilled for continuation is generic (a non zero determinant). However, there are topological constraints on the twist maps which force the appearance of bifurcations, where elliptic periodic orbits split into both elliptic and hyperbolic ones. We shall now show that such bifurcations necessarily occur in the restricted problem. Although many methods are available, we shall use a device proposed by Arnold (1976, appendix 9) and actually used by myself (Milani, 1985) because it appears to give a better insight into the geometry of such bifurcations.

Let us suppose we have a canonical map of a 2--dimensional space into itself:

$$\Theta_h : (\zeta, Z) \mapsto (q, p) \quad ; \quad q = Q(\zeta, Z) , \quad p = P(\zeta, Z)$$

then the matrix of partials:

$$D\Theta_h = \begin{pmatrix} Q_\zeta & Q_Z \\ P_\zeta & P_Z \end{pmatrix} = C$$

is symplectic: $Q_\zeta P_Z - Q_Z P_\zeta = 1$; if $Q_\zeta > 0$ and $P_Z > 0$ it can be represented by a generating function:

$$I(Z, q) = -Zq - S(Z, q)$$

in the implicit form:

$$\begin{cases} \zeta = -\dfrac{\partial I}{\partial Z} = q + S_Z(Z, q) \\ p = -\dfrac{\partial I}{\partial q} = Z + S_q(Z, q) \end{cases}$$

then it is possible to solve for q in the first equation, and the derivatives are given by the implicit function theorem:

$$Q_\zeta = (1 + S_{Zq})^{-1} \quad ; \quad Q_Z = -S_{ZZ} Q_\zeta$$

and by substituting into the second equation:

$$P_\zeta = S_{qq} Q_\zeta \quad ; \quad P_Z = 1 + S_{qq} Q_Z + S_{qZ}$$

Following Arnold, let us define the function:

$$F(\zeta, Z) = S(Z, Q(\zeta, Z))$$

Unlike the usual generating function, the Arnold function is defined as a function of the coordinates on one side only of the transformation; its derivatives can be directly computed :

$$\begin{cases} F_\zeta = S_q Q_\zeta \\ F_Z = S_Z + S_q Q_Z \end{cases}$$

then the critical points of F are the fixed points of Θ_h, that is $S_q = S_Z = 0$.

Let us now compute the hessian of F at some critical point: all the terms with the first derivatives are zero:

$$F_{\zeta\zeta} = S_{qq} Q_\zeta^2$$

$$F_{\zeta Z} = S_{qz} Q_\zeta + S_{qq} Q_\zeta Q_Z$$

$$F_{ZZ} = S_{ZZ} + 2 S_{Zq} Q_Z + S_{qq} Q_Z$$

then substitute inside the second derivatives of S as functions of C:

$$S_{qq} = P_\zeta Q_\zeta^{-1} \quad ; \quad S_{qz} = Q_\zeta^{-1} - 1 \quad ; \quad S_{ZZ} = -Q_Z Q_\zeta^{-1}$$

We obtain (with some algebra, and using $detC = 1$):

$$F_{\zeta\zeta} = P_\zeta Q_\zeta$$

$$F_{\zeta Z} = (P_Z - 1) Q_\zeta$$

$$F_{ZZ} = (P_Z - 2) Q_Z$$

$$det \frac{\partial^2 F}{\partial(\zeta, Z)^2} = Q_\zeta (2 - P_Z - Q_\zeta) \tag{2.1}$$

Equation (2.1) gives a relationship between the hessian determinant of the matrix of the second derivatives of the Arnold function F and the trace TrC of the derivative of the map at the fixed point. On the other hand the eigenvalues of the matrix C are solution of $x^2 - (TrC)x + 1 = 0$: $TrC > 2$ corresponds to real eigenvalues $\chi, 1/\chi$ and to an hyperbolic fixed point; $TrC < 2$ (taking into account that $TrC > 0$ always by the initial hypothesis on the map) coresponds to complex eigenvalues of modulus 1 and to an elliptic fixed point. Therefore :

Extrema of F $\iff$ elliptic fixed points

Saddles of F $\iff$ hyperbolic fixed points

Warning: the generating function is not intrinsic: a canonical change of coordinates changes the function, but the fixed points are intrinsic and the eigenvalues of the hessian of F are intrinsic.

Now let us apply this relationship between fixed points of a canonical map and critical points of the Arnold function to the reduced Poincaré map in the neigbourhood of a periodic orbit of the restricted problem : in the coordinates used to describe such a map as a twist:

$$p = Z + \mathcal{O}(\mu) \quad ; \quad q = \zeta + \theta(Z) + \mathcal{O}(\mu)$$

Then the generating function is given by:

$$S = -\int_0^Z \theta(Z')dZ' + \mathcal{O}(\mu) = S_0(Z) + \mathcal{O}(\mu) = F_0(Z) + \mathcal{O}(\mu)$$

$$\frac{\partial S_0}{\partial Z} > 0 \iff \theta(0) < 0 \iff Z = 0 \ \ minimum \ \ for \ \ F_0$$

Now let us look at $F_0(Z)$ as a function of η, ξ; it is smooth and symmetric; when

$$\theta(0) = -2\pi j/(n(0) - 1) - 2\pi m = 0$$

a maximum changes into a minimum; on one side of the bifurcation value there is a circle of degenerate extrema. $F(\eta, \xi)$ is a perturbation of F_0 and preserves all the non--degenerate extrema; the circle of degenerate extrema bifurcates into maxima, minima and saddles.

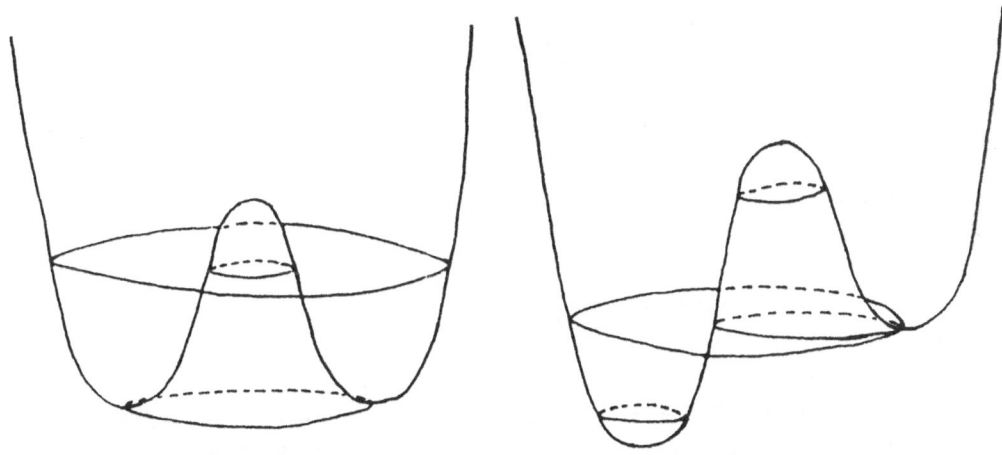

By elementary *Morse theory* the number of extrema bifurcating from the circle is equal to the number of saddles, and both numbers are $\neq 0$, provided the critical points are non--degenerate. Thus by crossing (for changing values h of the Hamiltonian) a resonance, an even number of elliptic and hyperbolic orbits is generated.

3 GLOBAL BEHAVIOUR OF THE STABLE AND UNSTABLE MANIFOLDS

3.1 AVERAGING

This section is dedicated to the purpose of describing the global behaviour of invariant stable and unstable manifold arising from an hyperbolic fixed point, in particular in our case, namely in the case of an hyperbolic periodic orbit bifurcated from a resonant orbit of the restricted problem as in section 2.4. Since the perturbation methods have been described in other lectures both in this and in previous meetings (see e.g. Message 1976, Milani 1988, Milani 1990, Henrard 1990), we shall describe them very briefly; the only essential point is the possibility to transfrom the dynamical situation in the neighbourhood of a resonance to some standard form plus a''small'' perturbation, and this can be done with a variety of techniques, the choice being to a large extent a matter of taste.

Let us start again from the restricted problem, presented in the same Delaunay variables used in section 2.3--2.4:

$$H = -\frac{k^4}{2\Lambda^2} - \Lambda + Z + \mu R \qquad (3.1)$$

$$\frac{d\lambda}{dt} = n - 1 + \mathcal{O}(\mu) \qquad ; \qquad \frac{d\zeta}{dt} = 1 + \mathcal{l}(\mu)$$

Resonance occurs whenever the mean motion n is a rational number:

$$j \cdot (n-1) + k \cdot 1 \simeq 0 \qquad ; \qquad j, k \in \mathcal{Z}$$

$$\frac{d}{dt}[j(\ell + \varpi - \ell') + k(\ell' - \varpi)] = \frac{d}{dt}[j\ell + (k-j)\ell' + (j-k)\varpi] \simeq 0$$

Then the slow moving combination angle $j\lambda + k\zeta = \sigma$ is a critical argument.

Unimodular transformation: are there integers p, q such that:

$$det \ A = det \begin{pmatrix} j & k \\ p & q \end{pmatrix} = jq - kp = 1$$

Answer: yes, if $MCD(j, k) = 1$. Then let us transform the angles by A, the actions by $[A']^{-1}$; this is a canonical transformation, and the transformed angles are still angular variables (that is, the Hamiltonian is still 2π periodic with respect to them):

$$\begin{pmatrix} \sigma \\ \gamma \end{pmatrix} = \begin{pmatrix} j & k \\ p & q \end{pmatrix} \begin{pmatrix} \lambda \\ \zeta \end{pmatrix}$$

$$\begin{pmatrix} \Sigma \\ \Gamma \end{pmatrix} = \begin{pmatrix} q & -p \\ -k & j \end{pmatrix} \begin{pmatrix} \Lambda \\ Z \end{pmatrix}$$

$$H_0 = -\frac{k^4}{2(j\Sigma + p\Gamma)^2} - (j\Sigma + p\Gamma) + (k\Sigma - q\Gamma)$$

$$\frac{d\sigma}{dt} = \frac{\partial H_0}{\partial \Sigma} + \mathcal{O}(\mu) = j(n-1) + k + \mathcal{O}(\mu) \qquad slow$$

$$\frac{d\gamma}{dt} = \frac{\partial H_0}{\partial \Gamma} + \mathcal{O}(\mu) = p(n-1) + q + \mathcal{O}(\mu) \qquad fast$$

We shall use now the transformation method, also called Lie series: use a *determining function* $W = W(\sigma, \gamma, \Sigma, \Gamma)$ to define an hamiltonian flow with independent variable s; the derivative of H along the flow of W is:

$$\frac{dH}{ds} = \{H, W\} = \frac{\partial H}{\partial \sigma}\frac{\partial W}{\partial \Sigma} + \frac{\partial H}{\partial \gamma}\frac{\partial W}{\partial \Gamma} - \frac{\partial H}{\partial \Sigma}\frac{\partial W}{\partial \sigma} - \frac{\partial H}{\partial \Gamma}\frac{\partial W}{\partial \gamma}$$

The effect of the flow for ''time'' s of W upon H is the composition of H with the inverse of the flow, that is the flow for $-s$; by Taylor formula:

$$T_W H = H - s\{H, W\} + \frac{s^2}{2}\{\{H, W\}, W\} + \cdots$$

Set $s = 1$ and chose W to be small: $W = \mu W_1 + \cdots$.

$$T_W H = H_0 + \mu[R - \{H_0, W_1\}] + \mathcal{O}(\mu^2)$$

To simplify the first order part of the Hamiltonian, we can chose W_1 in a suitable way: by using Fourier expansions:

$$R = \sum_{m,r} R_{mr}(\Sigma, \Gamma) \cos(m\sigma + r\gamma)$$

$$W_1 = \sum_{m,r} W_{mr}(\Sigma, \Gamma) \sin(m\sigma + r\gamma)$$

$$\{H_0, W_1\} = -\frac{\partial H_0}{\partial \Sigma}\frac{\partial W_1}{\partial \sigma} - \frac{\partial H_0}{\partial \Gamma}\frac{\partial W_1}{\partial \gamma}$$

$$\frac{\partial H_0}{\partial \Sigma} = \nu \qquad small$$

$$\frac{\partial H_0}{\partial \Gamma} = \delta \qquad large$$

$$W_{mr}(m\nu + r\delta) = R_{mr}$$

Since $\nu \ll \delta$ the small divisors are those with $r = 0$; we therefore chose W_1 in such a way that all the fast changing perturbation terms with $r \neq 0$ are removed:

$$W = \mu \sum_{r \neq 0} \frac{R_{mr}}{m\nu + r\delta} \sin(m\sigma + r\gamma)$$

and the transformation is such that:

$$T_W H = H_0 + \mu \sum_m R_{m0} \cos(m\sigma) + \mathcal{O}(\mu^2) = K(\sigma, \Sigma, \Gamma) + \mathcal{O}(\mu^2)$$

with K a much simpler Hamiltonian, indeed an integrable one:

$$\{\Gamma, K\} = 0 \quad ; \quad \Gamma = const \quad along \quad flow \quad of \quad K$$

and a perturbation of order in μ higher than the one we had before.

For some fixed Γ:

$$K(\sigma, \Sigma) = c(\Sigma - \Sigma_0)^2 + \mu \sum_m d_m(\Sigma) \cos(m\sigma) + \mathcal{O}(\Sigma - \Sigma_0)^3 \qquad (3.2)$$

If the term with $\cos(\sigma)$ is dominant:

$$K(\sigma, \Sigma) = c(\Sigma - \Sigma_0)^2 + \mu d_1(\Sigma_0) \cos(\sigma) + higher \ harmonics +$$

$$+ \mathcal{O}\big((\Sigma - \Sigma_0)^3, \mu(\Sigma - \Sigma_0)\big) \qquad (3.3)$$

In this way it is apparent that a first approximation to the problem of the dynamics near a resonance is the classical pendulum equation: $\Sigma_* = \Sigma - \Sigma_0$

$$K = c\Sigma_*^2 + \mu d_1 \cos(\sigma) + \ldots \qquad (3.4)$$

3.2 THEORY OF CONSEQUENTS

Let us now consider the qualitative properties of the situation near a resonance, as described in the previous section. Because of the standard properties of the pendulum Hamiltonian (3.4), the neigbourhood of exact resonance which has to be considered is of size $\mathcal{O}(\sqrt{\mu})$, therefore equation (3.3) can be described by saying that the transformed Hamiltonian is given by an integrable one --depending only upon one angle σ-- plus a remainder of order $\mu^{3/2}$:

$$K(\sigma, \gamma, \Sigma, \Gamma) = K_0(\sigma, \Sigma, \Gamma) + \mathcal{O}(\mu^{3/2})$$

We shall now consider the Poincaré map associated to the section $\gamma = 2j\pi$. Since γ is fast variable, the time spent by an orbit between

$\gamma = 0$ and $\gamma = 2\pi$ is of order 0 in μ, and the reduced Poincaré map is (for a fixed value of the Hamiltonian):

$$\Theta_\mu : (\sigma, \Sigma) \longrightarrow (\sigma', \Sigma')$$

a canonical mapping of the plane; for $\mu = 0$ this map is such that the Hamiltonian K_0 is preserved, that is, there are on the plane (σ, Σ) invariant level curves preserved by the map Θ_0 (the dynamical system with Hamiltonian K_0 is integrable). In particular, the level curve corresponding to the saddle (that is, through the hyperbolic fixed point) will contain the stable and the unstable manifold of the periodic orbit, and the two will be overlapping, as in the first figure of section 1.3. Then for a μ small but different from zero the stable and unstable manifold will split by a distance of order $\mu^{3/2}$ (this statement would require some work to be considered rigorously proved, but we cannot do this here).

We can now follow the argument given by Poincaré in the third volume of his *Méthodes nouvelles*, chapters XXVII and XXXIII. Let us draw the stable and unstable manifolds of the hyperbolic fixed point C on the surface of section (see figure below). Given a point A on the stable manifold, and a nearby point B on the unstable manifold, let D be the region of the surface of section bordered by some segment joining A with B, by the portion from C to A of the stable manifold, and by the portion from C to B of the unstable manifold. By the results of section 2.2 the reduced Poincaré map is symplectic, therefore area preserving (the determinant of the matrix of partial derivatives is equal to 1); thus the image $\Theta(D)$ has the same area as D. On the other hand $\Theta(D)$ is bordered by a curve joining $\Theta(A)$ and $\Theta(B)$, by the portion from C to $\Theta(A)$ of the stable manifold and by the portion from C to $\Theta(B)$ of the unstable manifold. Therefore the relative position of the four points A, B, $\Theta(A)$ and $\Theta(B)$ cannot be as in the figure below, because this would result in an area of $\Theta(D)$ larger than that of D.

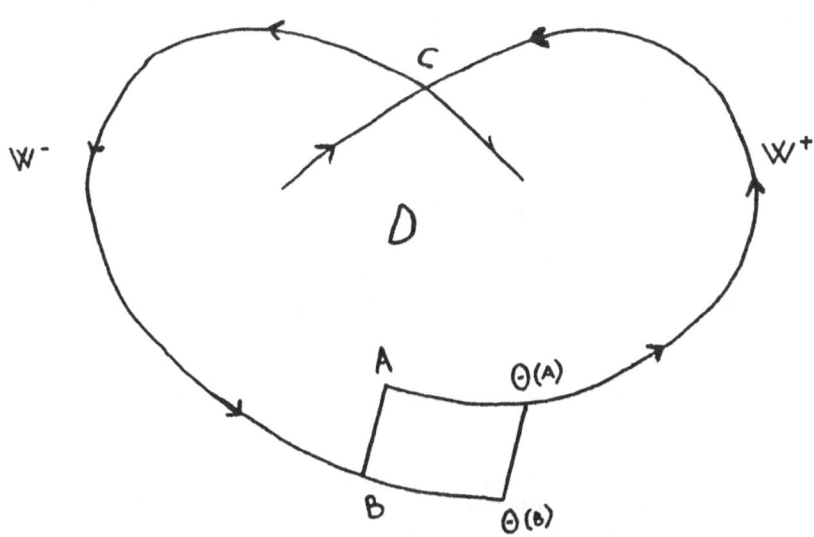

The two distances from A to B and from $\Theta(A)$ to $\Theta(B)$ are of order $\mu^{3/2}$, while the distances between A and $\Theta(A)$ and from B to $\Theta(B)$ are of order $\mu^{1/2}$ (from the pendulum equation (3.4)); it follows that the two portions of invariant manifold between A and $\Theta(A)$ and from B and $\Theta(B)$ must cross, for μ different from zero but small enough. The point at this necessary intersection of the stable and the unstable manifold is called by Poincaré homoclinic (the meaning of the word is that it tends toward the same limit, namely the hyperbolic fixed point, both for $t \longrightarrow +\infty$ and for $t \longrightarrow -\infty$). Thus the occurence of homoclinic points, as consequence of the existence of resonances and of the corresponding hyperbolic periodic orbits, is a necessary feature of the restricted 3--body problem, and indeed of any non--integrable Hamiltonian system.

3.3 CONCLUSION: CHAOS EXISTS

The behaviour of the stable and unstable manifolds of an hyperbolic fixed point, in presence of an homoclinic point, was qualitatively described by Poincaré but better understood much later by S. Smale (for a readable presentation, see Newhouse 1980). In short, the existence of only one homoclinic point (at which the stable and unstable manifold cross in a transversal way) can be described by a universal model based upon Smale's horsehoe map and the Bernouilli shift of symbolic dynamics. Without going into the details of all this, it can be shown that it follows the existence of an invariant closed set, which can be parametrised by a product of two Cantor sets, on which for every initial point there is a positive Lyapounov exponent. This means that there is a portion of the phase space on which the dynamical long term behaviour is critically sensitive to the initial conditions.

The theory of chaotic dynamics has developed significantly in the last decades, and there is no room here for any discussion even of the most interesting results. What I would like to stress as a conclusion is that the modern concept of chaos was found by Poincaré in his study of the 3--body problem, almost 100 years ago. Poincaré himself predicted that a long time would pass, before this kind of concept would be relevant for actual computations to be used in dynamical astronomy. Whether we like it or not, this time is now, and therefore the basic tools needed to build a theory of chaotic dynamics, including the ones briefly sketched here, are now becoming essential tools of our trade.

References

Abraham, R. and Marsden, J: 1967, 'Foundations of Mechanics', Benjamin, New York

Arnold, V.: 1976, *Méthodes Mathématiques de la Mécanique Classique*, MIR, Moscou

Froeschlé, C.: 1984, 'The Lyapounov Characteristic Exponents and Applications to the Dimension of the Invariant Manifolds of a Chaotic Attractor', in *Stability of the Solar System and Its Minor Natural and Artificial Bodies*, Szebehely, V. editor, Reidel, Dordrecht, 265--282

Hadjidemetriou, J. : 1982, 'A qualitative study of stabilizing and destabilizing factors in planetary and asteroidal orbits', in *Applications of modern dynamics to celestial mechanics and astrodynamics*, Szebehely, V., editor, Reidel, Dordrecht, 25--44

Hartmann, P.: 1964, *Ordinary differential equations*, J. Wiley and sons

Henrard, J: 1990, 'The adiabatic invariant in classical mechanics', *Dynamics Reported*, in press

Hirsch, F. and Smale, S.: 1974, *Differential equations, dynamical systems and linear algebra*, Academic press

Lefschetz, S.: 1963 ,*Differential equations: geometric theory*, Dover, New York (reprinted 1977)

Message, P.J.: 1976, 'Formal expressions for the motion of N planets in the plane, with the secular variations included, and an extension to Poisson's theorem', in *Long-Time Predictions in Dynamics*, Szebehely and Tapley eds., Reidel Pu. Co., Dordrecht, Holland, 279--293

Milani, A.: 1988, 'Secular perturbations of planetary orbits and their representation as series', in *Long Term Behaviour of Natural and Artificial N-Body Systems*, Roy, A.E., editor, Kluwer, Dordrecht, 73--108

Milani, A.: 1990, 'Perturbation methods in Celestial Mechanics', in Proceedings of the Goutelas Astronomy School, Froeschlé, C., ed., in press

Newhouse, S.E.: 1980, 'Lectures on Dynamical Systems', in *Dynamical Systems*, Marchioro, C., editor, liguori, Napoli, 209--311

Poincaré H.:, *Methodes Nouvelles de la Mechanique Celeste*, Vol. I, 1892; Vol. II, 1893; Vol. III, 1899; Gauthier-Villars, Paris (reprinted by Blanchard, Paris, 1987).

A NEW ROUTE TO CHAOS: GENERATION OF SPIRAL CHARACTERISTICS

G. Contopoulos

Astronomy Department
University of Florida
Gainesville, Florida, USA

and

Astronomy Department
University of Athens, Greece

1. INTRODUCTION

The best known route to chaos in Hamiltonian systems of two degrees
of freedom is the Feigenbaum sequence of period doubling bifurcations
(Feigenbaum 1978, Coullet and Tresser 1978). As one parameter h (e.g.
the energy) changes, a stable family of periodic orbits becomes unstable
(at $h=h_1$) and then a double period family bifurcates, which is stable.
The characteristic of this family, that gives the coordinate x (intersection of the periodic orbit with the x-axis) as a function of h, is
directed to the right, i.e. towards larger h. At a larger value of the
parameter ($h=h_2$) the double period family becomes unstable and generates
a period-4 family and so on. The intervals (h_1h_2), (h_2h_3)... decrease
approximately geometrically with universal ratio $\delta=8.72$ (Bennettin et al.
1980). Thus at the limit of the sequence h_1,h_2,h_3... we have an infinity
of unstable periodic orbits, that generate a large degree of chaos.

In some cases, as ε increases behond h_∞, an inverse Feigenbaum sequence is generated (Contopoulos 1983a), which produces an infinity of
characteristics directed towards smaller h, that join the characteristics
of the first Feigenbaum sequence. In this way, as ε increases, the
number of families of periodic orbits decreases until only the original
family remains which becomes again stable.

However in the restricted three-body problem a different phenomenon
was found (Pinotsis 1988). The short period family for $\mu=0.5$, becomes
unstable as the energy in the rotating frame h (Jacobi constant) increases
beyond h=2.2848 and generates a Feigenbaum sequence. At a larger value of
h (h=3.0538, where h=3 is the value of h corresponding to the Lagrangian
points L_4,L_5) this family becomes again stable and generates a new
Feigenbaum sequence of characteristics, directed towards smaller h.
However, these characteristics do not join the original characteristics.
Instead, the characteristics of both Feigenbaum sequences are of spiral
form, terminating at different focuses that have the same Jacobi constant
(h=3) but different values of x.

Thus the problem arises how these spirals are generated.

Predictability, Stability, and Chaos in N-Body Dynamical Systems
Edited by A.E. Roy, Plenum Press, New York, 1991

2. THE GENERATION OF THE MAIN SPIRAL CHARACTERISTICS

The generation of spiral characteristics starts when the Lagrangian points L_4, L_5 become unstable. This happens in the restricted three body problem when $\mu=\mu_1=0.03852$. In this paper we consider the rotating Hamiltonian system

$$H = \frac{v^2}{2} - [1+(r^2+1)^{\frac{1}{2}}]^{-1} + \varepsilon r^{\frac{1}{2}}(16-r)\cos 2\Theta - \frac{1}{2}\Omega^2 r^2 = h , \qquad (1)$$

that represents a barred galaxy. The coordinates (r,Θ) and the velocity v refer to a rotating system with angular velocity Ω. The axisymmetric backround $V_0 = -[1+(r^2+1)^{\frac{1}{2}}]^{-1}$ represents an isochrone model, while the perturbation is proportional to ε. The value of the energy in the rotating frame (Jacobi constant) is h.

This Hamiltonian has been studied in two previous papers (Contopoulos 1988 (paper I) and 1990 (paper II)). Here we describe the two main mechanisms that produce spiral characteristics. We give mainly schematic figures that emphasize these mechanisms.

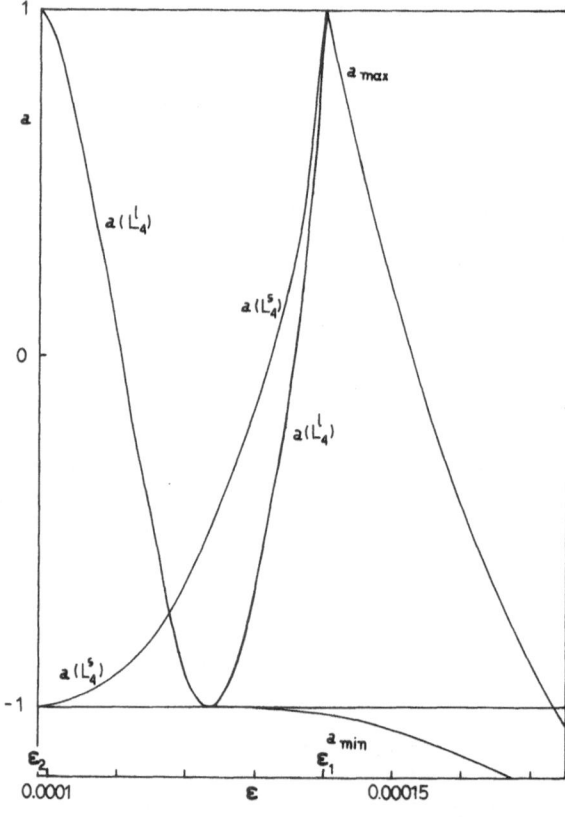

Fig. 1.   The Henon stability parameter of the Lagrangian point L considered as limit of short period orbits $[a(L_4^s)]$ or long period orbits $[a(L_4^\ell)]$. The curves $a(L_4^s)$ and $a(L_4^\ell)$ join into one for $\varepsilon \geqslant \varepsilon_1$. The maximum and minimum a of the LPO characteristics is marked.

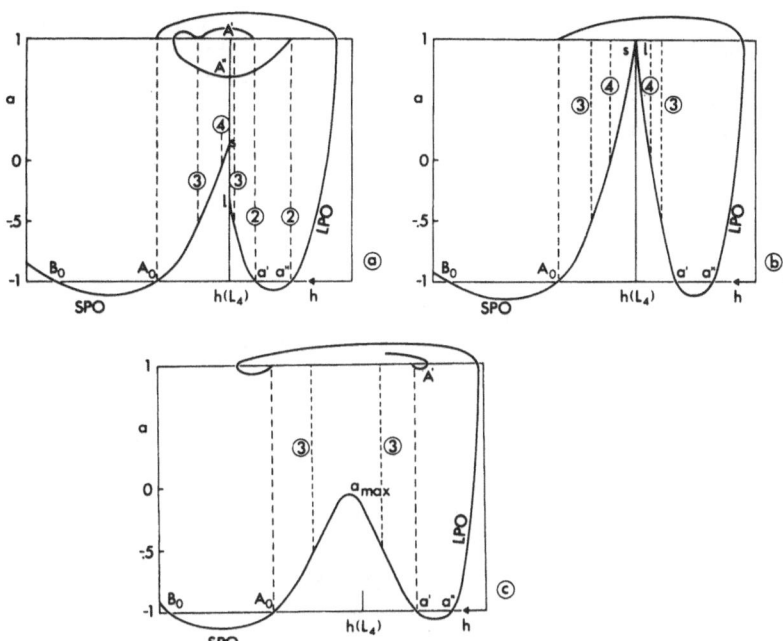

Fig. 2. Stability curves of the SPO and LPO families, giving the stability parameter a as a function of the energy in the rotating
frame h. (a) for $\varepsilon < \varepsilon_1$, (b) for $\varepsilon = \varepsilon_1$ and (c) for $\varepsilon > \varepsilon_1$. A few
bifurcating families are marked (schematically).

The critical value of ε beyond which the Lagrangian points L_4, L_5
are unstable is $\varepsilon = \varepsilon_1 = 0.000140$. When $\varepsilon < \varepsilon_1$ the Lagrangian point L_4 is
stable but its stability parameter (Hénon 1965) is different if it is
considered as the limit of short period orbits (SPO) of infinitesimal
dimensions, or as the corresponding limit of long period orbits (LPO)
(Fig. 1). There are, in fact, infinite LPO families starting and terminating at particular points of the SPO family (Henrard 1983, Contopoulos
1983b, 1988). The SPO families are called A,B,C and they are described
in detail by Contopoulos (1983b). The stability curves of the SPO and
LPO families join when $\varepsilon = \varepsilon_1$. For $\varepsilon > \varepsilon_1$ the SPO and LPO characteristics
join into one curve, as it is well known (Fig. 3 below). The
maximum $a = a_{max}$ along this joint characteristic is given in Fig. 1
beyond ε_1.

In Fig. 1 we give the curves $a(L_4^s)$ and $a(L_4^\ell)$ for $\varepsilon \geqslant \varepsilon_2 = 0.0000985$.
For the particular value $\varepsilon = \varepsilon_2$ the long period family shrinks to the
point L_4. Then the stability parameters are $a(L_4^s) = -1$ and $a(L_4^\ell) = +1$.

As ε increases $a(L_4^\ell)$ decreases to -1 and then increases again until
it reaches $+1$ together with $a(L_4^s)$ for $\varepsilon = \varepsilon_1$. When $a(L_4^\ell) = -1$ ($\varepsilon = \varepsilon' =$
0.000124) the stability curve of the long period family reaches a minimum
$a_{min} < -1$, and the LPO family has an unstable interval (Fig. 2a).

When $\varepsilon' < \varepsilon < \varepsilon_1$ the stability curve of the SPO has an unstable interval,
below $a = -1$, between B_0 and A_0 and then a increases to a maximum $a = a(L_4^s)$.
The stability curve of the LPO starts at $a = a(L_4^\ell)$, which is smaller than
$a(L_4^s)$, goes below $a = -1$ and then increases to $a = +1$ at a maximum h. Then
the LPO becomes unstable and returns to smaller h, until it joins the SPO
at its minimum h (point A_0 of Fig. 2a).

37

At the points a',a" starts and terminates a double period family, bifurcating from LPO, that we call family A'A" (Fig. 2a, Fig. 3). This family crosses the SPO at a point marked (3) in Fig. 2a, where a=+1, but the family A'A" is unstable on both sides of the crossing point which is marked as b_0 in Fig. 3.

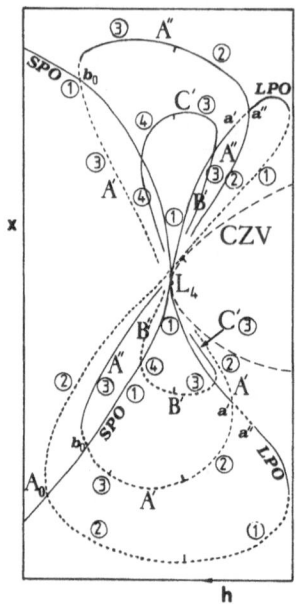

Fig. 3. The characteristics SPO, LPO, and the bifurcating families A'A" and B'B" joining the SPO and LPO, for $\varepsilon<\varepsilon_1$. (——) stable, (····) unstable families, (---) curve of zero velocity (CZV). The numbers indicate the multiplicity of each orbit (number of intersections of the x-axis upwards). At every tangent point with the CZV the multiplicity changes by one, and the corresponding orbits form a cusp on the CZV that develops into an extra loop. Ticks mark these transitions of multiplicity (schematically).

The family B'C' of Fig. 3 starts on the LPO as a period -3 family and terminates on the SPO as a period -4 family.

Whenever a family of period n bifurcates from the SPO or LPO its stability parameter is

$$a = \cos \frac{2\pi}{n}.$$ (2)

E.g. for n=1, a=1, for n=2, a=-1, for n=3, a=-0.5, for n=4, a=0 etc. The beginning and end of the family B'C' are marked as (3) and (4) in Fig. 2a. From Fig. 2a it is evident that there are only a finite number of families starting with a=cos2π/n on the LPO family and terminating with a=cos2π/(n+1) on the SPO family. In fact if a (L_4^ℓ) <cos2π/n_1 there are no bifurcations of multiplicities ≥n_1. As increases the maxima a(L_4^ℓ) and a(L_4^s) increases and new bifurcations appear whenever a(L_4^ℓ) = cos2π/n. Then it can be shown that a(L_4^s) = cos2π/(n+1).

In the case $\varepsilon=\varepsilon_1$ the two curves $a(L_4^s)$ and $a(L_4^\ell)$ join at $a=+1$. Then we have all higher order bifurcations of arbitrary high multiplicity n from LPO that reach the SPO as bifurcations of multiplicity (n+1) (Fig. 1b).

When $\varepsilon>\varepsilon_1$ the maximum a_{max} is below +1. If $a_{max}<\cos 2\pi/(n+1)$ the corresponding family of multiplicity (n+1) cannot bifurcate from the SPO family. What happens then can be seen by a comparison of Figs. 2b and 2c and Figs. 3a,b, and c. In Fig. 2c we have $a_{max}<0=\cos 2\pi/4$, therefore there are no bifurcations of multiplicity (4), while such bifurcations exist in Fig. 2b. It is obvious that as ε increases above ε_1 it reaches a value $\varepsilon=\varepsilon''$, at which $a_{max}=0=\cos 2\pi/4$, and then the two bifurcations (4) join with each other. The corresponding characteristics join on the curve SPO-LPO (below and above L_4; Fig. 4b). If $\varepsilon<\varepsilon''$ there are two bifurcations (4) from the characteristic SPO-LPO (Fig. 4a) and if $\varepsilon>\varepsilon''$ the period (4) families detach themselves from the SPO-LPO characteristics (Fig. 4c).

The characteristics of multiplicity 4 join all the higher order characteristics, of multiplicities $5,6,...\infty$, that have joined each other for smaller values of ε (but $\varepsilon>\varepsilon_1$) and they form a spiral. As ε increases further a_{max} decreases even further, and the families with n=3 (Fig. 3c) join the spiral. Namely B' joins A' below both branches of SPO-LPO and C' joins A'' above these two brances. Finally, as ε increases further a_{max} reaches -1. Then the points a' and A_0 of Fig. 2c join and for larger ε they disappear and we have $a_{max}<-1$. Therefore the LPO family itself cannot join the SPO family at A_0 but forms a spiral together with all the higher order families. (This evolution can be seen in Figs. 7 and 8 of Paper I).

The formation of spirals starts immediately after ε goes beyond the critical value ε_1, at which the orbits L_4,L_5 become unstable. For ε slightly larger than ε_1 the spirals affect only higher multiplicity bifurcations of order $n,n+1,n+2,..\infty$. As ε increases lower order families join, one by one, the higher order spirals, until the SPO-LPO family itself participates in the spiral.

As we move along the spiral towards its focus (I', or I'' in Fig. 4c) the corresponding orbits develop more and more loops, until at the focus itself they have infinite loops (see section IV below). Such are the asymptotic-periodic orbits of types I-IV in the Copenhagen problem (Szebehely 1967), which are heteroclinic orbits joining L_4 and L_5.

In paper II we describe how other families, bifurcating from SPO-LPO family form different spirals around the same focuses I' and I''. Such families bifurcate between a'' and the maximum h of the LPO (Fig. 2a,b,c).

In these cas ∍ also the spirals are formed by joining infinite families of higher orders. The only difference is that the sequence of families are not $n,n+1,n+2,...\infty$, as before but $n,n+2,n+4,...\infty$. The corresponding orbits develop pairs of loops, as we will describe in section 4.

3. SPIRALS WITH DIFFERENT FOCUSES

When ε is sufficiently larger than ε_1 we find many spirals with different focuses. All these focuses have the same h as L_4 but different x's. An example of the generation of such focuses is described in Fig. 5a,b,c,d. In Fig. 5a ($\varepsilon=0.00018$), we see the upper branch of the

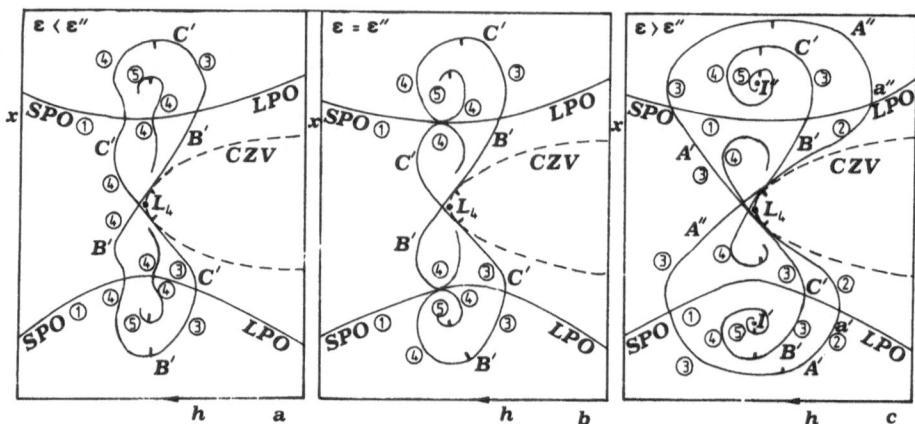

Fig. 4. Bifurcations from the SPO-LPO characteristic for ε close to ε''
(which corresponds to $a_{max}=0=\cos 2\pi/4$). (a) For $\varepsilon<\varepsilon''$ there are
2 bifurcating families of multiplicity 4. (b) For $\varepsilon=\varepsilon''$ the
two families (4) join on the characteristic SPO-LPO. (c) For
$\varepsilon>\varepsilon''$ the families (4) join each other, detaching themselves
from SPO-LPO. These families join higher order families and
form spiral characteristics with focuses I',I", that have
the same h as L_4. In Fig. 4c we have drawn also the family
A'A", which joins the B'C' for larger ε (schematically).

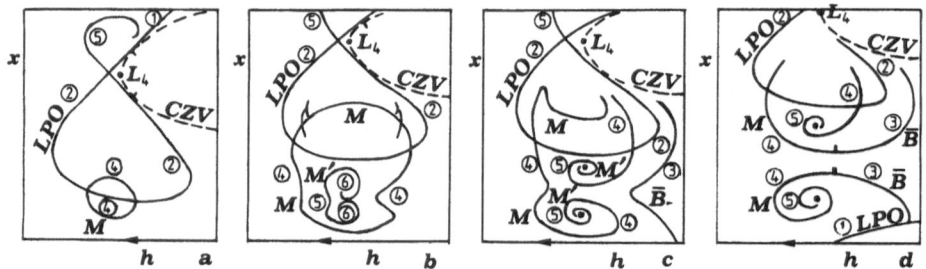

Fig. 5. Bifurcations from the LPO family and formation of new spirals
with different focuses. (a) The family LPO generates a bubble
M. (b) The bubble M forms an infinity of higher order bubbles
by successive period doubling bifurcations. At the same time a
new family M' is formed inside M which terminates at two spirals.
(c) The spirals M' join the characteristic M on the right which
is split into two parts. (d) The characteristics M and $\bar{B}$ are
split into two parts that join each other. Thus the lower part
of $\bar{B}$, that bifurcates from the lower part of LPO, participates
in the lower spiral (schematically).

LPO (a continuation of the SPO) which comes from a maximum h to the left and downwards and produces an infinity of loops above and below L_4 (the lower branch of LPO forms a spiral around the focus I' of Fig. 4c). The family LPO is originally of multiplicity 1, and it becomes successively of multiplicity 2,3,5,6... (Fig. 5a does not show all these details). The important fact is that at a certain $\varepsilon(>\varepsilon_1)$ LPO forms the bubble (closed characteristic) M of multiplicity 4.

As ε increases the bubble M becomes larger and produces higher order bubbles by successive period doublings (Fig. 5b). Thus an infinite sequence of bubbles is formed, as in other simple cases (Contopoulos 1983a).

At the same time a most important development takes place. Inside the lower part of the bubble M we see a closed characteristic M' that forms two spirals, containing orbits of multiplicities $5,6,...\infty$. In Fig. 5b we consider the case when the upper and the lower spiral are tangent and then close on the left, forming a continuous curve on the left, representing orbits of multiplicity 5. For a slightly smaller ε the curve (5) closes on itself, having the topology of a circle, while the spirals join inside this curve (5), forming again a double spiral with orbits of multiplicities $6,7,...\infty$. If ε is reduced still further this process seems to be repeated adinfinitum until the spirals disappear. In fact we found that as we approach the case of Fig. 5a, by decreasing ε, the bubble (5), that surrounds the double spiral, shrinks to a point and disappears.

The evolution of the characteristics beyond the case 5b is as follows. As ε increases to $\varepsilon=0.0002$ we have the case of Fig. 5c. In this case the characteristic M' has been split into two parts. At the same time the curve M on the right and below the LPO has also been split into two parts. The lower part has joined the lower part of M', forming a spiral of multiplicities $4,5...\infty$, while the upper part of the spiral M' joins the part of M closest and below the LPO.

In the same figure we see another family, $\bar{B}$, of multiplicity (3), which bifurcates from the lower branch of the LPO. As ε increases further ($\varepsilon=0.000205$, Fig. 5d) this family $\bar{B}$ is split into two and at the same time the left part of the characteristic M on the left and below the upper branch of the LPO is split into two parts. The lower branches of $\bar{B}$ M join and form a spiral of multiplicities $3,4,...\infty$, and the parts of $\bar{B}$ M above the splitting form a continuous curve.

This scenario shows that low order bifurcations from the family SPO-LPO can join higher order families to form new spirals with focuses different from the original focuses I' and I" of the previous paragraph.

As ε increases further the evolution of the family SPO-LPO is of further interest (Fig. 6a,b,c). We have already remarked that beyond Fig. 2c the maximum of the stability curve SPO-LPO goes below a=-1 and the points A_0a' disappear. In fact, the maximum $a=a_{max}$ disappears at some value of ε and then we have only a minimum $a=a_{min}$ between B_0 and a" (Fig. 6a, $\varepsilon=0.00025$). Beyond a" the stability curve goes to a minimum h and then it turns around towards larger h. After an unstable part (a>1) there is a stable interval, an unstable interval below a=-1, and a new stable interval until a maximum h'. The stability curve has infinite oscillations between maxima and minima h the characteristic it approaches asymptotically the focus I'.

As ε increases beyond $\varepsilon=0.00025$ the value of a_{min} increases and goes above a=-1 (Fig. 6b, $\varepsilon=0.0003$). At the same time the form of the stability curve near its minimum h has changed, i.e. the curve has first

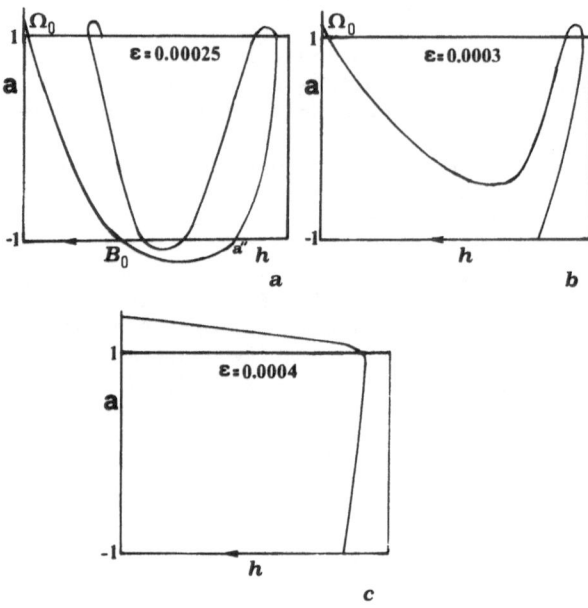

Fig. 6. Stability curves of the family SPO-LPO for large values of ε.
(a) $\varepsilon=0.00025$, (b) $\varepsilon=0.0003$, (c) 0.0004 (schematically).

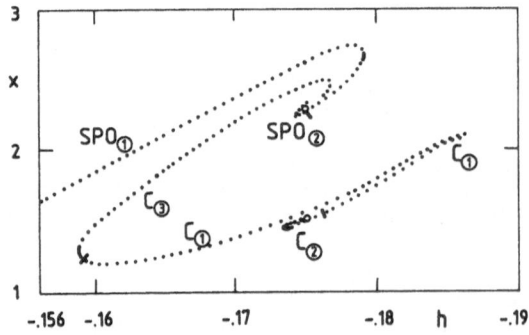

Fig. 7. The lower branches of the characteristics of the families SPO
and C for $\varepsilon=0.0004$. These families are mostly unstable (...),
with only small stable parts (——) near their maxima and minima.
The points o mark the focuses.

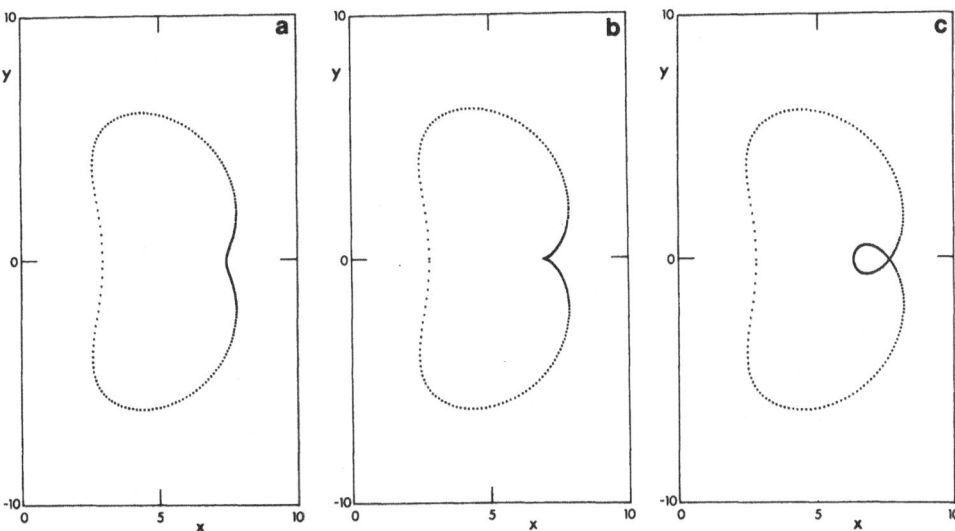

Fig. 8. An orbit of multiplicity 1 (a), develops a cusp (b), and forms
a loop (c) along the x-axis, and the orbit becomes of multiplicity 2.

an unstable part and then reaches its minimum h. As a_{min} increases,
the families bifurcating from SPO-LPO, before B_0 and a'' (up to a=+1),
join each other in pairs and form further spirals. Such are the families
B,C,... (Contopoulos 1983b) of multiplicities 3,4,...∞, that bifurcate
from the SPO between B_0 and Ω_0 (Fig. 6b). Detailed figures and des-
criptions of the cases ε=0.0002, ε=0.00025, and ε=0.0003 are given in
Paper II.

Finally as ε increases still further the whole family SPO up to its
maximum h becomes unstable (Fig. 6c, ε=0.0004) and all families that
were bifurcating from it for smaller ε, are now detached. Thus all
the families B,C,... of multiplicities 3,4,... form spirals that start
and terminate at particular focuses.

An example of such families for ε=0.0004 is shown in Fig. 7. In this
figure we give the lower branches of the characteristics of the families
SPO and C. The family C bifurcates from SPO for smaller ε, as a long
period family of multiplicity 3. In the present case the family C is
completely independent from SPO. Its lower branch starts and terminates
at two different focuses, one of them being the same as the focus of the
family SPO. (There are also other branches of the family not given in
Fig. 7).

In the same way all other long period families B,D,... are detached
from the family SPO in the case ε=0.0004.

4. ORBITS

Consider a simple orbit of multiplicity 1 (Fig. 8a). As the parameter
ε increases for some value of ε this orbit reaches the curve of zero
velocity (CZV). Then the orbit forms a cusp on the CZV (Fig. 8b) and
for larger ε the orbit forms a loop (Fig. 8c). Usually the cusp and the

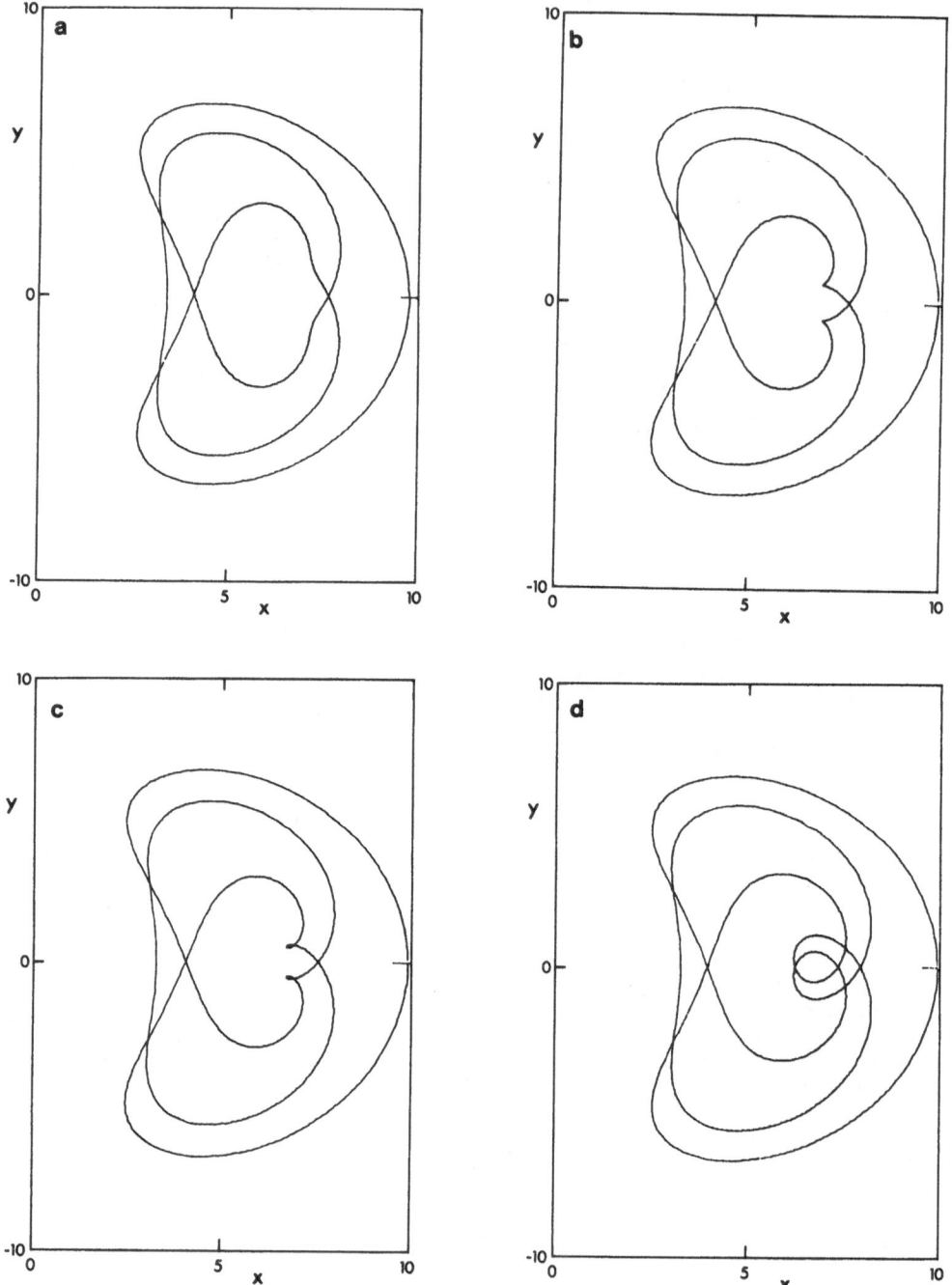

Fig. 9. An orbit of multiplicity 3 (a), develops two cusps (b) outside
the x-axis, which form 2 loops (c). When the loops grow and
intersect the x-axis (d) the orbit is of multiplicity 5.

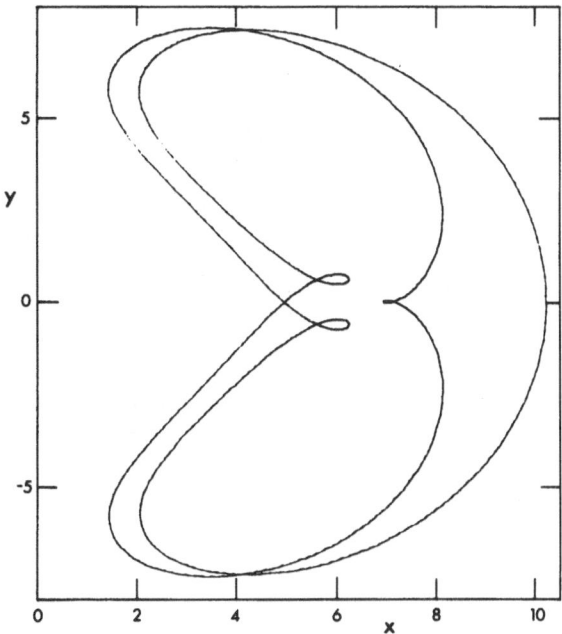

Fig. 10.   A multiplicity 3 orbit a small loop on the x-axis and 2 loops
outside this axis. For a slightly smaller ε the small loop
disappears and the orbit is of multiplicity 2. For larger ε
the two symmetric loops intersect the x-axis and the orbit
becomes of multiplicity 5.

loop are formed on the x-axis, therefore the multiplicity of the orbit
(number of crossings of the x-axis upwards) increases by 1.

However sometimes the orbit (Fig. 9a) reaches the CZV and forms a
cusp, outside the x-axis (Fig. 9b). Then, because of the symmetry of the
Hamiltonian (1) there is also a second cusp, symmetric to the first with
respect to the x-axis. As ε increases the cusps become loops (Fig. 9c).
As long as the loops do not intersect the x-axis the multiplicity of the
orbit does not change (in Fig. 9c the multiplicity is 3). However the
size of the loops increases with ε and beyond a certain value of ε the
loops intersect the x-axis. Then the multiplicity of the orbit increases
by 2 (Fig. 9d). As ε increases further, new cusps and loops are formed
in pairs. Thus we have a sequence of multiplicities along a family
n,n+2,n+4,... .

Sometimes an orbit has cusps or loops both on the x-axis and outside
this axis (Fig. 10).

Most of the periodic orbits close to L_4, L_5 for ε larger than $ε_1$ (when
the Lagrangian points L_4, L_5 become unstable), are highly unstable. Thus
the non-periodic orbits close to them are chaotic. This is particularly
true near the regions where we have the spirals. Then the stability
parameter of the orbits becomes large, sometimes of order a = ±1000, and
the chaotic behaviour of the non-periodic orbits is particularly evident.

When we have further spirals with different focuses a large part of
phase space becomes chaotic near corotation. This explains the observed
chaotic behaviour of the orbits near corotation in galactic models
(Contopoulos 1981).

5. REFERENCES

Benettin, G., Cercignani, Galgani, L. and Giorgilli, A., 1980, Lett. Nuovo Cim. 28, 1.

Contopoulos, G., 1981, Astron. Astrophys. 102, 265.

Contopoulos, G., 1983a, Lett. Nuovo Cim. 37, 149.

Contopoulos, G., 1983b, Celest. Mech. 31, 193.

Contopoulos, G., 1988, Celest. Mech. 43, 147 (Paper I).

Contopoulos, G., 1990, Celest. Mech. (in press) (Paper II).

Coullet, P. and Tresser, C., 1978, J. Phys. C. (Paris) 5, 25.

Feigenbaum, M.J., 1978, J. Stat. Phys. 19, 25.

Henon, M., 1965, Ann. Astrophys. 28, 992.

Henrard, J., 1983, Celest. Mech. 31, 115.

Pinotsis, A., 1988, in A. Roy (ed) "Long Term Behaviour of Natural and Artificial N-Body Systems", Reidel, Dordrecht, p. 465.

Szebehely, V., 1967, Theory of Orbits, Academic Press, New York.

CHAOS IN THE N-BODY PROBLEM OF STELLAR DYNAMICS

Douglas C. Heggie

Department of Mathematics
University of Edinburgh
Edinburgh, U.K.

1. THE N-BODY PROBLEM OF STELLAR DYNAMICS

Stellar dynamics uses several different models of stellar systems, but in this paper we consider the most fundamental, which is governed by the N-body equations:

$$\ddot{\mathbf{r}}_i = -G \sum_{j=1, j \neq i}^{N} m_j \frac{\mathbf{r}_i - \mathbf{r}_j}{|\mathbf{r}_i - \mathbf{r}_j|^3}. \tag{1}$$

Stellar dynamics shares these equations with celestial mechanics (a term which is used here to denote the study of the orbital dynamics of bodies in the solar system), but there are important differences of emphasis. In stellar dynamics all masses are comparable, whereas in celestial mechanics one mass tends to dominate (either the sun or a primary). This has an effect on the methods used and the types of motion which result. Approximate analytical methods are of immense value in celestial mechanics, but not in stellar dynamics, where numerical methods predominate. In celestial mechanics motions tend to be very nearly regular for long intervals of time, whereas in stellar systems motions are highly irregular.

This last point is one that is illustrated by graphical results from numerical integrations (Carnevali & Santangelo 1980), but in the present paper we analyse the irregular nature of motions in stellar systems from a more quantitative point of view; i.e. how sensitively the orbits of the stars depend on the initial conditions. We summarise both numerical data and theoretical investigations on this problem. In addition we consider the implications of these results for the validity and interpretation of numerical simulations.

2. SOME IDEAS OF STELLAR DYNAMICS

2.1 *Some Terminology*

For future reference we list here some useful standard notions of stellar dynamics (cf. Binney & Tremaine 1987), though the applications we have in mind are to star clusters. The kinetic and potential energies of N point masses are, as usual, $T = \frac{1}{2} \sum_{1}^{N} m_i \mathbf{v}_i^2$ and $W = -G \sum_{i<j} (m_i m_j / |\mathbf{r}_i - \mathbf{r}_j|)$, respectively. We define the *virial ratio* $q = 2T/|W|$. This is found to be of order unity in stellar systems in dynamic equilibrium, by the virial theorem (which is the extension of Lagrange's identity in the three-body problem to N bodies.) It is convenient to consider a time scale set by the *crossing time* $t_{cr} = 2R/v$,

where R is the size of the system (in a sense which may be made precise), and v is the mean stellar speed. Using estimates based on the magnitude of q it is then easy to deduce that $t_{cr} \sim \sqrt{(R^3/(GM))}$, where M is the total mass of the system.

2.2 A Summary of the Evolution of a Star Cluster

Later we shall present some numerical data whose interpretation depends partly on what is known about the way in which stellar systems evolve, and so here we summarise some ideas about this (cf. Spitzer 1987). If the system is not initially in dynamic equilibrium (e.g. it is collapsing) then it first settles down into dynamic equilibrium, on a time scale of order t_{cr}. The equilibrium is not exact, however, and the system continues to evolve, though much more slowly. Two-body interactions cause stars to be ejected out of the densest part of the cluster, which is called the *core*. The core shrinks (in size and mass); a process referred to as *core collapse*. It takes place on a *relaxation* time scale, which is of order $Nt_{cr}/\ln N$. When the density in the core has become high enough, three-body interactions become sufficiently common to eject substantial numbers of stars from the cluster. This loss of mass eventually reverses the collapse of the core, and causes overall expansion of the entire system.

The eventual outcome of this evolution is somewhat conjectural, but is thought to be an asymptotic state in which escaping single stars, and some escaping binaries, are spreading out to infinity.

3. CHECKING NUMERICAL ERRORS

As already mentioned, many results on the dynamical evolution of star clusters result from numerical computations, using eqs.(1). As always, it is desirable to control the errors in such computations, but this is clearly a difficult task here, as there exist no suitable non-trivial exact solutions against which to test the results. The following partial checks are available.

(i) *Conservation of integrals* In practice the total energy $E = T + W$ is the most useful. Usually it is considered sufficient if the relative numerical error is limited by $|\Delta E/E| < 10^{-4}$ per t_{cr}, or better, (Aarseth 1974). Incidentally, this might seem crude by the standards of celestial mechanics, but several important shortcuts are needed in order to obtain useful results for systems of any size, and accuracy is one of the sacrifices that must be made.

(ii) *Time-reversal* This test is rarely carried out on large systems. An example is the celebrated Burrau problem in the general three-body problem, where Szebehely & Peters (1967) reversed the velocities at the point where the triple system embarked on its final asymptotic motion (a single mass escaping hyperbolically from a binary), and were able to recover the initial coordinates to about three significant figures. This example also shows how unreliable it is to use conservation of energy as a test: even though there were such large errors in the coordinates the total energy was conserved to a relative accuracy better than 10^{-11}!

(iii) *Independent calculations* Numerical stellar dynamics is sometimes referred to as an experimental method, and, like all experimental techniques, should produce reproducible results. But it was shown by Lecar (1968) that different computations with the same initial conditions (using different computers, or even simply different algorithms on the same computer) produced widely divergent results, after the first crossing time or so.

This trial shows that the detailed positions of the stars in an N-body computation cannot, in general, be regarded as being even approximately correct. On the other hand the interpretation of the results is a little complicated, because we see not only the effects of truncation and rounding errors, but also the propagation of errors made at earlier stages of each calculation. One method of isolating the last mechanism (propagation of errors) is to study the evolution of N-body systems with (slightly) different initial conditions. We now discuss how this can be done in practice, and what the results are.

4. THE GROWTH OF NUMERICAL ERRORS

Consider two N-body systems I, II satisfying the same N-body equations but slightly different initial conditions, i.e.

$$\mathbf{r}_i^I(0) = \mathbf{r}_{i0}^I, \quad \dot{\mathbf{r}}_i^I(0) = \dot{\mathbf{r}}_{i0}^I,$$

and

$$\mathbf{r}_i^{II}(0) = \mathbf{r}_{i0}^{II}, \quad \dot{\mathbf{r}}_i^{II}(0) = \dot{\mathbf{r}}_{i0}^{II}.$$

Suppose we now integrate both systems as accurately as possible. How, we ask, does the difference between the solutions evolve?

The "difference" may be measured in several different ways. One possibility is to define

$$\Delta(t) = \left(\sum (\mathbf{r}_i^I - \mathbf{r}_i^{II})^2 + \sum (\dot{\mathbf{r}}_i^I - \dot{\mathbf{r}}_i^{II})^2\right)^{1/2}.$$

This choice was made in a classic investigation by Miller (1964), who found, for $4 \leq N \leq 32$, that Δ grows roughly as $\exp(\mu t)$, where μ is constant. There was much scatter, but typical values were given by $\mu t_{cr} \sim 2, 4, 20$ for $N = 8, 12, 32$, respectively. Very comparable results were obtained later by Standish (1968) and by Dejonghe & Hut (1986) for $N = 25$ and $N = 3$, respectively.

These studies immediately explain Lecar's results for $N = 25$; the growth of errors is very rapid on the time scales of interest. And Dejonghe & Hut found that initial errors in Burrau's Problem grow by a factor of order 10^9 up to the time at which Szebehely & Peters carried out time reversal; since errors also grow during the reverse integration by a comparable factor (which is a property of Hamiltonian systems), it is clear that Szebehely & Peters did well to recover the initial conditions even to three significant figures.

It also follows that it is impossible to predict the positions and velocities of the stars in a simulated star cluster for more than a few crossing times. But Miller's results are also puzzling. They suggest that the logarithmic rate at which the errors grow depends on N, and indeed increases roughly linearly with N. If his results could be extrapolated to a rich globular cluster, where $N \sim 10^6$ and $t_{cr} \sim 10^6 \text{yr}$, they suggest that the time scale for growth of errors is of order 1yr. In this time a star in a globular cluster moves a distance of order 1AU, whereas the mean distance between the stars is of order 10^4AU. What could cause such rapid growth of errors?

For these reasons the author, in collaboration with Goodman & Hut, has carried out further studies, in order to determine better the N-dependence of the growth of errors. In these new studies a standard alternative formulation, using the variational equations, has been employed. Let us abbreviate eqs.(1) as

$$\ddot{\mathbf{r}}_i = \mathbf{a}_i(\mathbf{r}_1, \mathbf{r}_2, ..., \mathbf{r}_N),$$

and let $\Delta \mathbf{r}_i = \mathbf{r}_i^{II} - \mathbf{r}_i^I$ be the difference between two neighbouring solutions. While this difference remains small, it satisfies approximately the linearised equations

$$\Delta \ddot{\mathbf{r}}_i = \sum_{j=1}^N \Delta \mathbf{r}_j . \nabla_{\mathbf{r}_j} \mathbf{a}_i, \tag{2}$$

and these equations are sufficient to determine the growing separation of the two solutions within the linear regime.

What follows is a summary of the numerical results reported in Goodman, Heggie & Hut (1990). The systems ranged in size from $N = 4$ to $N = 512$, in steps of a factor of 2. The components of the initial positions and velocities of the stars were chosen from a random distribution on (0,1), then rescaled to make the virial ratio $q = 1$. The components of Δr_i and $\Delta \dot{r}_i$ were assigned initial values equal to ± 1 at random. At each value of N ten cases were integrated (differing only in the initial conditions, selected as stated), except that only five cases were studied for N = 512. The integrations were continued up to $t = (5/\sqrt{2})t_{cr}$. To measure the growth of the variations Δr_i a different choice from Miller's was made. It turns out that the variations in the velocities $\Delta \dot{r}_i$ tend to have large 'spikes' during close approaches, whereas Δr_i is better behaved. To smooth the data further, a geometric mean over all the stars was taken, i.e. we define $\ln \Delta = (\sum \ln |\Delta r_i|)/N$.

Since the purpose of this investigation was to investigate the growth of errors in N-body integrations, it was desirable to use an integration algorithm allowing automatic step-size control based on the local truncation error. The fourth-order Runge-Kutta routines given in Press et al. (1986) proved suitable in this respect.

The results of these calculations confirmed that the growth of Δ is nearly exponential, especially for large N (Fig.1). Indeed the scatter in the results (across runs made with different random initial conditions) decreases as N increases. The N-dependence of the growth rate was one of the most interesting conclusions (Fig.2). Defining the growth rate as $\mu = d(\ln \Delta)/dt$, it was found that μ decreases as N increases in the range $4 \lesssim N \lesssim 32$, but then levels off up to the largest value we studied. Thus Miller's results, which stopped at $N = 32$, gave a similar N-dependence to what we have found, but clearly cannot be extrapolated to larger N. For large values of N our results are consistent with $\mu \simeq 6/t_{cr}$, with no dependence on N. Since our results are based on a geometric mean over all the stars in each system, it might be thought that the variations Δr_i might be large only for a small number of stars, but Fig.3 shows that this is not so.

These new results allow us to sharpen our conclusions about how quickly errors grow in large N-body computations. For example, if such a computation is carried out in double precision then all accuracy is lost at a time such that $\exp(6t/t_{cr}) \sim 10^{18}$, $\Rightarrow t \simeq 7t_{cr}$. By comparison, the time to the end of core collapse is of order $30t_{cr}$ for $N = 100$, and is approximately proportional to N.

5. LIAPOUNOV EXPONENTS

Before we pass on from these numerical studies to a theoretical analysis of error growth, it is worth attempting to clarify the link between this investigation and the standard language in which such studies are often expressed. Consider a finite dynamical system with state vector $\mathbf{q}$, and let $\Delta \mathbf{q}(t)$ be a solution of the corresponding variational equations. Define

$$\mu = \lim_{t \to \infty} \frac{\ln \|\Delta \mathbf{q}\|}{t}. \tag{3}$$

It can be shown that this limit exists if the system is tolerably smooth, and (for a given initial value $\mathbf{q}(0)$) takes one of a discrete set of values (depending on $\Delta \mathbf{q}(0)$), called *Liapounov characteristic exponents* (see, for example, Lichtenberg & Lieberman 1983). For almost all $\Delta \mathbf{q}(0)$ the value of μ is the *largest* exponent.

This is clearly related to the quantity μ discussed previously, but the present definition is of no value for our purposes, because of the requirement that $t \to \infty$. If the discussion of §2.2 (above) is correct, as $t \to \infty$ the stellar N-body problem is asymptotically integrable, and so $\mu = 0$. (In an integrable system neighbouring orbits deviate linearly with t, not exponentially.) Therefore the largest Liapounov Characteristic Exponent says nothing about loss of predictability, sensitive dependence on initial conditions, etc.

What is needed is a similar concept which measures the divergence of neighbour-

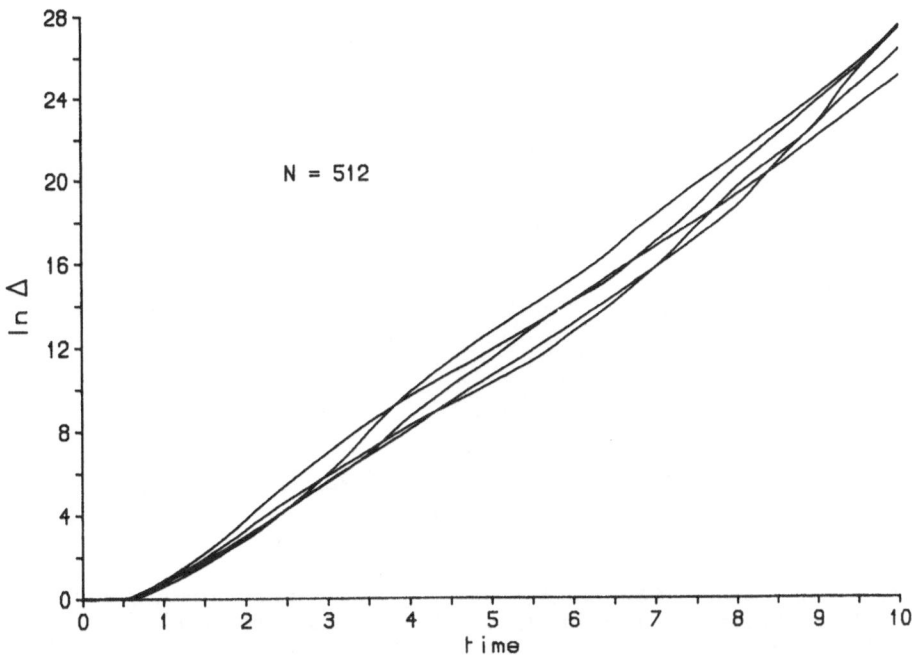

Fig.1 Growth of solutions of the variational equations, for 5 cases with $N = 512$. The ordinate is a logarithmic measure of the size of the solution, as defined in the text, and the abscissa is time.

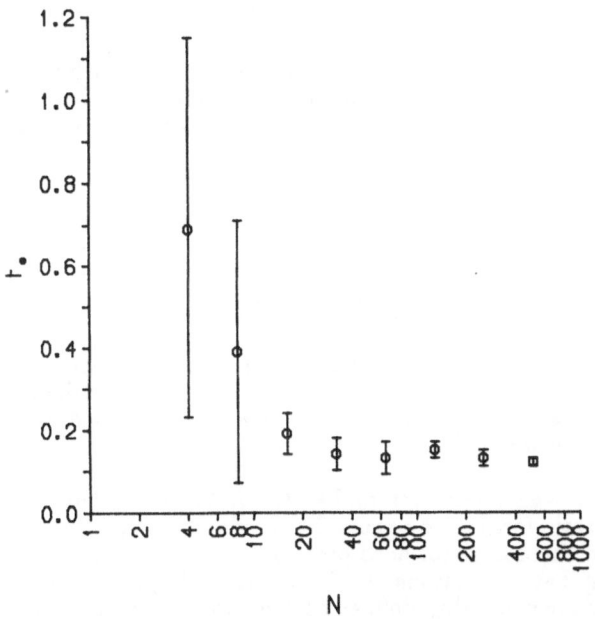

Fig.2 The time scale for the exponential divergence of neighbouring solutions of the N-body problem, as a function of N. The error bars give the standard deviation estimated from the 10 computations carried out for each value of N (except $N = 512$, for which there are only 5). The ordinate is expressed in units of the crossing time.

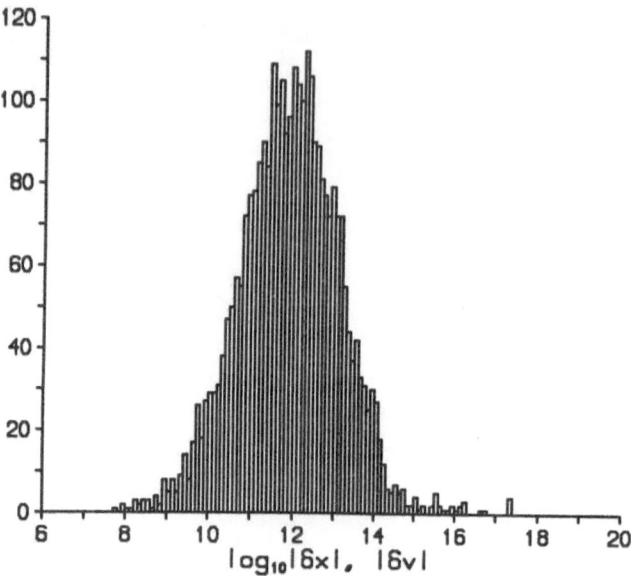

Fig.3 *Histogram of values of δx_i and δv_i, i.e. components of the variations in position and velocity, for 512 stars at $t = 10$. Thus 3072 data values are included. The initial values are of order 1.*

ing orbits over finite intervals of time which are of interest. If we modify eq.(3) and define μ by

$$\mu^{-1} = \frac{t}{\ln(\|\Delta(t)\|/\|\Delta(0)\|)},$$

then we have a definition of a time-dependent quantity, μ, which is sometimes conveniently referred to as a "Liapounov Characteristic Indicator". On the other hand we can emphasise its physical significance if we refer to μ^{-1} as the "time scale of instability", or "e-folding time".

6. THEORY OF THE EXPONENTIAL INSTABILITY

6.1 *Solutions of the N-Body Problem as Geodesics*

We now leave behind the numerical investigation of our problem, and turn to the theoretical approach. There are two lines of attack, one geometrical in nature, the other physical. Both date back to work of Krylov which was published in 1950 (Krylov 1979). In this section we consider the first of these approaches.

We consider a Lagrangian system, like the N-body problem, in which the kinetic energy T is a quadratic form in the generalized velocities $\dot{q}_i$, with coefficients depending on the q_i. Then there is a standard piece of theory (see, for example, Arnold 1978, p.247) which shows that all motions of a fixed energy h give orbits in configuration space which are geodesics for a suitably defined metric. In fact the distance between neighbouring points is given by $ds^2 = 2T(h - U)dt^2$, where U is the potential function. (By our assumption about the form of T, it can be seen that this is a quadratic form in the differentials dq_i, with coefficients which are functions of the generalized coordinates q_i.)

The advantage of casting a dynamical system in this form is that much is known about geodesic flows; and, while the conditions under which such results may be proved may not always apply to the N-body problem, they are strongly suggestive of behaviour to look for. Our interest is in the rate at which orbits diverge, and the rate of divergence of geodesics is described by Jacobi's equation (cf. Arnold 1978, Appendix 1). This is simply the variational equation for geodesic flow, but it is expressed in terms of *covariant* derivatives. The importance of these is that they have a significance independent of the particular coordinates in use; if a covariant derivative vanishes in one system then it vanishes in all.

Let us denote by v the unit tangent vector to a geodesic, and by ξ a vector which takes us from a point on this geodesic to a corresponding point on a neighbouring geodesic. We are interested in the rate of growth of $|\xi|$, and it is easily shown from Jacobi's equation (Gurzadyan & Savvidy 1986) that

$$\frac{d^2|\xi|^2}{ds^2} \geq -2K(\xi, v)|\xi|^2, \tag{4}$$

where K, the Riemannian curvature, can be easily computed in terms of the metric co-efficients. If K remained negative we would be able to show that $|\xi|$ eventually grows exponentially with time. Unfortunately, it turns out that K can have either sign in the classical gravitational N-body problem.

The next simplification that can be attempted is to average the right-hand side of eq.(4) over all possible directions of the vectors v and ξ. When this is done it is found that the average value of the Riemann curvature is

$$\langle K \rangle = \frac{2 - N}{4N} \frac{\sum_{i=1}^{N} m_i f_i^2}{T^3}, \tag{5}$$

where N is the number of stars, f_i is the acceleration of the ith star, and m_i is its mass. Interestingly, this vanishes when $N = 2$, and it is sometimes said that this corresponds to the integrability of the 2-body problem. Our interest is in larger N, and since then we have $\langle K \rangle < 0$, it is possible to argue (Gurzadyan & Savvidy 1986) that this implies exponential divergence of neighbouring orbits.

Furthermore, a theoretical estimate can be made of the rate of growth, which Gurzadyan & Savvidy attempted to do by estimating the average value of the right-hand side of eq.(5). The average is taken over a sensibly chosen, random distribution of the positions of the stars. Unfortunately, however, the average value of f_i^2 diverges because of the contribution of stars at small distances (on the assumption that the positions of neighbouring stars are uncorrelated). Because of the nature of this difficulty, these authors estimated the average value $\langle f_i^2 \rangle$ essentially by substituting the value of f_i^2 which would be contributed by a typical nearest neighbour. (In a system of radius R, the nearest neighbour is at a distance typically of order $RN^{-1/3}$.) In this way Gurzadyan & Savvidy estimated that the time scale for the divergence of neighbouring orbits in a stellar system is of order $N^{1/3}t_{cr}$.

As far as the author is aware, this was the first attempt to estimate the time scale for the divergence of orbits in the classical gravitational N-body problem. On the other hand the predicted N-dependence is inconsistent with the numerical results summarised in §4, even those published before the time of this theoretical study. It is not hard to see that the problem might lie with the divergence at small distances. While the typical interparticle distance is indeed of order $RN^{-1/3}$, it is not hard to show that, in each crossing time, each star has a close encounter at a distance of order $RN^{-1/2}$, where the rate of divergence of neighbouring orbits is larger than average. Although such encounters are short-lived it is not immediately clear whether a strong, brief effect will be dominated by a weaker, constant effect.

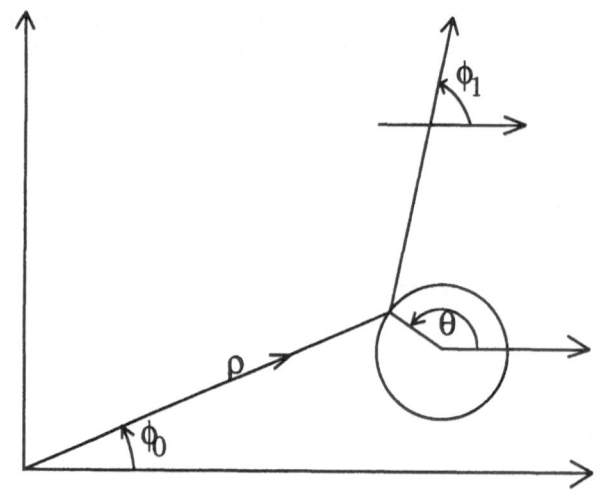

Fig.4 *Path of a particle which strikes, and is deflected by, a hard sphere. The notation is explained in the text.*

6.2 *The Growth of Errors in "Collisions"*

Perhaps it would be possible to adapt the geometric theory, just described, so as to estimate the effect of occasional close encounters. But there is another type of theory, also apparently due to Krylov, in which this may be attempted. He developed it in the context of the "hard-sphere gas", where the bodies have no effect on each other, except during collisions.

We shall outline the theory for a two-dimensional hard sphere gas first. Suppose a particle sets off from the origin, in a direction making an angle ϕ_0 with the x-axis (Fig.4). It strikes a sphere at a point on the sphere corresponding to the polar angle θ, and then sets off in the direction ϕ_1. Since the angles of incidence and reflection are $\pi - \theta + \phi_0$ and $\theta - \phi_1$, respectively, equality of these angles shows that

$$\phi_1 = \phi_0 + \pi - 2\theta. \tag{6}$$

Now suppose the particle had set off at a slightly different angle $\phi_0 + d\phi_0$. If the surface of the sphere is at a distance ρ from the origin, then the particle now has a transverse spatial displacement

$$dp_1 = \rho d\phi_0, \tag{7}$$

(where we use p to denote the minimum distance at which the particle would pass the centre of the sphere if it were not deflected.) Therefore the particle strikes the surface of the sphere at a point which is displaced by an amount $dp_1/\cos(\pi - \theta + \phi_0)$ clockwise round the surface of the sphere. Equating this to $-r_0 d\theta$, where r_0 is the radius of the sphere, the differential of eq.(6) leads to the result $d\phi_1 = d\phi_0(1 + 2\rho/[r_0 \cos(\pi - \theta + \phi_0)])$. The important point about this result is that errors in the direction of motion are magnified by a factor of order ρ/r_0, and we may imagine this can be large in a dilute gas.

Now we turn to the consideration of the gravitational N-body problem. Here simple estimates (cf. Heggie 1988) show that $\phi_1 \sim \phi_0 - Gm/(pv^2)$, where v is the speed of the particle. Hence

$$d\phi_1 \sim d\phi_0 + Gm(pv)^{-2} dp_1, \simeq d\phi_0(1 + Gm\rho/[p^2 v^2]). \tag{8}$$

A complication, to which we shall return, is that there is no length-scale in the gravitational problem which corresponds to the radius r_0 of a hard sphere. Instead, there are encounters at a wide range of distances p, and the distance travelled (ρ) depends on how close the encounters are which we are considering. What can be done (Heggie 1988) is to estimate the effect of those encounters which have the greatest magnifying effect on the errors. It turns out that the most effective encounters are those at a distance of order $RN^{-1/2}$, where, as before, R is a measure of the size of the system. Then the time scale for growth of errors is of order $1t_{cr}$. This is in agreement with the numerical results of Goodman et al. which were discussed in §4 above (and were obtained after the theoretical estimate!)

We now discuss some refinements of this theory which have been developed recently. An important consideration was advanced by Goodman and is described in more detail in Goodman et al. (1990). He pointed out that still closer encounters (at distances much less than $RN^{-1/2}$) affect only a few stars in each crossing time, but have a very large effect on the errors in the positions and velocities of these stars. These strongly affected stars can then "infect" the other stars in the system, because the errors in all the stars are coupled together, by the form of the terms on the right-hand side of the variational equations, eq.(2). (The physical reason for this coupling can also be illustrated by Fig.4. In the associated discussion we assumed that the position of the sphere was not subject to error. If, however, the error in its position is large, this will be what mainly determines the error in the direction of the particle after the collision.) It takes several crossing times for this "infection" to spread throughout the entire system, but the net effect is to shorten the time scale for exponential divergence by a factor of order $\ln N$.

It is worth pointing out that there is no numerical evidence for this factor. This may be because such logarithmic variations are rather small in practice, and may be too difficult to detect in the face of the considerable statistical uncertainty in our numerical estimates. On the other hand Goodman's theory makes it clear that the time needed for the "spread of the infection" (its incubation period?) increases with N, whereas our numerical experiments were concluded at a time which was chosen independent of N.

6.3 The Growth of Errors as a Stochastic Process

The theoretical estimates we have discussed simply yield orders of magnitude, and teach us how the rate of growth of errors depends on the number of stars, etc. It is much harder to use a result like eq.(8) to derive a numerical estimate of the time scale. The reason for this is that the magnification factor depends on quantities like ρ and p, which vary randomly from one encounter to the next. One solution is to simulate the process numerically, choosing values of these parameters from suitably chosen random distributions, and then multiplying together the effects of successive encounters. This has been done by Goodman (Goodman et al. 1990). (Note that this is a different kind of numerical simulation from those discussed in §4. Those were based on the full N-body equations, eq.(1). Here we are using an equation like eq.(8), which is already a great simplification from eq.(1).) In what follows we describe a framework for an analytical attack on the same problem. It is mainly due to my colleague A.M. Davie.

First we have to correct eq.(7), which assumes that the particle emerges from the origin with an error only in its initial direction. But if it emerges from its previous encounter, it already has a transverse error in its position, which we may denote by dp_0. Thus eq.(7) is to be replaced by

$$dp_1 = dp_0 + \rho d\phi_0,$$

and eq.(8) by the more fundamental form

$$d\phi_1 = d\phi_0 + Gm(pv)^{-2}dp_1.$$

(A more careful discussion would also be needed to determine the numerical value of the coefficient of the last term.)

These two equations define a map which is symplectic (since its Jacobian is unity), and stochastic (since the quantities ρ, p and v will differ from one encounter to another, and may be assumed to be random variables drawn from appropriate distributions.) Introducing a vector $\mathbf{x} = (dp, d\phi)^T$, we may write the iterated map as $\mathbf{x}_n = A_n \mathbf{x}_0$, where the matrices A_n are related by

$$A_{n+1} = \begin{pmatrix} 1 & \rho \\ \dfrac{Gm}{p^2 v^2} & 1 + \rho\dfrac{Gm}{p^2 v^2} \end{pmatrix} A_n.$$

(Note that the quantities ρ, p and v should be subscripted to show that they correspond to the nth encounter. These subscripts have been omitted for clarity of notation.) Expressing A_n explicitly by

$$A_n = \begin{pmatrix} a_n & b_n \\ c_n & d_n \end{pmatrix},$$

we deduce easily that

$$\begin{aligned}
a_{n+1} &= a_n + \rho c_n, \\
&= a_0 \Pi_{i=1}^n (1 + \rho_i c_i / a_i).
\end{aligned} \tag{9}$$

It is easy to see now that the average logarithmic divergence (per encounter) is given by the average value of $\ln(1 + \rho_i c_i / a_i)$, but it remains to determine how to take the average. In other words, what is the distribution of c_i / a_i? Let us define $z_i = c_i / a_i$. Then, by the kind of argument which led to eq.(9), we immediately obtain the relation $z_{n+1} = Gm(pv)^{-2} + z_n/(1 + \rho z_n)$. This defines a sequence of random variables, since the values of ρ, p and v are to be chosen from appropriate distributions. The probability density functions of this sequence can be obtained recursively from

$$\begin{aligned}
f_{n+1}(z_{n+1}) &= \int f_n(z_n) f_{\rho,p,v}(\rho, p, v) \delta(z_{n+1} - \frac{Gm}{p^2 v^2} - \frac{z_n}{1 + \rho z_n}) d\rho\, dp\, dv\, dz_n, \\
&= \int f_n(z_n) K(z_n, z_{n+1}) dz_n,
\end{aligned}$$

say. If we suppose that the sequence of density functions tends to a limit, i.e. $f_n \to f$, then f must be a solution of the integral equation

$$f(z') = \int f(z) K(z, z') dz. \tag{10}$$

Finally, the rate of exponential divergence of neighbouring orbits can now be computed as $\mu = \langle \ln(1 + \rho z) \rangle / \langle \rho/V \rangle$, where V is the speed of the star. In the computation of the first average, the distribution of z to be used is obtained (in principle) from the solution of the integral equation (10).

7. RELAXATION AND THE GROWTH OF ERRORS

In large stellar systems in dynamic equilibrium, the stars move around on r tively smooth orbits, only occasionally coming close enough to another star to be significantly deflected. Most of the time, they can be thought of as moving in a smooth, almost stationary gravitational field, in which the individual energy of each star is nearly constant. Because of the occasional close approach (and the cumulative effect of encounters which are not so close) its energy varies slightly, but it takes a time of order $N t_{cr} / \ln N$ (the Chandrasekhar relaxation time) to change by a significant amount (cf. Binney & Tremaine 1987).

Any process which changes the energy of individual stars is called a "relaxation" mechanism, and the main reason why Gurzadyan & Savvidy were interested in estimating the rate of divergence of orbits is that they assumed they were also calculating a relaxation rate. Furthermore they found, as we have seen, a time scale for the divergence of orbits which, for large systems at least, must be shorter than Chandrasekhar's time scale. (We saw in §6.1 that they estimated a time scale of order $N^{1/3} t_{cr}$ for the rate of divergence. In §6.2 this has been revised to a time scale of order t_{cr} or even $t_{cr} / \ln N$. This is smaller than Chandrasekhar's relaxation time scale by a factor of N.) Therefore they concluded that they had found a mechanism of relaxation which would cause stellar systems to evolve much more quickly than had been thought hitherto.

In hard-sphere gases it is true that the rate of divergence of orbits is comparable with the rate of relaxation: it takes only a few collisions to randomise the direction of motion of an atom, or to bring its kinetic energy in statistical equilibrium with the energies of the atoms around it. Stellar systems are different, as we now explain.

We have seen that the growth of errors in positions and velocities of stars in stellar systems is exponential, with an e-folding time scale t_e of order $t_{cr} / \ln N$. It is certainly true that errors in energy grow similarly, and on the same time scale. But this *exponential* growth of errors is obtained in a linear approximation (i.e. variational equations, or Jacobi's equation), and ceases to be valid when the linear approximation itself breaks down. It is easy to see that this will certainly happen when the error in the position of a star becomes comparable with the distance to its nearest neighbour.

In order to quantify this statement, let us denote by p and dp the distance of closest approach of two stars, and the error in this distance, respectively. We have seen (in §6.2) that the most effective encounters are those in which $p \sim R N^{-1/2}$, while exponential growth of errors on a time scale of order t_{cr} leads to $dp(t) \sim dp(0) \exp(t/t_{cr})$. Thus the linear approximation breaks down at a time when $dp(t) \sim R N^{-1/2}$, i.e. when $t \sim t_{cr} \ln(R/N^{1/2} dp(0))$. Thereafter the growth of errors in energy slows down, and there is no contradiction with the slow changes in energy predicted by Chandrasekhar's theory.

Actually the predictions of the two theories are quite consistent, in the following sense. The gravitational potential gradient (per unit mass) in a system of radius R, consisting of N stars of mass m each, is of order GNm/R. If the error in position of a star is of order $RN^{-1/2}$, it follows that the resulting error in its potential energy is of order $GmN^{1/2}/R$. Now Chandrasekhar's theory is based on a model in which the energy of a star varies in the fashion of a random walk, and so the change in energy varies with the square root of the time. Knowing the relaxation time given by Chandrasekhar's theory, we can easily estimate the predicted change in energy at the time (of order t_{cr}) when the exponential growth of errors slows down. It turns out to be the same (except for numerical factors) as our estimate of the error in energy at the same time.

There is one respect in which the estimate of the growth of errors in stellar systems teaches us something about galactic dynamics. It is usual to study galactic orbits by assuming, as mentioned above, that the stars move in a smooth potential, and that the graininess of the true potential does not matter over the time scales of interest. This leads to much interesting theory on whether the potentials are integrable, or whether motions are stochastic. In the latter case, it is possible to measure Liapounov Characteristic Exponents, again assuming that the potential is smooth. We now see, however, that the true rate at which neighbouring orbits diverge is of the order of a crossing time, and that encounters with individual stars are fundamental. Thus the approximation of a smooth potential is useful for studying orbits, but not for studying their divergence.

8. THE USE OF N-BODY SIMULATIONS

We now return to the issue which motivated this entire study: how reliable are the results of N-body simulations? We now know that the positions and velocities of individual stars are quite unreliable after a few crossing times, which is a period much shorter than that over which we would like to study the dynamics of star clusters. What sense, then, can be made of the results of simulations, which for many purposes seem to be the only way in which some problems of stellar dynamics can be approached? The usual answer given to this question is to assert that *statistical* results are still valid, even though the individual positions and velocities of stars are not known. For example, it is assumed that the average rate at which stars escape can be determined (with only statistical error), even though it is not usually possible to state which stars escape.

Despite the enormous numbers of N-body simulations which have been carried out, it is remarkable that very little has been done to test this assertion. It can also be looked at from another point of view: if all we can be interested in are *statistical* results, why should we bother to compute accurate orbits? How crudely (and cheaply) can we simulate stellar systems and still obtain statistically reliable results? There seems to be no theoretical understanding of these questions, and very little empirical evidence either.

One case which has been studied (Valtonen 1974) is the scattering problem for $N = 3$, i.e. the statistical study of the outcome of encounters between a single star and a binary. Valtonen conducted several series of experiments at different accuracy, and measured the statistical distribution of the final eccentricity of the binary. He found no dependence on the accuracy, judged by conservation of energy E, though the range of accuracies was rather limited: $0.005 < \langle (\Delta E/E)^2 \rangle^{1/2} < 0.03$.

Larger N-body systems ($N = 16$) have been studied by Smith (1977). The time step for different runs varied over a factor of 7, but he found no discernible differences in his statistical results except when the conservation of energy was grossly violated, i.e. unless $|\Delta E/E|$ exceeded $10 - 100$. His conclusion was that it is best to make large numbers of cheap runs, as long as the total energy is roughly conserved in each run.

Against this background, we now present some results of a new study involving somewhat larger systems ($N = 100$). Our aim will be able to test for any dependence of a number of statistical results on integration accuracy. The statistical data we test are chosen to correspond to a number of issues which are of special interest to stellar dynamicists (cf. §2.2).

The particular integration program used was the widely available program known as NBODY1 (Aarseth 1985, Binney & Tremaine 1987). In this program the local time-step Δt depends on a parameter η as $\sqrt{\eta}$, and in this study the chosen values of η were 0.1, 0.03 and 0.01. The value 0.03 is usually recommended, except when close encounters occur. Note that the integration method is of relatively high order (by the standards of stellar dynamics), the local truncation error in position varying as $(\Delta t)^7$. For the initial conditions a so-called Plummer model was selected (cf. Spitzer 1987). The results are scaled to units such that $G = 1, Nm = 1$ and $E = -1/4$; such values are rather standard in the field. For each value of η, 10 runs were carried out, differing in the random numbers used to generate the initial conditions.

It was found, as expected, that the error in the energy increased greatly around the time when the first close binary formed. Such an occurrence is to be expected on the basis of what is known of cluster evolution (cf. §2.2), and there are several refinements in more advanced N-body programs which greatly alleviate this difficulty. They are not present in the basic FORTRAN program NBODY1, and so the simulations were stopped at this point. To be precise, let us define $kT = \langle mv^2 \rangle /3$, i.e. as two-thirds the mean kinetic energy of a single star, in analogy with the appropriate definition of temperature in kinetic theory. Then we define an energetic binary as one whose energy exceeds $10kT$, and define t_{10} as the time at which the first such binary formed. Table 1 gives some statistical results on this quantity.

Table 1. Formation of binaries

η	median $\vert \Delta E/E \vert$	$\min(t_{10})$	$\mathrm{median}(t_{10})$	$\max(t_{10})$
0.01	0.0003	17	33	55
0.03	0.004	18	33	53
0.1	0.2	10	30	83
0.1*	0.14	10	30	34

* 7 cases with $\vert \Delta E/E \vert < 1$

Recalling that 10 cases were studied for each value of η, we see that there is very little evidence for a dependence on accuracy. Possibly the distribution of t_{10} is too wide at low accuracy, but the evidence is not compelling. Incidentally, the theory of core collapse (based on model equations valid for large N, i.e. the Fokker-Planck equation, cf. Spitzer 1987) predicts that this would be complete by about $t \simeq 32$ for $N = 100$. Formation of binaries is thought to be a signature of core collapse (§2.2).

Table 2 presents comparable data on the *half-mass radius* at the same time t_{10}. This is defined as the radius of an imaginary sphere, centred at the densest part of the system, which encloses half of its mass, and its evolution is often used as a simple measure of the evolution of the bulk of the cluster. Fokker-Planck theory predicts that this should increase from 0.77 at $t = 0$ to about 1.10 at the end of core collapse. The values found in these simulations are consistently smaller than this, which perhaps exposes the limitations of the Fokker-Planck model. The important point here is that the values are very consistent among themselves, and show no signs of any dependence on accuracy.

Table 2. Half-mass radius at t_{10}

η	smallest	median	largest
0.01	0.72	0.83	1.18
0.03	0.66	0.92	1.19
0.1	0.80	0.91	1.15

Table 3. Number of escapers (to time t_{10})

η	Total number	Total energy
0.01	13	0.0088
0.03	8	0.010
0.1	23	12.4
0.1*	14	0.25

* 7 cases with $\vert \Delta E/E \vert < 1$; data scaled to 10 cases.

Statistics on the number of stars which escape, and the total energy which they carry off, are shown on Table 3. Here at last is a result which clearly depends very sensitively on integration errors. Perhaps this is not surprising, because escapes tend to follow from relatively close encounters, and these are a major source of error. The last line of the table, in which the very worst of the low-accuracy runs are excluded, shows that the total energy of the escaping stars is even more sensitive to errors than their total number. Incidentally, the problem of escapers is an area where simulations are crucial, as the escape rate is notoriously difficult to predict theoretically (Wielen 1968).

The results also show that the savings achieved by sacrificing accuracy (i.e. setting $\eta = 0.1$ rather than 0.03) are less than a factor of 3 in the median number of integration steps required. Therefore there would be little to gain in carefully tuning η within this range to minimise the number of steps while preserving the statistical reliability of the results. Indeed the conclusion of this study is rather clear: use the recommended value (i.e. $\eta = 0.03$)!

It must be admitted that this brief study leaves many interesting questions untouched. For example, it would be interesting to extend it into the regime of dynamical evolution which follows core collapse, using a more refined N-body program dealing efficiently with close binaries. It would also be desirable to extend it to much larger values of N, values more in line with those which are customary in present-day simulations (e.g. $N \gtrsim 10^3$). The results of the above survey, along with a little theory, are indicative of what to expect, as the following argument shows.

Theory implies that the escape rate (number of escapers per crossing time) should be approximately independent of N, and that the mean energy of one escaper should be $\propto E/N$ approximately. Hence it follows that the rate at which escapers carry off energy is given approximately by $dE/dt \simeq 0.04E/(Nt_{cr})$, where a constant of proportionality has been estimated from the above series of 100-body simulations. Since the energy of escaping stars appears to be a sensitive indicator of integration accuracy, it seems desirable to ensure that the total energy is conserved to an accuracy better than this, which leads to the bound $|\Delta E/E| \ll 0.04t/(Nt_{cr})$ on the cumulative relative error in time t. For $N = 100$ up to the time t_{10} of formation of the first energetic binary, this leads to a limit of order 0.004 on the cumulative relative error (cf. Table 1).

All these considerations do nothing, however, to relieve one possible worry. It may be that numerical simulations of adequate accuracy give consistent results only because they are all equally inaccurate (by the standards which would be necessary to compute reliable positions and velocities for the individual stars.)

9. DESIRABLE PROPERTIES OF N-BODY SIMULATION PROGRAMS

If we are to rely on N-body simulations to produce consistent statistical results, we should consider how they ought to be designed in order to perform this task. A great deal is known about how to control the local truncation error in an integration routine, so that accurate positions and velocities can be ensured (at least for limited intervals of time). Very little seems to be known, by contrast, about how to ensure that the results are statistically valid.

We have already indicated that satisfactory energy conservation is a necessary condition for obtaining reliable statistical results on escaping stars, and we have been able to make this assertion approximately quantitative. Similar conditions can be obtained by requiring that the relaxation of stellar energies (cf. §7 above) should be simulated with sufficient accuracy.

The N-body equations have many other conservation properties, including conservation of the *Poincaré invariants* (cf. Arnold 1978). The first of these is $\omega^2 = \sum dp_i \wedge dq_i$, where each term is a 2×2 determinant, and q_i, p_i are cartesian coordinates and their conjugate momenta. The last invariant gives conservation of volume in the phase space

$R^{6N} = \{(q_1, \cdots, q_{3N}, p_1, \cdots, p_{3N})\}$, i.e. Liouville's theorem, which is fundamental to statistical mechanics. Since we are attempting to ensure the sound statistical behaviour of simulated N-body systems, preservation of the Poincaré invariants would seem to be a desirable property to require. Indeed it is guaranteed if the numerical scheme yields a symplectic map. An example is the familiar time-centred leapfrog, i.e.

$$v_{i+1/2} = v_{i-1/2} + a(r_i)\Delta t$$
$$r_{i+1} = r_i + v_{i+1/2}\Delta t,$$

where v_i and r_i are the velocity and position of a star at the ith step, respectively, a is the acceleration, and Δt is the time step, which here must be the same for all stars. On the other hand there appears to be no indication that symplectic schemes give better results when a system is chaotic (Channel & Scovel 1988).

10. INSTABILITIES AT THE STATISTICAL LEVEL

Finally, let us speculate on what we shall find when we study the results of N-body simulations, assuming that they are statistically reliable. Our best guide here are results based on simplified models for the evolution of stellar systems. We have already mentioned one of these models: the Fokker-Planck equation, which is a simplified form of the Boltzmann equation; and there are others.

These models show most interesting behaviour in the regime which follows core collapse, when binaries are sufficiently active to cause an overall expansion of the cluster. It turns out that the expansion can be unstable, depending on the number of stars in the cluster. Sometimes the expansion is modified by a regular oscillation, but there are values of N where the expansion looks quite chaotic (Heggie & Ramamani 1989). Indeed it is tempting to summarise the results by stating that, as N increases, the evolution exhibits the familiar period-doubling route to chaos (cf. Guckenheimer & Holmes 1983).

These intriguing observations have been greatly sharpened in recent work by Breeden et al. (1990), who have applied some of the standard techniques in chaos theory to study the irregular evolution which occurs for large values of N. For example they have measured positive "Liapounov exponents", and have shown that the solutions occupy an approximately two-dimensional submanifold, rather like the Rössler attractor. In this regard, it would be very interesting to produce a highly simplified model of the evolution of star clusters with three degrees of freedom (which seems to be the minimum needed to exhibit this kind of behaviour), and incorporating the essential physics of N-body systems. The "evaporative model" of stellar systems (cf. Spitzer 1987, §3.1) would seem to be a suitable starting point for such an investigation.

A number of very curious issues are raised by these results. For example they show that stellar systems are still unpredictable, even if we are able to ignore the instability of individual stellar orbits, and concern ourselves only with statistical results. Also they illustrate the old paradoxes by which a Hamiltonian system is able to exhibit behaviour in accordance with the laws of statistical mechanics: it is impossible for a Hamiltonian system to possess a low-dimensional attractor. Indeed these results suggest one other property which N-body simulations ought to possess, namely, correct statistical behaviour in the sense of the laws of thermodynamics.

In conclusion, it can be seen that the stellar N-body problem is an excellent and astrophysically important example of a system which, from every practical point of view, is highly chaotic. Indeed, chaotic systems appear to be as common in stellar dynamics as they are rare in celestial mechanics. Though it is difficult to study rigorously, it provides a concrete illustration of several of the important aspects of more abstract chaotic systems. Finally, it raises important questions about how such systems can be faithfully simulated.

11. REFERENCES

Aarseth, S.J., 1974, Dynamical Evolution of Simulated Star Clusters I. Isolated Models, *Astron. Astrophys.*, 35:237.

Aarseth, S.J., 1985, Direct Methods for *N*-Body Simulations, in: *Multiple Time Scales*, J.U. Brackbill & B.I. Cohen, eds., Academic Press, New York.

Arnold, V.I., 1978, *Mathematical Methods of Classical Mechanics*, Springer-Verlag, New York.

Binney, J. & Tremaine, S., 1987, *Galactic Dynamics*, Princeton University Press, Princeton.

Breeden, J.L., Packard, N.H. & Cohn, H., 1990, Chaos in Astrophysical Systems: Core Oscillations in Globular Clusters, preprint, CCSR-90-2 (Center for Complex Systems Research, Dept. of Physics, Beckman Institute, University of Illinois at Urbana-Champaign).

Carnevali, P. & Santangelo, P., 1980, Automated Graphical Plots for the Study of the Gravitational N-body Problem, *Mem.S.A.It.*, 51:529.

Channel, P.J. & Scovel, C., 1988, Symplectic Integration of Hamiltonian Systems, preprint, LA-UR-88-1828 (Los Alamos).

Dejonghe, H. & Hut, P., 1986, Round-Off Sensitivity in the N-Body Problem, in:*The Use of Supercomputers in Stellar Dynamics*, P. Hut & S. McMillan, eds., Springer-Verlag, Berlin.

Goodman, J., Heggie, D.C. & Hut, P., 1990, On the Exponential Instability of N-Body Systems, preprint.

Guckenheimer, J. & Holmes, P., 1983, *Nonlinear Oscillations, Dynamical Systems, and Bifurcations of Vector Fields*, Springer-Verlag, New York.

Gurzadyan, V.G. & Savvidy, G.K., 1986, Collective Relaxation of Stellar Systems, *Astron. Astrophys.*, 160:203.

Heggie, D.C., 1988, The *N*-Body Problem in Stellar Dynamics, in: *Long-Term Dynamical Behaviour of Natural and Artificial N-Body Systems*, A.E. Roy, ed., Kluwer, Dordrecht.

Heggie, D.C. & Ramamani, N., 1989, Evolution of Star Clusters After Core Collapse, *M.N.R.A.S.*, 237:757.

Krylov, N.S., 1979, *Works on the Foundations of Statistical Physics*, Princeton University Press, Princeton.

Lecar, M., 1968, A Comparison of Eleven Numerical Integrations of the Same Gravitational 25-Body Problem, *Bull. Astron.*, 3:91.

Lichtenberg, A.J. & Lieberman, M.A., 1983, *Regular and Stochastic Motion*, Springer-Verlag, New York.

Miller, R.H., 1964, Irreversibility in Small Stellar Dynamical Systems, *Ap.J.*, 140:250.

Press, W.H., Flannery, B.P., Teukolsky, S.A. & Vetterling, W.T., 1986, *Numerical Recipes*, Cambridge University Press, Cambridge.

Smith, H., Jr., 1977, The Validity of Statistical Results from N-Body Calculations, *Astron. Astrophys.*, 61:305.

Spitzer, L., Jr., 1987, *Dynamical Evolution of Globular Clusters*, Princeton University Press, Princeton.

Standish, E.M., 1968, *Numerical Studies of the Gravitational Problem of N Bodies*, Ph.D. Thesis, Yale University.

Szebeheley, V.G. & Peters, C.F., 1967, Complete Solution of a General Problem of Three Bodies, *A.J.*, 72:876.

Valtonen, M.J., 1974, Statistics of Three-Body Experiments, in: *The Stability of the Solar System and of Small Stellar Systems*, Y. Kozai, ed., Reidel, Dordrecht.

Wielen, R., 1968, On the Escape Rate of Stars from Clusters, *Bull. Astron.*, 3:127.

CHAOS, STABILITY AND PREDICTABILITY

IN NEWTONIAN DYNAMICS

Victor Szebehely

R.B. Curran Centennial Chair
Department of Aerospace Engineering
University of Texas, Austin, Texas, 78712

ABSTRACT

The entrance of the subjects of limited predictability and of chaos into the fields of celestial mechanics and gravitational n - body dynamics is treated in this paper. The non-integrability of the gravitational many-body problem (for three or more participating masses), when combined with errors in modelling and with the uncertain values of the initial conditions, leads to bundles of trajectories instead of single orbits for a given dynamical problem. The consequences of these realistic considerations are treated and their effects in celestial mechanics are discussed.

INTRODUCTION

The usually and generally accepted concept of Newtonian dynamics involve complete "solvability" and consequently, perfect predictability of the behavior of dynamical systems. This predictability is assumed to extend to arbitrary long times without any errors. It is to be emphasized that Newton himself did not accept such perfection, or in other words, he was not what today we call a Newtonian. Leibnitz on the other hand, as part of his various disagreements with Newton, accepted complete predictability, joining Laplace with his demon's superior abilities in dynamics. (Gleick, 1987).

After exorcising Laplace's fantazy (as Poincare likes to refer to the demon) we find ourselves in the real land of present day dynamics with its frustrating limitations, many unsolved problems and remarkable challenges.

FORMULATION OF PROBLEMS IN DYNAMICS

In this part of the paper the problems leading to finite predictability in dynamics are outlined.

(1) The differential equations of the motion are the results of our *modelling*, using approximations and simplifications. (Here we might wish to remember Einstein's comment according to which "Everything should be made as simple as possible, but not simpler.") Some of these approximations are arbitrary and others represent our limited knowledge concerning the "true" physical models and the "best" values of the parameters involved. These choices might seriously influence the description of the motion since they affect the structural stability of the solution. Advances in science can strongly

influence the choice of our model. Pre-Newtonian formulations changed considerably when the law of gravity entered the picture and further changes occurred when relativity effects were taken into consideration. This process does not stop and it might not be asymptotic.

Along theses lines attention is called to the importance of using the proper physical constants, the values of which, especially for new models are rather uncertain. The higher order gravitational coefficients of the Earth and of other bodies are determined by the observations of the orbits of artificial satellites and of space probes. It happens frequently that the values of some of these coefficients change with improved observations and/or with using different orbits of the satellites. Those who are responsible for the observations and data analysis claim that the new values are "better" than the old ones but sometimes one cannot help wondering why the Earth's gravitational field should depend on the orbit of the satellite which is used to obtain the data?

(2) The uncertainty of the *initial conditions* used leads to the dynamic stability of the solution. These initial conditions are often obtained by observations or are subject of arbitrary choices. When in text books and in scientific papers we see the expression "the following initial conditions are given," we must ask about the origin of these values. Under practical and operational circumstances the initial conditions are given only approximately with certain errors. (This fact does not stop our undergraduate text book writers to create the obviously wrong impression that precise initial conditions exist!)

Even when no observations are involved in establishing initial conditions, their exact values might be in question. An example for this is the study of motions originating at the triangular libration points of the circular restricted problem of three bodies (Szebehely, 1967). These points are located at the apices of equilateral triangles, the base-line of which is formed by the primaries participating in the problem. If these primaries are located at $x_1 = \mu$ and $x_2 = \mu-1$ on the rotating x axis (known as the line of syzygies), the coordinates of the apices are $x_3 = \mu-1/2$ and $y_3 = \pm \sqrt{3}/2$. Here m is the mass-parameter defined by

$$\mu = \frac{m_2}{m_1 + m_2},$$

where $m_1 \geq m_2$ are the masses of the primaries.

Even if the value of the physical parameter (μ) is known with arbitrary precision, the irrational value of y_3 introduces a non-zero error for the initial conditions.

(3) The *non-integrability* property of practically important dynamical problems represents one of the unavoidable reasons for limited predictability. It is to be emphasized that this difficulty can not be overcome by high speed super-computers and it is the inherent property of almost all dynamical systems of importance in celestial mechanics.

The original definition, of non-integrability given by Poincaré (1892) states that no analytical and globally valid integrals (invariant relations between the variables) exist besides the energy, momentum and center of mass integrals for the gravitational problem of $n \geq 3$ bodies. (See also Whittaker, 1904).

One important consequence of non-integrability is that results must be obtained by analytical or numerical perturbation methods using approximate techniques. It is to be remembered that the use of existing integrals can offer important qualitative informations but not complete solutions.

(4) The *approximate* analytical or numerical solutions lose their meaning as the time for making predictions increases. The general perturbation techniques result in divergent series and the special perturbation approaches (numerical integrations) accumulate errors due to truncation and round-off.

In conclusion, the acceptance of finite predictability is recommended for real and actual problems in dynamics. Individual trajectories lose their meanings and are replaced by bundles. Certain general and statistical properties might be still available and predictable.

DEFINITIONS OF CHAOTIC MOTION

In the previous part the reasons for instability and unpredictability were discussed. These same reasons lead to chaotic behavior which we define as non-periodic (irregular) behavior caused by the inherent non-linear nature of the system. The relation between instability and chaoticity is that the system shows high sensitivity to small changes of the initial conditions. This idea was expressed by Hadamard in 1898, by Poincare in.1908 and by Lorenz in 1963. Popularity came with Gleick (1987) and Stewart (1989).

A simple analytical way to express this idea, following Ruelle (1990) is to introduce the symbol $\delta x(0)$ representing the change in the initial conditions in the n-dimensional phase space. Let $\delta x(t)$ be the change of the trajectory at time t. One definition of chaotic motion is the exponential increase of $\delta x(0)$ or

$$\delta(t) = |\delta x(o)|e^{\lambda t},$$

where $\lambda \geq 0$ is often referred to as Lyapunov's (1892) characteristic exponent.

The possible large effect of small changes in meteorology is often referred to as the butterfly effect. The corresponding phenomenon in gravitational systems may be described as the rattle effect. The mass-distribution of the gravitational system is changed slightly when the baby (corresponding to the butterfly) throws his rattle (corresponding to the wings of the butterfly). The slight wind produced by the wings of the butterfly may produce (in case of instability) a storm. The change of the mass-distribution due to displacement of the rattle (in case of instability) may result in entirely different motion of the masses participating in a given dynamical system, part of which is the rattle.

A small change in initial conditions does not always result in unstable behavior and chaotic motion. As various regions of the phase space are investigated or as the physical parameters change chaoticity might be encountered. In several dynamical systems low levels of energy are associated with regular motion (consequently, with the appearance of Contopoulos' "Third Integral," 1963 and chaoticity appears only at higher values at the energy integral. Changes in the physical parameters are responsible for structural stability which means robustness of the phase portrait or topological equivalence of the behavior. Once again, there might appear regions of the physical parameters where this topological equivalence is displaced by chaoticity.

It is essential to note that neither random inputs nor complexity are required to produce chaos which is basically the outcome of the nonlinear and non-integrable nature of the dynamical system.

The measure of the degree of chaoticity is still an open area for research, especially in the field of gravitational n-body dynamics. Grassberger's (1986) and Maddox' (1990) papers offers several measures of complexity leaving open their possible applications to celestial mechanics. On the other hand if we turn to basic approaches, there seem to be several possibilities to evaluate the degree of chaoticity of gravitational systems. In the following these approaches will be listed with their dynamical applicabilities.

Poincaré's surface of section method allows a clear distinction between periodic, quasiperiodic and irregular or chaotic motion. Periodic orbits appear as single points, quasi-periodicity is represented by a line and chaoticity by irregularly distributed points in the surface of section plots. One disadvantage of this method is that it requires long-term numerical integration (since our systems are non-integrable). After sufficient time the numerical errors might result in chaoticity, without actual dynamical chaos. In order to circumvent this problem the time evolution for continuous time is often replaced by evolution for discrete time, or the original differential equation representing the dynamical system,

$$\frac{dx(t)}{dt} = X[x(t)]$$

is replaced by the difference equation,

$$x(t+1) = f[x(t)].$$

This replacement results in increased speed of obtaining points on the surface of section since no numerical integration is involved but presents the problem of relating the new system to the original dynamical system. It should be noted that examples are available in the literature (see for instance R.H. Miller, 1990) showing that using different methods of numerical integration can result in qualitatively different solutions, depending on the sensitivity of the dynamical system.

Another approach is to submit the result of the numerical integration to frequency analysis, i.e. to establish a power spectrum and observe the amplitude - frequency relation. Chaotic motion is associated with a flat distribution while periodicity shows up as peaks. Once again, long-time numerical integration is required, the result of which is subject to additional numerical analysis.

One of the popular methods of establishing a measure of chaoticity of a given orbit is finding the value of Lyapunov's characteristic exponent and associating chaos with positive values. Here the problem seems to be, in addition to the need of long-time numerical integration, that the definition of the characteristic number contains a follow up as $t \rightarrow \infty$.

It is not unreasonable to conclude that the measure as chaoticity in non-integrable dynamical systems represent a fundamental problem in principle. We wish to investigate the property of the solution of non-integrable dynamical systems when the solutions are available only for a limited time and with limited accuracy. The basic problem is to analyse a solution which is not known precisely and which is not available for arbitrary long time.

IRREVERSIBILITY, INTEGRALITY AND INTERMITTENCY

Since neither the future nor the past state of a dynamical system is defined under chaotic conditions, such motions can not be considered reversible. This is of interest since classical conservative dynamical systems have the property of reversibility, but at the same time they might also be performing chaotic motion, which results in irreversibility. The contradiction becomes more acceptable when it is realized that irreversibility is the basic element of our description of the physical word, according to Prigogine (1980). Furthermore, chaoticity eliminates the idea of a single trajectory which is replaced by a bundle, the elements of which deviate exponentially. Therefore, Hamiltonian systems become irreversible in case of chaos since if the future state cannot be predicted, a reversal of the system will not result in its original state.

Poincaré's non-integrability theorem (1892) states that for the gravitational n-body problem no analytically and globally valid integrals exist besides the energy, momentum and center of mass integrals when $n \geq 3$. This theorem is of importance for chaotic dynamics since chaoticity requires non-integrability.

In order to avoid a possible misunderstanding the definition of an integral of a dynamical system is offered here. The function $F(x_j, t) = C$ is an integral (or an invariant relation) of a dynamical system, represented by the differential equations of motion, $x_j =$

G (x_i, t) if along any orbit the following relation is satisfied:

$$\frac{dF}{dt} + \sum_i \frac{\partial F}{\partial x_i} G_i + \frac{\partial F}{\partial t} = 0 .$$

It is to be noted that other integrals might exist for systems of $n \geq 3$ in certain regions of the please space and therefore the global validity of an integral is an important part of the definition. For the various derivations of non-integrability of dynamical systems the reader is referred to Whittaker's book (1904). The proof of non-integrability of a given dynamical system is not at all trival, which statement might be supported by recalling that the non-integrability of the problem of satellite motion around an oblate planet has only very recently been established.

The importance of non-integrability becomes clear when Prigogine (1980) is quoted: "Poincaré's dictum of non-integrability was in a sense the point at which the development of classical dynamics ended."

Intermittency usually refers to the transition between chaotic and regular behavior. According to Poincaré periodic motions are dense in the phase space and so are quasi-periodic orbits. (Admittedly some of the periods might be very long.) On the other hand chaotic orbits cover densely certain ranges of the phase space, so these two types of orbits might alternate causing intermittency. A simple example is when a chaotic motion becomes quasi-periodic by "sticking" to the boundary of the chaotic zone. The reverse cannot happen except when a chaotic motion is generated from a quasi-periodic behavior due to numerical error accumulation. Thompson and Stewart (1986) formulate this law by stating that "there is no known quasi-periodic transition to chaos in mathematical dynamics - only in experimental dynamical systems." The idea of "order out of chaos" is emphasized by Prigogine (1979) and it has many interesting and important implications in thermodynamics and in dynamics. For transitions in hydrodynamics, see Swinney (1978).

With a slight freedom of thought the following example is offered using some recent results in classical dynamics. In the general problem of three bodies it has been known since Lagrange's (1760) publication of the equation, known today as the Lagrange-Jacobi equation, that when the total energy of the three-body system is zero or positive at least two of the three distances increase to infinity as $t \to \infty$. Recent investigations show that even with negative total energy the motion asymptotically approaches the separation of the three bodies into a binary and an escaping body as shown by Szebehely (1973). Prior to such a regular (orderly) motion (i.e. elliptic two-body motion, plus a hyperbolic escape) the three bodies perform a non-integrable bounded motion called interplay. It is to be mentioned that this simple, orderly outcome follows for any arbitrary set of initial conditions. The behavior of this dynamical system might be considered unpredictable originally because of the non-integrability of the three-body problem. As the time increases the motion becomes regular and predictable with the escape of the third body. This order out of chaos phenomenon of the problem of three bodies has been applied recently to ecomic and organizational problems.

If we reverse this problem and let a third body, coming from infinity, enter the field of the binary, we obtain another bounded three-body motion with eventual ejection of one of the bodies to infinity. The process will not be reversible unless the reentrance of the escaped third body satisfies precisely the previously established initial conditions.

Intermittency, of course is well known in fluid mechanics where Osborne Reynolds in 1883 found (deterministic) laminar flows becoming (chaotic) turbulent flows. Several papers dealing with intermittency in fluid dynamics are offered in the proceeding by Iooss, Helleman and Stora (1983).

Since transition to chaos requires instability, stability research might be considered the corner stone of chaotic dynamics (Ioos and Joseph, 1980).

NUMERICAL EXAMPLES

During the late fifties and early sixties Earth-to-Moon trajectories were designed using the model of the planar, circular restricted problem of three bodies. Most of these orbits were periodic with close approaches to the Earth and to the Moon (Egorov, 1957; Thüring, 1959; Broucke, 1962; Arenstorf, 1963 and Davidson, 1964). For details of these orbits see Szebehely (1967.) More recently Henon (1983), Broucke (1990), Érdi (1990), Smith (1990) and Szebehely (1990) established periodic, quasi-periodic and chaotic orbits in the restricted problem of three bodies. Some of the chaotic orbits were generated by slight modifications of the initial conditions which gave figure-eight Earth-to-Moon periodic orbits. The above references dated 1990 are presently prepared for publication.

For the sake of symmetry and simplicity the case of equal masses of the primaries is selected giving $\mu = 0.5$ for the value of the mass-parameter.

The curves of zero velocity give two identical, separate and bounded regions around m_1 and m_2 provided the value of the Jacobian constant satisfies the inequality $4.25 \leq C \leq \infty$. When $C = 4.25$ the two zero velocity curves intersects the axis at the origin ($x = 0$) and when $C < 4.25$ the third body is free to move between m_1 and m_2. This critical value of the Jacobian constant is denoted by C_2 since the figure-eight zero velocity curve intersects the x axis at the second collinear equilibrium point. As the value of the Jacobian constant decreases from $C_2 = 4.25$, the size of the opening at $x = 0$ increases. The next critical value of the Jacobian constant is reached when the singular points appear at $x_3 = x_1 = 1.198406$ giving $C_3 = C_1 = 3.706796$. If the value of the Jacobian constant is between C_3 and $C_4 = 3$ the zero velocity curves are symmetrically located above and below the x axis (without intersection with the x axis) and enclose only the triangular libration points.

The size of the opening of the zero velocity curve at $x = 0$ when $C_1 < C < C_2$ can be evaluated as follows.

The Jacobian constant is defined by the Jacobian integral which connects the potential function (Ω) of the restricted problem with the velocity of the third body (v):

$$v^2 = 2\Omega - C .$$

Here the function Ω is given by

$$\Omega = \frac{1}{2} [(1-\mu) r_1^2 + \mu r_2^2] + \frac{1-\mu}{r_1} + \frac{\mu}{r_2} ,$$

where r_1 and r_2 are the distances between the primaries and the third body.

The curves of zero velocity are given by $C = 2\Omega$. This equation for $\mu = 1/2$ and $x = 0$ gives

$$C = \frac{1}{4} + y_o^2 + \frac{2}{\sqrt{y_o^2 + \frac{1}{4}}} ,$$

where $\pm y_o$ are the ordinates of the points of intersections of the curves of zero velocity with the y axis. The size of the opening is $2y_o$. To determine the value of y_o for a given value of C, the above relation between C and y_o is to be solved which can be reduced to the solution of a cubic equation when

$$Y_o = \sqrt{y_o^2 + \frac{1}{4}}$$

is introduced as the new variable. (The physical meaning of Y_o is the distance between the primaries and the intersection of the curve of zero velocity with the y axis.) In this way the relation between the new variable Y_o and C becomes

$$C = Y_o^2 + \frac{2}{Y_o} \ ,$$

or

$$Y_o^3 - CY_o + 2 = 0 \ .$$

When $y_o = 0$, $Y_o = 1/2$ and $C = C_2 = 4.25$ and when $C = C_1 = C_3$, the opening becomes $2 \ y_o = 0.653$.

For small opening the relation between the Jacobian constant and the opening becomes

$$y_o = \sqrt{\frac{4.25 - C}{7}} \ .$$

The previously mentioned work by Henon (1983) presents results for $\mu = 1/2$, C = 4.75, 4.25, 3.75 and 3.25. (Note that Henon's definiton of the Jacobian constant (C_H) is related to our value of C by

$$C = C_H + \mu(1-\mu) \ ,$$

or $C = C_H + 0.25$.) The results show quasi-periodic behavior for C = 4.75 when the perturbation of the other primary (m_2) is small and the motion is restricted to the vicinity of m_1 since $C > C_2$ and the third particle is bounded by the curve of zero velocity around m_1. For the other values of the Jacobian constant Henon (1983) found chaotic motion even for C = 4.25 when the curve of zero velocity is not open and no communication is allowed between m_1 and m_2. The chaoticity in this case can be contributed to the strong perturbing effect of the other primary. For smaller values of the Jacobian constant, C = 3.75 and C = 3.25 the motion is strongly chaotic as shown by the corresponding surface of section plots.

Broucke's (1990) periodic orbit for C = 4.19, $\mu = 1/2$ shows an opening of $y_o \cong$ 0.1 of the curve of zero velocity. The orbit performs 5 revolutions around m_1, and then it moves to the vicinity of m_2 where it again makes 5 revolutions (in the synodic system) and it keeps repeating this periodic behavior. The previously mentioned results by Egorov (1957) and others, published during the period of 1957 to 1964 are similar to Broucke's 1990 results but with m = 0.0124 corresponding to the Earth and Moon as primaries.

Some recent results by Érdi, Smith and Szebehely (1990) might be mentioned concerning the motion for $\mu = 1/2$, when the opening of the curve of zero velocity is very small, i.e. for values of the Jacobian constant in the region $4 \le C \le 4.2491$. These results will be published in the near future in considerable more details but the preliminary result might be mentioned here, according to which small openings require long time for the third body to leave the vicinity of one of the primaries to move to the other. The motion is chaotic according to preliminary evaluations. The dynamical evolution might be described as follows. The third body performs a perturbed (chaotic) two-body motion around one of the primaries at the beginning. Sooner or later the location of the apo-apsis of the orbit will be close to the opening. The apocentric velocity is low (in the sidereal sense) and it will also be low in the synodic system near the opening, where the center of rotation is located. At this time the perturbing primary might be able to capture the third body which will pass through the opening and will start its motion around this primary. The phase and period relation between the synodic system and the motin of the third body is critical concerning the time it takes to pass through the opening.

An interesting, numerically produced effect is mentioned at this point, calling attention to the importance of careful control of numerical accuracy during such experiments.

Using $C = 4.2491$ and $\mu = 1/2$, the opening is very small ($y_0 = 0.0113$) which requires a long time for passing from the vicinity of one primary to the other. When this computer experiment was performed, the change from one primary to the other did not take place, even after an hour or so computer time. When the computation was stopped it was found that the error accumulation resulted in a change of the Jacobian constant from its initial value of $C = 4.2491$ to C slightly larger than 4.25. This value of the Jacobian constant, of course, does not allow penetration of the neighborhood of the other primary. So the expected switch of motion from one primary to the other did not occur because of numerical error accumulation.

CONCLUDING REMARKS

This lecture is concluded recalling some pre-Newtonian ideas concerning predictions. Without knowing the basic physical or dynamical principles and without understanding the basic mechanics, predictions were made for a short time following the behavior pattern. For chaotic systems we seem to be in a similar position since predictability falls off with time, i.e. meaningful predictions have limitations.

Various fields of science have recognized and their contributors have admitted the limitations of predictability long before the representatives of "Newtonian" dynamics and workers in celestial mechanics accepted such fundamental limitations. Celestial mechanics, inspite of Newton's announcement of its limitations, has been considered one of the successful areas of Laplace's demon's activities and playground. Present findings of chaoticity in the Solar System (Hyperion and the Kirkwood gaps, Wisdom, 1982), the still unsolved problem of its stability and the limits of predictability (10^8 years) suggest that Laplace's demon should be exorcised from its last stronghold of celestial mechanics which is, after all, based on the non-integrable gravitational problem of $n \geq 3$ bodies.

J. Ford's metaphore (1978) concerning integrability in dynamics may be of interest at this point. According to this, integrable dynamical systems correspond to integers on the real line and chaotic systems are as numerous as irrational numbers.

Prigogine (1980) associates unlimited predictability with "the founding myth of classical science."

References

Arenstorf, R.F., "Existence of Periodic Solutions Passing Near Both Masses of the Restricted Three-Body Problem," Am. Inst. of Aeronautics and Astronautics Journal, Vol. 1, p. 238, 1963.
Broucke, R., "Recherches d'orbites périodiques dans le problème restreint plan (système Terre-Lune), Univ. Louvain, 1962.
Broucke, R., Private communication, 1990.
Contopoulos, G., "On the Existence of a Third Integral of Motion," Astron. J., Vol. 68, p. 1, 1963.
Davidson, M.C., "Numerical Examples of Transition Orbits in the Restricted Three-body Problems." Astronautica Acta, Vol. X., pp. 308-319, 1964.

Egorov, W.A., "Certain Problems of Moon Flight Dynamics," Usp. Fiz. Nauk, Vol. 63, p. 73, 1957.

Erdi, B., Private communication, 1990.

Ford, J., "A Picture Book of Stochasticity," Am. Inst. of Physics, pp. 121-146, 1978.

Gleick, J., "Chaos," Viking Penguin Inc., New York, 1987.

Grassberger, P., "Toward a Quantitative Theory of Self-Generated Complexity," International Journal of Theoretical Physics, Vol. 25, No. 9, pp. 907-938, 1986.

Hadamard, J., J. Math. Pure and Applied, Vol. 4, pp. 27-73, 1898.

Helleman, R.H.G., (Editor), "Nonlinear Dynamics," New York Academy of Sciences, Vol. 357, 1980.

Henon, M., "Numerical Exploration of Hamiltonian Systems," in Chaotic Behavior of Deterministic Systems" (Editors: G. Iooss, R.H.G. Helleman and R. Stora), North Holland Publ. Co., Amsterdam, pp. 54-170, 1983.

Iooss, G. and D.D. Joseph, "Elementary Stability and Bifurcation Theory," Springer Publ., New York,) 1980.

Ioos, R., R.H.G. Helleman and R. Stora (Editors), "Chaotic Behavior of Deterministic Systems," North-Holland Publ. Co., Amsterdam, 1983.

Lagrange, J.L., "Miscellanea Taurinensia ou Melanges de Turin." Vol. II, 1760.

Lyapunov, A.A., "The General Problem of the Stability of Motion," Commun. Math. Soc. Krakow, Vol. 2, p. 1, 1892.

Lorenz, E., "Deterministic Nonperiodic Flow," Journal of the Atmospheric Sciences, Vol. 20, pp. 130-141, 1963.

Maddox, J., "Complicated Measures of Complexity," Nature, Vol. 344, p. 705, 1990.

Miller, R.H., "A Horror Story about Integration Methods" to appear in Celestial Mechanics, 1990.

Poincaré, H., "Les Methodes Nouvelles de la Mecanique Celeste," Gauthier - Villars, Paris, Vol. I, 1892; Vol. II, 1893; Vol. III, 1899.

Poincaré, H., "Science et Méthode," Flammardon, Paris, 1908.

Prigogine, I., "From Being to Becoming," W.H. Freeman Co., San Francisco, 1980.

Prigogine, I. and I. Stengers, "Order out of Chaos," Bantam Books, New York, 1984. Original "La nouvelle alliance," Gallimard, Paris, 1979.

Reynolds, O., Phil. Trans. Roy. Soc. London, Vol. 174, p. 935, 1883.

Ruelle, D., "Deterministic Chaos," Proc. Royal Soc. London, Vol. 427, pp. 241-248, 1990.

Smith, R., Private communication, 1990.

Stewart, I., "Does God Play Dice?" Penguin Books, London, 1989.

Swinney, H.L., and J.P. Gollub, "The Transition to Turbulence," Physics Today, Vol. 31, No. 8, pp. 41-49, 1978.

Szebehely, V., "Theory of Orbits." Academic Press, New York, 1967.

Szebehely, V., "The Problem of Three Bodies," Proc. NATO Advanced Study Institute in Dynamical Astronomy, D. Reidel Co. Holland, 1973.

Thompson, J.M.T. and H.B. Stewart, "Non-linear Dynamics and Chaos, " J. Wiley & Sons, Wichester, 1986.

Thüring, B., "Zwei spezieller Mondeinfang Bahnen in der Raumfahrt um Erde und Mond," Astronautica Acta, Vol. V., pp. 241-250, 1959.

Whittaker, E.T., "A Treatise on the Analytical Dynamics of Particles and Rigid Bodies," Cambridge University Press, 1904.

Wisdom, J., "The Origin of the Kirkwood Gaps," Astronomical Journal, Vol. 87, pp. 577-593, 1982.

PREDICTABILITY, STABILITY AND CHAOS IN DYNAMICAL SYSTEMS

Christian Marchal

D.E.S. - Onera
92320 Chatillon, France

ABSTRACT

Progress in the theory of stability is presented in the particular example of the Lagrangian motions of the three-body problem and is then generalized.

The notion of predictability is discussed and, surprisingly, in some cases the chaotic motions allow better predictions than the regular motions.

The Arnold diffusion conjecture gives a general picture of Hamiltonian dynamical systems, a picture in which chaotic motions have a major part.

All these recent advances have renewed the picture of Physics.

1. INTRODUCTION

The analysis of dynamical systems, and especially n-body dynamical systems, has known considerable improvements in recent years with a rare conjunction of numerical, analytical and topological progress. The "chaotic motions", the "strange attractors", the "fractals", the "Arnold tori", the "Arnold diffusion", the "effective stability" belong now to the usual vocabulary of mathematicians and mechanicians.

However, because of this progress, the present situation is very confusing. People are fighting about the meaning of such words as "Stability", "determination", "chaos", "temporary chaotic motions", etc... and a clear definition of a suitable vocabulary is necessary, with the confrontation of mathematical, mechanical and physical points of view.

Predictability depends very much on the considered scale. Many phenomena are apparently deterministic at a large scale (e.g. the tides at the decimetric scale) and become chaotic and "stochastic" at a smaller scale (e.g. the tides at the centimetric scale). This "chaoticity of small amplitude" plays a major part in the organization of the Universe... and in the present confusion of mind!

Predictability, Stability, and Chaos in N-Body Dynamical Systems
Edited by A.E. Roy, Plenum Press, New York, 1991

2. DYNAMICAL SYSTEMS

The analysis of a natural phenomenon is greatly simplified by the discovery of its main "laws" and by the construction of a corresponding suitable "mathematical model" even if, of course, this model has only a limited accuracy and if successive approximations are generally needed.

Thus the main laws of planetary motions are the law of inertia and the law of universal attraction. The corresponding mathematical model is thus of either punctual or spherical masses moving in their mutual gravitational fields. The successive approximations are that of oblate planets, tidal effects, relativistic effects, etc...

The dynamical systems correspond to mathematical models with a <u>finite</u> number of parameters x_1, x_2,...x_n :

$$\vec{X} = \text{state vector} = (x_1, x_2, ...,x_n) \tag{1}$$

The number of necessary parameters is related to the expression of the equation of evolution with respect to the time t : *

$$d\vec{X}/dt = f(\vec{X},t) \tag{2}$$

For instance in the planetary motions the main laws give the accelerations of planets and thus the state vector $\vec{X}$ must contain not only the positions of planets but also their velocities (6n parameters for the n-body problem).

There are of course many models with partial differential equations instead of ordinary differential equations (air or water flows, Navier-Stokes equations, etc...), but these distributed parameter systems will not be considered in this paper.

We will especially consider the three following types of dynamical systems :

(A) The autonomous dynamical systems:

$$d\vec{X}/dt = f(\vec{X}) \tag{3}$$

(B) The periodic dynamical systems (period T) :
$$d\vec{X}/dt = f(\vec{X},t) \quad ; \quad f(\vec{X},t) \equiv f(\vec{X},t+T) \tag{4}$$

(C) The mappings, i.e. the discrete version of the above systems:
$$\vec{X}_{n+1} = F(\vec{X}_n) \tag{5}$$

With the Poincaré section method these three types are essentially the same, for instance for the second type we obtain:

$$\left.\begin{array}{l}\text{Let us put } \vec{X}_n = \vec{X}(nT), \text{ hence along any given solution } \vec{X}(t) \\ \text{we can write: } \vec{X}_{n+1} = F(\vec{X}_n), \text{ this mapping } F(\vec{X}_n) \\ \text{corresponding to the integration of (4) over one period.}\end{array}\right\} (6)$$

----

*Or sometimes with respect to another suitable parameter of description

3. THE STABILITY

The word "stability" has many different meanings.

A trajectory $\vec{X}(t)$ is "Lagrange stable for the future" after some initial time t if and only if it remains bounded after t :

$$\rceil \text{ B bounded, such that: } t \geq t_o \text{ implies } \| \vec{X}(t) \| \leq B \qquad (7)$$

A trajectory $\vec{X}(t)$ is "Poisson stable for the future" if and only if it always come back in any neighbourhood of its past positions :

$$\lor \ \varepsilon > 0 \ ; \quad \forall (t_1, t_2) \ \rceil \ t_3 / \ \{ t_3 > t_2 \ ; \ \| \vec{X}(t_3) - \vec{X}(t_1) \| < \varepsilon \ \} \ (8)$$

The Liapounov stability is very famous, it is related to the behaviour of solutions in the vicinity of the solution $\vec{X}(t)$ of interest.

Let us call $\delta\vec{X}(t)$ the difference between the solution $\vec{X}(t)$ and an arbitrary neighbouring solution. This solution $\vec{X}(t)$ will be Liapounov stable after t_o if and only if :

$$\lor \ \varepsilon > 0 \ \rceil \ \delta > 0 \ / \ \{ \| \delta\vec{X}_o \| \ < \delta \ ; \ t > t_o \ \} \Rightarrow \| \delta\vec{X}(T) \| \ < \varepsilon \qquad (9)$$

In most cases simple continuity conditions make the Liapounov stability independent of t_o.

The periodic and the quasi-periodic solutions are both Lagrange and Poisson stable (for the past as well as for the future) but they are not necessarily Liapounov stable.

There are also the orbital or Poincaré stability, the Hill stability. the structural stability, the asymptotic stability, etc...

The difficulty of the demonstration of the Liapounov stability and the existence of the extremely slow phenomenon called "Arnold diffusion" has led to a new type of stability the "effective stability" (Ref. 1,2). This stability appears when an escape from the vicinity of the trajectory of interest is not strictly impossible but requires extremely long durations.

The effective stability is related to the "all-order stability" of the next section.

4. THE STABILITY OF LAGRANGIAN MOTIONS

The Lagrangian motions will give many examples of stability and . instability of different types.

These Lagrangian motions (Figures 1 and 2) are classical solutions of the three-body problem, their stability has been the subject of many studies (Ref. 1-11).

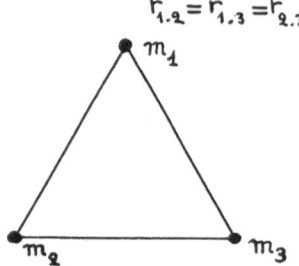

$r_{1.2} = r_{1.3} = r_{2.3}$

Fig. 1. The Lagrangian triangular
configuration.

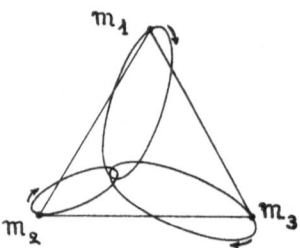

Fig. 2. An elliptic Lagrangian
motion.

The first-order study is sufficient for the demonstration of the exponential instability in most elliptic or circular Lagrangian motions (Figure 3) and especially in all cases where the largest of the three masses is less then 95.3% of the total mass.*

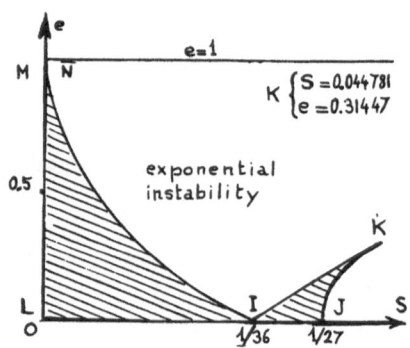

Fig. 3. First-order study of the
stability of Lagrangian
motions in terms of the
mass-ratio S and the eccen-
tricity e. $S = (m_1 m_2 + m_1 m_3 + m_2 m_3) / (m_1 + m_2 + m_3)^2$. The two shaded
regions correspond to the
"first-order stable cases" or
"critical cases". Equation
of the curve JK: $27S = 5 + e^2 - \sqrt{(16 - 8e^2 - 8e^4)}$.

In the two small remaining "critical zones" it is necessary to go beyond the first-order analysis and the following results have been obtained.

_____

*
Furthermore all "Eulerian motions" (similar to the Lagrangian motions, but using a collinear central configuration instead of a triangular one) are unstable.

4.1 Circular Restricted Three-Body Problem

In this case the three masses have a circular motion (hence $e = 0$ in Figure 3) and the small "third mass", m_3, is infinitesimal.

If, as usual, the angular velocity of the "primaries" m_1 and m_2 is chosen as unity the eigenpulsations ω_1, ω_2, ω_3 of motions in the vicinity of the Lagrangian motion of interest are the following ($\omega_1/2\pi$, $\omega_2/2\pi$, $\omega_3/2\pi$ being the corresponding eigenfrequencies):

$$S = m_1 m_2/(m_1 + m_2)^2$$

$$A = (1 - 27S)^{\frac{1}{2}} \tag{10}$$

$$\left.\begin{array}{l} \omega_1 = \{(1+A)/2\}^{\frac{1}{2}} \\[4pt] \omega_2 = -\{(1-A)/2\}^{\frac{1}{2}} \leqslant 0 \end{array}\right\} \quad \begin{array}{l} \text{for "in-plane" motions} \\ \omega_1^2 + \omega_2^2 = 1 \end{array}$$

$$\left.\begin{array}{l} \omega_3 = 1 \end{array}\right\} \quad \begin{array}{l} \text{for "out-of-plane"} \\ \text{motions.} \end{array}$$

Except for the three main resonances $\omega_2 = 0$; $\omega_1 + \omega_2 = 0$ and $\omega_1 + 2\omega_2 = 0$ the corresponding "normal form" of the Hamiltonian H of neighbouring motions is the following (and its three first terms define the signs of the three eigenpulsations ω_1, ω_2, ω_3):

$$\left.\begin{array}{l} H = 0.5\,(\omega_1 I_1 + \omega_2 I_2 + \omega_3 I_3) + D_{11} I_1^2 + D_{12} I_1 I_2 + D_{22} I_2^2 + D_{13} I_1 I_3 \\[4pt] \quad + D_{23} I_2 I_3 + D_{33} I_3^2 + E(p_1 + iq_1)(p_2 + iq_2)^3 + \overline{E}(p_1 - iq_1)(p_2 - iq_2)^3 + \\[4pt] \quad + O\{p_1^5, q_1^5, p_2^5, q_2^5, p_3^5, q_3^5\} \end{array}\right\} \tag{11}$$

with
$$\left\{\begin{array}{l} I_k = p_k^2 + q_k^2 \ ; \quad k = \{1,2,3\} \\[4pt] p_1, q_1, p_2, q_2 : \text{parameters for "in-plane motions";} \\[4pt] p_3, q_3 \qquad\quad : \text{parameters for "out-of-plane motions";} \end{array}\right.$$

and with (Ref. 6, 10) :

$$\left.\begin{array}{l} D_{11} = \omega_2^2\,(81 - 696\,\omega_1^2 + 124\,\omega_1^4)/576(2\omega_1^2 - 1)^2(1 - 5\omega_1^2) \\[4pt] D_{12} = \omega_1\,\omega_2(43 + 64\omega_1^2\,\omega_2^2)/24(2\omega_1^2 - 1)(2\omega_2^2 - 1)(1 - 5\omega_1^2)(1 - 5\omega_2^2) \\[4pt] D_{22} = \omega_1^2\,(81 - 696\,\omega_2^2 + 124\,\omega_2^4)/576\,(2\omega_2^2 - 1)^2(1 - 5\omega_2^2) \\[4pt] D_{13} = 2\omega_1\,\omega_2^2/3\,(4 - \omega_1^2)(2\omega_1^2 - 1) \qquad \text{Notice the symmetry of} \\[4pt] D_{23} = 2\omega_2\,\omega_1^2/3\,(4 - \omega_2^2)(2\omega_2^2 - 1) \qquad \text{subscripts 1 and 2} \\[4pt] D_{33} = -\omega_1^2\,\omega_2^2\,/(144 + 12\omega_1^2\,\omega_2^2) \\[4pt] E, \overline{E}\ \text{can always be reduced to zero except when } \omega_1 + 3\omega_2 = 0 \\[4pt] \text{case for which } E = 0.41323... - 0.37806\,i \end{array}\right\} \tag{12}$$

There are thus four main cases of resonance that correspond to the following if we assume $m_1 \geqslant m_2$:

Resonance		S	$m_2/(m_1+m_2)$	
$\omega_2 = 0$		0	0	These resonances destabi-
$\omega_1 + 3\omega_2 = 0$		1/75	0.013516	lize the motion
$\omega_1 + 2\omega_2 = 0$		16/675	0.024294	
$\omega_1 + \omega_2 = 0$		1/27	0.038521	The motion remains stable in spite of the resonance

$$(13$$

For the other mass ratios there is a major difference between the plane circular restricted three-body problem and the three-dimensional restricted circular three-body problem.

In the planar case, the Hamiltonian H has only two degrees of freedom and the "Arnold tori" enclose bounded parts of phase space, we can thus have the certainty of Liapounov stability out of the main resonances (Figure 4).

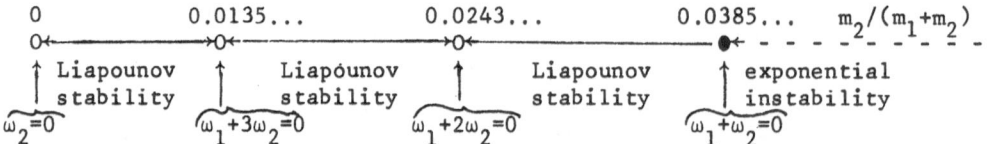

Fig. 4. Plane circular restricted three-body problem. Stability of the Lagrangian motion in terms of the primaries mass-ratio $m_2/(m_1+m_2)$. Unstable positive resonances for $\omega_2 = 0$; $\omega_1+3\omega_2 = 0$ and $\omega_1+2\omega_2 = 0$. Stable positive resonance for $\omega_1+\omega_2 = 0$.

In the three-dimensional case, the Hamiltonian H has three degrees of freedom and the Arnold tori cannot enclose bounded parts of phase space.

The integral of motion H and the quasi-integral (Ref.11, p206), the main term of which is $D_{11}I_1^2 + D_{12}I_1I_2 + D_{22}I_2^2 + D_{13}I_1I_3 + D_{23}I_2I_3 + D_{33}I_3^2$ allow us to draw Figure 5 and to conclude in many cases to the "all-order stability": only the very slow Arnold diffusion can then destroy the stability and this is the first step of the very long "effective stability".

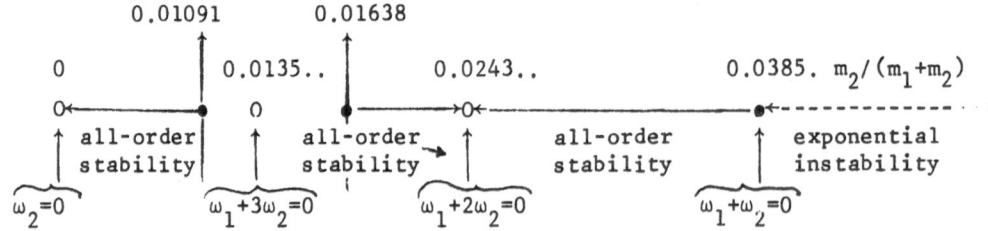

Fig. 5. Three-dimensional circular restricted three-body problem. Stability of Lagrangian motions in terms of the primaries mass-ratio $m_2/(m_1+m_2)$. There are three segments of "all-order stability" as well as the three limit cases $m_2/(m_1+m_2) = 0.0109..,0.0164..,$ 0.0385... .

The three resonances $\omega_2 = 0$; $\omega_1 + 3\omega_2 = 0$ and $\omega_1 + 2\omega_2 = 0$ destabilize the motion.

In the small range $0.0109136 < m_2/m_1 + m_2) < 0.0163767$ there is generally all-order stability but the "positive resonances" $a_1\omega_1 + a_2\omega_2 + a_3\omega_3 = 0$ with a_1, a_2, $a_3 \in \mathbb{N}$ can sometimes destabilize the motion.

The ratio $m_2/(m_1 + m_2) = 0.0109136$ is famous for a first Birkhoff invariant equal to zero and a non-zero second Birkhoff invariant.

4.2 Non Restricted and/or Elliptic Cases

Notice first a particularity of non-restricted cases. In the vicinity of a Lagrangian motion there exist always other Lagrangian motions with <u>different</u> periods. These Lagrangian motions cannot remain together and the Liapounov stability is impossible. We have then to consider an orbital or Poincaré stability.

In these more complex cases the Hamiltonian of neighbouring motions has a difficult normal form, the fourth-order terms of which have not been obtained. Thus the analysis is restricted to the information given by the eigenpulsations and by the general pulsation $\omega_o = 1$ of the elliptic cases (this general pulsation meaning that the Hamiltonian has the periodicity $2\pi/\omega_o$ that is here 2π).

In the "circular general case" (i.e. three non-infinitesimal masses rotating on circular orbits), the eigenpulsations ω_1, ω_2, ω_3 are given by (10) with the mass-ratio S having now the general expression already used in Figure 3 :

$$S = (m_1 m_2 + m_1 m_3 + m_2 m_3)/(m_1 + m_2 + m_3)^2 \qquad (14)$$

This case has one more eigenpulsation:

$$\omega_4 = 1. \qquad (15)$$

It has thus also one more degree of freedom (remaining after the reductions given by the usual integrals of motion).

The analysis of the circular general case in term of the mass-ratio S gives then the following (Ref. 10, p65)

$$S > 1/27 \quad \Rightarrow \quad \text{exponential instability}$$

$$S \leqslant 1/27 \quad \Rightarrow \quad \text{critical case with the eigenpulsations}$$

ω_1, ω_2, ω_3 (for 3-D motions) and ω_4:

$$\sqrt{0.5} \leqslant \omega_1 \leqslant 1 = \omega_3 = \omega_4 \; ; \quad -\sqrt{0.5} \leqslant \omega_2 \leqslant 0 \; ; \; \omega_1^2 + \omega_2^2 = 1 \; ; \; 4\omega_1^2 \omega_2^2 = 27S \qquad (16)$$

The resonances that are dangerous for the stability are the positive resonances (Ref.11, p208):

$$\vec{a} . \vec{\omega} = 0 , \text{ with } \vec{\omega} = (\omega_1, \omega_2, \omega_3, \omega_4); \; \vec{a} = (a_1, a_2, a_3, a) \neq 0 ; \text{ all } a_j \in \mathbb{N} \qquad (17)$$

In the general case (17) is impossible and the system has the "all-order stability".

If (17) is possible the system can be unstable but resonances of large orders are generally harmless.

The resonance $\omega_2 = 0$ leads to the instability.

The third-order resonances $\omega_1 + 2\omega_2 = 0$ and $2\omega_2 + \omega_4 = 0$ (i.e. S = 16/675 and S = 1/36) lead almost always to the instability.

The second and fourth-order resonances $\omega_1 + \omega_2 = 0$; $\omega_1 + 3\omega_2 = 0$; $3\omega_2 + \omega_4 = 0$ (i.e. S = 1/27 ; S = 1/75 ; S = 32/2187) are potentially dangerous: a complex inequality on the masses m_1, m_2, m_3 must be satisfied in order to avoid the instability.

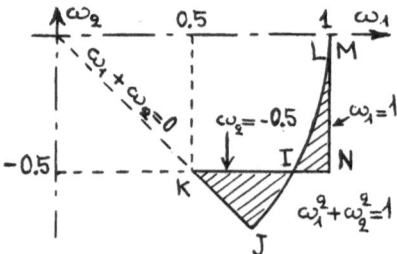

Fig. 6. The two domains of the
point w_1, w_2.

In the elliptic cases, as this of Figure 2, the analysis is similar.

The first-order analysis is presented in Figure 3 in terms of the mass-ratio S and the eccentricity e of the elliptic orbits.

In the critical cases the eigenpulsations ω_1 and ω_2 are functions of S and e, the eigenpulsation ω_3 for three-dimensional motions remains equal to one and the eigenpulsation ω_4 disappears, the general pulsation $\omega_0 = 1$ takes its place ($\omega_0 = 1$ means that the Hamiltonian of neighbouring motions is periodic in terms of the parameter of description, with the period 2π).

$\omega_1(S,e)$ and $\omega_2(S,e)$ can be developed in terms of S and e in the vicinity of S = 0 and/or of e = 0 (Ref. 10 page 64); they can also be computed numerically. The point (ω_1, ω_2) always corresponds to one of the two shaded domains of Figure 6, the letters IJKLMN giving the correspondence with Figure 3.

The eigenpulsations ω_1, ω_2, ω_3 and the general pulsation ω have slightly different functions and the positive resonances, the only resonances that are dangerous for the stability, are given by a slight modification of (17):

$$\vec{a}.\vec{\omega} = 0, \text{ with } \vec{\omega} = (\omega_o,\omega_1,\omega_2,\omega_3); \ \vec{a} = (a_o,a_1,a_2,a_3) \neq 0; \ a_o \in \mathbf{Z};$$
$$a_{1,2,3} \in \mathbb{N}$$
(18)

Since $\omega_o = \omega_3 = 1$ there is of course the positive resonance $\omega_3 - \omega_o = 0$ in the three-dimensional case (while the correspondonding resonance $\omega_3 - \omega_4 = 0$ of (17) is not positive). Hence in the elliptic three-dimensional cases a large part of the critical zones of Figure 3 can be destabilized by the fourth-order positive resonance $\omega_3 - \omega_o = 0$; but this analysis is difficult.

The limits JK, KI, IN, NM and ML of Figures 3, 6, 7 correspond to positive resonances of order one or two and there are the following (almost always destabilizing) positive resonances of order three:

$$3\omega_1 = 2 \qquad \text{curve FG}$$
$$2\omega_1 + \omega = 1 \qquad " \qquad \text{CE}$$
$$\omega_1 + 2\omega_2 = 0 \qquad " \qquad \text{BN}$$
$$3\omega_2 = -1 \qquad " \qquad \text{AN}$$
$$3\omega_2 = -2 \qquad " \qquad \text{DF}$$

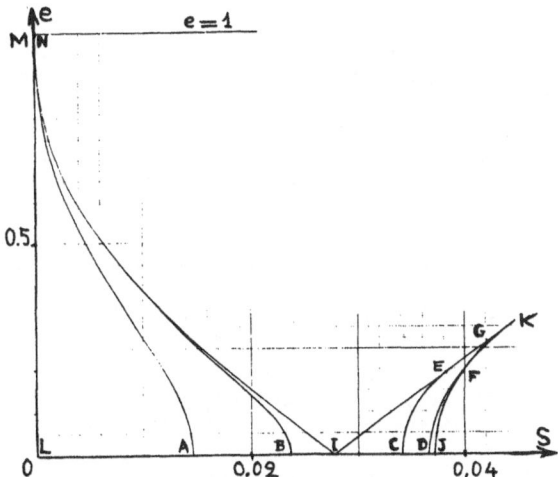

Fig. 7. (From Ref. 10). Elliptic Lagrangian motions. The zones IJK and ILMN of critical stability and the curves AN, BN, CE, DF, FG of third-order positive resonances in terms of the mass-ratio S and the eccentricity e. $S = (m_1m_2 + m_1m_3 + m_2m_3)/(m_1 + m_2 + m_3)^2$.

5. ANALYTIC HAMILTONIAN SYSTEMS

General considerations on the stability and the instability of periodic solutions

The Lagrangian motions of the previous section provide almost all necessary examples.

The difference between the Liapounov stability and the orbital or Poincaré stability is that in the former all parameters are considered while in the latter the analysis is concentrated on the stability of non-ignorable parameters and one or several ignorable parameters are "ignored" (the time in most usual cases of Poincaré stability, the time and the longitude of periastre in the elliptic general case of the previous section).

Notice that in case of Poincaré stability the "ignored parameters" have, at worst, a linear escape.

The motions in a small vicinity of the solution of interest can be analysed with suitable Hamiltonian systems. The "normal form", to some order, of these Hamiltonians is the simplest real Hamiltonian obtained after suitable canonical transformations.

In a critical case, when the eigenpulsations $\omega_1, \omega_2, \omega_3 \ldots$ are different, the second-order terms of the normal form are the following:

$$
\left.
\begin{aligned}
H = 0.5 \sum_{j=1}^{n} \omega_j \ (p_j^2 + q_j^2) + \text{higher order terms} \\
\text{with} \qquad n = \text{number of degrees of freedom} \\
p_j, q_j : \text{small Hamiltonian parameters (that becomes zero} \\
\text{for the periodic solution of interest).}
\end{aligned}
\right\} \quad (19)
$$

The first-order analysis of neighbouring motions gives the absolute values of the eigenpulsations and this expression (19) defines their signs.

The eigenpulsations are related to the eigenfrequencies f_1, f_2, f_3 by the usual relations $f_j = \omega_j/2\pi$ and their signs are obviously an essential element of the "positive resonances", the only resonances that are dangerous for the stability (Ref.11 p.208).

These positive resonances generalize (17) and (18) that is:

A) for autonomous Hamiltonians $H(\vec{p}, \vec{q})$ of neighbouring motions:

$$
\vec{a}.\vec{\omega} = 0; \text{ with } \vec{\omega} = (\omega_1, \omega_2, \ldots \omega_n); \ \vec{a} = (a_1, a_2, \ldots a_n) \neq 0; \text{ all } a_j \in \mathbb{N} \quad (20)
$$

Notice an obvious particular case: if all ω_j have the same sign the Hamiltonian H is locally either minimum (all ω_j positive) or maximum and since autonomous Hamiltonians are integrals of motions the solution of interest is obviously stable (Lejeune-Dirichlet theorem). This is in agreement with the absence of positive resonance when all ω_j have the same sign.

B) For periodic Hamiltonians $H(\vec{p}, \vec{q}, t)$ of neighbouring motions.

A periodic Hamiltonian is such that $H(\vec{p},\vec{q},t) \equiv H(\vec{p},\vec{q},t + T)$ and the normal form of Hamiltonian of neighbouring motion has generally the period T of the periodic solution of interest.

The positive resonances become then:

$$\vec{a}\,\vec{\omega} = 0, \text{ with } \vec{\omega} = (\omega_o,\omega_1,\ldots,\omega_n); \ \vec{a} = (a_o,a_1,\ldots a_n) \neq 0 \ ; \ a_o \in \mathbb{Z} \ ; \\ a_{1,2,\ldots n} \in \mathbb{N} \quad (21)$$

and with $\omega_o = 2\pi/T =$ general pulsation.

The order of a resonance is equal to the degree of the first corresponding term in the normal form of the Hamiltonian of neighbouring motions, for instance the E and $\overline{E}$ terms in (11) that correspond to the fourth-order resonance $\omega_1 + 3\omega_2 = 0*$. For positive resonances this order is generally $a_1 + a_2 + \ldots + a_n$ (if the set of integers a_o, a_1, $\ldots$ a_n is irreducible) but it can be larger if some terms are missing in the normal form. Thus the resonance $\omega_3 - \omega_o = 0$ of elliptic three-dimensional Lagrangian motions is only of fourth-order.**

The positive resonances of order two are generally at the limit of the zones of stability (Figure 6); they are very often destabilizing.

The positive resonances of order three are the most dangerous; they are destabilizing except when a suitable (and complicated) equality is satisfied.

The positive resonances of order four are also very dangerous; they are destabilizing when a suitable inequality is satisfied.

The positive resonances of order five, six, etc... are much less dangerous; the destabilization requires the satisfaction of one or several suitable equalities.

With a suitable ordinary "distance" δ to the solution of interest it is possible to give some quantitative elements.

The first-order instability is an "exponential instability". The escape velocity $d\delta/dt$ is of the order of δ and the time necessary to go from some infinitely small distance ε to some bounded distance A is of the order of Log $(1/\varepsilon)$.

A destabilizing resonance of order n leads to an escape velocity $d\delta/dt$ of the order of δ^m (with usually m = n - 1), it is a "power-m escape" and the duration between the distances ε and A is at least of the order $\varepsilon^{(1-m)}$, that is very large for large m and small ε.

The "all-order stability" gives either the true Liapounov stability or the true Poincaré stability (if some ignorable parameters are "ignored") or at least the Arnold diffusion with for small δ:

----

* the agreement on this definition of the order of a resonance is recent, the order was one unity less in Ref.11.

** The order of a resonance is at least two, it is even at least three if $a_o \neq 0$ and in the above case of $\omega_3 - \omega_o = 0$ the order is even because of the out-of-plane symmetry of the problem.

$$\forall \ m \ : \quad d\delta/dt = O(\delta^m) \quad\quad\quad\quad (22)$$

For instance, $d\delta/dt = O(\exp \{-1/\sqrt{\delta} \})$ in the Arnold example (Ref. 12).

The usual integrals of motion and the quasi-integrals (Ref.11 p.206) are very often sufficient for the deomonstration of the "all-order stability". For instance it is sufficient that the pulsations have no positive resonance, but the discrimination between the true stability and the Arnold diffusion seems very difficult.

This discrimination is sometimes purely theoretical: what is the physical meaning of an instability that requires millions of years to appear?

This situation has inspired the concept of "effective stability" to A. Giorgilli, C. Simo and their friends (Ref. 1-2), precisely in a problem of Lagrangian motions. They have delimited about the Sun-Jupiter Lagrangian point L_4 a small zone of stability "for at least twenty billions of years".... that is much more than the age of the Solar System.

This classification of exponential instability, power-m escape, Arnold diffusion and true (non-asymptotic) stability can be extended from the analytic Hamiltonian systems to all other analytic conservative systems. It can even be extended to analytic non-conservative systems if we add to the classification the two following types of asymptotic stability: the power-m stability (as in $dx/dt = -x^m$; m odd) and the exponential stability (as in $dx/dt = -x$).

If we compare the evolution for increasing time and the evolution for decreasing time it seems that, in an analytic conservative system, a periodic solution has always the same type of stability or instability for the past and for the future.

6. DETERMINISM, PREDICTABILITY, CHAOS

The concept of determinism was for centuries opposed to chaos and associated with stability and predictability. However, the definitions of mathematicians and physicists have recently diverged.

For a methematician a system such as the dynamical systems (2)-(5) is deterministic if "the conditions of existence and uniqueness of solutions are satisfied", for instance if in $d\vec{X}/dt = f (\vec{X},t)$ the function $f(\vec{X},t)$ is locally Lipschitz. This can be summarized by: "If we do again _exactly_ the same experiment we will obtain _exactly_ the same results".

This expression has of course no physical meaning because it is impossible to do twice exactly the same experiment and the physical expression of determinism can be "a phenomenon is deterministic if when doing again _almost exactly_ the same experiment we obtain _almost exactly_ the same results".

It is there that the notion of stability is essential.

We have seen in the previous section that in the vicinity of a solution with first-order instability the neighbouring solutions have an exponential divergence. The coefficients of these exponentials are called "Liapounov characteristic exponents" by the mathematicians and the physicists speak of "sensitivity to initial conditions".

The very fast increase of diverging exponentials is well known and the slightest initial difference, even at the quantum level, grows very rapidly to very large proportions. Each physical phenomenon has thus a "time of divergence" or an "horizon of prediction" (about some tenths of the inverse of its largest Liapounov characteristic exponent). Before the horizon deterministic predictions can be made and beyond the chaoticity dominates and statistical analysis gives the best results. For these large time-scales the choice of the initial conditions is as a pure random choice exactly as it happens in the game of head or tail.

It is a recent discovery, related to the progress of computers, that mathematically deterministic systems as the dynamical systems (2)-(5) can lead, even in very simple cases, to unstable phenomena and to chaotic motions (Ref.13-15).

Many "roads of chaos" have been discovered (Ref.16-18) and the chaotic motions as well as the "strange attractors" are now common features in most scientific and technical domains.*

Furthermore, as we will see in the next section, it must be emphasized that it is the existence of chaotic motions and strange attractors that give their importance to the statistical elements. Thus the temperature and the pressure are statistical properties of the motion of molecules for the physicists of the kinetic theory of gas and they are basic tools for the aerodynamicists.

7. A PREDICTABILITY IMPROVED BY THE CHAOS

What do we mean by predictability?

The mathematical predictability is a pure convention and is very artificial. When we "predict" that the sum 17 369 + 15 964 will be 33 333 we only apply some arithmetical rules of our decimal systems. The result is rigorous but it has few concrete meanings.**

The physical predictability always meets the questions of measurements, errors and accuracy. It is always confronted with the question of the certainty or even the probability of the result and we will need the following concepts.

----

*The mathematicians speak of "chaotic motions" when they consider conservative systems and "strange attractors" when they consider dissipative systems. These phenomena have in common a positive largest Liapounov characteristic exponent and the corresponding divergence of neighbouring solutions. Diplomatically the mathematicians and the physicists call all this "deterministic chaos".

** During the Cortina meeting Clause Froeschlé noted that accurate sums and especially differences can be very concrete: when they are on your bank account! But money itself is artificial.

7.1 The unique law of random

Emile Borel has written:

All laws of random are consequences of the unique following law:
"a remarkable event with too small a probability never happens".

What is a remarkable event? It is an event to which we think before it happens: No one will dare to buy the lottery ticket number 77 777 777, but the winner, 21 318 910 had the same probability to win.

For the events of the past (a scenario for the birth of the Earth...) a remarkable event is an event to which we think before finding its effects (in magnetism, in geology, etc...).

What is a "small probability" ? This depends on the phenomenon of interest and is related to the notion of "threshold of certainty".

7.2 The thresholds of certainty

In ordinary life many small probabilities are considered as negligible even in questions of life or death: for instance most people neglect the probability (a few 10^{-7}) of being killed in the traffic during the next weekend.

A prudent evaluation leads to the following thresholds of certainty:

In ordinary life: $1 - 10^{-8}$

In history of Earth: $1 - 10^{-30}$

In history of observable Universe: $1 - 10^{-200}$

7.3 Application of the unique law of random

All stochastic events can be expressed in terms of the unique law of random. For instance when we consider that the event E has a probability of 29% it means the following that correspond to the exclusion of remarkable events with too small a probability:

Over 10 000 tests the frequency of E will always be between 26% and 32%

Over 10^6 tests the frequency E will always be between 28,7% and 29.3% etc... } With the threshold of certainty of ordinary life

It is sometimes believed that 10^{-200} is so small that the threshold of certainty for the history of the observable Universe is meaningless (this allows hypotheses as "Life appears in the Universe by random") but this is not true and at this threshold, over 10^6 tests, a probability of 29% gives a frequency between 27.5% and 30.5%.

We can for instance write that an atom of Uranium has a probability of disintegration of 29% over a period Δ of about 7.4 billions of years. A microgram of Uranium contains about 2.5×10^{15} atoms and for all micrograms of Uranium of the observable Universe the proportion of disintegrated atoms during the last interval of time Δ is between 28.99997% and 29.00003%.

7.4 A definition of predictability

We can now define the predictability as the following property: "An event is predictable if the probability of its occurrence is above the threshold of certainty".

Of course the event of interest can be either individual or collective and in the latter case the existence of chaotic motions improve very much the possibility of predictions.

Consider for instance a closed vessel full of gas and a large crystal.

Let us define for these two objects a random procedure of choice of a cubic centimeter: C_1 in the vessel and C_2 in the crystal.

Let us now try to answer the following question: where is after one hour the centre of mass of the molecules of gas initially in C_1 and that of the atoms of the crystal initially in C_2?

The existence of the brownian chaotic motion allows a very accurate answer, even if the threshold of certainty is $1 - 10^{-200}$: the molecules initially in C_1 have their centre of mass in a very small zone about the centre of the inner volume of the vessel.

On the contrary the absence of chaotic motion in the crystal forbid us to improve the initial information with its large uncertainty.

Thus the sensitivity to initial conditions and the corresponding chaotic motions destabilize the individual motions but also stabilize the collective averaged elements that become independent from small scale initial conditions.

For instance, these statistical elements called "temperature", "pressure", "density", "velocity of the wind" can be considered to an excellent accuracy, as deterministic elements by the aerodynamists who can ignore phenomena at microscopic scale. Similarly, at a larger scale, the astronomers consider this averaged element called "centre of mass" of a planet and they can ignore the complexity of inner planetary motions.

The three major sources of errors in the study of a physical phenomenon are traditionally the errors on the model of this phenomena, the errors on initial conditions and the errors in numerical integrations. The existence of chaotic motions reduce very much the effects of the two last sources of errors on properly chosen averaged elements.

8.   THE ARNOLD DIFFUSION CONJECTURE AND TEMPORARY CHAOTIC MOTIONS

Many different models of chaotic motions and strange attractors have been presented and analysed (Ref.13-15), the fractal structure of strange attractors is now well known. On the other hand regular motions have several general properties, the best known, for Hamiltonian systems, being the structure in Arnold tori of quasi-periodic motions.

The existence of these tori was conjectured by Kolmogorov and demonstrated by Arnold (Ref.12) in the vicinity of periodic and "first-order stable" solutions.

These Arnold tori fill a set of positive measure in phase space but they are "nowhere dense" in non-integrable problems because of the abundance of resonances that forbid their formations: they have then a Cantor set structure looking like a fractal.

The "Arnold diffusion conjecture" (Ref.11, page 509) is applicable to autonomous (or periodic) analytic Hamiltonian systems when they have been simplified as much as possible using their integrals of motion and their eventual decomposability. This conjecture is the following:

In a part of phase-space with given values of the integrals of motion the Arnold tori can be considered as the backbone of the set of solutions and between them (Figure 8):

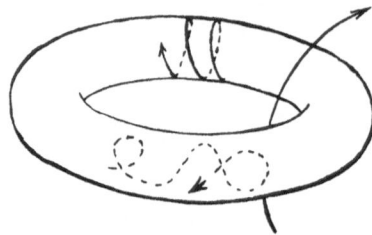

Fig. 8. Schematic representation of the Arnold diffusion conjecture with a torus of quasi-periodic solutions, an open solution and an enclosed chaotic solution.

A) The holes of infinite measure are almost everywhere filled with open solutions coming from infinity and going back to infinity,

B) The holes of finite measure are almost everywhere filled with chaotic solutions that are dense in their hole.
   The philosophy of this conjecture is thus that for non-quasi periodic solutions the Arnold diffusion, however slow, is strong enough to lead either to infinity or to all the attainable domain.

The "temporary chaotic solutions" are open solutions that lay in a hole with only very small apertures. Before finding these apertures the motion is bounded and looks like a chaotic motion.

The prototype of temporary chaotic motions is the cometary motion: a comet arrives from infinity, has a close approach to a planet and is sent on an elliptic orbit about the Sun: it will then undergo milleniums of erratic planetary perturbations before another very close planetary approach that will send it back to infinity.

In the general n-body problem the integrals of motion and the Arnold tori cannot enclose holes of finite measure; then the escape solutions are certainly everywhere dense and the Liapounov stability is impossible. This emphasizes the importance of the notion of "effective stability": very few changes during billions of years (Ref. 1-2).

The situation is qualitatively different when the integrals of motion can enclose parts of phase-space with bounded measure, as in the circular restricted three-body problem. It is also different when the number of degrees of freedom can be reduced to two: the Arnold tori enclose bounded parts of attainable phase-space (in the three-body problem: the collinear case, the Sitnikov case and the plane circular restricted case).

9. THE CHAOTICITY OF SMALL AMPLITUDE

Let us consider the sea level and the tides. At the decimetric scale, they have the regularity of astronomical motions and can be predicted well in advance. At the centimetric scale they inherit the chaoticity of meteorology: the passage of barometric depressions raises the sea level by a few centimeters.

Such phenomena happen very often in dissipative systems when a small chaotic effect is added to a large regular effect. But the same analysis is much more difficult for conservative systems and this "chaoticity of small amplitude" (that appears in the zones where predominate regular quasi-periodic motions as those of the Arnold tori, Ref.19) always threaten to be temporary chaotic and to lead finally to very large perturbations and even to infinity.

10. CONCLUSION

Recent years have seen tremendous progress in many domains: in numerical computations and numerical experiments, in the theory of stability and predictability, in the discussion of the concept of determinism, in the analysis of chaotic motions and strange attractors. This progress has renewed our picture of Physics and the Universe.

At each scale of nature: sub-atomic, atomic, microscopic, ordinary, geographical, astronomical, the presence of chaotic motions destabilizes and randomizes too small elements and stabilizes the corresponding mean statistical elements that become the significant elements of the scale of interest such as the temperature and the pressure at the ordinary scale.

The evolution of these significant elements remains essentially deterministic as long as the "divergence time" or "horizon of prediction" of the scale of interest is not reached, but becomes stochastic and statistical beyond this time. Mean statistical elements concerning larger sets then become significant and the phenomena are thus nested in one another.

The divergence time increases very much with the scale: extremely short at the sub-atomic and atomic levels (in agreement with the probabilistic and statistical character of quantum mechanics) it becomes about 15 days for meteorology (the "butterfly effect") and about 100 millions of years for the astronomy of the Solar System. Astronomy is thus the most deterministic of all sciences with the best possibilities of predictions.

For many years it was believed that the progress in measurements and scientific knowledge, the adoption of new models closer to the phenomena of interest and more powerful performing computation methods would substantially increase the divergence time. On the contrary, the numerical experiments, the existence, variety and generality of chaotic motions and strange attractors and a better understanding of the notions of stability and predictability have shown that there exists a natural barrier virtually impossible to break down.

Thus far from being mathematical curiosities the chaotic motions and the strange attractors are essential components of nature which they divide into many domains of different scales, passing stability, regularity and predictability from one to another according to the scale and duration considered. Their study and their understanding will be essential elements of future progress.

REFERENCES

1. A Giorgilli - A. Delshams - E. Fontich - L. Galgani - C. Simo - "Effective Stability for a Hamiltonian System near an Elliptic Equilibrium Point, with an Application to the Restricted Three Body Problem". Journal of Differential Equations - 77 p.167-198 (1989).

2. A. Celletti - A. Giorgilli, "On the stability of the Lagrangian points in the restricted problem of three bodies". CARR reports in Mathematical Physics - n19/90 - July 1990.

3. A Bennett - "Characteristics exponents of the Five Equilibrium Solutions in the Elliptically Restricted Problem" Icarus 4, p.177 (1965).

4. J.M.A. Danby - "Stability of Triangular Points in the Elliptic Restricted Problem of Three Bodies" Astronomical Journal, 69, p.165 (1964).

5. A.B. Leontovitch - "On the stability of the Lagrange periodic solution of the restricted three-body problem" Dokl-Akad. Nauk.SSSR. Vol.143 No.3-p. 525-528 (1962).

6. A. Deprit - A. Deprit-Bartholomé "Stability of the triangular Lagrangian point" Astronomical Journal, 72, No.2, p.173-179 (1967).

7. C. Marchal - "Etude de la stabilité des solutions de Lagrange du problèm des trois corps. Cas où l'excentricité et les trois masses sont quelconques". Séminaire du Bureau des Longitudes (7 mars 1968).

8. J. Tschauner - "Die Bewegung in der Nähe der Dreieckspunkte des elliptischen eingeschränkten Dreikörperproblems". Celestial Mechanics 3, p.189-196 (1971).

9. A. Alothman-Alragheb - "Influence des perturbations d'ordre élevé sur la stabilité des systèmes hamiltoniens (cas à deux degrés de liberté)". Thèse de Doctorat d'Etat. Observatoire de Paris - ler juillet 1986.

10. L. El Bakkali - "Voisinage et stabilité des solutions périodiques des systèmes hamiltoniens. Application aux solutions de Lagrange du problème des trois corps". Thèse de Doctorat d'Etat. Observatoire de Paris. 27 juin 1990.

11. C. Marchal - "The three-body problem". Elsevier Science Publishers B.V. Amsterdam (1990).

12. V.I. Arnold - A. Avez - "Problèmes ergodiques de la Mécanique classique" Gauthier-Villars, Paris (1967).

13. E. N. Lorenz - "Deterministic non-Periodic Flow". J. Atmos. Sci, Vol.20 p.130-141 (1963).

14. M. Hénon - "Numerical study of quadratic area-preserving mappings" Quarterly of Applied Mathematics - Vol.27 - No.3 p.291 (1969).

15. M. Hénon - "A two-dimensional mapping with a strange attractor" Communications in Mathematical Physics 50, p.69 (1976).

16. D. Ruelle - F. Takens - "On the nature of turbulence". (Commun. Math. Physics - Vol.20, p.167-192 (1971).

17. P. Collet - J.P. Eckmann- "Iterated maps on the interval as dynamical systems" Progress in Physics 1 - A. Jaffe - D. Ruelle editors - Birkhäuser (1983).

18. P. Bergé - Y. Pomeau - C. Vidal - "L'ordre dans le chaos". Hermann (1984).

19. J. Laskar - "Numerical experiment on the chaotic behaviour of the Solar System" Nature 338 - p.237-238 (1989).

ANALYTICAL FRAMEWORK IN POINCARE VARIABLES FOR

THE MOTION OF THE SOLAR SYSTEM

J. Laskar

SCMC du Bureau des Longitudes
UA 707 du CNRS
77 Avenue Denfert-Rochereau
75014, Paris
France

INTRODUCTION

The subject of this meeting is chaotic behaviour in celestial mechanics. It just means that we realize, one hundred years after Poincaré that the solutions of our equations are complicated. In order to better understand this complex behaviour of the solutions, a convenient framework is necessary. In the *Méthodes Nouvelles de la Mécanique Céleste* and *Leçons de Mécanique Céleste*, Poincaré sets up such a framework.

What Poincaré did not give in the *Méthodes Nouvelles* is the actual expansions of the planetary hamiltonian in this framework. For that he refers to LeVerrier. I have tried here to make these expansions appear more simple than in Le Verrier's *Recherches Astronomiques* (1855). Part of the present lecture results from the working group on the *Méthodes Nouvelles* which held in Paris during 1988–1990 and was already in (Laskar, 1989 and 1990a) although the exposition of the expansion of the hamiltonian was entirely revised.

I. HELIOCENTRIC REFERENCE FRAME

I.1. Usual non-canonical equations of second order

Ce changement de variables présente de graves inconvénients. Non seulement il altère la forme canonique des équation, mais il ne conserve pas la forme des intégrales des aires. Aussi n'en ferai-je aucun usage. C'est pour cette raison que je me borne à renvoyer le lecteur à l'Ouvrage de Tisserand. (H. Poincaré, Leçons de Mécanique Céleste, 1905, t.I, p.61)

Let $P_0, P_1, \ldots, P_n$ be $n + 1$ bodies of masses $m_0, m_1, \ldots, m_n$ in gravitational interaction, and let O be their center of mass. For every body P_i, we shall denote $\mathbf{u}_i = \overrightarrow{OP_i}$. In

Predictability, Stability, and Chaos in N-Body Dynamical Systems
Edited by A.E. Roy, Plenum Press, New York, 1991

the barycentric reference frame with origin O the Newton equations of motion form a differential system of order $6(n+1)$ and can be written for each i

$$\frac{d^2\mathbf{u}_i}{dt^2} = -G\sum_{j\neq i} m_j \frac{\mathbf{u}_i - \mathbf{u}_j}{\Delta_{ij}^3}. \tag{1}$$

where $\Delta_{ij} = \|\mathbf{u}_i - \mathbf{u}_j\|$, and G is the constant of gravitation. A canonical set of coordinates is $(\mathbf{u}_i, \tilde{\mathbf{u}}_i = m_i\dot{\mathbf{u}}_i)_{i=1,n}$ and in this set of coordinates the Hamiltonian will be

$$H = \frac{1}{2}\sum_{i=0}^{n} \frac{\|\tilde{\mathbf{u}}_i\|^2}{m_i} - G\sum_{0\leq i<j} \frac{m_i m_j}{\Delta_{ij}} \tag{2}$$

The reduction of the center of mass is achieved generally in the planetary case by using the heliocentric variables $(\mathbf{r}_i, \dot{\mathbf{r}}_i)$ with $\mathbf{r}_i = \mathbf{u}_i - \mathbf{u}_0$ and $\dot{\mathbf{r}}_i = d\mathbf{r}_i/dt$. We obtain then the usual differential system of order $6n$

$$\frac{d^2\mathbf{r}_i}{dt^2} = -\frac{G(m_0 + m_i)\mathbf{r}_i}{r_i^3} - \sum_{j\neq i} Gm_j\left(\frac{\mathbf{r}_j - \mathbf{r}_i}{\Delta_{ij}^3} - \frac{\mathbf{r}_j}{r_j^3}\right) \tag{3}$$

That is for each body $P_{i,i=1,n}$.

$$\frac{d^2\mathbf{r}_i}{dt^2} = \mathbf{grad}_{\mathbf{r}_i}\frac{G(m_0 + m_i)}{r_i} + \mathbf{grad}_{\mathbf{r}_i}\left(\sum_{j\neq i} Gm_j(\frac{1}{\Delta_{ij}} - \frac{\mathbf{r}_i \cdot \mathbf{r}_j}{r_j^3})\right) \tag{4}$$

These are the equations of the planetary motion in the heliocentric reference frame which have been used most of the time in planetary or satellite theories (except for the Moon where we have a single satellite) since the early works of Laplace. These equations are not the most convenient since they are not in an hamiltonian form. This is not very important when the planetary theory is of low order (1 or 2), but it will be a major drawback for higher orders as all the expressions will get more complicated.

In the *Méthodes Nouvelles*, Poincaré suggests to use Jacobi coordinates in order to express the equations in an hamiltonian form, but these coordinates are not well suited for planetary theories: the hamiltonian needs to be expanded with respect to the powers of the masses and it is difficult to add a new body in the middle of the system (see for more details Laskar, 1989, 1990a). In 1896, Poincaré introduced a new set of coordinates which are canonical and very natural for planetary theories (Poincaré, 1896, 1897, 1905). Despite these advantages, these coordinates which will be called the canonical heliocentric coordinates were never really adopted by the astronomers [1].

---

[1]While finishing this text, I realized that these coordinates have been used once by Yuasa and Hori (1979) who improperly refers to Charlier (1902) for the origin of these coordinates. Actually, it is true that Charlier's book constantly refers to Poincaré's work.

I.2. Canonical Heliocentric variables

The canonical heliocentric coordinates are very easy to introduce: Let $r_0 = u_0$. The change of coordinates to heliocentric coordinates is given by

$$
\begin{bmatrix}
r_0 \\
r_1 \\
r_2 \\
\vdots \\
r_{n-1} \\
r_n
\end{bmatrix}
=
\begin{bmatrix}
1 & 0 & 0 & \cdots & 0 & 0 \\
-1 & 1 & 0 & \cdots & 0 & 0 \\
-1 & 0 & 1 & \cdots & 0 & 0 \\
\vdots & \vdots & \vdots & \ddots & \vdots & \vdots \\
-1 & 0 & 0 & \cdots & 1 & 0 \\
-1 & 0 & 0 & \cdots & 0 & 1
\end{bmatrix}
\begin{bmatrix}
u_0 \\
u_1 \\
u_2 \\
\vdots \\
u_{n-1} \\
u_n
\end{bmatrix}
= A_H
\begin{bmatrix}
u_0 \\
u_1 \\
u_2 \\
\vdots \\
u_{n-1} \\
u_n
\end{bmatrix}
\tag{5}
$$

This linear transformation can be easily extended to a canonical transformation

$$
(u_i, \tilde{u}_i) \longrightarrow (r_i, \tilde{r}_i) = \left(A_H u_i, {}^t A_H^{-1} \tilde{u}_i \right)
\tag{6}
$$

that is

$$
\begin{bmatrix}
\tilde{r}_0 \\
\tilde{r}_1 \\
\tilde{r}_2 \\
\vdots \\
\tilde{r}_{n-1} \\
\tilde{r}_n
\end{bmatrix}
=
\begin{bmatrix}
1 & 1 & 1 & \cdots & 1 & 1 \\
0 & 1 & 0 & \cdots & 0 & 0 \\
0 & 0 & 1 & \cdots & 0 & 0 \\
\vdots & \vdots & \vdots & \ddots & \vdots & \vdots \\
0 & 0 & 0 & \cdots & 1 & 0 \\
0 & 0 & 0 & \cdots & 0 & 1
\end{bmatrix}
\begin{bmatrix}
\tilde{u}_0 \\
\tilde{u}_1 \\
\tilde{u}_2 \\
\vdots \\
\tilde{u}_{n-1} \\
\tilde{u}_n
\end{bmatrix}
= {}^t A_H^{-1}
\begin{bmatrix}
\tilde{u}_0 \\
\tilde{u}_1 \\
\tilde{u}_2 \\
\vdots \\
\tilde{u}_{n-1} \\
\tilde{u}_n
\end{bmatrix}
\tag{7}
$$

The reduction of the center of mass gives $\tilde{r}_0 = 0$. As was already noticed by Poincaré, the expression of the angular momentum is preserved by a canonical linear transformation $(u_i, \tilde{u}_i) \longrightarrow (r_i, \tilde{r}_i)$. We have

$$
G = \sum_{i=0}^{n} r_i \wedge \tilde{r}_i = \sum_{i=1}^{n} r_i \wedge \tilde{r}_i
\tag{8}
$$

and the total angular momentum is equal to the sum of the angular momentum of each bodies.

The expression of the kinetic energy is

$$
T = \frac{1}{2} \sum_{i=1}^{n} \frac{\|\tilde{r}_i\|^2}{m_i} + \frac{1}{2} \frac{\|\tilde{r}_1 + \tilde{r}_2 + \ldots + \tilde{r}_n\|^2}{m_0}
\tag{9}
$$

That is

$$
T = T_0 + T_1 \quad \text{with} \quad T_0 = \frac{1}{2} \sum_{i=1}^{n} \|\tilde{r}_i\|^2 \left[\frac{1}{m_i} + \frac{1}{m_0} \right] \quad \text{and} \quad T_1 = \sum_{0<i<j} \frac{\tilde{r}_i \cdot \tilde{r}_j}{m_0}
\tag{10}
$$

The computation of the potential is also simple and gives

$$U = U_0 + U_1 \quad \text{with} \quad U_0 = -G \sum_{i=1}^{n} \frac{m_0 m_i}{r_i} \quad \text{and} \quad U_1 = -G \sum_{0 < i < j} \frac{m_i m_j}{\Delta_{ij}} \tag{11}$$

with $\Delta_{ij} = \|\mathbf{r}_i - \mathbf{r}_j\|$

The full hamiltonian is on the form

$$H = H_0 + H_1; \quad \text{with} \quad H_1 = T_1 + U_1 \tag{12}$$

Due to the presence of the T_1 part, the generalized momentum variables ($\tilde{\mathbf{r}}_i$) are not proportional to the derivative of the positions ($\dot{\mathbf{r}}_i$)

$$\frac{d\mathbf{r}_i}{dt} = \mathbf{grad}_{\tilde{\mathbf{r}}_i} F = \mathbf{grad}_{\tilde{\mathbf{r}}_i} T = \tilde{\mathbf{r}}_i \left[\frac{1}{m_i} + \frac{1}{m_0} \right] + \sum_{j \neq i} \frac{\tilde{\mathbf{r}}_j}{m_0} \tag{13}$$

this is probably the reason why these coordinates were not very well accepted by the astronomers: the "osculating" ellipses are no longer tangent to the trajectories but are intersecting them. The equations of the unperturbed problem are defined with the integrable hamiltonian $H_0 = T_0 + U_0$ which gives

$$\frac{d^2 \mathbf{r}_i}{dt^2} = -G(m_0 + m_i) \frac{\mathbf{r}_i}{r_i^3} \tag{14}$$

This is the equation of a union of disjoint two body problems: the "planet" of mass m_i around the "sun" of mass m_0 (or of $\dfrac{m_0 m_i}{m_0 + m_i}$ around $m_0 + m_i$ fixed). It should be stressed that the expressions of the complete hamiltonian in these variables are very symetrical and simple.

I.3. Delaunay elliptical variables

Once the planetary problem is expressed as the perturbation of an integrable problem (here a union of two body problems) of hamiltonian H_0, the coordinates which are convenient are coordinates which form angle-action coordinates for the integrable unperturbed problem. In the present case, such coordinates are given by the Delaunay coordinates.
Let us consider the Keplerian motion of a body of mass m around a fixed center of mass μ

$$\frac{d^2 \mathbf{r}}{dt^2} = -\mu \frac{\mathbf{r}}{r^3} \tag{15}$$

The position vector and momentum vector $(\mathbf{r}, \tilde{\mathbf{r}})$ $(\tilde{\mathbf{r}} = m\dot{\mathbf{r}})$ give a set of canonical variables and the hamiltonian is

$$F = \frac{1}{2} \frac{\|\tilde{\mathbf{r}}\|^2}{m} - \frac{\mu m}{r} \tag{16}$$

With Hamilton-Jacobi's method, one can obtain a canonical change of variables (see for example Poincaré, 1905, or Chenciner, 1989)

$$(\mathbf{r}, \tilde{\mathbf{r}}) \longrightarrow (l, g, \theta, mL, mG, m\Theta) \tag{17}$$

where $L, G, \Theta, l, g, \theta$ are the Delaunay variables

$$
\begin{aligned}
L &= \sqrt{\mu a} \\
G &= \sqrt{\mu a(1 - e^2)} \\
\Theta &= G \cos i = \sqrt{\mu a(1 - e^2)} \cos i \\
l &= M \qquad &\text{(mean anomaly)} \\
g &= \omega \qquad &\text{(argument of perihelion)} \\
\theta &= \Omega \qquad &\text{(longitude of the node)}
\end{aligned}
\tag{18}
$$

(a is the semi major axis, e the excentricity and i the inclination). With these variables, the hamiltonian F becomes

$$F = -\frac{\mu m}{2a} = -\frac{\mu^2 m}{2L^2} \tag{19}$$

The hamiltonian depends only of L. All the variables are constants, except for the mean anomaly l, for which

$$\frac{dl}{dt} = -\frac{\partial F}{\partial L} = \frac{\mu^2 m}{L^3} \qquad \text{(Kepler's law)} \tag{20}$$

I.4. Poincaré elliptical variables

The Delaunay variables are not very well suited for planetary theories. In fact, the eccentricities and inclinations are small and we would like to make expansions with respect to quantities equivalent to these small quantities. This is why Poincaré makes a first linear change of coordinates to coordinates which are small for small eccentricities and inclinations.

$$
\begin{aligned}
\Lambda &= mL \\
H &= mL - mG \quad &= m\sqrt{\mu a}(1 - \sqrt{1 - e^2}) \\
Z &= mG - m\Theta \quad &= m\sqrt{\mu a(1 - e^2)}(1 - \cos i)
\end{aligned}
\tag{21}
$$

This linear change of the coordinates is extended to a canonical transformation

$$
\begin{aligned}
\lambda &= l + g + \theta \quad &&\text{(mean longitude)} \\
h &= -g - \theta \quad (= -\varpi) \quad &&(- \text{ longitude of perihelion}) \\
\zeta &= -\theta \quad (= -\Omega) \quad &&(- \text{ longitude of the node})
\end{aligned}
\tag{22}
$$

These new coordinates still have a disadvantage: they are not defined for zero eccentricity or inclination. A new change of variable is thus necessary which was introduced by Poincaré in the *Méthodes Nouvelles*. This change of variables is canonical and is similar to a change from polar to rectangular coordinates in the plane.

$$\Lambda = m\sqrt{\mu a}$$

$$\lambda$$

$$\xi = \sqrt{2H}\cos h = \sqrt{2\Lambda}\sqrt{1 - \sqrt{1 - e^2}}\cos\varpi$$

$$\eta = \sqrt{2H}\sin h = -\sqrt{2\Lambda}\sqrt{1 - \sqrt{1 - e^2}}\sin\varpi \qquad (23)$$

$$p = \sqrt{2Z}\cos\zeta = \sqrt{2\Lambda}\sqrt{\sqrt{1 - e^2}(1 - \cos i)}\cos\Omega$$

$$q = \sqrt{2Z}\sin\zeta = -\sqrt{2\Lambda}\sqrt{\sqrt{1 - e^2}(1 - \cos i)}\sin\Omega$$

$(\Lambda, \lambda, \xi, \eta, p, q)$ is a canonical system of variables, (ξ, η, p, q) are small for small eccentricities and inclinations, and the variables are regular for zero eccentricity and inclinations.

I.5. Poincaré variables vs classical elliptical elements

A set (among others) of classical non canonical elliptic variables is

$$z = \mathbf{k} + i\mathbf{h} = e\,E^{i\varpi}$$

$$\zeta = \mathbf{q} + i\mathbf{p} = \sin\frac{i}{2}\,E^{i\Omega} \qquad (24)$$

We can make a canonical transformation on the Poincaré variables to put them on the complex form $(x, -i\bar{x}, y, -i\bar{y})$ with

$$x = \frac{1}{\sqrt{2}}(\xi - i\eta) = \sqrt{\Lambda}\sqrt{1 - \sqrt{1 - e^2}}\,E^{i\varpi}$$

$$y = \frac{1}{\sqrt{2}}(p - iq) = \sqrt{\Lambda}\sqrt{\sqrt{1 - e^2}(1 - \cos i)}\,E^{i\Omega} \qquad (25)$$

We have the relation

$$z = x\sqrt{\frac{2}{\Lambda}}\left(1 - \frac{x\bar{x}}{2\Lambda}\right)^{1/2}$$

$$\zeta = y\frac{1}{\sqrt{2\Lambda}}\left(1 - \frac{x\bar{x}}{\Lambda}\right)^{-1/2} \qquad (26)$$

with the expansion up to order 4

$$z = \sqrt{\frac{2}{\Lambda}}\, x \left(1 - \frac{1}{4}\frac{x\bar{x}}{\Lambda} - \frac{1}{32}\frac{(x\bar{x})^2}{\Lambda^2} + o(e^4)\right)$$

$$\zeta = \frac{1}{\sqrt{2\Lambda}}\, y \left(1 + \frac{1}{2}\frac{x\bar{x}}{\Lambda} + \frac{3}{8}\frac{(x\bar{x})^2}{\Lambda^2} + o(e^4)\right)$$

(27)

The change of coordinates from complex elliptical elements (z, ζ) to complex Poincaré variables (x, y) is analytic for eccentricities smaller than 1. From the results of section (II.2), it can also be deduced that the change of coordinates from usual rectangular coordinates $(\mathbf{r}_i, \tilde{\mathbf{r}}_i)$ to complex elliptical elements is also analytic.

II. COMPUTATION OF THE HAMILTONIAN

I will present now a method for the computation of the expansion of the hamiltonian which I think is very natural and which can be implemented in a straightforward manner on a computer. It was actually completely implemented under the algebraic manipulator TRIP which is in development at the Bureau des Longitudes (Laskar, 1990b). The form of the hamiltonian in Poincaré canonical heliocentric variables is given by

$$H = H_0 + H_1; \qquad \text{with} \qquad H_1 = T_1 + U_1$$

and

$$H_0 = -\sum_{i=1}^{n} \frac{\mu_i^2 \beta_i^3}{2\Lambda_i^2}$$

$$U_1 = -G \sum_{0<i<j} \frac{m_i m_j}{\Delta_{ij}}$$

$$T_1 = \sum_{0<i<j} \frac{\tilde{\mathbf{r}}_i \cdot \tilde{\mathbf{r}}_j}{m_0}$$

(28)

with $\beta_i = \dfrac{m_i m_0}{m_i + m_0}$ and $\mu_i = G(m_0 + m_i)$. Notice that β_i and μ_i are the values of the masses to be used for the change of variables to elliptical Poincaré coordinates.

II.1 The direct part of the hamiltonian

The difficulty consists to compute the inverse of the distance of two planets (of radius vector $\mathbf{r}$ and $\mathbf{r}'$) $1/\Delta$ in term of the Poincaré variables $(\Lambda, \Lambda', \lambda, \lambda', x, x', \bar{x}, \bar{x}', y, y', \bar{y}, \bar{y}')$. We will make the expansion of this quantity in the vicinity of the circular planar problem (small excentricities and inclinations). We have

$$\Delta^2 = \mathbf{r}^2 + \mathbf{r}'^2 - 2rr' \cos S$$

(29)

where S is the angle beetween the two vectors $\mathbf{r}$ and $\mathbf{r}'$. With $\rho = r/r'$, and $\alpha = \dfrac{a}{a'}$ we obtain

$$\frac{a'}{\Delta} = \frac{a'}{r'}(1 + \rho^2 - 2\rho \cos S)^{-1/2}$$

$$= \frac{a'}{r'}\left(1 + \alpha^2 - 2\alpha \cos(\lambda - \lambda')\right) \tag{30}$$

$$+ \alpha^2\left(\left(\frac{\rho}{\alpha}\right)^2 - 1\right) + 2\alpha\left(\cos(\lambda - \lambda') - \frac{\rho}{\alpha}\cos S\right)\Big)^{-1/2}$$

We shall denote

$$A = 1 + \alpha^2 - 2\alpha \cos(\lambda - \lambda')$$

$$V_1 = \cos(\lambda - \lambda') - \frac{\rho}{\alpha}\cos S$$

$$V_2 = \left(\frac{\rho}{\alpha}\right)^2 - 1 \tag{31}$$

$$V = 2\alpha V_1 + \alpha^2 V_2$$

The computation of the inverse of the distance is thus reduced to

$$\frac{a'}{\Delta} = \frac{a'}{r'}(A + V)^{-1/2} \tag{32}$$

When V is smaller than A, that is when e and i are small and α not too large, we can make a Taylor expansion around A which then gives

$$\frac{a'}{\Delta} = \frac{a'}{r'}A^{-1/2} - \frac{1}{2}\frac{a'}{r'}VA^{-3/2} + \frac{3}{8}\frac{a'}{r'}V^2 A^{-5/2}$$

$$- \frac{5}{16}\frac{a'}{r'}V^3 A^{-7/2} + \frac{35}{128}\frac{a'}{r'}V^4 A^{-9/2} + \cdots \tag{33}$$

Let

$$U_i = \frac{a'}{r'}V^i = \frac{a'}{r'}(2\alpha V_1 + \alpha^2 V_2)^i \tag{34}$$

The quantities U_i can be computed once for all, as they contain a moderate number of terms for a given maximum degree in eccentricity and inclination. V_2 does not depend of the inclination and is of degree 1 at least in eccentricity. The computation of V_2 is straightforward from the two body usual expansions. V_1 is of degree 2 at least in inclination. Its computation is more complicated and will be explained in detail in the next section. If The hamiltonian is computed up to a given power of excentricity-inclination m, the quantities U_i need to be computed only for $i \le m$.

On the other hand, the expansions of the circular problem have a well-known expression (Laplace, 1798, Poincaré, 1905)

$$A^{-s} = \frac{1}{2} \sum_{k=-\infty}^{+\infty} b_s^{(k)}(\alpha) e^{ik(\lambda-\lambda')} \tag{35}$$

where $b_s^{(k)}(\alpha)$ are the usual Laplace coefficients. This expression is an infinite series which converges slowly for the values of α which are present in the planetary cases, but it depends only of two variables (α and $\lambda - \lambda'$). It can be computed in different ways (expansion in α, literal in the $b_s^{(k)}$, literal in the $b_1^{(0)}$ and $b_1^{(1)}$ with the use of recurrence formulas, expanded around a given value α_0 of α, etc ...) (See the annex 1.).

Once A^{-s} and the U_i are computed, the expression of any argument $k\lambda - k'\lambda'$ of the hamiltonian is very rapid. Let $c_I = k - k'$ be the characteristic of the argument of a term (characteristic of the inequality $k\lambda - k'\lambda'$). the characteristic of a product is the sum of the characteristics of the terms. All the terms of A^{-s} have characteristic zero. The term of the product $A^{-s} \times U_i$ with argument $k\lambda - k'\lambda'$ is thus the sum of the products of all the terms of characteristic $k - k'$ from U_i with the coresponding term of A^{-s}. Indeed, in TRIP, there is a special truncation function which allow to compute directly the coefficient of an argument in this manner, without computing the whole product. An example for the computation of the secular hamiltonian is given in the annex 2.

II.2. Computation of $\cos S$ and T_1

The computations of

$$\cos S = \frac{xx' + yy' + zz'}{rr'} \tag{36}$$

and of the complementary part of the hamiltonian

$$T_1 = \frac{\tilde{\mathbf{r}} \cdot \tilde{\mathbf{r}}'}{m_0} = \frac{mm'}{m_0}(\dot{x}\dot{x}' + \dot{y}\dot{y}' + \dot{z}\dot{z}') \tag{37}$$

are very similar and we will compute then simultaneously. Let

$$\mathcal{R}_1(\theta) = \begin{pmatrix} 1 & 0 & 0 \\ 0 & \cos\theta & -\sin\theta \\ 0 & \sin\theta & \cos\theta \end{pmatrix}$$

$$\mathcal{R}_3(\theta) = \begin{pmatrix} \cos\theta & -\sin\theta & 0 \\ \sin\theta & \cos\theta & 0 \\ 0 & 0 & 1 \end{pmatrix} \tag{38}$$

In the plane of the orbit, with origin at the pericenter, the position and velocity are given by

$$X = r \cos v; \qquad\qquad Y = r \sin v$$

$$\dot{X} = -\frac{na}{\sqrt{1-e^2}} \sin v; \qquad \dot{Y} = \frac{na}{\sqrt{1-e^2}}(e + \cos v) \tag{39}$$

where v is the true anomaly. Positions and velocities in the fixed reference frame are thus given by

$$
\begin{bmatrix} x \\ y \\ z \end{bmatrix} = \mathcal{R}_3(\Omega) \times \mathcal{R}_1(i) \times \mathcal{R}_3(\omega) \times \begin{bmatrix} X \\ Y \\ 0 \end{bmatrix}
$$
$$
\begin{bmatrix} \dot{x} \\ \dot{y} \\ \dot{z} \end{bmatrix} = \mathcal{R}_3(\Omega) \times \mathcal{R}_1(i) \times \mathcal{R}_3(\omega) \times \begin{bmatrix} \dot{X} \\ \dot{Y} \\ 0 \end{bmatrix} \tag{40}
$$

let

$$\mathcal{R} = \mathcal{R}_3(\Omega) \times \mathcal{R}_1(i) \times \mathcal{R}_3(-\Omega) \tag{41}$$

we have

$$
\begin{bmatrix} x \\ y \\ z \end{bmatrix} = \mathcal{R} \times \mathcal{R}_3(\varpi) \times \begin{bmatrix} X \\ Y \\ 0 \end{bmatrix}
$$
$$
\begin{bmatrix} \dot{x} \\ \dot{y} \\ \dot{z} \end{bmatrix} = \mathcal{R} \times \mathcal{R}_3(\varpi) \times \begin{bmatrix} \dot{X} \\ \dot{Y} \\ 0 \end{bmatrix} \tag{42}
$$

If we notice that $\mathcal{R}_1(i) = \mathcal{R}_1(i/2)\mathcal{R}_1(i/2)$, a straightforward computation gives

$$
\mathcal{R} = \begin{bmatrix} 1 - 2\mathbf{p}^2 & 2\mathbf{pq} & 2\mathbf{p}\cos(i/2) \\ 2\mathbf{pq} & 1 - 2\mathbf{q}^2 & -2\mathbf{q}\cos(i/2) \\ -2\mathbf{p}\cos(i/2) & 2\mathbf{q}\cos(i/2) & 1 - 2\mathbf{p}^2 - 2\mathbf{q}^2 \end{bmatrix} \tag{43}
$$

From now on we will use complex notations which will be more suited for these computations. We will need now a very simple remark which will make the remaining part of the computation much more easy. Let us notice that if $\eta = a + ib$ and $\eta' = a' + ib'$

$$(a' \quad b' \quad c') \times \begin{bmatrix} a \\ b \\ 0 \end{bmatrix} = aa' + bb' = \Re(\eta\bar{\eta}') \tag{44}$$

Using complex notations for the two first coordinates, we obtain

$$\mathcal{R}_3(\varpi) \times \begin{bmatrix} X \\ Y \\ 0 \end{bmatrix} = \mathcal{R}_3(\varpi) \times \begin{bmatrix} r\cos v \\ r\sin v \\ 0 \end{bmatrix} = r\begin{bmatrix} \cos w \\ \sin w \\ 0 \end{bmatrix} = \begin{bmatrix} re^{iw} \\ 0 \end{bmatrix} \tag{45}$$

and for the velocities

$$\mathcal{R}_3(\varpi) \times \begin{bmatrix} \dot{X} \\ \dot{Y} \\ 0 \end{bmatrix} = \frac{na}{\sqrt{1-e^2}} \mathcal{R}_3(\varpi) \begin{bmatrix} -\sin v \\ e + \cos v \\ 0 \end{bmatrix} = \frac{na}{\sqrt{1-e^2}} \begin{bmatrix} i(e^{iw}+z) \\ 0 \end{bmatrix} \qquad (46)$$

($w = \varpi + v$ is the true longitude and $z = e\exp i\varpi$). It is thus easy to obtain, using the small lemma (44), that

$$\begin{bmatrix} x \\ y \\ z \end{bmatrix} = \mathcal{R} \times \begin{bmatrix} re^{iw} \\ 0 \end{bmatrix} = \Re \left\{ re^{iw} \begin{bmatrix} 1 - 2\mathbf{p}^2 - 2i\mathbf{pq} \\ 2\mathbf{pq} - i(1 - 2\mathbf{q}^2) \\ -2\cos(i/2)(\mathbf{p}+i\mathbf{q}) \end{bmatrix} \right\}$$
$$= \Re \left\{ re^{iw} \begin{bmatrix} \cos^2(i/2) + \bar{\zeta}^2 \\ -i(\cos^2(i/2) - \bar{\zeta}^2) \\ -2i\cos(i/2)\bar{\zeta} \end{bmatrix} \right\} \qquad (47)$$

The lemma (44) allows us in some sense to commute the two terms of the product, making the results much more easy.

The computation of $\cos S = \dfrac{xx' + yy' + zz'}{rr'}$ is then immediate and gives[1]

$$\cos S = \Re\{ \quad e^{i(w+w')}(\bar{\zeta}\cos(i'/2) - \bar{\zeta}'\cos(i/2))^2$$
$$+ e^{i(w-w')}(\cos(i/2)\cos(i'/2) + \bar{\zeta}\zeta')^2 \} \qquad (48)$$

The computation of the indirect part of the disturbing function can be computed along the same lines. We obtain the same results as previously with

$$Z = \frac{na}{\sqrt{1-e^2}} i(e^{iw} + z) \qquad (49)$$

instead of e^{iw}. The complementary part of the hamiltonian is thus equal to

$$T_1 = \frac{mm'}{m_0} \Re\{ \quad ZZ'(\bar{\zeta}\cos(i'/2) - \bar{\zeta}'\cos(i/2))^2$$
$$+ Z\bar{Z}'(\cos(i/2)\cos(i'/2) + \bar{\zeta}\zeta')^2 \} \qquad (50)$$

In practice, on the computer, the two previous quantities can be computed at the same time.

---

[1] This expression was first established in a slightly different manner by Abu El Ata and Chapront (1975)

Annex 1. Expansions of the circular motion and Laplace coefficients

In the computation of the hamiltonian, we need to compute the powers of the inverse of the distance in the circular motion, that is

$$A^{-s} = (1 + \alpha^2 - 2\alpha \cos(\lambda - \lambda'))^{-s}$$

for $s = (2k+1)/2$. These expressions are usually given in term of Laplace coefficients $b_s^{(i)}(\alpha)$ (for more complete discussion, see Laplace, 1798, T.I, p 292, Poincaré, 1905, ch.XVIII, or Brouwer and Clemence, 1961, ch.XV, sec.7)(*). If we denote $z = e^{i(\lambda - \lambda')}$, we obtain

$$A^{-s} = (1 - \alpha z)^{-s}(1 - \alpha z^{-1})^{-s}$$

and the expansion of this quantity as a series in z gives the Laplace coefficients. We have

$$A^{-s} = \frac{1}{2} \sum_{k=-\infty}^{+\infty} b_s^{(k)}(\alpha) e^{ik(\lambda - \lambda')} = \frac{1}{2} b_s^{(0)}(\alpha) + \sum_{k=1}^{+\infty} b_s^{(k)}(\alpha) \cos k(\lambda - \lambda')$$

In these expressions, we can keep the Laplace coefficients in a symbolic way, but if the computations are conducted at a higher order than the first one, this will lead very quickly to a huge number of terms. A better choice is then to express these coefficients as polynomials of a small number of variables. The first thing which can be done is the use of the Taylor expansions of the Laplace coefficients, which can be useful for very small values of α

$$\frac{1}{2} b_s^{(j)}(\alpha) =$$

$$\frac{s(s+1)(s+2)\cdots(s+j-1)}{j!} \alpha^j F(s, s+j, j+1; \alpha^2) =$$

$$\frac{s(s+1)(s+2)\cdots(s+j-1)}{j!} \alpha^j \left[1 + \frac{s}{1} \frac{(s+j)}{(j+1)} \alpha^2 + \frac{s(s+1)}{2!} \frac{(s+j)(s+j+1)}{(j+1)(j+2)} \alpha^4 + \cdots \right]$$

If we are interested only for the values of the distance in the vicinity of fixed values of the semi-major axis, as in planetary theories, the Laplace coefficients can also be expanded in Taylor series around a fixed value α_0 of α (Abu El Ata and Chapront, 1975). But we can also search for exact expressions with a finite numbers of terms.

In the expansions of the disturbing function, we need the Laplace coefficients for the values $s = 2k + 1/2$. Laplace (1798) provided two recurrence relations which allow to express all the needed coefficients with respect to $b_{1/2}^{(0)}$ and $b_{1/2}^{(1)}$.

(*) The Laplace coefficients of Poincaré differ by a factor 2 from the other authors. Although Poincaré's choice can be considered as more natural, I will stick to the convention of Laplace, which is the traditional one and which was also used by Le Verrier

$$b_{s+1}^{(j)} = \frac{(s+j)(1+\alpha^2)b_s^{(j)} - 2(j-s+1)\alpha b_s^{(j+1)}}{s(1-\alpha^2)^2}$$

$$b_{s+1}^{(j+1)} = \frac{j}{j-s}(\alpha + \frac{1}{\alpha})b_{s+1}^{(j)} - \frac{j+s}{j-s}b_{s+1}^{(j-1)}$$

With these two relations, we can express all of the Laplace coefficients in the following form

$$b_s^{(j)} = P(\alpha, \frac{1}{\alpha}, \frac{1}{1-\alpha^2}) \, b_{1/2}^{(0)} + Q(\alpha, \frac{1}{\alpha}, \frac{1}{1-\alpha^2}) \, b_{1/2}^{(1)}$$

where P and Q are polynomial with rational coefficients.

The two coefficients $b_{1/2}^{(0)}$ and $b_{1/2}^{(1)}$ can then be computed directly from the series expansions, or can be expressed with respect to the the complete elliptical integrals of the first and second kinds (Charlier, 1927, Brouwer and Clemence, 1961)

$$K(\alpha^2) = \int_0^{\pi/2} \frac{d\theta}{\sqrt{1-\alpha^2\sin\theta}} \qquad E(\alpha^2) = \int_0^{\pi/2} \sqrt{1-\alpha^2\sin\theta}\, d\theta$$

as

$$b_{1/2}^{(0)} = \frac{4}{\pi}K(\alpha^2) \qquad b_{1/2}^{(1)} = \frac{4}{\pi}\frac{K(\alpha^2) - E(\alpha^2)}{\alpha}$$

It should be noticed that these analytical expressions are not extremely interesting since $K(\alpha^2)$ and $E(\alpha^2)$ are quantities of the order of unity which cancel in the substraction to give a quantity of order α^2.

Annex 2.Computation with TRIP

In the following, we will actually make the computation of the planetary hamiltonian in Poincaré variables following the previous discussion. The intermediate results will be given up to degree 2 with respect to excentricities and inclinations in order to save space and these series are the direct output of the algebraic manipulator TRIP. The final hamiltonian will be given up to degree 4 with respect to excentricities and inclinations. The variables are defined as follows (x, y are the Poincaré complex variables and are defined in Eq. (25))

$$
\begin{aligned}
\text{X} &= x/\sqrt{\Lambda} \\
\text{Xp} &= x'/\sqrt{\Lambda'} \\
\text{Xb} &= \bar{x}/\sqrt{\Lambda} \\
\text{Xbp} &= \bar{x}'/\sqrt{\Lambda'} \\
\text{Y} &= y/\sqrt{\Lambda} \\
\text{Yp} &= y'/\sqrt{\Lambda'} \\
\text{Yb} &= \bar{y}/\sqrt{\Lambda} \\
\text{Ybp} &= \bar{y}'/\sqrt{\Lambda'} \\
\text{sqr2} &= \sqrt{2} \\
\text{L} &= \exp(i\lambda) \\
\text{L'} &= \exp(i\lambda')
\end{aligned}
$$

The first quantities to compute are a/r and r/a. These quantities are usual expansions of the two body problem and are given as standard functions by the algebraic manipulator (see any textbook on celestial mechanics).

```
aSr(L,X,Xb,sqr2)  =
+                1*L**-2*X**2
+        1/2       *L**-1*X*sqr2
+                1
+        1/2        *L*Xb*sqr2
+                1*L**2*Xb**2


rSa(L,sqr2,X,Xb)  =
-        1/2       *L**-2*X**2
-        1/2       *L**-1*sqr2*X
+                1
+                1*X*Xb
-        1/2        *L*sqr2*Xb
-        1/2        *L**2*Xb**2
```

which allow to compute $\text{ro} = \rho/\alpha = r/r' \times a'/a$

```
ro(L,Lp,X,Xb,Xp,Xpb,sqr2)  =
-        1/2       *L**-2*X**2
-        1/2       *L**-1*Lp**-1*X*Xp
-        1/2       *L**-1*X*sqr2
-        1/2       *L**-1*Lp*X*Xpb
+                1*Lp**-2*Xp**2
+        1/2        *Lp**-1*Xp*sqr2
+                1
+                1*X*Xb
+        1/2        *Lp*Xpb*sqr2
+                1*Lp**2*Xpb**2
```

```
-       1/2     *L*Lp**-1*Xb*Xp
-       1/2     *L*Xb*sqr2
-       1/2     *L*Lp*Xb*Xpb
-       1/2     *L**2*Xb**2
```

Then we compute $\theta = \exp i(w - \lambda)$ which provides the transformation from true longitudes w to mean longitudes λ. This expression is derived from $\exp iv$ which is also a standard function of the two body problem.

```
TE(L,X,Xb,sqr2) =
-       1/4     *L**-2*X**2
-               1*L**-1*X*sqr2
+               1
-               2*X*Xb
+               1*L*Xb*sqr2
+       9/4     *L**2*Xb**2
```

This expression is necessary in the computation of $\cos(S)$ (48)

```
CosS(L,Lp,X,Xb,Xp,Xpb,Y,Yb,Yp,Ypb,sqr2) =
+       9/8     *L**-3*Lp*X**2
-               1*L**-2*X*Xp
+       1/2     *L**-2*Lp*X*sqr2
+               1*L**-2*Lp**2*X*Xpb
+       1/4     *L**-1*Lp**-1*Yp**2
-       1/2     *L**-1*Lp**-1*Y*Yp
+       1/4     *L**-1*Lp**-1*Y**2
-       1/8     *L**-1*Lp**-1*Xp**2
-       1/8     *L**-1*Lp**-1*X**2
-       1/2     *L**-1*Xp*sqr2
+       1/2     *L**-1*Lp
-       1/4     *L**-1*Lp*Yp*Ypb
+       1/2     *L**-1*Lp*Y*Ypb
-       1/4     *L**-1*Lp*Y*Yb
-               1*L**-1*Lp*Xp*Xpb
-               1*L**-1*Lp*X*Xb
+       1/2     *L**-1*Lp**2*Xpb*sqr2
+       9/8     *L**-1*Lp**3*Xpb**2
-               1*Lp**-2*X*Xp
-       1/2     *Lp**-1*X*sqr2
+               1*Xb*Xp
+               1*X*Xpb
-       1/2     *Lp*Xb*sqr2
-               1*Lp**2*Xb*Xpb
+       9/8     *L*Lp**-3*Xp**2
+       1/2     *L*Lp**-2*Xp*sqr2
+       1/2     *L*Lp**-1
-       1/4     *L*Lp**-1*Yp*Ypb
+       1/2     *L*Lp**-1*Yb*Yp
-       1/4     *L*Lp**-1*Y*Yb
-               1*L*Lp**-1*Xp*Xpb
-               1*L*Lp**-1*X*Xb
-       1/2     *L*Xpb*sqr2
+       1/4     *L*Lp*Ypb**2
-       1/2     *L*Lp*Yb*Ypb
+       1/4     *L*Lp*Yb**2
-       1/8     *L*Lp*Xpb**2
-       1/8     *L*Lp*Xb**2
+               1*L**2*Lp**-2*Xb*Xp
+       1/2     *L**2*Lp**-1*Xb*sqr2
-               1*L**2*Xb*Xpb
+       9/8     *L**3*Lp**-1*Xb**2
```

At the same time, we compute the complementary part of the hamiltonian. The computation of Z (49) is immediate (here $z = Z/(i\,na)$)

```
Z(L,X,Xb,sqr2) =
-      1/4      *L**-1*X**2
+               1*L
-               1*L*X*Xb
+               1*L**2*Xb*sqr2
+      9/4      *L**3*Xb**2
```

and gives $\mathrm{FI} = T_1 m_0/mnam'n'a'$ (50)

```
FI(L,Lp,X,Xb,Xp,Y,Yb,Yp,Ypb,sqr2) =
+      9/8      *L**-3*Lp*X**2
+      1/2      *L**-2*Lp*X*sqr2
+               1*L**-2*Lp**2*X*Xpb
-      1/4      *L**-1*Lp**-1*Yp**2
+      1/2      *L**-1*Lp**-1*Y*Yp
-      1/4      *L**-1*Lp**-1*Y**2
-      1/8      *L**-1*Lp**-1*Xp**2
-      1/8      *L**-1*Lp**-1*X**2
+      1/2      *L**-1*Lp
-      1/4      *L**-1*Lp*Yp*Ypb
+      1/2      *L**-1*Lp*Y*Ypb
-      1/4      *L**-1*Lp*Y*Yb
-      1/2      *L**-1*Lp*Xp*Xpb
-      1/2      *L**-1*Lp*X*Xb
+      1/2      *L**-1*Lp**2*Xpb*sqr2
+      9/8      *L**-1*Lp**3*Xpb**2
+      9/8      *L*Lp**-3*Xp**2
+      1/2      *L*Lp**-2*Xp*sqr2
+      1/2      *L*Lp**-1
-      1/4      *L*Lp**-1*Yp*Ypb
+      1/2      *L*Lp**-1*Yb*Yp
-      1/4      *L*Lp**-1*Y*Yb
-      1/2      *L*Lp**-1*Xp*Xpb
-      1/2      *L*Lp**-1*X*Xb
-      1/4      *L*Lp*Ypb**2
+      1/2      *L*Lp*Yb*Ypb
-      1/4      *L*Lp*Yb**2
-      1/8      *L*Lp*Xpb**2
-      1/8      *L*Lp*Xb**2
+      1/2      *L**2*Lp**-2*Xb*Xp*sqr2**2
+      1/2      *L**2*Lp**-1*Xb*sqr2
+      9/8      *L**3*Lp**-1*Xb**2
```

The other necessary quantities are V_1 and V_2 (31).

```
V1(L,Lp,X,Xb,Xp,Xpb,sqr2) =
-      1/2      *L**-2*X**2
-               2*L**-1*Lp**-1*X*Xp
-               1*L**-1*X*sqr2
-               2*L**-1*Lp*X*Xpb
+      5/2      *Lp**-2*Xp**2
+               1*Lp**-1*Xp*sqr2
+               1*Xp*Xpb
+               3*X*Xb
+               1*Lp*Xpb*sqr2
+      5/2      *Lp**2*Xpb**2
-               2*L*Lp**-1*Xb*Xp
-               1*L*Xb*sqr2
-               2*L*Lp*Xb*Xpb
-      1/2      *L**2*Xb**2
```

```
V2 (L, Lp, X, Xb, Xp, Xpb, Y, Yb, Yp, Ypb, sqr2) =
-      3/8        *L**-3*Lp*X**2
+      1/4        *L**-2*X*Xp
-      1/4        *L**-2*Lp*X*sqr2
-      3/4        *L**-2*Lp**2*X*Xpb
-      1/4        *L**-1*Lp**-1*Yp**2
+      1/2        *L**-1*Lp**-1*Y*Yp
-      1/4        *L**-1*Lp**-1*Y**2
+      1/8        *L**-1*Lp**-1*Xp**2
-      1/8        *L**-1*Lp**-1*X**2
+      1/4        *L**-1*Xp*sqr2
+      1/4        *L**-1*Lp*Yp*Ypb
-      1/2        *L**-1*Lp*Y*Ypb
+      1/4        *L**-1*Lp*Y*Yb
+             1*L**-1*Lp*Xp*Xpb
+      1/2        *L**-1*Lp*X*Xb
-      3/4        *L**-1*Lp**2*Xpb*sqr2
-     17/8        *L**-1*Lp**3*Xpb**2
+      9/4        *Lp**-2*X*Xp
+      3/4        *Lp**-1*X*sqr2
-      3/4        *Xb*Xp
-      3/4        *X*Xpb
+      3/4        *Lp*Xb*sqr2
+      9/4        *Lp**2*Xb*Xpb
-     17/8        *L*Lp**-3*Xp**2
-      3/4        *L*Lp**-2*Xp*sqr2
+      1/4        *L*Lp**-1*Yp*Ypb
-      1/2        *L*Lp**-1*Yb*Yp
+      1/4        *L*Lp**-1*Y*Yb
+             1*L*Lp**-1*Xp*Xpb
+      1/2        *L*Lp**-1*X*Xb
+      1/4        *L*Xpb*sqr2
-      1/4        *L*Lp*Ypb**2
+      1/2        *L*Lp*Yb*Ypb
-      1/4        *L*Lp*Yb**2
+      1/8        *L*Lp*Xpb**2
-      1/8        *L*Lp*Xb**2
-      3/4        *L**2*Lp**-2*Xb*Xp
-      1/4        *L**2*Lp**-1*Xb*sqr2
+      1/4        *L**2*Xb*Xpb
-      3/8        *L**3*Lp**-1*Xb**2
```

Secular Hamiltonian (first order)

Once all the previous quantities are computed, the first order hamiltonian can be obtained directly through equation (33). Depending on the way the distance in the circular problem A is expressed, we will obtain different forms for the hamiltonian. As an example, I present here the secular part of the hamiltonian of degree 4 in the variables excentricities and inclination. The first expression is the general expression with coefficients $C_i(\alpha)$. The values of these coefficients are given as Taylor series with respect to α, as exact expressions with respect to the Laplace coefficients $b_{1/2}^{(0)}$ and $b_{1/2}^{(1)}$, or as exact expressions with respect to the Laplace coefficients $b_{1/2}^{(0)}$, $b_{1/2}^{(1)}$, $b_{3/2}^{(0)}$, $b_{3/2}^{(1)}$, $b_{5/2}^{(0)}$, $b_{5/2}^{(1)}$ (these two last expressions will be discussed in a forthcoming paper). It should be noted that there are no secular terms in the complementary part of the hamiltonian. $X = x/\sqrt{\Lambda}, Y = y/\sqrt{\Lambda}$ (see Eq.(25)).

$$
\begin{aligned}
a'/\Delta = \ & C_1(\alpha) \\
+ \ & C_2(\alpha)\,(X\bar{X}' + \bar{X}X') \\
+ \ & C_3(\alpha)\,(X\bar{X} + \bar{X}'X' + Y\bar{Y}' + \bar{Y}Y' - Y\bar{Y} - \bar{Y}'Y') \\
+ \ & C_4(\alpha)\,(X\bar{X}'Y\bar{Y}' + \bar{X}X'\bar{Y}Y') \\
+ \ & C_5(\alpha)\,(X\bar{X}'^2X' + \bar{X}\bar{X}'X'^2) \\
+ \ & C_6(\alpha)\,(X^2\bar{X}\bar{X}' + X\bar{X}^2X') \\
+ \ & C_7(\alpha)\,(X\bar{X}'Y\bar{Y} + \bar{X}X'Y\bar{Y} + X\bar{X}'\bar{Y}'Y' + \bar{X}X'\bar{Y}'Y') \\
+ \ & C_8(\alpha)\,(2(XX'\bar{Y}\bar{Y}' + \bar{X}\bar{X}'YY' - \bar{X}X'Y\bar{Y}' - X\bar{X}'\bar{Y}Y') \\
& \qquad - \bar{X}\bar{X}'Y^2 - XX'\bar{Y}^2 - XX'\bar{Y}'^2 - \bar{X}\bar{X}'Y'^2) \\
+ \ & C_9(\alpha)\,(\bar{X}'^2Y^2 + X'^2\bar{Y}^2 + X'^2\bar{Y}'^2 + \bar{X}'^2Y'^2 - 2X'^2\bar{Y}\bar{Y}' - 2\bar{X}'^2YY') \\
+ \ & C_{10}(\alpha)\,(X^2\bar{X}'^2 + \bar{X}^2X'^2) \\
+ \ & C_{11}(\alpha)\,(X\bar{X}Y\bar{Y} + \bar{X}'X'\bar{Y}'Y') \\
+ \ & C_{12}(\alpha)\,(Y^2\bar{Y}\bar{Y}' + Y\bar{Y}^2Y' + Y\bar{Y}'^2Y' + \bar{Y}\bar{Y}'Y'^2) \\
+ \ & C_{13}(\alpha)\,(X^2\bar{X}^2) \\
+ \ & C_{14}(\alpha)\,(Y^2\bar{Y}^2 + \bar{Y}'^2Y'^2 + Y^2\bar{Y}'^2 + \bar{Y}^2Y'^2) \\
+ \ & C_{15}(\alpha)\,(X\bar{X}Y\bar{Y}' + \bar{X}'X'Y\bar{Y}' + X\bar{X}\bar{Y}Y' + \bar{X}'X'\bar{Y}Y') \\
+ \ & C_{16}(\alpha)\,(\bar{X}'^2X'^2) \\
+ \ & C_{17}(\alpha)\,(\bar{X}^2Y^2 + X^2\bar{Y}^2 + X^2\bar{Y}'^2 + \bar{X}^2Y'^2 - 2X^2\bar{Y}\bar{Y}' - 2\bar{X}^2YY') \\
+ \ & C_{18}(\alpha)\,(X\bar{X}\bar{X}'X' + Y\bar{Y}\bar{Y}'Y' - \bar{X}'X'Y\bar{Y} - X\bar{X}\bar{Y}'Y')
\end{aligned}
$$

$$C_1(\alpha) = 1 + \frac{1}{4}\alpha^2 + \frac{9}{64}\alpha^4 + \frac{25}{256}\alpha^6 + \frac{1225}{16384}\alpha^8 + \cdots$$

$$C_2(\alpha) = -\frac{15}{16}\alpha^3 - \frac{105}{64}\alpha^5 - \frac{4725}{2048}\alpha^7 + \cdots$$

$$C_3(\alpha) = \frac{3}{4}\alpha^2 + \frac{45}{32}\alpha^4 + \frac{525}{256}\alpha^6 + \frac{11025}{4096}\alpha^8 + \cdots$$

$$C_4(\alpha) = -\frac{75}{16}\alpha^3 - \frac{735}{32}\alpha^5 - \frac{127575}{2048}\alpha^7 + \cdots$$

$$C_5(\alpha) = -\frac{285}{64}\alpha^3 - \frac{4305}{256}\alpha^5 - \frac{335475}{8192}\alpha^7 + \cdots$$

$$C_6(\alpha) = -\frac{75}{64}\alpha^3 - \frac{1995}{256}\alpha^5 - \frac{193725}{8192}\alpha^7 + \cdots$$

$$C_7(\alpha) = \frac{165}{32}\alpha^3 + \frac{3045}{128}\alpha^5 + \frac{259875}{4096}\alpha^7 + \cdots$$

$$C_8(\alpha) = \frac{45}{16}\alpha^3 + \frac{1575}{128}\alpha^5 + \frac{33075}{1024}\alpha^7 + \cdots$$

$$C_9(\alpha) = \frac{135}{128}\alpha^4 + \frac{2625}{512}\alpha^6 + \frac{231525}{16384}\alpha^8 + \cdots$$

$$C_{10}(\alpha) = \frac{315}{128}\alpha^4 + \frac{4725}{512}\alpha^6 + \frac{363825}{16384}\alpha^8 + \cdots$$

$$C_{11}(\alpha) = -3\alpha^2 - \frac{495}{32}\alpha^4 - \frac{5775}{128}\alpha^6 - \frac{407925}{4096}\alpha^8 + \cdots$$

$$C_{12}(\alpha) = -\frac{15}{16}\alpha^2 - \frac{855}{128}\alpha^4 - \frac{21525}{1024}\alpha^6 - \frac{782775}{16384}\alpha^8 + \cdots$$

$$C_{13}(\alpha) = -\frac{3}{8}\alpha^2 + \frac{45}{128}\alpha^4 + \frac{525}{128}\alpha^6 + \frac{209475}{16384}\alpha^8 + \cdots$$

$$C_{14}(\alpha) = \frac{3}{8}\alpha^2 + \frac{405}{128}\alpha^4 + \frac{2625}{256}\alpha^6 + \frac{385875}{16384}\alpha^8 + \cdots$$

$$C_{15}(\alpha) = \frac{21}{8}\alpha^2 + \frac{945}{64}\alpha^4 + \frac{22575}{512}\alpha^6 + \frac{804825}{8192}\alpha^8 + \cdots$$

$$C_{16}(\alpha) = \frac{3}{2}\alpha^2 + \frac{855}{128}\alpha^4 + \frac{8925}{512}\alpha^6 + \frac{584325}{16384}\alpha^8 + \cdots$$

$$C_{17}(\alpha) = \frac{15}{8}\alpha^2 + \frac{945}{128}\alpha^4 + \frac{4725}{256}\alpha^6 + \frac{606375}{16384}\alpha^8 + \cdots$$

$$C_{18}(\alpha) = \frac{9}{4}\alpha^2 + \frac{225}{16}\alpha^4 + \frac{11025}{256}\alpha^6 + \frac{99225}{1024}\alpha^8 + \cdots$$

$$C_1(\alpha) = b_{1/2}^{(0)} \left(\frac{1}{2}\right)$$

$$(1-\alpha^2)^2 \, C_2(\alpha) = b_{1/2}^{(0)} \left(-\frac{1}{4}\alpha - \frac{1}{4}\alpha^3\right) \qquad\qquad + b_{1/2}^{(1)} \left(\frac{1}{2} - \frac{1}{2}\alpha^2 + \frac{1}{2}\alpha^4\right)$$

$$(1-\alpha^2)^2 \, C_3(\alpha) = b_{1/2}^{(0)} \left(\frac{1}{2}\alpha^2\right) \qquad\qquad + b_{1/2}^{(1)} \left(-\frac{1}{4}\alpha - \frac{1}{4}\alpha^3\right)$$

$$(1-\alpha^2)^4 \, C_4(\alpha) = b_{1/2}^{(0)} \left(\frac{1}{4}\alpha - \frac{13}{4}\alpha^3 - \frac{13}{4}\alpha^5 + \frac{1}{4}\alpha^7\right) \quad + b_{1/2}^{(1)} \left(-\frac{1}{2} + \frac{15}{8}\alpha^2 + \frac{13}{4}\alpha^4 + \frac{15}{8}\alpha^6 - \frac{1}{2}\alpha^8\right)$$

$$(1-\alpha^2)^4 \, C_5(\alpha) = b_{1/2}^{(0)} \left(\frac{1}{16}\alpha - \frac{43}{16}\alpha^3 - \frac{5}{16}\alpha^5 - \frac{1}{16}\alpha^7\right) \quad + b_{1/2}^{(1)} \left(-\frac{1}{8} + \frac{15}{16}\alpha^2 + \frac{21}{8}\alpha^4 - \frac{9}{16}\alpha^6 + \frac{1}{8}\alpha^8\right)$$

$$(1-\alpha^2)^4 \, C_6(\alpha) = b_{1/2}^{(0)} \left(-\frac{1}{16}\alpha - \frac{5}{16}\alpha^3 - \frac{43}{16}\alpha^5 + \frac{1}{16}\alpha^7\right) + b_{1/2}^{(1)} \left(\frac{1}{8} - \frac{9}{16}\alpha^2 + \frac{21}{8}\alpha^4 + \frac{15}{16}\alpha^6 - \frac{1}{8}\alpha^8\right)$$

$$(1-\alpha^2)^4 \, C_7(\alpha) = b_{1/2}^{(0)} \left(-\frac{1}{8}\alpha + \frac{25}{8}\alpha^3 + \frac{25}{8}\alpha^5 - \frac{1}{8}\alpha^7\right) \quad + b_{1/2}^{(1)} \left(\frac{1}{4} - \frac{9}{8}\alpha^2 - \frac{17}{4}\alpha^4 - \frac{9}{8}\alpha^6 + \frac{1}{4}\alpha^8\right)$$

$$(1-\alpha^2)^4 \, C_8(\alpha) = b_{1/2}^{(0)} \left(\frac{3}{2}\alpha^3 + \frac{3}{2}\alpha^5\right) \qquad\qquad + b_{1/2}^{(1)} \left(-\frac{3}{16}\alpha^2 - \frac{21}{8}\alpha^4 - \frac{3}{16}\alpha^6\right)$$

$$(1-\alpha^2)^4 \, C_9(\alpha) = b_{1/2}^{(0)} \left(\frac{1}{32}\alpha^2 + \frac{7}{16}\alpha^4 + \frac{33}{32}\alpha^6\right) \qquad + b_{1/2}^{(1)} \left(-\frac{1}{16}\alpha + \frac{3}{16}\alpha^3 - \frac{23}{16}\alpha^5 - \frac{3}{16}\alpha^7\right)$$

$$(1-\alpha^2)^4 \, C_{10}(\alpha) = b_{1/2}^{(0)} \left(-\frac{3}{32}\alpha^2 + \frac{27}{16}\alpha^4 - \frac{3}{32}\alpha^6\right) \qquad + b_{1/2}^{(1)} \left(\frac{3}{16}\alpha - \frac{15}{16}\alpha^3 - \frac{15}{16}\alpha^5 + \frac{3}{16}\alpha^7\right)$$

$$(1-\alpha^2)^4 \, C_{11}(\alpha) = b_{1/2}^{(0)} \left(-\frac{13}{8}\alpha^2 - \frac{11}{4}\alpha^4 - \frac{13}{8}\alpha^6\right) \qquad + b_{1/2}^{(1)} \left(\frac{1}{4}\alpha + \frac{11}{4}\alpha^3 + \frac{11}{4}\alpha^5 + \frac{1}{4}\alpha^7\right)$$

$$(1-\alpha^2)^4 \, C_{12}(\alpha) = b_{1/2}^{(0)} \left(-\frac{7}{16}\alpha^2 - \frac{17}{8}\alpha^4 - \frac{7}{16}\alpha^6\right) \qquad + b_{1/2}^{(1)} \left(-\frac{1}{16}\alpha + \frac{25}{16}\alpha^3 + \frac{25}{16}\alpha^5 - \frac{1}{16}\alpha^7\right)$$

$$(1-\alpha^2)^4 \, C_{13}(\alpha) = b_{1/2}^{(0)} \left(-\frac{7}{32}\alpha^2 + \frac{15}{16}\alpha^4 + \frac{25}{32}\alpha^6\right) \qquad + b_{1/2}^{(1)} \left(\frac{1}{16}\alpha + \frac{1}{16}\alpha^3 - \frac{25}{16}\alpha^5 - \frac{1}{16}\alpha^7\right)$$

$$(1-\alpha^2)^4 \, C_{14}(\alpha) = b_{1/2}^{(0)} \left(\frac{5}{32}\alpha^2 + \frac{19}{16}\alpha^4 + \frac{5}{32}\alpha^6\right) \qquad + b_{1/2}^{(1)} \left(\frac{1}{16}\alpha - \frac{13}{16}\alpha^3 - \frac{13}{16}\alpha^5 + \frac{1}{16}\alpha^7\right)$$

$$(1-\alpha^2)^4 \, C_{15}(\alpha) = b_{1/2}^{(0)} \left(\frac{11}{8}\alpha^2 + \frac{13}{4}\alpha^4 + \frac{11}{8}\alpha^6\right) \qquad + b_{1/2}^{(1)} \left(-\frac{1}{8}\alpha - \frac{23}{8}\alpha^3 - \frac{23}{8}\alpha^5 - \frac{1}{8}\alpha^7\right)$$

$$(1-\alpha^2)^4 \, C_{16}(\alpha) = b_{1/2}^{(0)} \left(\frac{25}{32}\alpha^2 + \frac{15}{16}\alpha^4 - \frac{7}{32}\alpha^6\right) \qquad + b_{1/2}^{(1)} \left(-\frac{1}{16}\alpha - \frac{25}{16}\alpha^3 + \frac{1}{16}\alpha^5 + \frac{1}{16}\alpha^7\right)$$

$$(1-\alpha^2)^4 \, C_{17}(\alpha) = b_{1/2}^{(0)} \left(\frac{33}{32}\alpha^2 + \frac{7}{16}\alpha^4 + \frac{1}{32}\alpha^6\right) \qquad + b_{1/2}^{(1)} \left(-\frac{3}{16}\alpha - \frac{23}{16}\alpha^3 + \frac{3}{16}\alpha^5 - \frac{1}{16}\alpha^7\right)$$

$$(1-\alpha^2)^4 \, C_{18}(\alpha) = b_{1/2}^{(0)} \left(\frac{9}{8}\alpha^2 + \frac{15}{4}\alpha^4 + \frac{9}{8}\alpha^6\right) \qquad + b_{1/2}^{(1)} \left(-3\alpha^3 - 3\alpha^5\right)$$

$$C_1(\alpha) = \frac{1}{2}b_{1/2}^{(0)}$$

$$C_2(\alpha) = \frac{3}{4}\alpha b_{3/2}^{(0)} - (\frac{1}{2} + \frac{1}{2}\alpha^2)b_{3/2}^{(1)}$$

$$C_3(\alpha) = \frac{1}{4}b_{3/2}^{(1)}\alpha$$

$$C_4(\alpha) = (-\frac{15}{4}\alpha - \frac{15}{4}\alpha^3)b_{5/2}^{(0)} + (\frac{3}{2} + \frac{27}{8}\alpha^2 + \frac{3}{2}\alpha^4)b_{5/2}^{(1)}$$

$$C_5(\alpha) = (-\frac{15}{16}\alpha + \frac{15}{16}\alpha^3)b_{5/2}^{(0)} + (\frac{3}{8} - \frac{9}{16}\alpha^2 - \frac{3}{8}\alpha^4)b_{5/2}^{(1)}$$

$$C_6(\alpha) = (\frac{15}{16}\alpha - \frac{15}{16}\alpha^3)b_{5/2}^{(0)} + (-\frac{3}{8} - \frac{9}{16}\alpha^2 + \frac{3}{8}\alpha^4)b_{5/2}^{(1)}$$

$$C_7(\alpha) = (\frac{15}{8}\alpha + \frac{15}{8}\alpha^3)b_{5/2}^{(0)} + (-\frac{3}{4} - \frac{9}{8}\alpha^2 - \frac{3}{4}\alpha^4)b_{5/2}^{(1)}$$

$$C_8(\alpha) = \frac{9}{16}\alpha^2 b_{5/2}^{(1)}$$

$$C_9(\alpha) = -\frac{15}{32}\alpha^2 b_{5/2}^{(0)} + (\frac{3}{16}\alpha + \frac{9}{16}\alpha^3)b_{5/2}^{(1)}$$

$$C_{10}(\alpha) = \frac{45}{32}\alpha^2 b_{5/2}^{(0)} + (-\frac{9}{16}\alpha - \frac{9}{16}\alpha^3)b_{5/2}^{(1)}$$

$$C_{11}(\alpha) = \frac{3}{8}\alpha^2 b_{5/2}^{(0)} + (-\frac{3}{4}\alpha - \frac{3}{4}\alpha^3)b_{5/2}^{(1)}$$

$$C_{12}(\alpha) = -\frac{15}{16}\alpha^2 b_{5/2}^{(0)} + (\frac{3}{16}\alpha + \frac{3}{16}\alpha^3)b_{5/2}^{(1)}$$

$$C_{13}(\alpha) = \frac{9}{32}\alpha^2 b_{5/2}^{(0)} + (-\frac{3}{16}\alpha + \frac{3}{16}\alpha^3)b_{5/2}^{(1)}$$

$$C_{14}(\alpha) = \frac{21}{32}\alpha^2 b_{5/2}^{(0)} + (-\frac{3}{16}\alpha - \frac{3}{16}\alpha^3)b_{5/2}^{(1)}$$

$$C_{15}(\alpha) = \frac{3}{8}\alpha^2 b_{5/2}^{(0)} + (\frac{3}{8}\alpha + \frac{3}{8}\alpha^3)b_{5/2}^{(1)}$$

$$C_{16}(\alpha) = \frac{9}{32}\alpha^2 b_{5/2}^{(0)} + (\frac{3}{16}\alpha - \frac{3}{16}\alpha^3)b_{5/2}^{(1)}$$

$$C_{17}(\alpha) = -\frac{15}{32}\alpha^2 b_{5/2}^{(0)} + (\frac{9}{16}\alpha + \frac{3}{16}\alpha^3)b_{5/2}^{(1)}$$

$$C_{18}(\alpha) = \frac{9}{8}\alpha^2 b_{5/2}^{(0)}$$

Acknowledgements. Many thanks to P. Robutel who checked the computer expansions and to B. Morando who carefully went through the draft version of this text.

REFERENCES

Abu-El-Ata, N., Chapront, J.: 1975 'Développements analytiques de l'inverse de la distance en mécanique céleste' *Astron. Astrophys.* **38,** 57–66.

Brouwer, D. and Clemence, G.M.: 1961, *Methods of Celestial Mechanics*, Academic Press, London and New York.

Charlier, C. L.: 1902, *Die Mechanik des Himmels*, Berlin, (russian translation Nauka, Moscow, 1966).

Chenciner, A.: 1989, 'Intégration du problème de Képler par la méthode de Hamilton-Jacobi: coordonnées "action-angles" de Delaunay', dans: *Groupe de travail sur la lecture des Méthodes Nouvelles de la Mécanique Céleste*. Notes scientifiques et techniques du Bureau des Longitudes **S 026**, Paris.

Laplace, P.S.: 1798, *Traité de Mécanique Céleste*, T.I, in *Oeuvres Complètes* T.I, Gauthier-Villars, 1878.

Laskar, J.: 1989, 'Les variables de Poincaré et le développement de la fonction perturbatrice.' in: *Groupe de travail sur la lecture des Méthodes Nouvelles de la Mécanique Céleste*, Notes scientifiques et techniques du Bureau des Longitudes **S 026**, Paris.

Laskar J.: 1990a, 'Systèmes de variables et éléments',in *Méthodes Modernes en Mécanique Céleste (Goutelas 89) D. Benest, C. Froeschlé (eds)*.

Laskar J.: 1990b, 'Manipulation des séries',in *Méthodes Modernes en Mécanique Céleste (Goutelas 89) D. Benest, C. Froeschlé (eds)*.

Le Verrier, U. J.: 1855, *Annales de l'Observatoire Impérial de Paris*, t.I, Mallet-Bachelier, Paris.

Poincaré, H.: 1892, *Méthodes Nouvelles de la Mécanique Céleste*, ch. I, t.I.

Poincaré, H.: 1896, 'Une forme nouvelle des équations du problème des trois corps ' *C.R.A.S.* **123**, 1031–1035.

Poincaré, H.: 1897, 'Une forme nouvelle des équations du problème des trois corps ' *Bull. Astron.* **14**, 53–67.

Poincaré, H.: 1905, *Leçons de Mécanique Céleste*, tome I.

Poincaré, H.: 1907, *Leçons de Mécanique Céleste*, tome II.

Yuasa, M. and Hori, G.: 1979, 'New appoach to the planetary theory', in *Dynamics of the Solar System, R.L. Duncombe (ed.)*.

ORIGIN OF CHAOS AND ORBITAL BEHAVIOUR

IN SLOWLY ROTATING TRIAXIAL MODELS

Stéphane Udry and Louis Martinet

Geneva Observatory
Ch. des Maillettes 51
CH-1290 Sauverny
Switzerland

INTRODUCTION

It is now well accepted that rotating triaxial models may provide good representations of stellar systems like elliptical galaxies, cD galaxies, spiral bulges or bars. These systems are known to possibly have very different structural and dynamical parameters such as axis ratios, figure rotation or central density slope. In order to understand the shape of such systems and with a future aim of constructing self-gravitating models, it is necessary to know the structure and stability of orbits in triaxial models. The main stable periodic orbits form the backbone around which the global structure is built, whereas unstable orbits can contribute to trigger chaos and evolution in the system. In particular cases, resonant periodic orbits occupying a large volume in phase space also may play a non-negligible role in the construction of the skeleton of the system (Miralda-Escudé and Schwarzschild, 1989; Pfenniger and de Zeeuw, 1989). Our purpose in the present paper is to give the main results of a systematic study of the periodic orbits, the resonances and the onset of chaos in models with various axis ratios, figure rotation and central mass concentration (For more details see Martinet and Udry, 1990; and Udry, 1991).

MODEL DESCRIPTION AND INVESTIGATION TOOLS

Our model representing triaxial systems is given by the following density law:

$$\rho = \frac{\rho_c}{(1+m^2)^{3/2}} \quad \text{where} \quad m^2 = \frac{x^2}{a^2} + \frac{y^2}{b^2} + \frac{z^2}{c^2} \quad \text{with} \quad a \geq b \geq c \quad (a \neq c). \quad (1)$$

This model corresponds to the ellipsoidal triaxial version of the modified *Hubble* profile. The associated potential (called $P_{3/2}$) and forces may be easily computed by a judicious use of mixed Cartesian and confocal ellipsoidal coordinates (Udry and Pfenniger, 1988). The constants a, b, c are free model parameters, they allow a complete study of the configurations ranging from oblate to prolate triaxial systems. Rotation around the minor axis is also introduced up to values of the considered pattern angular velocity Ω_p which do not exceed a value corresponding to a corotation radius equal to $\sim 10a$. In order to fix the models for numerical calculations, the length unit

Predictability, Stability, and Chaos in N-Body Dynamical Systems
Edited by A.E. Roy, Plenum Press, New York, 1991

has been set to 1 kpc and the gravitational constant G to 1. We also have chosen $M(m=2.06)=1$ and $a=8$.

In order to study the connection between periodic orbits, resonances and onset of chaos in the system, we use classical investigation tools. Surfaces of section allow to locate the stochastic zones. The position and stability of the periodic orbits are numerically calculated by means of the often used method described in Hadjidemetriou (1975), Magnenat (1982) or Pfenniger (1984). This method is based on the analysis in the plane of section of the eigenvalues of the linear transformation of the Poincaré map. The program supplies stability parameters which may be related to the rotation number of the considered orbit (Hénon, 1965, 1966). The evolution of the periodic orbits, found at $(x, \dot{x}) = (x, 0)$ in the surfaces of section, as of function of the Hamiltonian H is then displayed in characteristic diagrams (H, x).

BEHAVIOUR IN THE EQUATORIAL PLANE

Main Orbits

Much work has already been done in order to understand the orbital behaviour in the equatorial plane of triaxial systems (see e.g. Contopoulos and Papayannopoulos, 1980; Martinet and de Zeeuw, 1988). In the non-rotating case, four main periodic orbits are found: two axial orbits and one prograde and one retrograde ellipses bifurcating from the y-axis as the energy increases. When a figure rotation is added to the model, the axis orbits open themselves and become ellipses. So, in the notation of Contopoulos and Papayannopoulos (1980) we have: the stable x_1 direct orbit elongated along the major axis; the direct x_2 (stable) and x_3 (unstable) orbits, slightly elongated along the intermediate axis; and finally the x_4 retrograde orbit, nearly circular. In the (H, x) plane, the families x_2 and x_3 are connected and form a bubble. This bubble shrinks and disappears when the rotation increases.

Origin of Chaos

The method of surface of section lets us obtain information on the relative contribution of regular and irregular (chaotic) motions in the global orbital structure of a system. Two examples proper to our investigation are shown in Figs. 1a and 2a for models with axis ratios respectively equal to $8:5:4$ and $8:2:1$, in both cases we have $H = -0.15$ and $\Omega_p = 0.002$. In Fig. 1a the classical periodic orbits x_1, x_2, x_3, x_4 are clearly indicated. In particular, in our representation (Ox major-axis), x_1 and the invariant curves around it are very close to the frontier of the accessible region.

The unstable x_3 family, which is the prolongation for $\Omega_p \neq 0$ of the unstable part of the y-axis orbit, seems to play a determining role in the onset of chaos. Therefore, it is interesting to obtain a quantitative estimate of the instability of these orbits. It is given by their e-folding rate evaluated by the Poincaré characteristic exponents $\alpha = T^{-1} \ln \lambda$, where T is the orbital period and λ the modulus of the eigenvalue of the associated Jacobian matrix along the most dilating direction (see e.g. Heissler et al., 1982). These exponents correspond to the Lyapunov characteristic exponents in the case of unstable periodic orbits. They are given as a function of H in Fig. 3 for models with axis ratios equal to $8:7.18:6.55$ and $8:2:1$ for various values of rotation. The progressive reduction of α with increasing Ω_p and its dependence on the deviation from axisymmetry are clearly shown. Nevertheless, two remarks have to be pointed out. First, from the systematic exploration of surfaces of section, we infer that $\alpha_{\max}$ does not necessary coincide with the most extended chaotic region in the surfaces of section. Secondly, for H larger than the value corresponding to the right

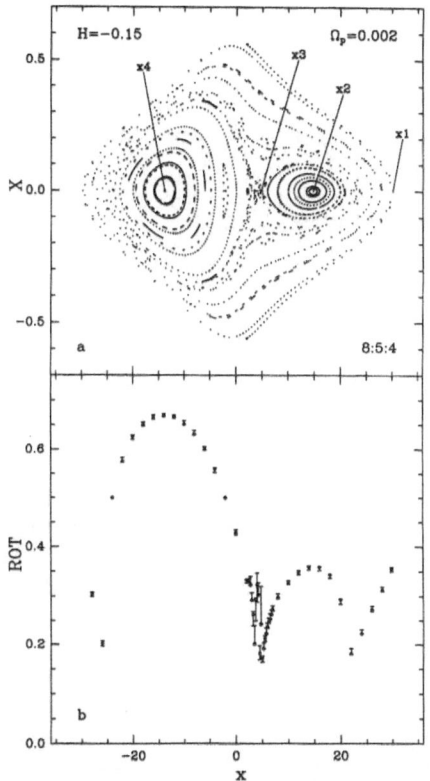

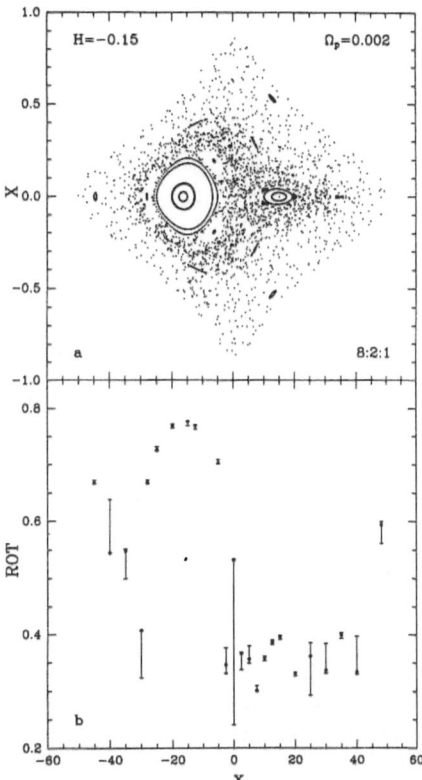

Fig. 1. a. Surface of section in a model with axis ratios equal to 8:5:4 for $H = -0.15$ and $\Omega_p = 0.002$.
b. The corresponding rotation numbers for points starting at $(x, 0)$ in the surface of section.

Fig. 2. Same as Fig. 1 in a model with axis ratios 8:2:1 for $H = -0.15$ and $\Omega_p = 0.002$.

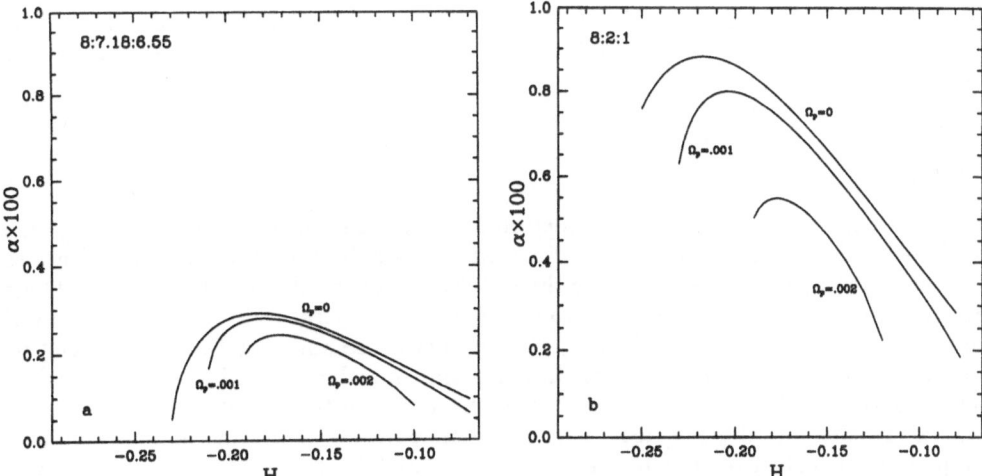

Fig. 3. e-folding rate α of the x_3 unstable familiy for models with axis ratios equal to 8:7.18:6.55 (a) and 8:2:1 (b) for different values of Ω_p.

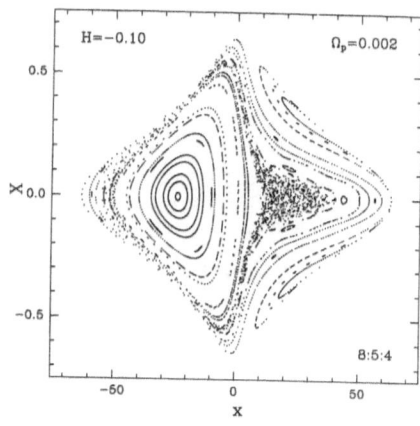

Fig. 4. *Surface of section in a model with axis ratios 8:5:4 for $H = -0.10$ and $\Omega_p = 0.002$. At this value of H the (x_2, x_3) bubble does not exist anymore.*

edge of the (x_2, x_3) bubble (Fig. 4) or for Ω_p larger than the critical value $(\Omega_p)_c$ of the disappearance of the bubble *i.e.* when the x_3 orbit no longer exists, some irregular behaviour subsists in the system. The previous remarks suggest that, apart from the presence of the unstable x_3 orbit, some higher order resonance interactions contribute to the propagation of chaos observed in our models.

In order to see what resonances are implicated in different cases, we will consider the rotation number $rot(x)$ corresponding to each invariant curves as a function of x, starting point of the orbit with $\dot{x}_0 = y_0 = 0$ at a given H. It is obvious from the definition of the rotation number that it is not possible to determine it unambiguously in a zone of irregular motions and its discontinuities allow to see what resonance interactions trigger significant chaos. These rotation numbers are given in Figs 1b and 2b for the surfaces of section given in Figs 1a and 2a. In the first case, the involved resonances are in the range (1/6, 1/3) whereas in the second one they are more numerous (between 1/5 and 2/3) and the stochastic zone is more developped.

Summing up, we can conclude that the possible chaotic behaviour in the equatorial plane of our triaxial systems generally spreads out from the x_3 unstable periodic orbit and that interactions with higher order resonances cause it to extend to a more or less important region of phase space depending on the figure rotation and on the axis ratios of the model since both of them determine the possible resonances in the system.

BOXLETS

As we have seen, unstable multiperiodic orbits are important for the propagation of chaos in the models. If multiperiodic orbits are stable, they also may contribute to matter trapping and consequently to the elaboration of the shape of the system. These orbits are related to resonances *i.e.* to rational frequency ratios in the system, we will thus call them resonant periodic orbits. Their existence and stability may depend on the values of the model parameters. In particular, Miralda-Escudé and Schwarzschild (1989) and Pfenniger and de Zeeuw (1989) have shown that in models with a sharp central density distribution, low order resonant periodic orbits like banana- or fish-shaped orbits, called boxlets by these authors, may fill a substantial volume in phase space and replace the boxes which can be destroyed if they approach the very central region too closely. The crucial role of these orbits in the construction of self-gravitating models is thus brought back into discussion.

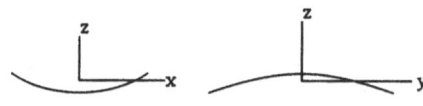

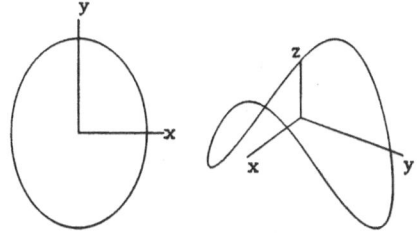

Fig. 5. *Example of the shape of the vertical bananas appearing in the models.*

In this work we are interested in triaxial rotating systems, the main involved frequencies are thus ω_x, ω_y, $\omega_z = \nu$ corresponding to the natural oscillations along each axis of the ellipsoid and $(\Omega \pm \Omega_p)$ the retrograde and direct relative revolution frequencies around the rotation axis. In view of their influence zone in the surface of section in the plane case and of their apparent importance in 3D N-body simulations (Pfenniger and Friedli, 1991), resonant periodic orbits of low order ($m = \omega_y/\omega_x \leq 2$ or $n = \nu/(\Omega \pm \Omega_p) \leq 2$) seem to be the main ones to consider. Because of their shape in configuration space, $m = 2$ ($n = 2$) orbits will be called (vertical) banana orbits. An example of such an orbit is displayed in Fig. 5. The resonant orbits will be found by following the stability parameters on the main stable periodic orbits in the equatorial plane which are related to the frequency ratios (m, n) of the bifurcating orbits.

THREE DIMENSIONAL ORBITAL BEHAVIOUR

In the third dimension, a great interest has been put into the study of the vertical instability strip on the plane retrograde orbit (Binney, 1981). It corresponds to a $n = 1$ resonance which gives rise to the so called stable and unstable anomalous orbits. The latter are inclined orbits respectively circling the x- and y-axis. Towards decreasing energies, they join the z-axis orbit or come back in the plane on the retrograde orbit depending on the central mass concentration, rotation or axis ratios of the model (Martinet and Pfenniger, 1987). For more information on the structure and stability of these orbits, see also Heissler et al. (1982) and Magnenat (1982).

Only few studies deal with other existing vertical periodic orbits (Mulder and Hooimeyer, 1984; Pfenniger, 1984; Cleary, 1989; Patsis and Zachilas, 1990). Although they provide a rather complete set of low order resonant orbits appearing in realistic triaxial models, these authors only consider a single or a few discrete values of axis ratios and a narrow range of rotation. Our purpose now is to find all the low order resonances ($n, m \leq 2$) involved in a realistic triaxial model, with varying axis ratios, figure rotation and central mass concentration.

Vertical Resonances on the Direct Orbits

No $n = 1$ resonance is found on the direct orbits. On the contrary, the $n = 2$ resonance appears on x_1 for a large range of model parameters. Fig. 6, called existence diagram, displays the set of axis ratios for which the resonant periodic orbit exists. The solid line represents the limit of the existence region ($\exists$) in the non-rotating case. On the limit, the bifurcation occurs at the minimum of the Hamiltonian $H_{\min}$, or very

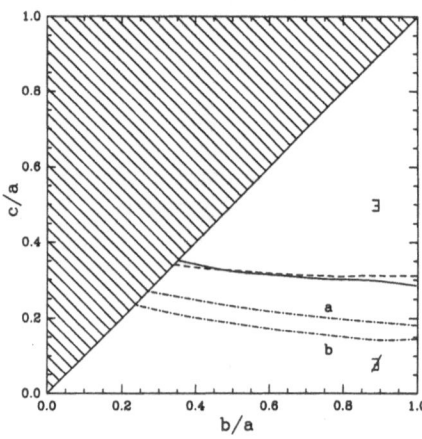

Fig. 6. *Existence diagram* $(b/a, c/a)$ *of the* $n = 2$ *stable vertical resonant orbit bifurcating from the* x_1 *direct orbit. In the non-rotating case the limit of the existence zone* $(\exists)$ *is represented by the solid line and by the dashed line for* $\Omega_p = 0.002$. *The dashed-dotted lines provide the limits of the existence zone as we add to the* $P_{3/2}$ *model a central Plummer potential with respective masses* $M_P = 0.001$ (a) *and* $M_P = 0.002$ (b), R_P *the extent of the Plummer sphere being equal to 0.1 and* $\Omega_p = 0$ *in both cases.*

near to it. As c/a increases, it appears then at higher and higher values of H. The limit depends weakly on the departure from axisymmetry (b/a value); for c/a greater than ~ 0.3 the vertical bananas bifurcate. Rotation also has no relevant influence on this limit. The existence region is only slightly widened towards the near axisymmetric models and shortened on the side of prolate models. The limit for the case $\Omega_p = 0.002$ is given by the dashed curve in the figure.

Observations have shown that giant ellipticals often present central cusps of brightness (Schweizer, 1979; Lauer, 1985). Central mass concentrations are also suspected to appear in the shape of black holes or at least of high density regions in many other galaxies (Lynden-Bell, 1969; Sargent et al., 1978; Rees, 1984). These facts lead us to introduce in the central part of our models, a steep-profile component which we model by a Plummer potential. The central mass concentration acts in the sense of an increase of the existence zone. This effect is illustrated in the figure by the dashed-dotted curves giving the limit of the existence zone for a Plummer mass $M_P = 0.001$ (a) and $M_P = 0.002$ (b). In both cases, the characteristic radius of the Plummer sphere is 0.1.

Among the vertical resonances of order 2, the bananas bifurcating from x_1 seem to be the most important ones. They occupy a wide part in the existence diagram, they are stable over a large range in parameter space and their shape, elongated in the x-direction, fit easily into the global structure of the system. They may thus have some dynamical influence and contribute to the 3-dimensional shape of the system. Furthermore, these orbits seem to be important in N-body simulations (Pfenniger and Friedli, 1991) which, by nature, select only the main resonances appearing in the system. Particularly, they seem to be crucial in the formation processes of box- and peanut-shaped galaxies (Combes et al., 1990).

Other Resonances of Order 2

Examining the bifurcations from the main families, other $n = 2$ vertical resonant periodic orbits exist in the models. They appear for small values of c/a (≤ 0.3) and are related to the x_4 retrograde, z-axis and x_2 direct orbits. One of them, connecting x_4 and Oz, presents the same stability behaviour as the anomalous orbits. It bifurcates backwards in energy and, depending on the central mass concentration, may come back on the retrograde orbit instead of reaching the z-axis orbit. The other vertical bananas (from x_2 and from the z-axis) are stable for small values of H and so may have some dynamical influence only in the central region of flattened models.

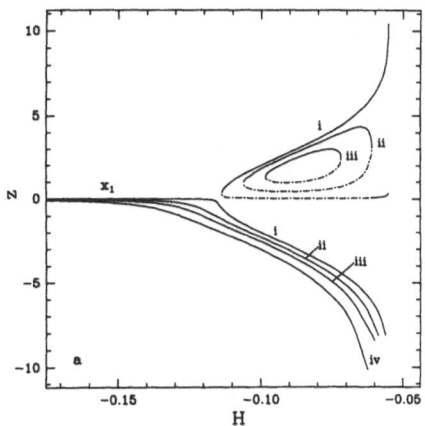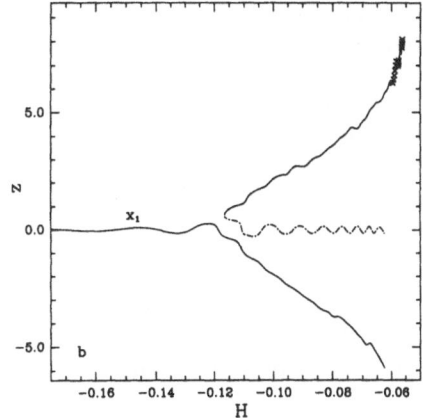

Fig. 7. *Characteristic diagrams of the vertical bananas bifurcating from x_1: a. as a Plummer sphere of varying mass M_P and extent $R_P = 1.0$ is placed at $(x_0, y_0, z_0) = (-20, 0, -20)$ below the plane in model with axis ratios 1:0.625:0.5 and $\Omega_p = 0.02$. The considered values for M_P are 0.01 (i), 0.25 (ii), 0.5 (iii) and 1.0 (iv). The stable parts of the curve are indicated by a solid line and the dashed dotted line corresponds to the vertically unstable parts of x_1. b. Same as a in the noisy $P_{3/2}$ potential. Complex instability is indicated by crosses.*

In the equatorial plane, in addition to the main orbits, bananas bifurcating from x_1 and x_4 are found. Their existence diagrams are similar but the zones of existence and of non-existence are inverted. Similarly to the vertical case, bananas from x_1 exist in a wide range of model parameters ($b/a \geq 0.4$). Bananas from x_4 appear only in very triaxial models ($c/a \leq 0.3$).

ASYMMETRY

Elliptical galaxies often present isophotes with departures from ellipticity or small irregularities. Similar features are also observed in *N*-body simulations (Pfenniger and Friedli, 1991). So, in order to compare our results with more realistic systems, we introduce asymmetry in our models. The latter manifests itself by global or local irregularities of the isodensity contours corresponding to low or high frequency spatial perturbations. It can thus be introduced in two different ways. The low frequency perturbation is modelled by a Plummer sphere outside the system whose tidal effect smoothly deforms the density distribution. The high frequency spatial perturbation (noise), representing the granularity effect of some system's components, is introduced by superposing a sinusoidal function to the main potential.

An important effect of asymmetry consists in a change in the connections between the orbits. Generally speaking, in a typical characteristic diagram of a symmetrical case, we have a "main" stable orbit which bifurcates into two stable branches at a point of change of stability. In the asymmetrical case, we find that a gap appears between the two stable branches, they are no longer connected: one prolongs the stable part of the "main" orbit and the other is related to its unstable part forming a bubble. This bubble shrinks and disappears as the amplitude of the perturbation increases. An example of such a behaviour is given in Fig. 7a for the vertical bananas bifurcating from x_1 as Plummer spheres of masses between 0.01 (i) and 1 (iv) are placed outside the system. A gap also appears for the same orbits when asymmetry is induced by a higher frequency perturbation. The high frequency even appears on the characteristic curve which then seems noisy (Fig. 7b). For more details see Udry (1991). Another example of such a behaviour has already been met with the introduction in the models of a figure rotation which destroys the symmetry between the direct x_2 and x_4 retrograde orbits triggering a gap between them in the characteristic diagram. The introduction of asymmetry let the main orbits transform themselves gradually taking a new shape instead of bifurcating abruptly and so in a natural way, emphasize the orbits which play some dynamical role in the system. For example the above mentioned x_1 direct orbit deviates from the plane and progressively becomes a vertical banana-shaped orbit thus showing its importance in establishing the 3-dimensional shape of the system. Asymmetry also increases stochasticity in the models. This effect grows with the amplitude and the frequencies of the perturbation.

CONCLUDING REMARK

Real galaxies may present radial variations of ellipticity of isophotes, asymmetries or central cusps of brightness. They are rather non-integrable systems. Stochasticity and resonance trapping may become important features in such systems. In particular, boxlets replacing box orbits may occupy a non-negligible volume in phase space and so have to be considered in the construction of self-gravitating models whereas they do not exist in integrable potentials.

REFERENCES

Binney, J. J.: 1981, *Monthly Notices Roy. Astron. Soc.* **196**, 455
Cleary, P. W.: 1989, *Astrophys. J.* **337**, 108
Combes, F., Debbash, F., Friedli, D., Pfenniger, D.: 1990, *Astron. Astrophys.* **233**, 82
Contopoulos, G., Papayannopoulos T.: 1980, *Astron. Astrophys.* **92**, 33
Hadjidemetriou, J.D.: 1975, *Celest. Mech.* **12**, 255
Hénon, M.: 1965, *Annales d'Astrophysique,* Paris, **28**, 992
Hénon, M.: 1966, *Bulletin astron. CNRS 3e série* **1**, 57
Heissler, J., Merritt, D.R., Schwarzschild, M.: 1982, *Astrophys. J.* **258**, 490
Lauer, T.R.: 1985, *Astrophys. J.* **292**, 104
Lynden-Bell, D.: 1969, *Nature* **223**, 690
Magnenat, P.: 1982, *Astron. Astrophys.* **108**, 89
Martinet, L., Pfenniger, D.: 1987, *Astron. Astrophys.* **173**, 81
Martinet, L., Udry, S.: 1990, *Astron. Astrophys.* **235**, 69
Martinet, L., de Zeeuw, T.: 1988, *Astron. Astrophys.* **206**, 269
Miralda-Escudé, J., Schwarzschild, M.: 1989, *Astrophys. J.* **339**, 752
Mulder, W. A., Hooimeyer, J. R. A.: 1984, *Astron. Astrophys.* **134**, 158
Patsis, P.A., Zachilas, L.: 1990, *Astron. Astrophys.* **227**, 37
Pfenniger, D.: 1984, *Astron. Astrophys.* **134**, 373
Pfenniger, D., Friedli, D.: 1991, in preparation
Pfenniger, D., de Zeeuw, T.: 1989, in *"Dynamics of Dense Stellar Systems"*, ed. D. Merritt, cambridge Univ. Press, Cambridge, p.81
Rees, M.: 1984, *Ann. Rev. Astr. Astrophys.* **22**, 471
Sargent, W.L., et al.: 1978, *Astrophys. J.* **221**, 731
Schweizer, F.: 1979, *Astrophys. J.* **233**, 23
Udry, S., Pfenniger, D.: 1988, *Astron. Astrophys.* **198**, 135
Udry, S.: 1991, submitted to *Astron. Astrophys*

PART II

DYNAMICS OF ASTEROIDS, COMETS AND METEORS

MODELLING: AN AIM AND A TOOL FOR THE STUDY OF THE CHAOTIC BEHAVIOUR OF ASTEROIDAL AND COMETARY ORBITS

Claude Froeschlé

Observatoire de la Côte d'Azur - B.P. 139
F-06003 NICE Cedex, France

ABSTRACT

Chaotic solutions of Newton equations are deeply rooted to both asteroidal and cometary dynamics. Great progresses have been made in the last decade using tools and results of the theory of dynamical systems. Both the existence of Kirkwood gaps and the transfer of comets into observable orbits will be related to chaos. Mapping and massive parallel computers are the main tools which will be discussed.

I. INTRODUCTION

For many years the motion of celestial bodies has been more or less implicitly thought of as akin to clockwork regularity and except for these mysterious and frightening objects called comets the efforts of the celestial mechanicians assumed that all problems are basically integrable. Therefore the game of classical perturbation theory using canonical transformations of the variables has been to successively push the non integrable part of the Hamiltonian to higher and higher orders. Such an approach lasted until the work of Poincaré who pionered a new approach to mechanics investigating the qualitative nature of the motion. He showed that most Hamiltonian systems do not possess the integrals required to reduce the solution to quadratrures. He also found that motions near unstable periodic orbits possess almost unimaginable complexity. He wrote "on sera frappé de la complexité de cette figure que je ne cherche même pas à tracer. Rien n'est plus propre à nous donner une idée de la complication du problème des trois corps et en général de tous problèmes dynamiques où il n'y a pas d'intégrale uniforme....". Poincaré's discoveries become fully appreciated in the sixties with the computer revolution. Indeed the seminal numerical studies of Hénon and Heiles (1964) show that complicated behaviour can result from simple Hamiltonian systems. Such irregular behaviour is called "chaotic" which first means that the so-called "chaotic orbits" look markedly more irregular than all others which look quasi-periodic. For conservative systems with two degrees of freedom chaotic behaviour may be distinguished from quasi-periodic behaviour using Poincaré's maps, i.e. looking at the set of points generated by the successive intersections of a trajectory with a surface of section which is generally some particular plane of the phase space. The successive points of a quasi-periodic trajectory lie on smooth curves, while the points generated by a chaotic trajectory seem to fill an area in an irregular way. For systems with more than two degrees of freedom, Poincaré's map technics are more problematic. Fortunately Lyapunov characteristic exponents (LCE) provide objective criteria of stochasticity (see Froeschlé 1984). Let us recall that the LCEs of a given trajectory characterize the mean exponential rate of divergence of a neighbouring trajectory and that for the trajectory of a dynamical system with any number of degrees of freedom they provide a quantitative measure of the degree of stochasticity. All these technics (surface of section and LCEs) have been used to study the stochasticity of both asteroidal and cometary orbits. Depending of the origin and the strengh of the stochasticity it is necessary to compute many orbits over time-spans which can range from 10^5 to 10^9 years and therefore either sophisticated computers or special methods (mappings) have been used.

In section II we revisit briefly the different tools used by dynamical system theory and some typical examples.

In section III we will review some examples of asteroidal orbital chaos and we will discuss the Chirikov-Wisdom mapping which has been so successful to explain the 3/1 resonance.

Section IV will be devoted to cometary orbits which are by essence chaotic.

We will finally discuss in section V a new general approach for building mappings.

II. THE PARADIGM OF THE DYNAMICAL SYSTEMS

1. Introduction

In recent years much work, not only numerical, has been devoted to the investigation of the ergodic properties of classical dynamical system (Hénon 1981). However, while the extreme cases of integrable and of ergodic systems are at least partially understood in a mathematical context the situation is different for many models of physical interest. Of course the now classical KAM theory has been the great break-through to prove the persistence in some regions of the phase space of uniform integrals. Another important break-through for understanding the behaviour of the so-called chaotic regions is the theory of Lyapunov exponents (hereafter called LCE) which measure the mean exponential rate of divergence of nearby trajectories. The use of such exponents dates back to Lyapunov, but in a form adapted to the theory of dynamical systems and to ergodic theory. It was only in 1968 that Oseledec (see Benettin et al. 1980) published his non-communitative Ergodic Theorem which provides a general and simple way to compute all the LCEs. The first numerical characterization of the stochasticity of a phase space trajectory in terms of the divergence of nearby trajectories was introduced by Hénon and Heiles and then further studied by Chirikov; Ford; Froeschlé; Froeschlé and Scheidecker who in order to give a precise quantitative definition of exponential divergence and thus of stochasticity have been led to considering the spectral properties of a linear operator. Studying directly the behaviour of the eigenvalues of linear tangential mapping for discrete dynamical systems they computed LCEs without mentioning the terms (Comme Monsieur Jourdain faisait de la prose sans le savoir).

The connection between LCEs and the preceding numerical works has been given and popularised by Benettin et al. (1980) who give also a simple procedure to compute all LCEs.

A review on the use of LCEs in celestial mechanics can be found in Froeschlé (1984). We will not present here the full theory but only recall some results showing the different kinds of diffusion in phase space. After a short survey of the tool-box of dynamical system we will go through the systems with one, two and three degrees of freedom.

2. The tool-box of Dynamical Systems

2.1. The theoretical tools

2.1.a. The integrable systems, perturbations theory

These were essentially the subject of all other Cortina meetings. Besides the proceedings of these meetings we recommend the proceedings of the Goutelas School (1988).

2.1.b. The hyperbolic systems

Besides the lectures by A. Milani and C. Simo, in this book we recommend the book on Chaotic behaviour of deterministic systems, Les Houches (1981), and the books by Lichtenberg and Lieberman (1983) and Guckenheimer and Holmes (1983).

2.2. The numerical tools

Here again, we cannot develop this very important topic. Let us emphasize that both the numerical method (constant or variable steps, predictor corrector methods, regularisation etc...) and the kinf of computer (CRAY, Connection Machine) used are very important for the validity of the results.

2.2.a. Modelling of the problem

Already in Newton equations there is an ideal situation like point mass approximation. Within this model we have to integrate the system

$$\frac{dx}{dt} = F(x, t)$$

A first approach consists in integrating the full system wich gives all the motion including the short period terms. Of course this kind of approach is very costly as far as computing time is concerned.

A further step in modelling is to average the equations either analytically or numerically. Unfortunately in order to eliminate the short periodic terms analytical averaging requiers series expansions followed by truncation. Therefore the procedure fails when the eccentricity is not reasonably small. Numerical averaging overcomes this problem and allows a factor ten on the computing time. However the problem remains of the validity of the averaging procedure, i.e. to which extend the solution of the average equations remains close to the averaged full solution.

Last but not least, the "nec plus ultra" in modelling is to replace the system of equations by an explicit mapping which maps one surface of section into itself. Of course the same questions than those about averaging remain.

2.2.b. Surface of section. Poincaré map - mappings

In this paragraph we will reproduce in extenso the very clear text of the lecture given by Hénon (1981) at Les Houches Summer School.

Since we are interested in the asymptotic behaviour of a trajectory, it is not really necessary to follow this trajectory in great detail; it might be sufficient to "sample" it from time to time. this is the idea behind the method of the surface of section. For the exposition, it will be convenient to consider first the case $N = 3$, so that phase space can be pictured as the familiar three-dimensional space. We select in this phase space a *surface of section* Σ which in the present case will be an ordinary two-dimensional surface; and we consider the successive intersections $Y_0, Y_1, Y_2, \ldots$ of the trajectory with Σ. Since we assume the trajectory to be recurrent, there will be an infinite sequence of such points. The sequence can also be extended towards the past: $\ldots Y_{-3}, Y_{-2}, Y_{-1}, Y_0, \ldots$.

This sequence has the fundamental property that if one of its points Y_i is given, then the next point Y_{i+1} can be deduced. This is quite obvious: all one has to do is to follow the trajectory from Y_i, by integrating the differential equations, until it intersects Σ again: this new intersection is Y_{i+1}. We have thus a mapping G of Σ on itself which is called the *Poincaré map*. Thus:

$$Y_{i+1} = G(Y_i). \tag{1}$$

More generally, one has

$$Y_{i+j} = G^j(Y_i). \tag{2}$$

Since the trajectory can be integrated towards both directions of time, the mapping is invertible: G^{-1} exists, and

$$Y_{i-1} = G^{-1}(Y_i). \tag{3}$$

More generally, Eq. (2) holds for any j, positive or negative.

The generalization to an arbitrary order N is obvious: we select, in the N-dimensional phase space, an $(N-1)$-dimensional subspace Σ, which should now be properly called a *space of section*, but is in practice frequently referred to as a "surface of section". We then consider the successive intersections... $Y_{-1}, Y_0, Y_1, Y_2,$ ... of the trajectory with Σ. We shall call $M = N - 1$ the dimension of the space of section. We introduce in this space a system of coordinates $y_1, \ldots, y_M$, and we represent the coordinates of the point Y_i by $y_{i,1}, \ldots y_{i,M}$. Then the mapping G can be written

$$y_{i+1,1} = g_1(y_{i,1}, \ldots y_{i,M}),$$
$$\vdots \tag{4}$$
$$y_{i+1M} = g_M(y_{i,1}, \ldots y_{i,M}).$$

When considering the mapping itself, rather than the sequence of points, it will often be convenient to adopt a simpler notation: two consecutive points Y_i and Y_{i+1} are represented by Y and Y', and eq. (4) becomes

$$y'_1 = g_1(y_1, \ldots y_M),$$
$$\vdots \tag{5}$$
$$y'_M = g_M(y_1, \ldots y_M).$$

Now comes the fundamental step: we decide to consider only the sequence of points Y_i from now on, and to forget about the rest of the trajectory, i.e. the detail of what happens between two intersections. The sequence is considered as being defined directly by the mapping G, and not any more by the system of differential equations, which will also be forgotten. The justifications for doing this are as follows:

(1) Experience shows that the essential properties of the differential system are reflected in equivalent properties of the mapping. A close correspondance exists between these two objects. Example: a simple periodic orbit of the differential

system, closing back upon itself after one revolution, corresponds to a fixed point of the mapping G; the periodic orbit is stable if and only if the fixed point is stable; and so on.

(2) The new problem is much simpler: we have only the ordinary equations (5) to consider instead of the differential equations. Also the dimension of the relevant space has been reduced by one unit: a dynamical system of order N gives rise to a mapping in an $(N - 1)$-dimensional space. Therefore the study is easier, both for theory and for numerical experiments.

(3) The essential properties of the system, related to the long-term behaviour, are more clearly seen because the irrelevant details of the short-term evolution have been eliminated. (Conversely, the surface-of-section would be of no use for a short-term problem, such as the trajectory of an interplanetary probe!)

(4) The graphical representation of the results is easier. For instance for $N = 3$, the surface of section has two dimensions and is much more easily represented than the three-dimensional phase space.

(5) Computing time at first is not improved by the introduction of a surface of section. If our problem is originally defined by a set of differential equations, and we introduce the surface of section afterwards, then in general it is not possible to obtain explicit equations for the mapping G; the only way to compute the image of a point Y_i is to go back to the differential equations and to integrate the trajectory until the next intersection is reached.

It is here, however, that the "philosophy" of numerical experiments comes into play in a crucial way. Remember that we are not interested in a particular dynamical system, but rather in the general properties of dynamical systems. Now the study of dynamical systems can be reduced to the study of iterated mappings, through the introduction of a surface of section. We may as well, then, attack this latter subject directly. Instead of defining the mapping G implicitly through the differential equations, we shall define it by giving explicitly the functions $g_1, \ldots, g_M$ in Eq. (5).

This reduces the computing time drastically (typically by a factor of the order of 1000), since all we have to do now is to evaluate expression instead of integrating differential equations. The accuracy is also much better; only round-off errors are left.

The chosen mapping G should be invertible, in order to correspond to the case of a dynamical system. This seems to be the only condition. Numerous experiments have shown that essentially the same properties are found in mappings defined implicity from dynamical systems and in mappings defined explicitly from given equations.

When studying a mapping, we shall sometimes refer to the infinite sequence of points Y_i as a "trajectory", or "orbit", since this is now the equivalent of the original trajectory in phase space.

3. System with one degree of freedom

This case is simple but nevertheless instructive.

A Hamiltonian system with one degree of freedom has always one integral, H; therefore *it is always integrable*. Phase space has only two dimensions (q_1, p_1) and can be easily represented. The subspaces H=constant are curves; therefore a trajectory essentially coincides with a curve H=constant. This means that the system is

ergodic. (It may happen, however, that H=constant corresponds to several discon-
nected curves; in that case, a trajectory will lie on one of the curve only).

The space of section has a dimension $2n - 2 = 0$; it reduces to a point, or a
finite sequence of points forming a cycle. This shows that the case of one degree of
freedom is in a sense trivial.

As an example let us consider the case of the Pendulum whose Hamiltonians is
given by

$$H(\psi, \ P) = \frac{P^2}{2} + \alpha^2 \ \cos \psi \tag{6}$$

Fig.1 shows in phase space the set of curves $H = h$. We see clearly the different kinds
of motion.

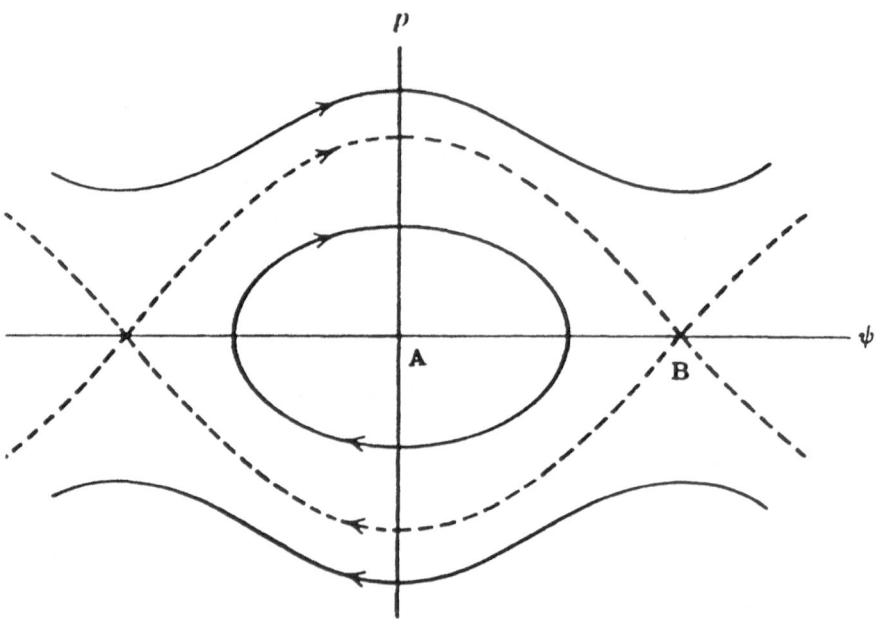

Fig.1 . Phase space diagram of the pendulum.

The fixed point $A(\psi = 0)$ is elliptic and corresponds to the minimum of the
potential

$$V = -\alpha^2 \ \cos \psi$$

The fixed point $B(\psi = \pi)$ is hyperbolic, i.e. unstable. The dashed curve is the
separatrix on which the point B is reached with an infinite time. This curve separates
two distinct modes of behaviour, oscillation and circulation, which can be separated
by infinitesimal differences in initial conditions. The celestial system has analogous
separatrixes, where behaviour can be grossly modified by the infinite number of small
sinusoidal components and/or by small differences in initial conditions. The result
is that over a significant range of initial conditions, long term behaviour may be
unpredictable or chaotic. The effect is exacerbated if changes in period during high-
amplitude oscillations render some short-period Fourier components to have long
period, i.e. resonant terms which create a resonance overlap.

4. System with two degrees of freedom

Here we come to the really interesting and non-trivial cases. For a Hamiltonian system with two degrees of freedom, in general, only one integral is known: the Hamiltonian $H(q_1, q_2, p_1, p_2)$. Therefore, the system is generally non-integrable. One cannot solve it analytically and write down an explicit general solution. However, the problem can be attacked by numerical computations. We will do it on the famous old planar restricted three body problem defined by the following restrictions. Again we will closely follow Hénon (1981).

(1) The third body M_3 has zero mass. Therefore it does not influence bodies M_1 and M_2, whose motions are given as a solution of a 2-body problem.

(2) A particular motion is chosen for M_1 and M_2: they describe circular orbits around their common center of mass.

(3) The motion of M_3 takes place in the orbital plane of M_1 and M_2.

The restricted problem is of interest both because it represents a useful first approximation to many real problems, and because it is the simplest unsolved case of the N-body problem.

The problem is reduced to a non-dimensional form as follows: (i) one takes the sum of the masses of M_1 and M_2 as unit of mass; these two masses are then called $1-\mu$ and μ respectively; (ii) one takes as unit of length the constant distance between M_1 and M_2; therefore the radii of their orbits are respectively μ and $1-\mu$; (iii) one chooses the unit of time in such a way that the gravitational constant $G = 1$. It follows from these choices that the angular velocity of M_1 and M_2 is also equal to 1. It is convenient to use a system of axes (x, y) which rotates with M_1 and M_2, with the x axix pointing towards M_2. In this system, M_1 has fixed coordinates $(-\mu, 0)$ and M_2 has fixed coordinates $(1 - \mu, 0)$. M_3 moves in the (x, y) plane, and its equations of motion are easily shown to be

$$\ddot{x} = 2\dot{y} + x - (1 - \mu)\frac{x + \mu}{r_1^3} - \mu\frac{x - 1 + \mu}{r_2^3},$$

$$\ddot{y} = -2\dot{x} + y - (1 - \mu)\frac{y}{r_1^3} - \mu\frac{y}{r_2^3}, \tag{7}$$

with

$$r_1 = [(x + \mu)^2 + y^2]^{1/2}, \qquad r_2 = [(x - 1 + \mu)^2 + y^2]^{1/2}. \tag{8}$$

This is brought into the form of a Hamiltonian system with two degrees of freedom by

$$q_1 = x, \ q_2 = y, \ p_1 = \dot{x} - y, \ p_2 = \dot{y} + x, \tag{9}$$

$$H = \frac{1}{2}(p_1^2 + p_2^2) + p_1 q_2 - p_2 q_1 - \frac{1 - \mu}{r_1} - \frac{\mu}{r_2}. \tag{10}$$

Therefore there exists an integral; it is customary to define it as $C = -2H$, and to call it the *Jacobi integral*. In the original variables, it is given by

$$C = x^2 + y^2 + \frac{2(1 - \mu)}{r_1} + \frac{2\mu}{r_2} - x^2 - y^2 \tag{11}$$

No other integral is known. We follow the usual procedure: we choose a particular value of C, and we define a surface of section by

$$y = 0, \qquad \dot{y} > 0. \tag{12}$$

The coordinate in the surface of section will be x and $\dot{x}$.

We will see the results for the case $\mu = 1/2$, i.e. equal masses for M_1 and M_2. Fig.2 (Hénon 1966a) represents the surface of section for $C = 4.5$. As before, points linked by a curve correspond to the same trajectory. The dashed lines represent the boundaries of the accessible region. The system seems to be integrable: the whole accessible region appears to be covered with curves. the successive points have been numbered in one of the sequences; and Fig.3 shows the corresponding orbit in the (x, y) physical plane, with the points of intersection with the surface of section Eq.(12), again represented. This orbit shows a great regularity, which is typical of quasi-periodic orbits in general. M_3 stays comparatively close to M_1, so that in a first approximation, the effect of M_2 can be neglected and the orbit is simply a two-body elliptical motion around M_1, seen in rotating axes. However, the symmetry around the origin is not perfect because of the perturbation from M_2.

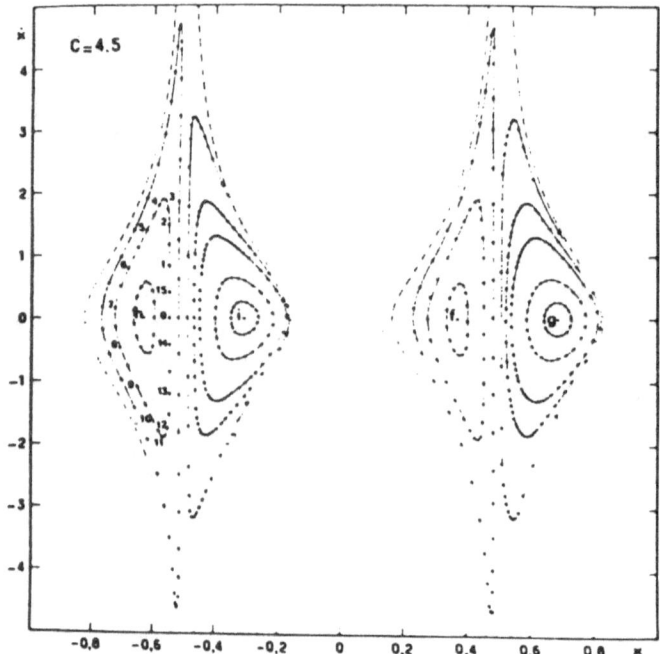

Fig.2 . Surface of section of the restricted problem
for C = 4.5: quasi integrable case.

We note also on Fig.3 that at a given point of the (x, y) plane, only two directions appear to be possible for the velocity. This again is typical of quasi-periodic orbits, and corresponds to the fact that the trajectory in phase space is a 2-dimensional torus.

We consider now a somewhat lower value: $C = 3.5$ (Fig.4), the chaotic region increases in extent. A number of successive points have been numbered on the trajectory which occupies that region, in an attempt to show how the points jump quasi-randomly from one place to another. A similar chaotic trajectory (for $C = 3$) is represented in the (x, y) plane on Fig.5, which should be contrasted with Fig.3. This orbit has a very disordered character. The third body M_3 describes a few loops

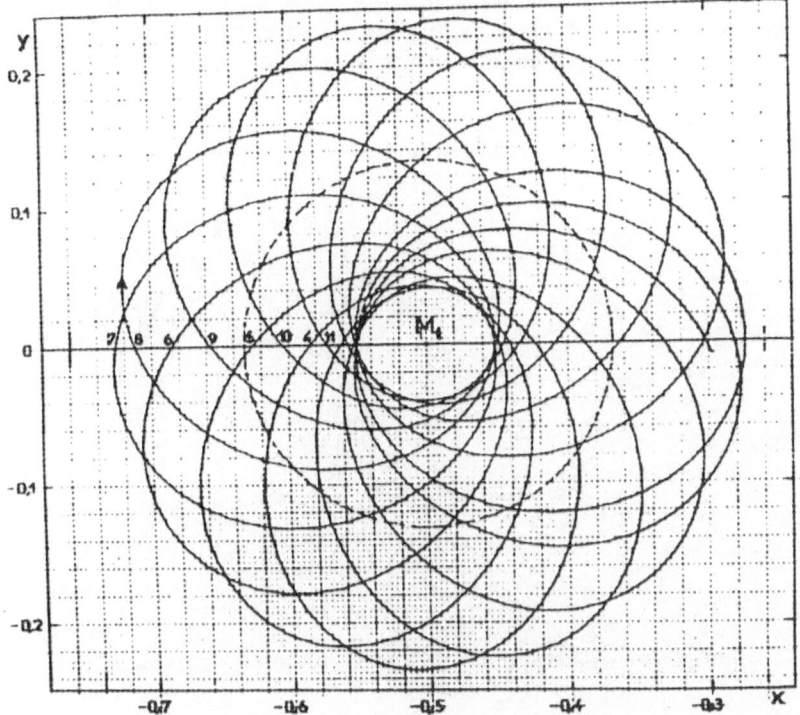

Fig.3 . Regular orbit corresponding to the numbered points of Fig.2.

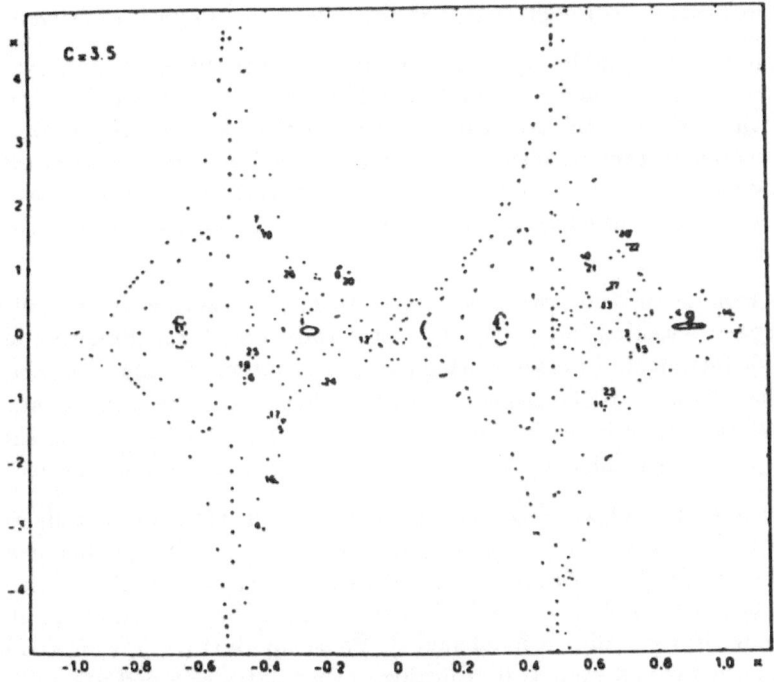

Fig.4 . Surface of section of the restricted problem
for C= 3.5: mixte case.

133

around M_1, then jumps over the vicinity of M_2, then back to M_1, in a highly irregular fashion. We note also that at a given point, the velocity can apparently have any direction. This figure is typical of chaotic orbits in general.

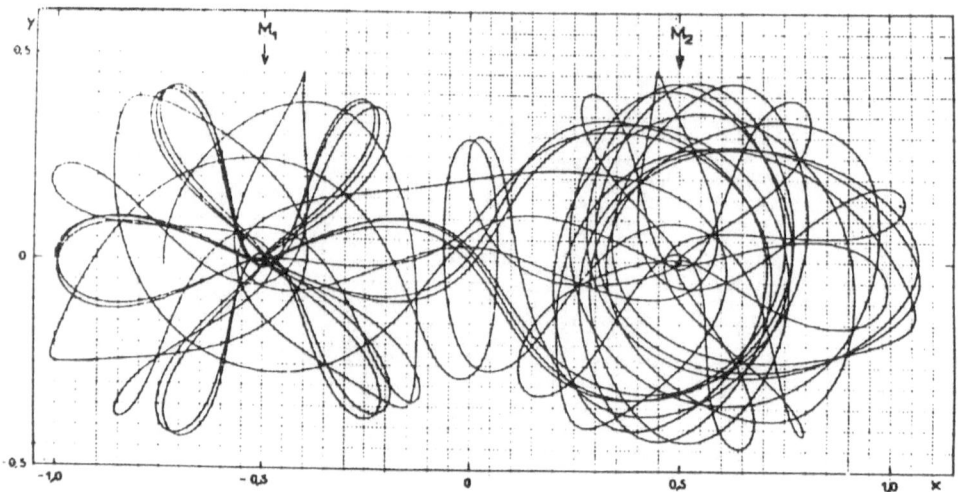

Fig.5 . Chaotic orbit corresponding to the numbered points of Fig.4.

5. System with three degrees of freedom

This has been much less studied than the case of two degrees of freedom. Phase space has now 6 dimensions; fixing $H = $ constant and taking a section, we obtain a 4-dimensional space of section, which is much more difficult to visualize and to study than the 2-dimensional surface of section of the case of two degrees of freedom.

If the system is integrable, there are, in addition to the Hamiltonian, two other integrals. Therefore the sequence of points should lie on a 2-dimensional subset of the space of section. If the system is non-integrable, but if we are in a region where the KAM theorem applies, then we also expect 2-dimensional tori. At the other extreme, for an ergodic system, points should fill the 4-dimensional space of section. Thus, the dimension of the manifold occupied by a sequence of points can range from 2 to 4.

A first technique of study consists in ignoring one coordinate: instead of the 4-dimensional space of section (q_1, q_2, p_1, p_2), we consider only the 3-dimensional space (q_1, q_2, p_1) (for instance). In essence, this is a projection of the space of section on the 3-dimensional space. If the sequence of points lies on a 2-dimensional subset in the space of section, then in projection the points also lie on a 2-dimensional subset on the space (q_1, q_2, p_1). This is easier to test. Two methods have been tried.

5.1. One can make stereoscopic projections in order to actually see the arrangement of the points in the 3-dimensional space. This technique has been pioneered by Froeschlé (1970a); Figures 6 and 7 show examples for the three-dimensional restricted problem (the motion M_1 and M_2 is still as described in section 4, but M_3 is now free to move in three-dimensional space). Phase space is $(x, y, z, \dot{x}, \dot{y}, \dot{z})$. The space of section is defined by $z = 0$. the three coordinates retained are x, y, $\dot{z}$. In other words, a point of coordinates $(z, y, \dot{z})$ is "plotted" at each intersection of the orbit with the plane $z = 0$. Actually, a pair of stereoscopic projections of this point on the (x, y) plane are plotted. With a stereoscopic viewer, the points of Fig.6 are seen to lie on a smooth surface, which resembles somewhat a cooling tower. This case therefore appears to be quasi-periodic orbit, with the points lying on a 2-dimensional torus in the space of section, and the orbit itself lying on a 3-dimensional torus in phase space.

For Fig.7, on the other hand, the stereoscope reveals that the points fill more or less a three-dimensional volume. The invariant tori have been destroyed. Beyond this, however, it is not possible to deduce from Fig.7 any information about the arrangement of the points in the true 4-dimensional region. This technique has also been used by Martinet and Magnenat (1981) in connection with the problem of a star in a galaxy without symmetry.

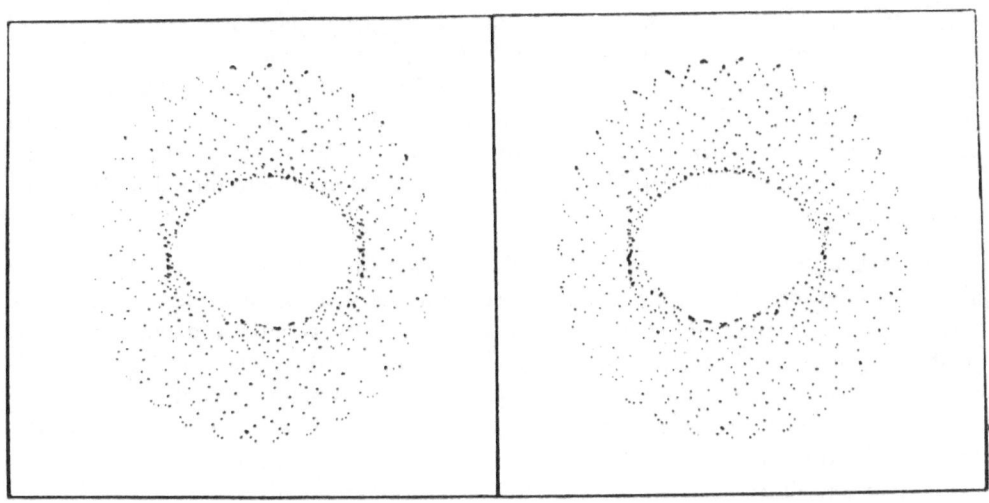

Fig.6 . Stereoscopic view of the subspace $xy \mid \dot{z} \mid$ for a quasi integrable case.

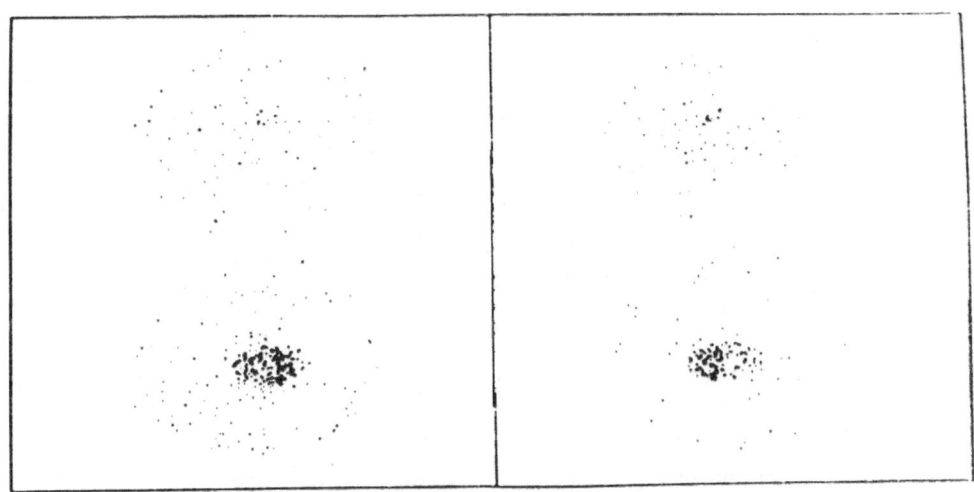

Fig.7. The same as for Fig.6, but for a quasi stochastic case.

5.2. We can also try to make a further section through the 3-dimensional space (q_1, q_2, p_1). However, this space is populated by points, not by a continuous trajectory; and the probability of a point falling on an arbitrarily chosen surface is zero. Therefore we must take not a true section, but rather a *slice*, having a small but finite thickness, in order to catch some points. Figures 8 and 9 are examples taken from Froeschlé (1972), for the 4-dimensional mapping

$$
\begin{aligned}
x' &= x + a_1 \sin(x + y) + b \sin(x + y + z + w), \\
y' &= x + y, \\
z' &= z + a_2 \sin(z + w) + b \sin(x + y + z + w), \\
w' &= z + w.
\end{aligned}
\tag{13}
$$

All variables are defined modulo 2π. A trajectory is a set of points in the (x, y, z, w) space. It is first projected on the 3-dimensional (x, y, z) space; and then a series of nine slices are taken, defined by $\mid z - z_0 \mid < 0.01$ with nine regularly spaced values for z_0. Parameter values are $a_1 = -1.3$, $a_2 = -1$ for both figures; $b = 0.15$ for Fig.8, and $b = 0.5$ for Fig.9. In the first case, the points fall on well-defined curves in each slice; this suggests that they fall on a 2-dimensional surface in (x, y, z) space. In the second case, the situation is rather different: the points appear to fill a 3-dimensional region in (x, y, z) space.

A rather different technique of study consists in determining numerically the Lyapunov characteristic exponents of the trajectories. This allows one to distinguish between quasi-periodic and chaotic orbits and to map the regular and chaotic regions in phase space. The question raised by Froeschlé (1970) on the number of quasi-integrals for a one parameter set of orbits of the non planar restricted three-body problem has been solved more easily, using the LCEs by Gonczi and Froeschlé (1981). The combination of the two methods (slice cutting and LCEs) has been very usefull to discover and illustrate a sticking-exploding phenomenon. However, this technique does not provide any information about the shape of an orbit in phase space, or of the set of points in the space of section; it gives just two numbers for each trajectory.

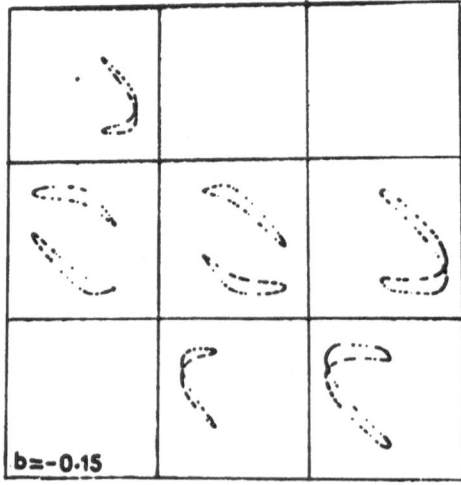

Fig.8 . Multisection of the subspace x, y, z for a regular case ($N = 40\,000$).

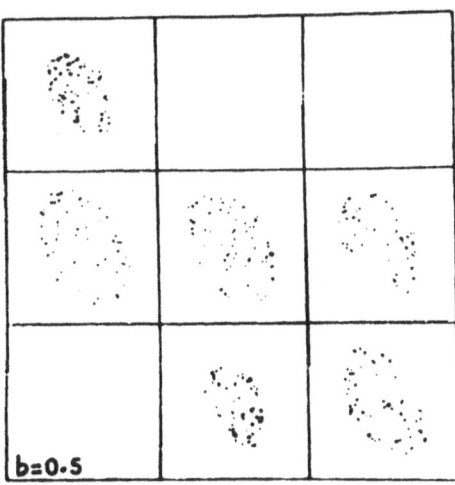

Fig.9 . Multisection of the subspace x, y, z for a chaotic case $N = (40\,000)$.

Figure 10a shows, on a log-log scale, the behaviour of the two positive γ_i^n (whose limits are identified with the Lyapunov characteristic numbers) as a function of time (i.e. the number n of iterations), for initial conditions taken in the ordered region, and with $b = 0.075$. Both γ_i^n appear to be decreasing functions of time. the largest shows very small variations around the expected linear behaviour.

On the other hand, Fig.10b shows a stochastic situation. As time increases, the two coefficients approach their strictly positive limit value $\chi_1 \approx 3.10^{-1}$ and $\chi_2 \approx 10^{-1}$.

Next we try to clarify the ambiguous situations. Up to $N = 4.10^4$ iterations, the orbits with $b = 0.15$ and $b = 0.275$ (that is case 1)) look the same: the points remain in a neighbourhood of an invariant surface. However, for $b = 0.15$, the LCNs exhibit (Fig.10c)) the same type of behaviour as for the clear integrable case of Fig.10a), but the slope of the line is much less accentuated. This suggests an integrable behaviour but with a manifold of very irregular shape. On the other hand, for $b = 0.275$ (Fig.10d)), after some fluctuations below $n = 1.5.10^5$, a sharp increase of the coefficients up to $\chi \approx 2.10^{-1}$ and $\chi_2 \approx 10^{-1}$ is observed. The jump of the γ_2^n 's occurs simultaneously with a sudden explosion of the points in the four-dimensional space.

We remark that, already for $n = 4.10^4$. the curves show obvious differences in shape and dispersion with respect to those of the case with $b = 0.15$ (recall that the cases were indistinguishable by the slice-cutting method).

In the second type of problem (case 2), that is in the case with $b = 0.3$, the slice cutting showed that the points were "ergodically" localized in a broad, well-bounded region. In this situation, we have obtained the curves of Fig.10e). We find the same type of behaviour as for $b = 0.275$, the only difference lying in the numerical values: the explosion of the y_i^n 's begins earlier (at $n = 10^5$ instead of $1.5.10^5$) and their values before the explosion are, on the average, rather larger (this latter statement can explain the two different results of slice cutting for $n = 0.275$ and $b = 0.3$ at $n = 4.10^4$).

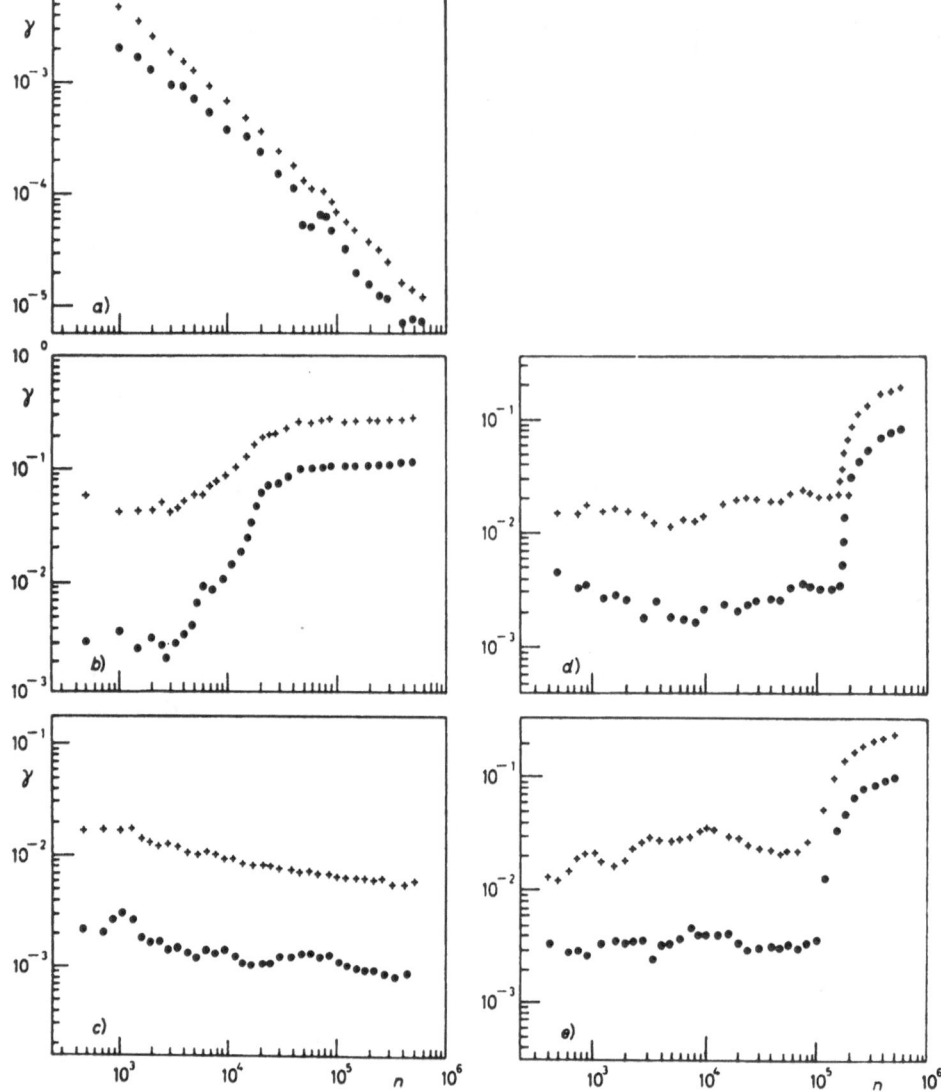

Fig.10 . Variation of the $\gamma_i^n(P_0)$ (whose limits are identified with the Lyapunov characteristic numbers) as functions of the number n of iterations for initial conditions: $a_1 = -1.3$, $a_2 = -1$, $P_0 = (1, 0, 0.5, 0, 5)$ and several values of the coupling parameter b: a) $b = 0.075$, b) $b = 0.425$, c) $b = 0.15$, d) $b = 0.275$, e) $b = 0.3$.

In order to visualize the evolution in time and show more concretely how the burst of the points occurs, we have plotted the projections on the (x, y) plane of the set of points P_n such that $| w | \leq \varepsilon$ with $\varepsilon = 0.01$ for $n = 40000$, (40 000), 360 000.

In the integrable case and even for $b = 0.015$ (Fig.11), we do not observe large variations in the shape or in the size of the picture: the system is quite stationary in time, as previously noted by Froeschlé (1972). On the other hand, an ergodic situation ($b = 0.275$) shows (fig.12) the expected explosion of the points starting at $n \approx 2.10^5$.

This sticking-exploding phenomenon has been emphasized recently by G. Contopoulos and B. Barnabis (1989). (See G. Contopoulos this book).

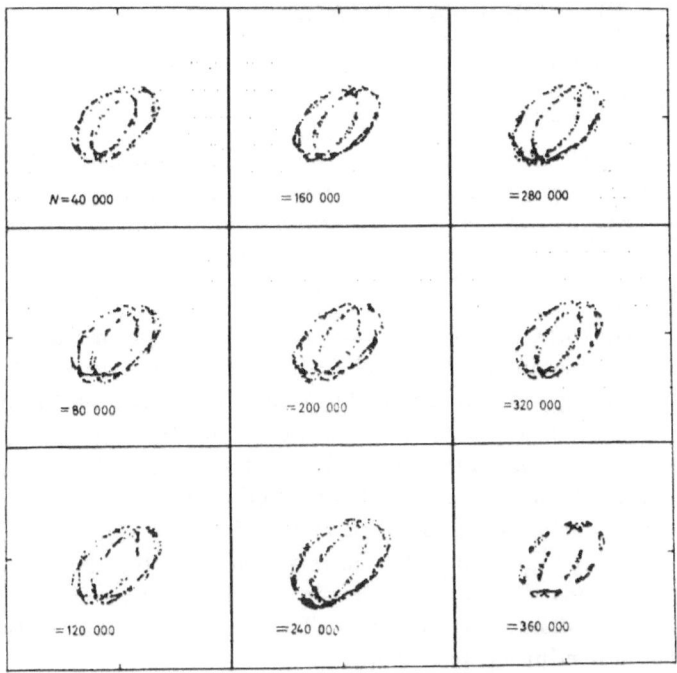

Fig.11 . Projection on the (x, y)-plane of the set of points $P_n = T(P_{n-1})$ for $n = 0$, $N = 0$, (40 000), 360 000. the initial conditions are the same as for Fig.10 $b = 0.15$.

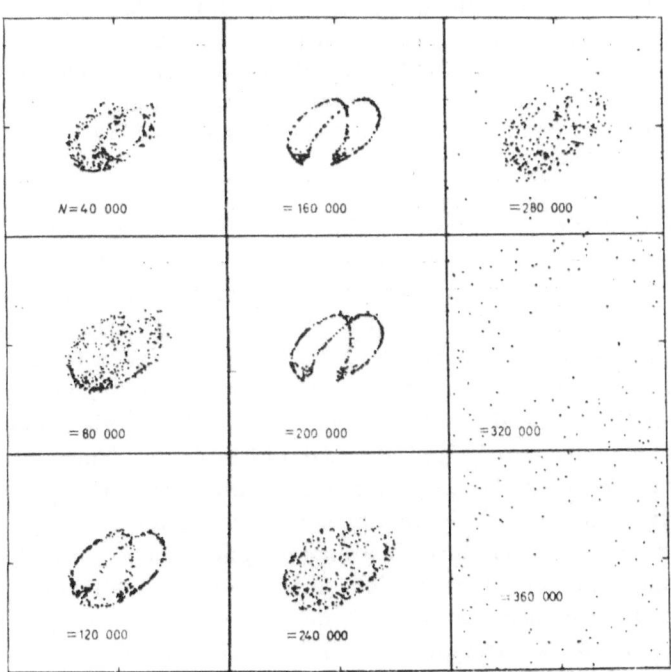

Fig.12 . Projection on the (x, y)-plane of the set of points $P_n = T(P_n - 1)$ for $n = 0$, $N = 1$ when $N = 0$, (40 000), 360 000; the initial conditions are the same for Fig.10b $b = 0.275$.

III. ASTEROIDAL ORBITAL CHAOS

1. Resonant motion

1.1. The Kirkwood gaps

The narrow gaps called Kirkwood gaps in the main asteroidal belt coincide with the location of mean motion resonances (see Froeschlé and Scholl 1983). Many attempts have been made in the last two decades to relate this depletion with intrinsic dynamical stochasticity of the elliptic restricted three body problem, i.e. within the framework of the so called gravitational hypothesis.

In order to get some feeling about the mean features of a stochastic trajectory let us consider the pendulum which has often been taken as the basic model of resonance (Lichtenberg and Lieberman 1983) since in celestial mechanics a resonance problem quite often is reduced to that of a pendulum through an expansion of the disturbing function and the removal of both the short periodic terms and the secular and resonant terms which are considered as negligible near resonance. This is the case for example when an asteroid's period is near 1/3 of Jupiter's, in which case, one term may have a very long period and thus have a much greater effect than the others. This effect corresponds to a periodic repetition of geometric configurations which enhances perturbations and introduces oscillations in orbital elements analogous to the behaviour of a pendulum. (see chapter II).

1.2. Full integration

The chaos which occurs for some orbits starting in the Kirkwood gaps is quite weak and occurs clearly only within timespans of the order of 10^8 years for orbits starting with low eccentricities. Even for resonant orbits starting with higher eccentricities timespans of the order of 10^5 years are necessary. This slow build up explains why chaos was evidenced only in the seventies by Giffen, Scholl and Froeschlé using Schubart's averaging program. However direct computations have recently been carried out over millions of year using a very fast computer "The Digital Orrery", specifically designed to study problems in celestial mechanics

1.3. Averaging

Analytical averaging is performed when short periodic terms are removed from the Fourier expansion of the disturbing function. Of course if further truncation are made in order to reduce the system to an integrable system there is no hope to exhibit chaotic behaviour. On the other hand if we are not able to integrate the system analytically, what is the gain since we are back to numerical integration? However following an idea of Poincaré, Schubart (1968) developed a different approach. Using a suitable change of canonical coordinates he modified Poincaré's canonical equations for the planar restricted Sun - Jupiter - asteroid problem. Then the Hamiltonian was further averaged over the corresponding commensurability period. This new Hamiltonian gives a new set of differential equations which is integrated numerically. For the circular case, this set reduces to a single implicit equation which can be solved without resorting to numerical integration. Orbits are represented by closed curves in $S - \sigma$ space with the conjugate variables S and σ given by : $S = \sqrt{a/a_J} \left(1 - \sqrt{1 - e^2}\right)$ and $\sigma = (p+q) \lambda_J - p\lambda - \tilde{\omega}$. The quantities a, λ and $\tilde{\omega}$ are respectively the semi-major axis, the mean longitude and the longitude of pericenter of the asteroid and quantities with the subscript J refer to Jupiter orbit . Besides the averaged Hamiltonian $\bar{H} = \frac{1}{2\pi(p+q)} \int_0^{2\pi(p+q)} H \, d\lambda$, the quantity $A = \sqrt{a/a_J} \left(2 - \sqrt{1 - e^2}\right)$ which is a variable in the elliptic model, is also an integral of motion in the circular case. These two integrals exhibit the topology of the second fundamental resonance problem. (see Fig.13 in the case of the 2/1 resonance).

In the circular averaged model with $e_J = 0$, orbits remain in their corresponding regions. In the more general model still with $e_J = 0$ but without averaging, or even with $e_J \neq 0$ and with averaging, an orbit can cross the bifurcation curve and consequently can change its behaviour. In particular, librators can become circulators and vice versa.

Schubart's averaging differs strongly from the so called averaging principle (see Arnold 1976): he does not drop short periodic terms and does not use a series expansion. Therefore with Schubart's averaging there is no restriction for eccentricities or resonance type.

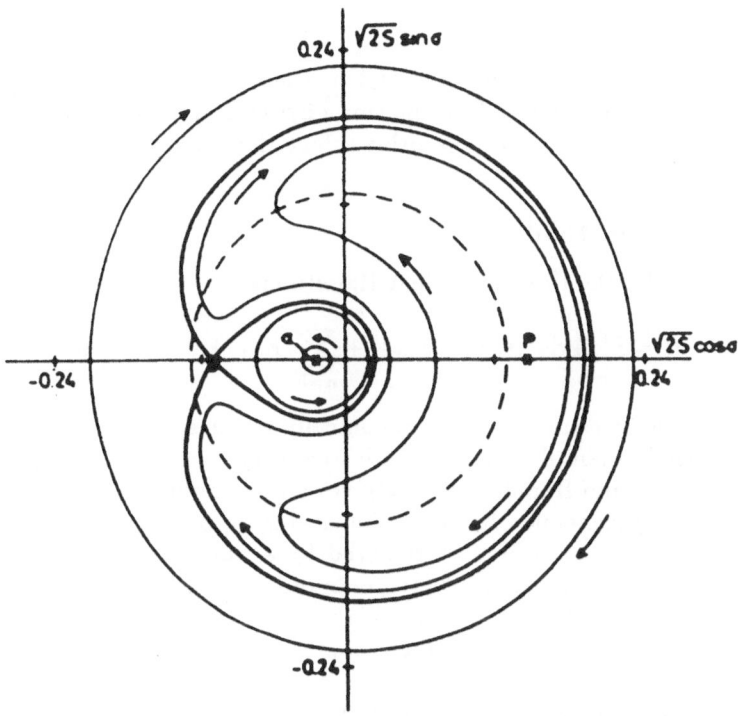

Fig.13 . Trajectories in σ versus $\sqrt{2S}$ space from Schubart's averaged circular model at the 2/1 commensurability for $A = 0.802$. The arrows indicate directions of motion in this space. The darker lines correspond to critical bifurcation trajectories. Paths immediatly around point a are apocentric librators; those about p are pericentric librators. The dashed circles correspond to the exact center of the resonance. From Greenberg and Scholl (1979).

It is clear that Schubart's topology as displayed in Fig.13 is only valid for the circular planar averaged model, i.e. an integrable model. The critical bifurcation point is called a homoclinic point in modern dynamics (Hénon, 1981). It is well known that integrable systems are not generic, i.e. as explained above for the pendulum small perturbations can destroy the integrability, and the separatrix or homoclinic orbit can cause wild regions with chaotic behaviour (Hénon, 1981). This peculiar behaviour for Schubart's topology was displayed by Froeschlé and Scholl (1977) in the elliptic averaged case.

Besides ellipticity, non-averaging as well as non-coplanarity destroys the integrability. For the case of non-coplanarity Schubart's topology displayed in Fig.12 remains valid to some extend and can be regarded as a good example for understanding and describing the behaviour of resonant orbits in the three-dimensional elliptic averaged case. Schubart (1978, 1979) has extended the planar model to deal with this more general case.

Using Schubart's model, Froeschlé and Scholl (1982) have performed a systematic exploration of the three dimensional asteroidal motion at the 2/1 resonance. Some orbits called alternators display a chaotic behaviour.

1.4. Mapping

The use of a mapping yields very fast method of numerical calculations. The basic idea of the method comes from plasma physics and is due to Chirikov (1979). Wisdom (1982) applied it to asteroid orbit calculations for the 3/1 resonance.

1.4.a. Chirikov's method

Lest us consider the time dependent Hamiltonian

$$H = I^2/4\pi + K_0/2\pi \, \cos V + \sum_{n \neq 0} K_n(I) \, \cos(V - nt)$$

Where I is the momentum and V its canonically conjugate coordinate. If the constants K_n are small then, droping the high frequency terms $\sum_{n \neq 0} K_n \cos(V - nt)$, the restricted pendulum Hamiltonian H_0 gives a good approximation to the system using the averaging principle. However this averaging procedure is no longer valid near the separatrix which is in fact replaced by a narrow chaotic band when the high frequency terms are present. Therefore in order to deal more properly with this problem Chirikov, instead of ignoring the high frequency terms included them all but in an approximated way, thus obtaining a new Hamiltonian

$$H_C = I^2/4\pi + K_0/2\pi \cos V + K_0/2\pi \sum_{n \neq} \cos(V - nt)$$

which can be considered closer to H than H_0, since chaos generating high frequency terms are present.

Using the Fourier transform of the Dirac δ function with period 2π, H_C becomes:

$$H_C = I^2/4\pi + K_0 \, \cos \, V \, \delta_{2\pi}(t)$$

Then by a straight forward integration, using the property that the delta function acts instantaneously, the standard map (see Lichtenberg and Lieberman 1983) is obtained:

$$\begin{cases} I' = I + K_0 \, \sin V \\ V' = V + I' \end{cases}$$

142

1.4.b. Wisdom's generalization

Wisdom has applied this method to the restricted elliptic three body problem for the 3/1 resonance. From the Hamiltonian derived through a second-order expansion of the perturbing function :

$$H = -\mu_1/2a + H_{secular}(a, e, i, \omega, \Omega, a_J, e_J, i_J, \omega_J, \Omega_J) + H_{resonant}(- - -, 3\lambda_J - \lambda, - - -) + H_{highfrequency}$$

he obtains the new Hamiltonian :

$$H_W = -\mu_1/2a + H_{sec} + H_{res}\ \delta_{2\pi}(t)$$

from which he derives his mapping. With this, orbits can be computed over millions of years. He found a surprising behaviour : a test particle starting in the gap could remain on a low eccentricity (< 0.05) orbit for one million years and then suddenly jumped to a large eccentricity (> 0.3) trajectory, thus becoming a Mars crosser (Wisdom, 1982).

The occurence of such a sudden jump is related to chaotic motion as Wisdom has shown by computing maximum Lyapunov characteristic exponents at the 3/1 resonance.

Abrupt changes in orbital behaviour have been known for some time, even before Wisdom work. Scholl and Froeschlé (1974, 1975) thought that these were exceptional cases, but Wisdom has shown that such changes are to be expected for any orbit near the 3/1 resonance if one follows an orbit for a sufficiently long period of time. Wisdom's other great achievement was to demonstrate that the observed width of the 3/1 Kirkwood gap coincides with the size of the chaotic region (Wisdom, 1983).

1.4.c. A semi analytical interpretation

In order to interpret these challenging numerical results Wisdom (1985) developed a semi-analytic perturbation theory for motion near the 3/1 commensurability in the planar restricted three body problem. Three natural time scales are considered : (i) the orbital period (a few years); (ii) the period of libration of the resonant argument (a few hundred years); (iii) the period of motion of the longitude of the perihelion (several thousands years), i.e. the time scale for the slow evolution of the "guiding center" of σ and e. Taking advantage of these well separated time scales Wisdom approximated analytically the fastest oscillations, i.e. only terms which contain σ in the disturbing function were considered. Also terms beyond second order in the eccentricity were ignored. Then the very-long-period behaviour was computed numerically under the assumption that the action of the motion on the intermediate time scale is adiabatically conserved during the slow evolution.

The predictions of the theory are in good agreement with the features found on surfaces of section as shown on Figure 14 for orbits generated with his mapping. This figure shows clearly two large chaotic zones. A trajectory in the chaotic zone surrounding the origin enters the narrow part of the chaotic zone which extends to high eccentricity at irregular intervals thus explaining the intermittent bursts of eccentricity. A new criterion for the existence of a large scale chaotic zone is presented and shows that the eccentricity of Jupiter's orbit is at the source of chaos, which confirms the results obtained by Froeschlé and Scholl (1977) using the Schubart averaging procedure.

Of course these results have been checked using different numerical methods. Murray and Fox (1984) have computed the motion of asteroids near the 3/1 resonance

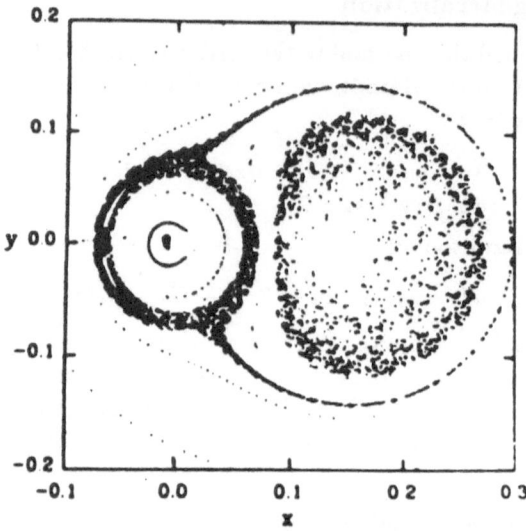

Fig.14 . Numerically generated surface of section computed with the Wisdom's mapping. Large chaotic zones appear. The narrow region generates high eccentricities at irregular intervals. From Wisdom (1985).

using three numerical methods : (a) integration of the full equations of motion, (b) integration of the analytically averaged equation of motion, (c) Wisdom's algebraic mapping. The agreement has been found to be good in the regular regions of phase space. However we have to remark that the higher the eccentricity becomes the less reliable the mapping becomes and therefore after the first increase of eccentricity the dynamical features of the real system may be significantly different from the dynamical feature of the mapping. No comparison has been made with Schubart's method.

1.4.d. Other mappings

What about the other gaps ? The same mapping-methods have been applied by Murray (1986) for the 2/1 and 3/2 Jovian resonances. He determined the chaotic region within these resonances by computing the largest Lyapunov exponent with the rescaling method. He found that both resonances have extensive chaotic regions.

Sidlichovsky and Melendo (1986) obtained similar mappings for the 5/2 resonance. They extended Scholl and Froeschlé (1975) orbits to much longer time intervals (millions of years instead of 38 000) for 96 orbits, and found 53 orbits instead of 33 (Scholl and Froeschlé 1975) for which the eccentricities go beyond 0.3. They thus reproduced the V - shaped nature of the gap.

Let us come back to Murray's results. The map he derived for the 2/1 and 3/2 resonances includes terms through second order following the earlier derivation of the 3/1 map. The large chaotic zones obtained in both cases are questionable. Indeed the very large eccentricity increases at the 2/1 and 3/2 chaotic zones are artifacts of the truncation of the disturbing function to terms of second order. Henrard and Lemaître (1987) have shown that a good representation of the motion is only obtained when terms of very high order are included. Actually the miracle of the 3/1 map arises primarily from the fact that as a second order resonance the largest terms in the disturbing function are proportional to two powers of the eccentricities and the next terms are fourth order. An other fortunate and purely combinatorial coincidence is also that the 3/1 resonance is far from other resonances which complicate the dynamics. Anyway it must be emphasized that the failures of the 2/1 and 3/2 mappings were not due to the mapping method itself but to the level of truncation which should not change too much the topology of phase space. Of course an other constraint is that the equations obtained through truncation have to be solvable analytically.

2. Planet crossing asteroids

2.1. A massive and extended numerical integration

Extended numerical computations have been performed by two groups (Hahn and Lagerkvist, 1988; Milani, Carpino, Hahn and Nobili, 1989) for hundreds of planet crossing asteroids. These orbits cannot be computed by the traditional tools of analytical celestial mechanics because they are strongly perturbed by close approaches and therefore can be described as unpredictable, i.e. chaotic. Indeed a small change in the initial conditions leads to an exponential divergence of computed orbits, i.e. to a largest Lyapunov characteristic exponent strictly positive. It follows that the details and behaviour of any single orbit is meaningless, except perhaps the computation of the LCEs. Therefore a large number of orbits and for a long enough time have to be integrated. Using the 15 th order Radau integrator Hahn and Lagerkvist computed 26 orbits of planet crossing asteroids of Apollo-Armor type. They discussed individual orbital evolutions of these asteroids. Milani et al. have computed 410 planet-crossing asteroids spanning 200 000 years. They made an attempt to define classes of dynamical behaviour taking into account graphic representation of the orbital evolution. Some of them like Oljato class show large scale chaotic effects.

We agree with the authors when they say that "the development of more and more efficient and affordable parallel computers, and the need of the methods of statistical mechanics to understand chaotic motions, both point to the need for this kind of work. This might only be the beginning of a difficult line of research, but we hope it will lead in the end to a better knowledge of the planet-crossing population of asteroids as well as to some understanding of the general properties of chaotic motions".

2.2. The close approach problem

In addition we want to emphasize a fundamental limitation of the parallel computing approach and therefore express some reserves about the results of Milani et al. Indeed it is not possible to handle properly the orbit of object having close encounters without loosing all the benefits of parallelism since the step size is the same for all objects. In a different context Froeschlé and Rickman (1980) have shown the importance of this phenomena by regularising the equations of the orbits of short period comets. Surprinsigly enough even statistical properties which are at the basis of their Monte Carlo method depend of the treatment of close approaches. And the shape of the statistical distribution of perturbations may govern the most basic outcome of the dynamical transfert (Stagg and Bailey 1989).

IV. COMETARY ORBITAL CHAOS

From the beginning of numerical computations of cometary orbits and especially for mechanism bringing them from their remote places of origin and storage towards observable orbits, chaotic routes are considered necessary. Like for planet crossing asteroids extensive numerical integrations have been performed but for a large sample of initial conditions we will not focus our attention on this approach and refer the reader to the recent review of C.R. Stagg and M.E. Bailey (1989) but rather to the mapping-methods since stochasticity is indeed at the root of the Monte Carlo modelling of cometary orbits. Actually for both long period and short period comets and using both extrinsic (exogeneous) or intrinsic (endogeneous) stochasticity, building of mapping appears as the main tool for the study of cometary dynamics. However we will distinguish two families of mappings. The first family supposes and uses extrinsic or intrinsic stochasticity, they will be called **stochastic mappings** and have already been reviewed and discussed recently by Froeschlé and Rickman (1988). Some new results by Baille and Froeschlé (see this book) will be quickly reviewed. But determinastic mappings have also been derived in the framework of cometary dynamics and exhibit large chaotic zones in the phase space. A first mapping modelizing the dynamics of long period comets has been devised by Petrosky and Broucke (1988)and applied to the study of Halley's comet by Chirikov and Vecheslavov (1986). Their results have been confirmed by Froeschlé and Gonczi (1988) who looked also on the values of Lyapunov characteristic numbers. Motivated by cosmogonic problems Duncan et al. (1989)have derived a mapping that approximates the restricted circular three-body problem for small eccentricities of the test particle and semi-major axis close to that of the planet.

1. Long period comets

1.1. Exogeneous stochastic mapping

While it has recently been realized that an important part of the dynamics of Oort cloud comets arises from regular motion in the Galactic tidal field (Heisler and Tremaine 1986), a decisive role is nonetheless played by individual stellar encounters. Given the physical parameters of such an encounter, its effect on the cometary motion is fully determined. However, the parameters of individual stellar encounters are unpredictable so the stellar perturbations impose a stochastic variation on the cometary orbits.

In Monte Carlo simulations of stellar perturbations (see, e.g., Weissman 1982; Remy and Mignard 1985) the dynamical evolution of a cloud of comets is studied as follows. At a given starting epoch each comet is initialized by choosing a set of orbital elements. These are perturbed by the gravitational effect of passing stars. The geometrical parameters of the stellar encounters are chosen at random. During the passage of a star a comet receives a heliocentric impulse through the interaction of the star with the comet and the Sun. This induces a change in the cometary orbital elements, so these are updated and the comet moves along a new Keplerian ellipse until the next encounter with a random star. In Weissman's procedure the impulse are taken in an even more simplified way from a pre-determined distribution. In other words a stochastic mapping is iterated where the perturbations caused by random stars impose a stochastic process on the cometary orbital elements, which therefore undergo a random walk. It is obvious that all orbits are chaotic and correspondingly the largest LCE is strictly positive. The stochasticity is exogeneous since the stellar encounters occur at random.

1.2. A deterministic mapping : The Keplerian Map

We know that in a neighborhood of any separatrix (i.e. the trajectory with zero frequency of the unperturbed motion) some chaotic motion has to be expected. Actually, the simplest example of separatrix is the parabolic trajectory of the two body problem which separates bounded and unbounded motions. These systems have very long characteristic time scale which allow Petrosky and Broucke (1988) to attack the usual difficulty of small denominators by a new method through Fourier analysis. The system with long time scale can be characterized by "continuous" Fourier spectrum and in this case such a "non integrable system" may become integrable by embedding the small denominator in an analytic function through a suitable analytic continuation. They apply this idea to construct a solution of the restricted three body problem in the case of nearly parabolic motion of the third body and then derive a two-dimensional canonical map which describes the dynamics of long period cometary motion.

Their map called "Keplerian map" reads

$$P_{n+1} = P_n + 2\sigma \mu s_1 \sin g_n$$

$$g_{n+1} = g_n - 2\pi \sigma/(-P_{n+1})^{3/2}$$

where g_n is the phase angle of Jupiter when the comet passes its perihelion and P_n the energy. μ is the mass of the small primary, $\sigma = \pm 1$ the parity of the direction of revolution of the particle and s_1 some scaling energy. Iterations of the Keplerian map show clearly chaotic region confined by KAM torus. Let us remark that the motion near the separatrix of a nonlinear pendulum is much less chaotic (the whisker map). A physical implication of such chaotic motion of comets concerns the structure of a cometary cloud surrounding the solar system: to each parabolic orbit is associated a cloud of comets revolving on elliptic orbits around the Sun with holes and (islands)in the corresponding surface of section. The actual cloud being a superposition of these planar clouds. Chirikov B.V. and Vecheslavov V.V. (1986) have used such a "Keplerian map" fitting the parameters to the observational datas to show the chaotic motion of comet Halley. They give estimates for the error growth in the extrapolation of comet's trajectory and show that the lifetime in the solar system crucially depends on weak non-gravitational forces.

A similar mapping related to Keplerian problem has also been proposed by Sagdeev and Zaslavsky (1987)

Let us remark that resonances rather than close approaches are at the origin of chaos. However close approaches play a non negligeable role for the stochasticity of comet Halley as it appears clearly with the numerical experiments performed by Froeschlé and Gonczi (1988) and Dvorak and Kribbel (1989).

2. Short-Period Comets

2.1. Endogeneous Stochastic Mapping

We now have to deal with the intrinsic stochasticity of a dynamical system which is usually approximated by a three-body problem : Sun - planet - comet. This stochasticity derives mainly from close encounters with the planet and during the intervals between such encounters the cometary motion is quasi-regular and predictable. Indeed the phase space domain of short period comets is similar to the phase space of Appolo-Amor asteroids and presents an intricate mixture of chaotic and ordered regions. This underlying stochasticity is at the basis of the mappings worked out by Rickman and Vaghi (1976); Froeschlé and Rickman (1980). In the framework of the elliptic restricted three body problem their mapping consists in regarding as a stochastic variable the perturbations of the orbital elements for one orbital revolution. More precisely different cells in phase space or one of its sub-spaces separate perturbation distribution are calculated. The orbital evolution is then simulated as a sequence of perturbations randomly chosen according to the calculated distribution. The main questions raised by this approach concern : the treatment of close approaches (why regularize ?); the dynamical homogeneity, i.e. the extent to which the average perturbation distributions are really applicable throughtout the respective regions (how many cells are needed) and the topological homogeneity, i.e. the extent to which the system is stochastic.

The validity of such procedures have been studied recently by Froeschlé and Rickman (1988); Baille and Froeschlé (see this book) have shown using the standard mapping as a model problem the connection between the Kolmogorov entropy and the efficiency of such Monte Carlo mappings.

2.2. A deterministic mapping

Again in the framework of the restricted circular three-body problem (Sun - planet - test particle) a mapping has been derived by Duncan, Quinn and Tremaine (1988) in order to study the long-term evolution of the planetesimal like orbits in the solar system. As usual the aim is to replace the Hamiltonian dynamical system by an area-preserving mapping that captures much of the relevant physics of the dynamical system. The approximation they use is based on the fact that planets have small eccentricities and the fractional difference in semi-major axes between a near circular orbit of a test particle and the nearest planet is small. Then the perturbations to the motion of a test particle on a near circular orbit are localized near conjunction with the planet then the first-order perturbations to the orbital element can be computed analytically using Hill's equations (Hénon and Petit 1986) or equivalently Gauss' or Lagrange's equations. Away from conjunction the perturbating effect of the planet is assumed to be negligible and hence the orbital elements after conjunction remain unchanged until the next conjunction.

Finally they get a map

$$
\begin{cases}
Z_{n+1} = Z_n + \frac{ig\, exp(i\lambda_n)}{\varepsilon_1^2} sgn(\varepsilon_1)\, \frac{m}{M_\odot} \\
\varepsilon_{n+1} = \varepsilon_n \sqrt{1 + \frac{4(|Z_{n+1}|^2 - |Z_n|^2)}{3\,\varepsilon_n^2}} \\
\lambda_{n+1} = \lambda_n + 2\,\pi f(\varepsilon_{n+1})
\end{cases}
$$

where $Z = e\, exp(i\tilde{\omega})$

$$f(\varepsilon) = |\, (1+\varepsilon)^{-3/2} - 1\,|^{-1}$$

e being the eccentricity, λ_n the longitude of the nth conjunction and $\tilde{\omega}$ the longitude of periastre and $\varepsilon = \frac{|a - a_P|}{a}$ with a and a_P the semi major axes of the particle and of the nearest planet.

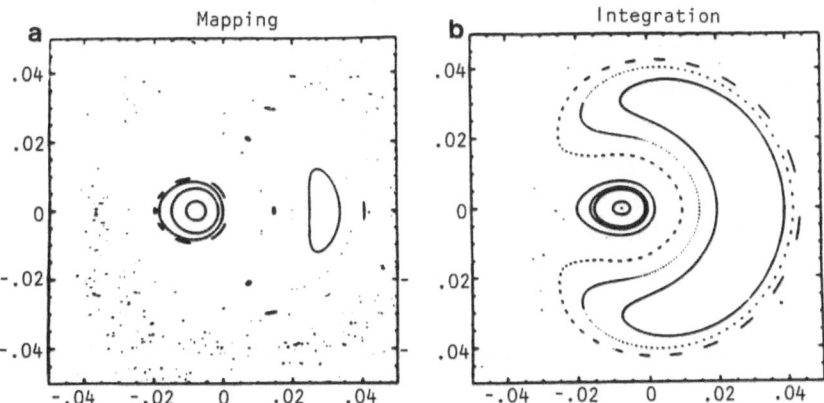

Fig.15 . (a) Trajectories of the one-planet map in the complex y planes, for mass ration $m/M_{\odot} = 5.178 \times 10^{-5}$ (Neptune's mass) and dimensionless Jacobi constant $\gamma = 8.0694 \times 10^{-5}$. (b) Surface of section for the restricted three-body problem with the same mass ratio and Jacobi constant chosen so that the longitude change of the periodic orbit is the same as in part (a).

Let us introduce the variable

$$y_n = Z_n \, exp(-i\,\theta_n)$$

where $\theta_n = \lambda - \pi \, f(\varepsilon_n) + \pi$, i.e. the mean longitude of the test particle at the opposition preceding the nth conjunction. Then Fig.3 shows orbits of the mapping in the complex y plane and the corresponding surface of section through direct integration.

As the authors state the map correctly reproduces the approximate location, shape and size of the zone of regular orbits near zero eccentricity as well as the existence of a larger, second zone of regular orbits centered near $z = 0.03$. The map underestimates the size of the second zone but this is not surprising since the approximation $\mid z \mid \ll \varepsilon$ is no longer accurate once $\mid z \mid \gtrsim 0.03$. The qualitative similarity of the two parts of Figure 15 suggests that the map provides a useful representation of the restricted three-body problem in this parameter regime. They both suggest that there may still be stable bands in the outer solar system containing surviving planetesimals. Let us emphasize that such a model does not take into account secular resonances which certainly play a crucial role for such a problem.

V. A NEW GENERAL APPROACH FOR BUILDING MAPPINGS

1. A semi-analytical approach

This approach has been pionnered by J. Hadjidemetriou. He develops a new method for the construction of an algebraic mapping that approximates a dynamical system which is close to an integrable one. Instead of using the usual method of solving approximately the actual system of differential equations to obtain a discrete system, he starts directly from the integrable system and obtains analytically the corresponding mapping on a suitable surface of section. Then he perturbs this mapping in such a way that it includes all the main features of the non-integrable systems, which are known by analytic and numerical studies.

Since there is a chapter in this book devoted to this method, we refer the reader to this chapter written by J. Hadjidemetriou himself.

2. A synthetic approach

All the mappings described above are ad hoc mappings valid only in some regions of the phase space and for specific purposes. C. Froeschlé and J.M. Petit (1990) build a mapping valid everywhere in phase space following an idea already used by Varosi et al. (1987) but in the framework of non hamiltonian systems, i.e. systems where attractors do exist.

The basic idea of the synthetic mapping is to interpolate the image of a point, given the images of a set of points, located at the vertices of a grid. The simplest way to achieve this, is to use a linear interpolation (Varosi et al., 1987). Unfortunately, when attractors do not exist, this requires a rather fine graining of the phase space, i.e. we have to compute a lot of points on the grid. We tried to find more accurate methods. The more commonly used methods of higher order are the Taylor and Spline interpolations.

Taylor interpolations appear to be more precise. There exist in any case two key parameters: the number of divisions in each direction $N = $ (total number of cells)$^{(1/D)}$, where D ($D = 2$ in their paper) is the dimension of the surface of section and the order M of the Taylor expansion. In order to explore the validity of the synthetic approach they applied their method for two cases:

1) An algebric area preserving mapping for which the computation of orbits is very fast. This allows to calculate a lot of orbits and to perform enough iterations for a meaningful comparison.

2) A special case of the restricted three body problem studied by Duncan et al. (1989) (see IV. 2.2.)

First the well-known standard mapping has been used. (Froeschlé, 1970; Chirikov, 1979; Lichtenberg and Lieberman, 1983)

$$x^{(n+1)} = x^{(n)} + a \, \sin(x^{(n)} + y^{(n)}) \qquad\qquad (mod \ 2\Pi)$$
$$y^{(n+1)} = x^{(n)} + y^{(n)} \qquad\qquad (mod \ 2\Pi)$$

Fig.16a shows orbits of the standard mapping for $a = -1.3$, indeed such a mapping exhibits all the characteristics and well-known features of problems with two degrees of freedom such as invariant curves, islands and stochastic zones where the points wander in a chaotic way. Figs.16b and 16c are magnifications of the small boxes indicated in Figs. 16a and 16b respectively. At this magnification details like second order islands become evident and the approximation qualities of the synthetics maps are easily visualized. Figs. 16g, 16h and 16i correspond respectively to the same orbits and same magnifications as 16a, 16b and 16c for the Taylor interpolation mapping in order $M = 5$ (T5) where the grid is characterized by $N = 40$. We obtain results which look the same as for the original map. When the grid is such as $N = 20$ (Figs. 16d, 16e and 16f), it is interesting to notice the agreement of the details in magnifications e and f. This agreement is remarkable since the dimension of the cell (0.31) is much bigger than the dimension of the smallest islands seen in Fig. 16f (0.005). Of course the invariant curve shows some thickness which disappears for $N = 40$ but would reappear with a further magnification. This phenomenon is due to the fact that synthetic mapping like T5 does conserve areas only to order 6.

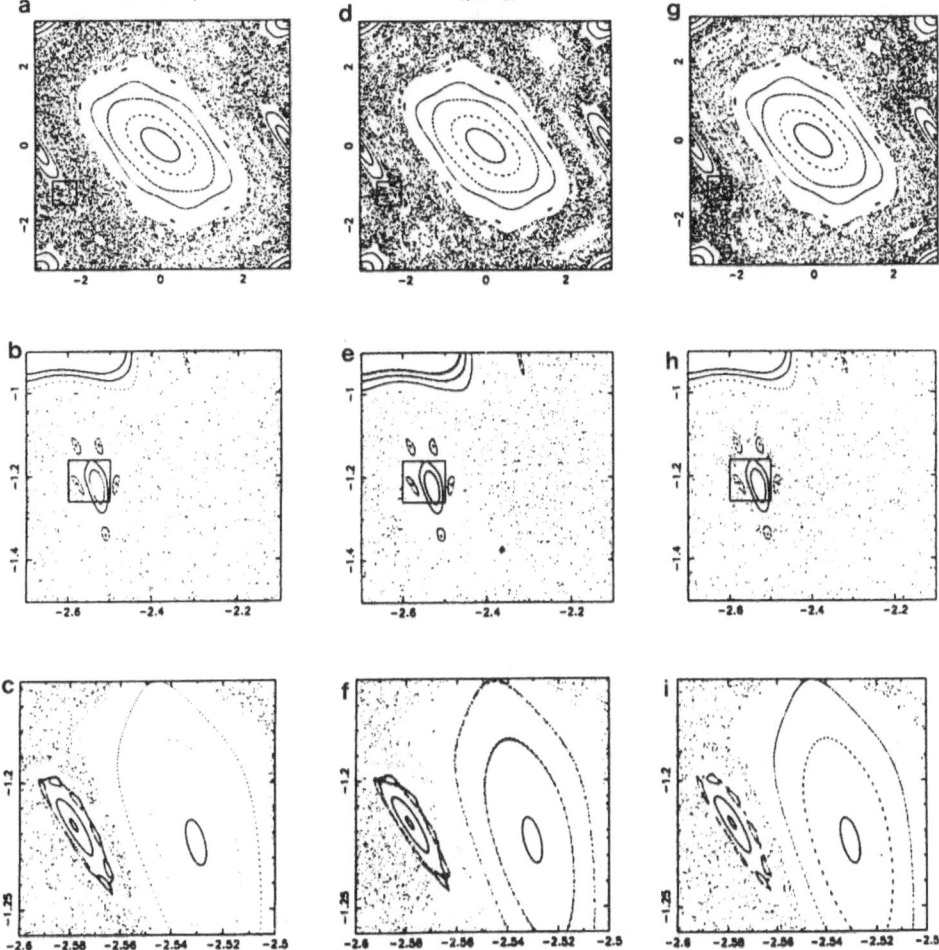

Figs. 16a, b, c . (a) Standard mapping picture for $a = -1.3$ (b) and (c) are enlargements of the small boxes shown in Figs. 16a and 16b respectively.
Figs. 16d, e, f : Same as Figs. 16a, b, c for the synthetic mapping T5 and $N = 20$.
Figs. 16g, h, i : Same as Figs. 16d, e, f for $N = 40$.

We have also tested the method on a special case of the restricted three body problem for which Duncan et al. have developed a special mapping. Fig. 17a exhibits orbits of the Poincaré map taking as surface of section the plane defined by the eccentricity and the mean longitude as polar coordinates, when the particle is in conjunction with the planet (i.e. in the rotating frame, when $y = 0$ and $y' > 0$). On Figs. 17b, c, d are plotted the corresponding orbits for synthetic mapping T1, T3, T5 for a grid with $N = 100$. If linear mapping T1 shows only poor qualitative similarities with the Poincaré map, the map T3 correctly reproduces the locations, shapes and sizes of zones of regular orbits. Of course the map T5 is even better. Let us point out that to obtain the same accuracy we needed a smallest grid size (i.e. more points) than for the standard map since the functions which have to be interpolated are less regular. In this case, about 80 minutes were needed to compute the Poincaré map by integrating the equations of motion (Fig.17a) and only 2 seconds for T5 (6d). However this should be tempered by the time needed to compute the grid.

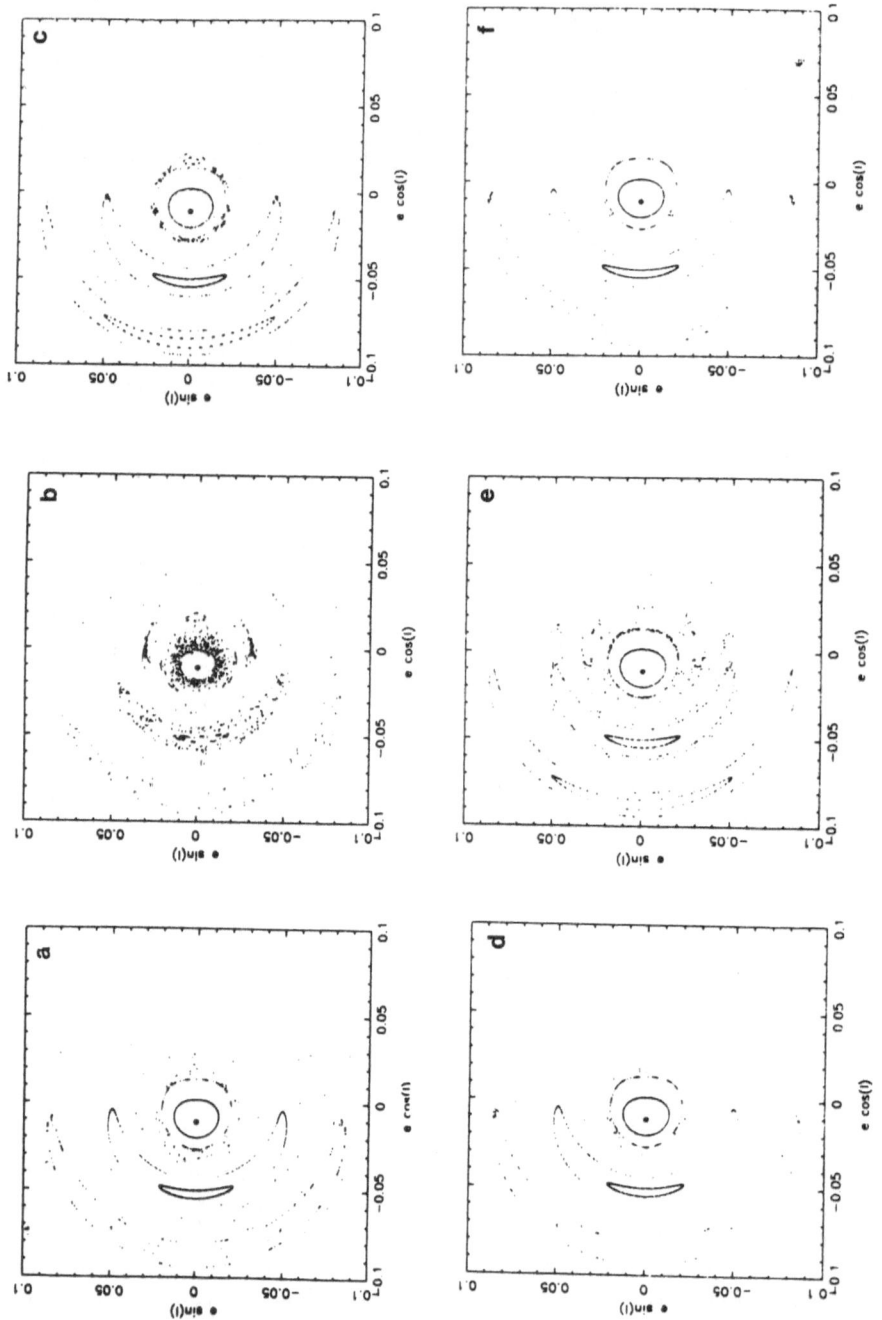

Fig. 17. (a) Trajectories of the Poincaré map of the restricted three body problem, in the plane (eccentricity, mean longitude) at conjunction with the planet, for the Neptune-Sun mass ratio $m/M_\odot = 5.178 \; 10^{-5}$ and Jacobi constant 3.0080694. (b,c,d) Same as (a) for the synthetic maps T1, T3 and T5. (e,f) Same as (b,c) for a grid with $N = 400$ obtained with T5.

Let us remark that the same accuracy than the one obtained by the map T5 can be obtained even for the map T1 by decreasing drastically the grid cell size (i.e. increasing N). Actually for such an increase of the number of vertices, the price to pay is very low since the additional Poincaré return map values at the vertices can be computed using the more precise mapping T5. Figs. 17e, 17f, show the mapping T1 and T3 for a grid with $N = 400$ where the 15/16 of the N^2 map values at the vertices have been computed using T5 through the already computed grid with $N = 100$. Indeed the accuracy of T1 is drastically improved. Of course the same holds for T3. This combination of the order of interpolation and of different coarse graining, i.e. of different mappings appears very promising for four dimensional mappings, where the computational time of T5 is far to be negligible.

VI. CONCLUSION

The "new paradigm" (Wisdom 1987) provided by the study of dynamical systems has allowed considerable progress in the field of chaotic dynamics of the solar system and especially for asteroidal and cometary orbits. Indeed chaotic routes are essential to bring either asteroidal material or comets to the vicinity of Earth. Mappings to study the extent of chaotic motion are used more often. Of course we have to understand the limits of the numerical methods involved not only for mappings but also for averaging and numerical integrations of the full set of equations. Much care should be given to the treatment of close approaches especially when massive parallel computers are used.

"Everything should be made as simple as possible but not simpler"

A. Einstein

REFERENCES

Arnold, V.I. (1976). *Méthodes Mathématiques de la Mécanique Céleste* , Ed. de Moscou

Baille, Ph., and Froeschlé, C. (1989). in *Asteroids Comets Meteors III*, Eds. Lagerkvist. Rickman, Lindblad, Lindgren, Uppsala Univ., pp. 231-234

Benettin, G., Galgani, L., Giorgilli, A., Strelcy, J.M. (1980) *Meccanica* p. 9-20, 21-30

Chirikov, B.V. (1979). *Phys. Rep.* **52**, 263-379

Chirikov, B.V., Vecheslavov, V.V. (1986). preprint

Ducan, M., Quinn, T., Temaine, S. (1989). preprint

Dvorak, R., Kribbel, J. (1989). preprint

Froeschlé, C. (1970a). *Astron. Astrophys.* **4**, p. 115

Froeschlé, C. (1970b). it Astron. Astrophys. **9**, 15

Froeschlé, C. (1972). *Astron Astrophys.* **16**, 172

Froeschlé, C., Scholl, H. (1977). *Astron. Astrophys.* **57**, 33-39

Froesché, C., Scholl, H. (1982). *Astron. Astrophys.* **111**, 346-356

Froeschlé, C. Scholl, H. (1983). in *Asteroids Comets Meteors*. Eds. C.I. Lagerkvist and H. Rickman pp. 115-125

Froeschlé, C. (1984). *Celestial Mech.*, **34**, 95-115

Froeschlé, C., Rickman, H. (1980). *Astron. Astrophys.* **82**, 183-194

Froeschlé, C., and Rickman, H. (1988). *Celestial Mech.* **43**, 265-284

Froeschlé, C., and Gonczi, R. (1988). *Celestial Mech.* **43**, 325-330

Froeschlé, C., and Greenberg, R. (1989). to appear in *Asteroid II*

Froeschlé. C., Petit, J.M. *Astron. Astrophys.* to appear

Greenberg, R., and Scholl, H. (1979) in *Asteroids*. Ed. T. Gehrels, Tucson, Univ. of Arizona Press pp. 310-333

Gonczi, R., and Froeschlé C. (1981) *Cel. Mech.* **25**, 271-280

Guckenheimer, J., Holmes, P. (1983). *Nonlinear Oscillations, Dynamical Systems, and Bifurcations of Vector Fields.* Springer Verlag

Hahn, G. and C.I. Lagerkvist (1988). *Celest. Mech.* **43**, 285-302

Heisler, J., Tremaine, S. (1986). *Icarus* **65**, 13-26

Hénon, M., and Heiles C. (1964). *Astron. J.* **69** 73-79

Hénon, M. (1966a). *Bull. Astron. (3) 1, fasc 1*, p. 57

Hénon, M. (1981). in *Cours des Houches XXXVI North Holland 1981*

Hénon, M., and Petit, J.M. (1986). *Celestial Mech.* **38**, 67-100

Henrard, J., Lemaître, A. (1987). *Icarus* **69**, pp. 266-279

Lichtenberg, A.J., Lieberman, M.A. (1983). *Regular and Stochastic Motion* Springer-Verlag

Martinet, L. and Magnenat, P. (1981). *Astron. Astrophys.* **96**, 68

Milani, A., Carpino, M., Hahn, G., and Nobili, A.M. (1989). *Icarus* **78**, 212-269

Murray, C.D., and Fox, K. (1984). *Icarus* **59**, 221-223

Murray, C.D. (1986). *Icarus* **65**, 70-82

Petrosky, T.Y. and Broucke, R. (1988). *Celest. Mech.*. **42**, 53-79

Remy, F., Mignard, F. (1985). *Icarus* **63**, 1-19

Rickman, H., Vaghi, S. (1976). *Astron. Astrophys.* **51**, 327-342

Sagdeev, R.Z. and Zaslavsky, G.M. (1987). *il Nuovo Cimento*, Vol **97**, BN2

Scholl, H., and Froeschlé, C. (1974). *Astron. Astrophys.* **33**, 455-458

Scholl, H., and Froeschlé, C. (1975). *Astron. Astrophys.* **42**, 457-463

Schubart. J. (1968). *Astron. J.* **73**, 99-103

Schubart, J. (1978). in *Dynamics of Planets of Satellites and theories of their motion.* Szebehely V. ed., pp. 137-143

Schubart, J. (1979). in *Dynamics of the Solar System.* Duncombe R.L. ed., pp. 207-215

Sidlichosky, M., and Melando, B. (1986). *Bull. Astron. Inst. Czech.* **37**, 66-80

Stagg, C.R. and Bailey, M.E. (1989). To appear in *Monthly Notices*

Varosi F., Grebogi, C., Yorke J.A. (1987). *Phys. Lett A* **124**, 59

Weissman, P.R. (1982). in *Comets* (ed. L.L. Wilkening), Univ. Arizona Press, Tucson, pp. 637-658

Wisdom, J. (1982). *Astron. J.* **87**, 577-593

Wisdom, J. (1983). *Icarus* **56**, 51-74

Wisdon, J. (1985). *Icarus* **63**, 272-289

Wisdom, J. (1987). *Icarus* **72**, 241-275

MAPPING MODELS FOR HAMILTONIAN SYSTEMS

WITH APPLICATION TO RESONANT ASTEROID MOTION

John D. Hadjidemetriou

Department of Theoretical Mechanics
University of Thessaloniki
GR-540 06 Thessaloniki, Greece

ABSTRACT. A systematic method is presented to construct a mapping model for a Hamiltonian system. We start with two degrees of freedom and extend the method to three degrees of freedom. The basic notions of the averaging method are presented first, as the method of constructing the mapping is based on the averaged Hamiltonian. A mechanism of generation of chaos will be described, through the interaction of two degrees of freedom in systems with three degrees of freedom.Finally, the method is applied to the 3:1 resonance in asteroid motion for nonzero eccentricity of Jupiter, where chaotic behaviour of the eccentricity of the asteroid is found.

INTRODUCTION

The evolution of a dynamical system for a long time is a very interesting problem with many applications. However,since almost all dynamical systems are nonintegrable, a rigorous analytic study of such a system is not possible. On the other hand the numerical integration of the differential equations which govern the evolution of the system cannot be carried out for long time intervals, both because it is costly and also because the accumulation of numerical errors makes the results unreliable,especially if we are far from an ordered region. We may note at this point that one cannot draw conclusions for the long term evolution of a dynamical system by extrapolating results obtained (numerically or otherwise) for a short time, as it may happen that irregular motion may appear after a very long time,preceeded by a rather regular motion. Finally,we may note that an analytic study of the dynamical system by a perturbation method by series expansions is not very helpful for long time predictions because of the problem of small divisors.

One way to study the general behaviour of a nonlinear dynamical system is to represent it by a mapping. It is obvious that a mapping has many advantages over the numerical integration both with respect to computing time and with respect to accuracy.Such a mapping has been found by Wisdom, based on a method proposed by Chiricov, who used it in a series of papers (1982,1983,1985) for the study of asteroid motion near a resonance. A different method to construct a mapping for the study of asteroid motion near a resonance

Predictability, Stability, and Chaos in N-Body Dynamical Systems
Edited by A.E. Roy, Plenum Press, New York, 1991

has been used by Hadjidemetriou(1986,1988,1990). Also mapping models have been used by Froeschle and Petit (1990), Petrosky and Broucke (1989) and Liu and Yi-Sui Sun (1990), for the study of dynamical systems.

The construction of a mapping for a dynamical system is not equivalent to a method of numerical integration. Its usefulness is due to the fact that many dynamical systems behave in a similar way and thus it is expected that the evolution obtained from the study of the mapping will be essentially the same as that of the actual dynamical system. Hence, the method of constructing the mapping is crucial for its reliability as a useful substitute for the actual system. It is essential that the topology of the phase space of the actual dynamical system and of the mapping coincide, at least in their basic features.

In what follows we shall develop a systematic way to obtain a mapping model for a nearly integrable Hamiltonian system.We shall start with two degrees of freedom and we shall extend the method to three degrees of freedom. In the transition from two to three degrees of freedom we shall see a mechanism of generation of chaos through the interaction between the degrees of freedom.This mechanism has been described by Henrard and Lemaitre(1986) in a different context. In the construction of the mapping we shall require that the basic features of the actual system are preserved in the mapping, so that the topology of the phase space is the same,not only qualitatively but also quantitatively.

THE BASIC IDEA

Let the Hamiltonian of a system with two degrees of freedom be of the form

$$H=H_0(p_1,p_2,q_1,q_2)+ \varepsilon H_1(p_1,p_2,q_1,q_2) \tag{1}$$

where H_0 is the integrable part and ε a small parameter. We shall use action-angle variables of the unperturbed problem and the Hamiltonian takes the form

$$H=H_0(J_1,J_2)+ \varepsilon H_1(J_1,J_2,\theta_1,\theta_2) . \tag{2}$$

We shall study first the unperturbed system defined by the Hamiltonian $H_0(J_1,J_2)$. It is clear that the unperturbed motion is motion on a 2-torus (Fig.1) with fixed radii J_1,J_2 and constant frequencies $n_1(J_1,J_2),n_2(J_1,J_2)$ given by

$$n_1=\frac{\partial H_0}{\partial J_1}, \quad n_2=\frac{\partial H_0}{\partial J_2} . \tag{3}$$

We define now a mapping on a surface of section ($H_0=h,\theta_2=0$) or ($J_2=J_{20},\theta_2=0$). This mapping takes the form

$$\left.\begin{array}{c} J_1 \rightarrow J_1 \\ \theta_1 \rightarrow \theta_1+ 2\pi\dfrac{n_1}{n_2} \end{array}\right\} \tag{4}$$

where the change of the angle

$$\Delta\theta_1=2\pi n_1/n_2 \tag{5}$$

is a function of J_1 (and of the parameter of the mapping, h, or J_{20} depending on the definition of the surface of section). Hence, this is a twist mapping.It is more convenient to use Cartesian-like coordinates

$$X=\sqrt{2J_1}\ \cos(\theta_1),\quad Y=\sqrt{2J_1}\ \sin(\theta_1) \tag{6}$$

and in these variables the invariant curves are all circles (Fig.2).

Along each invariant curve the ratio n_1/n_2 is constant but it varies continously as the radius ρ varies. Consequently, for some values of ρ it is rational, equal to s/r, and the change of the angle is now $\Delta\theta_1 = 2\pi s/r$. The values of ρ for which n_1/n_2 is rational are dense, but only those corresponding to small integers s,r are of practical importance. On a rational invariant curve all points are r-multiple fixed points, i.e. we come to the starting point after r iterations of the mapping. It is easy to see that a point on a rational invariant curve corresponds to the initial conditions of a periodic motion on the 2-torus during which the angle θ_1 changes by $\Delta\theta_1 = 2\pi s$ and the angle θ_2 by $\Delta\theta_2 = 2\pi r$. Note that one iteration of the mapping corresponds to one revolution on the 2-torus along the θ_2 angle.

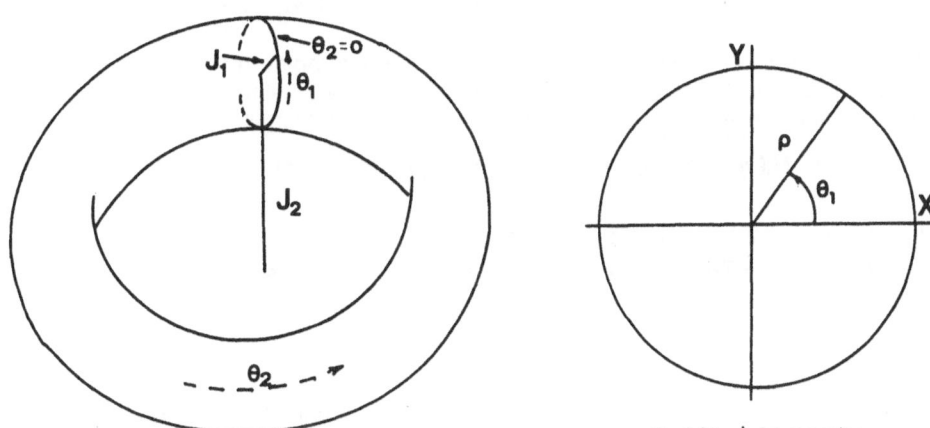

Fig.1.Motion on a torus

Fig.2.The twist mapping

Let us see now what happens when the Hamiltonian H_0 is perturbed.The new Hamiltonian is of the form (2) and,for small ε,the corresponding mapping is a perturbed twist mapping.Thus the mapping is area preserving and we consider it as being generated by the generating function

$$F=J_{1n+1}\theta_{1n}+\ G_0(J_{1n+1})+\ \varepsilon G_1(J_{1n+1},\theta_{1n}) \tag{7}$$

through the transformation

$$J_{1n}=\frac{\partial F}{\partial\theta_{1n}},\quad \theta_{1n+1}=\frac{\partial F}{\partial J_{1n+1}} \tag{8}$$

which gives the perturbed mapping

$$J_{1n+1} = J_{1n} - \varepsilon \frac{\partial G_1}{\partial \theta_{1n}}, \quad \theta_{1n+1} = \theta_{1n} + \frac{\partial G_0}{\partial J_{1n+1}} + \varepsilon \frac{\partial G_1}{\partial J_{1n+1}} \ . \tag{9}$$

Note that for $\varepsilon = 0$ the mapping (9) coincides with the unperturbed mapping (4), if G_0 is properly defined. We may also note that the functions G_0, G_1 depend also on the parameter J_{20} (or h) of the mapping. Our main problem is to find the functions G_0, G_1 in the generating function (7) so that the perturbed mapping model (9) has all the main features of the actual Hamiltonian system.

The above ideas were presented for a system with two degrees of freedom, but it is clear that this method can be generalized to three or more degrees of freedom.

As we shall explain in the sequence, the method of constructing the mapping rests heavily on the averaged Hamiltonian $\overline{H}$ of the original Hamiltonian (2) near a resonance. For this reason we shall present in the next section briefly the method of averaging and we shall find the relation between the averaged and the original, nonaveraged, system.

THE RELATION BETWEEN THE AVERAGED AND THE ORIGINAL SYSTEM

We shall describe the averaging method by a simple example. In this way we shall be able to present all the main features of this method without being lost in unnecessary details. The example we shall use is the system defined by the Hamiltonian

$$H = H_0(J) + \varepsilon H_{00}(J) + \varepsilon H_{11}(J)\cos(\theta_1 - \theta_2) + \varepsilon H_{rs}(J)\cos(r\theta_1 - s\theta_2) \ . \tag{10}$$

We shall consider initial conditions near the s/r resonance, i.e. the frequencies n_1, n_2 obey the relation $n_1/n_2 \simeq s/r$. Thus $H_{rs}\cos(r\theta_1 - s\theta_2)$ is the resonant term and $H_{11}\cos(\theta_1 - \theta_2)$ is the nonresonant term.

Resonant angles

In order to study the motion near the s/r resonance we perform a canonical change of variables from J, θ to $\hat{J}, \hat{\theta}$ by the generating function

$$F = (\theta_1 - \frac{s}{r}\theta_2)\hat{J}_1 + \theta_2 \hat{J}_2 \ . \tag{11}$$

The corresponding canonical transformation which, in a sense, is a transformation to a rotating frame, is

$$\hat{J}_1 = J_1, \qquad \hat{J}_2 = \frac{s}{r}J_1 + J_2$$

$$\hat{\theta}_1 = \theta_1 - \frac{s}{r}\theta_2, \qquad \hat{\theta}_2 = \theta_2 \tag{12}$$

and the corresponding new Hamiltonian is

$$\hat{H}=\hat{H}_0(\hat{J})+\varepsilon[\hat{H}_{00}(\hat{J})+\hat{H}_{rs}(\hat{J})\cos(r\,\hat{\theta}_1)]+\varepsilon\hat{H}_{11}(\hat{J})\cos(\hat{\theta}_1+\frac{s-r}{r}\hat{\theta}_2)\ . \tag{13}$$

The term $\hat{H}_0$ is related to H_0 by the relation

$$\hat{H}_0=H_0(J_1,-\frac{s}{r}J_1+J_2) \tag{14}$$

and the new unperturbed frequencies of the system are

$$\hat{n}_1=\partial\hat{H}_0\,/\partial\hat{J}_1=n_1-(s/r)n_2\quad:small \tag{15}$$
$$\hat{n}_2=\partial\hat{H}_0\,/\partial\hat{J}_2=n_2\ .$$

We see that we have a slow angle, $\hat{\theta}_1$ and a fast angle, $\hat{\theta}_2$.

<u>The method of averaging</u>

We shall try next to find a new canonical change of variables, from $\hat{J},\hat{\theta}$ to $\bar{J},\bar{\theta}$, such that the new Hamiltonian, $\bar{H}$, does not depend on one of the angles. Let

$$F=\bar{J}_1\hat{\theta}_1+\bar{J}_2\hat{\theta}_2+\varepsilon S(\bar{J},\hat{\theta}) \tag{16}$$

be the generating function, where S is a function to be determined. The corresponding canonical change is given by

$$J_i=\bar{J}_i+\varepsilon\partial S/\partial\hat{\theta}_i,\quad \hat{\theta}_i=\bar{\theta}_i-\varepsilon\partial S/\partial\bar{J}_i \tag{17}$$

and the new Hamiltonian is, to $O(\varepsilon)$,

$$\bar{H}=\hat{H}_0(\bar{J})+\varepsilon\frac{\partial\hat{H}_0}{\partial\bar{J}_1}\frac{\partial S}{\partial\hat{\theta}_1}+\varepsilon\langle\hat{H}_1\rangle_{\theta_2}+\varepsilon\left[\frac{\partial H_0}{\partial\bar{J}_2}\frac{\partial S}{\partial\hat{\theta}_2}+\{\hat{H}_1\}\right] \tag{18}$$
$$where \quad \langle\hat{H}_1\rangle_{\theta_2}=\hat{H}_{00}+\hat{H}_{rs}\cos(r\bar{\theta}_1)$$
$$\{\hat{H}_1\}=\hat{H}_{11}\cos(\bar{\theta}_1+\frac{s-r}{r}\bar{\theta}_2)\ .$$

Note that the third term in the right hand side is the averaged part of the Hamiltonian (10) and the fourth term is the only term which contains the fast angle $\hat{\theta}_2$. Also the second term is of the second order in ε, when we are close to resonance, as we can see from the first equation (15). Thus, if we define the function S in such a way that the fourth term in (18) is eliminated, we shall obtain from (17), for this function S a new Hamiltonian, up to second order in ε,

$$\bar{H}=\hat{H}_0(\bar{J}_1,\bar{J}_2)+\varepsilon\hat{H}_{00}(\bar{J}_1,\bar{J}_2)+\varepsilon\hat{H}_{rs}(\bar{J}_1,\bar{J}_2)\cos(r\bar{\theta}_1) \tag{19}$$

in the new canonical variables $\bar{J},\bar{\theta}$, where the angle $\bar{\theta}_2$ is ignorable and consequently $\bar{J}_2=\bar{J}_{20}=$constant. For the example (10)

$$S=-\frac{1}{n_2}\hat{H}_{11}(\bar{J})\frac{r}{s-r}\sin(\bar{\theta}_1+\frac{s-r}{r}\bar{\theta}_2) \ . \tag{20}$$

The Hamiltonian $\bar{H}$ is called the <u>averaged Hamiltonian</u>. We see that (19) can be interpreted as the Hamiltonian of a one degree of freedom system, with $\bar{J}_{20}$ as a parameter. However, we shall also see it as the Hamiltonian of a <u>two</u> degrees of freedom system with one ignorable coordinate.

Using (12) and (17),(20) we find the following transformation from the averaged variables $\bar{J},\bar{\theta}$ to the original variables J,θ:

$$J_1=\bar{J}_1-\text{E } \hat{H}_{11}(\bar{J})COS$$

$$\theta_1=\bar{\theta}_1+\frac{s}{r}\bar{\theta}_2+ \text{E }\left(\frac{\partial\hat{H}_{11}}{\partial\bar{J}_1}+\frac{s}{r}\frac{\partial\hat{H}_{11}}{\partial\bar{J}_2}\right)SIN$$

$$J_2=-\frac{s}{r}\bar{J}_1+\bar{J}_2+\text{E }\hat{H}_{11}(\bar{J})COS \tag{21}$$

$$\theta_2=\bar{\theta}_2+\text{E }\frac{\partial\hat{H}_{11}}{\partial\bar{J}_2}SIN$$

where we abbreviated

$$\text{E}=\frac{\varepsilon}{n_2}\frac{r}{s-r}, \ COS=\cos(\bar{\theta}_1+\frac{s-r}{r}\bar{\theta}_2), \ SIN=\sin(\bar{\theta}_1+\frac{s-r}{r}\bar{\theta}_2) \ . \tag{22}$$

<u>Differential equations and fixed points in $\bar{J},\bar{\theta}$</u>

The differential equations in $\bar{J},\bar{\theta}$, obtained from (19) are

$$\dot{\bar{J}}_1=\varepsilon r\hat{H}_{rs}(\bar{J})\sin(r\,\bar{\theta}_1)$$

$$\dot{\bar{\theta}}_1=\frac{\partial}{\partial\bar{J}_1}(\hat{H}_0+\varepsilon\hat{H}_{00})+\varepsilon\frac{\partial\hat{H}_{rs}}{\partial\bar{J}_1}\cos(r\,\bar{\theta}_1)$$

$$\dot{\bar{\theta}}_2=\frac{\partial}{\partial\bar{J}_2}(\hat{H}_0+\varepsilon\hat{H}_{00})+\varepsilon\frac{\partial\hat{H}_{rs}}{\partial\bar{J}_2}\cos(r\,\bar{\theta}_1) \tag{23}$$

$$\dot{\bar{J}}_2=0 \quad \rightarrow \quad \bar{J}_2=\bar{J}_{20} \ .$$

Note that the first two equations of (23) can be studied separately, as they depend on $\bar{J}_1,\bar{\theta}_1$ only with $\bar{J}_2=\bar{J}_{20}$ a parameter. Note also that only the averaged part of the Hamiltonian (10) enters in (23). This part is the resonant part. The nonresonant terms enter through the transformation equations (21).

The fixed points of (23) in $\bar{J}_1,\bar{\theta}_1$ are obtained if we set $\dot{\bar{J}}_1=0,\dot{\bar{\theta}}_1=0$.The corresponding solution of the complete system (23) is

$$\bar{J}_1=\bar{J}_{10}, \ \bar{\theta}_1=\bar{\theta}_{10}, \ \bar{J}_2=\bar{J}_{20}, \ \dot{\bar{\theta}}_2=\bar{n}_{20}=constant, \tag{24}$$

where

$$\dot{\theta}_2 = n_2 + \varepsilon \frac{\partial \hat{H}_{00}}{\partial \bar{J}_2} + \varepsilon \frac{\partial \hat{H}_{rs}}{\partial \bar{J}_2} c_0 \equiv \bar{n}_{20} \quad . \tag{25}$$

The solution (24),(25) can be considered as a T-periodic solution.

(a) <u>Resonant fixed points</u>. From the first equation (23) we obtain, for $\dot{\bar{J}}_1 = 0$,

$$\sin(r\bar{\theta}_{10}) = 0 \quad \rightarrow \quad \bar{\theta}_{10} = k\frac{\pi}{r}, \quad (k=0,1,2,...) \tag{26}$$

Then $\bar{J}_{10}$ is obtained from the second equation (23), for $\dot{\bar{\theta}}_1 = 0$,

$$\frac{\partial}{\partial \bar{J}_1}(\hat{H}_0 + \varepsilon \hat{H}_{00}) + c_0 \frac{\partial \hat{H}_{rs}}{\partial \bar{J}_1} = 0 \tag{27}$$

$$where \quad c_0 \equiv \cos(r\, \bar{\theta}_{10}).$$

Note that there are 2r fixed points of this kind,belonging to two groups:for k=even and for k=odd,respectively. For the first group (k=even) we have $c_0 = +1$ and for the second (k=odd) we have $c_0 = -1$.Each group has the same value of $\bar{J}_{20}$, as it is clear from the equation (27).

(b) <u>Nonresonant fixed points</u>. From the first equation (23) we can also have $\dot{\bar{J}}_1 = 0$ if $H_{rs}(\bar{J}_1,\bar{J}_{20}) = 0$. Then $\bar{\theta}_{10}$ is given from the second equation (23) for $\dot{\bar{\theta}}_1 = 0$. In the following we shall study the resonant fixed points, where it will become clear why they are related to a resonance.

<u>Relation between the fixed points in $\bar{J}_1, \bar{\theta}_1$ and the periodic orbits in J,θ</u>

Consider a fixed point solution of (23)

$$\bar{J}_1 = \bar{J}_{10}, \; \bar{J}_2 = \bar{J}_{20}, \; \bar{\theta}_1 = \bar{\theta}_{10}, \; \bar{\theta}_2 = \bar{n}_{20}t + \bar{\theta}_{20} \quad . \tag{28}$$

Using the transformation (21) we find that this solution, expressed in the original variables, is

$$J_1 = \bar{J}_{10} - E\, \hat{H}_{11}(\bar{J}_0)\cos\left(\bar{\theta}_{10} + (s-r)\frac{\bar{n}_{20}}{r}t\right)$$

$$J_2 = -\frac{s}{r}\bar{J}_{10} + \bar{J}_{20} + E\, \hat{H}_{11}(\bar{J}_0)\cos\left(\bar{\theta}_{10} + (s-r)\frac{\bar{n}_{20}}{r}t\right)$$

$$\theta_1 = \bar{\theta}_{10} + \frac{s}{r}\bar{n}_{20}t + E\left(\frac{\partial \hat{H}_{11}}{\partial \bar{J}_{10}} + \frac{s}{r}\frac{\partial \hat{H}_{11}}{\partial \bar{J}_{20}}\right)\sin\left(\bar{\theta}_{10} + (s-r)\frac{\bar{n}_{20}}{r}t\right)$$

$$\theta_2 = \bar{n}_{20}t + E \ \frac{\partial \bar{H}_{11}}{\partial \bar{J}_{20}} \sin\left(\bar{\theta}_{10} + (s-r)\frac{\bar{n}_{20}}{r}t\right) \qquad (29)$$

where E is given by (22). The solution (29) is T-periodic with

$$T = \frac{2\pi}{\bar{n}_{20}} r \ . \qquad (30)$$

Note that during one period we have $\Delta\theta_1 = 2\pi s$ and $\Delta\theta_2 = 2\pi r$. This method to study periodic orbits has been used by Message(1966).

As we mentioned before, there are 2r fixed points, divided into two groups, for k= even and k=odd respectively. Thus, one might think that there are 2r periodic solutions of the form (29). We shall prove now that each of the above groups corresponds to the same r-multiple periodic solution: Consider for example the first group (k=2v, v=0,1,2,..r-1). We have r initial conditions for a periodic orbit, one for each fixed point. For a particular value v_0 of v the initial conditions of the periodic orbit are of the form

$$J_1(0) = A_0 + \varepsilon F_{10} \cos(\frac{2v}{r}\pi)$$

$$J_2(0) = B_0 + \varepsilon F_{20} \cos(\frac{2v}{r}\pi)$$

$$\theta_1(0) = \frac{2v}{r}\pi + \varepsilon F_{30} \sin(\frac{2v}{r}\pi) \qquad (31)$$

$$\theta_2(0) = \qquad \varepsilon F_{40} \sin(\frac{2v}{r}\pi) \qquad .$$

Take now the values of the solution (31) at the time $\tau = T/r = 2\pi/\bar{n}_{20}$. We find

$$J_1(\tau) = A_0 + \varepsilon \ F_{10} \cos(\frac{2v_0'}{r}\pi) \qquad (32)$$

$$\cdots\cdots$$

where $v'_0 = v_0 + s$. It is clear that these are the initial conditions of another periodic orbit belonging to the same group. Thus, finally, we come to the conclusion that there exist two periodic orbits in J,θ with multiplicity r which correspond to the s/r resonance. In the next section we shall prove that one of them is stable and the other is unstable.

Stability

We shall consider the averaged system (23) as a system with two degrees of freedom, because we wish to compare it with the nonaveraged system. Let us define, for reasons of brevity, the following functions:

$$g(\bar{J})\equiv\tilde{H}_{rs},\;\; f_l(\bar{J})\equiv\frac{\partial}{\partial J_l}(\tilde{H}_0+\varepsilon\tilde{H}_{00}),\;\; g_l(\bar{J})\equiv\frac{\partial\tilde{H}_{rs}}{\partial J_l},\;\; l=1,2 \;\; . \tag{33}$$

Then the system of differential equations takes the form

$$\begin{aligned}
\dot{\bar{\theta}}_1 &= f_1(\bar{J}) + \varepsilon g_1(\bar{J})\cos(r\,\bar{\theta}_1)\\
\dot{\bar{J}}_1 &= \qquad\;\; \varepsilon rg(\bar{J})\sin(r\,\bar{\theta}_1)\\
\dot{\bar{\theta}}_2 &= f_2(\bar{J}) + \varepsilon g_2(\bar{J})\cos(r\,\bar{e}_1)\\
\dot{\bar{J}}_2 &= 0
\end{aligned} \tag{34}$$

The fixed point solution (24) is

$$\bar{\theta}_{10}=k\pi/r,\;\; \bar{J}_{10},\;\bar{J}_{20},\;\bar{\theta}_2=\bar{n}_{20}t+\bar{\theta}_{20} \tag{35}$$

and the corresponding variational equations are

$$\dot{\xi} = A\xi,\;\;\; \xi=(\Delta\bar{\theta}_1,\Delta\bar{J}_1,\Delta\bar{\theta}_2-\bar{n}_{20}t,\Delta\bar{J}_2)^t \tag{36}$$

where

$$A = \begin{pmatrix} 0 & \alpha_1 & 0 & \alpha_2\\ \gamma & 0 & 0 & 0\\ 0 & \beta_1 & 0 & \beta_2\\ 0 & 0 & 0 & 0 \end{pmatrix} \tag{37}$$

and

$$\gamma=\varepsilon r^2 g_0 c_0,\;\; \alpha_l=\frac{\partial}{\partial J_l}(f_1+\varepsilon g_1 c_0),\;\; \beta_l=\frac{\partial}{\partial J_l}(f_2+\varepsilon g_2 c_0)\;\; . \tag{38}$$

The eigenvalues of A are

$$\lambda_{1,2}=\pm\sqrt{\alpha_1\gamma},\;\;\; \lambda_3=\lambda_4=0 \;\; . \tag{39}$$

Since $c_0=\cos(r\,\bar{\theta}_{10})$ is equal to $+1$ or -1, according to whether r is even or odd, respectively, we come to the conclusion that the T-periodic orbit in the original variables corresponding to the one group of fixed points is stable ($\lambda_{1,2}$:imaginary) and the other is unstable($\lambda_{1,2}$:real). The fixed point solution (35) corresponds to the T-periodic solution (29) in J,θ. Thus we can also see (35) as a T-periodic solution and the monodromy matrix corresponding to it is

$$M = e^{AT}. \tag{40}$$

The stability index is defined as $K=\text{trace}(M)-2$ and using the fact that $\text{trace}(M) = \Sigma e^{\lambda i T}$ we obtain from (39)

$$K = 2 + T^2 \alpha_1 \gamma + O(\varepsilon^2). \tag{41}$$

A MAPPING MODEL FOR THE TWO DEGREES OF FREEDOM SYSTEM (2)

The perturbed model

The mapping (4) corresponds to the unperturbed motion on the 2-torus, defined by the Hamiltonian H_0. When this Hamiltonian is perturbed to $H = H_0 + \varepsilon H_1$, for small ε, the motion takes place on a perturbed torus and the corresponding mapping is a perturbed twist mapping of the form (9) determined by the generating function (7). In constructing the perturbed mapping we shall require that it has the following properties: (a) it is area preserving (b) it has the same fixed points as the actual system and (c) the fixed points have the correct stability index. Then the basic topological structure of the phase space is preserved to the mapping not only qualitatively but also quantitatively.

The area preserving property is automatically fulfilled since the mapping is obtained by a generating function. We shall prove that the properties (b) and (c) will also be fulfilled if the averaged Hamiltonian is used to define the generating function F:

$$F = \bar{J}_{1n+1}\bar{\theta}_{1n} + \tau \hat{H}_0(\bar{J}_{1n+1}, \bar{J}_{20}) + \tau\varepsilon \hat{H}_{00}(\bar{J}_{1n+1}, \bar{J}_{20}) + \\ \tau\varepsilon \hat{H}_{rs}(\bar{J}_{1n+1}, \bar{J}_{20})\cos(r\,\bar{\theta}_{1n}) \tag{42}$$

where τ is a parameter to be defined later and $\bar{J}_{20}$ is a fixed parameter. From (42) we obtain the mapping

$$\left.\begin{array}{l}
\bar{J}_{1n+1} = \bar{J}_{1n} + \tau\varepsilon r \hat{H}_{rs}(\bar{J}_{1n+1}, \bar{J}_{20})\sin(r\,\bar{\theta}_{1n}) \\[2mm]
\bar{\theta}_{1n+1} = \bar{\theta}_{1n} + \tau\dfrac{\partial}{\partial \bar{J}_{1n+1}}(\hat{H}_0 + \varepsilon\hat{H}_{00}) + \tau\varepsilon\dfrac{\partial \hat{H}_{rs}}{\partial \bar{J}_{1n+1}}\cos(r\,\bar{\theta}_{1n})
\end{array}\right\} \tag{43}$$

or, using the notation (33),

$$\bar{J}_{1n+1} = \bar{J}_{1n} + \tau\varepsilon rg(\bar{J}_{1n+1}, \bar{J}_{20})\sin(r\,\bar{\theta}_{1n}) \\
\bar{\theta}_{1n+1} = \bar{\theta}_{1n} + \tau f_1(\bar{J}_{1n+1}, \bar{J}_{20}) + \tau\varepsilon g_1(\bar{J}_{1n+1}, \bar{J}_{20})\cos(r\,\bar{\theta}_{1n}). \tag{44}$$

Note that if we take $\varepsilon = 0$ and use the transformation (21) we obtain the mapping, for $\tau = T$ given by (30),

$$J_{1n+1} = J_{1n}$$

$$\left. \theta_{1n+1} = \theta_{1n} + 2\pi \frac{n_1}{n_2} r \right\} \qquad (45)$$

This is the mapping (4) applied r times.

Fixed points of the mapping

The fixed points of the mapping (44) are $\bar{J}_{1n} = \bar{J}_{1n+1} = \bar{J}_{10}$, $\bar{\theta}_{1n} = \bar{\theta}_{1n+1} = \bar{\theta}_{10}$ and satisfy the relations

$$g(\bar{J}_{10}, \bar{J}_{20}) \sin(r\,\bar{\theta}_{10}) = 0$$

$$f_1(\bar{J}_{10}, \bar{J}_{20}) + \varepsilon g_1(\bar{J}_{10}, \bar{J}_{20}) \cos(r\,\bar{\theta}_{10}) = 0 \quad . \qquad (46)$$

These are identical with the first two equations (23) for $\dot{\bar{J}}_1 = \dot{\bar{\theta}}_1 = 0$. Thus, we proved that the mapping (43) has the same fixed points as the averaged system.

Stability of the fixed points of the mapping

Let us define the variational variables $(\xi_0, \eta_0), (\xi, \eta)$ by

$$\bar{J}_{1n} = \bar{J}_{10} + \xi_0, \qquad \bar{\theta}_{1n} = \bar{\theta}_{10} + \eta_0$$

$$\bar{J}_{1n+1} = \bar{J}_{10} + \xi, \qquad \bar{\theta}_{1+n1} = \bar{\theta}_{10} + \eta \quad . \qquad (47)$$

Linearizing the mapping (44) near the fixed point $\bar{J}_{10}, \bar{\theta}_{10}$ we obtain the variational equations

$$\begin{pmatrix} \xi \\ \eta \end{pmatrix} = A \begin{pmatrix} \xi_0 \\ \eta_0 \end{pmatrix}, \qquad A = \begin{pmatrix} 1 & \tau\gamma \\ \tau\alpha_1 & 1 + \tau^2 \alpha_1 \gamma \end{pmatrix} \qquad (48)$$

where α_1, γ are defined by (38). The stability index is given by $K = \text{trace}(A)$, i.e.

$$K = 2 + \tau^2 \alpha_1 \gamma \quad . \qquad (49)$$

Consequently, if we select τ as the period of the resonant periodic orbit, given by (30), the stability index (49) coincides, to $O(\varepsilon^2)$, with that of the corresponding periodic orbit, as we can see from (41). Consequently, the mapping (43) with $\tau = T$, is a mapping model which has all the required properties.

Invariant curves of the mapping (43) near the s/r resonance

As we mentioned before, the invariant curves of the unperturbed mapping (4) in X,Y variables, defined by (6), are circles. When the mapping is perturbed, the invariant curves that correspond to $n_1/n_2 =$ rational survive the perturbation (for a sufficiently small perturbation) as distorted circles, according to the KAM theorem. On the contrary, the invariant curves for which $n_1/n_2 = s/r$ (rational) are destroyed: Out of the infinite number of r-multiple fixed point only a finite number survive, half of them stable and half unstable, according to the Poincaré-Birkhoff fixed point theorem (Lichtenberg and Lieberman, 1983). In general, only two fixed points survive the perturbation, one stable and the other unstable, unless there are special symmetries in the system. Thus, the topology of the phase space near a rational invariant curve changes qualitatively, in contrast to the .vicinity of an irrational invariant curve (Fig.3). Note however that in general the structure of the phase space near a rational invariant curve is apparent only for small values or s,r. When these values are large, the corresponding invariant curve behaves practically as an irrational invariant curve.

Fig.3. Invariant curves of the perturbed mapping. The resonant invariant curve gives rise to a set of stable and unstable fixed points

We remark that the level curves of the averaged Hamiltonian $\bar{H}$, obtained from

$$\bar{H} = h \tag{50}$$

are similar to the invariant curves of Fig.3, as we have the same fixed points with the same stability characteristics. This however is true only when the system behaves as integrable, i.e. when we are in an ordered region of phase space. If chaos develops, and this is a generic property of all nonintegrable dynamical systems, some of the invariant curves in Fig.3 dissolve, starting from the vicinity of the unstable fixed points. On the other hand, the level curves remain always smooth curves, as the averaged system is integrable, since it corresponds to a system with only one degree of freedom. Thus, the averaged Hamiltonian can be used in regions of ordered motion only.

THREE DEGREES OF FREEDOM (OR TIME DEPENDENT HAMILTONIAN)

We shall extend now our study by increasing from two to three the degrees of freedom. Here also we shall present the main ideas by an example, in order to make the presentation clear. Let the <u>averaged</u> Hamiltonian of a system with three degrees of freedom be

$$\bar{H} = \bar{H}_0(\bar{J}) + \varepsilon \bar{H}_{00} + \varepsilon \bar{H}_{rs}(\bar{J})\cos(r\bar{\theta}_1) + \varepsilon' \bar{H}_{11}(\bar{J})\cos(\bar{\theta}_1 + \bar{\theta}_2) \ . \tag{51}$$

Note that we have two averaged variables $\bar{\theta}_1, \bar{\theta}_2$. The mapping is in a four dimensional space $\bar{J}_1, \bar{\theta}_1, \bar{J}_2, \bar{\theta}_2$ and can be obtained from the generating function

$$F = \bar{J}_{1n+1}\bar{\theta}_{1n} + \bar{J}_{2n+1}\bar{\theta}_{2n} + T\hat{H}_0(\bar{J}_{1n+1},\bar{J}_{2n+1}) + T\varepsilon\hat{H}_{00}(\bar{J}_{1n+1},\bar{J}_{2n+1}) +$$
$$+ T\varepsilon\hat{H}_{rs}(\bar{J}_{1n+1},\bar{J}_{2n+1})\cos(r\,\bar{\theta}_{1n}) + \tag{52}$$
$$+ T\varepsilon'\hat{H}_{11}(\bar{J}_{1n+1},\bar{J}_{2n+1})\cos(\bar{\theta}_{1n}+\bar{\theta}_{2n})$$

by formulas similar to (8). The obtained mapping is

$$\left.\begin{array}{l} \bar{J}_{1n+1} = \bar{J}_{1n} + T\varepsilon r\hat{H}_{rs}(\bar{J}_{1n+1},\bar{J}_{2n+1})\sin(r\,\bar{\theta}_{1n}) + T\varepsilon'\hat{H}_{11}\sin(\bar{\theta}_{1n}+\bar{\theta}_{2n}) \\[2mm] \bar{J}_{2n+1} = \bar{J}_{2n} + T\varepsilon'\hat{H}_{11}\sin(\bar{\theta}_{1n}+\bar{\theta}_{2n}) \\[2mm] \bar{\theta}_{1n+1} = \bar{\theta}_{1n} + T\varepsilon\dfrac{\partial}{\partial\bar{J}_{1n+1}}(\hat{H}_0 + \varepsilon\hat{H}_{00}) + T\varepsilon\dfrac{\partial\hat{H}_{rs}}{\partial\bar{J}_{1n+1}}\cos(r\,\bar{\theta}_{1n}) + \\[4mm] \qquad + T\varepsilon'\dfrac{\partial\hat{H}_{11}}{\partial\bar{J}_{1n+1}}\cos(\bar{\theta}_{1n}+\bar{\theta}_{2n}) \\[4mm] \bar{\theta}_{2n+1} = \bar{\theta}_{2n} + T\dfrac{\partial}{\partial\bar{J}_{2n+1}}(\hat{H}_0 + \varepsilon\hat{H}_{00}) + T\varepsilon\dfrac{\partial\hat{H}_{rs}}{\partial\bar{J}_{1n+1}}\cos(r\,\bar{\theta}_{1n}) + \\[4mm] \qquad + T\varepsilon'\dfrac{\partial\hat{H}_{11}}{\partial\bar{J}_{2n+1}}\cos(\bar{\theta}_{1n}+\bar{\theta}_{2n}) \end{array}\right\} \tag{53}$$

For $\varepsilon'=0$ this mapping coincides with the mapping (43). If $\varepsilon'\neq0$ then there is a coupling between the two degrees of freedom that exist in the averaged Hamiltonian (after the elimination of the "fast" angle). Now $\bar{J}_{20}$ is no longer a constant but varies slowly. This coupling is especially important when we are near a critical invariant curve of the mapping (43), i.e the mapping (53) for $\varepsilon'=0$. As we will see in the example that follows, chaos is generated at these regions of phase space because the variation of $\bar{J}_2$ makes the points of the mapping to jump from one side of the critical curve to the other. This chaotic phenomenon is more evident if in the two dimensional mapping in $\bar{J}_1,\bar{\theta}_1$ already exists a chaotic region around the critical curve.

APPLICATION:ASTEROID MOTION NEAR THE 3:1 RESONANCE

As an example to the method described, we shall study the planar motion of an asteroid near the 3:1 resonance. Only some basic aspects will be given here.A complete study is in preparation. The planar motion of an asteroid in the gravitational field of the Sun and Jupiter is described by the well known restricted three-body problem. We shall start with action-angle variables (Goldstein,1980, Hadjidemetriou 1990) given in terms of the elements of the orbit by the relations

$$J_1 = L - G, \qquad \vartheta_1 = \lambda - \omega$$
$$J_2 = G \quad , \qquad \vartheta_2 = \lambda - \lambda' \tag{54}$$
$$\text{where} \quad L = \sqrt{(1-\mu)a}, \quad G = L\sqrt{1-e^2}$$

λ is the mean longitude, ω the longitude of pericenter and the primed quantities refer to Jupiter. We have $n_1 : n_2 = \dot{\vartheta}_1 : \dot{\vartheta}_2 = n : (n-n')$. For motion near the 3:1 resonance we have $n : n' = 3 : 1$, i.e. $n_2 = 2$ and also $r = 2$.

For the study of motion near this resonance we will use the resonance variables σ, v, S, N defined by the relations (Henrard, 1988)

$$\sigma = (3\lambda' - \lambda - 2\omega)/2$$
$$v = -(3\lambda' - \lambda - 2\omega')/2$$
$$S = L - G = \sqrt{(1-\mu)a}\left(1 - \sqrt{1-e^2}\right) \tag{55}$$
$$N = 3L - G = \sqrt{(1-\mu)a}\left(3 - \sqrt{1-e^2}\right) \quad .$$

Next we go from the resonant variables to the averaged variables. The Hamiltonian in these variables is (Henrard, 1988, Henrard and Caranicolas 1989, Lemaitre, 1984a, Wisdom, 1985)

$$\bar{H} = -\frac{2(1-\mu)^2}{(N-S)^2} - \frac{3}{2}(N-S) + \mu\left[2FS - \frac{2.4}{N}S\cos(2\sigma)\right] +$$
$$+ \mu e'\left[G\sqrt{2S}\cos(\sigma+v) + D\sqrt{2S}\cos(\sigma-v)\right] \tag{56}$$

where $\mu = 0.001$, $e' = 0.048$ and

$$F = -0.205, \quad C = -0.863, \quad G = 0.199, \quad D = 2.656, \quad E = -0.363 \tag{57}$$

To make the notation simpler we have not used the bar to denote the averaged variables, but we must bear in mind that the variables in the averaged Hamiltonian are not identical with (55) but differ by quantities of the order of μ, by formulas similar to (17). Note that the Hamiltonian (56) is of the form (51) and corresponds to an original system with two degrees of freedom whose Hamiltonian is time dependent (through periodic terms, which in fact were eliminated by the method of averaging).

Circular motion of Jupiter

We will start with a circular orbit of Jupiter, i.e. we will use the model of the circular planar restricted three-body problem. In this case we have $e' = 0$ and the Hamiltonian (56) is a Hamiltonian for a one degree of freedom system, with N as a parameter. Using the method described before, we find the generating function

$$F = S_{n+1}\sigma_n + T \frac{2(1-\mu)^2}{(N-S_{n+1})^2} - \frac{3}{2}T(N-S_{n+1}) +$$

$$+2\mu TFS_{n+1} - \mu T\frac{2.4}{N} S_{n+1} \cos(2\sigma_n) \tag{58}$$

which gives the mapping

$$S_{n+1} = \frac{S_n}{1 + \mu T\frac{4.8}{N} \sin(2\sigma_n)}$$

$$\sigma_{n+1} = \sigma_n - \frac{4T(1-\mu)^2}{(N-S_{n+1})^3} + \frac{3}{2}T + 2\mu TF - \mu T\frac{2.4}{N} \cos(2\sigma_n) \tag{59}$$

The parameter T is the period of the resonant 3:1 periodic orbit, given by (30). In our case where r=2, and $n_2=2$ we have $T \approxeq 2\pi$.

In Fig.4 we present the invariant curves of the mapping (59) for different values of the parameter N, in X,Y variables defined by (6). Note that for a range of values of the parameter N the central fixed point is unstable. This corresponds to a circular periodic orbit of the asteroid at the 3:1 resonance. As the value of N increases the central fixed point becomes stable but a pair of doubly symmetric resonant 3:1 fixed points appear, one stable and the other unstable. These are resonant elliptic fixed points at this resonance (Hadjidemetriou and Ictiaroglou,1984). No chaotic motion is apparent in this region of phase space in the mapping. This is also true in the actual case, as we have seen from the numerical integrations.

The invariant curves of Fig.4, Obtained from the mapping (59) coincide with the level curves of the averaged Hamiltonian, obtained by Lemaiter (1984a).

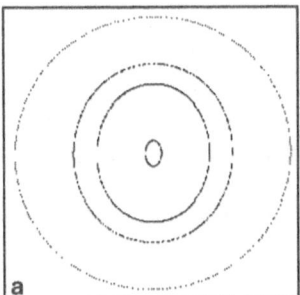 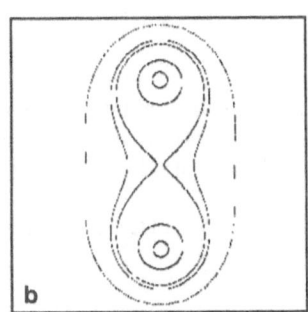 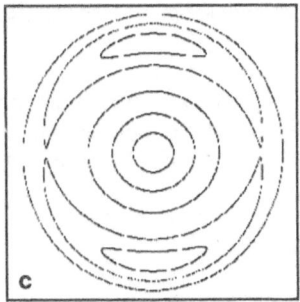

Figs.4a,b,c.Invariant curves of the mapping (59) in the XY plane.The values of N are 1.385, 1.386, 1.40 respectively. The dimensions are ±0.07 for a,b and ±0.25 for c.

Elliptic motion of Jupiter

Let us now assume that $e' \neq 0$ in (56): Then a four dimensional mapping in σ,v,S,N is obtained from the generating function

$$F = S_{n+1}\sigma_n + N_{n+1}v_n + TH_c(S_{n+1}, N_{n+1}, \sigma_n) +$$
$$+ \mu\theta' \sqrt{2s_{n+1}} \left[G \cos(\sigma_n + v_n) + D \cos(\sigma_n - v_n) \right] \tag{60}$$

where H_c is the part of the generating function corresponding to the Hamiltonian of the circular restricted problem (i.e. the part in the right hand side of (58)). The mapping obtained from (60) is

$$
\left.
\begin{aligned}
S_{n+1} &= S_n - \mu T \frac{4.8}{N_{n+1}} S_{n+1} \sin(2\sigma_n) + \\
&\quad + \mu\theta' T \sqrt{2S_{n+1}} \left[G \sin(\sigma_n + v_n) + D \sin(\sigma_n - v_n) \right] \\
N_{n+1} &= N_n + \mu\theta' T \sqrt{2S_{n+1}} \left[G \sin(\sigma_n + v_n) - D \sin(\sigma_n - v_n) \right] \\
\sigma_{n+1} &= \sigma_n - \frac{4T(1-\mu)^2}{(N_{n+1} - S_{n+1})^3} + \frac{3}{2}T + 2\mu TF - \mu T \frac{2.4}{N_{n+1}} \cos(2\sigma_n) + \\
&\quad + \frac{\mu\theta' T}{\sqrt{2S_{n+1}}} \left[G \cos(\sigma_n + v_n) + D \cos(\sigma_n - v_n) \right] \\
v_{n+1} &= v_n + \frac{4T(1-\mu)^2}{(N_{n+1} - S_{n+1})^3} - \frac{3}{2}T + \mu T \frac{2.4}{N_{n+1}^2} S_{n+1} \cos(2\sigma_n)
\end{aligned}
\right\} \tag{61}
$$

It is clear that there is a coupling between the two degrees of freedom (averaged variables σ, v). In particular, if we start with initial conditions near those corresponding to a critical invariant curve of the mapping (59), the variation of N (which was constant in (59)) will generate chaotic motion.

Since (61) is a four dimensional mapping, it is difficult to have a clear geometric presentation. For this reason we will define a "mapping" of the mapping (61) which will be two dimensional and consequently easy to present.

Note that in the elliptic model we have three time scales in the motion of the asteroid: (1) The orbital period of the asteroid. This is eliminated by the averaging process to obtain the Hamiltonian (56). (2) The motion of the resonant angle σ. This appears on the X,Y plane in the mapping (59), Fig.4. (3) The motion of the longitude of the pericenter, $\omega = -(\sigma + v)$.

The motion of σ can be eliminated by considering in (61) a mapping on the surface of section

$$\sigma = \pi/2, \quad \Delta\sigma > 0 \ . \tag{62}$$

Note that since the mapping is discreet, an interpolation will be needed to obtain the exact position on the above surface of section. We have now a two dimensional mapping in the variables $S, \omega = -(\sigma + v)$. The results will be presented in the Cartesian-like coordinates ξ, η defined by

$$\xi = \sqrt{2S}\ \cos(\sigma + v), \eta = \sqrt{2S}\ \sin(\sigma + v) \tag{63}$$

In Figs.5a,b we have two chaotic orbits in the ξη plane, for the initial conditions in S,σ corresponding to the unstable fixed point of Fig.4c. Note the effect of the initial value of the angle v on the orbit. In Figs.6a,b we have the corresponding mapping on the surface of section (62).Also in Fig.7 we have the mapping for the same initial conditions as in Figs.5,6 but for zero eccentricity of Jupiter. The effect of the nonzero eccentricity of Jupiter on the generation of chaos is evident.

The interesting thing is the evolution of the eccentricity for a chaotic orbit. An example is shown in Fig.8 for the initial conditions of the unstable fixed point of Fig.6b: For a long time

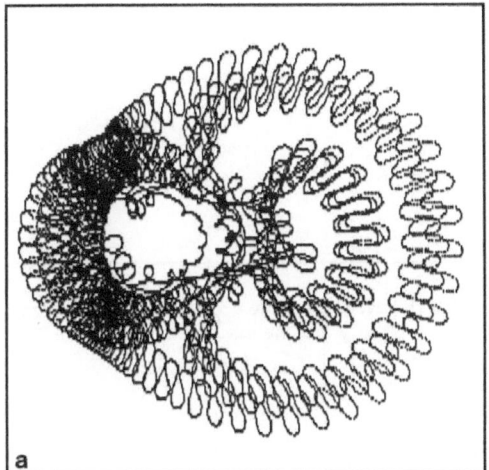 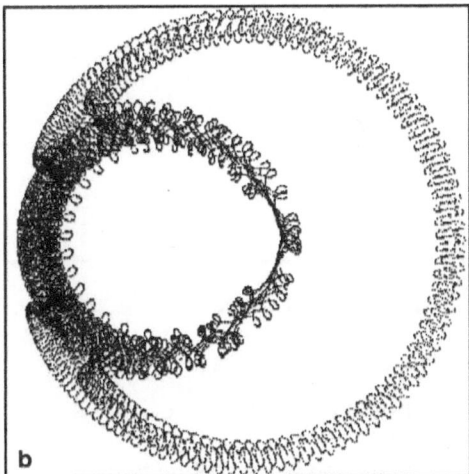

Fig.5a,b.Two chaotic orbits in the ξ,η plane corresponding to the same initial conditions: J = 0.013547, σ = 0, N = 1.4 and different values of v: v = 0 and v = 1.57 rad,respectively.

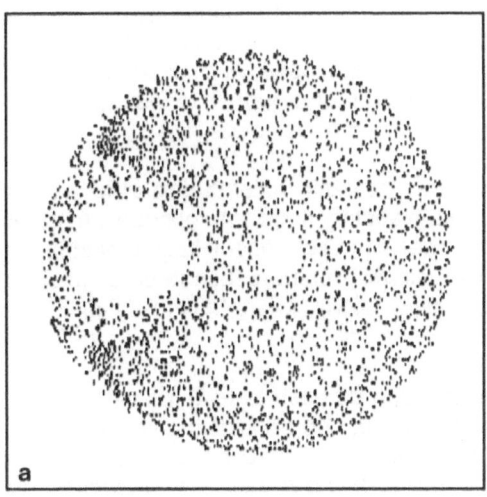 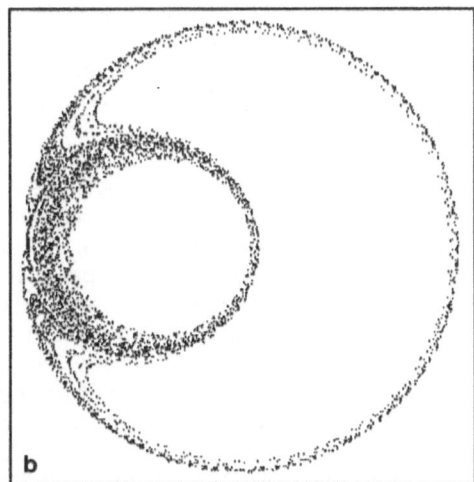

Figs.6a,b.The mapping on the ξ,η plane for the same initial conditions as in figs.5a,b,respectively.

the eccentricity oscillates at low values and suddenly it jumps up to high values. This property has been found by Wisdom, using a different mapping,for the motion of an asteroid at the 3:1 resonance, and in this way he explained the gap in the distribution of the asteroids at this resonance.

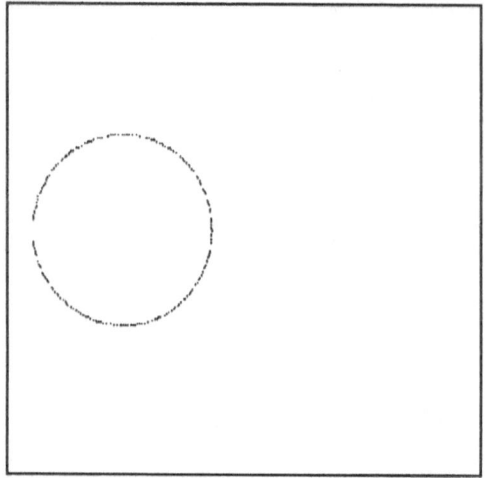

Fig.7.The case of zero eccentricity of Jupiter.

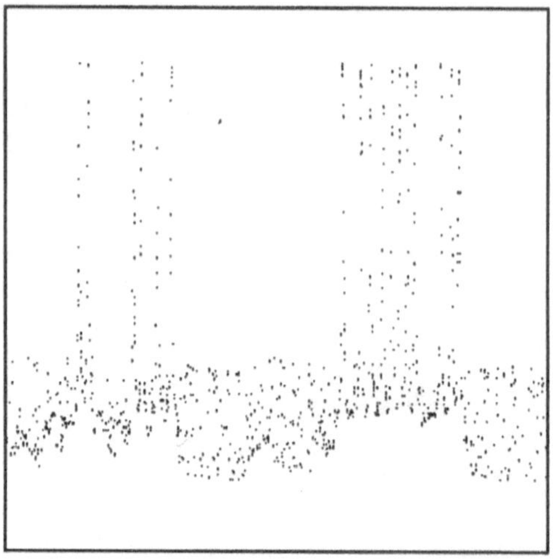

Fig.8.The evolution of the eccentricity. The total time is $3*10^6$y and the max eccentricity is 0.6.

DISCUSSION

The method we described to obtain a mapping model for a Hamiltonian system with two and three degrees of freedom is very simple and can be obtained easily from the averaged Hamiltonian. It is a good model for the original system because the fixed points of the mapping coincide with the periodic orbits of the original system and also they have, to $O(\varepsilon^2)$, the same stability indices. Consequently, the topology of the phase space of the mapping coincides with that of an accurate mapping of the original system obtained by the method of section of Poincaré. We may also note that this method ,applied to a famous problem, the motion of an asteroid near the 3:1 resonance, gave very good results which in fact coicide with similar results obtained by different methods (Wisdom,1985 who used a different mapping model, Henrard and Caranicolas, 1990, who used a further averaging to the averaged Hamiltonian(56)).

The above method of constructing a mapping can be easily extended to other similar problems, for example the satellites of major planets, planetary rings, and of course to other resonant cases of asteroid motion, for example the 2:1 resonance (Hadjidemetriou, in preparation). Also this method can be extended to systems with more than three degrees of freedom, for example motion of an asteroid in three dimensions, where the inclination of the orbit of the asteroid will be one more variable.

The mapping model constructed as shown above can be used to test, in an easy way, several hypotheses for the evolution of the asteroid,for e'≠0, and considering variable mass of the Sun and/or Jupiter, variable e', a slow removal of an accretion disc (studied by Henrard and Lemaitre,1983, Lemaitre,1984b for the circular case) e.t.c. It is also useful to compare similar studies of dynamical systems by mapping models and see what properties are common in all models. This is what we did here in the example of the 3:1 resonant asteroid motion and we found that two very different mapping models for this resonant asteroid motion,the one presented here and Wisdoms' mapping, behave in a very similar way.

REFERENCES

Froeschle,C. and Petit.J.M.:1990, Polynomial approximations of Poincaré maps for Hamiltonian Systems, Astron.Astrophys. (to appear).

Goldstein,H.:1980, Classical Mechanics,(2nd ed.), Addison-Wesley.

Hadjidemetriou,J.D.:1986, A hyperbolic twist mapping model for the study of resonant asteroid motion near the 3:1 resonance,J.Appl.Math.Phys. 37,776-796.

Hadjidemetriou,J.D.,1988: Algebraic mappings near a resonance with an application to asteroid motion in Long term behaviour of Natural and Artificial N-body Systems,A.E.Roy (ed),257-276,Kluwer Publ.

Hadjidemetriou,J.D.:1990, Mapping models for the motion of asteroids near a resonance,proceedings of the Vars Meeting,C.Froeschle (ed).

Hadjidemetriou,J.D. and Ichtiaroglou,S.:1984, A qualitative study of the Kirkwood gaps in the asteroids, Astron.Astrophys. 131, 20-32.

Henrard,J.:1988, Resonances in the planar elliptic restricted problem,in Long-Term Dynamical Behaviour of Natural and Artificial N-body Systems, Roy(ed), 405-425,Kluwer Publ.

Henrard,J. and Lemaitre,A.:1983, A mechanism of formation of the Kirkwood gaps, Icarus55, 482-494.

Henrard J. and Lemaitre,A.:1986, A perturbation method for problems with two critical arguments,Celes.Mech. 39, 213-238.

Henrard,J.and Caranicolas,D.:1990, Motion near the 3:1 resonance of the planar elliptic restricted three body problem,Celes.Mech. 47, 99-121.

Lemaitre,A.:1984a, Higher order resonances in the restricted three-body problem Celes.Mech 32,109-126.

Lemaitre,A.:1984b, Formation of the Kirkwood gaps in the asteroid belt,Celes.Mech.34, 329-341.

Lichtenberg,A.J. and Liebermann,M.A.:1983, Regular and Stochastic Motion, Springer-Verlag.

Liu,J. and Yi-Sui Sun: 1990, On Sitnikov Problem, Celes.Mech.

Message,P.M.:1966, On nearly-commensurable periods in the restricted three body problem in Proceedings of IAU Symposium No 25, Contopoulos (ed).

Petrosky, T.Y. and Broucke, R.:1989, Area -preserving mappings and deterministic chaos for nearly papabolic motion.

Wisdom,J.:1982, The origin of the Kirkwood gaps,Astron.J. 87,577-593.

Wisdom,J.:1983, Chaotic behavior and the origin of the 3/1 Kirkwood gap, Icarus 56,51-74.

Wisdom,J.:1985, A perturbative treatment of motion near the 3/1 commensurability,Icarus 63,272-289.

A MODEL FOR THE STUDY OF VERY-HIGH-ECCENTRICITY ASTEROIDAL

MOTION: THE 3:1 RESONANCE

S.Ferraz-Mello and J.C.Klafke

Instituto Astronômico e Geofísico
Universidade de São Paulo
São Paulo

INTRODUCTION

Recent work by one of the authors (Ferraz-Mello, 1989, 1990 a,b) and by Morbidelli and Giorgilli (1990) has shown the existence of very-high-eccentricity stable and unstable equilibrium solutions (corotation centers) in the averaged Sun-Jupiter-asteroid planar problem, when a secular resonance and a resonance of periods occur simultaneously. They correspond to stationary motions in which the orbits of the asteroid and Jupiter share the same apsidal line. In the case of a 3:1 resonance of periods, two of these corotation centers form a stable-unstable pair at $e = 0.812$ and $e = 0.788$, respectively. The stable corotation center corresponds to a maximum of the energy ($E_S = -1.775728$ in astronomical units). For values close to this maximum the motions are regular oscillations in the neighbourhood of the corotation center (as those shown in the left-hand side of fig. 2).

The study of the phase portrait of such dynamical system may be done by means of purely numerical techniques, but, generally, the CPU times involved are large and limit the possible exploration of the phase space. More extended analyses become possible if an analytical averaging of the equations is made before the numerical integration. However, the classical technique for the expansion of the potential of the disturbing forces due to Jupiter, in terms of Keplerian elements, may be used only for small eccentricities; indeed, as shown by Sundmann (Silva, 1909; Hagihara, 1971), the convergence of the series is limited to $e \approx 0.33$ in the resonance $3 : 1$, to $e \approx 0.18$ in the resonance $2 : 1$ and only to $e \approx 0.09$ in the resonance $3 : 2$, values which are less than the observed ones (the original proof by Sundmann was restricted to orbits with circulating perihelia – but this is generally the case). In order to circumvent this difficulty, we may use asymmetric expansions (Ferraz-Mello, 1987; Ferraz-Mello and Sato, 1989); they are Taylor series about generic points in the phase space (with $e_0 \neq 0$) and may represent the disturbing potential for large values of e. However, there is no reason for which these series may have a better convergence than the classical, symmetric one – on the contrary, the convergence must become poor when $e_0 \to 1$. Thus, analytically, the asymmetric series can only be used for the study of librations and corotations of small amplitude, when the motion remains in the neighbourhood of the point about which the expansion is done.

Predictability, Stability, and Chaos in N-Body Dynamical Systems
Edited by A.E. Roy, Plenum Press, New York, 1991

THE MODEL

In the planar model presented in this paper, in order to cope with this difficulty, the phase space is divided into a given number of cells and the analytical averaging of the disturbing potential is done locally, in the center of each cell. The averaged equations are, then, integrated numerically and, at every point in the integration, the coefficients of the series representation of the disturbing function are taken as computed in the center of the corresponding cell.

The adopted equations are non-canonical. The reason for this choice is that the asymmetric expansion of Ferraz-Mello and Sato is done using Keplerian elements. These equations are obtained straightforwardly from Lagrange's equations for the variation of the elements:

$$\frac{da}{dt} = \frac{2r}{na}[h\frac{\partial F}{\partial k} - k\frac{\partial F}{\partial h} - \frac{\partial F}{\partial \sigma_1}] \tag{1}$$

$$\frac{d\sigma_1}{dt} = (r+1)n_1 - rn + \frac{2r}{na}\frac{\partial F}{\partial a} - \frac{r\beta}{na^2(1+\beta)}[k\frac{\partial F}{\partial k} + h\frac{\partial F}{\partial h}] \tag{2}$$

$$\frac{dk}{dt} = \frac{\beta}{na^2}\frac{\partial F}{\partial h} - h\frac{d\sigma_1}{dt} - \frac{k\beta}{2a(1+\beta)}\frac{da}{dt} \tag{3}$$

$$\frac{dh}{dt} = -\frac{\beta}{na^2}\frac{\partial F}{\partial k} + k\frac{d\sigma_1}{dt} - \frac{h\beta}{2a(1+\beta)}\frac{da}{dt}. \tag{4}$$

The variables are the semi-major axis a and the parameters k, h, σ_1 defined through the equations

$$\begin{aligned}
k &= e\cos\sigma \\
h &= e\sin\sigma \\
\sigma &= \phi - \varpi \\
\sigma_1 &= \phi - \varpi_1 \\
\phi &= (r+1)\lambda_1 - r\lambda.
\end{aligned} \tag{5}$$

ϕ is the critical angle associated with the qth-order resonance $(p+q):p$ (or $(r+1):r$ with $r = p/q$). e, e_1 are the eccentricities, λ, λ_1 the mean longitudes, ϖ, ϖ_1 the longitudes of the perihelia, n, n_1 the mean motions and $\beta = \sqrt{1-e^2}$; the subscript 1 refers to Jupiter. The fifth equation, in the derivative of the mean synodic longitude $Q = \lambda - \lambda_1$, is not considered, since the function F is averaged over q times the mean synodic period and $\frac{\partial F}{\partial Q} = 0$. The orbital elements of Jupiter are assumed constant.

The function F is the averaged disturbing function:

$$F = \frac{\mu}{a_1}[A_0 + A_1\delta k + A_2\delta h + \frac{1}{2}A_3\delta k^2 + \frac{1}{2}A_4\delta h^2 + A_5\delta k\delta h$$

$$+(A_6 + A_8\delta k + A_{10}\delta h)e_1\cos\sigma_1 + (A_7 + A_9\delta k + A_{11}\delta h)e_1\sin\sigma_1$$

$$+ \frac{1}{2}A_{12}e_1^2 + \frac{1}{2}A_{13}e_1^2\cos 2\sigma_1 + \frac{1}{2}A_{14}e_1^2\sin 2\sigma_1] \tag{6}$$

expanded about one center k_0, h_0 up to the second power of the differences $\delta k = k - k_0, \delta h = h - h_0$ (see Ferraz-Mello and Sato, 1989). The coefficients A_j are functions of the semi-major axis a and of the chosen center k_0, h_0.

The Lagrangian equations for a, k, h, σ_1 have the energy-like first integral

$$-E = \frac{\mu}{2a} + \frac{p+q}{p} n_1 n a^2 + F. \tag{7}$$

The value of the coefficients A_j and their derivatives with respect to a are calculated previously in a set of points k_0, h_0. These coefficients are assumed to allow a good representation of the function F in a small domain around k_0, h_0, in the neighbourhood of the value of the semi-major axis a characteristic of the given resonance. The plane k, h is then divided into a finite number of square cells. During the numerical integration of the equations, at each point, the expansion about the center of the cell where the point is found, is selected. The dependence of the coefficients with a is given by a linear approximation.

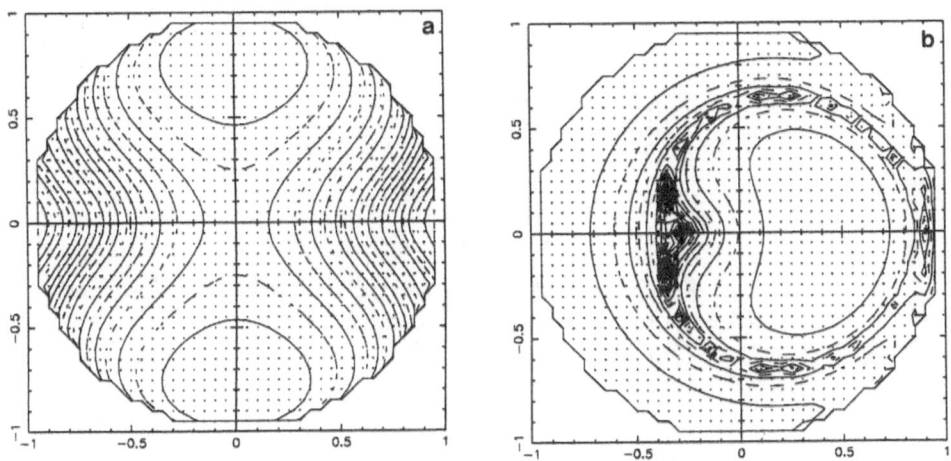

Fig. 1 – Level curves $A_0 = const.$ in the plane k, h. (a) Resonance 3:1. (b) Resonance 3:2. The dots are the centers used for the local expansions.

Figure 1 shows the level curves $A_0 = const.$ in the plane k, h, in two cases: the resonances 3:1 and 3:2. Some results for the resonance 3:1 are discussed in this paper. The level curves for the resonance 3:2 show some difficulties we face in this resonance, but not in the resonance 3:1. In the case 3:2 we may see a *reef* along the line corresponding to averages over orbits going through a collision with Jupiter. The infinite values of A_0 on this line, in fact, are not seen in the figure, since the function is sampled only in the points of a finite grid. In both cases the calculations were done in the center of a net of squares sized 0.05×0.05.

THE RESONANCE 3:1

In this section, some preliminary results concerning the resonance 3:1, obtained with this method, are given. Figures 2 and 3 show surfaces of section defined by $\sigma = \pi/2$ (with $\dot{\sigma} < 0$) and the energies (referred to the maximum) $\Delta E = -4.3 \times 10^{-5} E_S$ and $\Delta E = -4.7 \times 10^{-5} E_S$, respectively. These figures extend to high eccentricities results previously found by Wisdom (1983,1985) and by Henrard and Caranicolas (1990) (They are somewhat larger than those obtained using the classical Laplacian expansion). In both figures there is a stable periodic solution, near the origin, surrounded by

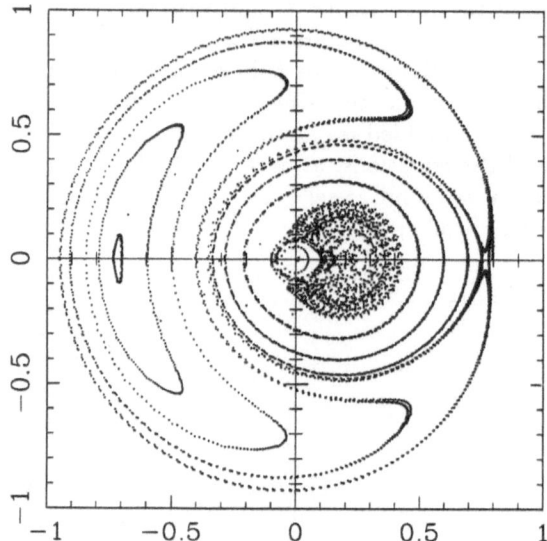

Fig. 2 – Surface of section for $\Delta E = -4.3 \times 10^{-5} E_S$. The axes are $\epsilon.\cos(\varpi - \varpi_1)$ and $\epsilon.\sin(\varpi - \varpi_1)$.

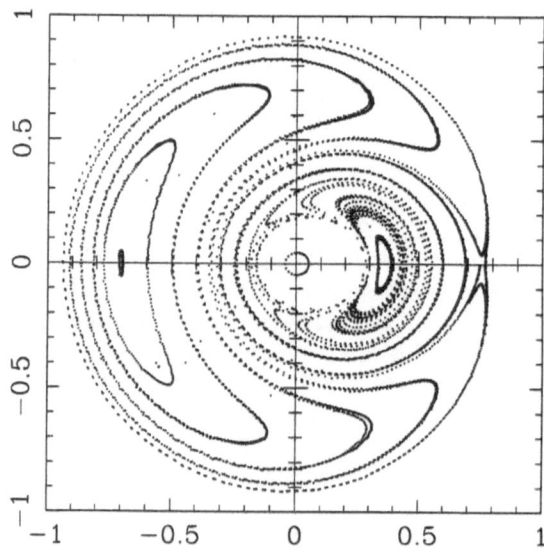

Fig. 3 – Surface of section for $\Delta E = -4.7 \times 10^{-5} E_S$. Axes as in fig. 2.

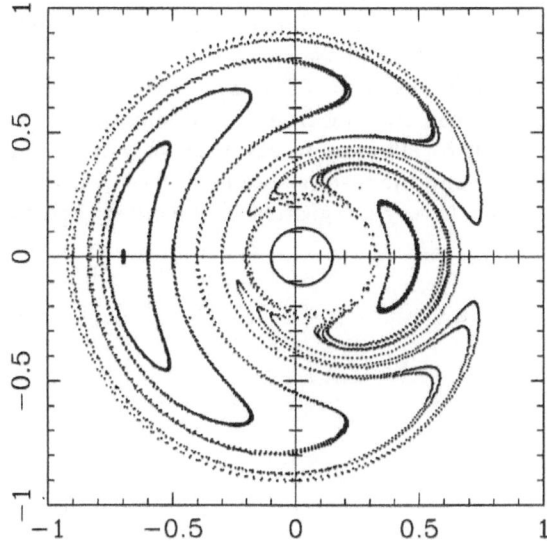

Fig. 4 – Surface of section for $\Delta E = -5.05 \times 10^{-5} E_S$. Axes as in fig. 2.

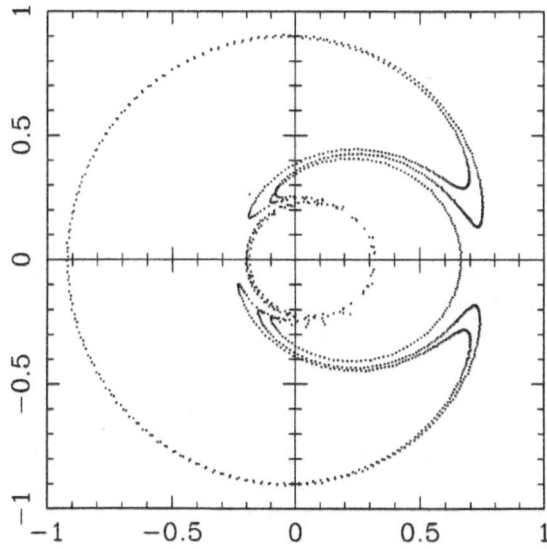

Fig. 5 – Detail of fig. 4. The chaotic solution making the communication between the inner and outer part of the plane of section.

regular motions. The boundary of this region (called *zone of uncertainty*, by Wisdom) is the inner limit of the chaotic region emanating from the saddle in the horizontal axis, at $k \approx -0.1$ (fig. 2) and $k \approx -0.2$ (fig. 3). In fig. 3 another center is seen, at $k \approx 0.35$, as well as some seemingly regular curves around it. In fig. 2 a similar center has not been found. In both cases, the whole inner region is enveloped by a bunch of regular curves. In the outer part of the phase space the topology is governed by the center at $k = -0.7$ and the saddle near $k = +0.8$. The motions about the center are regular. Some irregularities are seen starting from the saddle, but the precision of the calculations is not good enough to allow us to say that they mean chaos. Anyway, chaotic motion is expected there and the results obtained serve to say that the diffusion in such region must be very slow (every integration in the outer region covers 300,000 years).

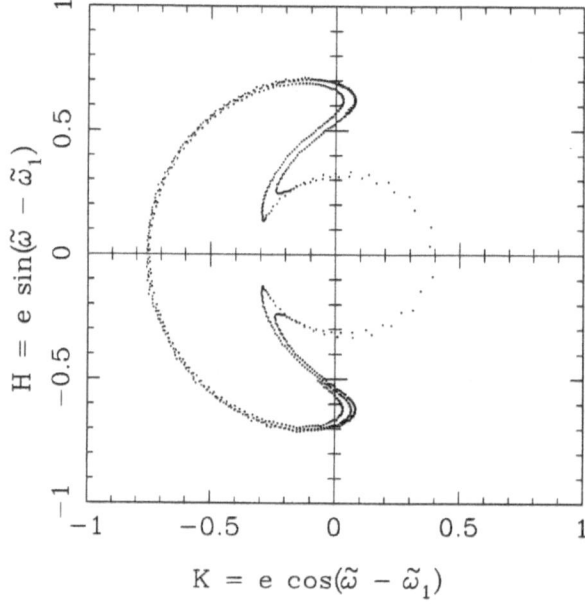

Fig. 6 – Solution corresponding to the present orbital elements of (887) *Alinda* for 300,000 years.

In the inner region of these figures we may see orbits allowing the eccentricity to vary between $0.1 - 0.2$ and $0.4 - 0.5$; in the outer one the allowed variation is between 0.3 and 0.9. However, these two regions, for these energies, do not communicate.

Figure 4 shows a similar surface of section for $\Delta E = -5.05 \times 10^{-5} E_S$. Now, the inner and outer regions communicate; the outer stable and unstable manifolds of the innermost saddle and the inner manifolds of the outermost one entangle. While the chaotic regions of figures 2 and 3 are associated with the existence of homoclinic points, the communication between the two regions is associated with heteroclinic points. One solution leading from $e < 0.2$ to $e = 0.9$ is, now, possible. The evolution shown in fig. 5 corresponds to 7×10^5 years. One asteroid (or meteoroid), in such a motion, would keep its eccentricity higher than 0.6 for periods of $\approx 10^5$ years, one

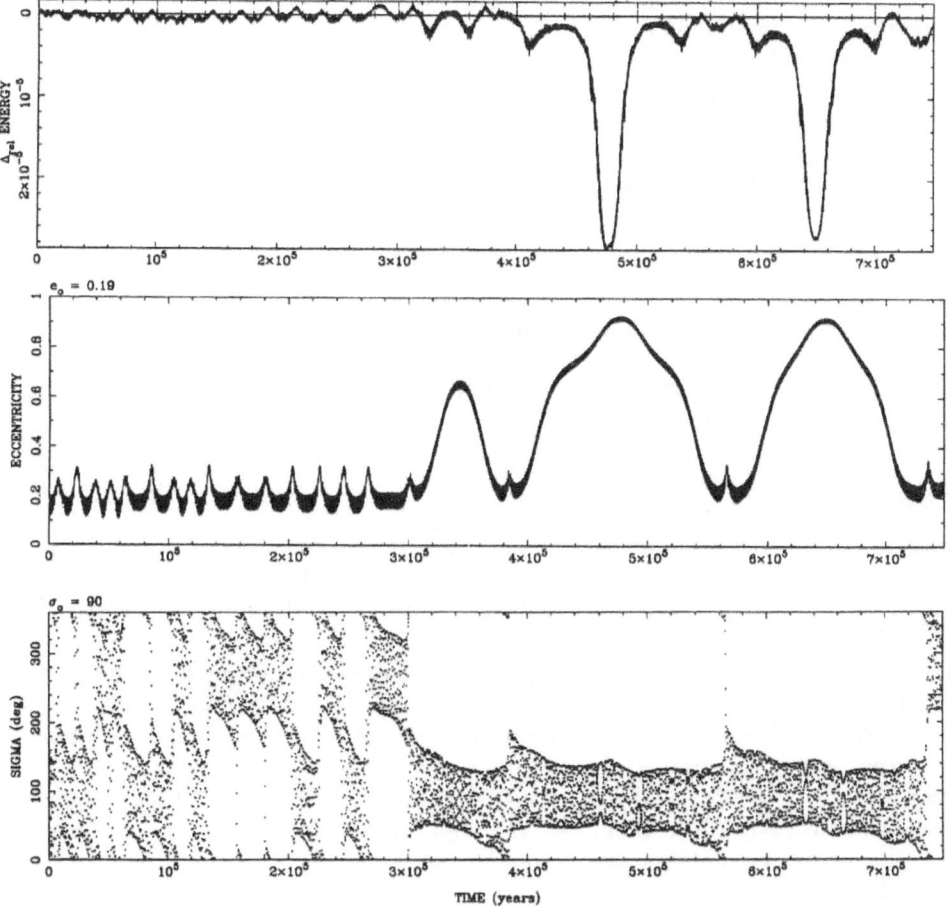

Fig. 7 – Variations of the energy (in units of E_S), the eccentricity and the critical angle σ, for $750,000$ years, in the solution whose section is shown in fig. 5.

time long enough to allow him to be strongly perturbed by the Earth or Mars and to become, perhaps, a permanent Apollo.

It is worth mentioning that the asteroid (887) *Alinda* ($\Delta E = -6.6 \times 10^{-5} E_S$) is moving in the chaotic zone and may pass from one region to the other, undergoing variations of the eccentricity from 0.25 to 0.75 in 6×10^4 years. The secular evolution of its eccentricity and perihelion for 3×10^5 years is shown in figure 6 (the numerical integration of the exact equations of the motion shows variations in the range $0.35 - 0.76$ since the capture of this asteroid in the resonance 3:1; see Milani *et al.* 1969). We emphasize the extreme sensitivity of this solution with respect to initial conditions. Very small changes are sufficient to give a crescent-like curve not including the origin.

The integrations were checked mainly by looking at the energy E. Inside a cell, the precision obtained is very good – ten figures, or more, for instance – but in the jump from one cell to the next, the results are impaired by the difference of the computed values of the function F in the two sides of the border. Figure 7 shows the behaviour of the energy of the solution shown in fig. 5, over 750,000 years. This figure also shows the evolution of the eccentricity and the angle σ, in the same time. The variation in the energy is clearly related to large eccentricities and large amplitudes

183

of the libration of σ. When these factors occur, the regions of faster variation of A_0 are reached (see fig. 1(a)) and the border errors become larger. This result shows that improvements are still necessary. However, the succession of similar sections guarantees the results, at least under a qualitative point of view. The large variations in the energy also serve to show the importance of having a good representation of the averaged potential of the disturbing forces–F over the phase space. Integrations with a continuous representation of F may give good internal accuracy hiding, in this way, the large errors due to the poor representation of that function.

Other checks were done making the eccentricity of Jupiter equal to zero, in which case the dynamical system is completely integrable. The surfaces of section thus obtained show only well marked-invariant curves, even at high eccentricities.

Acknowledgements. This research was partly sponsored by the Research Foundation of São Paulo, grant FAPESP 89/0964-9.

REFERENCES

Ferraz-Mello, S.: 1987, "Expansion of the disturbing force-function for the study of high-eccentricity librations", *Astron. Astrophys.* **183**, 397-402.

Ferraz-Mello, S.: 1988, "The high-eccentricity libration of the Hildas", *Astron. J.* **96**, 400-408.

Ferraz-Mello, S.: 1989, "A semi-numerical expansion of the averaged disturbing function for some very-high-eccentricity orbits", *Celest. Mech.* **45**, 65-68.

Ferraz-Mello, S.: 1990a, "Regular Motions of Resonant Asteroids", *Rev. Mexicana Astron. Astrof.* (in press).

Ferraz-Mello, S.: 1990b, "A theory of planar planetary corotations" (in preparation).

Ferraz-Mello, S., Sato, M.: 1989, "A very-high-eccentricity asymmetric expansion of the disturbing function near resonances of any order", *Astron. Astrophys.* **225**, 541-547.

Hagihara, Y.: 1971, *Celestial Mechanics*, Vol.II, Part I, M.I.T.Press, Cambridge.

Milani, A., Carpino, M., Hahn, G., Nobili, A.M.: 1989, "Project Spaceguard: Dynamics of planet- crossing asteroids. Classes of Orbital Behaviour", *Icarus* **78**, 212-269.

Morbidelli, A., Giorgilli, A.: 1990, "On the Dynamics of the Asteroidal Belt", *Celest. Mech. Dyn. Astron.* **47**, 145-204.

Silva, G.: 1909, "Sur les limites de convergence du développement de la fonction perturbatrice", *Bulletin Astron.* **26**, 49-75, 97-114.

Wisdom, J.: 1983, "Chaotic behaviour and the origin of the 3/1 Kirkwood gap", *Icarus* **56**, 51-74.

Wisdom, J.: 1985, "A perturbative treatment of motion near the 3/1 commensurability", *Icarus* **63**, 272-289.

THE LOCATION OF SECULAR RESONANCES

Alessandro Morbidelli

Dep. of Math., F.U.N.D.P.
Rempart de la Vierge 8
B–5000 Namur, Belgium

INTRODUCTION

The importance of the secular resonances for the dynamics in the asteroid belt is well known since Tisserand (1882) and Poincaré (1892). Indeed a Lagrangian linear theory for the secular motion of the planets shows that their elements $e, \overline{\omega}, i, \vartheta$ (eccentricity, longitude of perihelion, inclination and longitude of the ascending node, using the notations of Poincaré) are not constants of motion, as they are in the Keplerian problem, but vary with the following law (see Bretagnon (1974)) :

$$
\begin{aligned}
e_k \cos \overline{\omega}_k = \sum_{j=1,8} M_{k,j} \cos \left(g_j t + \alpha_j\right) , \qquad & e_k \sin \overline{\omega}_k = \sum_{j=1,8} M_{k,j} \sin \left(g_j t + \alpha_j\right) \\
\sin \frac{i_k}{2} \cos \vartheta_k = \sum_{j=1,8} M_{k,j} \cos \left(s_j t + \beta_j\right) , \qquad & \sin \frac{i_k}{2} \sin \vartheta_k = \sum_{j=1,8} M_{k,j} \sin \left(s_j t + \beta_j\right) .
\end{aligned}
\tag{1}
$$

Here the indexes j, k refer to each planet, from Mercury (1), to Neptune (8). If one considers only the system formed by the Sun, Jupiter, and Saturn, only the terms with $k = 5, 6$ and $j = 5, 6$ survive.

The right hand side of (1) plays the role of a forcing term in the linearized averaged equations for the motion of the elements of an asteroid; like the linearized equations of a forced pendulum, these equations have a singular divergent solution in case of resonance, namely when the frequency of the longitude of perihelion or of the longitude of node of the asteroid is equal to one of the fundamental frequencies of the solar system g_j or s_j respectively; these resonances are usually called *secular*.

The problems connected to the secular resonances are essentially two : the location of the resonances themselves and the exploration of the resonant dynamics. This is strictly relied to the problem of the determination of the regions of the Solar System where chaotic or quasi–periodic motion is present, in the hope to get an explanation for the observed puzzling distribution of the asteroids. To determine the location of the resonances is easy in principle : one should just evaluate the frequencies of the perihelion and of the node of the asteroid; however, such frequencies are not constant, and so one has to define somehow two normalized constant frequencies, called "proper frequencies". In general this is done by expanding the Hamiltonian of the problem up to a quite low order in the eccentricity and in the inclination of the asteroid, and averaging it with respect to the mean anomalies and to the argument of perihelion; this is an approximated method, whose accuracy must evidently become worse and worse with increasing eccentricity and/or inclination. In 1969 Williams developed a method where he avoided the expansion of the perturbation with respect to the elements of the asteroid. Unfortunately his method was described only briefly in his Ph.D. dissertation, and the results, published after a revision by Faulkner (1981), were never obtained independently by any other author. Moreover Williams and Faulkner limited themselves to locate the resonances, and never extended their theory to study the resonant dynamics.

For what regards the theoretical attempts of exploration of resonant dynamics, the most interesting results have been obtained by Yoshikawa (1987) for the ν_6 resonance ($g_0 = g_6$), and Nakai and Kinoshita (1985) for the ν_{16} resonance ($s_0 = s_6$). In their works, the authors average the perturbation (expanded in terms of the eccentricity and inclination up to order 4 in Yoshikawa) with respect to the mean anomalies and over the argument of perihelion. These studies have not been extended to the ν_5 resonance ($g_0 = g_5$) as its high inclination ($\sim 30°$) gives many problems to the accuracy of these approximated models.

In two recent papers (1990a, 1990b), Morbidelli and Henrard developed a general theory which can be easily applied to locate the secular resonances as well as to study the resonant dynamics. The precision of the results is not limited to a small portion of the phase–space; this is obtained by avoiding any expansion of the Hamiltonian in terms of the eccentricity and of the inclination of the asteroid and by introducing suitable action–angle variables. In this brief paper, my aim is to outline the basic ideas of these our works, and discuss the results obtained on the location of secular resonances.

THE SECULAR PROBLEM

The problem of the secular resonances is usually studied in the framework of a massless body (asteroid), attracted by the Sun and perturbed by the planets (essentially Jupiter and Saturn). The Hamiltonian of the problem can be easily expressed in the form

$$K = -\frac{1}{2\Lambda^2} + \varepsilon^2 K^p(\Lambda, H, Z, \lambda, h, \zeta, \lambda_j, \xi_j, \eta_j, p_j, q_j) \,, \tag{2}$$

where we have adopted for the asteroid the modified Delaunay variables which, in the notations of Poincaré are :

$$\begin{aligned}
\Lambda &= L \,, & \lambda &= l + g + \vartheta \\
H &= L - G \,, & h &= -g - \vartheta = -\overline{\omega} \\
Z &= G - \Theta \,, & \zeta &= -\vartheta \,;
\end{aligned} \tag{3}$$

moreover $j(1 \le j \le 8)$ is an index which refers to the planets, λ_j are the mean longitudes of the planets, and ξ_j, η_j, p_j, q_j are related to the elliptic elements of the planets by the relations

$$\xi_j = e_j \cos \overline{\omega}_j, \quad \eta_j = -e_j \sin \overline{\omega}_j, \quad p_j = \sin \frac{i_j}{2} \cos \vartheta_j, \quad q_j = -\sin \frac{i_j}{2} \sin \vartheta_j \,. \tag{4}$$

In formula (2) one easily recognizes the Hamiltonian of the two-body problem $-1/(2\Lambda^2)$ and a perturbation $\varepsilon^2 K^p$ which is globally of order of the ratio between the mass of Jupiter and that of the Sun ($\sim 10^{-3}$); then $\varepsilon^2 = 10^{-3}$. The perturbation K^p is time–dependent through the variables $\lambda_j, \xi_j, \eta_j, p_j, q_j$, which are considered as given functions of time with proper fixed frequencies.

The first step of our analysis of the Hamiltonian (2) is to expand it in Taylor series around 0, with respect to the variables of the planets ξ_j, η_j, p_j, q_j; as the secular theory for these variables (see Bretagnon, 1974) shows that they are always of order ε, one gets an expansion in orders of ε of the Hamiltonian, of the kind :

$$K = -\frac{1}{2\Lambda^2} + \varepsilon^2 K_0 + \varepsilon^3 K_1 + \varepsilon^4 K_2 + \ldots \tag{5}$$

where the index m of K_m refers to the degree of the corresponding polynomial in ξ_j, η_j, p_j, q_j. Then K_0 is the perturbation given by the planets considered on coplanar circular orbits, and thus it is dependent only on $\Lambda, H, Z, \lambda, h, \zeta, \lambda_j$.

We suppose now to be away from a mean motion resonance; in this case we can normalize the Hamiltonian with respect to the mean anomalies λ and λ_j of the asteroid and of

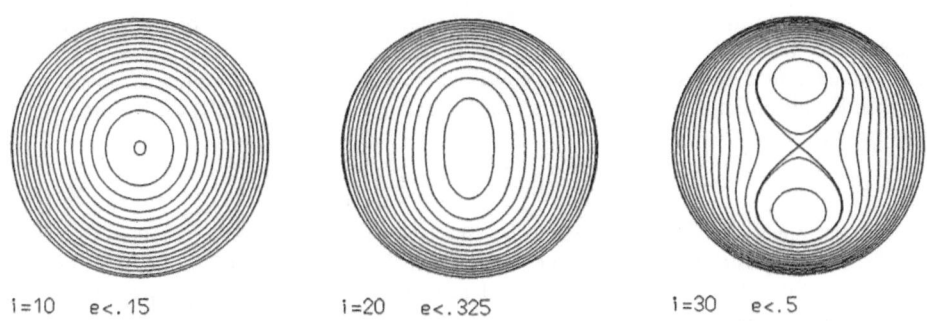

| $i=10$ $e<.15$ | $i=20$ $e<.325$ | $i=30$ $e<.5$ |

Fig. 1.   Secular motion of the asteroid described by $\overline{K}_0$; each circle represents a surface $L - \Theta$ =const. in polar coordinates e and g; below each picture are reported the inclination of the circular orbit and the maximal eccentricity of the orbits on the plane.

the planets respectively. This can be done via the well known Lie's algorithm, and one gets, up to order 4, the following Hamiltonian

$$K_A = -\frac{1}{2\Lambda^2} + \varepsilon^2 \overline{K}_0 + \varepsilon^3 \overline{K}_1 + \varepsilon^4 \overline{K}_2 + \frac{\varepsilon^4}{2} \overline{\{\chi_2, K_0\}} \,, \tag{6}$$

where $\{.,.\}$ denotes the Poisson bracket, and χ_2 is the generating function of lowest order of the Lie transform. Here, with the bar over each term, we wish to indicate that an average over λ and λ_j has been performed; note however that, as we have not expanded the functions with respect to the elements of the asteroid, the operation of averaging can not be done by simply dropping the harmonics in λ and λ_j, but requires a numerical procedure. This is the price to pay to carry on a study, which must be valid all over the phase–space. The term $\frac{\varepsilon^4}{2}\overline{\{\chi_2, K_0\}}$ is usually addressed to as "quadratic term in the masses" since its nominal order in ε is 4. However, as it contains the so called *small denominators*, given by an integer combination of the mean motions of the asteroid and of the perturbing planet, it's size is expected to become relevant approaching a mean–motion resonance.

As a first approximation, we neglect the quadratic term in the masses; this is consistent if we are far from a mean–motion commensurability. In this case, the term $\overline{K}_0$ is the one of predominant size in (6). Such a term has a precious propriety : considered as an isolated Hamiltonian, it is an integrable one. Indeed one can prove that it depends only on $\zeta - h$, namely on the argument of perihelion g, and so has two constants of motion, which are $H + Z$ and $\overline{K}_0$ itself. The motion described by such Hamiltonian has been studied in detail by Kozai in 1962, and we summarize here his main results. We consider a given surface of constant $H + Z = L - \Theta$ (we remind here that, for small e and i, $L - \Theta \sim e^2 + i^2$) and we describe the motion on it in polar coordinates e and g. In the first picture of Fig. 1. we show the motion on a surface given by a small value of $L - \Theta$ (namely for a small value of the inclination of the circular orbit on the surface). As one sees, the origin ($e = 0$), at the center of the figure, is a stable equilibrium point, and around it there are almost circular cycles, where g is a circulating variable. The second picture of Fig. 1. corresponds to a higher value of $L - \Theta$; the origin is still a stable equilibrium point, but the cycles around it are no more circular. Finally the last picture of Fig 1. is drawn for a value of $L - \Theta$ greater than a critical value; this critical value is actually a function of the semi-major axis, which can be found in Kozai, but, more or less, in the asteroid belt corresponds to a circular orbit with an inclination of $\sim 30°$. In Fig 1c the circular orbit is an unstable equilibrium point, and two new stable stationary points appear at $g = 90°$ and $g = 270°$, surrounded by a separatrix (bold line). Inside of the separatrix, g is a librating variable, while outside g is circulating on a cycle which is very far from a circular one. The three drawn pictures are just qualitative, as they have been obtained by using the first terms reported in Yuasa's paper (1973) for a semi–major axis equal to .58 times that of Jupiter, but they reproduce well the essential features of the real motion.

We want now to compute the free frequencies of the longitudes of perihelion and of the node of an asteroid. These are defined as the frequencies induced by the perturbations of the planets considered on coplanar circular orbits, namely as the frequencies given by the integrable Hamiltonian $\overline{K}_0$, which is the most relevant term of the full Hamiltonian (6). So we have simply

$$\varepsilon^2 \nu_h = \varepsilon^2 \frac{\partial \overline{K}_0}{\partial H}(2g) \,, \quad \varepsilon^2 \nu_\zeta = \varepsilon^2 \frac{\partial \overline{K}_0}{\partial Z}(2g) \,, \tag{7}$$

which are not constants, but functions of g; this is very bad to build a perturbative theory and to locate the position of the secular resonances. A first possibility, which is followed by many authors, like Yoshikawa and Nakai and Kinoshita is to average these frequencies over the argument of perihelion g. Even from a formal point of view, this is correct only if we can write

$$\nu_h(2g) = \overline{\nu}_h + \varepsilon f_h(2g) \,, \quad \text{where} \quad \overline{\nu}_h = \frac{1}{\pi}\int_0^\pi \nu_h(2g)\,dg$$

$$\nu_\zeta(2g) = \overline{\nu}_\zeta + \varepsilon f_\zeta(2g) \,, \quad \text{where} \quad \overline{\nu}_\zeta = \frac{1}{\pi}\int_0^\pi \nu_\zeta(2g)\,dg \tag{8}$$

with f_h and f_ζ of order of the unity, in order to reject the part dependent on g in the perturbation of higher order. It's easy to check numerically that (8) is true only in a small neighbourhood of the planar circular orbit.

As we wish to develop a theory which is valid all over the phase–space, we then propose a different general method. First of all we perform the change of variables

$$\begin{aligned} P &= H + Z = L - \Theta \,, & h &= h = -\overline{\omega} \\ Z &= Z = G - \Theta \,, & g &= \zeta - h \,, \end{aligned} \tag{9}$$

in order to have an action, Z, conjugated to g. Now $\overline{K}_0$ is a function of P, Z and g only, and so it appears as a very simple separable Hamiltonian. Following Henrard (1990), we then introduce the action–angle variables

$$\begin{aligned} P &= P \,, & h &= h' + \varrho(\psi) \\ Z &= \mathbf{P}(P, J, \psi) \,, & g &= \mathbf{Q}(P, J, \psi) \,, \end{aligned}$$

such that the Hamiltonian $\overline{K}_0$ in the new variables is a function of the actions P and J only. The system has now two constant characteristic frequencies, $\nu_{h'}$ and ν_ψ, defined by

$$\varepsilon^2 \nu_{h'} = \varepsilon^2 \frac{\partial \overline{K}_0}{\partial P} \,, \quad \varepsilon^2 \nu_\psi = \varepsilon^2 \frac{\partial \overline{K}_0}{\partial J} \,, \tag{10}$$

which can be used at the base for a general perturbative theory all over the phase–space, even in the regions where the argument of perihelion g is a librating variable. Let us call $-\varepsilon^2 \nu_{h'}$ as the *proper free frequency* of the longitude of perihelion and $-\varepsilon^2(\nu_{h'} + \nu_\psi) \stackrel{\text{def}}{=} -\varepsilon^2 \nu_{\zeta'}$ as the *proper free frequency* of the longitude of node. One can easily prove that, if g is circulating,

$$\nu_{h'} = \frac{1}{T}\int_\gamma \frac{\partial \overline{K}_0}{\partial P}(P, Z(g(t)), g(t))\,dt \,, \text{and} \quad \varepsilon^2 \nu_\psi = \frac{2\pi}{T} \,,$$

where $\gamma \equiv (i(g(t)), e(g(t)), g(t))$ is the cycle described by $\overline{K}_0$, whose period is denoted by T.

THE LOCATION OF SECULAR RESONANCES

We have used the frequencies computed as in (10) to locate the main secular resonances. This can be simply done by confronting the above computed proper frequencies with the characteristic frequencies of the perturbing system, namely g_5, g_6 and s_6. We call "secular

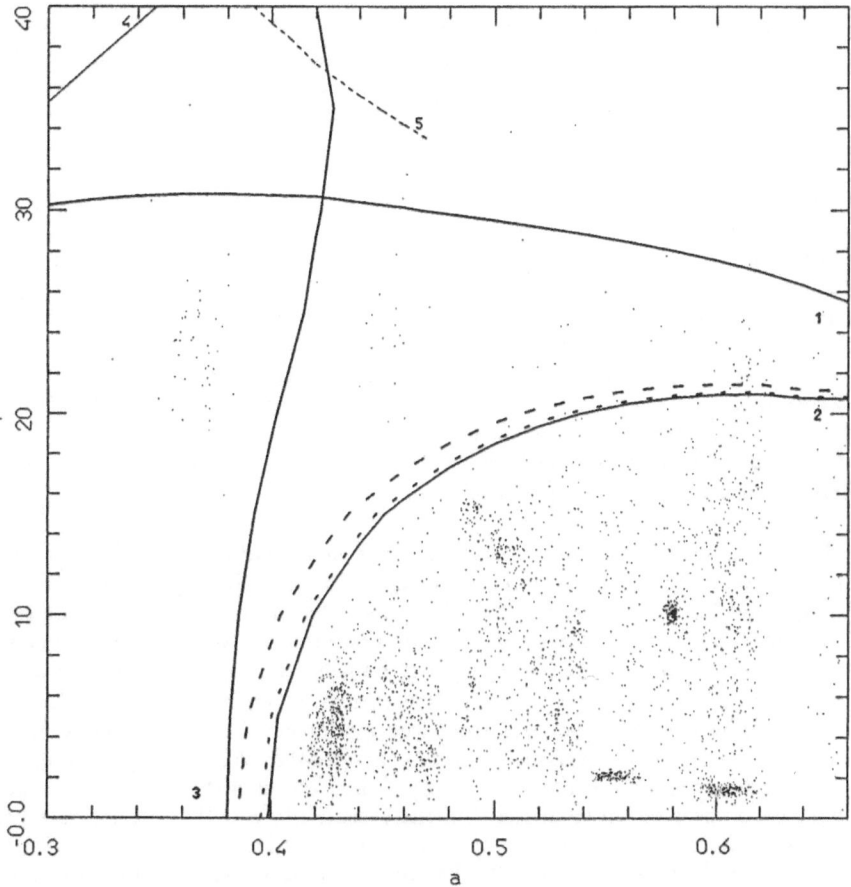

Fig. 2. The location of the main secular resonances for $e = .1$. The numbers refer to the following resonances : **1** ν_5; **2** ν_6 assuming for g_6 the values given by Nobili, Brouwers and Bretagnon (from the right to the left); **3** ν_{16}; **4** $\varepsilon^2(\nu_{h'} - 2\nu_\psi) = -g_6$. The dots represent the observed asteroids.

resonances of order 1" the commensurabilities among these frequencies which correspond to harmonics that are present in the term $\varepsilon^3 \overline{K}_1$. As this is the term of greatest size in the perturbation of $\overline{K}_0$, (see (6)), these are expected to be the strongest secular resonances. Analogously we call "secular resonances of order m" the commensurabilities which correspond to harmonics that are present in $\overline{K}_m$, and not at any previous order.

In Figure 2 we report the location of the secular resonances of order 1 in case of rotating argument of perihelion g. The results are expressed as a function of the semi–major axis and of the *proper inclination* for a given value of the *proper eccentricity* equal to 0.1. These last are defined in Williams (1969) as the values of i and e reached when $g = 0$, and so correspond to the maximal inclination and the minimal eccentricity assumed over the cycle γ. We have chosen to adopt this definition as, in this way, the computation of the action variables J, P is straightforward.

In figure 2 bold lines with numbers 1, 2, 3 refer respectively to the resonances ν_5, ν_6 and ν_{16}. These are the common names of the resonances that occur when the proper free frequency of the longitude of perihelion is equal to g_5 or g_6, or when the proper free frequency of the longitude of node is equal to s_6. The values assumed for the frequencies g_5, g_6 and s_6 are those given by the Longstop project and published by Nobili et al.(1988), which are in very good agreement with those obtained by Laskar (1988); the bold dashed line and the bold dotted line, however, designate the location of the ν_6 resonance if one considers for g_6 the value assumed by Bretagnon (1974) or by Brouwers and Van Woerkom (1950). For the values of g_5 and s_6 the agreement among the authors is much better, and so a similar comparison among the different positions of the relative secular resonances is not necessary.

Line (4) in the upper left corner locates the resonance $-\varepsilon^2(\nu_{h'} - 2\nu_\psi) = g_6$, which is the only secular resonance of order 1 involving ν_ψ to be present in the portion of the phase–space shown in Fig. 2. Finally line (5) reports the location of the ν_{16} resonance when the argument of perihelion g is in one stable equilibrium point at 90 or 270 degrees. In this case, however, the proper inclination is just the inclination of the equilibrium point, and the eccentricity is a function of the semi–major axis and of the inclination itself.

On the same graphic, for comparison, the position of the asteroids with $e < .2$ is shown; anyway such a comparison has to be considered only as qualitative, because the asteroids are represented through their osculating variables.

The results are in perfect agreement with those of William and Faulkner (1981), also for different values of the proper eccentricity.

The main limit of the results reported above is that, having neglected the terms which come from the Lie's algorithm of normalization of the Hamiltonian with respect to the mean motions, we have lost all the information about the influence of nearby mean–motion resonances on the location of the secular ones. So, in order to have a more realistic portrait, we have extimated the correction of the proper free frequencies given by the quadratic term in the masses $\frac{\varepsilon^4}{2}\overline{\{\chi_2, K_0\}}$, which has the same structure of $\overline{K}_0$ itself (namely it is function of g only). This computation has been performed through the classical series expansion of K_0 in terms of the eccentricity and the inclination of the asteroid, thus leading to an easy calculation of the generating function χ_2, and of the secular terms of their Poisson bracket. The result has been obtained up to degree 4 in the eccentricity and the inclination separately, thus improving the existing results by Yuasa (1973) and Milani and Knežević (1990) where only the terms in e^2 and i^2 are retained. We are still far, however, to obtain a computation correct for any eccentricity and inclination, which could be achieved only by regarding the compact form of the original Hamiltonian with respect to e and i. Indeed this involves so many technical difficulties that seems to us impractical up to now.

Our results are reported in figure 3a and 3b. The difference between the two figures is that in Fig. 3a we do not consider the contribution of the term $e^2 i^4$ of $\frac{\varepsilon^4}{2}\overline{\{\chi_2, K_0\}}$. This is done to give an idea of the localizations of the regions where the computation is very sensitive to the limit order of truncation of the series. As one sees, this happens in the upper right corner of the graphic, where the inclination is very high and we are close to strong mean motion resonances. In that case $i \sim 30°$ (which is equivalent to $1/2$ in radians), and consequently the convergence of the series in terms of i^2 is very slow. On the other hand the coefficients of this powers are very big, thanks to the nearby presence of mean motion resonances, and, at least for the first few orders, their size becomes bigger with growing powers of i^2. In a case like this, we are not allowed to conclude that Fig. 3b is a better approximation of Fig. 3a. A similar phenomenon was largely discussed by Lemaître and Henrard (1990) studying the 2/1 mean motion resonance.

The three dashed regions in Fig. 3a and 3b refer to the blank zones for the theory associated to the mean motion resonances 3/1, 5/2 and 2/1 (from left to right). Indeed the Lie's algorithm of normalization with respect to both the mean longitudes of the asteroid and of the perturbing planet is singular in correspondence of mean motion commensurabilities, and so the vicinity of these should be studied with different techniques. The width of the dashed regions depends on the geometrical size of the grid adopted for the computations and so does not have a direct dynamical meaning. Other mean motion resonances, like the 4/1 or 7/3, for example, have a too narrow region of influence to be detected by our grid, and so we have ignored them drawing continuous lines.

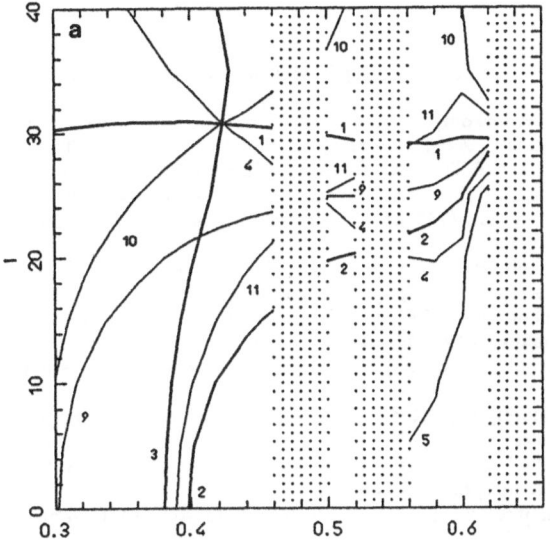

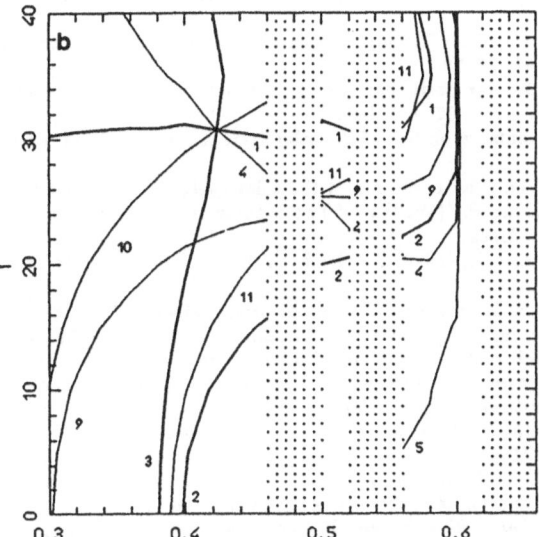

Fig. 3. The location of the main secular resonances considering also the quadratic term in the masses, truncated in the inclination at order 2 (Fig. a top) and at order 4 (Fig. b bottom). The enumeration of the secular resonaces is the following : **1** $\varepsilon^2 \nu_{h'} = -g_5$; **2** $\varepsilon^2 \nu_{h'} = -g_6$; **3** $\varepsilon^2 \nu_{\zeta'} = s_6$; **4** $\varepsilon^2(\nu_{h'} + \nu_{\zeta'}) = -(g_5 + s_6)$; **5** $\varepsilon^2(\nu_{h'} + \nu_{\zeta'}) = -(g_6 + s_6)$; **9** $2\varepsilon^2 \nu_{h'} = -(g_5 + g_6)$; **10** $\varepsilon^2(\nu_{h'} - \nu_{\zeta'}) = -(g_5 - s_6)$; **11** $\varepsilon^2(\nu_{h'} - \nu_{\zeta'}) = -(g_6 - s_6)$. See text for comment.

In Fig. 3a and 3b the secular resonances of order 1, ν_5, ν_6 and ν_{16} (bold lines) and the main secular resonances of order 2 are shown. The enumeration is that given by Milani and Knežević (1990).

A particular remark must be made about the resonance (5), namely $\varepsilon^2(\nu_{h'} + \nu_{\zeta'}) = -(g_6 + s_6)$, which seems to be related to the existence of the Eos family (see Milani and Knežević, 1990) : if we considered only the term $\overline{K}_0$ to compute the frequencies $\nu_{h'}$ and $\nu_{\zeta'}$ such a resonance would not exist. Indeed in that case $\nu_{h'} + \nu_{\zeta'}$ would be always positive, while $-(g_6 + s_6)$ is negative.

References

Bretagnon, P., (1974) : "terms à longues périodes dans le système solaire.", *Astron. Astrophys.*, **30**, 141–154.

Brouwers, D. and Van Woerkom, A., J., J., (1950) : "The secular variations of the orbital elements of the principal planets.", *Astr. Papers. U.S. Naval Obs.*, **13**, 85–107.

Carpino, M., Milani, A., Nobili, A.M., (1987) : "Long–term numerical integrations and synthetic theories for the motion of the outher planets.", *Astr. Astrophys.*, **181**, 182–194.

Froeschlé, Ch., and Scholl, H., (1986a) : "The secular resonance ν_6 in the asteroidal belt.", *Astron. Astrophys.*, **166**, 326–332.

Froeschlé, Ch., and Scholl, H., (1986b) : "The effects of the secular resonances ν_{16} and ν_5 on the asteroidal orbits.", *Astron. Astrophys.*, **170**, 138–144.

Froeschlé, Ch., and Scholl, H., (1987) : "Orbital evolution of asteroids near the secular resonance ν_6.", *Astron. Astrophys.*, **179**, 294–303.

Froeschlé, Ch., and Scholl, H., (1988) : "Secular resonances : new results.", *Celest. Mech.*, **43**, 113–117.

Froeschlé, Ch., and Scholl, H., (1989) : "The three principal resonances ν_5, ν_6 and ν_{16} in the asteroidal belt.", *Celest. Mech.*, **46**, 231–251.

Froeschlé, Ch., and Scholl, H., (1990) : "Orbital evolution of known asteroids in the ν_5 resonance region.", *Astron. Astrophys.*, **227**, 255–263.

Henrard, J., and Lemaître, A., (1983) : "A second fundamental model for resonance.", *Celest. Mech.*, **30**, 197–218.

Henrard, J., (1989) : "The adiabatic invariant in celestial mechanics.", submitted to *Dynamics reported*.

Henrard, J., (1990) : " A semi–numerical perturbation method for separable hamiltonian systems.", preprint.

Heppenheimer, T., A., (1980) : "Secular resonances and the origin of eccentricities of Mars and the asteroids.", *Icarus*, **41**, 76–88.

Kozai, Y., (1962) : "Secular perturbations of Asteroids with high inclination and eccentricities.", *The Astronomical Journal*, **67**, 591–598.

Laskar, J., (1988) : "Secular evolution of the solar system over 10 million years.", *Astron. Astrophys.*, **198**, 341–362.

Laskar, J., (1990) : "The chaotic motion of the solar system.", submitted to *Icarus*.

Lemaître, A., and Dubru, P., (1990) : "The secular resonances in the primitive solar nebula." submitted to *Celest. Mech.*

Lemaître, A., and Henrard, J., (1990) : "Origin of the cahotic behaviour in the 2/1 Kirkwood gap.", *Icarus*, **83**, 391–409

Milani, A., and Knežević, Z., (1990) : "Asteroid proper elements.", submitted to *Celest. Mech.*

Morbidelli, A., and Henrard, J., (1990a) : "Secular resonances in the asteroid belt : theoretical perturbation approach and the problem of their location." submitted to *Celest. Mech.*

Morbidelli, A., and Henrard, J., (1990b) : "The main secular resonances ν_6, ν_5 and ν_{16} in the asteroid belt." submitted to *Celest. Mech.*

Nakai, H., and Kinoshita, H., (1985) : "Secular perturbations of asteroids in secular resonances.", *Celest. Mech.*, **36**, 391–407.

Nobili, A.M., Milani, A., and Carpino, M.. (1989) : "Fundamental frequencies and small divisors in the orbits of the outer planets.", *Astron. Astrophys.*, **210**, 313–336.

Poincaré, H., (1892) : *Les méthodes nouvelles de la mécanique céleste*, Gauthier–Villars, Paris.

Tisserand, M., F., (1882) : *Ann. Obs. Paris*, **16**, E1.

Williams, J., D. (1969) : "Secular perturbations in the solar system.", *Ph.D. dissertation*, University of California, Los Angeles.

Williams, J., G., and Faulkner, J., (1981) : "The position of secular resonance surfaces.", *Icarus*, **46**, 390–399.

Yoshikawa, M., (1987) : " A simple analytical model for the ν_6 resonance.", *Celest. Mech.*, **40**, 233–272.

Yuasa, M., (1973) : "Theory of secular perturbations of asteroids including terms of higher order and higher degree.", *Pubbl. Astr. Soc. Japan*, **25**, 399–445.

TEMPORARY CAPTURE INTO RESONANCE

Département de Mathématique FUNDP
8, Rempart de la Vierge, 5000 Namur, Belgique

Abstract. It seems likely that some pairs of satellites of Uranus have been temporarily captured into resonance in the past. In order to analyze these temporary captures, one must modify the model constructed for the capture into resonance of the satellites of Jupiter and Saturn. The key factor is the value of the oblateness of Uranus which is smaller than the corresponding value for Jupiter or Saturn. The smaller value allows some overlap of nearby resonances producing chaos and secondary resonances. The secondary resonances are instrumental in dragging the captured orbit back to the chaotic layer surrounding the primary resonance from which it can escape in the regular region outside the resonance.

1. Mechanism of Capture into Resonance

Since Goldreich (1965) demonstrated that pairs of satellites could be captured into resonance by tidal effect, the mechanism of such a capture has been investigated and well understood at least for the satellites of Jupiter and Saturn that are now in resonance (see for instance Peale 1986). Basically this mechanism is as follows. Let us take for instance a 3:1 commensurability between the mean motions of two satellites. In the vicinity of this commensurability, the restricted planar averaged three body problem can be approximated by the following Hamiltonian (see for instance Henrard and Sato 1990):

$$H = C(N - S)^2 + AS + 2DS \cos 2\sigma + 2Fe'^2 \cos 2\nu$$
$$+ e'\sqrt{2S}[E \cos(\sigma - \nu) + G \cos(\sigma + \nu)], \qquad (1)$$

where (S, N) are the momenta conjugated to the angular variables (σ, ν) and where the coefficient C is of the order of unity, the coefficient A of the order of the oblateness of the primary and the other coefficients, D, F, E, G of the order of the mass of the secondary. At the zeroth approximation (when $D = G = E = F = 0$) the frequencies of the periodic terms in (1) are:

$$\left[\frac{d}{dt}(-2\nu)\right]_0 = -2C(N - S) \quad,$$

$$\left[\frac{d}{dt}(\sigma - \nu)\right]_0 = -2C(N - S) + A \quad,$$

$$\left[\frac{d}{dt}(2\sigma)\right]_0 = -2C(N - S) + 2A \quad,$$

$$\left[\frac{d}{dt}(\sigma + \nu)\right]_0 = A \quad. \qquad (2)$$

The first three variables can be "resonant", i.e. for some value of $(N-S)$ their frequency can vanish. But, if the oblateness of the planet (i.e. the coefficient A) is large enough, these resonances are well separated (i.e. they take place for well separated values of $(N-S)$). In the vicinity of such a resonance (let us take for instance the resonance corresponding to 2σ), one can consider that the frequencies of the other angular variables are fast enough that they can be averaged out. The motion can then be approximated by the integrable system corresponding to the Hamiltonian :

$$H_\sigma = C(N-S)^2 + AS + 2DS\cos 2\sigma \quad . \tag{3}$$

The principal effect of the tides raised by the satellite on the planet is to make the "constant" N grows slowly with the time. The evolution of the system (3) can then be evaluated by means of the adiabatic invariant theory (see Henrard 1982 and Lemaître 1984). The conclusions are as follows. If the pair of satellites go through the resonance characterized by $(N-S)$ close to zero, it can be captured by it; $(N-S)$ remains close to zero while N and S grow. The ratio of the semi-major axis remains constant and the eccentricity grows. It is also possible (and one can evaluate a probability for such a event) that the pair of satellite "jumps" over the resonance. In this case the ratio of the semi-major axis continues to grow and the eccentricity, after having experienced a "jump" at the time of passage through the resonance, stays constant.

In any case, the pair of satellites follows one or the other path but cannot be captured for a significant time interval and then escape again.

Such an evolution repeats itself for each of the three possible resonances $(2\nu, \sigma - \nu, 2\sigma)$, as shown schematically in Figure 1.

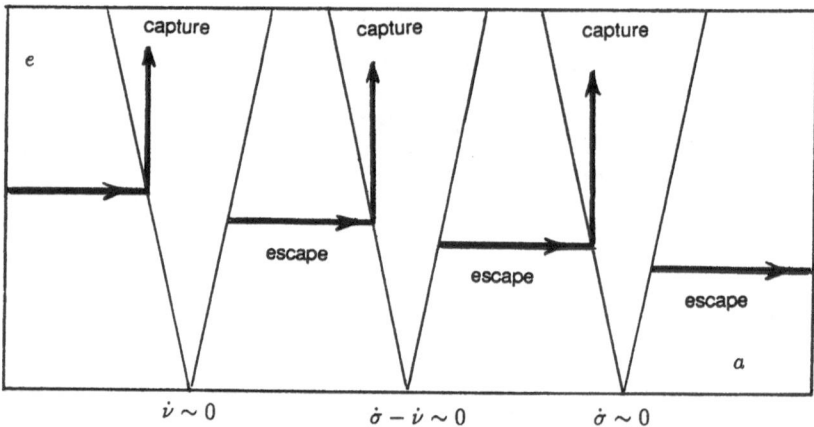

Fig. 1. Schematic evolution due to tidal effect close to a 3:1 commensurability when the oblateness of the primary is large enough.

2. The Case of the Satellites of Uranus

Recently, Tittermore and Wisdom (1988,1989), Dermott *et al.* (1988), Malhotra and Dermott (1990) and Malhotra (1990), have investigated the passage through resonance for some pairs of satellites of Uranus. These satellites are not at present captured into resonance, but a temporary capture in the past would help explain their thermal history which seems difficult to explain otherwise. This could also help in explaining the relatively high value of the inclination of Miranda.

In the case of Uranus, as pointed out by Dermott (1984), the smaller value of the oblateness (and thus of the coefficient A in (1)) allows some overlapping of the three main resonances. This interaction between the primary resonances is responsible for a broadening of the separatrix of (3) into a macroscopic chaotic layer and for the apparition of macroscopic secondary resonances inside the primary resonances.

The above mentioned authors have established numerically that in these circumstances the most probable evolution, in the Miranda-Umbriel 3:1 commensurability, is as follows.

The pair of satellites is captured into one of the primary resonances when the eccentricity is still small. As for small values of the eccentricity the primary resonances are well separated enough, the previous model can be used to evaluate the probability of capture and the subsequent evolution.

Once captured, the eccentricity of Miranda starts to grow which makes it encounter the secondary resonances (see Figure 2) where it can also be captured.

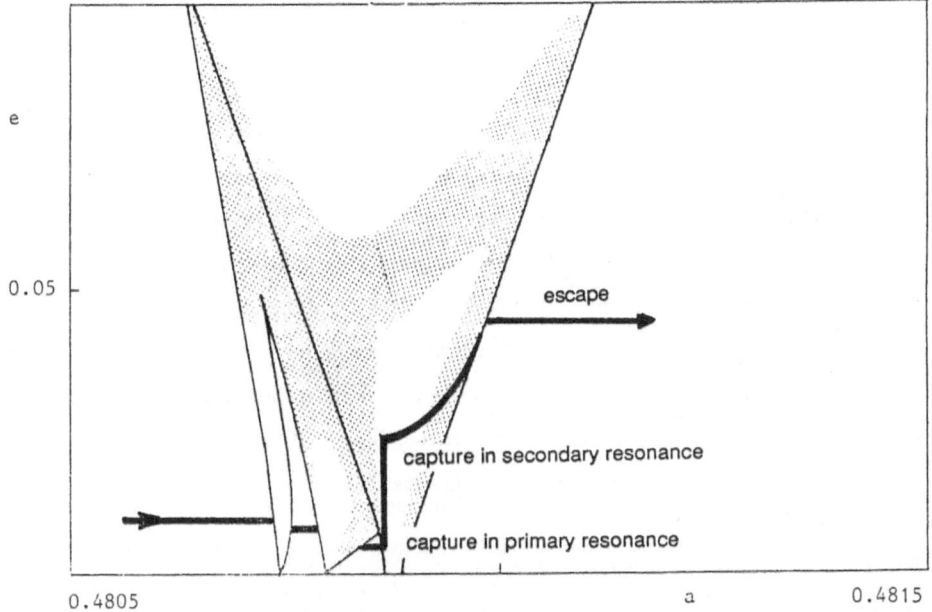

Fig. 2. Schematic view of the evolution of the pair Miranda-Umbriel through the commensurability 3:1. The shaded areas correspond to layers where the motion is chaotic

If it is captured in one of them (for instance the 2:1 secondary resonance in Figure 2), it is dragged by it back to the chaotic layer surrounding the primary resonance. When it is there, it seems very likely that it escape quickly both the secondary and the primary resonance.

3. Analytical Estimates

In order to understand better this scenario and to evaluate its sensitivity with respect to the parameters of the problem, we need to be able to:

(1) estimate the probability of capture in a secondary resonance,

(2) extend the theory of the adiabatic invariant not only to a crossing of a separatrix (Henrard 1982) but to a passage through the macroscopic chaotic layer which borders the primary resonance.

The first of these questions has already received an answer, at least qualitatively, in a not yet published paper by Malhotra (1990). This estimate is based upon the extension of the adiabatic invariant to separatrix crossing (Henrard 1982) applied to a simplified model of secondary resonances. The principal result is that the probability of capture is small for high order resonances but becomes significant for the secondary resonances 3:1 and 2:1.

The second question necessitates a new approach. According to Tittermore and Wisdom (1988), the presence of even a very small chaotic layer changes completely the mechanism of capture, so much so that the predictions based upon the extension of the adiabatic invariant mentioned above, have no relevance at all. We believe that it is not so and that a smooth extension of the adiabatic invariant to "chaotic layer crossing" is possible. Preliminary results along these lines show that indeed the probability of capture computed from the extension of the adiabatic invariant theory to separatrix crossing can be corrected by terms taking into account the size of the chaotic layer. A new feature is that the transition is no longer instantaneous but that the mean time spent in the chaotic layer depends critically upon the rate of increase of the area covered by the layer. We hope to be able to report progresses along these lines in a future contribution.

References

Dermott, S.F.: 1984, "Origin and evolution of the Uranian and Neptunian satellites: some dynamical considerations", in *Uranus and Neptune* (J Bergstrahl Ed.), Nasa Conf. Pub. 2330, pp 377-404.

Dermott, S.F., Malhotra, R. and Murray C.D.: 1988, "Dynamics of the Uranian and Saturnian satellite systems : A chaotic route to melting Miranda ?" *Icarus*, **76**, 295-334.

Goldreich, P.: 1965, "An explanation of the frequent occurence of commensurable mean motions in the Solar system", *M.N.R.A.S.*, **130**, 159-181.

Henrard, J.: 1982, "Capture into resonance : An extension of the use of the adiabatic invariants", *Celest. Mech.*, **27**, 3-22.

Henrard, J. and Lemaître, A.: 1983, "A second fundamental model for resonance", *Celest. Mech.*, **30**, 197-218.

Henrard, J., and Sato, M.: 1990, "The origin of chaotic behaviour in the Miranda-Umbriel 3/1 resonance". *Celest. Mech.*, submitted.

Lemaître, A.: 1984, "High order resonance in the restricted three body problem", *Celest. Mech.*, **32**, 109-126.

Malhotra, R.: 1990, "Capture probabilities for secondary resonances". *Icarus*, in print.

Malhotra, R. and Dermott, S.F.: 1989, "The role of secondary resonances in the orbital history of Miranda", *Icarus*, **85**, 444-480.

Peale, S.J.: 1986, "Orbital resonance, unusual configurations and exotic rotation states", in *Satellites* (J. Burns and M. Matthews eds.), Univ. of Arizona Press, 159-223.

Tittermore, W.C. and Wisdom, J.: 1988, "Tidal evolution of the Uranian satellites. I. Passage of Ariel and Umbriel through the 5:3 mean-motion commensurability", *Icarus*, **74**, 172-230.

Tittermore, W.C. and Wisdom, J.: 1989, "Tidal evolution of the Uranian satellites. II. An explanation of the anomalously high orbital inclination of Miranda", *Icarus*, **78**, 63-89.

APPLICATIONS OF THE RESTRICTED MANY-BODY PROBLEM TO BINARY ASTEROIDS

Victor Szebehely

Professor R.B. Curran Centennial Chair
University of Texas at Austin

Joseph R. Pojman

Graduate Research Associate
University of Texas at Austin

ABSTRACT

    This paper presents a general approach to the problem of stability of
binary configurations in solar system dynamics. As a special application
of the analytical approach the possible existence of binary asteroids is
investigated by numerical approaches. The analytical results are based on
the hierarchical model of the restricted four-body problem, in which the
primaries are the Sun and Jupiter and the "small masses" are the asteroid
and its satellite. The numerical results offer a parametric study considering the effects of the initial orbital parameters of the binary on
its stability.

1. ANALYTICAL FORMULATION OF THE PROBLEM

    The dynamical system known as the restricted problem of three bodies
consists of two major bodies, the primaries, which govern the motion of
a third body of much smaller mass. This third body does not influence
the motion of the primaries and this is where the restricted problem of
three bodies differs from the general problem. We arrive at the many-
body restricted problem by straight-forward generalization and by introducing the idea of hierarchy of the participation masses. Since this
concept is discussed in several papers available in the literature, (see
for instance Pojman and Szebehely, 1988), at this point only the problem
known as the 2 + 2 restricted problem is treated which is the model
directly applicable to the binary asteroid problem. (For the observational aspects see for instance Van Flandern, 1980).

    Consider two bodies with masses of m_1 and m_2 which influence gravi-
tationally two much smaller bodies of masses, m_3 and m_4:

$$m_1 \geq m_2 \gg m_3 \geq m_4.$$

    In the special case of interest in this paper m_1 is the mass of the
Sun, m_2 is the mass of Jupiter, m_3 is the mass of the asteroid and m_4 is
the mass of the satellite of the asteroid. The motion of the primaries

Predictability, Stability, and Chaos in N-Body Dynamical Systems
Edited by A.E. Roy, Plenum Press, New York, 1991

m_1 and m_2 is assumed to be known, and the problem is to present the behaviour of m_3 and m_4. If no gravitational forces act between m_3 and m_4 we are faced with two restricted problems of three bodies. In the model accepted in this paper, we assume gravitational attraction between m_3 and m_4. Consequently, the motions of m_3 and m_4 are influenced by each other and by m_1 and m_2. The motions of m_1 and m_2 are only influenced by each other but not by m_3 and m_4. The classical restricted problem assumes circular orbits of m_1 and m_2 but any other solution of the corresponding two-body problem is acceptable as model, such as the elliptic restricted problem. This is mentioned here since some recent numerical results offered in the literature considering the problem of satellites of asteroids (Zhang and Innanen, 1988) use elliptic orbits of the primaries.

The equations of motions of m_3 and m_4 are presented first. In these equations we use an inertial system with the origin located at the center of mass of m_1 and m_2. Note that at this point we do not use the conventional rotational (synodic) system (Szebehely and Whipple, 1984). We include the gravitational forces acting between m_1 and the minor bodies (m_3 and m_4), the forces between m_2 and the minor bodies (m_3 and m_4), and the force between m_3 and m_4.

The equations of motion of m_3 and m_4 are

$$m_3 \ddot{\bar{r}}_3 = G \frac{m_3 m_4}{|\bar{r}_{34}|^3} \bar{r}_{34} - G \frac{m_1 m_3}{|\bar{R}_{13}|^3} \bar{R}_{13} - G \frac{m_2 m_3}{|\bar{R}_{23}|^3} \bar{R}_{23} \qquad (1)$$

and

$$m_4 \ddot{\bar{r}}_4 = -G \frac{m_3 m_4}{|\bar{r}_{34}|^3} \bar{r}_{34} - G \frac{m_1 m_4}{|\bar{R}_{14}|^3} \bar{R}_{14} - G \frac{m_2 m_4}{|\bar{R}_{24}|^3} \bar{R}_{24} . \qquad (2)$$

Here $\bar{r}_3$ and $\bar{r}_4$ are the position vectors of the minor bodies with masses m_3 and m_4. The vector $\bar{r}_{34} = \bar{r}_4 - \bar{r}_3$ is the relative position vector of m_4 with respect to m_3. The vectors $\bar{R}_{13}$ and $\bar{R}_{14}$ represent the relative position vectors of m_3 and m_4 with respect to m_1 or $\bar{R}_{13} = \bar{r}_3 - \bar{R}_1$ and $\bar{R}_{14} = \bar{r}_4 - \bar{R}_1$ where $\bar{R}_1$ is the position vector of m_1. Similarly $\bar{R}_{23} = \bar{r}_3 - \bar{R}_2$ and $\bar{R}_{24} = \bar{r}_4 - \bar{R}_2$ are the relative position vectors of m_3 and m_4 with respect to m_2 and $\bar{R}_2$ is the position vector of m_2 (see Figure 1).

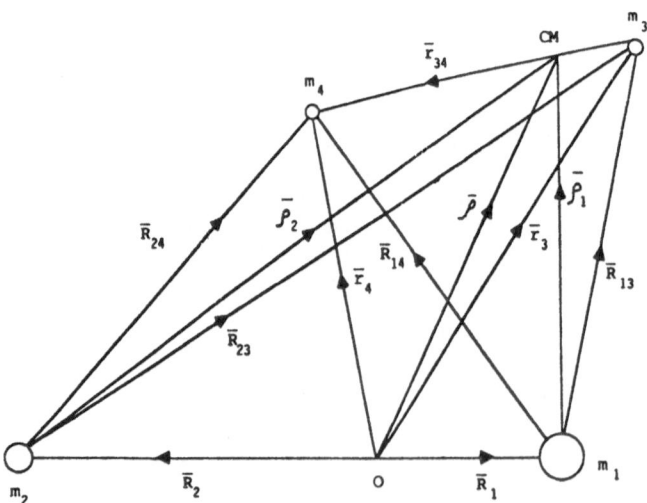

Figure 1. Position and relative position vectors of the participating bodies.

The relative motion of m_4 with respect to m_3 is described by $\bar{r}_{34}$. This equation is obtained from Equation (1) and (2) in the usual way, i.e. subtracting Equation (1) from Equation (2):

$$\ddot{\bar{r}}_{34} = -G \frac{m_3 + m_4}{|\bar{r}_{34}|^3} \bar{r}_{34} + \bar{F}_1 + \bar{F}_2 , \tag{3}$$

where

$$\bar{F}_1 = Gm_1 \left(\frac{\bar{R}_{13}}{|\bar{R}_{13}|^3} - \frac{\bar{R}_{14}}{|\bar{R}_{14}|^3} \right) \tag{4}$$

and

$$\bar{F}_2 = Gm_2 \left(\frac{\bar{R}_{23}}{|\bar{R}_{23}|^3} - \frac{\bar{R}_{24}}{|\bar{R}_{24}|^3} \right) . \tag{5}$$

In Equation (3) the first term on the right hand side represents the usual two-body effect between m_3 and m_4, while $\bar{F}_1$ and $\bar{F}_2$ represent the forces acting on the binary by the primaries, m_1 and m_2.

The corresponding equations in the synodic system are also available in the literature but Equation (3) serves our purpose better at this point.

Using again the well known approach, we compute the scalar product of both sides of Equation (3) by $\dot{\bar{r}}_{34}$ and obtain

$$\frac{d}{dt} \left[\frac{1}{2} \dot{\bar{r}}_{34}^2 - G \frac{m_3 + m_4}{|\bar{r}_{34}|} \right] = \dot{\bar{r}}_{34} \cdot (\bar{F}_1 + \bar{F}_2) . \tag{6}$$

Integrating both sides, we have

$$h = \int (\bar{F}_1 + \bar{F}_2) \cdot d\bar{r}_{34} , \tag{7}$$

where h is the total energy of the two-body system formed by m_3 and m_4.

At this point various approximations can be made to evaluate the variation of the total energy. The essential idea is that the initial value of $h(0) = h_o$ is negative since the initial values of $\bar{r}_{34}$ and $\dot{\bar{r}}_{34}$ are selected so that this condition is satisfied. If $h(t)$ stays negative the binary is stable, i.e. the masses m_1 and m_2 perform perturbed elliptic orbits. If $h(t) \geq 0$ the motion is considered unstable and the binary breaks up. The numerical integration concentrates on this criterion of stability and it evaluates the value of $h(t)$. If the initial value of the two-body energy is large, i.e. if h_o has a large negative value, then even if $h > 0$, it might take a long time before instability sets in. If $h \leq 0$, the binary might be considered stable. The initial value of h is therefore an important parameter which becomes

$$h_o = - G \frac{m_3 + m_4}{|\bar{r}_{34}|} + \frac{\dot{\bar{r}}_{34}^2}{2} .$$

Therefore, for close binary systems, stability can be expected for a longer time than for binaries with large initial separation.

The right side of Equations (6) and (7) can be approximated by considering that the distances between the primaries and the binary are much larger than the separation between the masses m_3 and m_4. This approximation might be expressed as follows.

Since $\bar{R}_{14} = \bar{R}_{13} + \bar{r}_{34}$, $\bar{F}_1 = Gm_1 \left(\dfrac{\bar{R}_{14} - \bar{r}_{34}}{|\bar{R}_{14} - \bar{r}_{34}|^3} - \dfrac{\bar{R}_{14}}{|\bar{R}_{14}|^3} \right)$

and
$$\bar{F}_1 \cong \frac{-Gm_1 \bar{r}_{34}}{|\bar{R}_{14}|^3} \cong - Gm_1 \frac{\bar{r}_{34}}{|\bar{\rho}_1|^3} . \tag{8}$$

Similarly, $\quad \bar{F}_2 \cong - Gm_2 \dfrac{\bar{r}_{34}}{|\bar{\rho}_2|^3} . \tag{9}$

Here $\bar{\rho}_1$ and $\bar{\rho}_2$ are the relative position vectors of the center of mass of the binary with respect to m_1 and m_2 or

$$\bar{\rho}_1 = \frac{m_3 \bar{r}_3 + m_4 \bar{r}_4}{m_3 + m_4} - \bar{R}_1 \tag{10}$$

and

$$\bar{\rho}_2 = \frac{m_3 \bar{r}_3 + m_4 \bar{r}_4}{m_3 + m_4} - \bar{R}_2 . \tag{11}$$

The approximate form of Equation (6) becomes

$$\frac{dh}{dt} \cong - \frac{G}{2} \left(\frac{m_1}{|\bar{\rho}_1|^3} + \frac{m_2}{|\bar{\rho}_2|^3} \right) \frac{d}{dt} \bar{r}_{34}^2 . \tag{12}$$

If the term $m_2/|\bar{\rho}_2|^3$ can be neglected in comparison to the term $m_1/|\bar{\rho}_1|^3$ and if in addition $|\bar{\rho}_1| = $ constant, we have

$$\frac{dh}{dt} \cong - \frac{G}{2} \frac{m_1}{|\bar{\rho}_1|^3} \frac{d}{dt} \bar{r}_{34}^2. \tag{13}$$

The physical interpretation of the above approximation can be made clearer if the Sun-Jupiter system is considered as the primaries, $m_1 = m_\odot$ and $m_2 = m_4$. If the center of mass of the binary moves on a circle around the Sun and Jupiter's effect is neglected, Equation (13) is obtained.

2. RESULTS OF NUMERICAL INTEGRATION

The dynamical system considered in this paper consists of the Sun, Jupiter and a binary asteroid. The parameters of the examples might be separated in two groups. Those which do not change for the various examples are the semi-major axis of the center of mass of the binary with respect to the Sun, $a_{cm} = 2 \cdot 8$ AU, the eccentricity of the orbit of the center of mass ($e_{cm} = 0.1$), the argument of perihelion $\omega_{cm} = 90^\circ$ and the ascending node, $\Omega_{cm} = 90^\circ$. The time of perihelion passage of the center of mass of the binary is given by $t=0$, corresponding to zero value of the true anomaly.

The values of the masses of the participating bodies are
$m_1 = 1.9891 \times 10^{33}$g, $m_2 = 1.8996 \times 10^{30}$ g, $m_3 = (m_1 + m_2)10^{-13}$ and $m_4 = (m_1 + m_2)10^{-17}$.

The initial inclination of the orbit of the centre of mass (i_{cm}) was varied between 0 and 180°.

The varied initial conditions for the members of the binary were $0 \leq i_{34} \leq 180^{\circ}$, $a_{34} = (4,6,8$ and $10) \times 10^3$ km or 4000km $\leq a_{34} \leq$ 10000km; while, $e_{34} = 0.01$, $\Omega_{34} = 90^{\circ}$, $\omega_{34} = 90^{\circ}$ and $f_{34} = 0$ were kept constant. Here f_{34} represents the true anomaly of the binary.

The numerical integration terminated at 1000 years for stable orbits and at a shorter time if instability set in before 1000 year. As mentioned before, instability of the binary was defined by $h \geq 0$ and a binary was considered stable if h was negative for 1000 years.

The regions of stability are represented in Figure 2. The initial value of the inclination of the binary's orbit (i_{34}) is plotted against the initial inclination of the orbit of the centre of mass (i_{cm}).

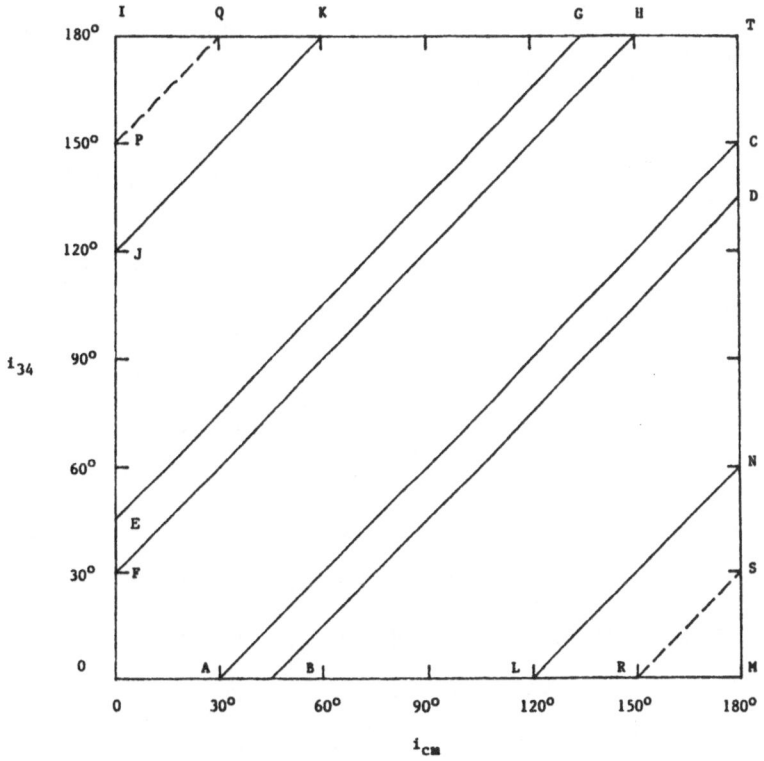

Figure 2. Regions of Stability.

If the initial value of the semi-major axis of the binary is 4000km, the region of stability is represented by the area of the square OMTI, i.e. all orbits appeared to be stable.

If the initial value of the semi-major axis of the binary is a_{34}=6000km, the regions of stability are ABDC, EFHG, JKI AND LMN.

If the initial value of the semi-major axis of the binary is $8000km \leq a_{34} \leq 10,000km$, the regions of stability are IPQ and RMS.

The boundaries can be represented by the relations :

PQ: $i_{34} = 150 + i_{cm}$, where $0 \leq i_{cm} \leq 30$

JK: $i_{34} = 120 + i_{cm}$, $0 \leq i_{cm} \leq 60$

EG: $i_{34} = 45 + i_{cm}$, $0 \leq i_{cm} \leq 135$

FH: $i_{34} = 30 + i_{cm}$, $0 \leq i_{cm} \leq 150$

AC: $i_{34} = -30 + i_{cm}$, $30 \leq i_{cm} \leq 180$

BD: $i_{34} = -45 + i_{cm}$, $45 \leq i_{cm} \leq 180$

LN: $i_{34} = -120 + i_{cm}$, $120 \leq i_{cm} \leq 180$

RS: $i_{34} = -150 + i_{cm}$, $150 \leq i_{cm} \leq 180$

As shown on Figure 3 the regions of stability were obtained using 15° increments for i_{cm} and i_{34}. If the scale of the increments is refined to 1° some fuzziness appears along the lines of stability, possibly due to numerical instability, dynamical uncertainties, chaotic behaviour, etc. Figure 4 shows the irregularity at the boundary in the region $108^{\circ} \leq i_{34} \leq 117^{\circ}$ and $0 \leq i_{cm} \leq 9^{\circ}$ for $a_{34} = 6000$ km when $\Delta i = 1^{\circ}$ is used. The detailed results of the investigations concerning the behaviour at the boundaries of the stability regions will be reported elsewhere.

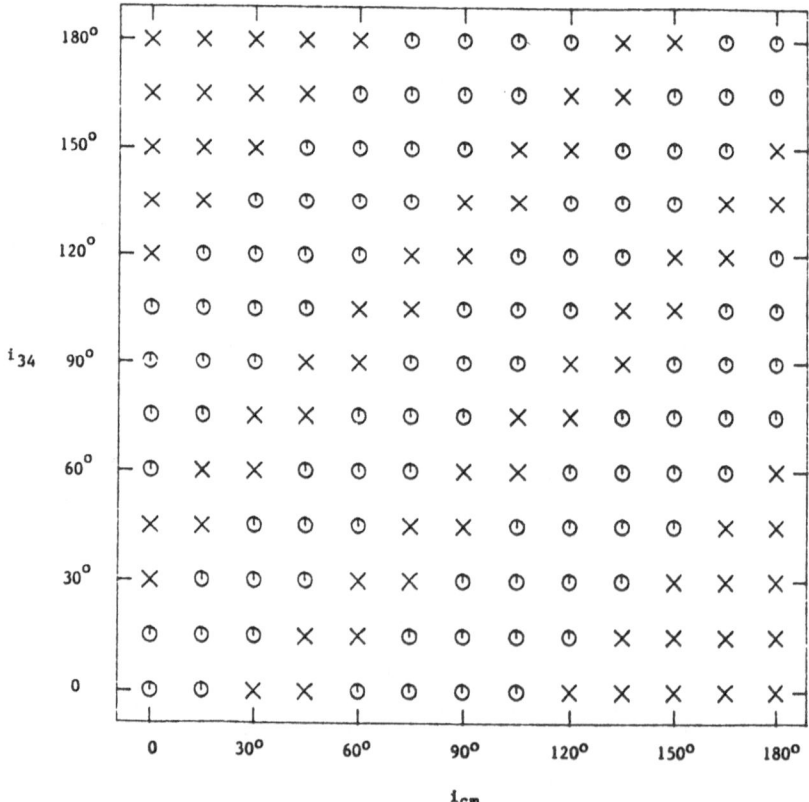

Figure 3. Regions of stability for $a_{34} = 6000$ km showing the initial conditions used; X = stable, ⊙ = unstable binaries.

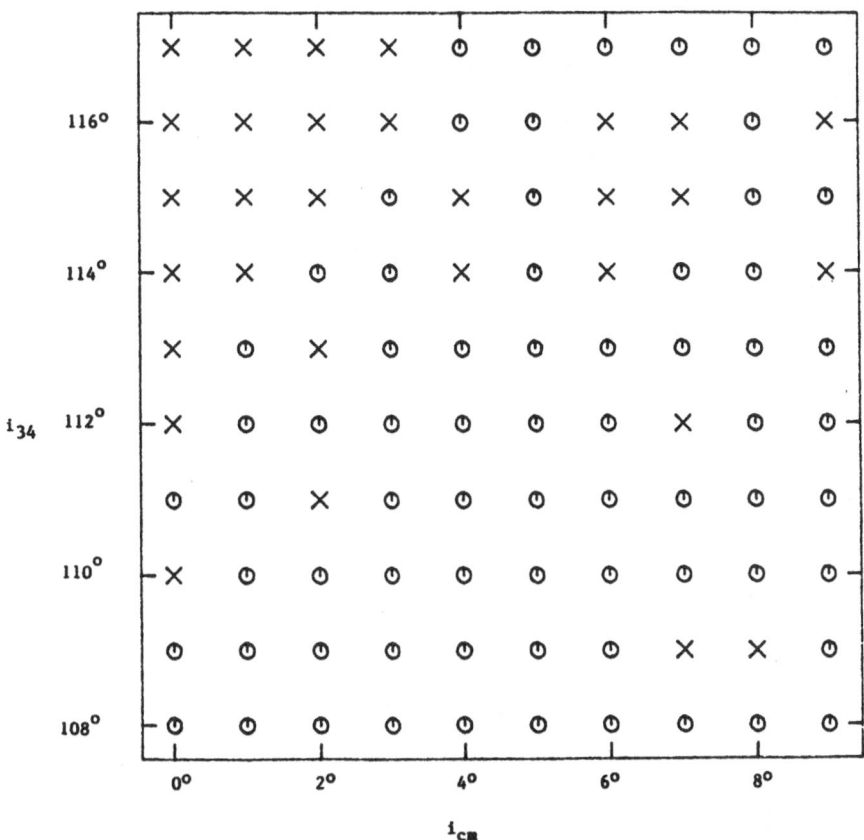

Figure 4.   Irregularity at the stability boundary for a_{34} = 6000 km.

3. CONCLUSIONS

The model of the generalized restricted many body problem is applicable
to the study of the dynamical behaviour of binary asteroids in the region
of the asteroid belt. The numerical results indicated stability (i.e.
negative value of the energy of the binary) for 1000 years when the participating asteroids started with an elliptic two-body motion for which the
semi-major axis was 4000 km.   As the initial value of the semi-major
axis increased, the regions of stability decreased. One of the unexpected
results of this investigation were that high angles of inclinations do not
necessarily introduce instability. For instance for the range of
8000 km $\leq a_{34} \leq$ 10000 km, low inclinations of the orbit of the centre of
mass required near 180° inclinations of the orbit of the binary (i.e.
retrograde orbit) and high values of i_{cm} were associated with low values of
i_{34}.   For smaller initial values of the semi-major axis (a_{34}=6000 km)
high inclinations of the binary (i_{34}=90°) required high inclinations of the
orbit of the center of mass (i_{cm}=50° or 130°) for stability.

4. ACKNOWLEDGEMENTS

Both authors wish to express their appreciation for the support
received from the R. B. Curran Centennial Chair of the University of Texas,
from the National Science Foundation and from the Scientific Affairs
Division of NATO.

5. REFERENCES

Szebehely V. and Whipple A.L. (1984). "Generalization of the Restricted Problem of Three Bodies". The Big Bang and George Lemaitre, pp. 195-205, D. Reidel Publ. Co.

Pojman J. and Szebehely V. (1988). "The Hierarchical Restricted Problem with Applications". Long-Term Dynamical Behaviour of Natural and Artificial N-body Systems. (Ed. A.E. Roy). pp. 277-288, Kluwer, Dordrecht.

Van Flandern T.C. (1980). "Satellites of Minor Planets: A New Frontier for Celestial Mechanics". Celestial Mechanics, 22, pp. 79-80.

Zhang S.P. and Innanen K.A. (1988). "The Stable Region of Satellites of Large Asteroids". Icarus, 75, pp. 105-112.

THE WAVELET TRANSFORM AS CLUSTERING TOOL FOR
THE DETERMINATION OF ASTEROID FAMILIES

Ph. Bendjoya, E. Slezak, and Cl. Froeschlé

Observatoire de la Côte d'Azur
B.P. 139
F-06003 Nice cedex (France)

INTRODUCTION

The determination of asteroid families is a long standing problem already present within their distribution with respect to their semi major axis. Such an histogram exhibits indeed structures, and, whereas holes resulted in the identification of the so-called Kirkwood gaps, peaks suggested the idea that asteroids might be distributed into families. This latter concept is supported by the break up theory. Within this framework the collision-induced burst of an asteroid leads in fact to what is named a family, that is, the set of fragments whose relative speeds are greater than the speed of ejection (see Housen and Holsapple, 1990 and references therein).

Since the pioneering work of Hirayama (1918) who first worked out families, several methods have been developed in order to perform in a more or less objective way their identification. Unfortunately the agreement between the results is usually pretty poor, except for large well-known families, such as Koronis ,Eos and Themis. The renewed interrest for a reliable identification of these families is due to two recent works: i) clarification about subtle issues concerning the long-term dynamical evolution of asteroids orbits (for reviews see Froeschlé et al., 1988, Valsechi et al., 1989), ii) a new method for proper elements computation (Milani and Knezevic, 1990).

A new way to carry out cluster analysis has been recently developed by Slezak et al. (1990) in order to achieve the objective detection of structures from galaxy catalogues. It is based on the wavelet transform (Goupillaud et al., 1984). Unlike the window Fourier transform (Gabor, 1946), this transform has indeed the great advantage to perform a space-scale analysis which takes into account the characteristic scale of the studied signal. The results obtained, as well as other successes related to many topics, such as speech recognition (Kronland-Martinet et al., 1987), fractal structures (Holschneider, 1988) or image analysis (Mallat, 1989), lead us to consider with great interest this new tool for the determination of asteroïds families in the main belt.

Predictability, Stability, and Chaos in N-Body Dynamical Systems
Edited by A.E. Roy, Plenum Press, New York, 1991

So we carried out such a wavelet analysis on the new proper elements recently computed by Milani and Knezevic (1990). Not only these proper elements are the most reliable ones up to now, but also they have already been used by Zappala et al. (1990), which will allow us to compare the results. It is indeed of great importance to test the quality of the result, that is to explore how families differ with respect to different methods.

The plan of this paper is as follows. The wavelet transform is described in section 2 and section 3 presents the data used in our analysis. The identification of families is described in section 4, and comparative results are given in section 5 before the conclusion.

THE WAVELET TRANSFORM

Introduced in 1983 by J. Morlet to study seismic data, the wavelet transform of a $1D$ function $f(x)$ with respect to the analyzing wavelet $\psi(x)$ is the $2D$ function

$$C(a,b) = K(a) \int_{-\infty}^{+\infty} f(x) \, \psi(\frac{x-b}{a}) \, dx$$

where a is a strictly positive scale variable and $K(a)$ a constant equal to $a^{-1/2}$ for normalization in energy. The so-called wavelet coefficients $C(a,b)$ describe the data both in space and scale, especially if the localization properties of the function $\psi(x)$ chosen as basic wavelet are good enough. Consequently, whereas Fourier transforms are inadequate for the study of phenomena with localised events, this new transform enables a good description of such structures. According to its definition, the wavelet transform can be viewed indeed as a filtering with a set of pass-band filters $\psi(a\omega)$ where thinner and thinner details are extracted when smaller and smaller scales are investigated. So the unfolding of $f(x)$ in the half-plane defined by its wavelet coefficients results in a multi-scale analysis characterized by its ability to detect and to follow singularities. One can therefore go beyond a global description of the scaling properties (Argoul et al., 1989).

An admissibility condition must be added for reconstruction purposes to the smoothness properties of $\psi(x)$ (Grossmann and Morlet, 1985). It implies that, for differentiable functions, the integral of $\psi(x)$ has to be nul. Since a Gaussian law ensures a good compromise between spatial and scale resolutions, a popular analysing wavelet is then its second derivative named the *mexican hat* :

$$\psi(x) = (2 - \frac{x^2}{\sigma^2}) \, e^{-\frac{1}{2} \, x^2/\sigma^2}$$

The scale parameter a and the translation parameter b have to be discretized for computer applications, and it has been shown that this discrete wavelet transform relies on a regular sampling of the space axis and on an uniform sampling of the frequencies on a logarithmic scale. I. Daubechies (1990) studied in particular the completeness and stability of the discrete wavelet transform through the induced errors on the restored function. For orthogonal wavelets (see Meyer 1989 and references therein) she showed that the sampling parameters can be chosen so that no error occurs. For wavelets which are not orthonormal bases of $L^2(R)$, however there always exists an error which depends on the sampling lattice and on the wavelet itself. It is the case for the mexican hat.

The multi-scale analysis concept developed by S. Mallat (1989a) has greatly improved the understanding of orthogonal wavelet bases. A systematic way to construct such bases with compact support can indeed be deduced from this framework where functions are regarded as the limit of a hierarchy of approximations. The wavelet coefficients appear in fact to measure the signal associated with the difference between two successive approximations and thus to describe the details at different scales. All these concepts can be extended quite easily to multidimensional signals with or without distinguishing any spatial orientation (Mallat, 1989b).

Apart from functional analysis, the wavelet transform, and especially its multiresolution description, is very well adapted for applications in signal and image processing. For instance, a new way to process astronomical images can be derived where intricate objects are decomposed into features of different scales, each one being detected at a given resolution (Bijaoui et al.,1989). The wavelet transform can also be viewed as a new powerful clustering method for data analysis (Slezak et al.,1989). Its main advantage over former algorithms is to provide a description of structures with no assumption about a hierarchical feature or about an initial partition of the set of points. The next section presents in detail such an application to the space of asteroid proper elements.

THE DATA

If there were only the Sun in the solar system the determination of asteroid families would not be a problem since all the members of the same family would have an elliptic motion very close to the one of the parent as the ejection speed is small compared to the orbital velocity of the parent. But because of the secular perturbations (mainly coming from Jupiter and Saturn) all the fragments have been scattered.They are now on absolutely decorrelated orbits.

However, there is a mathematical space in which some track of a break up is kept. This space is the space of *proper elements*.

A proper element is a first integral of the motion, that is a quantity unchanged during the motion. There are three proper elements : the proper semi-major axis (a') , the proper inclination (i') and the proper eccentricity (e') .

As the relative ejection velocities of the fragments are small compared to the parent's speed, it can be expected that all the proper elements of each member at break up time are very close to those of the parent. Thus it is pretty obvious that the mathematical $3D$ space of proper elements is the right space for searching families. The proper elements have been computed by the self consistent code written by Milani and Knezevic (1990). It is based on a Lie series development of the Hamiltonian up to fourth order terms for the m/M parameter (where m is the asteroid's mass and M is Jupiter's mass) and up to second order terms for eccentricity. It takes also into account Jupiter's and Saturn's perturbations, and gives moreover an estimation of the degree of accuracy of the computed proper element.

An asteroid family has a characteristic size. This implies therefore that a **metric** has to be defined in the proper element space. A natural choice which relies on the break-up theory is to define a metric whose dimension is a velocity. As shown by Brouwer (1951), the substitution of the osculating elements by the proper elements in the Gauss equations enables one to define such a metric. The Gauss equations (see Brouwer 1951 and Zappala et al. 1984) give indeed the components of the relative speed between two asteroids from the difference of their osculating elements. By changing the osculating elements by the proper ones, it is then possible to express the relative speed between asteroids as function of the proper elements.

Let n, a', e', and i' be respectively the mean motion, the proper semimajor axis, the proper eccentricity and the proper inclination of the asteroid, and let δ represent the operator of difference between the considered asteroid and the moving point of reference that will be defined in the fourth section. The modulus of a relative speed in the 3D proper element space is then :

$$\delta v = na' \sqrt{k_1 (\frac{\delta a'}{a'})^2 + k_2 \delta e'^2 + k_3 \delta i'^2}$$

The three coefficients (k_1, k_2, k_3) are not free and their choice is somewhat arbitrary. We have chosen the same values for these coefficients as Zappala et al.(1990) in order to have the possibility to compare results obtained by this method with theirs : $k_1 = 5/4$, $k_2 = 2$, $k_3 = 2$. The analysis has also been made with $k_1 = 3/2$, $k_2 = 2$, $k_3 = 4$, so as to estimate the influence of these coefficients on the family determination.

THE WAVELET CLUSTER ANALYSIS

Since the wavelet transform of a $3D$ function is at least a $4D$ one, we preferred for evident practical reasons to perform $2D$ transforms on the 3 planes (a', e'), (a', i') and (e', i'). Real structures are then obtained through the intersection of structures discovered in each of the planes.

The discrete wavelet transform is carried out on each of these planes of projection using a regular lattice where the wavelet coefficients are computed at each node. Dealing with a two dimensional set of diracs, their weighted sum inside a circle whose radius is such the integral of the wavelet is less than 0.001. Each weight is the value at the corresponding point of the analyzing wavelet centered at the current node.

Using the isotropic radial mexican hat as basic wavelet, it is quite easy to see that, because of its properties, a group of points spread on an area more or less equal to the studied scale gives a positive coefficient. On the contrary a hole of about this size gives a negative value. At last a coefficient close to zero will be obtained for a uniform distributed area. Then one has to extract the position and the value of the maxima of the wavelet transform in order to detect and model clusters of points (see Slezak et al. 1990).

The proper element space is divided in seven zones (cf. Table I) because of the mean motion resonances in the vicinity of which proper elements are not defined. The analysis is made on the 3 projected planes of each zone for scales between 60m/s and 900 m/s with a resolution of the frame from 512 *points* × 512 *lines* to 32 *points* × 32 *lines*; two successive scales and resolution are $\sqrt{2}$ rated in order to minimize the loss of information following Daubechies'(1990) computation.

Since it is necessary to consider the restrictions of the metric to each of the 3 planes, there is a discrepancie between a scale directly defined in the $3D$ space and the equivalent scale obtained from the intersection of the 3 planes. However, a structure

Table I . zones defined by the mean motion resonances.

zone	s.major axis	resonances
1	.... - 2.065	.... - 1/4
2	2.065 - 2.300	1/4 - .....
3	2.300 - 2.501	.... - 1/3
4	2.501 - 2.825	1/3 - 2/5
5	2.825 - 2.958	2/5 - 3/7
6	2.958 - 3.030	3/7 - 4/9
7	3.030 - 3.278	4/9 - 1/2
8	3.278 -	1/2 - .....

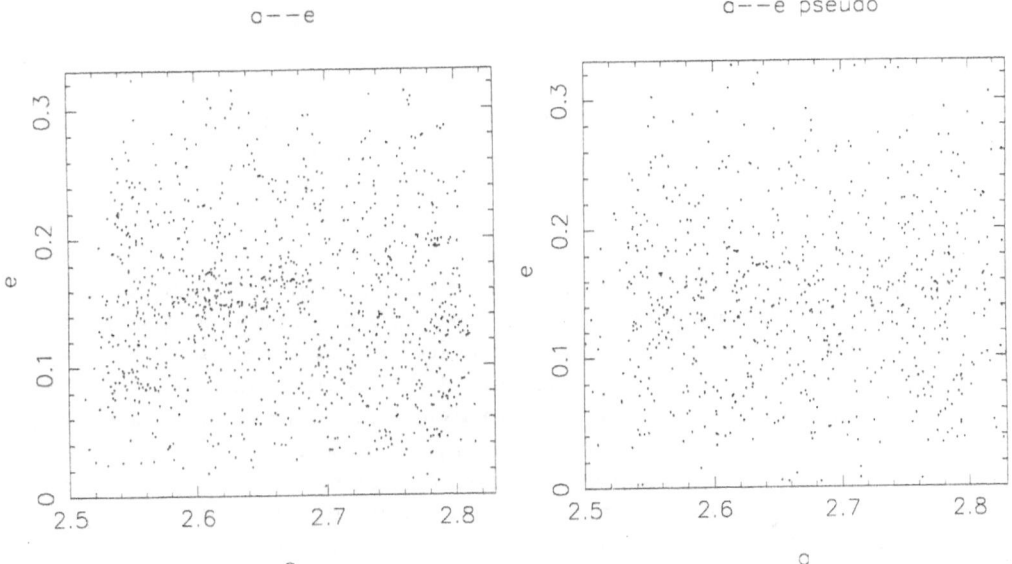

fig.1a
real distribution

fig.1b
pseudo random distribution

Asteroid distribution in the (a',e') proper elements plane

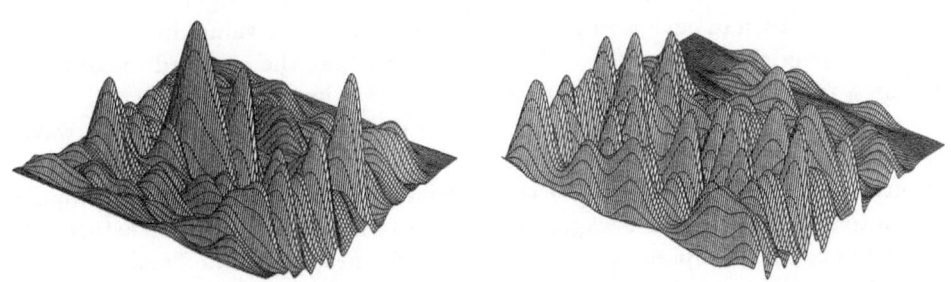

fig.1c   Maps of the wavelet coefficients   fig.1d
real distribution   pseudo random distribution

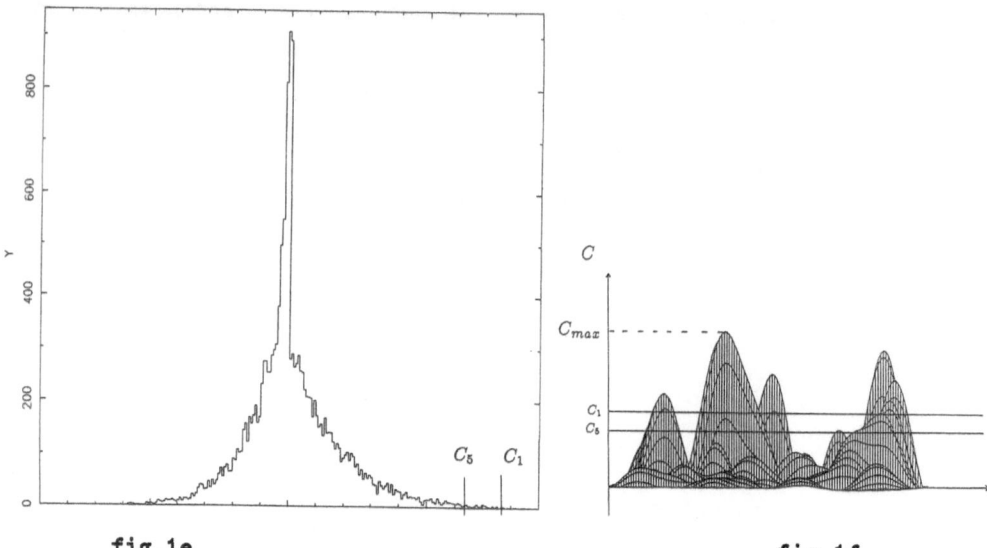

fig.1e

Histogram of the wavelet coefficient
of the pseudo random distribution.
C_5 threshold at $\frac{5}{1000}$, C_1 threshold at $\frac{1}{1000}$

fig.1f

Thresholds reported on the map
of the real distribution.

defined at σ_0 is modeled by an ellips taking into account the local background. The semi major axis of this modeling ellips is automaticaly fitted between σ_0 and $2 \times \sigma_0$. Thus, not only because of the use of the wavelet transform but also by the way of modeling structures from the first moments of the distribution, it is obvious that the study at a scale σ_0 contains also information about some greater scales, which considerably reduces the above discrepancie. A great advantage of this mathematical method is the possibility to quantify the statistical significance level of the detection with respect to a Poissonian distribution (private com. Bijaoui 1990). However, the analysis is performed in the mathematical proper element space where data do not follow this kind of distribution. Therefore, in order to evaluate the chances for the structures to be due to randomness, it is necessary to construct a random distribution of points based on the real one. The same analysis than descibed in Zappala's paper (1990) is then done on this pseudo-random distribution (cf. figs. 1a – 1d). The value of the coefficient corresponding to a 1000 to 5 probability to find by chance a higher value is computed. It is obtained using the histogram of values of wavelet coefficients from the pseudo-random distribution (fig. 1e). In that way a significance threshold is defined. It is used in such a way as to keep only the higher values in the coefficient map of the true distribution (fig. 1f). This means that the retained maxima have less than a thousand-to-five chance to get their values by coincidence.

The next step is to extract around each significant maximum of the wavelet map built from the "real" distribution all the points which have contributed to this computation. The first moments of the restricted distribution are then computed, taking into account the local background in order to fit ellipses to the structures. The intersection of these ellipses in the three planes defines finally the "real" structures in the $3D$ space.

One of the most important problems in any cluster analysis is to cut the hierarchy so as to change from structures to physical entities. In our specific application we need to find the charateristic scale at which the analysis has to be cut to define the family.

A criterion based on the philosophy of the wavelet transform, that is an increasing knowledge of the signal by thinner and thinner details, has been defined. To this end the number of added asteroids from one scale to the successive one is examined. For a given zone the highest scale to consider will be the one for which this number is a minimum, or the scale which detects the drastic change in the density. With a system for labelling the different structures it is then possible to follow each asteroid from scale to scale and to determine which family it belongs to. It is evident that all the families are defined with a degree of sinificance equal or better than 99.5%.

RESULTS

Twenty one families have been identified at the end of this analysis. Table II gives all these families : they are indexed by a number of 2 digits and the name of the lowest numbered asteroid of the family. The first digit is for the zone, the second one classifies the family in the zone by increasing number of the asteroid naming the family. This table gives also the number of members of each family. The similarities and dicrepancies in the identification of families defined from Zappala's method and ours are also described in this table.The comparison has not been done for the very populous zone 2 where results are too much different.

Let 31: Vesta be the sample for describing table 1. The first digit is 3 because of the localisation of this family in the third zone. The second digit is 1 because the number of the naming asteroid (Vesta) is the lowest one from the six numbers of the naming family asteroids in this zone. The third column of this table indicates that 5 asteroids have been attribuated to this family. The next column shows that the hierarchical analysis, has been cut at 110 m/s according to our criterion defined in order to cut the multi-scale study. The fifth column gives the percentage of identic results betwen both compared method. The last column indicates that, 31: Vesta is embeded in the Zappala's Vesta family definition, since no asteroid has been added to this definiton by the wavelet method.

The main characteristic of this table is to show that not only the well-defined families are cross identified, but also that there is a good agreement between even the small families.

Two different methods i.e the single linkage hierarchical clustering algorithm (see Zappala et al 1990) and the method based on the wavelet transform analysis, give pratically identical results. So the physical reality of the asteroid families is assess in a stronger way.

CONCLUSION

Succesful applications of the wavelet transform as clustering tool in observational cosmology, and the improvements of the accuracy on proper elements due to a new computation, made us decide to apply this transform to the determination of asteroids families. The first results obtained in celestial mechanics are very promising. Well-known families have been rediscovered and a better agreement than before has been established between two different methods even for small and less well defined families.

Table II . Asteroid families identification and comparisons.

zones:	Families	number of members	scale (m/s)	comparison idem (%)	number of added asteroids
zone 2	21: Lucretia	6	110		
	22: Berolina	114	450		
	23: 700	26	230		
	24: Iduberga	5	110		
zone 3	31: Vesta	5	110	71.4%	0
	32: Amalasuntha	6	80	75.%	0
	33: Leonce	5	80	100%	0
	34: Appenzela	20	230	0%	20
	35: 1821	19	320	63%	7
	36: Tinchen	7	110	50%	1
zone 4	41a: Eunomia	114	450	95.6%	31
	41b: Adeona	17	160	100%	2
	42: Leto	37	450	100%	15
	43: Lydia	22	320	100%	16
	44: Maria	34	320	94%	4
	45: Dora	18	320	100%	2
	46: Agnia	12	320	86%	6
zone 5	51: Koronis	125	320	91%	1
zone 6	61: Eos	172	230	87%	0
zone 7	71: Themis	117	160	51%	0
	72: Veritas	6	110	86%	0

These encouraging results make us confident about the work in progress, namely the introduction of other analizing wavelets and the development of a real $3D$ analysis in order to improve the detection of families by a direct analysis in the $3D$ proper element space. The characteristic scale of a family is however determined through physical processes. It is then of importance to model the physical mechanisms of break-up. We plan thus to carried out a wavelet analysis on artificial families for which the physical mechanisms are known so as to perform a feedback between models and family detection.

References

Argoul,F.,Arnéodo,A.,Elezgaray,J.,Grasseau,G.,Murenzi,R.: 1989, *Phys. Letters A* **135,327**
Wavelet transform of fractal aggregates.
Bijaoui,A.,Slezak,E.,Mars,G.,Giuducelli,M.: 1989, 12ème colloque du GRETSI sur le *Traitement du signal et des images.*, 1,209–211

Détection d'objets faibles dans les images célestes à l'aide de la transformée en ondelettes.

Brouwer,D.: 1951 *Astron. J.* **56**,9

Daubechies,I.: 1990, *I.E.E.E. Trans. on Information Theory*
   The wavelet transform, time frequency localization and signal analysis.

Gabor,D.: 1946, *J. I.E.E.E.* **93**,429–441
   Theory of communication.

Goupillaud,P.,Grossmann,A.,Morlet,J.: 1984, *Géoexploration* **23**,85–102
   Cycle-octave and related transforms in seismic signal analysis.

Froeschlé,Cl.,Farinella,P.,Carpino,M.,Froeschlé,Ch.,Gonzi,R.,Paolicchi,P.,Zappalà,V:
1988, in *The Few Body Problem* edited by M.J. Valtonen (Kluver, Amsterdam),pp
101–106

Grossmann,A.,Morlet,J.: 1985, in *Mathematics+Physics, Lectures on recent results*,
L.Streit (Ed.), World Scientific Publishing
   *Decomposition of functions into wavelets of constant shape, and
   related transforms.*

Holschneider,M.: 1988, *J. Stat. Phys.* **50**,953–993
   On the wavelet transform of fractal objects.

Housen,K.R, Holsapple,K.A.: 1990, *Icarus* **84**,226–253

Kronland-Martinet,R.,Morlet,J.,Grossmann,A.: 1987, *Int. J. Pattern Recognition
and Artificial Intelligence* **1**,273–302
   Analysis of sound patterns through wavelet transforms.

Mallat,S.: 1989a, *I.E.E.E. Trans. on Pattern Analysis and Machine Intelligence*
   A theory for multiresolution signal decomposition : the wavelet representation.

Mallat,S.: 1989b, *I.E.E.E. Trans. on Acoustic Speech and Signal Processing*
   *Review of multifrequency channel decompositions of images and
   wavelet models.*

Meyer,Y.: 1989, p. 21–37 in *Wavelets, time-frequency methods and phase space*,
J.M. Combes, A.Grossmann, Ph. Tchamitchian (Eds.), Springer-Verlag
   Orthonormal wavelets.

Milani,A. and Knezevic,Z. : 1990, *Celestial Mechanic* submitted

Slezak,E.,Bijaoui,A.,Mars,G.: 1990, *Astron. Astrophys.* **227**,301–316
   Identification of structures from galaxy counts : use of the wavelet transform

Valsechi,G.B.,Knezevic,Z.,Williams,J.G.: 1989, in *Asteroids II* ed. T. Gehrels (Univ.
of Arizona,Tucson), pp 1073–1089

Zappalà,V.,Farinella,P.,Knezevic,Z.,Paolicchi,P.: 1984, *Icarus* **59** ,261–285

Zappalà,V.,Cellino,A.,Farinella,P.,Knezevic,Z.: 1990, *Astron. J.* submitted
   Asteroid Families.

DELIVERY OF METEORITES FROM THE ν₆ SECULAR RESONANCE REGION NEAR 2 *AU*

Ch. Froeschlé and H. Scholl

Observatoire de la Côte d'Azur
Laboratoire G.D. Cassini
CNRS URA 1362
B.P. 139, F-06003 Nice Cedex, France

Abstract

Numerical integrations in the frame of Sun–Mars–Jupiter–Saturn model over 1Myr have been performed in order to investigate the orbital evolution of asteroid fragments produced in the innermost asteroid belt $(2.07 - 2.13 AU)$. Fragments injected in the vicinity of the ν_6 secular resonance enhance their eccentricities and become Mars-crossers. Close encounters to Mars will then lead to a random walk in semi-major axes. Two different mechanisms may occur to produce Earth-crossers. In the first case, the fragment enters the 4/1 mean motion resonance and becomes an Earth-crosser within at least $2.6 * 10^5$ years. In the second case, which involves only the secular resonance ν_6, the shortest timescale for deriving meteorites is of the order of $5.6 * 10^5$ years.

Key words: Asteroids, Celestial Mechanics, Meteorites

1 Introduction

Most asteroids probably come from the asteroid belt. Data obtained from meteorites provide strong constraints on the properties of the asteroids' parent bodies raising the following questions:

Can we understand the relationship between the various types of meteorites and the various classes of asteroids?

What are the different delivery mechanisms from the source regions to the Earth, which are the possible source regions in the asteroid belt, and what are the timescales for the delivery?

Predictability, Stability, and Chaos in N-Body Dynamical Systems
Edited by A.E. Roy, Plenum Press, New York, 1991

For instance, the most abundant recovered meteorites, the ordinary chondrites have no generally accepted spectrally identified parent bodies in the asteroid belt. Fierberg et al.(1982) claim that S asteroids, the most common class of asteroids might be parent bodies for ordinary chondrites. Ordinary chondrites have typically time exposures of the order of millions of years. Wetherill (1985, 1987) showed that the Kirkwood gap associated with the 3/1 mean motion resonance with Jupiter should be the most important source of ordinary chondrites. Wisdom (1983) namely found that the orbital element space (semimajor axis a, eccentricity e, inclination i) between 2.48 and 2.52 AU constitutes a "chaotic zone", centered on the 3/1 Kirkwood gap at 2.50 AU. Using this result, Wetherill (1985, 987) showed that asteroid collisions debris produced at the borders of the 3/1 resonance with ejection velocities between 50 to 200 m/s have a high probability to be placed into the chaotic region. Then, on a timescale of $t \leq 10^6$ years, the eccentricity pumps up to a value of ~ 0.6 causing a perihelion distance of $\sim 1AU$. Due to close encounters with the Earth, the semimajor axis will be removed from the resonant region, and the fragment will be "stranded" with practically constant eccentricity ~ 0.6 and with a perihelion distance of $1.AU$.

Quantitative studies based on a Monte Carlo method (Wetherill 1985,1987) are in good agreement with the observed ordinary chondrites falls.

Besides the 3/1 Kirkwood gap as a possible source to derive ordinary chondrites, Wetherill also suggested the innermost asteroid belt between 2.0 and 2.25 AU via the secular resonance ν_6. Wetherill showed by considering only the effect of the ν_6 resonance that injection velocities of $800 m/s$ are required. Since the fraction of the ejecta with such high value is small $\sim 0.3\%$ (Gault et al.1963), relativly few meteorites can be expected by this mechanism. Fragments with injection velocities of $100 m/s$ can suffer close encounters with Mars, and the debris is random walked in the vicinity of the ν_6 resonance increasing the debris' eccentricity. The fragments may become Earth-crossers after $10^8 - 10^9$, a timescale which corresponds to exposure tiem of differentiated meteorites.

However, in the region $2.0 - 2.2 AU$, three resonances are located: the secular resonances ν_6 and ν_{16}, and the mean motion resonance 4/1 with Jupiter at $2.065 AU$. Froeschlé and Scholl (1986) found that fragments injected at low inclination into the ν_6 resonance region with semimajor axes between $2.04 - 2.06 AU$ become Earth-crossers within 10^6 years. Yoshikawa (1989) performed numerical integrations of fictituous bodies located in the 4/1 resonance with starting eccentricities ranging from $0.02 \leq e \leq 0.1$. The maximum eccentricity of these bodies reaches values between $0.6 \leq e \leq 1.0$. Since the 4/1 resonance is situated very close the secular resonance ν_6, Mars encounters may move bodies in the 4/1 mean motion resonance. Then, a substantial broadening of the Earth-crossing region may be expected, and the delivery timescale to the Earth may be shorter.

In the following, we present numerical integrations which show that the combined effect of the two resonances ν_6 and 4/1 is an efficient mechanism to produce meteorites on timescales less than 10^6 years.

2　Numerical experiments in the region $2.07-2.13 AU$

Twenty orbits of test fragments were integrated in the frame of the Sun–Mars–Jupiter–Saturn model. All the fragments have a starting inclination of $2.°$ with respect to ecliptic 1950.0. Two initial eccentricities 0.1 and 0.05 are taken and ten values of initial semimajor axes in the range $2.07 - 2.13 AU$ were chosen. We used two sets of fragments, namely Meteor 1 for e= 0.1 (Table 1) and Meteor 2 for e= 0.05 (Table 2).

The orbits were integrated using the DVDQ code (Krogh 1970). This code is a predictor–corrector method allowing variable stepsize and variable order for differences. The integration time was 10^6 years at the most. We stopped the integration when the fragment remained at least $2 * 10^4$ years an Earth-crosser. Since the Earth was not included in our model, our results ,of course, are not realistic after crossing the Earth's orbit. By planet crossing we mean that the orbit of a small body intersects the ecliptic at a heliocentric distance smaller than the corresponding planets' mean distance from the Sun.

According to Tables 1 and 2, which summarize the results, 7 fragments with semimajor axis $a \leq 2.1 AU$ become Earth-crossers. Two different mechanisms may occur Earth-crossers.

One mechanism involves the two resonances ν_6 and $4/1$, the second is based only on the secular resonance ν_6. In the first case -which concerns six orbits- a fragment is at the beginnig situated in the secular resonance ν_6: the secular argument $\varpi - \varpi_s$ librates around 180° (fig.1); consequently, the eccentricity increases. When the eccentricity is ~ 0.27, close encounters with Mars occur, the fragment's semimajor axis starts a random walk (fig.2). Then after about some 10^5 years, (Tables 1 and 2) the fragment is locked in the $4/1$ mean motion resonance (fig.2). The resonance argument σ librates. The eccentricity starts to increase strongly and the threshold eccentricity $e = 0.52$ to become Earth-crosser is reached (fig.3). Then close encounters with Mars may eject the fragment from the $4/1$ resonance, injecting it eventually into the secular resonance ν_6 again, see Table 1. Some of the bodies (tables 1 and 2) with $e \geq 0.65$, become even Venus crossers (fig.3). Further calculations including the Earth in our model should verify this result.

The second mechanism which involves only the secular resonance ν_6, is illustrated by body 7 (table 1). The fragment is located in the secular resonance ν_6 during all the time of integration (fig.4). The eccentricity starts to increase and the body becomes a Mars-crosser. Then close encounters with Mars (fig.5) place the semimajor axis deep in the secular region, the eccentricity continues to increase (fig.6), and after $6 * 10^5$ years, the fragment becomes Earth-crosser and even Venus-crosser (fig.7).

As stated on tables 1 and 2, the bodies with semimajor axis between 2.105 - 2.120 AU have several close encounters with Mars. All of them are located in the ν_6 resonance. The secular resonance arguments $\varpi - \varpi_S$ seem to librate about 180° (fig. 8). However, their eccentricities do not exceed 0.32, (tables 1 and 2).

It appears that the starting value $2.105 AU$ for the semimajor axis is a boundary value to produce Earth-crossers on a timescale of 1 Myr.

As it is shown on Tables 1 and 2, fragments with initial semimajor axes $a \geq 2.125 UA$ do not suffer close encounters during at least 1 Myr.

3 Conclusion

These numerical experiments show clearly that ordinary chondrites can be derived from the region $2.07 - 2.105 AU$ on timescales less than 1Myr. Two mechanisms for producing meteorites are found. One, which is the most efficient, involves the

Table 1. Meteors 1. $e_{starting} = 0.1,$ $i_{starting}(ecliptic = 1950.0) = 2°$

Body Number	$a_{starting}$ (AU)	time spent in ν_6 resonance $(10^5 yrs)$	time spent in 4/1 resonance $(10^5 yrs)$	first Mars encounter $(10^5 yrs)$	Earth crosser after $(10^5 yrs)$	e_{max}
4	2.070	$0 \leq t < 2.3;$ $t > 3.2$	$2.3 \leq t \leq 3.2$	0.8	2.6	0.61
5	2.080	$0 \leq t < 2.5;$ $t > 6.7$	$2.6 \leq t \leq 6.7$	0.8	3	0.77
6	2.090	$0 \leq t \leq 2.8$	$t \geq 3.0$	2.4	3.2	0.89
7	2.100	always	no	2.7	5.6	0.84
8	2.105	always	no	2.5	no	0.26
9	2.110	always	no	1.8	no	0.30
10	2.115	always	no	1.5	no	0.29
11	2.120	always	no	1.8	no	0.28
12	2.125	always	no	no	no	0.25
13	2.130	always	no	no	no	0.23

Table 2. Meteors 2. $e_{starting} = 0.05,$ $i_{starting}(ecliptic = 1950.0) = 2°$

Body Number	$a_{starting}$ (AU)	time spent in ν_6 resonance $(10^5 yrs)$	time spent in 4/1 resonance $(10^5 yrs)$	first Mars encounter $(10^5 yrs)$	Earth crosser after $(10^5 yrs)$	e_{max}
4	2.070	always	no	1.5	no	0.32
5	2.080	$t \leq 3.6$	$t > 3.5$	2.5	3.9	0.63
6	2.090	$t \leq 4.0$	$t > 4.0$	1.8	4.4	~ 1.0
7	2.100	$t \leq 4.0$	$t > 4.0$	2.5	5.5	0.55
8	2.105	always	no	2.6	no	0.26
9	2.110	always	no	2.7	no	0.26
10	2.115	always	no	8.8	no	0.26
11	2.120	always	no	8.7	no	0.24
12	2.125	always	no	no	no	0.23
13	2.130	always	no	no	no	0.21

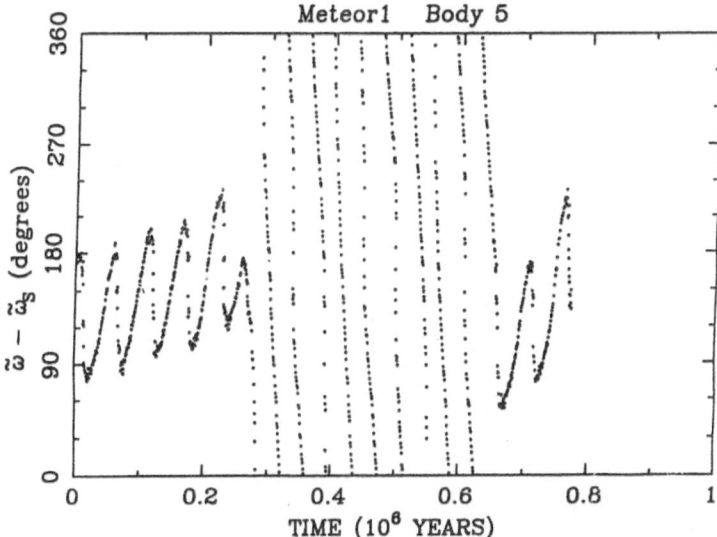

Figure 1. The ν_6 secular resonance argument $\varpi - \varpi_S$ versus time. Until $3*10^5$ yrs, the body is located in the secular resonance ν_6; $\varpi - \varpi_S$ librates about $180°$; then, the body is ejected from the ν_6 secular resonance. $\varpi - \varpi_S$ circulates until $\leq 6.7*10^5$ yrs, and again $\varpi - \varpi_S$ librates which indicates that the body is again situated in the secular resonance ν_6.

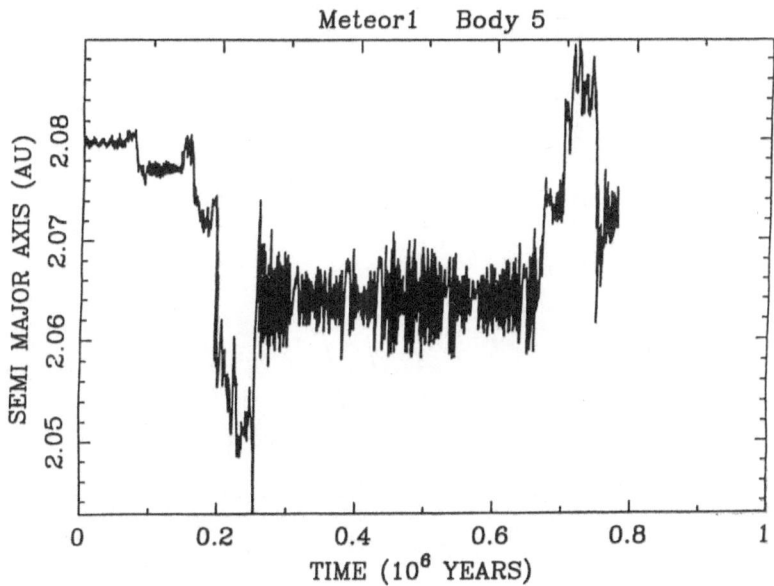

Figure 2. The semimajor axis of the body is trapped in the 4/1 resonance after $2.5*10^5$ until $6.7*10^5$ years.

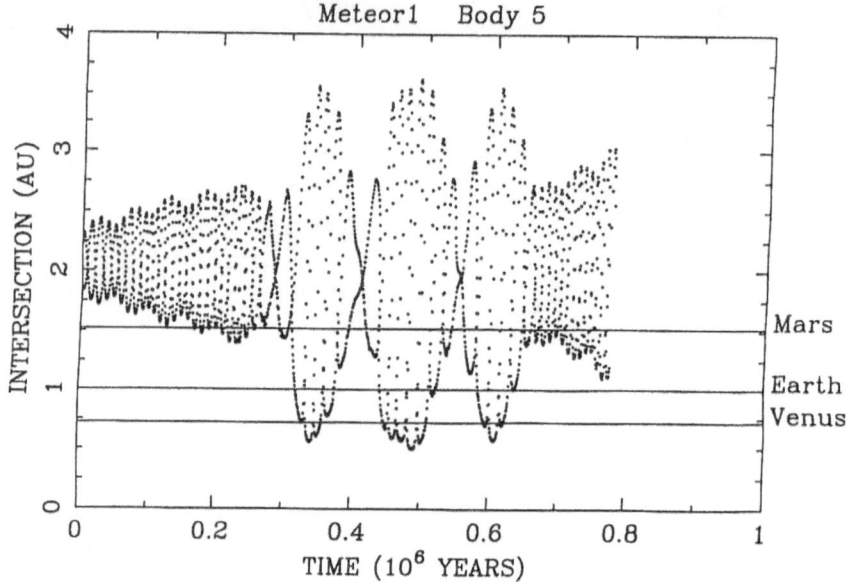

Figure 3. Time evolution of intersections between the orbit of Body 5 (starting values $i = 2°$, $e = 0.1$, $a = 2.08$ AU) and the ecliptic. The position of the semimajor axes of Mars, Earth, and Venus are indicated by corresponding lines.

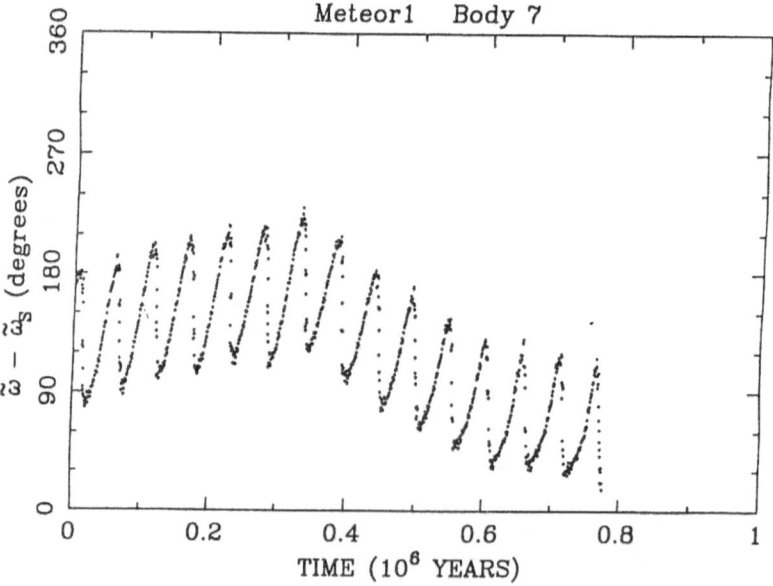

Figure 4. The ν_6 secular resonance argument $\varpi - \varpi_S$ versus time for body 7 (starting values $i = 2°$, $e = 0.1$, $a = 2.10$ AU). The body is located in the secular resonance ν_6.

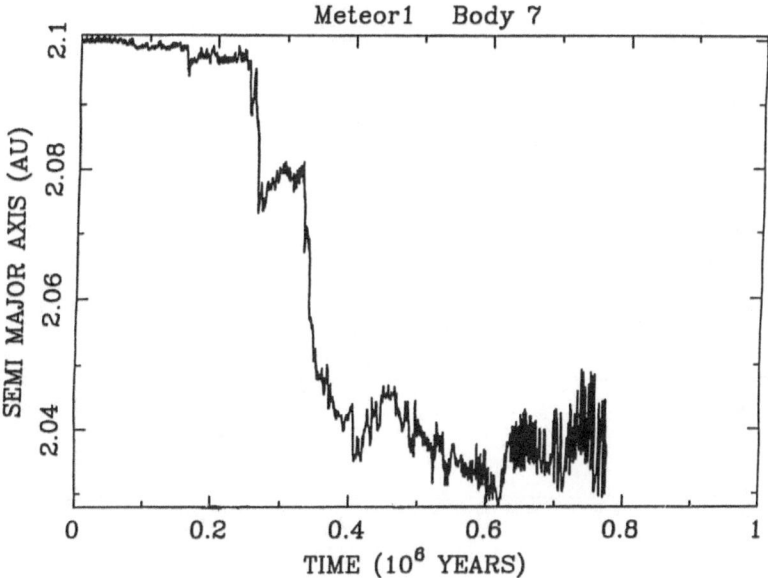

Figure 5. Time evolution of the semimajor axis of body 7 (starting values $i = 2°$, $e = 0.1$, $a = 2.10$ AU)

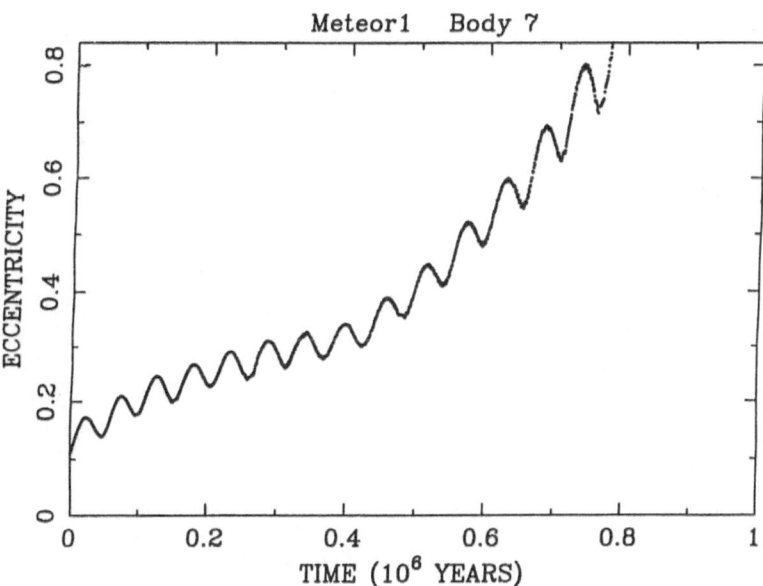

Figure 6. Eccentricity e versus time of body 7. The eccentricity increases due to the secular resonance ν_6.

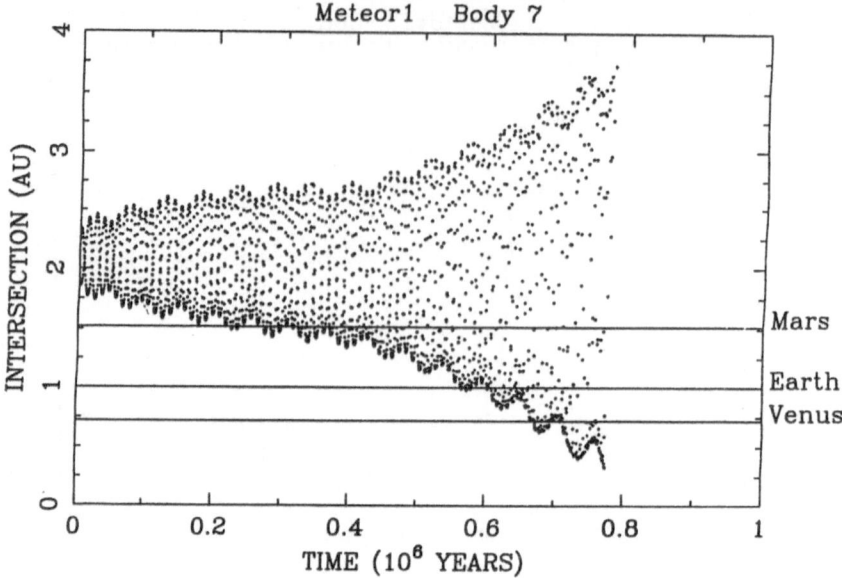

Figure 7. Time evolution of the intersections between the orbit of body 7 and the ecliptic. The body becomes Earth-crosser at $\leq 5.6 * 10^5$ yrs.

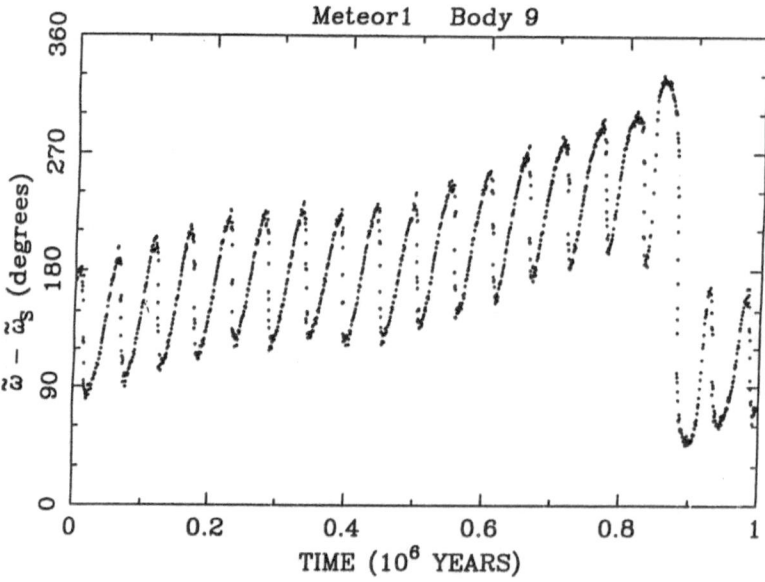

Figure 8. The ν_6 resonance argument $\varpi - \varpi_S$ circulates slowly due to the proximity to the ν_6 resonance.

two resonances ν_6 and 4/1. The second one is only due to the secular resonance ν_6. The eccentricity of a body injected in the ν_6 resonance will become high enough to permit deep Mars crossing. Due to close encounters with Mars, the semimajor axis starts a random walk. The body may be either ejected from the ν_6 resonance and trapped in the 4/1 mean resonance, or may be placed deeper in the ν_6 resonance. When the fragment is trapped in the 4/1 resonance, the threshold eccentricity to become Earth-crosser is reached in only some 10^4 years. While the fragment stays in the secular resonance ν_6, the eccentricity increases more slowly.

These results show also, that the proximity of the ν_6 and the 4/1 resonances plays an important role in the delivery of meteorites. In order to study the efficiency of the inner Flora region for production of meteoritic material, more investigations including realistic dynamical features must be done.

References

[1] Feierberg, M. A., Larson, H. P., and Chapman, C. R., 1982, Spectroscopic evidence for differentiated S–type asteroids, Astrophys. J., 257:361.

[2] Froeschlé, Ch., Scholl, H., 1986, The secular resonance ν_6 in the asteroid belt, Astron. Astrophys.,166:326.

[3] Krogh, F. T., 1982, JPL Technical Utilization Document, No. CP–,238.

[4] Wetherill, G. W., 1985, Asteroidal source of ordinary chondrites, Meteoritics, 20:1.

[5] Wetherill, G. W., 1987, Dynamical relation between asteroids, meteorites and Apollo-Amor objects, Phil. Trans. Roy. Soc. Lond., A 323:323.

[6] Wisdom, J., 1983, Chaotic behaviour and the 3/1 Kirkwood gap, Icarus, 56:51.

[7] Yoshikawa, M., 1989, The survey of the motions of asteroids in the commensurabilities with Jupiter, Astron. Astrophys., 213:436.

THE DYNAMICS OF METEOROID STREAMS

I.P.Williams and Zidian Wu

Astronomy Unit
Queen Mary and Westfield College
Mile End Rd
London, E1 4NS

SUMMARY

Meteor showers, seen at regular and frequent intervals on Earth, are caused by the interaction of meteoroids (that is small dust grains) in a coherent stream, all moving on similar heliocentric orbits, with the atmosphere of the Earth. The formation and dynamical evolution of such streams is discussed and techniques, including numerical integration, for following their evolution is described. In some cases the evolution is very critically dependent on initial conditions and the evolution may be chaotic. In addition to general considerations, some specific streams, all with individual areas of interest, will be discussed.

1.INTRODUCTION

Numerous meteors, or shooting stars, can be seen on any clear night. Regular observation will however show that the appearance of the meteors is not totally random but that above a random or sporadic background, many meteors can be seen at certain times of the year, the first few days of January or the second week in August being good examples. At these periods of high activity, it will also be evident that there is a pattern to the spatial distribution of the meteors with many of the trails appearing to originate from a single point on the sky, called the radiant. This point is however variable from shower to shower and it is the custom to name the shower of high activity after the constellation in which the radiant is located. Thus the high activity period in the second week of August is called the Perseid meteor shower. The explanation for these observed phenomenae is that in interplanetary space there exists sets, or streams, of small solid particles which we shall call meteoroids moving on nearly identical heliocentric orbits and that a meteor shower is seen whenever the Earth passes through such a stream, the visible meteor being caused by the meteoroid burning up in the upper atmosphere. The Earth must, by definition of the node, pass through each stream at the node of the stream and if there is no dynamical evolution of the stream, the shower will be seen at exactly the same time each year. Conversely, any observed change in the time of appearance of the shower maximum gives a very good determination for the rate of retrogression or progression of the node.

The data for the main known meteor showers is given in Table 1. Here, Z.H.R stands for the Zenith Hour Rate, that is the number of meteors that would have been seen per hour if the radiant had been at the Zenith and seeing conditions perfect. Naturally the number of meteors actually seen is somewhat less than this number. However, there are now agreed procedures for dealing with the correction for a non-zenith radiant and so the Z.H.R does give a reliable guide to the relative meteoroid population densities in the various streams.

Table 1. Data on some important Meteor Streams.

STREAM NAME	GEOCENTRIC RADIANT			DATE	MAXIMUM ZHR	
	λ_0	R.A.	Dec.		Visual	Radar
Quadrantids	282.7	230.1	+48.5	Jan 1–4	140	
Lyrids	31.7	271.4	+33.6	Apr 20–23	20	
η Aquarids	42.4	335.6	−1.9	Apr 21–May 12	30	
Daytime Arietids	77	44	+23	May 29–Jun 19		60
Daytime ζ Perseids	78	62	+23	Jun 1–17		40
Daytime β Taurids	96	86	+19	Jun 24–Jul 6		30
Phoenicids	109.6	31.1	−47.9	Jul 3–18		30
δ Aquarids	125	333.1	−16.5	Jul 21–Aug 29	30	
α Capricornids	127	307	−10	Jul 15–Aug 10	30	
Perseids	139	46.2	+57.4	Jul 23–Aug 23	70	
Daytime Sextantids	183.6	152	0	Sep 24–Oct 5		30
Orionids	208	94.5	+15.8	Oct 2–Nov 7	30	
Leonids	234.5	152.3	+22	Nov 14–20	10	
Geminids	261.0	112.3	+32.5	Dec 4–16	70	
Ursids	270.7	217.1	+75.9	Dec 17–24	20	

The possible association of comets and meteoroid streams appears to have first been suggested by G.V.Schaparelli in 1866 (see Oliver 1925 for this and other references to early work on meteors and meteoroid streams). This suggestion of Schapparelli was based on the similarity of the orbits of the Perseid stream and comet 1862ii (now known as periodic comet Swift-Tuttle) and of the Leonid stream with comet 1866i (now called periodic comet Tempel-Tuttle). The activities of comet Biela must also have played an important part in the formulation of the comet-meteoroid stream hypothesis. In 1832, comet Biela appeared as a perfectly normal comet with a period of about six and a half years. Atits next predicted return in 1839, observing conditions were not favourable and the comet was not seen. However in 1845 the comet was again recovered but was seen to have a faint companion accompanying it. At the 1852 return both comets were very faint and have never been seen since that date despite repeated and intensive searching. It was calculated that in 1872 the Earth actually crossed the orbit of the comet moderately close to where the comet would have been and a very strong meteor shower was seen at that time. This phenomenon was repeated in 1885, 1892 and 1899. Today, this association of meteoroid streams with comets is widely accepted and cometary companions have been identified for most of the major streams. One outstanding exception is the Quadrantid meteoroid stream and we will discuss the dynamics of this interesting stream later in the paper. Though the Geminid stream appears to have a clear parent, it is asteroid 3200 Phaethon, and this raises the question of whether some dead comets are in fact miss-identified as asteroids.

In the next section we give a simple overview of the process of dust ejection from comets and in section 3 proceed to apply this to the formation of meteoroid streams. The evolution of streams under the influence of planetary perturbations is discussed in section 4, which also includes some remarks on the numerical integration techniques which might be applied. In section 5 we discuss in more detail problems associated with an individual stream, the Quadrantid stream where some elements of Chaos may be present.

2. THE EMISSION OF DUST FROM COMETARY NUCLEII

The basic process of dust emission from cometary nucleii is now generally agreed, solar radiation heats up the predominantly icy nucleus, eventually causing sublimation which results in an outflow of gas, expanding out from the nucleus at a speed close to the sound speed. This outflow of gas drags with it many of the embedded small solid grains and accelerates them until such a time as the drag forces become smaller than other forces and the grains respond to these new forces for most grains either solar radiation pressure or solar gravity and it is this latter set that holds the possibility of forming meteoroid streams (see for example Whipple 1985). In order to gain some insight into the type of grains that might be found in meteoroid streams, let us take a very simplistic view of this whole process. The drag on a spherical grain of diameter d moving with relative speed v through a gas of density ρ is given for example by Baines et al (1965) as

$$3F_D = \pi d^2 \rho v W \qquad\qquad 2.1$$

where W is the mean molecular speed in the gas (roughly the same as the sound speed). Of course, the formulae applies equally when the gas is streaming out with a speed W and the grain is moving slower, the grain speed relative to the nucleus then being (W-v) , so that initially, when the grain is stationary on the nucleus, v = W. In theory, the grain could be accelerated until v = 0, that is its actual outward speed is W but this is not likely as other forces take over.

The gas density, ρ, is difficult to estimate but the mass loss rate of comets, $\dot{M}$, is easier to estimate from ground based observations, or even determine with a fly past of a spacecraft as for example in the case of comet Halley. Using the equation of continuity, we have

$$\dot{M} = 4\pi f\rho \; WR^2, \qquad\qquad 2.2$$

at a distance R from the nucleus, $4\pi f$ being the solid angle into which gas is ejected, or roughly, f is the fraction of the surface that is active. Eliminating ρ from equation 2.1 gives

$$F_D = (d^2 v\dot{M})/(12fR^2). \qquad\qquad 2.3$$

The only force attempting to prevent the grain from escaping is the gravity of the nucleus and is given by

$$F_N = (GM_N\pi d^3\sigma)/(6R^2), \qquad\qquad 2.4$$

where σ is the bulk density of the grains.

Escape will not be possible if $F_N > F_D$, that is if

$$d > (v\dot{M})/(2\pi\sigma fGM_N), \qquad\qquad 2.5$$

and v = W as the grain must start from rest on the surface of the nucleus.

For comet Halley, the nucleus mass M_N was found to be in the range 5 - 13 x 10^{16}g (Whipple, 1987), while a typical mass loss rate $\dot{M}$ was 2×10^{14}gy^{-1} and f at about 15%. With these values, and assuming a mean molecular weight of 18 for the escaping gas, equation 2.5 shows that the maximum size of grain that can be carried away from the nucleus by gas drag is about 12cm. This estimate is in good agreement with the largest size of particles actually found in meteoroid streams and is also consistent with the detection of a 1g grain by Giotto near the Halley nucleus (McDonnell et al 1987).

As the grains move out from the nucleus, other forces become important and eventually the motion will be completely determined by the combined effects of solar gravity and solar radiation pressure.

The ratio of solar radiation pressure effect to solar gravity is well known, (see for example Fox et al 1982) and is given by

$$\beta = 2.9 \times 10^{-5}/(d\sigma), \qquad\qquad 2.6$$

though, for very small grains ($d < 10^{-5}$cm) the expression is modified due to the efficiency factor for radiation pressure being a function of grain size and wavelength of the incident radiation (A full discussion is given in Simpson et al 1979)

If β is greater than one, the grains will clearly be driven out of the solar system by radiation pressure and so no grains with such a β value can be found in meteoroid streams. This limit implies grains with $d < 2.4 \times 10^{-4}$cm. In fact grains with β less than unity can escape from the system as their total energy is positive and (Kresak 1974) showed that escape is possible for all

$$\beta > (1-e)/2.$$

With typical values of e at 0.9, this gives a β value of .05, and increases the minimum grain radius to be found in meteoroid streams by a factor of 20 to about 5×10^{-3}cm, again very consistent with what is detected.

We now consider the speed with which a grain is likely to leave the influence of the gas drag and move under gravity, assuming its dimensions are in the correct range. An approximate equation of motion is obtained from equation 2.3 as

$$m\dot{R} = (d^2 W \dot{M})/(12fR^2) \qquad\qquad 2.7$$

which can be integrated to give

$$R^2 = (W\dot{M})/(\pi\sigma df R_C) \qquad\qquad 2.8$$

In fact, Whipple (1951) has considered this problem in some detail and produced a similar expression, namely

$$R = 464\,(n\sigma d r^{2.25} R_C)^{-0.5}\, \text{cms}^{-1} \qquad\qquad 2.9$$

where n includes both f and an estimate for the amount of energy available for ejection. Here R_C is in km and r in AU. Insertion of typical numerical values shows that the escape speed is typically a few hundred to a thousand cms^{-1}, and so is very much less than the orbital speed. Seen in a heliocentric frame these grains will appear to drift very slowly away from the nucleus and follow orbits that are very similar to that of the parent comet. However, as we shall see later, these small changes can be important as they can change the circumstances in any close encounter with a planet.

3. THE FORMATION OF A METEOROID STREAM

The picture we have then is of small grains leaving the nucleus with speeds up tp $1000cms^{-1}$ and moving primarily under the influence of solar gravity. If the speed relative to the nucleus is denoted by v and making an angle θ with the direction of the orbital velocity of the nucleus, denoted by V, then the new energy per unit mass of the grain E' is given by

$$2E' = V^2 + v^2 + 2vV\cos\theta - GM_o(1-\beta)/r, \qquad 3.1$$

while the energy per unit mass of the nucleus E is given by

$$2E = -GM_o/a = V^2 - 2GM_o/r. \qquad 3.2$$

Equation 3.1 can be rewritten, using equation 3.2 as

$$2E' = -GM_o/a + v^2 + 2vV\cos\theta + 2GM_o\beta/r, \qquad 3.3$$

and it can be seen that the energy is positive if $\beta > r/2a$ which was previously used with $r = a(1-e)$ to determine the minimum size of grains present.

The energy difference between the grain and the comet is given by

$$2\Delta E = 2E' - 2E = v^2 + 2vV\cos\theta + 2GM_o\beta/r. \qquad 3.4$$

From Kepler's third law we have

$$\frac{\Delta E}{E} = \frac{-\Delta a}{a} = \frac{-2\Delta P}{3P} \qquad 3.5$$

Inserting typical numerical values shows that $\Delta E/E$ is generally small and so the changes in a and P will also be small. However, as the period is slightly different grains will arrive back at perihelion at slightly different times from the comet with this delay, or advancement repeated every orbit so that, in time, the grains are spread like a doughnut around the cometary orbit. Fig 3.1 illustrates this in a schematic way. This doughnut is clearly a reasonable representation of a meteoroid stream. However both the grains and the comet are moving within the planetary system and are subject to gravitational perturbations by the planets. These perturbations may cause significant changes in the orbits and will be discussed next.

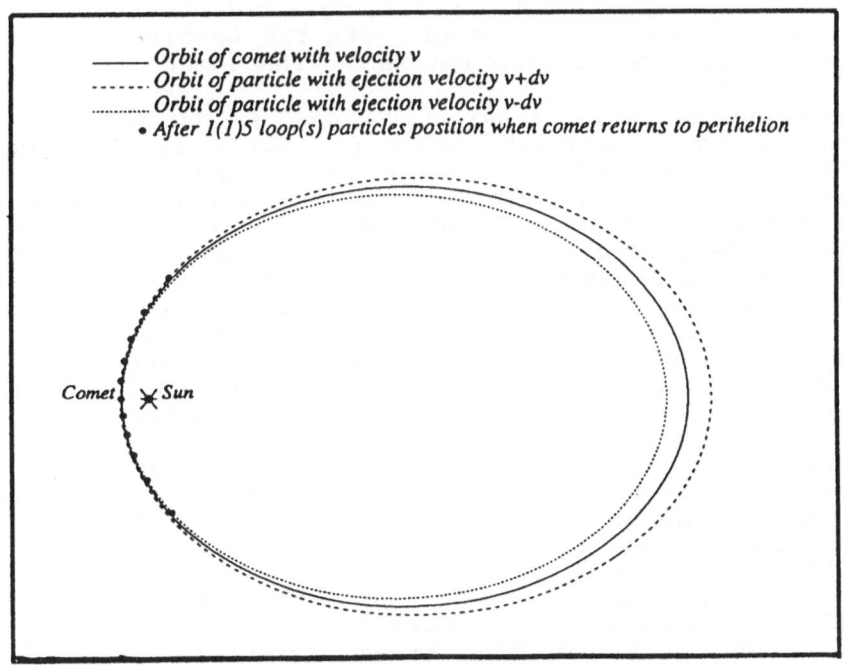

Figure 3.1 Formation of meteor stream

4. THE EVOLUTION OF METEOROID STREAMS

The physics of the situation is easy to understand. Each particle is subject to the gravitational forces of each of the planets as well as that of the Sun. Mutual particle - particle gravitation is negligible.Thus, if positions for the particles and planets are known, the accelerations can easily be obtained from Newton's laws. However, dealing with the effects of these accelerations is more problematic. The simplest of all models that can be used is a one particle model, that is, it is assumed that the motion of the parent comet accurately reflects the motion of the stream. Early success in predicting the behaviour of the Leonid stream was had with such a model (see Lovell, 1954). Yeomans(1981) has investigated the orbital evolution of comet Tempel-Tuttle over the period 1366 to 1965, including the non-gravitational terms due to outgassing and found that most of the impressive displays associated with the Leonid stream were caused by dust located outside and behind the comet.

An advancement on this simplistic idea is to average the perturbations over a complete orbit so that only the secular perturbation terms remain. Brouwer (1947) generated such a method which worked well even for highly eccentric orbits and this was used by Whipple and Hamid(1950) to show that 4700 years ago some of the Taurid meteors had orbits similar to Encke's comet. Plavec (1950) used the Gauss-Hill method to study the Geminid stream and found that it was evolving very rapidly. However, the most popular method in current use is the Gauss-Halphen-Goryachev method (see Hagihara, 1972) and has been used by Babadzhanov and Obrubov (1983) to study the long term evolution of both the Geminid and Quadrantid streams.

These secular perturbation methods are more suitable when the long term evolution of a stream is considered since they do not require vast computing capabilities using as they do the averaged perturbation equations. Two drawbacks are that they tell nothing about the behaviour of individual meteoroids or even groups as they use the mean stream. They will therefor miss the peculiar behaviour of the Quadrantid stream (see later) where different parts of the stream appear to be evolving differently and secondly, they are not that useful in dealing with the short term evolution of a stream again since in the short term, not all the meteoroids will experience the mean perturbation. In consequence of the above they are also highly inaccurate close to a resonance.

The alternative is to represent the stream by a set of test particles and to numerically integrate the equations of motion of all the test particles. With a reasonable number of test particles a reasonable representation of the real stream can be obtained. The main drawback is that such methods require a large amount of computer time and generally have not been successfully used with a sensibly large number of particles for a long period of integration. Levin et al (1972) used a program developed by Sherbaum (1970) to show that the Jovian perturbations considerably increase the width of meteoroid streams, a results not easily obtainable from secular perturbation techniques.

Numerical models for the Quadrantid stream were produced by Hughes et al (1979) using the Runge-Kutta method and the Geminid stream by Fox et al (1982) using the Gauss-Jackson method, while Runge-Kutta methods were also used by Jones and McIntosh (1986) to discuss the h-Aquarid stream and its connection with comet Halley. As numerical approaches are very useful and common it is worthwhile mentioning the main methods in use.

There are three main types of integration techniques in use,

(i) Taylor series methods,

(ii) Single step methods,

(iii) Predictor-Corrector methods.

In the Taylor series method, if the differential equation is

$$\frac{dy}{dx} = f(x,y), \qquad\qquad 4.1$$

then $y = y_0 + hf_0 + h^2/2\, f'_0 + \ldots ,$

where $h = x - x_0$ and as many terms as is necessary for accuracy can be included provided the differentiation of the function f can be carried out.

The single step methods such as the Runge-Kutta use some procedure for evaluating y from the values of y_0 and the differential equation. In its simplest, first order form, this would be

$$y_1 = y_0 + hf_0, \qquad\qquad 4.2$$

the procedure is then repeated for an other step using the values at y_1 only and thus getting to y_2 and to continue in this fashion. One great advantage of these methods is that the step length h does not have to be constant throughout, large h can be used when the motion is slow. Higher order methods modify the form of equation 4.2, but the philosophy is the same. The fourth order method (see for example Khabaza, 1969) is used in many different contexts and is generally available in most subroutine libraries.

A variant on this method, is the Runge-Kutta-Nystrom methods (Dormand and Prince, 1978, Dormand et al, 1987), where expansions for two orders are produced and the difference used to judge the error and adjust the step length.

The final method is the predictor-corrector method and many such as the Adams-Bashforth or the Milne-Simpson methods (see Khabaza, 1969) are in common use. Here a predictor formula is used, based on the information available up to point x_i to generate the predicted value at y_{i+1}. The old information plus the new prediction is then used to iteratively correct the prediction in a more accurate formula. Description of other such methods can be found in Brewer and Clements (1961). Though all methods have their advantages in specific situations, the second group are probably the most useful for general use (Fox, 1984).

By the use of such methods there is a fairly general understanding existing of the way in which meteoroid streams evolve.The Geminid stream seems to have come in for particular attention, together with the h Aquarid stream, the latter not being surprising in view of its connection with Halley's comet. Fox et al (1983), showed that the observed rate profile could be obtained from the ejection mechanisms described above, while Hunt et al (1985) investigate detailed planetary perturbations and Fox and Williams(1985) demonstrated that the fireballs observed in the 11th century were not related to the Geminid stream but to the more minor Monocerotid stream. Jones(1985) and Jones and Hawkes (1986) also numerically investigated the effects of planetary perturbations on the Geminids and came to roughly the same conclusion as earlier papers. Gustafson(1989) integrated backwards the orbits of a set of actual Geminid meteors and claimed to determine the ejection epoch and velocity. McIntosh and Hajduk (1983) proposed a model for the meteoroid streams associated with comet Halley and Jones and Macintosh (1986) used numerical modeling to further investigate the structure of the Halley meteoroid stream. The agreement between theory and observation in most cases is good enough for it to be clear that the major process of evolution are as we have described them. Recent reviews of the topic are by Hughes (1985), Williams (1989). We now turn to a specific stream where further comment may be useful.

5. THE QUADRANTID STREAM, AN EXAMPLE OF CHAOS ?

Though the general picture of meteoroid stream evolution is very satisfactory, there are instances where surprises still remain, the most interesting being the Quadrantid stream. The first indication that something unusual was occuring in the Quadrantid stream came when it was found that the occurrence of the maximum in the visual meteor rate was two days different from that found by radar. This phenomenon was investigated by Hughes et al (1981) who found that test meteoroids with the aphelia of their orbit between 5 and 6 AU suffered very irregular retrogression of the nodes, each particle behaving somewhat differently. Froeschle and Scholl (1985) pointed out that the 2 : 1 resonance region for orbits with perihelion close to the Earth had an aphelion around 5.5 AU and investigate the motion of stream particles close to this resonance. They concluded that test particles exhibited two distinct evolutionary behaviour, the stream was breaking up into two arcs. In fact, the situation is more complex than described above and Table 2 shows that a number of strong resonances with Jupiter exist in the region occupied by the Quadrantid stream.

The authors have reinvestigated the Quadrantid situation, but rather than using test particles to represent the stream, we

Table 2. The ressonance with Jupiter in the region of the Quadrantid Meteoroid Stream

Aphelion AU	Semi-major axis AU	Period Years	Ressonance
3.98	2.49	3.93	3 : 1
4.62	2.81	4.72	5 : 2
5.52	3.26	5.89	2 : 1
6.36	3.68	7.07	5 : 3
6.90	3.95	7.85	3 : 2

have used actual observed meteors. We find three very distinct type of behaviour. These three types can be seen in Figs 5.1 and 5.2 for perihelion distance and eccentricity using ten real orbits. One set (meteors 9945 9953 9955 9974 9985) shows a clear, almost sinusoidal oscillatory behaviour, while the other two sets (meteors 9902 9952 9997 and 9907, 9996) both show a near monotic change but the change being very different for the two sets. The numbering system used is the trail number from the Harvard Photographic Catalogue

Figs 5.3, 5.4 and 5.5 show a spatial representation of the evolution of these three different type of meteoroid behaviour, the diagrams representing meteors 9985, 9902 and 9907 respectively and cover the time interval from 1954AD (observation time) to 3000BC. The lack of evolution in one set and the very different orbital parameters of this set might

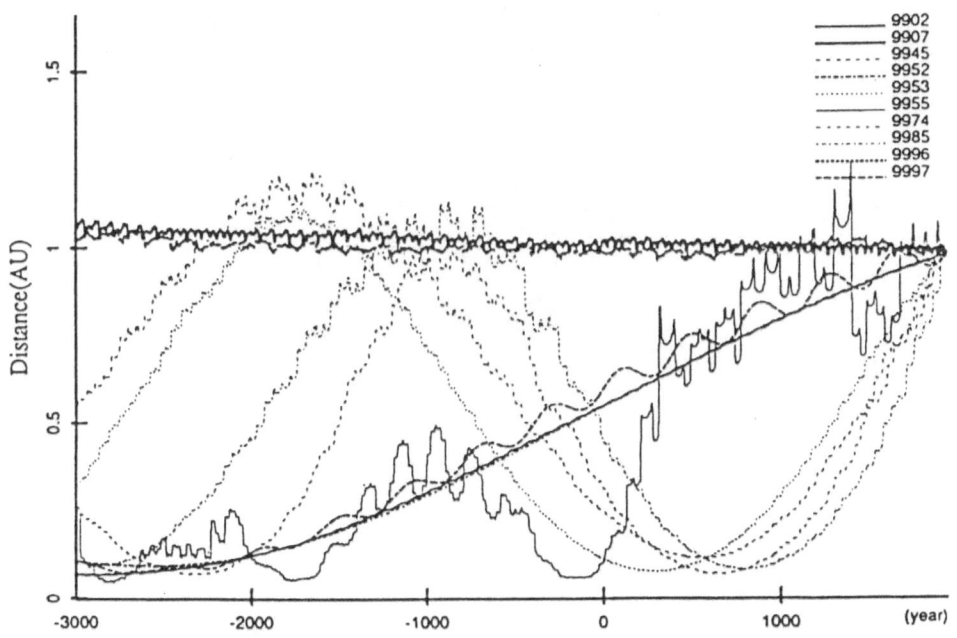

Figure 5.1 Perihelion Distance of the Orbits of the Quadrantids

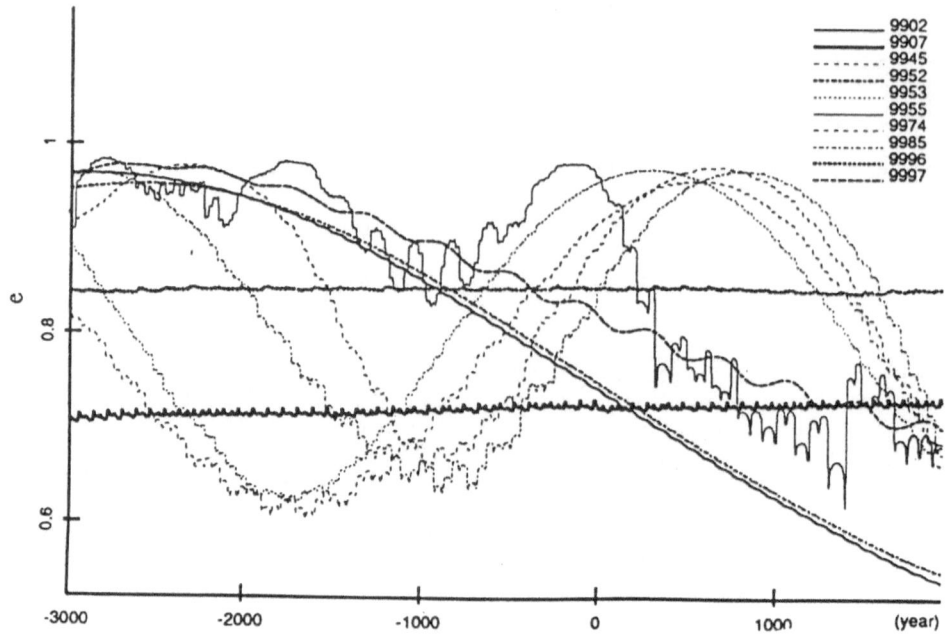

Figure 5.2 Eccentricity of the Orbits of the Quadrantids

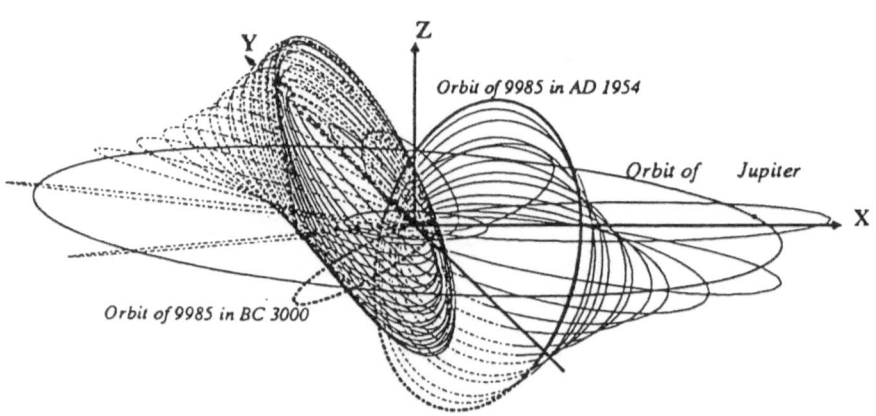

Figure 5.3 Orbital Motion of Quadrantid 9985 from BC 3000 to AD 1954

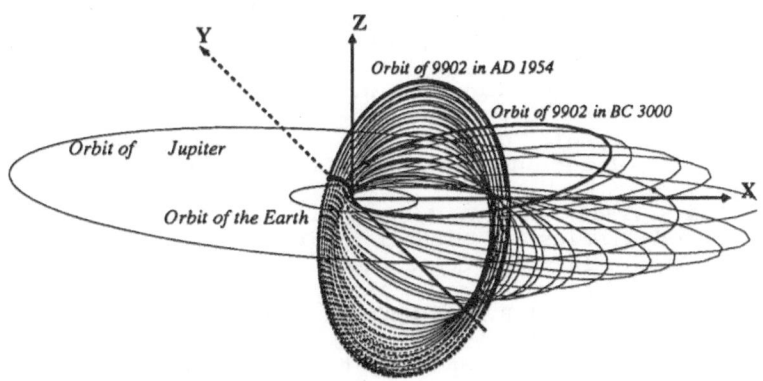

Figure 5.4 Orbital Motion of Quadrantid 9902 from BC 3000 to AD 1954

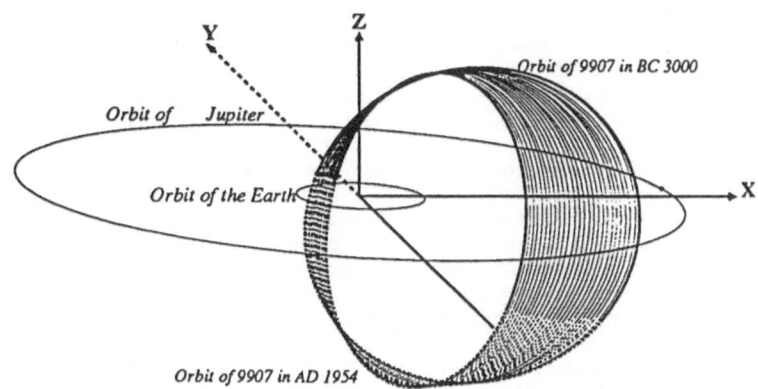

Figure 5.5 Orbital Motion of Quadrantid 9907 from BC 3000 to AD 1954

suggest that the meteors forming trails 9907 and 9996 may not actually be true Quadrantids but part of some other stream or sporadic background. This very discrepant behaviour of meteors whose parameters on impact with the Earth are very similar are strong indicators that chaos is present in this system and the whole system deserves further investigation in this light. It is also possible that by change, these chaotic changes have brought together at this time two different streams, with two different parents and different evolutionary histories.

REFERENCES

Babadzhanov, P. D. and Obrubov, Y.Y.:1983. *Highlights in Astronomy vol 6*, ed West, R.M., D.Reidel Pub Co.
Baines, M.J.,Williams, I.P. and Asebiomo, A.S.:1965. *Mon. Not. R.ast. Soc.*, **130**, 63.
Brouwer, D.:1947, *Astr. Jl.*, **52**, 190.
Brouwer, D. and Clemence, G.M.:1961. *Methods of Celestial Mechanics*, Academic Press, London.
Dormand, J. R. and Prince, P.J.:1978. *Celestial Mechanics*, **18**, 223.
Dormand, J.R., El-Mikkai, M.E.A. and Prince, P.J.:1987, *J. Numer. Anal.*, **7**, 423.

Fox, K.:1984. *Celestial Mechanics.*, **33**, 127.

Fox, K.,Williams, I.P. and Hughes, D.W.:1982. *Mon. Not. R. astr. Soc.*, **200**, 313.

Fox, K.,Williams, I.P. and Hughes, D.W.:1983. *Mon. Not. R. astr. Soc.*, **205**, 1155.

Froeschle, Cl. and Scholl, H.:1985. *Asteroids Comets Meteors II, Ed Lagerkvist, C.-I., Lindblad, B. A.,Lundstedt, H. and Rickman, H.* ,Uppsala Univ. Press.

Gustafson, B.A.S.:1989. *Astron. Astrophys.*, **225**, 533.

Hagihara, Y.: 1972. *Celestial Mechanics vol 2.* MIT, London.

Hughes, D.W.: 1985. *Asteroids Comets Meteors II, Ed Lagerkvist, C.-I., Lindblad, B. A.,Lundstedt, H. and Rickman, H.* ,Uppsala Univ. Press.

Hughes, D. W., Williams, I.P. and Murray, C. D.: 1979, *Mon. Not. R. astr. Soc.*, **189**, 493.

Hughes, D. W., Williams, I.P. and Fox, K.: 1981, *Mon. Not. R. astr. Soc.*, **195**, 625.

Hunt, J., Williams, I.P. and Fox, K.: 1985, *Mon. Not. R. astr. Soc.*, **217**, 533.

Jones, J.:1985. Mon. Not. R.astr. Soc., 217, 523.

Jones, J. and Hawkes, R.:1986. Mon. Not. R. astr. Soc.

Jones, J. and McIntosh, B. A.:1986, Exploration of Comet Halley, 233, ESA SP-250.

Khabaza, I. M.: 1969. *Numerical Analysis*, Pergamon Press, London.

Kresak, L.: 1974. *Bull. astr. Inst. Csl.*, **13** ,176.

Levin, B. Y., Simonenko, A. N. and Sherbaum, L. M.:1972. in *The Motion, Evolution of Orbits and Origin of Comets, ed Chebotarev, G. A. et al.*, D. Reidel Pub. Co.

Lovell, A. C. B.:1954. *Meteor Astronomy*, Oxford Univ. Press.

Plavec, M.:1950. *Nature*, **165**, 362.

Sherbaum, L. M.:1970. *Vestun. Kiev Un-ta. Ser. Astron.*, **12**, 42

McDonnell, J. A. M., Alexander, W. M., Burton, W. M., Bussoleti, E., Evans, G. C., Evans, S. T., Firth, J. G., Grard, R. J. L., Green, S. F., Grun, E., Hanner, M. S., Hughes, D. W., Igenberg, E., Kissel, J., Kuczera, H., Lindblad, B. A., Langevin, Y., Mandeville, J. C., Nappo, S., Pankiewicz, G. S. A., Perry, C. H., Schwehm, G. H., Sekanina, Z., Stevenson, T. J., Turner, R. F., Weishaupt, U., Wallis, M. K.and Zarnecki, J. C.:1987. *Astron. Astrophys*, **187**, 719.

McIntosh, B. A. and Hajduk, A.: 1983. *Mon. Not. R. astr. Soc.*, **205**, 931.

Oliver,C. P.:1925. *Meteors*, Williams and Wilkis Pub Co.

Simpson, I. C., Simons, S. and Williams, I. P.:1979. *Astrophys. Space Science*, **71**, 3.

Whipple, F. L.:1951. *Astrophys. Jl.*, **113**, 464.

Whipple, F. L.:1985. *The Mystery of Comets*, Cambridge Univ. Press.

Whipple, F. L.:1987. *Astron. Astrophys.*, **187**, 852.

Whipple, F. L. and Hamid, S. E. D.:1950, *Sky and Telescope*, **9**, 248.

Williams, I. P.:1989. *Asteroids Comets Meteors III, Ed Lagerkvist, C.-I., H.Rickman, Lindblad, B. A. and Lindgren, M.*, Uppsala Univ. Press.

Yeomans, D. K.:1981. *Icarus*, **47**, 492.

PERTURBATION THEORY, RESONANCE, LIBRATIONS, CHAOS, AND HALLEY'S COMET

P.J. Message

Department of Applied Mathematics and Theoretical Physics
University of Liverpool
United Kingdom

INTRODUCTION

The lectures given at the Advanced Study Institute began with a brief survey of quasi-ergodicity, wildness, and chaos-type phenomena in celestial mechanics, continued with a very brief outline of the development of solar system perturbation theory (itself very briefly indicated here), showing how resonance in orbital period leads to transitions between types of motion, so that, since rational values of the ratio of two orbital periods are everywhere dense, the motions in the solar system will show a complexity in which the eventual character of a particular orbit of the system may be expected to depend very finely on the initial conditions. The lectures ended with a description of some numerical investigations of resonant librations in the orbit of Halley's comet.

WILDNESS AND QUASI-ERGODICITY IN CELESTIAL MECHANICS

Features of dynamical systems, like those now being called "chaos", have been known in celestial mechanics since Hénon and Heiles' (1964) exploratory numerical integrations of orbits in an axially-symmetric gravitational field modelling a stellar system, when interest centred on the possible existence of a third integral of motion in addition to those of energy and angular momentum about the axis if symmetry. On surface of section plots in phase space, some orbits remained confined each to an invariant curve, as would be expected of all orbits if a third integral did exist. Let us call such orbits "regular". However, other orbits of the same dynamical system were found to intersect the surface of section in an apparently random scatter of points throughout that part of the surface not occupied by the invariant curves followed by the regular orbits. Orbits of this latter type came to be called "wild", or "quasi-ergodic". (They cannot be truly ergodic if they do not enter the part of the phase space occupied by the regular orbits.) These results appeared to indicate the existence of the sought-after third integral for some orbits, but not for others, in the same dynamical system, a surprising feature of the space of solutions of a Hamiltonian dynamical system. Similar features have been found in the restricted three-body problem (see, e.g., Wisdom (1985)) in motions corresponding to minor planets near to small-integer commensurability of orbital period with Jupiter, usually for larger values of the eccentricity. Numerical exploration of simplified planetary systems of two planets, whose

Predictability, Stability, and Chaos in N-Body Dynamical Systems
Edited by A.E. Roy, Plenum Press, New York, 1991

masses are taken larger in terms of that of the primary than those of the actual planets Jupiter and Saturn, to accelerate the interaction, not only show the expected acceleration of the mutual perturbations, but also, on passing an apparently well-marked ratio of enlargement of the masses, a transition to wilder behaviour, usually leading to the eventual loss of one of the planets from the system. (Nacozy, 1976.)

For the main planetary system of eight planets moving in nearly-circular, nearly-coplanar orbits of well separated and very nearly constant major axes (which we will call a "regular" planetary system), however, the methods of perturbation theory give expressions for the parameters of the instantaneous Kepler ellipses as multiple Fourier series with arguments which are linear functions of time, which indicate that, within their validity, the planetary orbits will continue to possess these characteristics, that is, will continue to be "regular" in this sense.   These expressions are however asymptotic, but not uniformly convergent (in common of course with the only analytic representation of solutions available for practically all non-linear differential equation systems, that is, for the equations which arise in just about all real situations encountered in physical science) and strict mathematical analysis cannot give realistic error bounds for them for more than very short time intervals.   However, the impressive agreement of predictions based on use of these expressions, not only with observed positions of the planets over some centuries, but also with numerical integrations of the equations of motion of the planets covering several millions of years (see for example, Cohen, Hubbard, and Oesterwinter, (1972 & 1973), and D.Richardson and Walker (1988)) shows that the expressions give correct predictions of the orbital behaviour for very much longer time intervals than those which strict analysis assures, and has led to confidence in the expectation of continued regularity of the planetary orbits, if not indefinitely, then at least for many hundreds of millions of years, and also to the further expectation that "wildness", or chaotic-type behaviour would not be found in a system as regular as that of the major planets.   Testing of this belief by numerical integration, however, meets the difficulty that the unavoidable accumulation of error eventually degrades the solution to the point where no real information about the solution on which the integration began is actually being gained.   The integration of the full equations of motion for the nine planet system by Richardson and Walker (1988) for two million years is probably close to the limit attainable for that system.   For the simplified system comprising the four largest planets and Pluto only, the orbital periods are longer, so that a longer time step may be used, and the integration may be continued for longer before the accumulation of error degrades the solution.   Thus the "Longstop" integration (Roy et al., 1988) covered one hundred million years, and from it Fourier analysis of filtered values of the parameters of the instantaneous Kepler elliptic orbits led to expressions in generally very good agreement with those arising from analytical secular perturbation theories (Carpino et al., 1987; see also Laskar 1988).   There is, however, an indication of "bunching" of frequencies around that of one of the modes of secular variation, that associated with the eccentricity vector of Saturn, showing some frequencies (not yet all identified) appearing to arise from near resonances involving combinations of the secular frequencies, which would give long-periodic terms corresponding to such resonant combinations larger amplitudes than they would have if they were not resonant, and so to appear unexpectedly at the precision being worked to. Of course these are precisely the circumstances, according to one conjectured view, in which "wild" or "chaotic" behavior sets in.   But that is indeed yet conjecture.

The recent results of Laskar (1988 & 1989), which include also the four inner planets, indicate that their addition to the model brings the suggestion of "wildness" in the main planetary system nearer than had been imagined possible.   His earlier work (1984, 1985 & 1986) on the extension to higher orders than previously of the analytical theory of the secular variations of the eccentricity vectors and the vectors giving the orientations of the orbit planes, found that there were so many near-resonances enlarging the terms of high order as apparently to present an impenetrable barrier to carrying the work to smaller than a certain precision.   More recently, he has carried out a numerical integration covering 200 million years, not of the complete equations for the nine planets, but of the averaged equations for the secular variations of the eccentricity and plane-orientation vectors, and finds that his results cannot be fitted to a single multiple Fourier series for more than a small fraction of this time interval.   He also calculates the largest Lyapounoff characteristic exponent, which gives the time scale of the enlargement by a factor e of the separation of initially close solutions, finding this to be about five million years.   This all indicates that there is something like "wildness", at some scale, in the orbits of the inner planets over this time scale, and suggests that the time interval for which a numerical integration of the complete equations of motion of the whole planetary system is carried out could not be increased from that of Richardson and Walker by too great a factor (supposing that the technical difficulties could be overcome) without encountering similar symptoms of possible "wild" behaviour.

This would still leave the breadth of any chaotic zone in the space of appropriate orbital parameters to be determined.   We know that in some special idealised dynamical systems, for example in the context of minor planets in near orbital resonance with Jupiter or other planets, that it can be very narrow, which would not imperil predictability except to precisions finer than its width.   The evidence of the present regular nature of the planetary orbital system suggests that there has been no major upheaval in the planetary orbits so recently as in a time scale of a few million years in the past (a still, quiet pond is not one into which a large stone has just recently been thrown, as Archie Roy has said in a different, but related context, and also, the climate on Earth has clearly not changed that much over hundreds of millions of years), and that in turn suggests that any "wild" motion is confined to a quite narrow belt in the orbital parameter space.

We are however again in the realm of conjecture, but must abandon the confidence of just a few years ago that we really understood all the important phenomena that there were to find in those parts of solar system dynamics of most significance practically , and that, apart from a few special cases where the effect of resonance was obvious, as in the orbits of some minor planets and satellites of Jupiter and Saturn, where qualitatively new phenomena might be found, the task of celestial mechanics, at least as far as the main planetary system was concerned, was essentially the securing of successively greater precision in prediction.   On the contrary, the effects of higher resonances may reach further into the main features of the planetary motions than we thought, over scores or hundreds of millions of years, and a new field in the study of dynamical systems is opening up.

Let us however first turn to the context in solar system perturbation theory in which some of the newly discovered phenomena have arisen, with a very brief review of some of the methods that have in fact proved so very successful in the prediction of planetary movements.

OUTLINE OF SOLAR SYSTEM PERTURBATION THEORY

The position and velocity of a planet or comet at a given instant may be specified by the parameters of the "instantaneous" Kepler ellipse, that is, the ellipse in the unperturbed inverse-square central force for which they are the initial conditions at that instant. A suitable set of such parameters is: a, the major semi-axis, e, the eccentricity, i, the inclination of the orbit plane to the reference plane, Ω, the longitude of the ascending node, ϖ, the longitude of the perihelion, and λ, the mean longitude in the orbit. In the presence of a perturbing force additional to that central force, these parameters will of course change with time, their rates of change being given by Gauss' planetary equations:

$$\frac{da}{dt} = \frac{2}{n}\left(\frac{ae \sin f}{b} S + \frac{b}{r} T\right) ,$$

$$\frac{de}{dt} = \frac{b \sin f}{na^2} S + \frac{r}{nab}\left(2 \cos f + e \cos^2 f + e\right) T,$$

$$\frac{di}{dt} = \frac{r \cos u}{nab} W,$$

$$\frac{d\Omega}{dt} = \frac{r \sin u}{nab} W,$$

$$\frac{d\varpi}{dt} = -\frac{b \cos f}{na^2 e} S + \frac{r \sin f (2 + \cos f)}{nabe} T + \frac{r \sin u \tan (i/2)}{nab} W,$$

and $\quad \dfrac{d\lambda}{dt} = n - \dfrac{1}{na^2}\left[2r + \left(1 - \dfrac{b}{a}\right)\dfrac{b \cos f}{e}\right] S$

$$+ \frac{1}{nabe}\left(1 - \frac{b}{a}\right) r \sin f (2 + e \cos f) T$$

$$+ \frac{r \sin u \tan (i/2)}{nab} W,$$

where S and T are the components of the perturbing force in the radial and transverse directions in the instantaneous orbit plane, that is, the plane defined by the radius vector and velocity, and W is the component of the force at right angles to that plane. Also r is the radial distance from the primary, f is the true anomaly (angle between the near apse direction and the radial direction), $b = a \sqrt{(1 - e^2)}$, n is the mean motion in longitude, (satisfying Kepler's third law $\mu = n^2 a^3$), and $u = \varpi + f$.

If the perturbing force can be expressed as the gradient of a function R (the "disturbing function"), then these rates of change are also given by Lagrange's planetary equations:

$$\frac{da}{dt} = \frac{2}{na} \frac{\partial R}{\partial \lambda} ,$$

$$\frac{de}{dt} = \frac{b(b - a)}{na^4 e} \frac{\partial R}{\partial \lambda} - \frac{b}{na^3 e} \frac{\partial R}{\partial \varpi} ,$$

$$\frac{di}{dt} = -\frac{1}{nab \sin i} \frac{\partial R}{\partial \Omega} - \frac{\tan (i/2)}{nab}\left(\frac{\partial R}{\partial \lambda} + \frac{\partial R}{\partial \omega}\right) ,$$

$$\frac{d\Omega}{dt} = \frac{1}{nab \sin i} \frac{\partial R}{\partial i} ,$$

$$\frac{d\omega}{dt} = \frac{b}{na^3 b} \frac{\partial R}{\partial e} + \frac{\tan (i/2)}{nab} \frac{\partial R}{\partial i} ,$$

and

$$\frac{d\lambda}{dt} = n - \frac{2}{na}\frac{\partial R}{\partial a} + \frac{b(a - b)}{na^4 e}\frac{\partial R}{\partial e} + \frac{\tan (i/2)}{nab}\frac{\partial R}{\partial i}.$$

(The form of these equations shows that they are closely related to a set of canonical equations of Hamiltonian type, and if λ, ϖ, and Ω are taken as canonical co-ordinates, their conjugate momenta are, respectively, $\Lambda = \sqrt{(\mu a)}$, $\Pi = \Lambda\{\sqrt{(1 - e^2)} - 1\}$, and $N = (\Lambda + \Pi)(\cos i - 1)$, with Hamiltonian function $- \mu^2/(2\Lambda^2) - R$.)

Where the perturbation is due to the gravitational attraction of a planet, the disturbing function R may be expressed as a multiple Fourier series in the six angular arguments λ, ϖ, Ω, λ', ϖ', and Ω' (primed symbols denoting Keplerian elements of the perturbing planet), of the form:

$$R = \mu m' \sum K_j \cos N_j$$

where m' is the ratio of the mass of the perturbing planet to that of the primary, $N_j = j_1\lambda + j_2\varpi + j_3\Omega + j_4\lambda' + j_5\varpi' + j_6\Omega'$, and the summation is over all sets $j = (j_1, j_2, j_3, j_4, j_5, j_6)$ with $j_1 \geq 0$ and $j_1 + j_2 + j_3 + j_4 + j_5 + j_6 = 0$. The co-efficients K_j are functions of a, e, and i, as well as of a', e', and i', and their form reflects the d'Alembert property, that R is expressible in terms of a and positive powers of $e \sin \varpi$, $e \cos \varpi$, $\sin (i/2) \sin \Omega$, and $\sin (i/2) \cos \Omega$, and their counterparts $e' \sin \varpi'$, $e' \cos \varpi'$, $\sin (i'/2) \sin \Omega'$, and $\sin (i'/2) \cos \Omega'$, which implies that K_j has the factors $e^{|j_2|}$, $e'^{|j_5|}$, $\sin^{|j_3|}(i/2)$, and $\sin^{|j_3|}(i'/2)$, the remainder of the dependence on e, e', i, and i' being through positive even powers of e, e', $\sin (i/2)$, and $\sin (i'/2)$ only.

The "first-order perturbations" are derived by integrating the Lagrange planetary equations after having, on the right-hand sides of the first five, and all terms but the first "n" in the sixth, substituted for the Keplerian elements (both where they appear explicitly, and where they are implicit through dependence on them of R through K_j and N_j) unperturbed expressions (say a_0 for a, &ce.), that is, constant values for all except λ, for which the unperturbed expression is $n_0 t + \varepsilon$, with n_0 and ε constant. Before integrating the equation for λ, the first term "n" is replaced by its first-order approximation

$$n_0 - \frac{3n_0}{2a_0}\,\delta a,$$

where δa is the first-order perturbation in a. Those terms in the expansion of R for which j_1 and j_4 are not both zero give rise to terms in the first-order perturbations which are periodic, and have denominator $\nu = j_1 n_0 + j_4 n'$ (or its square in the case of part of λ). Those terms which have $j_1 = j_4 = 0$ are called "secular", and are linear in t. It is clear that there are no such terms in the first-order perturbations in the major semi-axis, a. Substitution of the first-order perturbations into the right-hand sides of the Lagrange equations, and retaining terms up to second order in m' leads on integration to the "second-order" perturbations. Further substitution of these leads to the "third-order" perturbations, and so on. This method of solution of the equations for the perturbations in series in powers of m' has come to be called "Poisson's method". "Poisson's Theorem", that the expressions for the first-order and second-order perturbations in the major semi-axes of the planets' orbits contain no secular term, is now known to be true for a system of parameters for which the equations for the whole planetary system can be given by a single Hamiltonian function, for example if Jacobi's system of relative position vectors is used. (See Duriez 1978). It is now also known that the entire formal solution for the perturbations of the Kepler elements can be expressed entirely in periodic terms, some of the periods being of the order of millions of years, as arise in the theory of the secular variations. (Message (1982a).)

NEAR COMMENSURABILITY AND RESONANCE

The appearance of the denominator $\nu = j_1 n + j_4 n'$ in periodic terms indicates that Poisson's method may not be straightforward in the case of resonance between the mean motions n and n'. If $n:n' \simeq p:p+q$, where p and q are integers, then, for all terms with $j_1 = kp$ and $j_4 = k(p+q)$, where k is also an integer, this denominator is small, and the corresponding terms in the first-order perturbations are large (and, in perturbations of successively higher order, successively larger). The most notable example amongst the major planets involves Jupiter and Saturn, whose orbital periods are near in ratio to 2:5, and so whose mean motions are near in ratio to 5:2. The corresponding large perturbations, largest in the longitudes because of the denominators ν^2, are called the "great inequalities". The near-commensurability is not however in this case so close as to prevent Poisson's method from giving a serviceable representation of the perturbations. This situation we call "shallow resonance".

In the case of the mutual perturbation of two planets, only those near commensurabilities in which the integers p and q are small are in practice of importance, for the following reason. The eccentricities, e, and orbital inclinations, i, of the planetary orbits are all small. The co-efficient K_j has, from the d'Alembert property, the factor

$$e^{|j_2|} e'^{|j_5|} \sin^{|j_3|}(i/2) \sin^{|j_6|}(i'/2),$$

and this is smaller than $\kappa^{|j_2|+|j_5|+|j_3|+|j_6|}$, where κ is the greatest of e, e', $|\sin(i/2)|$, and $|\sin(i'/2)|$. Also

$$|j_2|+|j_5|+|j_3|+|j_6| \geq |j_2+j_3+j_5+j_6| = |j_1+j_4|,$$

since $j_1+j_2+j_3+j_4+j_5+j_6 = 0$. So although any ratio n/n' may be approximated arbitrarily closely by a ratio of integers, if in fact the corresponding integers j_1 and j_4 are large, then the co-efficient K_j will have a correspondingly very small factor. Consequently in practice difficulties in the use of Poisson's method to represent the perturbations in planetary orbits only arise in cases of near commensurability of orbital period in which the integers p and q are small. However, in the case of cometary orbits, for which the eccentricity is not usually small, this mechanism will not be so effective, and near commensurabilities of orbital period involving larger integers might well prove to be significant.

The case in which the near-commensurability of orbital periods is so close that Poisson's method fails to provide a representation of the actual motion we shall call "deep resonance", and other methods are necessary to study the motion. The angular argument N_j of any term for which ν is small is called a "critical argument". The "long-period" problem is one in which there have been removed from the disturbing function, by a Lie series or equivalent formal averaging transformation given by a generating function in the form of an asymptotic series, those terms ("short-periodic" terms) whose arguments contain at least one of λ and λ' and are not critical arguments (see, e.g., Message, 1988).   Consider first the simplified system known as the restricted problem of three bodies, that is, the study of the motion of a body of negligible mass moving under the perturbation of just one other planet, used to model the perturbation by Jupiter (or another single planet) of a minor planet or comet.   Simplified still further is the circular restricted problem, in which the perturbing planet moves in a circular orbit, and further simplified still is the planar restricted problem in which all the motion is confined to a single plane.   In the latter problem, a near commensurability $n{:}n{+}1{\approx}p{+}q{:}p$ will give rise to a single sequence of critical arguments, which are all multiples of a single basic critical argument $\vartheta = (p{+}q)\lambda - pl' - q\tilde{\omega}$.   In the long-period problem, ϑ is the only non-ignorable co-ordinate, so that this problem is then essentially of one degree of freedom only, and its phase-space is a plane, in which the solution curves may be classified, just as in the problem of the simple pendulum, as either "circulating", that is, in which ϑ either always increases or always decreases, or as "librating", in which ϑ oscillates about a mean value.   The fixed points in this phase plane about which the libration of ϑ takes place correspond to periodic solutions of the full problem before averaging.   (See, e.g., Message, 1966.)   But in a more realistic problem, there will be more than one basic critical argument, even for the same pair of integers p and q (there are two in the planar elliptic restricted problem, in which the disturbing planet moves in an ellipse, and in the general three-body problem, with two planets, each of finite mass, there are, in the plane case, two different critical arguments, $(p{+}q)\lambda - p\lambda' - q\tilde{\omega}$ and $(p{+}q)\lambda - p\lambda' - q\tilde{\omega}'$). There are still periodic solutions, in which each critical argument takes a fixed value in the long-period problem, and there are also solutions in which the critical arguments oscillate about these fixed values. However, classification of such solutions into circulating or librating for each critical argument is not possible, since a single solution may have intervals of time during which a particular critical argument is oscillating, and other intervals during which it is increasing or decreasing through complete revolutions.   (See Message, 1982b and 1984.)

As already noted, in the case of the orbit of a comet, since the eccentricity is usually not particularly small, the d'Alembert property does not prevent terms in the disturbing function, which correspond to quite large values of p and q, being significant.   Consequently, in the case of near commensurability of orbital period, there can be significant critical arguments corresponding to quite large values of p and q.   This makes it more likely that, at any one time, there maybe more than one critical argument playing a significant part in the motion.   This is relevant to the investigations of some features of the motion of Halley's comet, about to be described.

RESONANCE AND LIBRATIONS IN HALLEY'S COMET

The orbital period of Halley's comet has fluctuated in the vicinity of about 75 years since identified observations of it have been made, that is, since 240 B.C., and, for most of that time, the orbital period has appeared to oscillate, the cycle of oscillation having a period of between 700 and 800 years. (An attempt to predict the date of the 1910 return using harmonic analysis of the dates of previous returns was made by Angstrom (1862).)   It was shown by Kiang (1973) that this oscillation is related to the 13:2 near commensurability of orbital period with Jupiter (and not to perturbation by any undiscovered planet).   In fact the critical argument $\vartheta = 13\lambda - 2\lambda' - 11\tilde{\omega}$ was librating between a time near the beginning of the A.D. era and the mid 17th century, since when it has been increasing.   The oscillatory behaviour of the orbital period ceased at about the same time at which the critical argument ϑ ceased to librate.   The close approach of the comet to Jupiter in 1681 was noted by Halley as likely to cause perturbation of the date of the next return (see review in the introduction to the paper by Yeomans (1977)). Chirikov and Vecheslavov (1989) and Dvorak and Kribbel (1990), using a simplified mapping, have investigated chaotic-type behavior over several hundreds of thousands of years.   Let us concern ourselves now, however, with behaviour over a much shorter time scale, using a more detailed and realistic model of the motion.

A numerical integration of a model of the orbit from 313 B.C. to 2420 A.D. has been carried out, including the perturbations of the planets from Venus to Neptune (their orbits being represented by power series in the time, as given by Seidelmann, Dogett, and DeLuccia (1974)), to investigate the perturbations of the Kepler elements, and the behaviour of ϑ and other critical arguments of the form $(p+q)\lambda - p\lambda' - q\bar{\omega}$.   As expected, the larger changes in the elements occur near the perihelion passages, in response to close encounters with planets.   The argument with $p = 3$ and $q = 16$ begins to librate near to 1681 and continues to do so throughout the remainder of the integration; that with $p = 5$ and $q = 27$ librates between 1455 and 1835; that with $p = 5$ and $q = 28$ from 1066 to 1301; that with $p = 7$ and $q = 38$ between 684 and 989; that with $p = 7$ and $q = 39$ between 296 and 607, and again from 1066 to 1455; that with $p = 9$ and $q = 49$ between 530 and 1066; and that with $p = 9$ and $q = 50$ between 1066 and 1145.   Each of these dates of transition corresponds to an encounter with Jupiter to within about one astronomical unit, except 607, when there is an encounter with Venus to within 0.08 a.u., and 1145, when there is an encounter with the Earth to within 0.22 a.u.   It is seen that there are intervals of time when more than one critical argument is librating.

Further integrations were carried out, with slightly different initial conditions, to investigate the stability of the libration of ϑ against such changes.   It is found that increasing the initial major semi-axis by only 0.005 a.u., leaving the other elements unaltered, changes the motion so that the libration of ϑ, while beginning at the same time as in the actual motion, does not however cease in the 17th century, but continues throughout the integration, while decreasing the major semi-axis by 0.02 a.u. leads to the libration not even beginning.   Increasing the initial eccentricity by 0.00025 also leads to a motion in which the libration begins as in the real motion, but continues throughout the integration, while an increase of 0.001 causes it to end once more at about the same time as in the actual motion. Decreasing the initial eccentricity by 0.001 gives a motion in which the libration does not occur.   Moving the apse from which the integration starts backwards through 0.1 degrees gives a motion in which the libration starts as in reality, and indeed ends in about 1681, but is followed by retrograde circulation of the argument ϑ, that is, circulation in the opposite sense to that followed before the interval of libration.   These investigations will be described in more detail in a further paper.

Since the transitions between circulation and libration of many critical arguments, and the consequent changes in the pattern of the perturbations of the orbital period and other orbital parameters, are each very sensitive to the initial conditions of the motion, and since even slight changes in the orbital period in particular will greatly alter the circumstances of close approaches to planets, and hence the consequent perturbations and subsequent motion, it is clear that the predictability of the motion after only a few millennia will be very severely limited by uncertainties in the initial conditions, and also that the pattern of the motion will be subject to frequent transitions of type.   These causes will lead to the motion displaying some of the phenomena of "chaos", even over this time scale.

REFERENCES

Ångstrom, A.J., 1862, Actes de la Soc. Roy. de Sci. d'Uppsala, Ser.III, t.IV, pp.1–10.

Carpino, M., Milani, A., and Nobili, A.M., 1987, Astron. & Astrophys., vol.181, pp.182–194.

Chirikov, B.V., and Vecheslavov, V.V., 1989, Astron. & Astrophys., vol.221, pp.146–154.

Cohen, C.J., Hubbard, E.C., and Oesterwinter, C., 1972, Astronomical Papers for the American Ephemeris, vol.22, Part 1.

Cohen, C.J., Hubbard, E.C., and Oesterwinter, C., 1973, Celest. Mech., vol.7, pp.438–448.

Duriez, L., 1978, Astron. & Astrophys., vol.68, pp.199–216.

Dvorak, R., and Kribbel, J., 1990, Astron. & Astrophys., vol.227, pp.264–270.

Hénon, M, and Heiles, C., 1964, Astron. J., vol.69, pp.73–79.

Kiang, T., 1973, M.N.R.A.S., vol.163, pp.271–287.

Laskar, J., 1984, "Théorie générale planétaire: éléments orbitaux des planètes sur un million d'années", Thèse de troisième cycle, Observatoire de Paris.

Laskar, J., 1985, Astron. & Astrophys., vol.144, pp.133–146.

Laskar, J., 1986, Astron. & Astrophys., vol.157, pp.59–70.

Laskar, J., 1988, Astron. & Astrophys., vol.198, pp.341–362.

Laskar, J., 1989, Nature, vol.338, pp.237–238.

Message, P.J., 1966, Proc. I.A.U.Symp. 25, pp.197–222.

Message, P.J., 1982a, Celest. Mech., vol.26, pp.25–39.

Message, P.J., 1982b, in "Applications of Modern Dynamics to Celestial Mechanics and Astrodynamics" (ed. V.Szebehely, Reidel), pp.77–101.

Message, P.J., 1985, in "Stability of the Solar System and its Minor Natural and Artificial Bodies" (ed. V.Szebehely, Reidel), pp.193–199.

Message, P.J., 1988, in "Long-Term Dynamical Behaviour of Natural and Artificial N-Body Systems" (ed. A.E. Roy, Kluwer), pp.47–72.

Milani, A., Nobili, A.M., and Carpino, M., 1987, Astron. & Astrophys., vol.172, pp. 265–279.

Nacozy, P., 1976, Astron. Journ., vol.81, pp.787–791.

Richardson, D.L., and Walker, C.F., 1988 Bulletin of Amer. Astron. Soc., vol.20, p.901, and "Astrodynamics 1987" (proc. of AAS/AIAA Astrodynamics Conference at Kalispell, Montana, August 1987), pp.1473–1495.

Roy, A.E., Walker, I.W., Macdonald, A.J., Williams, I.P., Fox, K., Murray, C.D., Milani, A., Nobili, A.N., Message, P.J., Sinclair,A.T., and Carpino, M., 1988, Vistas in Astronomy, vol.32. pp.95–116.

Seidelmann, P.K., Doggett, L.E., and DeLuccia, M.R., 1974, Astron. Journ. vol.79, pp.57–60.

Wisdom, J., 1985, Icarus, vol.63, pp.272–289.

Yeomans, D., 1977, Astron. Journ., vol.82, pp.435–440.

ROTATIONAL BEHAVIOUR OF COMET NUCLEI

P. Oberti[1], E. Bois[2], and C. Frœschlé[1]

Observatoire de la Côte d'Azur
[1] *Le Mont Gros, B.P. 139, F-06003 Nice Cedex, France*
[2] *Av. N. Copernic, F-06130 Grasse, France*

Abstract. Numerical experiments of the rotational behaviour of comet nuclei have been performed, including the Sun and Jupiter's disturbing torques in the models. In a stable configuration, the solar torque induces great librations that remain unchanged along the orbit. A close approach with Jupiter can result in great changes on the rotational pattern because of the motion sensitivity to initial conditions. The unstable configuration is characterized by great librations of the nutation angle, and the existence of a possibly large chaotic zone in the phase space.

1. Introduction

Comet orbital motions are often observed, at least partially. Concerning their rotation motions, the only available data come from comet Halley's last appearance (Festou *et al.* 1987, Abergel and Bertaux 1990). Indications of the comet oscillating behaviour appear in the literature (Abergel and Bertaux 1990, Belton 1990), and a numerical model of its rotation has been built by Peale and Lissauer (1989). These new data and results provide some realistic information for a general dynamical qualitative study of the rotational motion for cometary-type bodies.

Forces that create torques may be split in two groups, gravitational and non-gravitational. Only gravitational perturbing effects are concerned here. However, nucleus rotation is assumed to greatly influence solar exposure and thus gas ejection, or more generally, non-gravitational forces (Peale and Lissauer 1989). Non-gravitational effects have to be taken into account for a complete description of the motion.

The comet is modeled by an ellipsoid with three semi-axes $a \times b \times c$. Equations of the rotational and orbital motions are numerically integrated. Two different angle sequences (3-1-3 and 1-2-3) are used, when integrating the motion, to prevent singularities (Bois 1986). The results are shown using Euler angles ψ, θ, and ϕ. For every figure, the ϕ-angle is cleared out of its mean rotation. In the stable case, two scenarios are shown: at first, Halley's orbit is taken as a test comet orbit. Then, a particular orbit encompassing a close approach with Jupiter is used.

Predictability, Stability, and Chaos in N-Body Dynamical Systems
Edited by A.E. Roy, Plenum Press, New York, 1991

2. A Halley-like comet

Different parameters (comet shape or orientation in space) influence the rotational motion properties. Halley's orbital parameters are used for these experiments (Peale and Lissauer 1989). The asymmetry of the comet determines the solar torque action, creating oscillations around the mean rotations. In the well known case of a symmetrical comet nucleus, the physical librations present only one short period of modulation. When the asymmetry increases, a second libration modulation appears. Its amplitude and period depend on the comet shape (Bois *et al.* 1991). Librations also depend on the comet initial orientation. With $\psi_0 = \theta_0 = \phi_0 = 0.1$ deg, $P_{rot.} = 2.2$ days, $P_{nut.} = 0$, $P_{prec.} = 7.4$ days, and the comet shape given by $8 \times 4 \times 3.4$ km, the three libration periods are equal to the initial proper rotation period (2.2 days). The amplitudes are of order 13 deg for ψ and ϕ, and near 0 deg for θ. When θ_0 increases, the amplitudes increase to a maximum that is reached for $\theta_0 = 90$ deg: 15 deg for ψ and ϕ, and now 12 deg for θ. For larger values of θ_0, the amplitudes decrease but the periods keep on increasing up to a maximum value, where they show about twice the initial value (4.4 days), for $\theta_0 = 180$ deg. For this value, the libration amplitudes for θ are 0 deg again (Bois *et al.* 1991).

For comet Halley's case (Peale and Lissauer 1989), initial conditions are the same as above, except $\psi_0 = -22.5$ deg, $\theta_0 = 20$ deg. The physical librations present a double modulation. Their amplitudes remain unchanged along the orbit, of order 15 deg for ψ and ϕ, and 3 deg for θ. (fig. 1).

3. Jovian close approach

The comet is assumed to be moving on an inner orbit in Jupiter's mean plane. Such comets can be found, their existence being explained by an aphelion – perihelion shift mechanism (Rickman and Frœschlé 1988). A close approach is a crucial phenomenon for comet diffusion process, rendering a comet visible, or ejecting it. Comet parameters are chosen in such a way that the close encounter occur near the aphelion of both the comet and Jupiter, for different minimum distances between the two bodies (Bois *et al.* 1991).

The most interesting results appear when the encounter distance is planned to occur at 0.01 AU. Here, the comet is ejected from an almost keplerian orbit two days before the anticipated aphelion. The orbital motion undergoes a brief but strong impulse. Figures 2 and 3 show two different resulting motions, depending on the particular comet orientation at the encounter moment. The variations are limited compared to the changes on the orbital motion, but the second case shows that the rotational pattern can be largely modified. After the close encounter, the comet rotational motion keeps the new features, like new initial conditions after an impact. In fact, the motion is greatly sensitive to particular sets of initial conditions. Then, Jupiter naturally plays the trigger role.

4. Unstable case: the rocking comet

The test comet is on a Halley-type orbit. The rotation is initially applied around the intermediate axis. The comet shape is given by $8 \times 1 \times 4$ km. As shown on figures 4 and 5, the comet presents a very large libration of its nutation angle. For θ_0 near 0 deg, the libration amplitude is 180 deg, leading to successive turnovers for the comet (fig. 4). For $\theta_0 = 120$ deg, the libration amplitude is smaller,

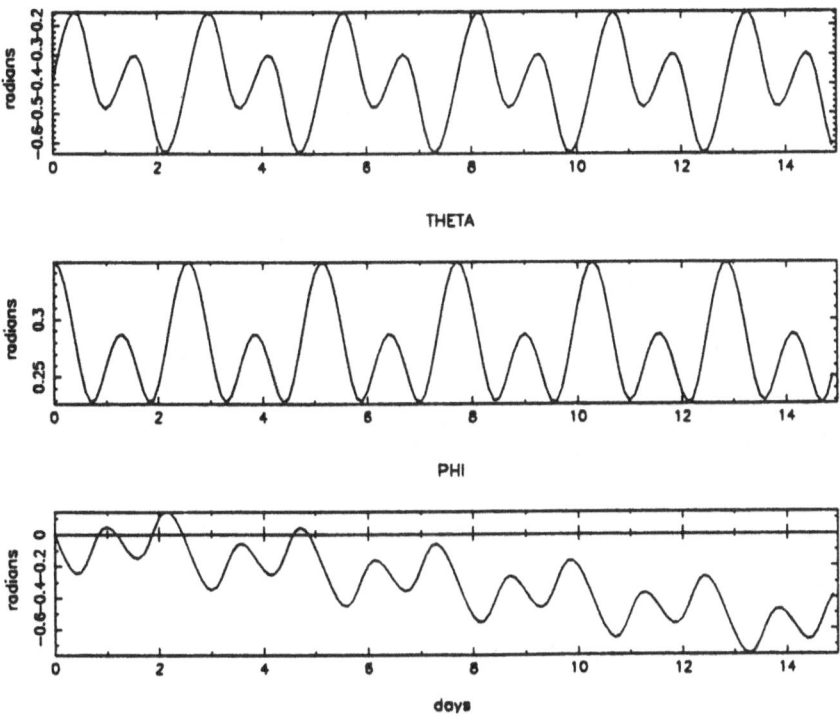

Fig. 1 . Comet Halley

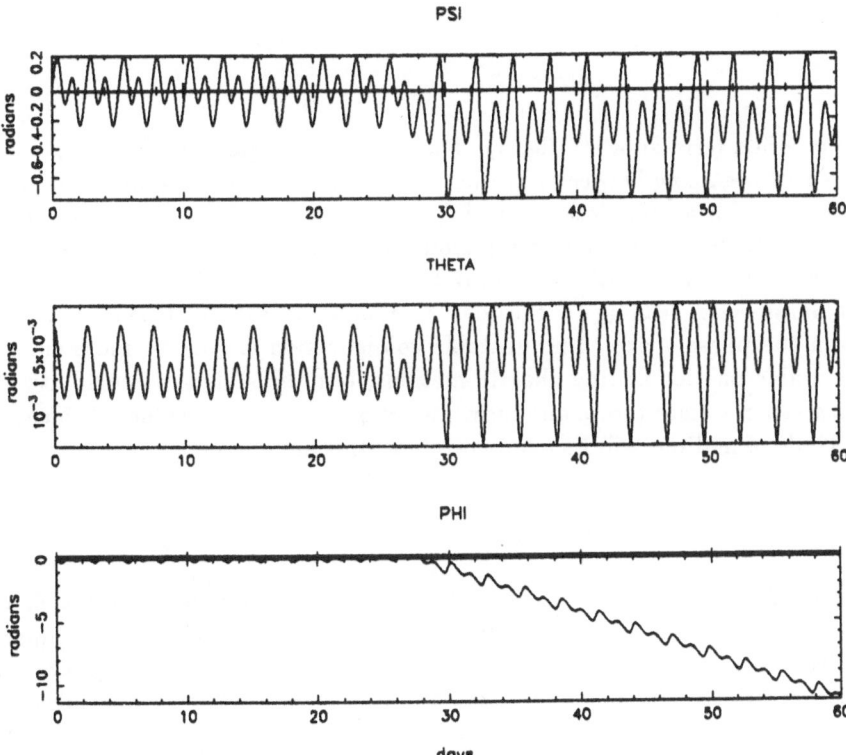

Fig. 2 . Jovian close approach (30th day)

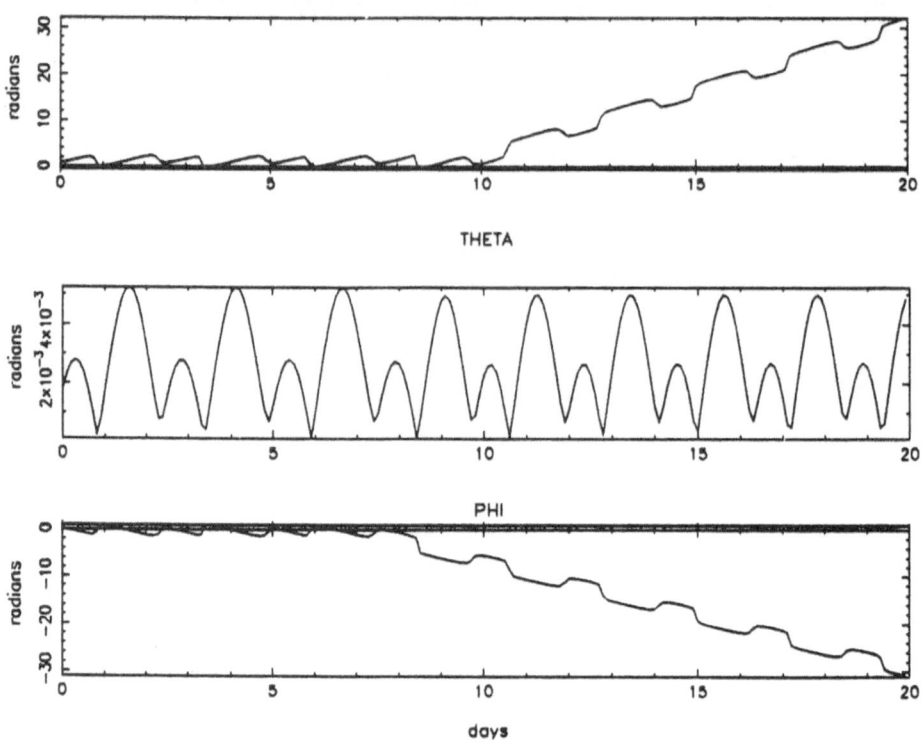

Fig. 3 . Jovian *exciting* effect (10th day)

but a long period term appears (fig. 5). The precession angle shows a very peculiar pattern, a sort of mixing between libration and circulation modes, that could pretty well be chaotic. For a very close value of θ_0, for instance 120.1 deg, the numbers of successive jumps during the circulation modes are different.

Such curves, with high sensitivity to initial conditions, can be found for many sets of input parameters, leading to the idea of a possibly extended zone of chaotic motions in the phase space. The unstable configuration is may be more suitable than the stable one for finding chaotic motions, but the stable case exhibits also curves deserving a more thorough study. Investigations on a possible stochasticity of these kinds of motions will be exposed in a forthcoming paper.

5. Conclusion

In stable position, the solar torque induces great librations on comet nuclei, and Jupiter's close approach leads to a limited change on the rotational motion compared to the orbital one. But, depending on initial conditions, the rotational features can be largely perturbed. The unstable case shows very large libration amplitudes, and the comet rotational behaviour could be chaotic for particular initial conditions. For a more realistic model, non-gravitational effects have to be inserted. This point and the stochasticity of the motion are still under study.

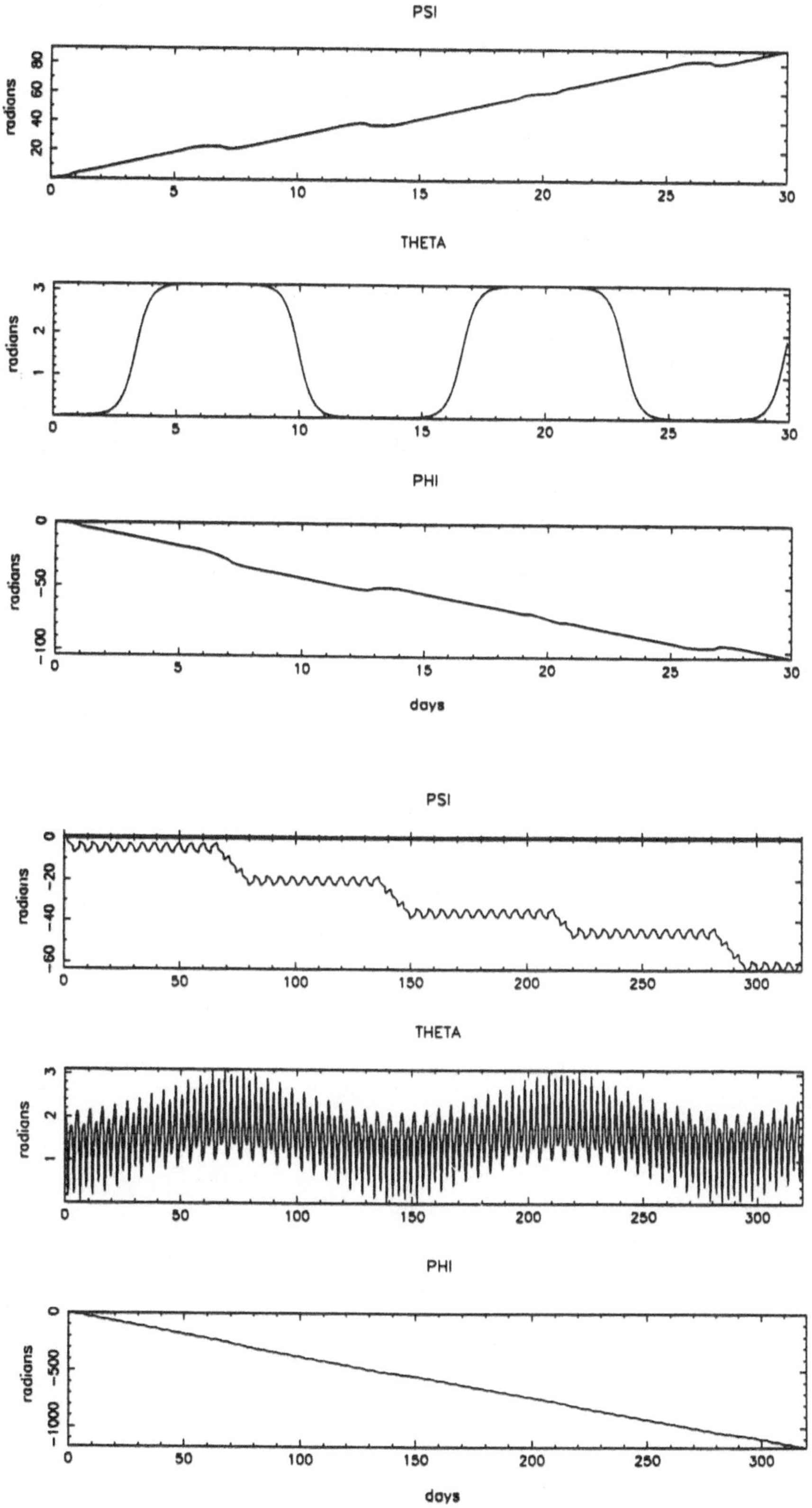

PSI

THETA

PHI

days

PSI

THETA

PHI

days

Fig. 5 . A possibly chaotic pattern

References

Abergel, A., Bertaux, J.L., 1990, "Rotation states of the nucleus of comet Halley compatible with spacecraft images", *Icarus* **86**, 21-29.

Belton, M.J.S., 1990, "Rationalization of comet Halley's periods", *Icarus* **86**, 30-51.

Bois, E., 1986, "First-order theory of satellite attitude motion - Application to Hipparcos", *Celestial Mech.* **39**, 309-327.

Bois, E., Oberti, P., Frœschlé, C., 1991, "Physical librations of comet nuclei", *Celestial Mech.*, submitted.

Festou, M.C., Drossart, P., Lecacheux, J., Encrenaz, T., Puel, F., Kohl-Moreira, J.L., 1987, "Periodicities in the light curve of P/Halley and the rotation of its nucleus", *Astron. Astrophys.* **187**, 575-580.

Peale, S.J., Lissauer, J.J., 1989, "Rotation of Halley's comet", *Icarus* **79**, 396-430.

Rickman, H., Frœschlé, C., 1988, "Cometary dynamics", Review paper, *Celestial Mech.* **43**, 243-263.

PART III

DYNAMICS OF NATURAL AND
ARTIFICIAL SATELLITES

THE MOON'S PHYSICAL LIBRATIONS
Part I: Direct Gravitational Perturbations

I. Wytrzyszczak[1] and E. Bois[2]

[1]ASTRONOMICAL OBSERVATORY OF A. MICKIEWICZ UNIVERSITY
UL. SLONECZNA 36, 60-286 POZNAŃ, POLAND

[2]OBSERVATOIRE DE LA CÔTE D'AZUR, DÉPARTEMENT CERGA
AVENUE COPERNIC, 06130 GRASSE, FRANCE

Abstract: An accurate Moon's rotation model has been performed by numerical integration. Direct gravitational perturbations on the Moon's rotational motion have been analysed. Their resulting librations are presented in this paper including complete physical librations, planetary effects and Earth-Moon figure-figure interactions.

1. Introduction

The great accuracy and accumulation of Earth-Moon's distance measurements collected by the Lunar Laser Ranging Stations permit to improve the Earth-Moon system knowledge, i.e. dynamics and physics of the Moon, Earth's rotation parameters and so on. However, in order to analyse these measures, to discorrelate the dynamical mechanisms they intrinsically contain, to understand and explain these mechanisms, to delimit the internal structure of the Moon etc., a theoretical model is needed, i.e. Dynamical Ephemeris of the Moon conform to the accuracy of the observations. Such a model is currently constructed in the Observatoire de la Côte d'Azur for the purposes of the Lunar Laser Ranging Station of Grasse (Veillet et al., 1989). In this paper only the equations of the Moon's rotational motion are numerically integrated while the positions of the bodies, Earth, Sun and planets are stemming from the JPL DE301 ephemeris. The model is in accordance with the requirements of observational accuracy and controled in function of different perturbations and physical parameters taken into account.

The lunar laser technique permits to measure the distance between a terrestrial station and a lunar reflector with a present range uncertainty of 3 cm accuracy for the three last years of observations and 1 cm on one night. The smallest amplitudes of lunar librations that can be then detected are respectively 0.01 and 0.001 seconds of arc (Veillet and Bois, 1990). So, all perturbations whose libration effects are at least of order 10^{-4} seconds of arc in amplitude have been included in the model. The analyses of lunar laser ranging data indeed keep requiring very accurate calculations of the lunar physical librations (Williams et al., 1973).

Forces that create torques acting on the rotational motion of the Moon are generally split into two groups, gravitational and non-gravitational. Gravitational disturbing effects can be produced by the Earth, the Sun and the planets, as point sources, acting on the gravity field of the Moon, or by the interactions between the figures of the Earth and Moon. Other effects, gravitational and non-gravitational, are caused by the fact that the Moon is non-rigid, and therefore, in connexion with its internal structure, different kinds of deformations have to be taken into account.

2. Conventions

A rotational motion in space can be described by three basic angular rotations used to locate a body-fixed rotating system of coordinates $(Oxyz)$ with respect to a fixed reference frame $(OXYZ)$, both with the origin at the center of mass. All curves are presented here in the classical sequence 313 of Euler angles according to the following decomposition and notation: the precession ψ (around a fixed axis OZ, from a reference axis OX), the nutation θ (around an intermediary axis pointing towards the ascending node and representing the inclination of the Oz body-fixed axis with respect to the OZ-axis), and the rotation ϕ (around Oz), generally called the *proper* rotation, conventionally understood as the rotation of greatest energy (Bois *et al.*, 1990, 1991); the axis of inertia around which it is applied representing then the axis of *figure*. However, in order to avoid singularities in the equations, another type of sequence of angles is also used in the computational process. It is the matter of the 1-2-3 sequence; numbers 1,2 and 3 refer respectively to the axes x or X, y or Y, z or Z. Shifting rules can be found in Bois (1986).

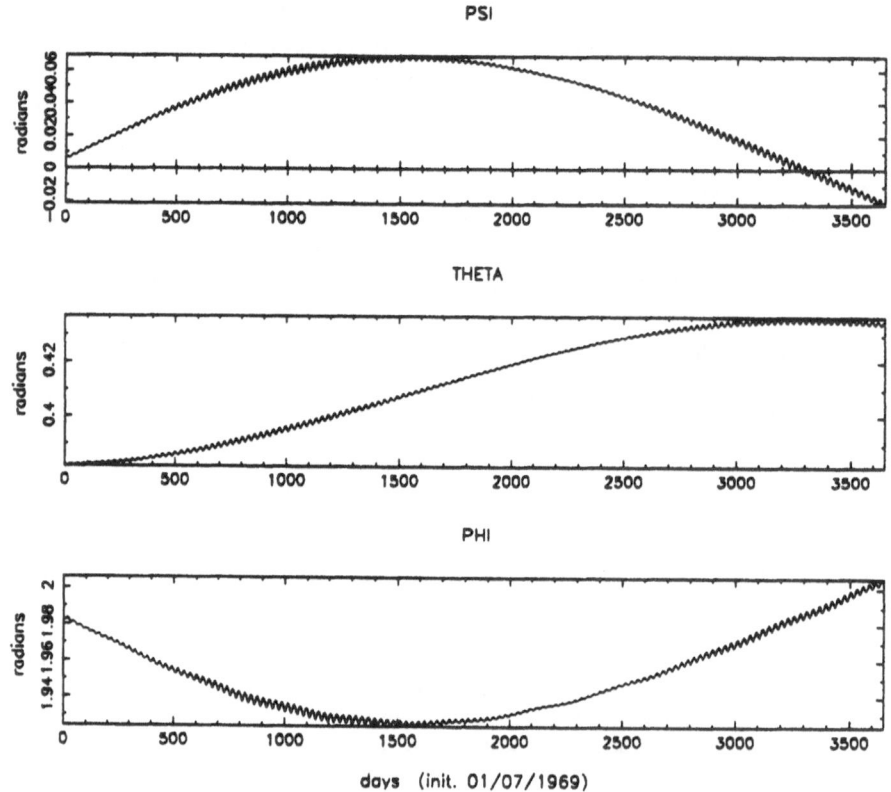

Fig. 1. Moon's physical librations

3. Physical librations

Figure 1 presents the Moon's physical librations plotted on a period of 10 years. Regarding the short periods, we see increasing and decreasing oscillations around a mean fictitious motion defined by the empirical laws of Cassini; the ϕ angle has been cleared out of its mean rotation. Their libration amplitudes are of the order of 500 seconds of arc in ψ and ϕ, of the order of 250 seconds of arc in θ, and that with the short period of 27.3 days. The main part of the disturbing torque is of course produced by a point source Earth acting on the Moon's gravity field reduced to three oblateness coefficients. The magnitude of the torque due to the gravitational action of the Sun is about 100 times smaller. The amplitudes of the long period librations are then of the order of 10 seconds of arc (fig. 2). On figure 1 too, in the ϕ angle curve, it is possible to discern an additional slight modulation of one year period due to the Sun.

4. Direct planetary effects

The effect of planets is almost negligible. The largest one is due to Venus (fig. 3), but only in moments of particular positions the torque is large enough to induce librations of the order of 10^{-3} seconds of arc. Action of Jupiter alone is negligible from the point of view of the present accuracy of observations, but the sum of such effects with, for example, several planets in particular conjunction is significant for long periods of time. The resulting librations are then of the order of 10^{-3} seconds of arc (fig. 4) where we infer by comparison to figure 3 other planetary interventions.

5. Earth-Moon figure-figure interactions

Let us now consider the librations directly connected to the figure of the Moon. In a first step, we assume Earth as a point mass while the gravity field of the Moon is developed through a spherical harmonic representation. Such expansions in term of torques and up to the fourth order can be found in Eckhardt (1981). The main disturbing effect is naturally connected to the spherical harmonics of second order; nevertheless the effects generated by the third order harmonics ("Moon3", fig. 5) are yet very important. For example, the mean libration amplitude for the ϕ angle is of about 200 seconds of arc, but with a 1050 days period i.e. about 2.9 years. The same behaviour as in "Moon3" occurs in "Moon4" (fig. 6), but the effect is a 100 times smaller; the amplitude in ϕ is now of the order of 2 seconds of arc, still with the same period of 2.9 years. Taking into account the magnitude of the ratio "Moon3/Moon4", permits to deduce that harmonics of fifth order ("Moon5") could have a significant action on the rotational motion. Expanding the Moon's potential to the fifth order we see that the libration amplitude in ϕ is of the order of 0.02 seconds of arc with the same period of 2.9 years (Bois and Wytrzyszczak, 1991). Figure 7 presents effects of "Moon5" for a period of 10 years.
In a second step the Earth is no longer considered as a point mass and we study the effects of the resulting mutual gravitational potential. Starting from the general and rigorous expansion of Borderies (1978) for the mutual potential energy of N solid bodies, the torques are developing for the case of a rotating Moon perturbed by gravitational attraction of a flat Earth (Bois and Wytrzyszczak, 1991). Figure 8 shows librations generated by interactions between the J_2 zonal harmonic of the Earth and all the Moon's spherical harmonics of second order ("E2M2"). The order of magnitude of these effects is more or less the same as for the effects of "Moon5", but the global behaviour is rather similar to "Moon2" (fig. 1). Interactions between the J_2 of the Earth and all the lunar spherical harmonics of third order ("E2M3", fig. 9) produce effects of amplitude 10 times smaller in ψ and ϕ and a 100 times smaller in θ than in the last case with "E2M2". However, the general behaviour is rather close to that of "Moon3", "Moon4" and "Moon5" and we find again the same period of 2.9 years in the ϕ angle. Let us notice the frequent presence of periods near 2.89 years, i.e near the adopted 2.8912 year resonance mentioned by Eckhardt (1982) for the free libration in longitude. It is the ordinary resonant frequency for physical librations in longitude (Newhall et al., 1983).

6. Conclusion

Because of the requirements of the observational accuracy some improvements have to be added to the theoretical models. First, the torques due to the figure of the Moon are included up to the fifth order harmonics. Second, mutual potential effects between the figure of the Moon and the figure of the Earth are expanded farther up, even to the fifth order ("E2M3"). It is also necessary to take into account the direct action of planets, whose effects are very small but not always negligible.

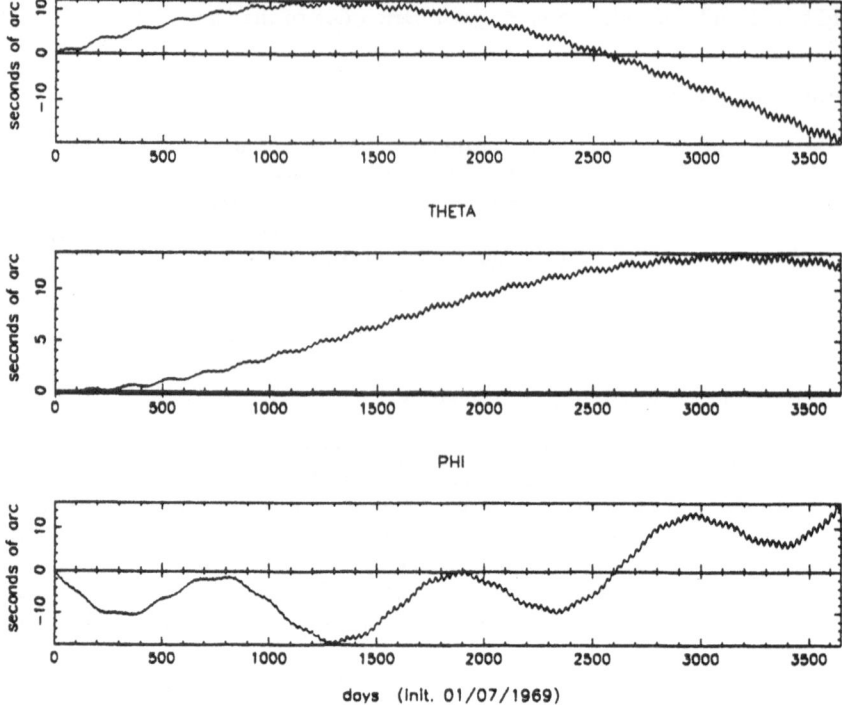

Fig. 2. Effect of the Sun

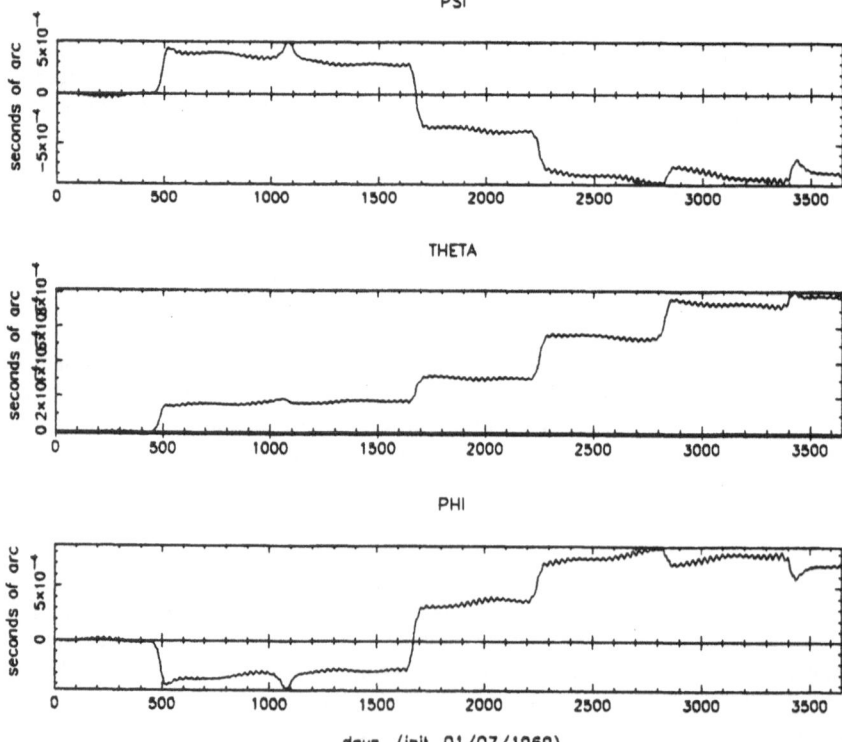

Fig. 3. Effect of Venus

260

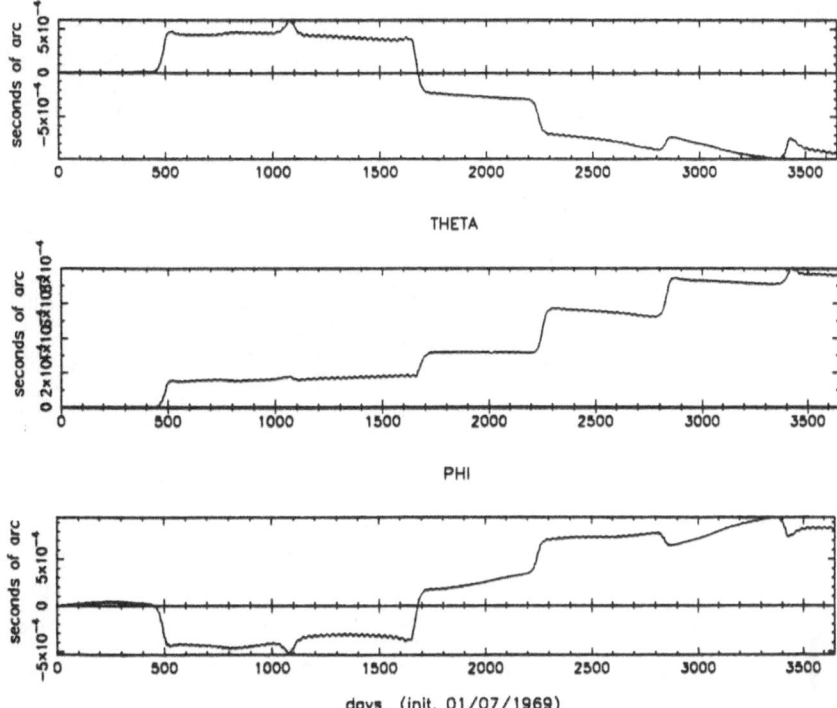

Fig. 4: Effect of the planets

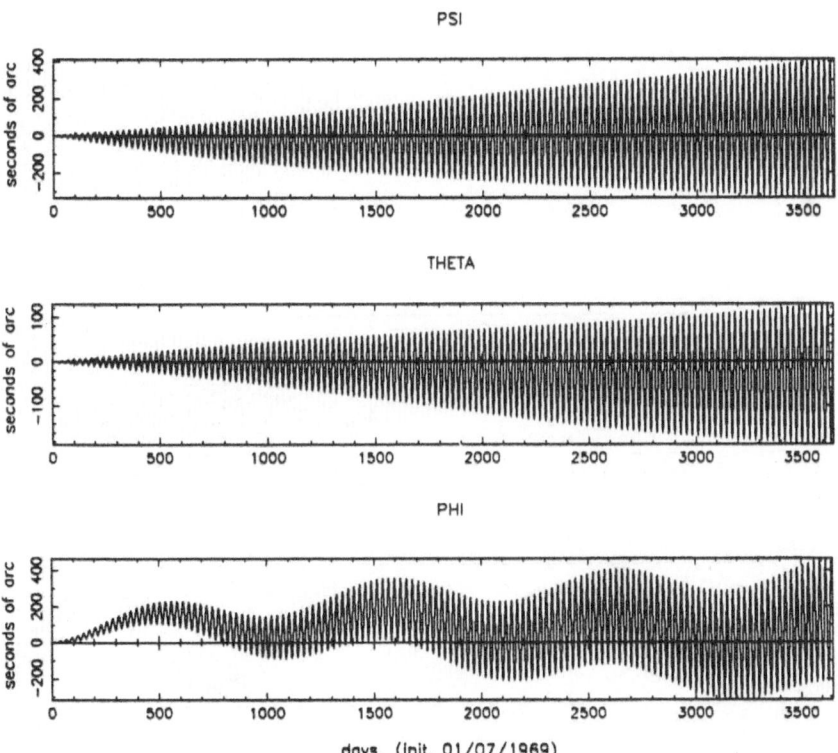

Fig. 5. Effect of "Moon3"

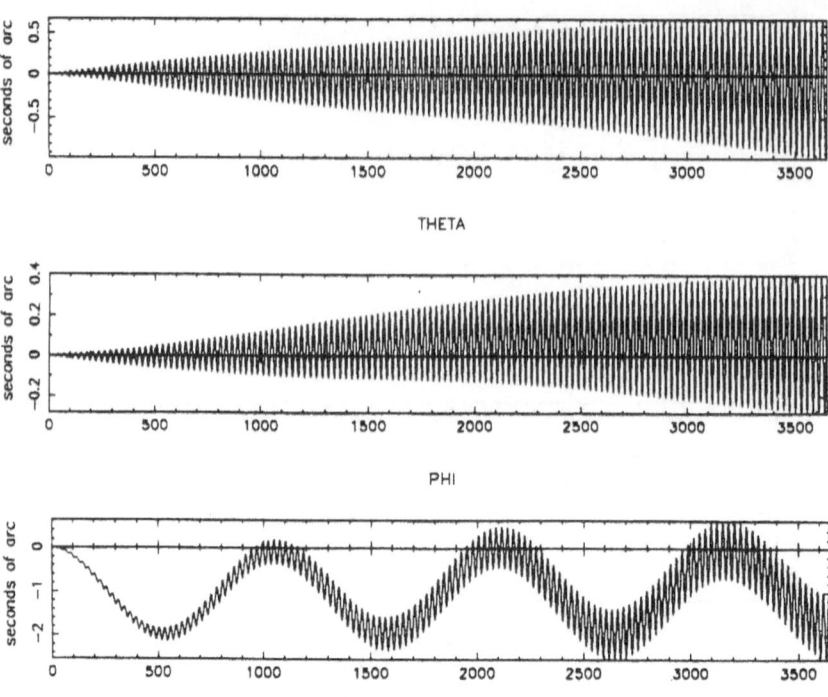

Fig. 6. Effect of "Moon4"

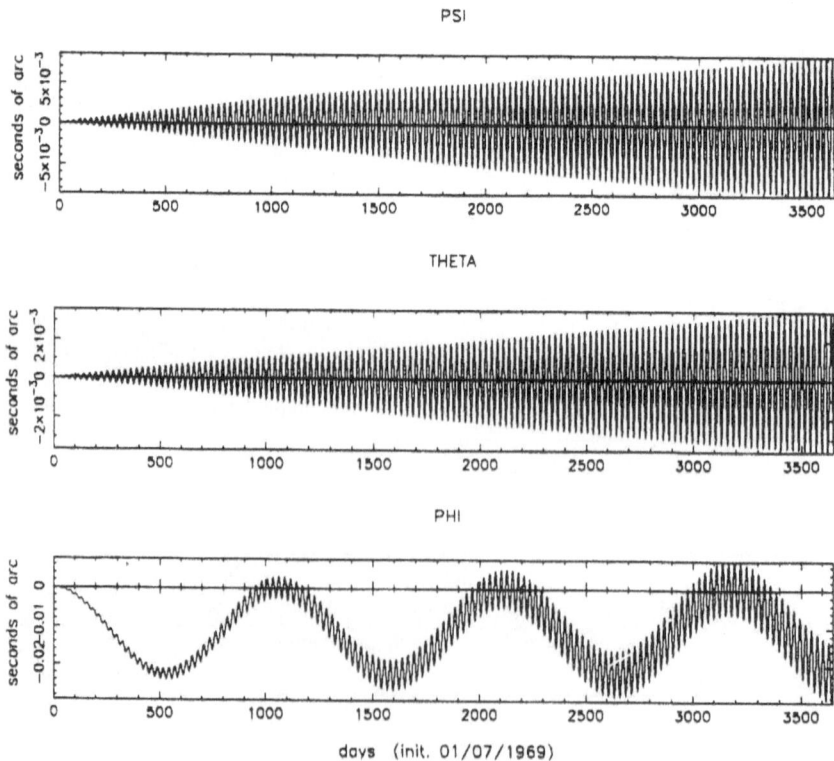

Fig. 7. Effect of "Moon5"

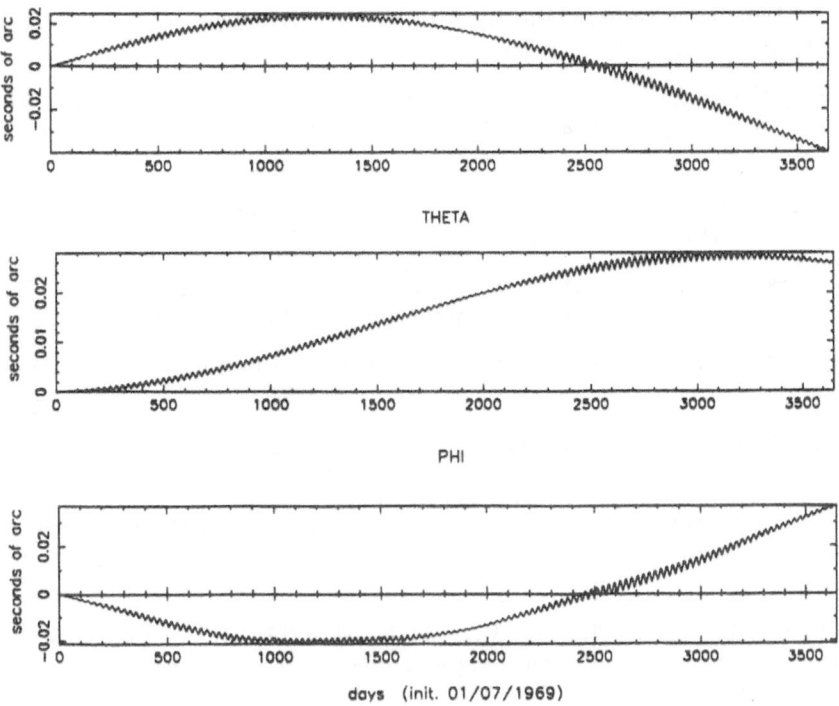

Fig. 8. Effect of "E2M2"

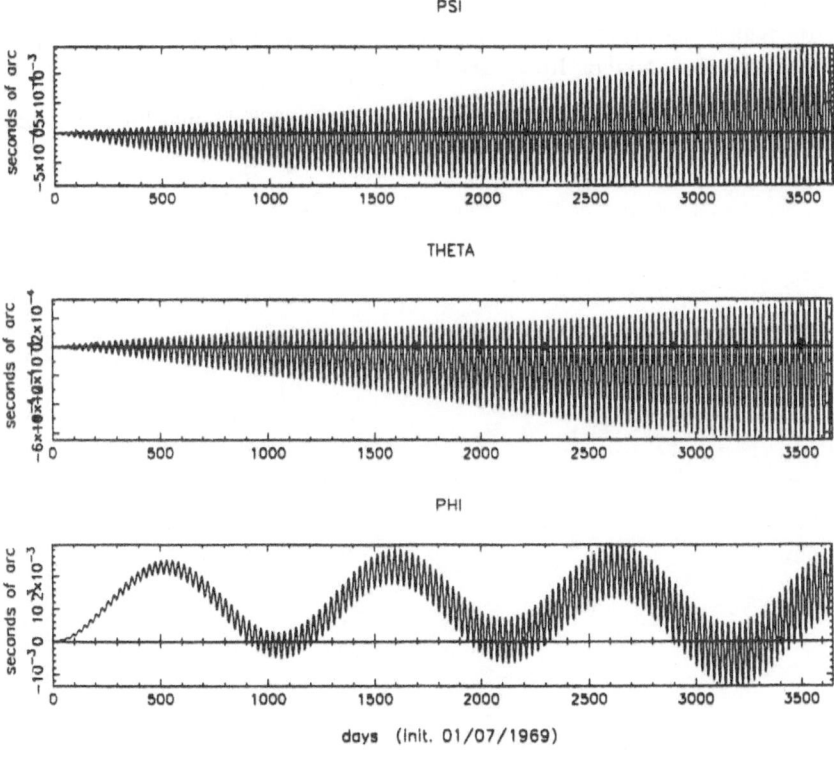

Fig. 9. Effect of "E2M3"

7. References

Bois, E., 1986, "First-Order Theory of Satellite Attitude Motion - Application to HIPPARCOS", *Celestial Mechanics* **39**, No. 4, pp. 309-327.

Bois, E., Oberti, P., Frœschlé, C., 1990, "Etude dynamique du mouvement de rotation des noyaux cométaires", *Seconde Table Ronde de Planétologie Dynamique*, Vars, Mars 90, proceedings to be published.

Bois, E., Oberti, P., Frœschlé, C., 1991, "Physical Librations of Comet Nuclei", paper in preparation for *Celestial Mechanics*.

Bois, E., Wytrzyszczak, I., 1991, "Direct Effects of Planets and Figure-figure Interactions on the Moon's Rotational Motion", paper in preparation for *Celestial Mechanics*.

Borderies, N., 1978, "Mutual Gravitational Potential of N Solid Bodies", *Celestial Mechanics* **18**, pp. 295-307.

Eckhardt, D.H., 1981, "Theory of the Libration of the Moon", *The Moon and the Planets* **25**, pp. 3-49.

Eckhardt, D.H., 1982, "Planetary and Earth Figure Perturbations in the Librations of the Moon", *High-Precision Earth Rotation and Earth-Moon Dynamics*, O. Calame (ed.), pp. 193-198.

Newhall, X.X., Standish, E.M., Williams, J.G. and Jr., 1983, "DE 102: A Numerically Integrated Ephemeris of the Moon and Planets Spanning Forty-four Centuries", *Astronomy and Astrophysics* **125**, pp. 150-167.

Veillet, C., Bois, E., 1990, "Sélénodésie par télémétrie laser depuis la Terre", *Journée d'information sur la Lune*, CNES, Toulouse, Avril 90, proceedings to be published.

Veillet, C., Bois, E., Chabaudie, J.E., Feraudy D., Glentzlin, M., Mangin, J.F., Pham-Van, J. and Torre, J.M., 1989, "The CERGA Lunar Laser Ranging Station", *7th International Workshop on Laser Instrumentation*, Matera, Italie, Oct. 89.

Williams, J.G., Slade, M.A., Eckhardt, D.H. and Kaula, W.M., 1973, "Lunar Physical Librations and Laser Ranging", *The Moon* **8**, pp. 469-483.

THE MOON'S PHYSICAL LIBRATIONS

Part II: Non-Rigid Moon and
Direct Non-Gravitational Perturbations

E. Bois[1] and I. Wytrzyszczak[2]

[1]Observatoire de la Côte d'Azur, Département Cerga
Avenue Copernic, 06130 Grasse, France

[2]Astronomical Observatory of A. Mickiewicz University
ul. Sloneczna 36, 60-286 Poznań, Poland

Abstract: Some relations between the non-rigidity of the Moon and its physical librations are described in this second part, including librations due to the tides and others due to the rotational motion. Starting from their nature, their cause and their behaviour, the different families of physical librations are presented here in a clear and compact classification.

1. Introduction

Following the study of the gravitational perturbations detailed in part I (Wytrzyszczak and Bois, 1990), this second part concerns the rotational behaviour of a non-rigid Moon. A non-rigid Moon model has been built with two possibilities:
- purely elastic deformations alone,
- both elastic and anelastic deformations,
including the other perturbations mentioned in table 1, except for the present time, modelisation of a liquid core. The present paper dealing with nature and classification of the physical librations, only curves stemming from an elastic Moon model are given here. The corresponding anelastic effects do not qualitatively change the aspects and features treated here. Presently, the current uncertainty on the determination of the k lunar Love number is 0.003, while the one on the kT lunar rotational dissipation coefficient has a value of 0.0048 ± 0.002 days (Veillet and Bois, 1990). It is not quite sufficient in particular for the value of the phase shift of the Love number. Anyway, we may use a realistic range of values for these parameters to begin the study of the viscosity effects on the Moon's rotational motion.

2. Viscosity effects

First of all, let us specify that there exist two kinds of deformations. Because the tensor of inertia is no longer constant in the case of a non-rigid Moon and contains two parts, some deformations are due to the tides (tidal deformations) and other deformations are due to the rotational motion (rotational deformations) that are not gravitational pertubations (table 1). In order to construct a non-rigid Moon's model, the following assumption (Eckhardt, 1981) has been made: the sum of deformations between the two poles of the Moon remains equal to zero;

these deformations cancel each other out along the z-axis. Starting from that, figure 1 represents the effects of purely elastic deformations on the three angles of librations.

Questions of the internal structure and the one of the "free" librations are obviously connected. Let us consider the librations due to the rotational deformations only. That means, we consider Moon alone in space, from the gravitational point of view. Figure 2 represents then the effects of the rotational deformations contained in the librations of figure 1. These rotational deformations are so faint that figure 1 represents in fact the librations due to the tidal deformations, with the present model of purely elastic deformations. That confirms the studies and expectations of Migus (1977). Amplitudes of these tidal librations are of the order of a few thousandth seconds of arc. The ϕ angle of figure 1, with its three modulations of libration expresses formation of an equatorial bulge in direction of the Earth. The great amplitude of the long period permits, with respect to the input value of the Love number, to bring evidence on the general behaviour of the lunar tides. But at this level, in order to precise the amplitude of these lunar tides, it is absolutely necessary to adjust the theoritical model, including anelastic deformations, to the observed tides. But unfortunately, in spite of the very good accuracy of the lunar laser technology, reduction of the observations can not give the value of the tide amplitude. The reason of that is simply connected with the places of the Moon where the five reflectors have been set down. Three of them are american reflectors deposited by the Apollo missions. The two others are french reflectors deposited by the russian automatic space probes. The highest in latitude reflector does not work; consequently the greatest difference in latitude between two reflectors is of about 25 degrees and it is not sufficient to reach the amplitude of the tides. However, a new spatial lunar program is provided by the NASA. The CNES of Toulouse is connected to this program and we have proposed to put very much higher in latitude one or two new reflectors on the Moon. We think that about 60° in latitude should be convenient. Then, with this new differentiation in latitude, a sensible progress could be obtained in many questions relative to selenophysics. Observing the tides implies then a better adjustment of the theoritical model, therefore a better determination of the lunar Love number and its phase shift, therefore a better determination of the dissipation coefficient of the Moon which characterises the damping of the "free" librations. Consequently, with the so improved model, this would lead to a better knowledge of the internal structure of the Moon.

3. Towards another terminology

Libration phenomena have been historically introduced step by step but presently, with the required abundance of superposed motions, it would be justified to question ourselves on the signification of the "free" librations and more generally on the classification and terminology of the physical librations. Are the qualifications of "forced" and "free" really suitable for the Moon's physical librations ? What are the free librations physically ? Does "free" mean unperturbed or free in space ?
- The free librations do not square with the oscillations of the Euler-Poinsot motion; in this case, the period would be 144 years (Migus, 1977). The periods of the observed "free" librations are (Calame, 1976, Migus, 1977) : 2.9 years in longitude, 27.3 days and 75 years in the two modes of latitude, let be 24 years in ψ, 75 years in θ, 2.9 years in ϕ; and in ϕ we find again this same period as in Part I for the figure librations. It is a difficulty to study dynamics of the Moon for the reason that, apparently, some librations having close periods would be of different nature. There could be also some confusion on the nature, in term of classification, of these librations. The Euler-Poinsot oscillations are given for a rigid body by the Eulerian differential equations without the right hand side (Goldstein, 1964, Landau and Lifchitz, 1969). They signify purely geometric oscillations without physical causes, neither gravitational nor nongravitational, only initial conditions. The effect of the non-rigidity included then into the tensor of inertia produces some added oscillations in the solution of these equations without the right hand side. The oscillations of figure 2 can also be obtained this way. Oscillations obtained by the homogeneous parts of the equations of motion are often called free librations (for instance, Calame and Mulholland, 1989) in the sense that the Moon is free in space but :
1/ Can these librations, plotted on a large interval of time, reach the observed amplitudes of

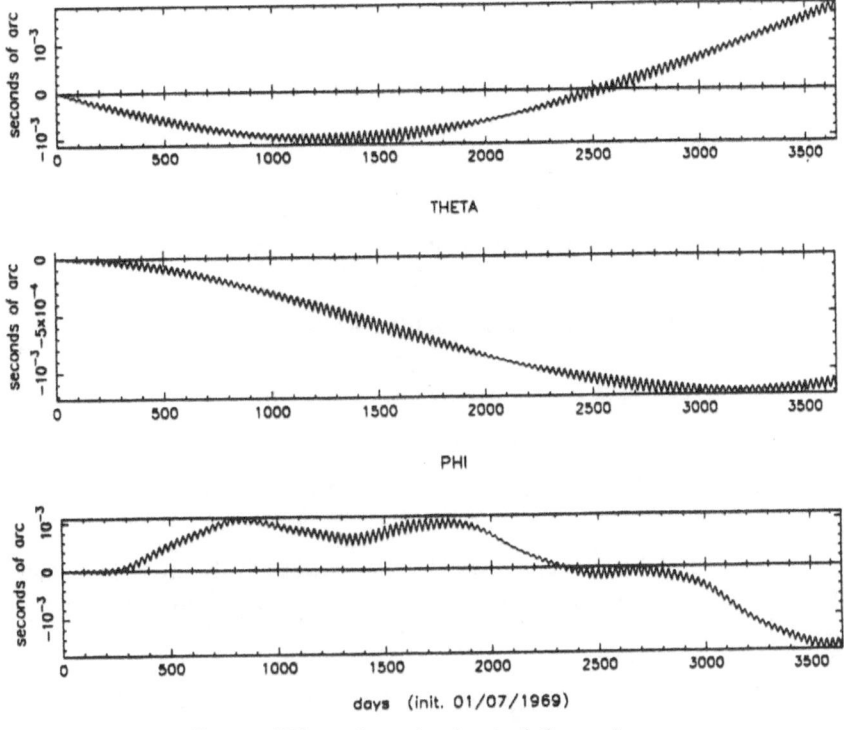

Fig. 1. Effect of purely elastic deformations

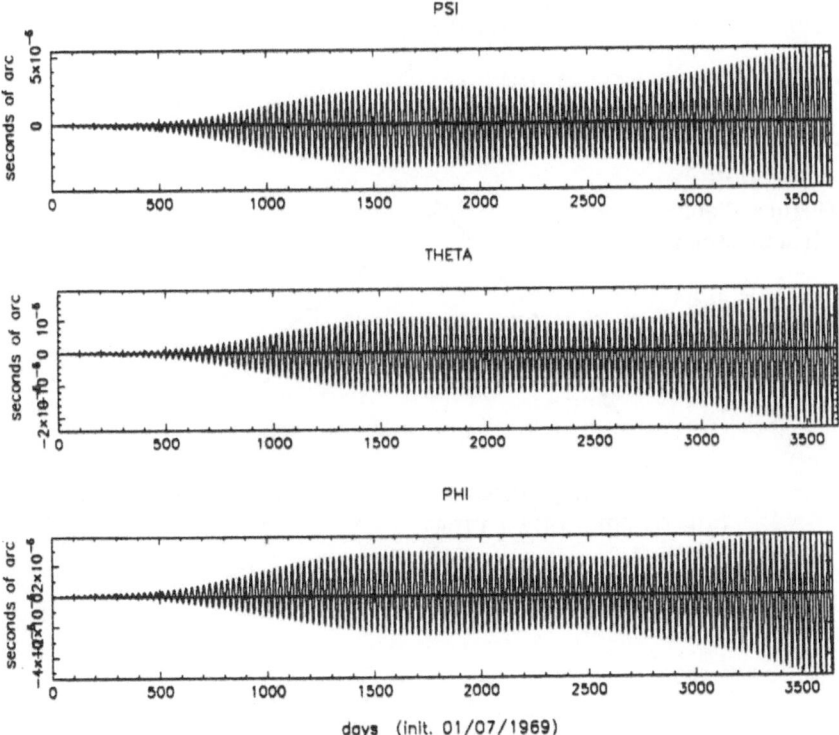

Fig. 2. Centrifugal librations with an elastic Moon model

Table 1. Model of Moon's rotational motion

MOON'S ROTATIONAL MOTION

DIRECT GRAVITATIONAL
PERTURBATIONS

DIRECT NON-GRAVITATIONAL
PERTURBATIONS

* Sun

* Planets

* Figure of the Moon

* Figure-Figure
  (Earth-Moon)

* Earth Tides
  (Equatorial bulge)

* Internal structure :

elastic Moon
elastic + anelastic
liquid or viscous core

DEFORMATIONS
DUE TO THE LUNAR TIDES

DEFORMATIONS DUE
TO THE ROTATION

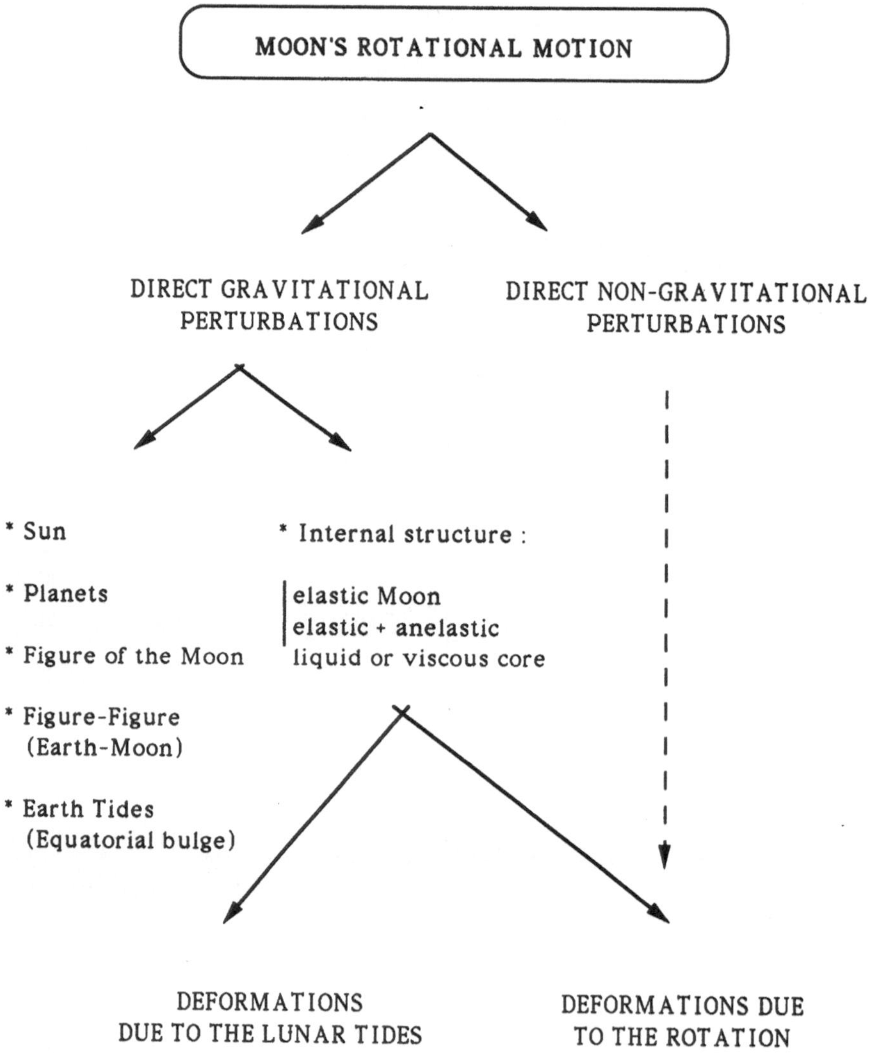

Table 2. The libration, cause and nature,
in the Moon's rotational motion

MOON'S PHYSICAL LIBRATIONS

GRAVITATIONAL
PERTURBATIONS

NON-GRAVITATIONAL
PERTURBATIONS

POTENTIAL LIBRATIONS
(Forced motion)
(Forced librations)

KINETICAL LIBRATIONS
(Free motion)
(Free librations)

* Classical potential

* Mutual potential
  (figure-figure interactions)

* Tidal deformations
  (non-rigid Moon)

  --› *Tidal libration*

* Rotational deformations
  (viscosity of the Moon)
  (liquid core and fluid
  turbulence)

  --› *Centrifugal libration*

* Viscous fluid friction ?
  (core-mantle interface)

* Meteorite impacts ?
  (possibly recent)

* ...

the so-called free librations ? Well, these determined amplitudes arise, according to Calame (1976, 1977), to 1.8 seconds of arc in longitude, 0.4 and 7.8 seconds of arc in latitude, with the respective periods of 2.9 years, 27.3 days and 75 years.

2/ The free librations can not be reduced to the above definition. Other causes may exist. In 1981, Yoder has shown that both a very recent impact (Calame and Mulholland, 1978) or fluid turbulence in the lunar core are plausible mechanisms for generating the observed free large librations.

3/ The internal structure of the Moon can be considered as disturbing torques acting on the rotational motion.

- Finally, a rotational deformation is as much "forced" as a tidal deformation. Free librations are physical librations. Family of free librations of an actual Moon can have many different actual causes, internal or external. Nevertheless, the most reasonable criterion is right to split the librations according to the gravitational aspect or not.

Let us try going towards a more suitable terminology. Aristotle defined the "being" as "act" and "power". But act is not act as the act II of a theater play, but act as existing, efficient, that what is, that what is independently of particular determinations. Power is not power as the power but as in power, that what is potentially. From that, stem the two usual qualifications for the physical energy: respectively the kinetic energy, that what is at present in motion, and the potential energy for that what is caused by, a potential for example and justly. Whence, the possibility to keep this meaning and this formulation for the physical libration families :

⋆ Kinetical librations : librations with non-gravitational causes and directly connected to the existing motion that produces them. A suitable example in this family is : the centrifugal libration, where it is clear that we are speaking of librations due to rotational deformations including viscosity of the Moon, a liquid core or yet a turbulent friction (Ferrari *et al.*, 1980, Yoder, 1981).

⋆ Potential librations : librations with gravitational causes where we can find for example the main-figure libration, the figure-figure interaction libration or the tidal libration (Wytrzyszczak and Bois, 1990).

Table 2 presents a summary of these considerations : "the libration, cause and nature, in the Moon's rotational motion", where it is clear that to each cause, its effect, its libration.

4. Conclusion

An accurate Moon's rotation model has been built up by numerical integration. Nature, cause and behaviour of the Moon's physical librations have been described, discorrelated, classified. The effects of purely elastic deformations are presented in this part II, including considerations about the "free" librations, a general physical libration classification and the search of a more clear and suitable vocabulary.

5. References

Calame, O., 1976, "Free Librations of the Moon Determined by an Analysis of Laser Range Measurements", *The Moon* **15**, pp. 343-352.

Calame, O., 1977, "Free Librations of the Moon from Lunar Laser Ranging", in *Scientific Applications of Lunar Laser Ranging*, J.D. Mulholland Ed., Reidel, Dordrecht, pp. 53-63.

Calame, O., and Mulholland, J.D., 1978, "Lunar Crater Giordano Bruno : A.D. 1178 Impact Observations Consistent with Laser Ranging Results", *Science* **199**, pp. 875-877.

Calame, O., and Mulholland, J.D., 1989, "The Problem of the Eulerian Oscillations: a weakness of numerical versus analytical methods", *Celestial Mechanics* **45**, pp. 275-280.

Eckhardt, D.H., 1981, "Theory of the Libration of the Moon", *The Moon and the Planets* **25**, pp. 3-49.

Ferrari, A.J., Sinclair, W.S., Sjogren, W.L., Williams, J.G., and Yoder, C.F., 1980, "Geophysical Parameters of the Earth-Moon System", *Journal of Geophysical Research* **85**, No. B7, pp. 3939-3951.

Goldstein, H., 1964, *Mécanique Classique*, Presses Universitaires de France, p. 173.

Landau, L., and Lifchitz, E., 1969, *Mécanique*, MIR Moscou, p. 158.

Migus, A., 1977, "Thèorie analytique de la libration physique de la Lune", *physical doctoral thesis*, Paris VI.

Veillet, C., Bois, E., 1990, "Sélénodésie par télémétrie laser depuis la Terre", *Journée d'information sur la Lune*, CNES, Toulouse, Avril 90, proceedings to be published.

Wytrzyszczak, I., Bois, E., 1990, "The Moon's Physical Librations - Part I: Direct Gravitational Perturbations", *Proceedings of the NATO Advanced Study Institute on: "Predictability, Stability and Chaos in N-Body Dynamical Systems"*, held in Cortina d'Ampezzo, Italy, Aug. 90, Edited by Archie E. Roy.

Yoder, C.F., 1981, "The Free Librations of a Dissipative Moon", *P.T.R.S.L.* **303**, 327-338.

SIGNIFICANT HIGH NUMBER COMMENSURABILITIES IN THE MAIN LUNAR PROBLEM:

A POSTSCRIPT TO A DISCOVERY OF THE ANCIENT CHALDEANS

A.E. Roy[1], B.A. Steves[2], G.B. Valsecchi[3] and E. Perozzi[4]

[1] University of Glasgow, U.K.
[2] Queen Mary College, University of London, U.K.
[3] Istituto di Astrofisica Spaziale, Roma, Italy
[4] Telespazio s.p.a., Roma, Italy

LUNAR CYCLES AND THE SAROS

Since ancient times the knowledge of several "lunar cycles" helped mankind to predict lunar phases and eclipses. These cycles owe their existence to high-number commensurabilities between the mean motions of the Sun and the Moon and the lunar nodical (or draconitic) and anomalistic months (Deslambre, 1817); some of them are reported in Table 1. The Metonic cycle ensures that, if a full Moon or new Moon occurs on a particular date, a full Moon or new Moon will occur on the same date 19 years later, allowing easy calibration of the lunar phase to the solar calendar. The Saros period of slightly more than 18 years gives the basic time span for eclipse prediction. Hipparcus (circa 140 B.C.) introduced three additional cycles, the shortest of which is reported in Table 1, involving the synodic, anomalistic and nodical months of the Moon, in an attempt to improve the predictability of events in the Earth-Moon-Sun system.

Table 1. Some long-known lunar cycles

Cycle		Commensurability in			Repeatability of			
Name	Length	T_S	T_N	T_A	C_d	E_{ms}	σ_m	σ_s
Meton	19.000	235			Y	–	–	Y
Saros	18.030	223	241.999	238.992	–	Y	Y	Y
Hipparchus	20.294	251		269.000	–	–	Y	–

Legend: T_S = synodic month
T_N = nodical month
T_A = anomalistic month
C_d = calendar date
E_{ms} = lunar and solar eclipses
σ_m = apparent semi-diameter of the Moon
σ_s = apparent semi-diameter of the Sun
Y = Yes

Predictability, Stability, and Chaos in N-Body Dynamical Systems
Edited by A.E. Roy, Plenum Press, New York, 1991

The discovery of the Saros, the subject of this paper, is attributed to the Babylonians. It is essentially a period of about 6585.321 days, or approximately 18 years and 10 or 11 days, depending on the number of leap years in the interval. The relative dynamical geometry of the Earth-Moon-Sun system during eclipses repeats itself very closely over one Saros period, so closely that the sequence of eclipses occurring within a Saros can be predicted very accurately (Roy, 1988). This is surprising considering that the eccentricities of the orbits may cause the Sun and the Moon to be respectively up to 2° and 5° off from their mean positions. As an example, in Table 2 (first row) are listed the relative positions and velocities for the Earth-Moon-Sun system during the partial lunar eclipse of 11 February 1952. We can see that the distance of the Moon r_m, the difference between the Moon and the Sun's geocentric longitudes $(\lambda_m - \lambda_s)$ and latitudes $(\beta_m - \beta_s)$, and the daily rates of change of these coordinates, $\dot{r}_m$, $(\dot{\lambda}_m - \dot{\lambda}_s)$ and $(\dot{\beta}_m - \dot{\beta}_s)$, return to much the same values after one Saros.

The repetition of eclipses is a consequence of the set of high integer near commensurabilities existing between the Moon's synodic, anomalistic and nodical months. Their mean values are:

Synodic (T_S) = 29.530589 days

Anomalistic (T_A) = 27.554551   "   (1)

Nodical (T_N) = 27.212220   "

These mean values remain steady over many centuries to within one second even though the actual values of these different months may vary considerably in any one revolution of the Moon, due to solar perturbations. Then, as is well known:

223 T_S = 6585.3213 days

239 T_A = 6585.5375   "   (2)

242 T_N = 6585.3575   "

This close agreement ensures that the mean geometry of the Earth-Moon-Sun system at the beginning of a Saros is almost exactly repeated at the end. Since there is no commensurability between the mean motions of the Moon and the Sun, the Sun's geocentric radius vector is about 10° from its former position.

The close similarity of eclipses separated by a Saros period also means that over any Saros cycle, the perturbations of the Sun on the Earth-Moon system, and in particular the large disturbances in the Moon's semimajor axis, eccentricity and inclination, are almost completely cancelled. The basic mechanism by which this can be explained relies on the repeated occurrence of mirror configuration (hereafter indicated by MC). It is well-known that if a system of n gravitating point masses (n≥2) enters a configuration where every radius vector from the centre of mass of the system is perpendicular to every velocity vector, then the behaviour of each body after that epoch will be a mirror repetition of its history before it (Roy and Ovenden, 1955). There are two and only two types of configurations possible: collinear, with the bodies on a line with all their velocity vectors perpendicular to that line, and coplanar, with the bodies on a plane with the velocity vectors perpendicular to that plane. A corollary to the mirror theorem states that if a dynamical system passes through two MC's then the system is periodic, its period being twice the time interval between the two MC's. It can be shown (Perozzi et al., 1991) that in each Saros there are indeed a minimum of two near MC's.

Table 2

Saros Period (days)	Beginning End	r_m (km)	$\dot{r}_m$ (km/day)	$\lambda_m-\lambda_s$ (deg)	$\dot{\lambda}_m-\dot{\lambda}_s$ (deg/day)	$\beta_m-\beta_s$ (deg)	$\dot{\beta}_m-\dot{\beta}_s$ (deg/day)	Q	Type
6585.32738	1952 2 11.02729	403657.9	-1749.7	180.075	10.942	0.848	-1.091		Observed
	1970 2 21.35467	404162.3	-1592.1	180.081	10.917	0.863	-1.087		Eclipse
6585.32696	1952 2 11.02729	403657.9	-1749.7	180.075	10.942	0.848	-1.091	0.015	Q
	1970 2 21.35425	404162.9	-1591.9	180.076	10.916	0.864	-1.087		Eclipse
6585.32751	1958 10 20.89109	392136.2	4829.7	105.436	11.633	4.980	-0.401	0.023	Q
	1976 10 31.21860	391372.4	4877.1	105.439	11.677	4.990	-0.397		
6585.32358	1988 6 7.71649	370856.4	1595.6	275.906	13.086	0.380	1.234	0.014	Q
	2006 6 19.04007	370542.8	1308.2	275.907	13.112	0.305	1.237		

The first two cases correspond to a partial lunar eclipse; in the second one, the second row refers to the epoch that minimizes Q. The epochs of the remaining two cases, listed for comparison, have been chosen far from eclipses.

THE EARTH-MOON-SUN SYSTEM AS A NEARLY PERIODIC ORBIT

The Saros cycle was discovered thanks to the repetition of easily observable dynamical configurations, the eclipses; but if the Earth-Moon-Sun system has recurrent configurations, then it may be nearly periodic, with period equal to one Saros.

We have devised a way to test this hypothesis, using three different ephemerides of the Moon:

a)   JPL high-precision planetary and lunar ephemerides (Standish et al, 1976);

b)   numerical integration of the restricted elliptic 3-body problem Earth-Moon-Sun, with starting elements taken from the JPL ephemeris at JD=2434000.5;

c)   same as b) but with starting elements at JD=2433000.5.

The following procedure was adopted: at an epoch t_1, chosen randomly, but avoiding the time of a solar or lunar eclipse, the relative geocentric position and velocity coordinates of the Moon and the Sun are found using the three ephemerides described above in turn. Assuming an approximate Saros length T = 6585.3 days, we then search through the ephemerides within the time interval t_1+T±0.5 days for the time t_2 at which the quantity

$$Q = \sqrt{\sum_{i=1}^{6} \left[\frac{X_{2i}-X_{1i}}{\delta X_{maxi}} \right]^2} \qquad (3)$$

has a minimum. In (3) X_{1i} and X_{2i}, with i=1 to 6, are the values of the six relative coordinates r_m, $(\lambda_m-\lambda_s)$, $(\beta_m-\beta_s)$, $\dot{r}_m$, $(\dot{\lambda}_m-\dot{\lambda}_s)$, $(\dot{\beta}_m-\dot{\beta}_s)$ at epochs t_1 and t_2 respectively, and the quantity δX_{maxi} is the maximum possible difference between any two values of a given coordinate. If the relative dynamical geometry of the Earth-Moon-Sun system is accurately repeated after one Saros period, the minimum should occur after a time span approximately equal to the Saros period and Q should also be very small.

Once the time t_2 near t_1+T which minimizes Q is found, we define the quantity $T^*=t_2-t_1$ as the "osculating" value of the Saros period for that particular epoch t_1. This procedure is repeated for values of t_1 which span two Saros periods, from 1952 to 1988. In total, one hundred values of t_1 were tested for each ephemeris. In Table 2 two examples are listed, while in Figure 1 the whole set of values of T^* found is plotted as a function of starting epoch t_1. The values appear rather regularly distributed around the Saros periods computed directly from the corresponding synodic months T_{saros} = 223 T_S, indicated by solid lines. Delaunay's formulae (Delaunay, 1872) have been used to compute the mean synodic months for the restricted 3-body systems. More details about the whole procedure can be found in Perozzi et al. (1991).

As Figure 1 shows, irregardless of the initial time chosen, the relative positions and velocities of the Moon and the Sun are best repeated after an interval of time close to the classical Saros period. The Saros period given by 223 mean synodic months is 6585.321347 days, and we find a value remarkably close to this one by the minimization of Q. This suggests that the synodic month plays the dominant role in driving the system towards the repetition of any particular configuration. Moreover, these considerations hold for the "real" Moon (i.e. the high-precision JPL

276

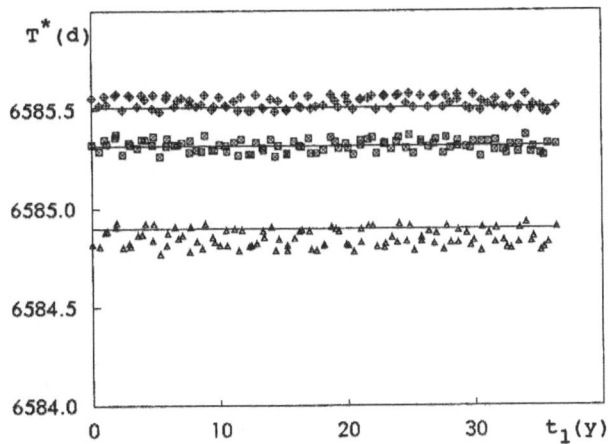

Fig.1. Osculating Saros periods T^* versus
starting date t_1, in years from the
partial lunar eclipse of 11 Feb. 1952
(JD 2434053.5); squares: JPL DE 118
ephemeris; diamonds: numerical integration of restricted elliptic 3-body
problem, starting on JD 2434000.5;
triangles: same integration, starting on
JD 2433000.5. Horizontal lines: Saros
lengths T_{saros} corresponding to 223 mean
synodic months.

ephemeris) as well as for the elliptic restricted 3-body integrations,
thus indicating that the Saros is basically a 3-body mechanism.

The repetivity of the orbit of the Moon after one Saros is further
illustrated in Figures 2, 3 and 4. In them some geocentric orbital elements of the Moon are plotted for a time span of 200 days centred on a
near-MC; the data are taken from a restricted elliptic 3-body integration
and cover two successive Saroses. The element plots have been superposed
using a time delay T_{saros}, computed from Delaunay's theory, as explained
above. We can easily see in Figure 2 that the eccentricity behaviour is
very well reproduced after one Saros, as is the behaviour of the semimajor
axis shown in Figure 3. Indeed so closely repetitive are the behaviours
of e and a that it seems necessary to emphasize that in each of Figures 2
and 3 two plots are superposed on each other, viz. one running from (t-
100) to (t+100) days, the other from (t+T_{saros}-100) to (t+T_{saros}+100)
days. In the case of the argument of perigee, shown in Figure 4, we can
see the effect of the non-exactness of the relationships (2), as there is
a small residual time shift between the lines corresponding to the two
successive Saroses.

We can conclude that the relative dynamical geometry of the Earth-
Moon-Sun system over one Saros period is closely repeated at any osculating phase of the period, not just in the mean geometry reference frame,
nor at the occurrence of certain particular events (e.g. eclipses). In
other words, the perturbations of the Sun on the Earth-Moon system, particularly the large disturbances in the Moon's semimajor axis, eccentricity and inclination, are almost completely cancelled out over a Saros, no
matter where it is taken to start.

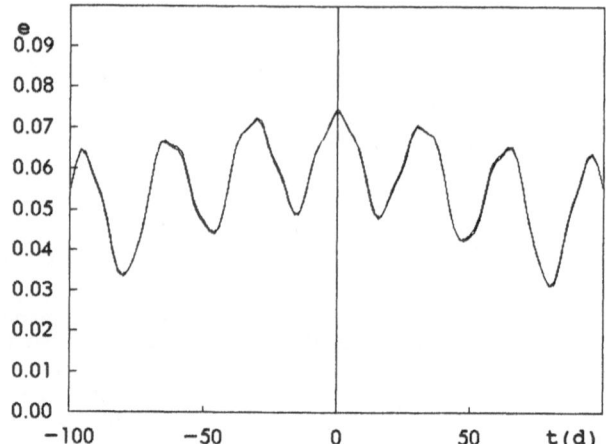

Fig.2. Time evolution of the geocentric osculating eccentricity of the Moon about a
near-MC; data taken from a restricted
elliptic 3-body integration, covering
two successive Saroses.

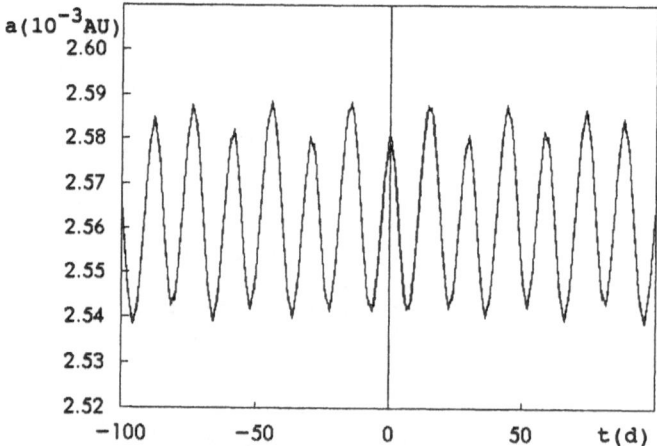

Fig.3. Time evolution of the geocentric osculating semimajor axis of the Moon about
a near-MC; data taken from a restricted
elliptic 3-body integration, covering
two successive Saroses.

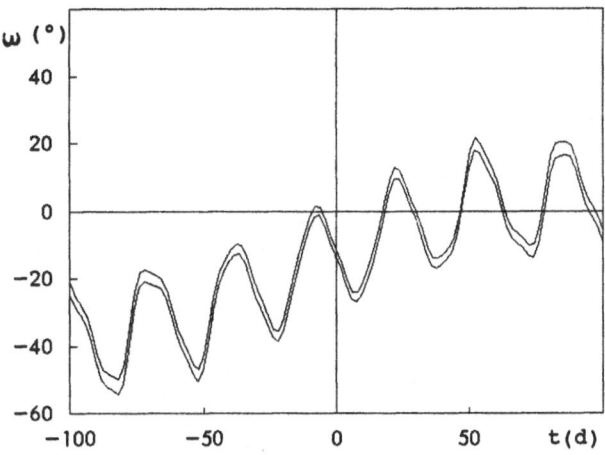

Fig.4. Time evolution of the geocentric osculating argument of perigee of the Moon
about a near-MC; data taken from a
restricted elliptic 3-body integration,
covering two successive Saroses.

THE PERIODIC ORBIT OF THE MOON

The almost exact repetivity of the orbital elements of the Moon after
223 synodic months raises the question of the possible existence of a
periodic orbit near (in phase space) that of the Moon. We cannot however
have such an orbit in the restricted elliptic 3-body problem, because 223
synodic months do not correspond to an integer number of years, and thus
the Earth-Sun distance would not come to exactly the same value after one
Saros; this difficulty, however, does not arise in the circular restricted
3-body problem.

In order to mimic well the behaviour of the real lunar orbit, which
passes very close to MC's (Perozzi et al., 1991), the periodic orbit
sought for should also pass, and exactly, through MC's. In the restricted
circular 3-body problem there are 16 possible MC's in all, listed in Table
3. Of them, 8 are of the collinear type, and the other 8 are of the
coplanar type. It is easy to see that they can be characterized by the
values of three angles: the argument of perigee and the true anomaly of
the Moon, and the difference of geocentric longitude between Moon and Sun;
each of these angles must be either 0° or 180° for the 8 collinear configurations, whereas to obtain the 8 coplanar configurations we must have the
argument of perigee equal to either 90° or 270°, the other angles still
being either 0° or 180°. We can unambiguously identify each MC by speci-
fying a three-digit code representing the values of the argument of perigee, true anomaly and longitude difference as multiples of 90° (see Table
3).

Other characteristics of the periodic orbit can be guessed by noting
that in it there should be exactly two MC's, because of the mirror theorem, and that they must obviously be different from each other (otherwise
the orbit would have a period of half a Saros!). This means that there
are 8 periodic orbits possible, and that they "connect" pairs of MC's in a
way that is dictated by the number of revolutions that each of the three
angles we are considering makes during a Saros. These numbers are 3 for

ω, 239 for f and 223 for $\lambda_m-\lambda_s$; since they are all odd, and since the two MC's must be exactly half a Saros apart in time, it follows that a periodic orbit must contain MC's in which ω, f and $\lambda_m-\lambda_s$ differ by 180°. Therefore, there should be 8 periodic orbits each containing two MC's, in the arrangement indicated by Table 3, whose last column lists these orbits as 1 to 8.

Table 3. Mirror configurations in the
circular restricted 3-body problem

Configuration identifier	ω	f	$\lambda_m-\lambda_s$	Orbit
000	0°	0°	0°	1
222	180°	180°	180°	1
200	180°	0°	0°	2
022	0°	180°	180°	2
002	0°	0°	180°	3
220	180°	180°	0°	3
020	0°	180°	0°	4
202	180°	0°	180°	4
100	90°	0°	0°	5
322	270°	180°	180°	5
300	270°	0°	0°	6
122	90°	180°	180°	6
102	90°	0°	180°	7
320	270°	180°	0°	7
120	90°	180°	0°	8
302	270°	0°	180°	8

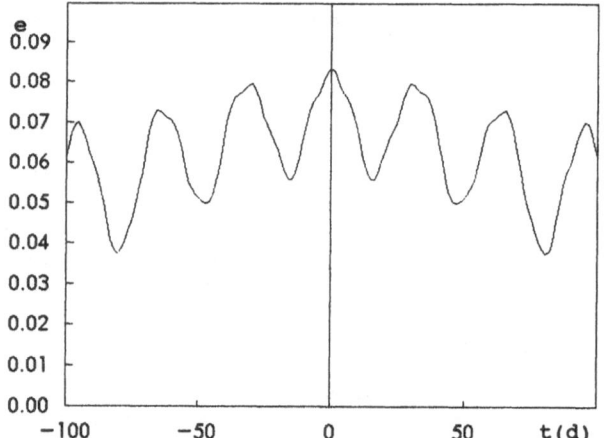

Fig.5. Time evolution of the geocentric osculating eccentricity of the Moon about a MC of the periodic orbit described in the text.

We have numerically verified the existence of 8 periodic orbits with characteristics corresponding to those shown in Table 3, of period 6585.321347d, exactly the same as 223 synodic months of the Moon (Valsecchi et al., 1991); in Figures 5, 6 and 7 we present the time behaviour of lunar eccentricity, semimajor axis and argument of perigee, for the same time span, and about the corresponding MC of Figures 2, 3 and 4, for the orbit labeled 1 in Table 3. The similarity of the time behaviour of the elements shown is striking; in comparing Figures 2 and 5 it is possible to recognize a slight systematic shift between the osculating lunar eccentricity and that of the periodic orbit. A similar shift exists also in the case of the inclination, and it is part of the reason for the half-day "inexactness" of the Saros given by (2).

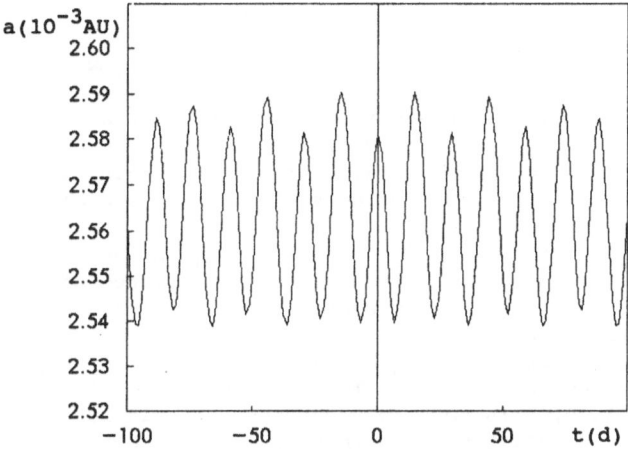

Fig.6. Time evolution of the geocentric osculating semimajor axis of the Moon about a MC of the periodic orbit described in the text.

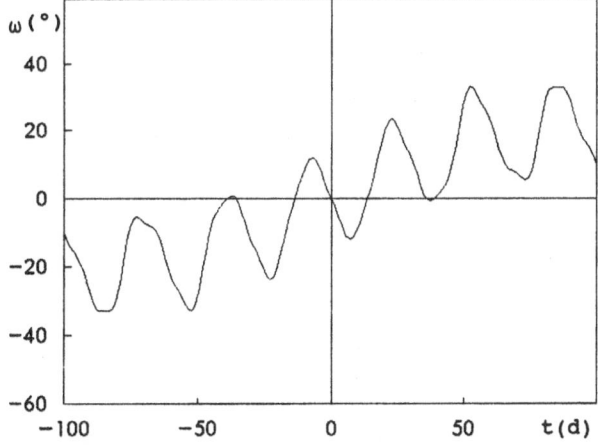

Fig.7. Time evolution of the geocentric osculating argument of perigee of the Moon about a MC of the periodic orbit described in the text.

CONCLUSIONS

The existence of the Saros implies a near repetition of the orbital elements of the Moon not only at eclipses, but at any other time during the Saros; both the JPL ephemeris and a numerical integration of the elliptic restricted 3-body problem confirm this finding. As a consequence, the Moon moves in a nearly periodic orbit of period equal to the Saros. Moreover, in the circular restricted 3-body problem it is possible to find periodic orbits, with period equal to the Saros, for which the behaviour of the osculating orbital elements in time is strikingly similar to that of the real Moon.

REFERENCES

Delaunay, Ch., 1872, Note sur le mouvements du perigee et du noeud de la Lune, Compt. rend. hebdom. Acad. Sci., 74:17.

Deslambre, J.B.J., 1817, "Histoire del l'Astronomie Ancienne", Paris.

Perozzi, E., Roy, A.E., Steves, B.A, and Valsecchi, G.B., 1991, Significant high number commensurabilities in the main lunar problem. I: the Saros as a near periodicity of the Moon's orbit, Celest. Mech., submitted.

Roy, A.E., 1988, "Orbital Motion", Adam Hilger, Bristol.

Roy, A.E., and Ovenden, M.W., 1955, On the occurrence of commensurable mean motions in the solar system. II. The mirror theorem, Mon. Not. Roy. Astron. Soc. 115:296.

Standish, E.M., Keesey, M.S.W., Newhall, X X, 1976, "JPL Development Ephemeris Number 96". NASA Technical Report 32-1603.

Steves, B.A., 1990, "Finite-Time Stability Criteria for Sun-Perturbed Planetary Satellites." PhD Thesis, Univ. of Glasgow.

Valsecchi, G.B., Steves, B.A, Perozzi, E., and Roy, A.E., 1991, Periodic orbits close to that of the Moon, Nature, submitted.

MOON'S INFLUENCE ON THE TRANSFER FROM THE EARTH TO A HALO ORBIT AROUND L_1

Gerard Gómez[1], Àngel Jorba[2],
Josep Masdemont[2] and Carles Simó[1]

(1) Departament de Matemàtica Aplicada i Anàlisi, Universitat de Barcelona, Gran Via 585, 08007 Barcelona, Spain

(2) Departament de Matemàtica Aplicada I, ETSEIB, Universitat Politècnica de Catalunya, Diagonal 647, 08028 Barcelona, Spain

Abstract

The influence of the Moon in the transfer of a satellite from the Earth to a Halo orbit around L_1 in the Earth+Moon-Sun system, is analysed by means of a simple bicircular model for the motion of the Earth and the Moon. The results suggest that using the stable manifold of the Halo orbit, slightly bent by the Moon, it is possible to carry out the transfer avoiding the insertion maneouvre in the Halo orbit.

1 Introduction

Taking the Restricted Three Body Problem (RTBP) as starting point, with the Sun as one primary and the Earth+Moon barycentre as the other one, the collinear Lagrangian equilibrium point L_1 lies between the two primaries aproximately at 0.01 au (1500000 km) from the Earth.

This point is of great importance in order to study the Sun, because a satellite located on it is able to monitor continuously the solar phenomena and at the same time to send the information to the Earth (see [1], [2]). The problem is that as seen from the Earth the satellite would appear in front of the solar disk, and due to the solar noise it would be impossible to receive the radio signal sent from the satellite. An exclusion zone subtending an angle of seven degrees from the Earth is necessary in order to avoid this problem.

There are two possible types of orbits around L_1 suitable for avoiding the exclusion zone: Lissajous orbits and Halo orbits. With Lissajous orbits of moderate size it is not possible to avoid the exclusion zone for more than roughly four years (see [5]). Halo orbits form a family of periodic orbits which bifurcate from the planar Liapunov family of orbits when the in-plane amplitude is big enough (654276 km.). At that point the in-plane and out-of-plane frequencies are equal and the family of planar orbits bifurcates to two families of three dimensional periodic orbits, which are the ones called Halo orbits (see [3], [4]).

On August 12, 1978 a scientific spacecraft called International Sun-Earth Explorer-3 (ISEE-3) was targeted to a Halo orbit of 120000 km of out-of-plane amplitude and 666672 km of in-plane amplitude. In computing the transfer trajectory two ways were considered, the slow transfer, and the fast transfer. The first one was chosen because the delta-v requirements were lesser, in spite of the fact that the transfer is somewhat slower (see [1]).

The ISEE-3 was injected in the transfer orbit at 6564.1 km from the centre of the Earth and its velocity relative to the Earth was 10.990 km/sec. The ΔV requirements for the Halo insertion were 36.95 m/sec in-plane magnitude and 0.03 m/sec out-of-plane magnitude.

The computation of the transfer paths between the Earth and the Halo orbits were done by means of an optimization procedure. The procedure searches for the solution by changing the parameters in the parking orbit near the Earth and varying the insertion point in the target Halo orbit in order to minimize the ΔV spent by the manoeuvres. Our approach is a geometric one and it has revealed that the transfer orbit followed by the ISEE-3 is in fact very near to an orbit of the stable manifold of the target Halo orbit. Therefore the process of globalization of the stable manifold is the natural way for getting the transfer trajectories.

The first numerical experiments with the RTBP showed that the closest distance from the stable manifold of the ISEE-3 Halo orbit to the centre of the Earth was about 10000 km. At the same time the distance decreased as the out-of-plane amplitude of the Halo orbit increased, reaching, at some value of the amplitude, collision with the Earth.

As we have said, our RTBP has the Earth+Moon barycentre as the small primary. This fact together with the displacement of the Earth when we separate the two bodies revolving around their centre of masses, and with the possibility that the Moon could perturb and bend the manifold, suggest the importance of studying what happens with the transfer orbits only taking into account the main perturbing body, the Moon.

The way we are going to study this influence is the simplest one, by means of a bicircular model.

2   The equations of motion

We will write the equations of motion as a perturbation of the RTBP ones:

$$\begin{aligned}
\ddot{x} - 2\dot{y} &= \Omega_x, \\
\ddot{y} + 2\dot{x} &= \Omega_y, \\
\ddot{z} &= \Omega_z,
\end{aligned}$$

with

$$\Omega = \frac{1}{2}(x^2 + y^2) + \frac{1-\mu}{r_1} + \frac{\mu}{r_2} + \mu(1-\mu),$$

$$r_1^2 = (x - \mu)^2 + y^2 + z^2, \qquad r_2^2 = (x - \mu - 1)^2 + y^2 + z^2,$$

where the Sun is supposed to be at $(\mu, 0, 0)$ with mass $m_1 = 1 - \mu$ and the Earth+Moon at $(\mu - 1, 0, 0)$ with mass μ.

As we want to take into account the different forces due to the Earth and Moon, we must split the μ/r_2 term in two parts. We shall consider the Earth and the Moon moving in circular orbits around their centre of masses.

Let θ be the angle, counted counterclockwise, between the vectors $\overrightarrow{Earth + Moon\,barycentre, Earth}$ and $\overrightarrow{Earth + Moon\,barycentre, Sun}$.

If n_l is the mean motion of the Moon around the Earth in RTBP units,

$$n_l = \frac{mean\,motion\,Moon\,around\,the\,Earth}{mean\,motion\,Earth + Moon\,barycentre\,around\,the\,Sun},$$

we have $\theta = n_l t + \theta_0$, where θ_0 is an initial phase.

We introduce also $\ell = \dfrac{mean\,Earth - Moon\,distance}{mean\,Earth + Moon\,barycentre - Sun\,distance}$,

and $\delta = \dfrac{mass\,of\,the\,Moon}{mass\,of\,the\,Moon + mass\,of\,the\,Earth}$.

Therefore the position of the Earth and Moon are:

$$Earth: \quad (\mu - 1 + \ell\delta\cos\theta, \ell\delta\sin\theta, 0) := (x_E, y_E, 0),$$
$$Moon: \quad (\mu - 1 - \ell(1 - \delta)\cos\theta, -\ell(1 - \delta)\sin\theta, 0) := (x_M, y_M, 0),$$

and we can split the term $\frac{\mu}{r_2}$ as:

$$\frac{\mu\delta}{r_3} + \frac{\mu(1 - \delta)}{r_4} = \mu\left(\frac{\delta}{r_3} + \frac{1 - \delta}{r_4}\right),$$

with:

$$r_3^2 = (x - x_M)^2 + (y - y_M)^2 + z^2, \qquad r_4^2 = (x - x_E)^2 + (y - y_E)^2 + z^2.$$

The equations of the bicircular model are the same ones as the RTBP but with the new Ω:

$$\Omega = \frac{1}{2}(x^2 + y^2) + \frac{1 - \mu}{r_1} + \mu\left(\frac{\delta}{r_3} + \frac{1 - \delta}{r_4}\right) + \mu(1 - \mu).$$

The following data summarize the values taken in the computations:

$\mu = 3.040357143 \times 10^{-6}$,   Unit of distance= $1.49597871411 \times 10^8$ km,
$\delta = 0.012150298$,   Unit of velocity= 29784.7358 m/s,
$\ell = 2.57245638 \times 10^{-3}$,   Unit of time= 58.1301004 days,
$n_l = 13.36411007$.

We note that the bicircular model is no longer an autonomous system as the RTBP was. It is also slightly non coherent: Earth, Moon and Sun motions do not satisfy Newton's equations.

3   The QPO obtained by parallel shooting

Due to the non autonomous character of the bicircular model, periodic Halo orbits do not longer exist. They must be substituted by quasiperiodic orbits (QPO). The aim of this section is to show how to compute these quasiperiodic orbits or, at least, how to compute a numerical orbit of the model which behaves like a quasiperiodic one moving near a Halo orbit of the RTBP during a time span of several years.

ntrst of all we define the Poincaré map, $\mathcal{P}$, associated with a half revolution with the surface of section $y = 0$ following the flow when t increases:

$$\mathcal{P}(t, x, z, \dot{x}, \dot{y}, \dot{z}) = (\bar{t}, \bar{x}, \bar{z}, \dot{\bar{x}}, \dot{\bar{y}}, \dot{\bar{z}}) :=$$
$$:= (\mathcal{P}_1(t, x, z, \dot{x}, \dot{y}, \dot{z}), \ldots, \mathcal{P}_6(t, x, z, \dot{x}, \dot{y}, \dot{z})).$$

As the system of differential equations is non autonomous, time must to be kept in the representation of an initial or final point.

We denote by $Q_i = (t_i, x_i, z_i, \dot{x}_i, \dot{y}_i, \dot{z}_i)$ the i-th component of a certain initial vector Q for the parallel shooting procedure which refers to the i-th point in the partition of the full time interval. If the parallel shooting is splitted in N subintervals then $i = 0, \ldots, N$.

Then in principle we have $6(N+1)$ free variables. We fix t_0, initial epoch, and z_0, initial z-amplitude, at the beginning of the parallel shooting. So in this way the number of free variables (the components of Q) is $6N + 4$. Therefore we need $6N + 4$ equations in order to fit those variables.

Any parallel shooting must satisfy the matching conditions:

$$F_{6i+j}(Q) = \mathcal{P}_j(Q_i) - (Q_{i+1})_j = 0, \qquad j = 1, \ldots, 6, \qquad i = 0, \ldots, N-1,$$

where F is the vector of $6N + 4$ equations to be satisfied by the variable Q.

The matching conditions give us $6N$ equations. The remaining four ones can be chosen in several ways obtaining different approaches to the QPO.

The ones which we have taken are:

$$
\begin{aligned}
F_{6N+1} &= x_0 - x_N &&= 0, \\
F_{6N+2} &= \dot{x}_{\text{halo}} - \dot{x}_N &&= 0, \\
F_{6N+3} &= \dot{y}_{\text{halo}} - \dot{y}_N &&= 0, \\
F_{6N+4} &= \dot{z}_{\text{halo}} - \dot{z}_N &&= 0.
\end{aligned}
$$

Summarizing, the system of equations to be solved is:

$$
\begin{aligned}
F_1(Q) &= \mathcal{P}_1(t_0, x_0, z_0, \dot{x}_0, \dot{y}_0, \dot{z}_0) - t_1 = 0, \\
F_2(Q) &= \mathcal{P}_2(t_0, x_0, z_0, \dot{x}_0, \dot{y}_0, \dot{z}_0) - x_1 = 0, \\
&\cdots \quad \cdots\cdots \\
F_6(Q) &= \mathcal{P}_6(t_0, x_0, z_0, \dot{x}_0, \dot{y}_0, \dot{z}_0) - \dot{z}_1 = 0, \\
F_7(Q) &= \mathcal{P}_1(t_1, x_1, z_1, \dot{x}_1, \dot{y}_1, \dot{z}_1) - t_2 = 0, \\
&\cdots \quad \cdots\cdots \\
F_{6N}(Q) &= \mathcal{P}_6(t_{N-1}, x_{N-1}, z_{N-1}, \dot{x}_{N-1}, \dot{y}_{N-1}, \dot{z}_{N-1}) - \dot{z}_N = 0, \\
F_{6N+1}(Q) &= x_0 - x_N = 0, \\
F_{6N+2}(Q) &= \dot{x}_{\text{halo}} - \dot{x}_N = 0, \\
F_{6N+3}(Q) &= \dot{y}_{\text{halo}} - \dot{y}_N = 0, \\
F_{6N+4}(Q) &= \dot{z}_{\text{halo}} - \dot{z}_N = 0,
\end{aligned}
$$

where

$$Q = (x_0, \dot{x}_0, \dot{y}_0, \dot{z}_0, t_1, x_1, z_1, \dot{x}_1, \dot{y}_1, \dot{z}_1, \cdots, t_N, x_N, z_N, \dot{x}_N, \dot{y}_N, \dot{z}_N).$$

Denoting the above equations by $F(Q) = 0$ and taking $Q^{(0)}$ as initial value of the variables (we take the positions and the velocities from the Halo orbit and $t_i = iP/2$ where P is the period of the halo), the improved values are obtained by means of a Newton procedure:

$$DF(Q^{(j)})(Q^{(j+1)} - Q^{(j)}) = -F(Q^{(j)}).$$

4　The stable manifold

In the case of periodic Halo orbits one can compute the eigenvalues of the monodromy matrix. For a quasiperiodic orbit, as the initial and final points are different, the monodromy matrix properly does not exist, nor exist the associated eigenvalues. But, in fact, the QPO is not too far from the Halo orbit. So we have used the variational matrix as if it were the monodromy matrix in order to compute the stable manifold.

Once a revolution is selected, we can compute the stable manifold in this way:

Let η be the eigenvalue of the Poincaré map related to the stable manifold and $\omega = (\omega_0, \omega_1, \omega_2, \omega_3, \omega_4, \omega_5)$ its eigenvector whose components are associated with $t, x, z, \dot{x}, \dot{y}, \dot{z}$ respectively.

Let $X0 = (t_0, x_0, 0, z_0, \dot{x}_0, \dot{y}_0, \dot{z}_0)$ be the initial conditions of the QPO in the selected revolution, FAC a shifting factor and $v = (\omega_0, \omega_1, 0, \omega_2, \omega_3, \omega_4, \omega_5)$.

Initial conditions $XW = (t_\omega, x_\omega, y_\omega, z_\omega, \dot{x}_\omega, \dot{y}_\omega, \dot{z}_\omega)$ approximating the linear part of the manifold can be obtained by means of:

$$XW = X0 + FAC * (\eta)^\lambda * v \qquad \text{with} \quad \lambda \in [0, 1].$$

Due to the fact that the manifold approaches exponentially to the QPO, equally spaced points of λ in [0,1] produce roughly "equally spaced points in the manifold".

The shifting factor FAC must be chosen small enough, in absolute value, in order to get the initial conditions not far from the QPO, where the linear part is a good aproximation, but also not very small since μ is quite small. We have taken $FAC = \pm 10^{-4}$, the sign being of course the one which produces approaches to the Earth.

Next plots represent the minimum distance from the Earth (in km) to the orbits of the stable manifold, for the QPO with normalized z-amplitude 0.08 (120.000 km), for certain revolutions of the QPO and several initial Earth-Moon positions. The minimum of each plot represents the minimum distance from the W^s to the Earth (the parameter XLA stands for λ).

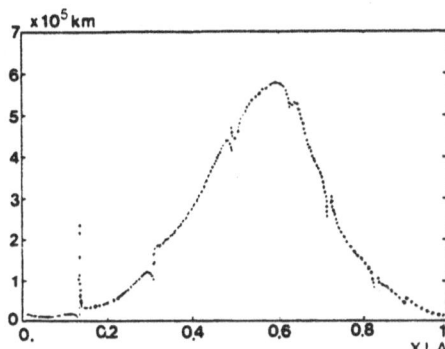

Figure 1. Distance in km from the stable manifold to the centre of the Earth for the revolution 5 of the QPO with normalized z-amplitude 0.08 and $\theta_0 = 30$ degrees.

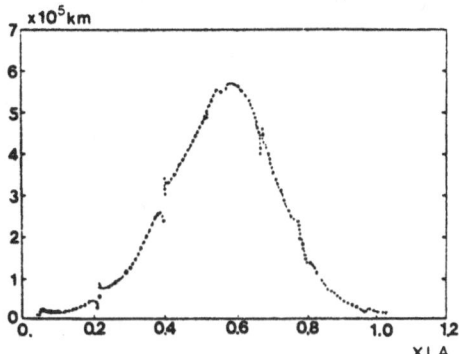

Figure 2. Distance in km from the stable manifold to the centre of the Earth for the revolution 5 of the QPO with normalized z-amplitude 0.08 and $\theta_0 = 200$ degrees.

It is easy to notice the Moon's influence bending the plot at several points adding a sharp oscilation. This is the main difference between the bicircular model and the RTBP. Some orbits pass near the Moon and the manifold bends. We note that the plot only has a discontinuouity if the manifold collides with the Moon.

Some of the bends, as we can see in the next enlargement, are close to the "RTBP minimum" and its displacement in some cases is big enough to reach, and even collide, with the Earth's surface in an acceptable range of the λ parameter, and at the same time they do not pass too close to the Moon.

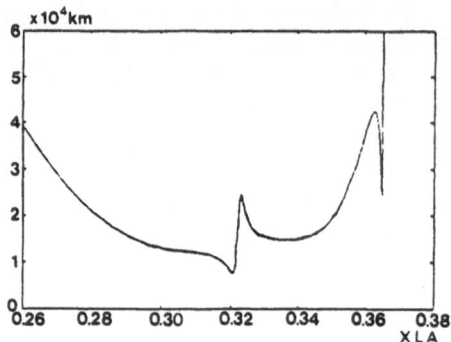 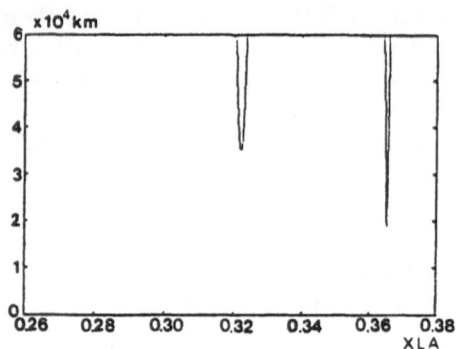

Figure 3. Enlargement of figure 2.

Figure 4. Distance from the stable manifold to the Moon for figure 3.

During the explorations, some other bends have been found far from the RTBP minimum giving also collision in a narrow range of the λ parameter. These parts of the manifold have been discarded because the strong bends come from a too close approach to the Moon (in fact most of them collide with the Moon's surface). From now on, all the results we shall show, come from the study of the manifold near the RTBP minimum.

The displacement and position of the bends depend on the angle θ_0 between the Earth and Moon as seen from the Sun at the initial epoch at which the QPO was computed.

Next plots show the change of several magnitudes when the angle θ_0 varies from 0 to 360 degrees. Letting aside translations, the behaviour is the same for all the revolutions.

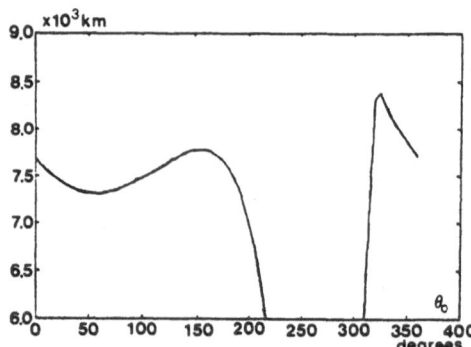 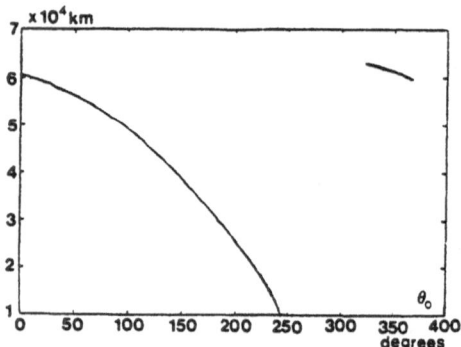

Figure 5. Minimun distance from the stable manifold to the centre of the Earth for the revolution 4 of the QPOs with normalized z-amplitude 0.08 .

Figure 6. Distance from the manifold to the centre of the Moon for the QPOs described in figure 5.

5 Orbital elements near the Earth

In order to demonstrate the feasibility of accomplishing the transfer using some of these orbits perturbed by the Moon, the orbital elements as well as the distances from the Moon and from the Earth and the velocity relative to the Earth have been computed.

Close enough to the Earth is it possible to obtain information of the orbit by means of its orbital elements since they hardly change.

Some of the results obtained are presented below. They correspond to the revolution number 5 of an orbit of normalized z-amplitude 0.08 (120000 km) and 70 degrees for the initial Sun-Earth-Moon angle. All the angles are measured in degrees, the distances in km and the velocities in m/s. The inclinations are mesured from the ecliptical plane.

(1) Parameter XLA (λ).

(2) Distance from centre of the Earth (not necessary minimum distance).

(3) Velocity relative to the Earth.

(4) Eccentricity.

(5) Inclination with respect to the ecliptical plane.

(6) Distance from the centre of the Moon.

(1)	(2)	(3)	(4)	(5)	(6)
0.148507	7270.765	10419.2	0.980	24.8	.29D+05
0.148614	6751.616	10815.4	0.981	25.7	.27D+05
0.148722	6554.471	10977.6	0.982	26.8	.26D+05
0.148829	6414.460	11097.0	0.983	28.2	.25D+05
0.148936	6462.957	11053.9	0.985	29.7	.23D+05
0.149044	6199.372	11287.7	0.986	31.7	.22D+05
0.149151	6495.623	11023.7	0.988	34.0	.21D+05
0.149258	6423.329	11084.9	0.989	37.1	.20D+05
0.149366	6492.712	11023.5	0.990	41.0	.18D+05
0.149473	6722.119	10830.3	0.992	46.2	.17D+05
0.149581	6372.335	11125.1	0.993	53.4	.16D+05
0.149688	6573.512	10949.8	0.994	63.6	.14D+05
0.149795	6521.865	10991.5	0.995	77.8	.13D+05
0.149903	6603.527	10920.1	0.995	95.9	.12D+05
0.150010	6418.619	11075.5	0.994	115.	.10D+05
0.150117	6374.898	11110.9	0.992	132.	.94D+04
0.150225	6434.840	11054.9	0.986	145.	.82D+04
0.150332	6359.881	11116.9	0.977	154.	.69D+04

ACKNOWLEDGEMENTS: The second and third authors have been partially supported by a GPC fund from MEC (Spain). The first and the fourth ones have been supported by CICYT Grant PB86-0527. Computer facilities were provided by CIRIT Grants. The methods presented here are being used in ESA project 8625/89/D/MD(SC).

References

[1] Farquhar, R.W., and Muhonen, D.P.: Mission Design for a Halo Orbiter of the Earth. *J. Spacecraft* **14**, 3 (1977), 170–177.

[2] Farquhar, R.W., Muhonen D.P., Newman, C.R., and Heuberger, H.S., "Trajectories and Orbital Maneuvers for the First Libration-Point Satellite", *J. Guidance and Control*, **3**, 6 (1980), 549–554.

[3] Farquhar, R.W., and Kamel, A.A., "Quasi-Periodic Orbits about the Translunar Libration Point", *Celestial Mechanics*, **7**, June 1973, 458–473.

[4] Hénon, M., "Vertical Stability of Periodic Orbits in the Restricted Problem I. Equal masses", *Astron, Astroph.*, **28** (1973), 415–426.

[5] Simó C., Gómez G., Jorba A., Masdemont J.: Invariant Unstable Tori Computed by Lindstedt-Poincaré Method. Reduction to the Central Manifold and Applications to Space Flight Dynamics, Proceedings of the NATO/ASI held at Cortina d'Ampezzo, 1990.

FIRST ORDER THEORY OF PERTURBED CIRCULAR MOTION:
AN APPLICATION TO ARTIFICIAL SATELLITES

E. Bois[1] and I. Wytrzyszczak[2]

[1]OBSERVATOIRE DE LA CÔTE D'AZUR, DÉPARTEMENT CERGA
AVENUE COPERNIC, 06130 GRASSE, FRANCE

[2]ASTRONOMICAL OBSERVATORY OF A. MICKIEWICZ UNIVERSITY
UL. SLONECZNA 36, 60-286 POZNAŃ, POLAND

Abstract: This paper describes briefly the particularities of an analytical theory of perturbed circular motion. The main advantage of the solution, expanded in Fourier series and in nonsingular variables, is the presence of iterative formation laws for its coefficients. The theory is then indeed particularly accurate and suitable whatever are the perturbations, their nature and their number. An application to the case of the geosynchronous satellite and the comparison of the results with a numerical integration show the degree of accuracy of the first-order solution.

1. Introduction

The present paper summarizes a new approach of analytical representation well suitable for geosynchronous satellite motion and more generally for circular orbits. The method's idea derives from the analytical theory of perturbed rotational motion that has led to confirming and generalizing the hypotheses and the algorithm set up in the numerical processes of calculation of the HIPPARCOS attitude motion by the FAST consortium in charge of the data reduction of this satellite (Bois, 1987; Kovalevsky and Bois, 1986). First results for the geosynchronous satellite have been presented at the third COGEOS meeting held in Brussels (1990) (Bois, 1990). The present paper is only devoted to a brief description of the particularities of the theory whereas the demonstration and the analytical developments can be found in Bois (1991).
The analytical theories initially developed by Kaula (1966) and Kozai in function of the keplerian classical elements are not suitable for any satellite, in particular in the cases of orbits with small eccentricity or small inclination, i.e. for almost circular or almost equatorial orbital motions and obviously not more for thoses wholly circular and equatorial. It is due to the presence of singularities in the Lagrange equations where eccentricity and inclination appear in denominators. Giacaglia (1977), Nacozy and Dallas (1977), Wytrzyszczak (1986) have then carried out regularisations of the singularities in the disturbing function but only for the geopotential. The so obtained Lagrange equations have a very complicated form and in fact none of these works gives an analytical solution. Wnuk presents however, in 1988, an analytical solution which is valid for orbits with a small eccentricity but contains singularities for zero inclinations. In 1990, Wytrzyszczak finds an analytical solution wholly regular in eccentricity and inclination. Using still the Hori's perturbation method, the author has expanded her theory in the nonsingular elements of Lagrange that are function of those of Kepler. However, the solution is very com-

plicated and uses a lot of series. Consequently, it needs a lot of computation time, much longer than with a numerical integration.

2. Generality on the theory

The present theory is directly devoted to perturbed circular motions. Starting from the equations of motion written in spherical coordinates, the resolution method gives a literal solution directly resolved in these nonsingular variables and suitable for weakly or strongly perturbed orbital motion, circular or quasi-circular, and suitable for any inclination. A formal solution is then expanded in the first order according to the powers of a small parameter characteristic of the order of magnitude of the disturbing forces and this theory is suitable whatever the forces. These forces are expanded in Fourier series and the theory applies whatever the length of these series (Bois, 1986). The coefficients of the solution are indeed given by iterative formation laws.

3. Advantages of the theory

The main advantage of the theory is the presence of iterative formation laws for the coefficients of the solution. That gives more generality and more precision for the applications. The theory is indeed suitable whatever the perturbations, their nature and their number. In many cases, directly using several perturbations improves the validity of the solution. For instance, in the case of the geosynchronous satellite, it can be particularly useful not to reduce the disturbing function to the geopotential but to extend it to potentials of the Moon and Sun and to the Solar radiation pressure. Moreover, because the algorithms of the solution use only few parameters, there is a real saving of computation time, several orders of magnitude with respect to numerical integration.

4. Accuracy of the theory

The principle of the method (minimum of approximation according to the theorem of Poincaré and a literal solution at each order) and the iterative character of its solution permit to obtain a good precision (*cf.* the analogous theory for rotational motions in Bois, 1986 and 1988, and in Bois and Kovalevsky, 1990); the first or second order being relative to the requirements in accuracy. Different tests of applications have been performed on geosynchronous satellites. Let us notice here the case of one of them submitted to the geopential and set in a mean resonance. Difference between the mean sideral motion of the Earth and the mean motion of the satellite is 6.10^{-4} rev./day. Figure 1 represents the differences between the first-order solution and a numerical integration plotted on one orbit without fitting the initial conditions, respectively in r, ϕ, λ. $\Delta r \leq 5$ m; $\Delta \phi \leq 0.9$ seconds of arc, $\Delta \lambda \leq 0.04$ seconds of arc. These differences could be better when fitting the initial conditions. Figure 2 represents the same curves but plotted on 5 orbits still without fitting the initial conditions. $\Delta r \leq 5$ m; $\Delta \phi \leq 0.9$ seconds of arc, $\Delta \lambda \leq 0.12$ seconds of arc. The accuracy is stable in r and ϕ while the appeerence of a slope in λ refers to a circulation of the variable signifing, at the first order, the not complete representation of the libration in longitude.

5. Conclusion

A formal solution is expanded in the first-order according to the powers of a small parameter characteristic of the order of magnitude of the disturbing forces that are themselves expanded in Fourier series. The theory can be applied whatever the length of these series because the coefficients of the solution are given by iterative formation laws. The comparison of the results with a numerical integration is convincing. Many applications to artificial satellites could be interesting, maybe advantageous, at least in computation time.

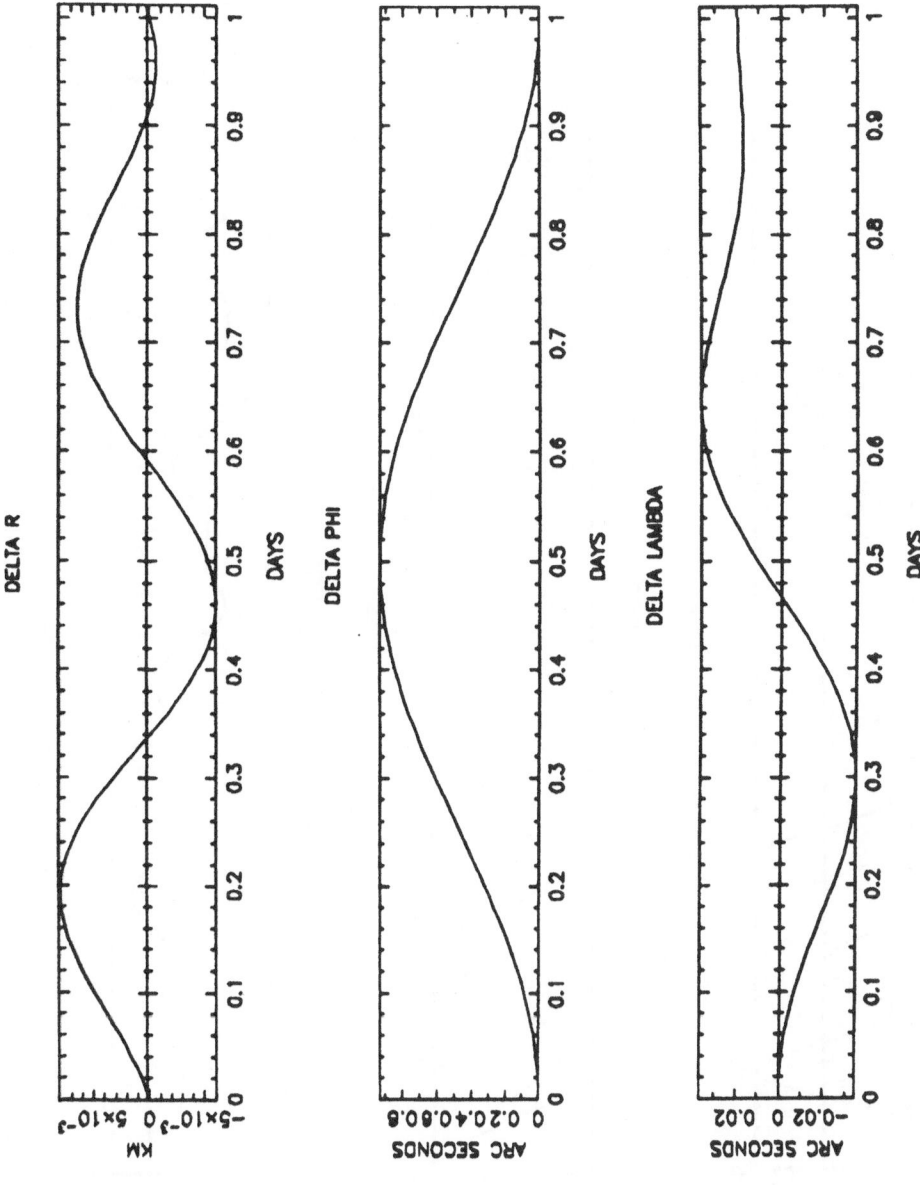

Fig. 1. Differences between the analytical solution and the numerical integration on one orbit.

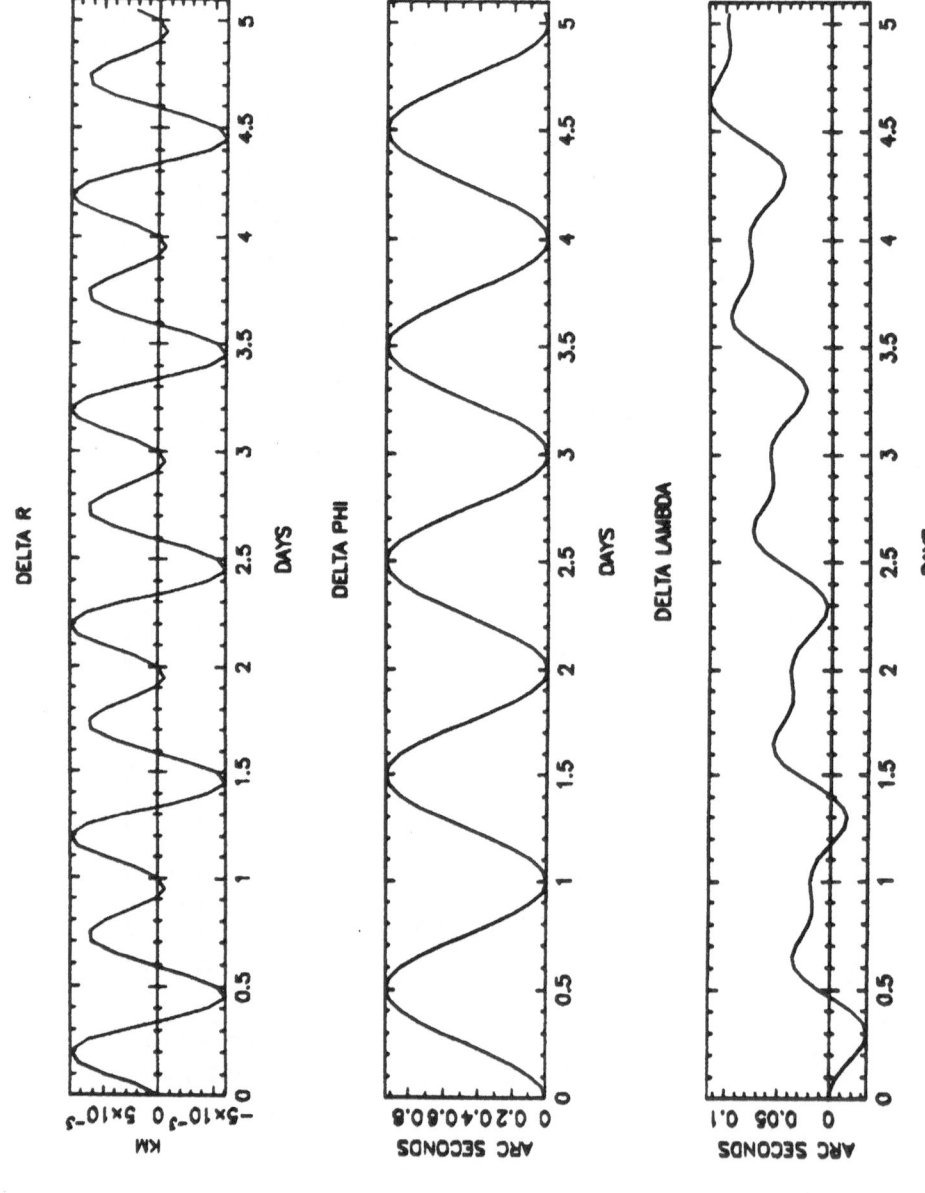

Fig. 2. Differences between the analytical solution and the numerical integration on 5 orbits.

6. References

Bois, E., 1986, "First-Order Theory of Satellite Attitude Motion - Application to HIPPARCOS", *Celestial Mechanics* **39**, No. 4, pp. 309-327.

Bois, E., 1987, "Analytical Theory of the Rotation of an Artificial Satellite", *Proceedings of the NATO Advanced Study Institute on: "Long-Term Dynamical Behaviour of Natural and Artificial N-Body Systems"*, held in Cortina d'Ampezzo, Italy, Aug. 87, Edited by Archie E. Roy, NATO ASI Series, Series C: Mathematical and Physical Sciences - Vol. 246, pp. 149-154.

Bois, E., 1988, "Second-Order Theory of the Rotation of an Artificial Satellite ", *Celestial Mechanics* **42**, Nos. 1-4, pp. 141-168.

Bois, E., 1990, "Analytical Theory of Perturbed Circular Orbital Motion", *Third COGEOS Workshop*, Bruxelles, Juin 90.

Bois, E., 1991, "First-Order Theory of Perturbed Circular Motion", paper in preparation for *Celestial Mechanics*.

Bois, E. and Kovalevsky, J., 1990, "Analytical Model of the Rotation of an Artificial Satellite", *Journal of Guidance, Control, and Dynamics* **13**, No. 4, Publication of the American Institute of Aeronautics and Astronautics, pp. 638-643.

Giacaglia, G.E.O., 1977, "The Equations of Motion of an Artificial Satellite in Nonsingular Variables", *Celestial Mechanics* **15**, p. 191.

Kaula, W., 1966, *Theory of Satellite Geodesy*, Blaisdell, Waltham, Mass..

Kovalevsky, J. and Bois, E., 1986, "Attitude Determination of the HIPPARCOS Satellite", in K.B. Bhatnagar (ed.), *Space Dynamics and Celestial Mechanics*, pp. 345-354.

Nacozy, P.E., Dallas, S.S., 1977, "The Geopotential in Nonsingular Orbital Elements", *Celestial Mechanics* **15**, p. 453.

Wnuk, E., 1988, "Tesseral Harmonic Perturbations for High Order and Degree Harmonics", *Celestial Mechanics* **44**, p. 179.

Wytrzyszczak, I., 1986, "Nonsingular Elements in Description of the Motion of Small Eccentricity and Inclination Satellites", *Celestial Mechanics* **38**, p. 101.

Wytrzyszczak, I., 1990, "Petite excentricité et petite inclinaison dans les problèmes orbitaux de satellites artificiels de la Terre", *Seconde Table Ronde de Planétologie Dynamique*, Vars, Mars 1990, Proceedings to be published.

Wytrzyszczak, I., 1990, "First Order Perturbations in Spherical Coordinates r, φ, λ", *Third COGEOS Workshop*, Bruxelles, June 1990.

POINCARÉ–SIMILAR VARIABLES INCLUDING J_2–SECULAR EFFECTS

Luis Floría and José M. Ferrándiz

Departamento de Matemática Aplicada a la Ingeniería
E.T.S. de Ingenieros Industriales
Universidad de Valladolid, 47011 Valladolid, Spain

ABSTRACT

In a previous paper we had defined a set of eight generalized canonical Delaunay–Similar (GDS) variables incorporating the first–order secular effects present in the Main Problem in the Theory of Earth's Artificial Satellite. The new GDS set was derived by means of a canonical transformation whose generating function is inspired by Deprit's radial intermediary and can be considered as a generalization of a canonical set introduced by Scheifele. When applied to Deprit's intermediary, the proposed variables lead to a simple solution, the momenta being constant and the co-ordinates being either a constant or a linear function of the independent variable. As a further step, a set of generalized Poincaré–Similar (PS) canonical variables corresponding to the aforesaid DS ones is now constructed; the new GPS set also exhibits the same feature of containing the whole first–order secular contribution of the J_2 zonal harmonic of the Earth's potential and is free from singularities.

1. INTRODUCTION

Within the framework of the Analytical Satellite Theories in the extended phase space attached to the polar–nodal Hill variables, we developed [5] a set of eight canonical variables that can be considered as a generalization of the Delaunay–Similar (DS) variables introduced by Scheifele ([10], [11], [12]). The new independent variable we used could be understood as a "generalized true anomaly".

Our DS–type set was derived by means of a canonical transformation whose generating function depends on the old co–ordinates and the new momenta, and was inspired by Deprit's radial intermediary ([3]). This fact allowed us to incorporate the whole first–order secular perturbation due to the J_2 zonal harmonic into the defining relations for the new variables. Moreover, when expressed in terms of this GDS set, Deprit's intermediary takes the same form as an unperturbed two–body problem written in Scheifele variables with the true anomaly as the independent variable ([12, p. 3]) , and therefore we are led to a simple solution to the intermediary problem: all four momenta and two angular co–ordinates are constant, while the remaining two coordinates behave like fast angular variables linearly dependent on our fictitious time.

Accordingly, the variables so defined do contain "per se" the whole first–order secular contribution of the potential corresponding to the Main Problem in Satellite Theory, and the first–order terms in the Hamiltonian function of the Main Problem are quasiperiodic in two angular variables.

However the redundant GDS canonical set that we constructed from the polar–nodal Hill variables is not, in certain cases, well suited for the calculation of perturbations, because it is not free from the appearance of virtual singularities ([4], [6]) corresponding to small values of eccentricity and inclination; as a matter of fact, when perturbation methods are applied and partial derivatives of the first–order Hamiltonian with respect to the new momenta are considered, there can appear small divisors causing singularities or indeterminate forms in the perturbation formulae.

We next aim to obtain a new canonical set which enables us to set up the perturbation equations in a nonsingular form. Classical, modified and generalized singularity–free Poincaré variables have proved ([1], [7], [8], [12]) to work successfully not only in analytical developments but also in circumventing numerical difficulties related to the use of sets of polar–like variables, although other resorts (e.g., the application of the Lie transform technique, [4], [6]) have also been devised in order to avoid virtual singularities in Perturbation Theory.

As a further development of the GDS set previously introduced, we propose an associate Poincaré–Similar (GPS) canonical set that shares with the former the property of incorporating in their own definition the whole first–order J_2–secular perturbation appearing in the Main Problem. It is expected that the application of these sets of variables should improve long–term predictions of the dynamical behaviour in satellite motion.

In Section 2 we shall summarize a description of the DS set from which the construction of the corresponding PS variables will be carried out in Section 3.

2. REVIEWING PREVIOUS GDS VARIABLES

In this Section we restrict ourselves to writing down the expressions for our DS variables, whose derivation was obtained in [5], and we apply them to Deprit's intermediary and to the Main Problem.

Starting from Hill variables in extended phase space $(r, \theta, \nu, t; P_r, P_\theta, P_\nu, P_0)$, where P_0 (the negative of the total energy in the problem to which the variables are applied) is the canonically conjugated momentum corresponding to the physical time t, we shall consider the transition to a set of canonical variables $(q_\Phi, q_G, q_N, q_L; \Phi, G, N, L)$ implicitly defined by the canonical transformation

$$P_r = \sqrt{Q} \ , \qquad q_\Phi = Cf \ ,$$

$$P_\theta = G \ , \qquad q_G = \theta - Cf = \theta - q_\Phi \ ,$$

$$P_\nu = N \ , \qquad q_N = \nu - \frac{3\varepsilon N}{\gamma^4} \frac{f}{\Gamma} \ ,$$

$$P_0 = L \ , \qquad q_L = t - \frac{\mu}{(2L)^{3/2}} [E - e \sin E - q_\Phi] \ ,$$

where the following notations and useful abbreviations have been introduced:

$$\gamma = G - \Phi + \frac{\mu}{\sqrt{2L}} \ , \qquad Q = \frac{2\mu}{r} - 2L - \frac{\gamma^2}{r^2} - \frac{\varepsilon}{r^2\gamma^2}\left[\frac{3N^2}{\gamma^2} - 1\right]$$

$$\Gamma^2 = \gamma^2 + \frac{\varepsilon}{\gamma^2}\left[\frac{3N^2}{\gamma^2} - 1\right] \ , \quad \sigma = \frac{6N^2}{\gamma^2} - 1 \ , \quad C = \frac{\gamma}{\Gamma} - \frac{\varepsilon}{\gamma^3}\frac{\sigma}{\Gamma} \ ,$$

and the two auxiliary variables f and E behave like true and eccentric anomalies for a Keplerian motion with semi–latus rectum, semi–major axis and eccentricity given by

$$p = \frac{\Gamma^2}{\mu} \ , \quad a = \frac{\mu}{2L} \ , \quad e = \sqrt{1 - \frac{2L}{\mu}p} = \sqrt{1 - \frac{2L\Gamma^2}{\mu^2}} \ ,$$

that is,

$$r = \frac{p}{1 + e\cos f} \ , \quad r = a(1 - e\cos E) \ .$$

Finally, μ stands for the gravitational parameter of the Newtonian central field of the Earth, and

$$\varepsilon = -\frac{\mu^2 R_E^2 J_2}{2}$$

denotes a small parameter related to the geophysical constant J_2 (Earth's oblateness parameter), R_E being the mean equatorial radius of the Earth.

Let us point out that the above canonical transformation can be derived from the generating function

$$S \equiv S(r,\theta,\nu,t;\Phi,G,N,L)$$
$$= \theta G + \nu N + tL + \int_{r_0}^{r} \sqrt{Q}\,dr \ ,$$

r_0 being a positive root of the equation $Q(r;\Phi,G,N,L) = 0$.

The Main Problem in extended Hill variables is known to be described by the Hamiltonian

$$H_{MAIN} = \frac{1}{2}\left[P_r^2 + \frac{P_\theta^2}{r^2}\right] - \frac{\mu}{r} + \varepsilon\,\frac{(3c^2 - 1)}{2\mu r^3} + \varepsilon\,\frac{3s^2}{2\mu r^3}\cos 2\theta + P_0 \ ,$$

where

$$c = \cos I = \frac{P_\nu}{P_\theta} \ , \quad s = \sin I \ ,$$

I being the inclination of the orbital plane. For the sake of brevity we put

$$\mathcal{G} = (G + \gamma)\left\{1 - \frac{\varepsilon}{G^2\gamma^2}\left[\frac{3N^2(G^2 + \gamma^2)}{G^2\gamma^2} - 1\right]\right\} \ .$$

After performing a change of time variable given by the differential relation

$$dt = \frac{2r^2}{\mathcal{G}}\,d\tau \ ,$$

we obtain the new homogeneous Hamiltonian in the GDS set as

$$K_{MAIN} = K_0 + \varepsilon\,K_1 \ ,$$

with

$$K_0 = \Phi - \frac{\mu}{\sqrt{2L}} \ ,$$

$$K_1 = \frac{1}{\mathcal{G}\Gamma^2} \left\{ e(3c^2 - 1) \cos\left(\frac{q_\Phi}{C}\right) + 3s^2 \cos(2q_\Phi + 2q_G) \right\}$$

$$+ \frac{1}{\mathcal{G}\Gamma^2} \frac{3es^2}{2} \left\{ \cos\left[\left(2 + \frac{1}{C}\right) q_\Phi + 2q_G\right] + \cos\left[\left(2 - \frac{1}{C}\right) q_\Phi + 2q_G\right] \right\} \ .$$

Obviously no secular term appears in the first–order part of K_{MAIN} . In view of this result, we conclude that these GDS variables indeed contain the whole first–order secular contribution of the potential defining the Main Problem, and K_1 is quasiperiodic in the two angular co–ordinates q_Φ and q_G.

Notice that this expression for K_{MAIN} is similar to the corresponding one in DS variables ([12, p. 5]), although somewhat more involved.

As a final remark concerning these variables, observe that from Deprit's radial intermediary R_0 ([3, p.138, formula (57)]) we construct the homogeneous Hamiltonian

$$H_{DEPRIT} = \frac{1}{2}\left[P_r^2 + \frac{P_\theta^2}{r^2}\right] - \frac{\mu}{r} + \varepsilon \frac{(3c^2 - 1)}{2r^2 P_\theta^2} + P_0 \ ,$$

that, after the aforesaid transformation, takes the form

$$\tilde{H}_{DEPRIT} = \frac{\mathcal{G}}{2r^2}(G - \gamma) \ ,$$

and the final expression for this Hamiltonian corresponding to the new indepedent variable τ is

$$K_{DEPRIT} = G - \gamma = \Phi - \frac{\mu}{\sqrt{2L}} \ .$$

These considerations show that, in the GDS variables, K_0 (the zero–order part in K_{MAIN}) exhibits the same form as K_{DEPRIT}, this form being also the same as that of the Hamiltonian of an unperturbed two–body problem expressed in Scheifele variables with the true anomaly as the independent variable. The canonical equations of motion for K_{DEPRIT} result in a simple solution: the co–ordinates q_Φ and q_L are linear functions of the independent variable τ, and the remaining GDS variables are constant with respect to τ.

3. DERIVATION OF THE GPS VARIABLES

In order to overcome the difficulties due to singularities introduced by the use of the Delaunay–like GDS variables, we shall next develop an adequate set of generalized Poincaré–Similar canonical variables that are regular in behaviour at small and vanishing values of eccentricity or inclination.

According to the usual practice ([2, pp.538–540]; [14, Bd.2., p. 183, pp.236–238]), in this Section we shall obtain our new GPS variables from the GDS ones by means of the composition of two successive canonical transformations.

Irrespective of the dynamical problem to which it might be applied, the set

$$(q_\Phi, q_G, q_N, q_L; \Phi, G, N, L)$$

of generalized DS variables is taken as the starting point of the process of construction. In what follows we will adopt Bond's derivation ([1, pp. 289–290]) of the PS variables.

We first consider the transition to an intermediate canonical set

$$(y_1, y_2, y_3, y_4; Y_1, Y_2, Y_3, Y_4)$$

implicitly defined via the generating function

$$S_1 = S_1(q_\Phi, q_G, q_N, q_L; Y_1, Y_2, Y_3, Y_4)$$

$$= q_\Phi Y_1 + q_G(Y_1 - Y_2) + q_N(Y_1 - Y_2 - Y_3) + q_L Y_4$$

depending on the old angular co-ordinates and the new momenta. The equations of transformation are then derived from the relations

$$y_1 = \frac{\partial S_1}{\partial Y_1} = q_\Phi + q_G + q_N , \qquad Y_1 = \Phi ,$$

$$y_2 = \frac{\partial S_1}{\partial Y_2} = -q_G - q_N , \qquad Y_2 = \Phi - G ,$$

$$y_3 = \frac{\partial S_1}{\partial Y_3} = -q_N , \qquad Y_3 = G - N ,$$

$$y_4 = \frac{\partial S_1}{\partial Y_4} = q_L , \qquad Y_4 = L ,$$

since

$$\Phi = \frac{\partial S_1}{\partial q_\Phi} = y_1 , \qquad G = \frac{\partial S_1}{\partial q_G} = Y_1 - Y_2 = \Phi - Y_2 ,$$

$$N = \frac{\partial S_1}{\partial q_N} = Y_1 - Y_2 - Y_3 = G - Y_3 , \qquad L = \frac{\partial S_1}{\partial q_L} = Y_4 .$$

We next perform a canonical transformation from the preceding intermediate set to the desired Poincaré–Similar variables $(\sigma_1, \sigma_2, \sigma_3, \sigma_4; \varrho_1, \varrho_2, \varrho_3, \varrho_4)$; it is achieved through the generating function ([9, p. 71])

$$S_2 = S_2(Y_1, Y_2, Y_3, Y_4; \sigma_1, \sigma_2, \sigma_3, \sigma_4)$$

$$= \sigma_1 Y_1 + \frac{1}{2}\left[\sigma_2\sqrt{2Y_2 - \sigma_2^2} + 2Y_2 \arcsin\left(\frac{\sigma_2}{\sqrt{2Y_2}}\right)\right]$$

$$+ \frac{1}{2}\left[\sigma_3\sqrt{2Y_3 - \sigma_3^2} + 2Y_3 \arcsin\left(\frac{\sigma_3}{\sqrt{2Y_3}}\right)\right] + \sigma_4 Y_4 ,$$

that clearly involves the old momenta and the new co-ordinates. The formulae defining the transformation are now

$$y_i = \frac{\partial S_2}{\partial Y_i} , \qquad \varrho_i = \frac{\partial S_2}{\partial \sigma_i} , \qquad i = 1, 2, 3, 4 .$$

After completing the required calculations we get

$$\sigma_1 = y_1 , \qquad\qquad \varrho_1 = Y_1 ,$$

$$\sigma_2 = \sqrt{2Y_2}\sin y_2 , \qquad \varrho_2 = \sqrt{2Y_2}\cos y_2 ,$$

$$\sigma_3 = \sqrt{2Y_3}\sin y_3 , \qquad \varrho_3 = \sqrt{2Y_3}\cos y_3 ,$$

$$\sigma_4 = y_4 , \qquad\qquad \varrho_4 = Y_4 .$$

By considering the composition of both canonical transformations we finally obtain the relation between our DS and PS variables:

$$\sigma_1 = y_1 = q_\Phi + q_G + q_N \;,$$

$$\sigma_2 = \sqrt{2(\Phi - G)}\sin(-q_G - q_N) = -\sqrt{2(\Phi - G)}\sin(q_G + q_N) \;,$$

$$\sigma_3 = \sqrt{2(G - N)}\sin(-q_N) = -\sqrt{2(G - N)}\sin q_N \;,$$

$$\sigma_4 = y_4 = q_L \;,$$

$$\varrho_1 = Y_1 = \Phi \;,$$

$$\varrho_2 = \sqrt{2(\Phi - G)}\cos(-q_G - q_N) = \sqrt{2(\Phi - G)}\cos(q_G + q_N) \;,$$

$$\varrho_3 = \sqrt{2(G - N)}\cos(-q_N) = \sqrt{2(G - N)}\cos q_N \;,$$

$$\varrho_4 = Y_4 = L \;.$$

The set of formulae corresponding to the inverse transformation will then read as follows

$$\Phi = \varrho_1 \;,$$

$$L = \varrho_4 \;,$$

$$q_L = \sigma_4 \;,$$

$$q_N = -\arctan\left(\frac{\sigma_3}{\varrho_3}\right) \;,$$

$$q_G = -\arctan\left(\frac{\sigma_2}{\varrho_2}\right) + \arctan\left(\frac{\sigma_3}{\varrho_3}\right) \;,$$

$$q_\Phi = \sigma_1 + \arctan\left(\frac{\sigma_2}{\varrho_2}\right) \;,$$

$$G = \varrho_1 - \frac{\varrho_2^2 + \sigma_2^2}{2} \;,$$

$$N = \varrho_1 - \frac{\varrho_2^2 + \sigma_2^2 + \varrho_3^2 + \sigma_3^2}{2} \;.$$

Inserting the above relations into the Hamiltonian K_{MAIN} would lead to the expression for the Main Problem in GPS formulation. This is a lengthy but straightforward task; we shall not display here the final form of K_{MAIN} subjected to that transformation, but will restrict ourselves to showing that corresponding to Deprit's intermediary in GPS variables , namely:

$$K_{DEPRIT}(\sigma_i, \varrho_i) = \varrho_1 - \frac{\mu}{\sqrt{2\varrho_4}} \;,$$

which allows the simple solution

$$\sigma_1 = \tau + \text{const.} \;, \qquad \sigma_4 = \frac{\mu}{(2\varrho_4)^{3/2}}\,\tau + \text{const.} \;,$$

the remaining variables behaving like constants with respect to τ.

Further particulars and more exhaustive research concerning DS and PS canonical sets, possible extensions (e.g., analogous DS and PS sets incorporating higher-order secular effects), modifications and their application to analytical and numerical integration in Earth's Satellite Theory will appear in the first author's Doctoral Dissertation.

ACKNOWLEDGEMENT

The authors acknowledge the financial support received from CICYT of Spain under Project ESP.88–0541.

References

[1] Bond, V.R.: 1976, *Celest. Mech.* **13**, 287–311.

[2] Brouwer, D. and Clemence, G.M.: 1961, *Methods of Celestial Mechanics*, Academic Press, New York. Ch. XVII, Section 5.

[3] Deprit, A.: 1981, *Celest. Mech.* **24**, 111–153.

[4] Deprit, A. and Rom, A.: 1970, *Celest. Mech.* **2**, 166–206.

[5] Ferrándiz, J.M. and Floría, L.: *Una Generalización de los Elementos DS*. Actas XIII Jornadas Hispano–Lusas de Matemáticas, Valladolid, September 1988. (In Press).

[6] Henrard, J.: 1974, *Celest. Mech.* **10**, 437–449.

[7] Lyddane, R.H.: 1963, *Astron. J.* **68**, 555–558.

[8] Palacios, M.P.: *Sistemas canónicos con un número superabundante de variables. Equivalencia de métodos de integración. Aplicaciones.* Doctoral Dissertation. Univ. Zaragoza (Spain). Available from Instituto Geográfico y Catastral (Madrid, Spain).

[9] Pollard, H.: 1966, *Mathematical Introduction to Celestial Mechanics*, Prentice Hall, Englewood Cliffs, New Jersey.

[10] Scheifele, G.: 1970a, *Celest. Mech.* **2**, 296–310.

[11] Scheifele, G.: 1970b, *Compt. Rend. Acad. Sc. Paris* **271**, 729–732.

[12] Scheifele, G. and Graf, O.: 1974, *Analytical Satellite Theories Based on a New Set of Canonical Elements*, AIAA Paper No.74–838.

[13] Stiefel, E.L. and Scheifele, G.: 1971, *Linear and Regular Celestial Mechanics*, Springer–Verlag.

[14] Stumpff, K.: 1959, 1965, *Himmelsmechanik* I, II. VEB Deutscher Verlag der Wissenschaften, Berlin.

MEASURING THE LACK OF INTEGRABILITY OF THE J_2 PROBLEM FOR EARTH'S SATELLITES

Carles Simó

Dept. de Matemàtica Aplicada i Anàlisi, Universitat de Barcelona
Gran Via 585, 08007 Barcelona, Spain

Abstract. We consider the motion around an oblate primary, keeping only the J_2 term in the expansion of the potential in spherical harmonics. The problem has cylindrical symmetry. It has been suspected for a long time, due to numerical evidences, that the problem is non integrable. This has been proved recently by Irigoyen (1990). However, even if the system is non integrable, the size of the stochastic zones can be so small that they can be neglected for all practical purposes. This is what we study here, and we show that for the case of the Earth and considering possible real orbits, i.e., non colliding with the Earth, the effect of the non integrability can be completely neglected.

1. Introduction

Let $q = (x, y, z)^T$ the coordinates of a particle around an oblate planet and p the corresponding momenta. The Hamiltonian of the system is

$$(1) \qquad H = \frac{1}{2}(p, p) - \frac{\mu}{r} + J_2 \frac{\mu a_e^2}{r^3} P_2\left(\frac{z}{r}\right),$$

where μ is the gravitational constant, $r = \|q\|$, a_e the equatorial radius of the attracting body, P_2 the Legendre polynomial of second degree and J_2 the coefficient of the zonal harmonic of order two. For the Earth these quantities are, approximately, $a_e = 6378\,\text{km}$, $\mu = 398600\,\text{km}^3\,\text{sec}^{-2}$, $J_2 = 1082 \cdot 10^{-6}$. Due to the cyclicity of the longitude in the Hamiltonian, we can use cylindrical coordinates (ρ, θ, z) and then (1) is expressed as

$$(2) \qquad H = \frac{1}{2}\left(P_z^2 + P_\rho^2\right) - \frac{\mu}{r} + \frac{c^2/2}{\rho^2} + \frac{\tilde{J}_2}{r^3}\left(\frac{3}{2}\left(\frac{z}{r}\right)^2 - \frac{1}{2}\right),$$

where c is the component in the z direction of the angular momentum and $\tilde{J}_2 = J_2 a_e^2 \mu$. This is a two degrees of freedom Hamiltonian. The dynamics of (2) can be studied by using an analytical approach or a numerical one. Probably the best thing is to combine both of them. The system obtained from (2) depends on the parameters $\mu, c, \tilde{J}_2$ and on the value of the energy, h.

On the (ρ, p_ρ) plane the available domain is defined by $p_\rho^2 \le 2h + \frac{2\mu}{\rho} - \frac{c^2}{\rho^2} + \frac{\tilde{J}_2}{\rho^3}$. The boundary $(p_z = 0)$ is an orbit of the Hamiltonian on the equatorial plane. Let $V(\rho) = -\frac{\mu}{\rho} + \frac{c^2/2}{\rho^2} - \frac{\tilde{J}_2/2}{\rho^3}$.

Predictability, Stability, and Chaos in N-Body Dynamical Systems
Edited by A.E. Roy, Plenum Press, New York, 1991

Then this orbits is written as $p_\rho^2 = 2(h - V(\rho))$. The function $V(\rho)$ has two extrema provided $c^4 > 6\mu\tilde{J}_2$. We denote them by ρ_1, ρ_2 with $0 < \rho_1 < \rho_2$. Assume $0 > h > V(\rho_2)$. Then ρ can range in two intervals, one of them containing $\rho = 0$ and being skipped in the case of the Earth because it is fully contained inside the Earth. We also assume c not too small to have $V(\rho_1) > 0$, i.e. $c^4 > 8\mu\tilde{J}_2$ (see figure 1).

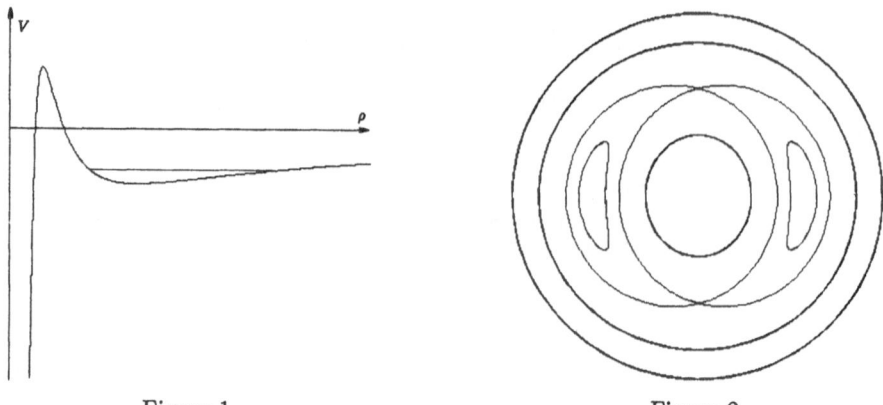

Figure 1   Figure 2

We can study a representation of the motion by using the Poincaré section through $z = 0$. The section is the interior of the available domain described before. If $J_2 = 0$ and $c \neq 0$ the Poincaré map is the identity inside the boundary, and the boundary is an orbit of the flow. Compactifying the Poincaré section (topologically an open disk) to a sphere (by adding one point which represents the boundary orbit) we have the identity on S^2. As J_2 is small and under the previous constrains on c we shall have a near the identity map for the case of the Earth. This map has, at least, two fixed points of elliptic type. There are also invariant curves and chaotic (or stochastic) zones associated to the resonances. The largest chaotic zone is related to the 1 to 1 resonance known to occur near the critical inclination, as it can be seen in Hagihara (1972). Figure 2 shows a qualitative picture of the Poincaré map when the two symmetric hyperbolic points are present. In fact the heteroclinic orbits between them do not agree. They create a very narrow stochastic layer whose size we shall bound.

2. The normal form. Hyperbolic fixed points

If J_2 is small and r is bounded away from zero we can compute a useful normal form of (1). As the system has, essentially, two degrees of freedom, as displayed by (2), the normal form is integrable. We learn from Coffey et al. (1990) that the Hamiltonian can be approximated, to the second order in J_2, by

$$
\begin{aligned}
\text{(3)} \quad N = &-\frac{\mu^2}{2L^2} + J_2 n (a_e \mu)^2 \left(-\frac{1}{4G^3} + \frac{3H^2}{4G^5} \right) \\
&+ \frac{1}{2} J_2^2 n (a_2 \mu)^4 \left\{ \frac{15}{64G^7} - \frac{15H^2}{32G^9} - \frac{105H^4}{64G^{11}} + \frac{1}{L}\left(-\frac{3}{16G^6} + \frac{9H^2}{8G^8} - \frac{27H^4}{16G^{10}} \right) \right. \\
&+ \frac{1}{L^2}\left(-\frac{15}{64G^5} + \frac{27H^2}{32G^7} - \frac{15H^4}{64G^9} \right) \\
&\left. + \cos(2g)\left[-\frac{39}{32G^7} + \frac{33H^2}{4G^9} - \frac{225H^4}{32G^{11}} + \frac{G^2 + LG + L^2}{L(L+G)}\left(\frac{3}{4G^7} - \frac{9H^2}{2G^9} + \frac{15H^4}{4G^{11}} \right) \right. \right.
\end{aligned}
$$

306

$$+ \frac{1}{L^2}\left(\frac{3}{32G^5} - \frac{3H^2}{2G^7} + \frac{45H^4}{32G^9}\right)\bigg]\bigg\}\,,$$

where L, G and H are the Delaunay momenta and g is the angle canonically conjugated to G. The variables L, G and H are related to the Keplerian semimajor axis, eccentricity and inclination by

(4) $$L(\mu a)^{1/2}, \quad G = L(1 - e^2)^{1/2}, \quad H = G\cos I,$$

and the problem is considered as an $O(J_2)$ perturbation of the Keplerian motion with mean motion n. We remark that G and H are nothing else then the total angular momentum and the z–component of the angular momentum. Hence $H = c$. The variables L and H being invariant we have that (3) is, essentially, a one degree of freedom Hamiltonian. The fixed points are obtained by setting $\frac{\partial N}{\partial G} = 0$, $\frac{\partial N}{\partial g} = 0$. In particular one obtains hyperbolic points for

$$g = \pm\frac{\pi}{L}, \; G = \sqrt{5}H\left[1 + \frac{J_2}{500}\frac{\mu^2 a_e^2}{H^4}\left(-13 + 35\frac{H^2}{L^2} + 40\frac{H^2/L^2}{1 + \sqrt{5}H/L}\right) + O(J_2^2)\right].$$

It is also possible to obtain the eigenvalues at those points. The hyperbolic points of (3) (concerning the (G, g) couple) are related to the hyperbolic points of the Poincaré section mentioned in section 1. By using (4) and the relation between a and h for the Kepler problem one obtains for the eigenvalues of the hyperbolic points:

(5) $$\log\lambda_{\text{Poincaré}} = \frac{2\pi(a_e\mu)^3}{10c^5 5^{5/2}}(3J_2)^{1/2}\left(\frac{3}{c^2} + \frac{30h}{\mu^2}\right)^{1/2}J_2(1 + O(J_2)),$$

provided the factor $\left(3c^{-2} + 30h\mu^{-2}\right)^{1/2}$ is not too small (then the higher order terms in J_2 become important).

As said in section 1, the separatrices of figure 2 are, in fact, split. In the next section we shall give bounds on the size of the splitting. The expression (5) will play a very important role. It agrees quite well with direct numerical computations of the Poincaré map.

3. The maximal width of the stochastic zone

As shown by Lazutkin (1984) for the standard map the width of the stochastic zone is exponentially small with respect to the parameter of that map. In Fontich and Simó (1990) it is proved that the size of the stochastic zone is bounded by expressions of the type $M\exp\left(-\frac{2\pi(\delta - \eta)}{\log\lambda}\right)$, where λ is the eigenvalue at the homoclinic or heteroclinic points of the analytic near the identity area preserving map, η is any positive quantity (to be choosen small), M is a constant which only depends on η (and not on the small parameter, J_2 in our case) and δ is the minimum distance to the real axis of the singularities of the separatrix of some Hamiltonian planar flow (see Fontich and Simó (1990) for the details). In our case the planar flow is essentially a pendulum as it follows from (3) using the fact that G changes by an small amount along the separatrix. So, with the required scalings one has $\delta = \pi/2$.

The basic idea now is that to have the largest possible values of the stochastic zone we had to use the largest possible value of λ as given by (5). Let D the minimum value allowed to the perigeon distance of an artificial Earth satellite. Typically D can be taken as 6600 km. Then (5) should be maximized under the constrain $q = a(1 - e) \geq D$. By using

$$G = \sqrt{5}c\left(1 + O(J_2)\right), \quad a = -\frac{\mu}{2h}\left(1 + O(J_2)\right) \quad \text{and} \quad 1 - e^2 = \left(\frac{G}{L}\right)^2,$$

one obtains the equivalent constrain (skipping, from now on, the terms $O(J_2)$ when they appear as $1 + O(J_2)$)

(6)
$$\left[1 - \left(1 + \frac{10c^2h}{\mu^2}\right)^{1/2}\right]\left(\frac{\mu}{-2h}\right) \geq D.$$

To maximize (5) it is enough to maximize the factor depending on c and h, i.e. $c^{-5}\left(\frac{3}{c^2} + \frac{30h}{\mu^2}\right)^{1/2}$. By using (6) this factor is bounded by

(7)
$$\frac{\left(1 + \frac{2Dh}{\mu}\right)h^3}{\left[\left(1 + \frac{2Dh}{\mu}\right)^2 - 1\right]^3},$$

where we have skipped numerical factors or factors depending only on μ. Let $w = 2Dh\mu^{-1}$. The maximum of (7) is obtained for $w = -1/2$ and this implies $h = -\frac{\mu}{4D}$, $c = \left(\frac{3\mu D}{10}\right)^{1/2}$ and $e = \frac{1}{2}$. Hence

(8)
$$\max \log \lambda_{\text{Poincaré}} = \frac{4\pi}{9\sqrt{5}}\left(\frac{a_e}{D}\right)^3 J_2^{3/2}.$$

Using $D = 6600$ km this amounts to $2 \cdot 10^{-5}$. As $\delta = \pi/2$ we can take $\eta = \frac{\pi^2 - 9}{2\pi} \simeq 0.1384$ to have the simple value $2\pi(\delta - \eta) = 9$. It remains to estimate M to have the desired upper bound. But this is irrelevant. Indeed one can use the constructive method given in Fontich and Simó (1990) or simply we can compute numerically for much larger values of J_2. Numerically one can estimate M to be of the order of units, but even a relative error by a factor of 10^{100} is irrelevant because the dominant term is $\exp\left(-\frac{9}{\log \lambda}\right) \leq \exp\left(-4.5 \cdot 10^5\right)$.

4. Conclusion

It has been obtained that the J_2 problem for a feasible artificial Earth satellite, even being non integrable, can be considered as integrable for all practical purposes. This behaviour is shared by many other problems (for instance, the Hénon–Heiles problem for energies less then 0.04). We remark that despite the practical integrable character it is a hard task to obtain, in general, rather good analytical approximations to the solutions. Normal forms up to high order can be very useful for this purpose.

Acknowledgements. This work has been partially supported by a CICYT Grant PB 86–527. The computing facilites were provided by a CIRIT Grant.

References

Coffey, S., Déprit, A., Déprit, E., Healy, L., 1990, Painting the Phase Space Portrait of an Integrable Dynamical System, <u>Science</u>, 247:833.

Fontich, E., Simó, C., 1990, The splitting of separatrices for analytic diffeomorphisms, <u>Ergod. Th. & Dynam. Sys.</u>, 10:295.

Hagihara, Y., 1972, "Celestial Mechanics", Vol. 2, Part 1, p.422, MIT Press, Boston.

Irigoyen, M., 1990, Non integrability of the J_2 problem, preprint, Univ. de Paris 2.

Lazutkin, V. F., 1984, Splitting of separatrices for the Chirikov's standard map, Preprint VINITI 6372/84, Leningrad.

THE EFFECTS OF THE J_3-HARMONIC (PEAR SHAPE) ON THE ORBITS OF A SATELLITE

R.A. Broucke

University of Texas, Austin, Texas 78712

INTRODUCTION

The object of the present article is a detailed numerical investigation of the perturbation on the orbit of a satellite, caused by the pear-shape or J_3-Harmonic of the central body. We principally use concepts from the general theory of periodic orbits, such as Poincaré surfaces of section, stability theory, characteristic exponents and bifurcations.

The principal orbital perturbations due to the J_3-Harmonic were first published simultaneously by Brouwer (page 390), Kozai (page 375) and Garfinkel (page 366) in the famous issue of the Astronomical Journal (November 1959) which contains fundamental theoretical articles on the Main Problem of Artificial Satellite Theory. The results given by Brouwer, Garfinkel and Kozai are actually identical. They all give first-order effects only and they average the disturbing function in order to eliminate the short period terms. They show that the principal first-order perturbations due to J_3 are long-term effects on the four classical elements e, i, ω, Ω while the semi major axis a remains constant. There are no first-order secular perturbations caused by J_3. These Brouwer-Garfinkel-Kozai formulas have been used as such by many different people later, for instance in the context of the so-called frozen orbits.

The principal justification of our undertaking a new detailed study of the J_3-perturbations is the fact that J_3 has a strong effect on the critical inclination ($\sin^2 i = 4/5$). It is well known that the average rate of the perigee motion, $\dot{\omega}$ contains the factor $(5 \sin^2 i - 4)J_2$, so that due to $J2$ alone, there will be virtually no motion of the perigee near $i = 63°.4$).

However it is not as well-known but it is a true fact that the averaged J_3-disturbing function also contains the same factor $(5 \sin^2 i - 4)$. This factor also turns up in the equations for $\dot{e}$ and $\dot{i}$. On the other hand the same factor is not present for all the other higher order harmonics such as J_4, J_5, ... or even the tesseral harmonics.

The consequence of this situation is that the fundamental properties of the orbital perturbations near the critical inclination are essentially due to two and only two harmonics in the expansion of the Earth's gravity field. Several satellites which have been launched (the Soviet Molnya's) in orbits at or near the critical inclination as well as much theoretical research on the critical inclination shows the extreme complexity of the perturbations in this region. For a list of references on the critical inclination, we refer the reader to A. Jupp's review article (1987).

It is because of the reasons and we started our present research on the simultaneous effects of two harmonics J_2 and J_3 on the properties of the orbits.

The principal conclusion of our studies of the $(J_2 + J_3)$ - effects is the existence of a complex sequence of bifurcations of periodic orbits at and just above the usual critical inclination of $63^\circ.4$. It is mainly in order to understand this sequence of bifurcations that we found it necessary to also make separate investigation of the J_2- effects alone (Broucke and Kim, 1990) and the J_3-effects alone.

We agree that the J_3-harmonic alone is a rather unrealistic abstract situation, but we feel that it is necessary to know some of the basic properties of each harmonic by itself in order to understand the global combined effects. We already published a study of the effects of the J_2-harmonic by itself. We are also preparing some other articles on the combined effects, both with numerical integrations and with the use of averaged analytical formulas.

We used canonical units in all of the present work. The total energy of all the orbits is -1/2 (corresponding to a mean semi-major axis $\bar{a} = +1$ and a Delaunay variable L = 1). We also assume that the central body has GM and radius both equal to unity. The two principal variable parameters in our study are $C_z^2 = H^2$, (Z-Component of angular momentum = C_z) and J_3 itself. We made integrations with rather large values of J_3, equal to ±0.01 for and ±0.001 for instance. The sign of J_3 is irrelevant here because it only interchanges the Northern and Southern Hemisphere properties of the orbits.

THE EQUATIONS OF MOTION

The potential function for the gravitational field of the Earth has a well-known expansion in orthogonal functions which are called Spherical Harmonics. We will only be concerned with the so-called zonal harmonics. The first three terms of the expansion are then

$$U = \frac{GM}{r} + \frac{GMR^2}{r^3} C_{20}P_2(\sin \phi) + \frac{GMR^3}{r^4} C_{30}P_3(\sin \phi), \qquad (2.1)$$

where μ = GM is the central mass gravity coefficient and R the equatorial radius of the central body. The two dimensionless constants C_{20} and C_{30} determine the size of the perturbations. The constant $J_2 = -C_{20}$ is associated with the flattening at the poles of the central body while the last constant $J_3 = -C_{30}$ is related to the pear shape. The two relevant coordinates of the satellite are here the radius-vector f and the latitude ϕ. In other words, we have a description with spherical coordinates, but the longitude λ is absent and ignorable because we assume cylindrical symmetry of the central body, around the z-axis. Finally, the two functions of sin ϕ, P_2 and P_3 are the Legendre Polynomials:

$$P_2 (\sin \phi) = \frac{1}{2} (3 \sin^2 \phi - 1), \tag{2.2}$$

$$P_3 (\sin \phi) = \frac{1}{2} (5 \sin^3 \phi - 3 \sin \phi). \tag{2.3}$$

We note that the rectangular coordinates (ρ, z) of the satellite in the co-rotating meridian plane will often be used. These are related to the three cartesian coordinates (x, y, z) by

$$x = \rho \cos \lambda ; \qquad \rho = r \cos \phi ,$$
$$y = \rho \sin \lambda ; \qquad z = r \sin \phi , \tag{2.4}$$

so that we also have

$$r^2 = x^2 + y^2 + z^2 = \rho^2 + z^2 \tag{2.5}$$

In the present paper we intend to concentrate ourselves on a detailed study of the J_3 - or Pearshape - harmonic of the central body and from now on we will set J_2 equal to zero.

The Hamiltonian of our problem can now be written as

$$K = \frac{1}{2} (\dot{r}^2 + \frac{p_\phi^2}{r^2}) + \frac{H^2}{2\rho^2} - \frac{\mu}{r} + \frac{\mu R^3 J_3}{2r^4} (5 \sin^3\phi - 3 \sin \phi), \tag{2.6}$$

or, if we express it in the variables ρ and z:

$$K = \frac{1}{2} (\dot{\rho}^2 + \dot{z}^2) + \frac{H^2}{2\rho^2} = \frac{\mu}{r} + \frac{\mu R^3 J_3}{2r^4}(5 \frac{z^3}{r^3} - 3 \frac{z}{r}), \tag{2.7}$$

where r should be thought of as a function of c and z, according to (5), and where we have the two conjugate momenta:

$$\dot{\rho} = \dot{r} \cos \phi - r \dot{\phi} \sin \phi ,$$
$$\dot{z} = \dot{r} \sin \phi - r \dot{\phi} \cos \phi . \tag{2.8}$$

In the Hamiltonian (7), we have also introduced the constant H, which is the z-component of the angular momentum, one of the two integrals of the problem:

$$r^2 \cos^2 \phi . \dot{\lambda} = \rho^2 \dot{\lambda} = H = G_z . \tag{2.9}$$

The other integral is of course the energy K itself. Our calculations for the present paper will be based on the two canonical variables (ρ, z). We will choose the units of length and time in such a way that $R = \mu = 1$.

The equations of motion in these variables can be written as:

$$\ddot{\rho} = + \frac{H^2}{\rho^3} + \frac{\partial U}{\partial \rho} = + \frac{H^2}{\rho^3} - \frac{\rho}{r^3} + \frac{5 J_3 \rho z}{2r^7} (7 \frac{z^2}{r^2} - 3),$$
$$\ddot{z} = \frac{\partial U}{\partial z} = - \frac{z}{r^3} + \frac{J_3}{2r^5} [3 - 30 \frac{z^2}{r^2} + 35 \frac{24}{r^4}] . \tag{2.10}$$

Most of our calculations in the J_3 - problem have been done with the same values for the different constants involved in this problem. We take the μ = GM as well as the radius of the Earth R equal to unity. We also take J_3 = 0.01 in most of the numerical work and we limit ourselves to isoenergetic orbits, with energy E = -0.5. In the perturbed Kepler problem, at this energy value, the circular orbits of unit radius also have a unit linear velocity and a period 2π, which gives a convenient scaling of the solutions, with no loss of generality.

INTEGRATION WITH EQUINOTIAL ELEMENTS OR WITH HILL VARIABLES

In the type of applications we are mostly interested in, which is the generation of Poincaré surfaces of sections, it is useful to integrate some orbital elements rather than coordinates. In this formulation, we need the three orthogonal components (R, S, W: Radial, Transverse, Normal) of the J_3 - perturbation:

$$R = \frac{4\mu J_3 R_e^3}{r^5} P_3(\sin \phi),$$

$$S = \frac{-4\mu J_3 R_e^3}{r^5} \sin i \cos u \cdot P_3'(\sin \phi), \qquad (3.1)$$

$$W = \frac{-\mu J_3 R_e^3}{r^5} \cos i \cdot P_3'(\sin \phi),$$

where the symbol P_3' represents the derivative of the Legendre Polynomial:

$$P_3' (\sin \phi) = \frac{3}{2} (5 \sin^2 \phi - 1). \qquad (3.2)$$

We notice the obvious relation between the S and W-components

$$S \cos i = W \sin i \cos u, \qquad (3.3)$$

which is just another way of expressing the cylindrical symmetry of the problem around the z-axis.

A remarkable set of variables selected for our study is the canonical set of two-dimensional Hill variables (also called Whittaker variables), defined as follows (Hill, 1913) (Aksnes, 1970):

r: the radius.
$\dot{r}$:   the radial velocity component.
u: the argument of latitude = $\omega + v$.
G: the magnitude of the angular momentum.

The Transformation between Meridian variables and Hill variables might begin with the simple relations:

$$r^2 = \rho^2 + z^2,$$
$$r\dot{r} = \rho\dot{\rho} + z\dot{z}. \qquad (3.4)$$

G can be obtained from the following relation between angular momenta:

$$G^2 = P_\phi^2 + \frac{H^2}{\cos^2\phi} \quad , \tag{3.5}$$

where P_ϕ is the conjugate momentum of ϕ given by

$$P_\phi = r^2\dot\phi = \rho\dot z - \rho\dot z . \tag{3.6}$$

To construct the equations for the transformation of u which is the angle between the ascending node and the radius vector, let's consider the spherical triangle formed by the equatorial plane, the meridian plane and the orbital plane.

We know the basic relation:

$$\sin i \, \sin u = \sin\phi \tag{3.7}$$

We also know the $|\phi|\leq\pi/2$ but u is a 2π-periodic variable so (4.5) is not sufficient to determine u. Taking the time - derivative of (4.5) and using the relations $G = r^2\dot u$ and $p_\phi = r^2\dot\phi$, we obtain:

$$G \sin i \, \cos u = P_\phi \cos\phi. \tag{3.8}$$

Now, we can summarize the two-way conversion formulas between Meridian variables and Hill variables

$$\left.\begin{array}{ll}
r = (\rho^2 + z^2)^{\frac{1}{2}} & \rho = r\cos\phi \\[2mm]
\dot r = (\rho\dot\rho + z\dot z)/r & \dot\rho = \dot r\cos\phi - \dfrac{P_\phi}{r}\sin\phi \\[2mm]
G = (P_\phi^2 + \dfrac{r^2}{\rho^2} H^2)^{\frac{1}{2}} & z = r\sin\phi \\[2mm]
\sin u = \dfrac{G}{\sqrt{G^2 - H^2}}\sin\phi & \dot z = \dot r\sin\phi + \dfrac{P_\phi}{r}\cos\phi \\[3mm]
\cos u = \dfrac{P_\phi}{\sqrt{G^2 - H^2}}\cos\phi &
\end{array}\right\} \tag{3.9}$$

where

$$\sin\phi = \frac{z}{r} = (1 - \frac{H^2}{G^2})^{\frac{1}{2}}\sin u,$$

$$\cos\phi = \frac{\rho}{r} = (\cos^2 u + \frac{H^2}{G^2}\sin^2 u)^{\frac{1}{2}} , \tag{3.10}$$

$$P_\phi = \rho\dot z - \rho\dot z = \frac{\cos u}{\cos\phi}(G^2 - H^2)^{\frac{1}{2}} .$$

Using the canonicity of the Hill variables, the equations of motion of the J_3 - problem can be derived from the Hamiltonian:

$$\ddot r = -\frac{\mu}{r^2} + \frac{G^2}{r^3} + R,$$

$$\dot u = \frac{G}{r^2} + \frac{r\sin u \cos i}{G\sin i} W, \tag{3.11}$$

$$\dot G = rS.$$

The efficiency of the Hill variables in satellite theory has been demonstrated principally by Iszak, Aksnes (1970) and Deprit (1981).

The set of canonical Hill variables $(r, \dot{r}, u, G)$ consists of three fast variables $(r, \dot{r}, u)$ and one slowly varying variable G. We will show here that it may be advantageous to introduce a new set of variables, where three of them will be slow variables and only one is a fast one. This is achieved by replacing the radius vector r and its rate of change $\dot{r}$ by two new slow variables: the equinoctial variables $h = e \sin \omega$ and $k = e \cos \omega$. The resulting set (h, k, u, G), however, is not a canonical set. These new variables have been called semi-equinoctial variables (Konopliv, 1986).

Let us first show that the transformation between $(r, \dot{r})$ and (h,k) is a rather simple one, (if one keeps u and G unchanged). We know the following Keplerian relations:

$$r = \frac{p}{1+e \cos v} = \frac{G^2/\mu}{1+e \cos v} \ , \qquad \dot{r} = \frac{\mu}{G} e \sin v, \tag{3.12}$$

where $v = u-\omega$ is the true anomaly. We conclude from these equations that

$$e \cos v = k \cos u + h \sin u = \frac{G^2}{\mu r} - 1,$$
$$e \sin v = k \sin u - h \cos u = \frac{G\dot{r}}{\mu} \ , \tag{3.13}$$

and consequently also:

$$h = \frac{-G\dot{r}}{\mu} \cos u + (\frac{G^2}{\mu r} -1) \sin u \ ,$$
$$k = \frac{G\dot{r}}{\mu} \sin u + (\frac{G^2}{\mu r} - 1) \cos u. \tag{3.14}$$

We see that only the constant μ and the four Hill variables are present on the right-hand side of these last two equations.

As for the rate of change of the two equinoctial elements (h,k), we usually start from the derivatives $\dot{e}$ and $\dot{\omega}$.

This gives us the final result, (Battin, 1987 pages 492-493):

$$\dot{h} = \frac{r}{G} \ [-A \cos u \ R + (B \sin u + h)S - Ck \ W],$$
$$\dot{k} = \frac{r}{G} \ [+A \sin u \ R + (B \cos u + k)S + Ck \ W], \tag{3.15}$$

where

$$A = 1 + e \cos v = \frac{p}{r} \ ; \quad B = 1+A; \quad C = \frac{\sin u \cos i}{\sin i} \ . \tag{3.16}$$

Of course the two remaining equations of motion are the usual Hill equations (for G and u):

$$G = \dot{r}S; \quad \dot{u} = \frac{G}{r^2} - \frac{r}{G} CW , \qquad (3.17)$$

where r needs to be expressed as a function of the four variables (h,k,G,u) with the use of the previous equations (3.12) and (3.13). The equations (3.15), (3.16) and (3.17) form the final system of new equations of motion that need to be integrated. The inclination i is a function of H and G, (cosi = H/G).

It is of course thanks to the cylindrical symmetry and the ensuing ignorable variable λ, that we are able to reduce this three-dimensional problem from the sixth to the fourth order. However we will end this section with the important remark that, for the purpose of constructing the Poincaré surfaces of section we can reduce the order of the above system still further: from 4 to 3. To do this, we eliminate the time and we use the angle u as the new independent variable. In other words, we integrate the system of three equations in (r, ṙ, G) as functions of u. We only need to divide the three equations for (ṙ, r̈, Ġ) by the equation for u̇. This even simplifies the construction of the Poincaré sections as well: we integrate with constant stepsize up to u = 2π. In other words the iterations are no longer needed in order to obtain an accurate intersection with the equatorial plane.

THE ALLOWABLE REGIONS OF MOTION IN THE J_3-PROBLEM

As in any conservative dynamical system, the Energy equation can be used to define equipotential lines, (zero-velocity curves) and allowable (or forbidden) regions of motion. In the present J_3-Problem we define the allowable or forbidden regions of motion in the rotating meridian plane (x,z). The energy equation $2(E+U(x,z)) = V^2$ defines the allowable regions of motion $E+U(x,z) \geq 0$. The detailed study and classification of all possible situations is rather complex, although it is elementary, because it all derives from the discussion of roots of quartic equations, exactly as the discussion of the zero-velocity curves of the J_2-Problem reduces to discussing the roots of cubic equations (Zare, 1983).

We will limit the present discussion to a cursory enumeration of the most important events occurring as a result of the variation of H^2, from 1.0 (the equatorial case) down to 0.0 (the polar orbits), keeping all other parameters constant: $J_2 = 0$, $J_3 = -0.01$, with the total energy E equal to -0.5, (which corresponds to circular orbits with unit radius in the unperturbed Keplerian case).

We will see that the available area available for the motion in the rotating meridian plane is minimum at $H^2 = 1.0$ and maximum for the polar case $H^2 = 0.0$. Near the equatorial configuration, only nearly circular orbit orbits (with radius close to one) are possible. In three dimensions, the allowable space is a thin torus which is slightly biased and shifted down below the equator. This is similar to the J_2-situation, although there, the torus is symmetric with respect to the equation. When the parameter H^2 decreases, the width of the torus increased up to the point $H^2 = 0.0$, where the available space is practically the whole sphere of radius 2 (which in the unperturbed Keplerian case would correspond to the orbits with a=1 and all possible eccentricities e from 0 to 1, corresponding to an apogee of 2). Figures 4.1 and 4.2 show the examples of forbidden regions (the shaded areas). The scales are the same for both figures (± 2 horizontal and ± 1 vertical). The values for H^2 are 0.80 and 0.16.

In the rotating meridian plane, in the case of H^2 close to 1.0, we have thus two small oval allowable regions near $x = \pm 1$ and slightly below the equator. As was said above, only nearly circular orbits are possible and there is a definite upper limit on the eccentricity of the orbits. These orbits cannot escape to infinity or collide with the origin.

We must say however that for all values of H^2, there is a very small region of allowable motions very near the origin. This is the region where the radius r has small enough values so that the perturbation term (in $1/r^4$) is actually the dominant term, while the Keplerian term (in $1/r$) is really a small perturbation now. A similar such region exists in the J_2-Problem, as well as in the theory of general relativity, (the region inside the so-called Schwarschild radius).

In the rotating meridian plane this inner allowable region takes the form of a 4-leaf clover which is symmetric with respect to the z-axis but not with respect to the horizontal x-axis, the two lower leaves being slightly larger than the two upper leaves.

When H^2 decreases, i.e. when the inclination increases, the two allowable oval regions near $x = \pm 1$ increase very fast in size, although the inner 4-leaf clover stays at roughly the same size.

We observe that in the course of the evolution of the parameter H^2, we pass through two important bottleneck openings of the zero-velocity curves (actually three if the polar case $H^2 = 0.0$ is included). These values of H^2 are thus critical points of the zero velocity curves. They occur near 0.50 and near 0.1365, for the present values of J_3 of -0.01. Note that critical H^2-values are highly dependent on the value of J_3 and the energy E. We did not explore the variation of these quantities.

At the first critical point ($H^2 = 0.50$), the outer ovals make contact with the inner 4 leaf-clover, at two points below the equator (with the two larger leaves). The opening of the bottleneck is below the equator. Past this value of H^2, the inner zone and the outer allowable zones are now connected, (only below the equator) and collision orbits will thus be possible from now on.

At $H^2 = 0.1365$, there is a new opening of a bottleneck: the two upper leaves (above the equator) of the inner allowable region now connect with the two outer oval zones. Past this value of H_2 (fow lower values) the orbits can reach the origin in two ways: above the equator and below the equator. There still are two small forbidden leaves, just above the equator as well as the two narrow forbidden tubes around the z-axis. This gets narrower when H^2 decreases.

Figure 4.3 shows the allowable and forbidden regions of motion (shaded = forbidden) for the value 0.50 of H^2, just before the opening of the two lower bottlenecks. Figure 4.4 shows the similar situation of $H^2 = 0.137$ which is close to the second critical value: the opening of the upper bottlenecks. The 2 figures have been magnified in comparison with the previous figures 4.1 and 4.2: the horizontal scale is here only ± 0.40.

318

For $H^2 = 0.0$, we have the polar orbits, and the allowable region
is maximum in size. The orbits can now cross over the North and South
poles. The allowable region is very nearly circular, with radius = 2,
except for the small forbidden region near the origin, in the shape of an
ordinary three-leaf clover, with one leaf below the equator. It is only
at the exact value $H^2 = 1.0$, that we have the last bottleneck opening,
when the bridge between the inner and outer forbidden regions collapses.
Figure 4.5 shows the regions of motion for the value H = 0.0001. The
forbidden three-leaf clover is clearly visible. The vertical forbidden
area around the z-axis is also clearly visible.

To end this section, we will make a few remarks about the nature of
the central collision singularity at (0.0). The zero-velocity curves
show that the central singularity cannot be reached from all arbitrary
directions (in the rotating meridian plane). The approximate allowed
collision directions are the angles 60° to 120°, (for H^2 near 0.0), as
well as 180° to 240° and 300° to 360°. The forbidden approach directions
are thus 0° to 60°, 120° to 180° and 240° to 300°, for the larger values
of H^2, the origin can be reached through 4 narrow cones near 10°, 170°,
260° and 280°. In particular, the Earth's spin axis (angles of 90° or
270°) belong to the forbidden approach directions.

Figure 4.6 shows a magnification of the central area of the regions
of motion of figure 4.5. The values of H^2 are the same (0.0001), but
figure 4.6 gives a clearer display of the possible approach directions
to the origin (0.0). The horizontal scale on this figure is ±0.2. We
intend to publish more results about the collision orbits later.

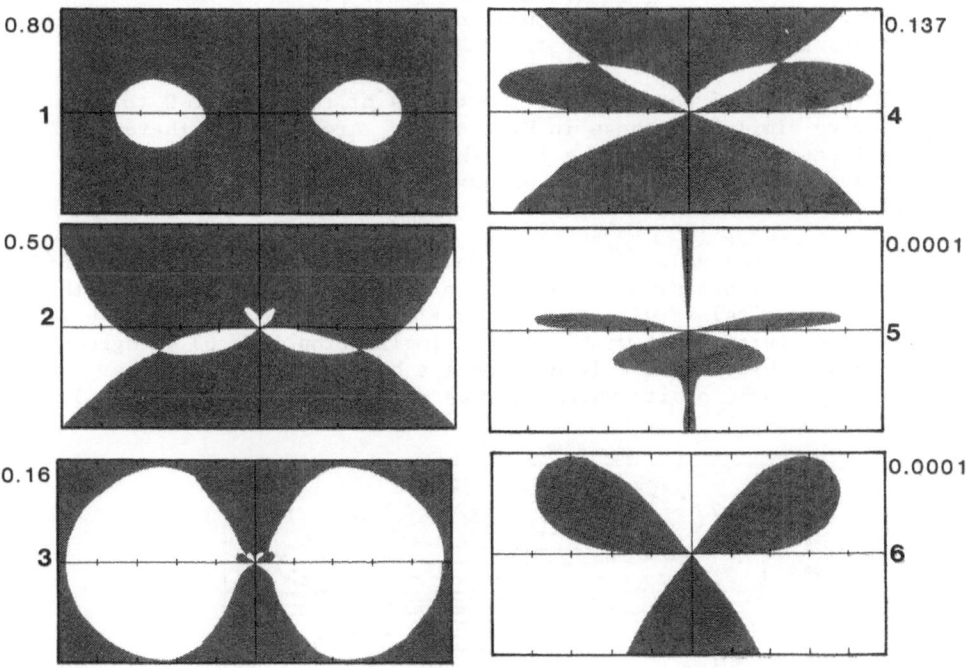

Figures 4.1 to 4.6

We will describe here some of the several Poincaré surfaces of section that were made in our initial exploration of the J_3-Problem. This has been our principal tool for the classification of the different types of orbits. In general, we used the equatorial plane z = 0 as our surface of section and plot the ascending points only, where z > 0. At each such intersection we plot thus the pair of canonically conjugate variables $(\rho, \dot{\rho})$ often designated by the alternate symbols $(x, \dot{x})$. We typically integrate each orbit for about 400 revolutions.

Figure 5.1 shows one of the typical surfaces of section corresponding to a very low inclination ($H^2 = 0.95$). The value of J_3 was 0.01. This surface of section is rather simple: it shows the two important elliptic fixed points corresponding to the two stable periodic orbits B_1 and B_2 (see Section 6). These two fixed points are surrounded by the two large regions of quasi-periodic orbits. There is no sign of chaos or other features relative to non-integrability. However, the different invariant curves each have a different structure and winding number for instance. A more detailed exploration would probably uncover several chains of numerous small islands.

We will mention only one unusual feature of the invariant curves for $H^2 = 0.95$. We clearly see a kind of separative curve just below the horizontal x-axis, which seems to wrap around B_1 on the right-side and around B_2 on the left side of Figure 5.1 In order to verify and illustrate this phenomenon in more detail, we made some refined calculations in this area. The results are shown in Figure 5.2. It is seen that 12 invariant curves are present on this figure, of which the 4 middle ones are of this special type which connects the B_1 group and the B_2-group. We do not know if this is an intrinsic feature of the problem or if it is related to our choice of variables for instance.

All the Poincaré sections for the values of the H^2 from 0 to about 0.50 are very similar to those in Figure 5.1. Around 0.50, there are some new features related to the fact that some orbits (even periodic orbits) never cross the equator. We have not yet explored all these properties in detail.

However, at the value $H^2 = 0.20$, (assuming decreasing values of H^2, corresponding to increasing values of the inclination, at least for the near-circular orbits) we witness the first important surprise: the birth of a nearly circular orbit with critical inclination (i = 63.4 degrees. This circular orbit immediately undergoes a bifurcation into three families of periodic orbits which will be described later (A, A_1, A_2).

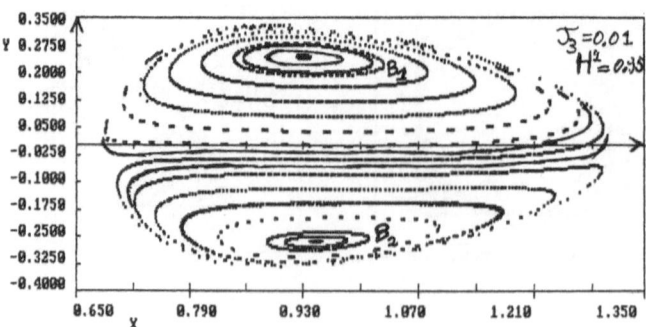

Figure 5.1

The Poincaré section is shown in Figure 5.3. We see the beginning of the bulge appearing near $x_o = 1$ and $\dot{x}_o = 0$, which is the fixed point corresponding to the circular periodic orbit. This is the circular orbit which undergoes the bifurcation. In Figure 5.4, we see the two new stable fixed points A_1 and A_2 which correspond to two new periodic orbits (described in Section 6, below). We also see the two hyperbolic points A which corresponds to a new family of periodic orbits, which is of course stable. These all correspond to $H^2 = 0.19$. In Figure 5.5, we have the invariant curves corresponding to $H^2 = 0.12$. The two stable fixed points A_1 and A_2 are still visible. In fact the region of influence of these two points (filled with the corresponding quasi-periodic orbits) is so large that it occupies about the whole admissible space.

We expect some homoclinic curves originating at the two hyperbolic points A. However, the angle of intersection of the two curves, which is a certain measure of the non-integrability, is expected to be extremely shallow here. This angle is also the basis of the Melnikov theory.

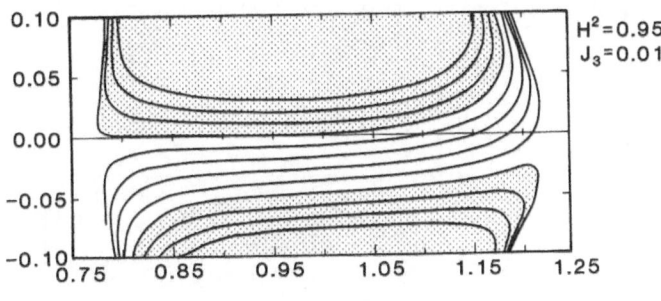

Figure 5.2

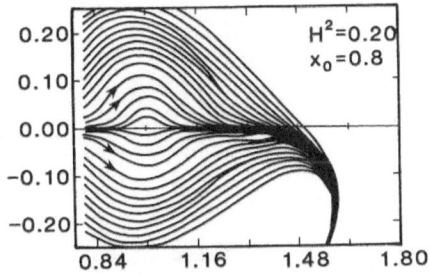

Figure 5.3

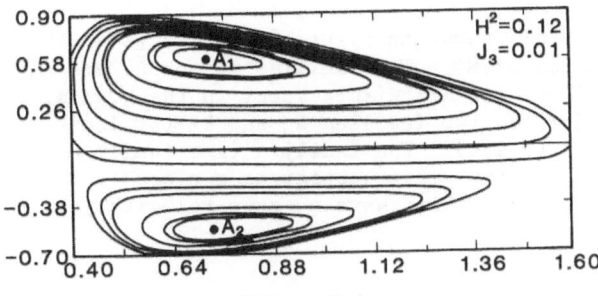

Figure 5.4

6.1 Summary

We were able to find several new families of periodic orbits in the J_3-Problem, but we limit our detailed classification to five principal families. These orbits are all periodic only in the rotating meridian plane, which contains the satellite. They are <u>not</u> symmetric with respect to the equator, (as was usually the case in the J_2-Problem). We used the value J_3 = 0.01.

The principal reason we limit ourselves to these five families is because the Poincaré sections show that these families have very large regions of influence. By this we mean that the stable periodic orbits are surrounded by very large regions of quasi-periodic orbits. Another reason we cnsider these families of periodic orbits as important is because they exist and they can be justified on the basis of an integrable averaged approximation of the J_3-Problem. In fact, the important connection and bifurcation between three of the families can be modelled with the averaged Hamiltonian, as we will show later.

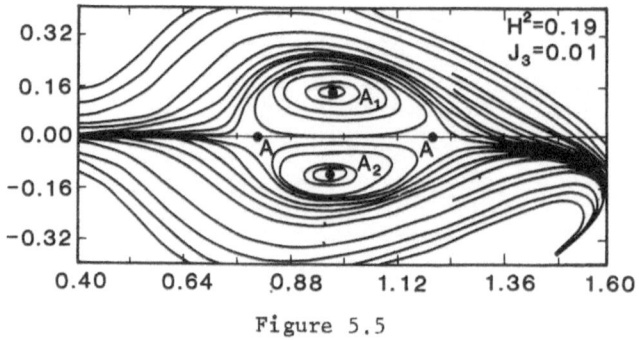

Figure 5.5

The five families have been called A, A_1, A_2, B_1 and B_2. The periodic orbits of the family A are unstable while the orbits in the other families A_1, A_2, B_1 and B_2, are all stable as is seen in the Poincaré sections. We actually verified this by computing the characteristic exponents, obtained by integration of the linearized equations of motion (see Section 11).

All five are all shown in Figure 6.1 which gives the initial value of x_o as a function of H^2. The families A_1 and A_2 are more or less mirror images of one another with respect to the equator. The same property holds for the two families B_1 and B_2 (at least for the large values of H^2 (0.5 to 1.0). The family A is a family of periodic orbits of the closed type: the orbits appear as a closed curve in the rotating meridian plane. These orbits have no zero-velocity points, (Figure 6.2).

322

On the other hand, the orbits in the families A_1, A_2, B_1, B_2 are all of the open type: in the rotating meridian frame, they appear as a single line, with zero-velocity points at each end.

We also mention that all these periodic orbits, in all five families are highly elliptic. An exception is the equatorial circular orbit at $H^2 = 1.0$. The only other nearly circular periodic orbits that we found are near the so-called critical inclination at $H^2 = 0.20$.

Table 1 which is at the end of this section gives a summary of the initial conditions of 32 typical periodic orbits, belonging all to the five major families. These orbits were all obtained by an automated computer program which iteratively determines the two initial parameters x_o, $\dot{x}_o$ leading to periodicity. No symmetry conditions were used as a sufficient periodicity criterion, in this problem.

6.2 The family A of Periodic Orbits

The family A is a family of periodic orbits which could be called "closed" Orbits, according to a terminology that has been used in the context of the restricted-three-body problem. The orbits are closed curves in the rotating meridian plane and they have no points of zero-velocity (relative to the rotating meridian).

In this family of periodic orbits the satellite seems to spend most of the time in the Southern Hemisphere, i.e. below the equator (Figure 6.2). There is no mirror image family of periodic orbits where the satellite would be mostly above the equator. Another important feature of the family A of periodic orbits is that this family exists only for the small values of H^2, from 0 to 0.2, i.e. for large inclinations, above the critical inclination. The family A terminates at the critical inclination ($H^2 = 0.20$) at a bifurcation where the three families A, A_1 and A_2 coalesce. The common orbit, which connects the three families A,A_1 and A_2 is very nearly circular. The orbits of the family A are all unstable.

The companion plots (e cos ω, e sin ω) given in Figure 6 show that, although the satellite is mostly below the equator, the perigee is generally in the Northern part at a latitude of about 20° above the equator. Some of the orbits of this family, with high inclination and eccentricity have their osculating perigee below the equator for a short time however.

6.3 The Family A$_1$ of Periodic Solutions

This is a family of stable periodic orbits which exists only for inclinations above the critical inclination i.e. for values of H^2 from 0.20 (critical inclination) to 0.0 (polar). The family originates at $H^2 = 0.20$ through a bifurcation of the pseudo-circular orbits, simultaneously with the other stable family A_2 and the unstable family A. The satellite spends most of its time above the equator: the perigee is near 270 degrees while the apogee is in the Northern Hemisphere, near 90 degrees. The orbits have the vertical symmetry axis (cos ω = 0), in the (k,h) - plane, (Figure 6.4).

6.4 The Family A₂ of Periodic Solutions

This is again a family of stable periodic solutions which exists only for the high inclination, above the critical one. More precisely, it corresponds to the values of H^2 from 0.20 to 0.0. The family A_2 is born out of the circular solutions at $H^2 = 0.20$, simultaneously with its stable near-mirror image A_1 and the unstable family A. The satellite spends most of the time below the equator where it has its apogee. The perigee is thus generally in libration around $\omega = 90$ degrees. For some orbits however, the perigee circulates, but the eccentricity is very small when the ω passes through 270 degrees (it actually has its minimum there). The whole family A_2 also has the vertical axis ($\cos \omega = 0$) as a symmetry axis in the (k,h) - plane, (Figure 6.5).

Although the table with initial conditions at the end of this section does not give the stability index k (the sum of the two non-trivial eigenvalues λ, λ^{-1} of the monodromy matrix) this number has been computed for all our periodic orbits. We observed that in a large number of cases, we see values of k very near + 2.0 (corresponding to near-unit eigenvalues). This seems thus to be a sign of near-integrability to the present dynamical system. To give an example taken from the family A_2 of periodic orbits, we mention that the values of k corresponding respectively to $H^2 = 0.16$, 0.12, 0.08, 0.04 and 0.01 are 1.99955, 1.99775, 1.98216, 1.8337 and 1.4668, (for $J_3 = 0.01$).

6.5 The Family B₁

Family B_1 is a family of stable periodic orbits which exists for all values of H^2 from 0.0 (polar) to 1.0 (equatorial). The companion orbits (Figure 6.6) show that the perigee rotates around the central body, with most of the time spent around 270°, for some of the orbits. For some other orbits, the curve is entirely below the horizontal axis, which seems that the perigee librates around the value $\omega = 270$ degrees. The vertical axis ($\omega = 90$ or 270 degrees) is a symmetry axis for the family B_1 of periodic orbits.

6.6 The Family B₂

Family B_2 begins as a bifurcation out of the near-equatorial, near-circular orbits. The periodic orbits of the family B_2 have the perigee above the equator and the apogee below the equator. At $H^2 = 1.0$, the value of x is about 1.02. (Note that for the family B_1, this x_o is only about 0.98).

The family B_2 continues and evolves without incident from $H^2 = 1.0$ (the equatorial orbits) up to the higher inclination at about $H^2 = 0.50$. The further evolution of the family is described in Section 7 below.

Figure 6.7 shows a typical companion orbit ($e \cos \omega$, $e \sin \omega$) for a member of the family B_2. We see that the perigee oscillates around the mean value of $\omega = 90^\circ$, (the vertical axis). In fact the orbit is clearly symmetric with respect to this axis. This kind of symmetry property would not have been noticed in the usual (x,z) - coordinate system. All the orbits of the family B_2 are stable.

The exact termination of the two families B_1 and B_2 has not been studied so far.

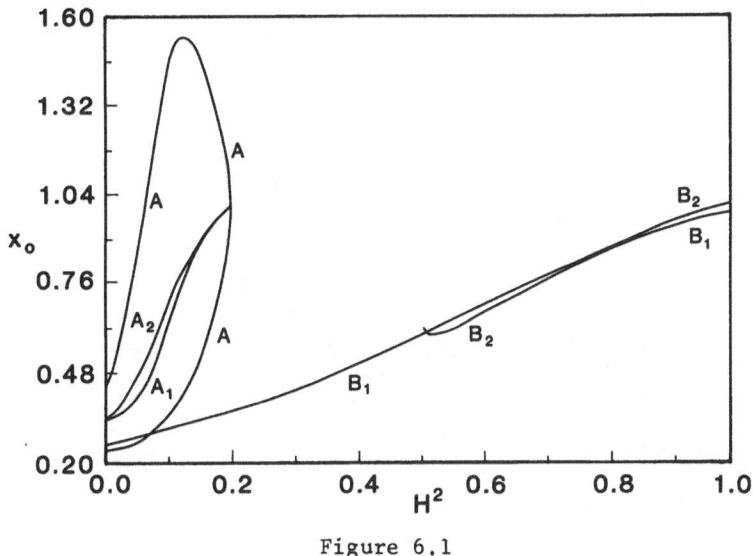

Figure 6.1

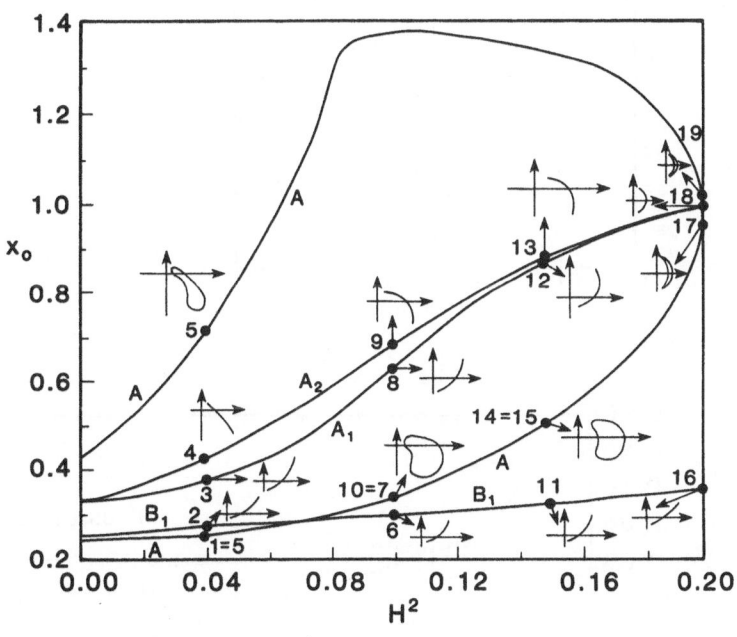

Figure 6.2

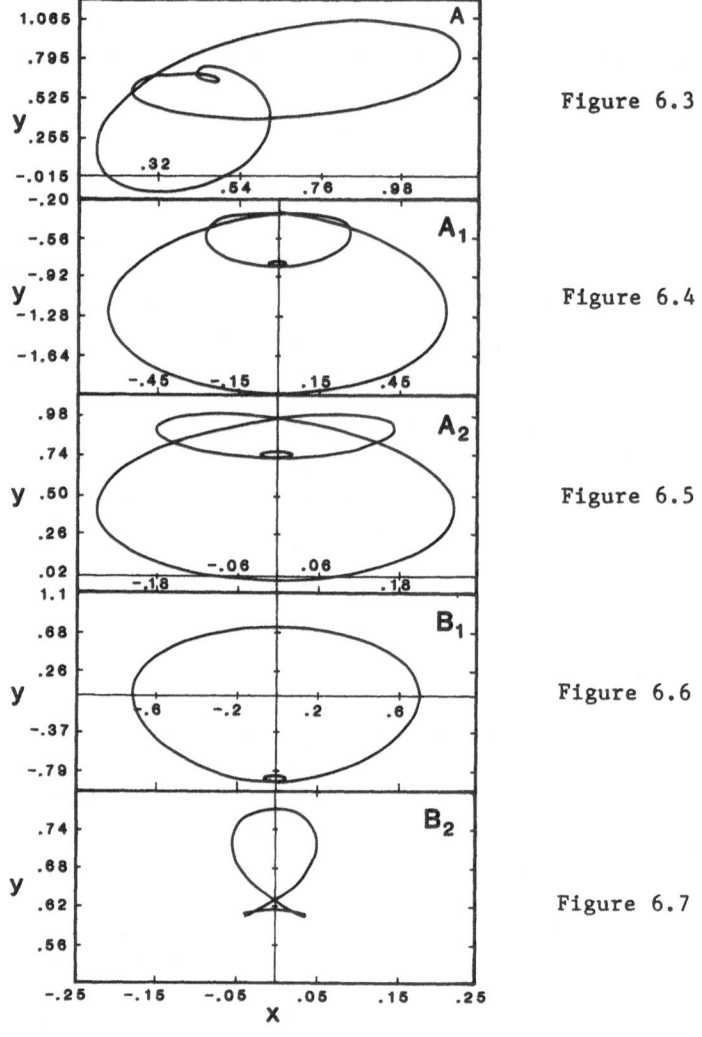

A

Figure 6.3

A₁

Figure 6.4

A₂

Figure 6.5

B₁

Figure 6.6

B₂

Figure 6.7

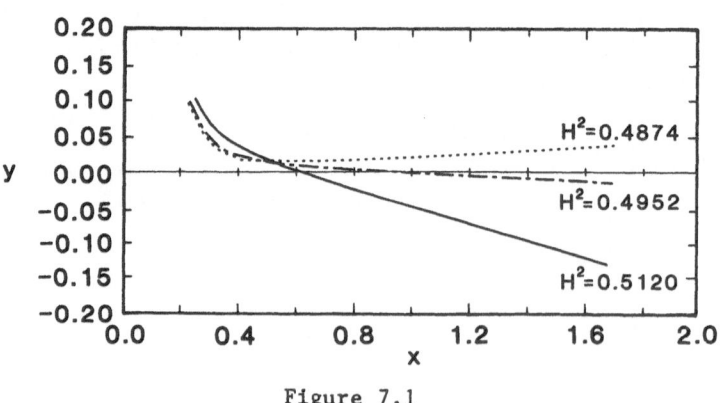

Figure 7.1

Table 1. Initial conditions for some typical periodic orbits, among the five principal families. The value of z_o was zero for all these orbits. The energy was -0.5 and the value of J_3 was always 0.01. The velocity component is not given because this was computed internally with the use of the energy integral. The orbits are numbered from 1 to 32. For the family A_1 two initial points are given, the left- and the right-side points. The correspondence between left- and right-side points is indicated in the table.

	H^2	X_o	$\dot{x}_o$	Family	Stability
1	0.04	0.25285	-0.60547	A (=5)	Unstable
2	0.04	0.26780	+1.87967	B1	Stable
3	0.04	0.37401	+1.37203	A1	Stable
4	0.04	0.42225	-1.11331	A2	Stable
5	0.04	0.71237	-1.02296	A (=1)	Unstable
6	0.10	0.29685	1.70262	B1	Stable
7	0.10	0.33243	-0.16214	A (=10)	Unstable
8	0.10	0.62777	0.78618	A1	Stable
9	0.10	0.67615	-0.67892	A2	Stable
10	0.10	1.42876	-0.36588	A (=7)	Unstable
11	0.15	0.32330	1.56484	B1	Stable
12	0.15	0.87145	0.39095	A1	Stable
13	0.15	0.88126	-0.36024	A2	Stable
14	0.15	0.50727	-0.032604	A (=15)	Unstable
15	0.15	1.47875	-0.081632	A (=14)	Unstable
16	0.20	0.47875	1.43635	B1	Stable
17	0.19948	0.95752	-0.0085900	A (=19)	Unstable
18	0.20	1.00	0.0	A1,A2,A	Bifurcation
19	0.1997	1.04202	-0.0096014	A (=17)	Unstable
20	0.30	0.42028	1.20619	B1	Stable
21	0.40	0.50018	1.00729	B1	Stable
22	0.50	0.58888	0.83276	B1	Stable
23	0.60	0.65618	-0.76040	B2	Stable
24	0.60	0.68181	0.67565	B1	Stable
25	0.70	0.76159	-0.58283	B2	Stable
26	0.70	0.77417	0.52947	B1	Stable
27	0.80	0.86104	-0.42404	B2	Stable
28	0.80	0.86038	0.38800	B1	Stable
29	0.90	0.94792	-0.26594	B2	Stable
30	0.90	0.93324	0.24290	B1	Stable
31	0.99	1.01034	-0.074182	B2	Stable
32	0.99	0.98173	0.0677575	B1	Stable

PERIODIC ORBITS ABOVE OR BELOW THE EQUATOR

In the previous section, in the description of the family B_2 of periodic orbits we already mentioned that the termination of this family is at this point unknown to us, but we described the evolution for H^2 going from 1.0 down to 0.50.

Then we see the unusual phenomenon that some of the orbits will actually be completely above the equator. Some of the last orbits to cross the equator are shown in Figures 7.1. The initial conditions for three such orbits are:

$$H^2 = 0.512 \; ; \; x_o = 0.603822 \; ; \; \dot{x}_o = -0.943946, \quad \text{(Figure 7.1)}$$
$$H^2 = 0.500 \; ; \; x_o = 0.680592 \; ; \; \dot{x}_o = -0.925132$$
$$H^2 = 0.495 \; ; \; x_o = 0.880607 \; ; \; \dot{x}_o = -0.795301$$

All three orbits also have $y_o = 0.0$ and $E = -0.5$ and correspond to $J_2 = 0$ and $J_3 = 0.01$. The following Orbit 7.3 is an example of a Periodic Orbit which remains permanently above the equator. It has the initial conditions:

$$H^2 = 0.4874 \; ; \; x_o = 0.297939 \; ; \; y_o = 0.04 \; ; \; \dot{x}_o = -716057 \; ; \; E = -0.5.$$

We did not follow this family of unusual periodic orbits further than the value $H_2 = 0.47$. It should be noted that by changing the sign of J_3, we interchange the "below" and "above" the equator properties. We also note that these orbits should not be completely surprising, as similar phenomena have been reported in the literature before, (Hall, 1962).

THE BIFURCATION AT THE CRITICAL INCLINATION

It is well known that the critical inclination $i_c (\cos^2 i_c = 1/5 = 0.20)$, ($i_c = 63^{\circ}.4$) plays a very special role in the J_2-Problem. It is a kind of a boundary between two major groups or orbits with very different behaviour, the so-called equatorial group and the polar group (Broucke and Kim, 1990). The critical inclination is known in the theory of the J_2-Problem since the work of the Soviet Scientist Orlov.

It turns out that the critical inclination plays also a very special role in the J_3-perturbation. We find that the three families A_1, A_2 and A all join in a single bifurcation at $H^2 = 0.2$, where the periodic orbit is actually circular. In other words, these three families only exist for the relatively high inclinations: for H^2 from 0.0 (Polar Case) to $H^2 = 0.20$ (the critical inclination). The junction point ($x_o = 1.0$, $H^2 = 0.20$) as well as the complete evolution of the 3 families is clearly shown on Figure 6.2. This special type of bifurcation between the unstable family A which is represented by two hyperbolic fixed points on the Poincaré sections and the two stable families A_1 and A_2 corresponding to an elliptic fixed point each. This unusual quadruplet of fixed point (two hyperbolic and two elliptic) can actually be modelled in a very simple way with polynomials as will be shown in a later article.

Our results show that both in the J_2-Problem and in the J_3-Problem the special properties of the critical inclination orbits and especially the resulting bifurcations are fundamentally related to the appearance of additional pairs of eigenvalues of the monodromy matrix of the quasi-circular orbits. In the J_2-Problem this is a rapid succession of two consecutive bifurcations due to the appearance and disappearance of a pair of eigenvalues, (Deprit 1990). This results in the existence of three families of periodic orbits, two of which are stable and one unstable. On the other hand in the J_3-Problem, there is the simultaneous birth, at $H^2 = 0.20$, of three new families, two of them Stable (A_1 and A_2) and one of them unstable (A). The common orbit at $H^2 = 0.20$ is very nearly circular. However for the smaller values of H^2, all orbits have higher eccentricities. In Figure 8.1 we compare the two bifurcation diagrams for the J_2 and J_3-Problems. These diagrams give one of the initial conditions (x_o) of the periodic orbits as a function of H^2. In Figure 8.2 we give two sketches which illustrate the same information in a Poincaré section type of form. The axes are not to scale but they can be thought of being ($e\cos\omega$, $e\sin\omega$) or (x_o, $\dot{x}_o$). We notice a major difference: in the J_2-Problem there are 5 fixed points, while in the J_3-Problem there are only 4 of them. The central stable fixed point is absent, corresponding to the absence of the near-circular periodic orbits. The two sketches in Figure 8.2 would correspond to a value of H^2 about 0.18, thus with inclination a little higher than the critical one. In Figure 8.1, the letter "S and U" stand for "Stable and "Unstable", respectively.

In a future article, we will describe the simultaneous interaction between the J_2- and J_3-harmonics.

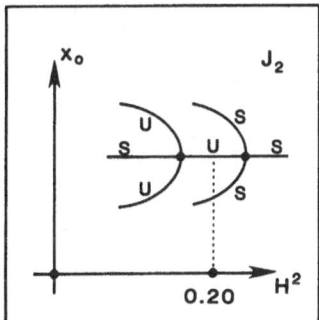 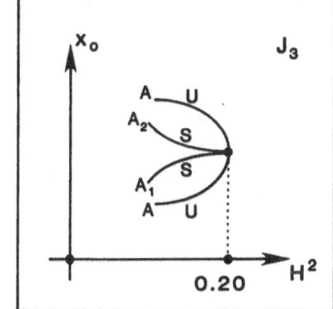

Figure 8.1

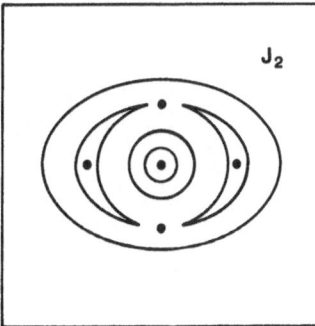 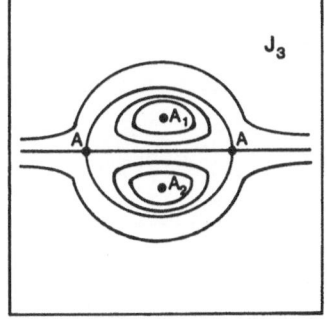

Figure 8.2

We found that in the case of energy E = -0.5 and J_3 = 0.01 (J_2= 0) the allowable region of motion is essentially the circular region with radius R = 2.0 around the origin. Therefore <u>no</u> escape orbits are possible at this energy level.

However we have three small <u>forbidden</u> regions around the centre (0,0) in the form of a clover leaf.

Figure 9.1 shows the central forbidden area in the case of polar orbits (H_2= 0.0) with energy E = -0.5. The boundaries of this figure are -0.10 to + 0.25 vertically and + 0.20 horizontally. The forbidden area is seen to be in the form of a clover leaf. The collision with the centre (0,0) will thus be possible through three different openings. The first opening is from the bottom along the negative increasing values of z. The two other openings are symmetric with respect to the z-axis, on either side with decreasing z and ρ. We discovered that actually a large number of orbits end up in collision at the centre (0,0).

Figure 9.2 shows the evolution of an initially circular polar orbit corresponding to the constants H^2 = 0, J_2 = 0, J_3 = 0.01. It is seen that the orbit is rapidly becoming more and more eccentric, with the eccentricity continuously increasing, resulting in a collision at the centre, after about 45 half revolutions.

The initial conditions of this polar orbit were x_o = 1.0, y_o = 0.0 $\dot{x}_o$ = 0 and Energy E = -0.5. Figure 9.1 is a magnification of the central part of Figure 9.2, showing parts of the last 16 revolutions, just before the central collision in between the two lower leaves of the clover.

Figure 9.3 actually shows another collision orbit, entering the centre from the top right side. It has the initial conditions x_o = 0.15 ; y_o = 0.05 ; $\dot{x}_o$ = 0.0, with $\dot{y}_o$ > 0 such that E = 0.5. We see that the satellite rapidly moves away from the centre towards the outer zero - velocity curve (at a radius 2.0) before the final return and collision at the centre. The constants for this orbit are again E = -0.5, J_2 = 0.0, J_3 = 0.01.

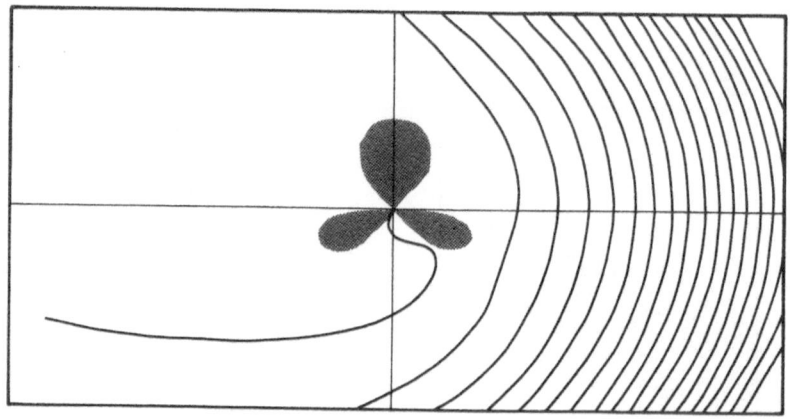

Figure 9.1

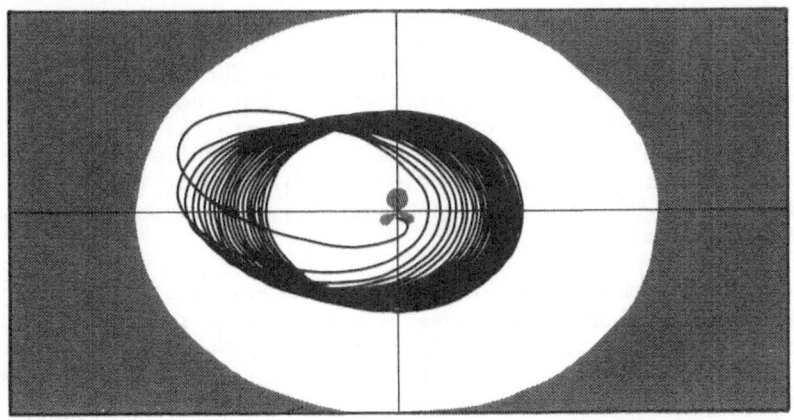

Figure 9.2

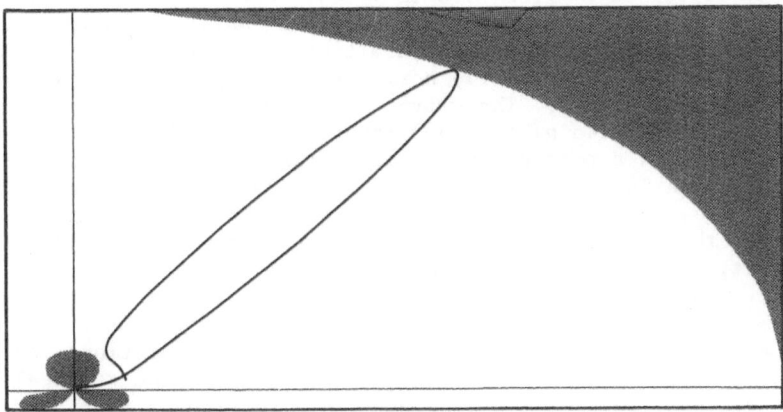

Figure 9.3

SOME TYPICAL ORBITS FOR THE POLAR J_3-PROBLEM

We have already said that the polar problem is a very special case of the general zonal harmonic problem. It is essentially an independent inbedded two-dimensional dynamical system, in a fixed (rather than rotating) meridian plane. We discovered several unusual types of orbits. We already mentioned in the previous section that many orbits end in the central collision in between one of the three leaves of the forbidden clover. We will describe a few more types of orbits here, in particular periodic orbits, without going in much detail, because these orbits are not of much practical or astronomical interest.

We first mention a few periodic orbits that were discovered. These are just a few simple cases among hundreds of periodic orbits that seem to exist with all different complicated shapes. These orbits arise essentially because of the many possibilities for the particle to bounce off the three leaves of the forbidden clover. The table that follows gives the initial conditions and stability index (k) for nine simple periodic orbits. Figure 10.1 shows four of them. It is seen that all these orbits are symmetric with respect to the z-axis. We found no periodic orbits which cross the equator (x-axis) without crossing the vertical z-axis, (in the polar case, $H^2 = 0$). We must remember that the polar case is the only case where the orbits can cross through the z-axis; also the fact that the z-axis is a symmetry axis in this problem, while the x-axis is not. We used values of -0.005 and -0.02 for J_3, keeping the energy at -0.5 (and $x_o = y_o = 0$). Changing the sign of J_3 gives us similar but mirror image forms of orbits.

We also notice that two of the nine orbits are stable ($|k| < 2$). We can thus expect the standard types of bifurcations that exist in all conservative Hamiltonian systems. In particular, the fifth orbit in the table is related to the fourth through a bifurcation with the same period, at the value $k = +2$ (corresponding to a pair of unit eigenvalues).

We finally show an example of a Poincaré surface of section that was made, using the equator (y = 0) as a plane of section. We used the value $J_3 = +0.001$. As initial conditions we used E = -0.5, $x_o = 1.6$, with several values of $\dot{x}_o$ from -1.0 to + 1.0. Most of these orbits end up in collision: The eccentricity is increasing while the perigee angle ω is moving only very little. The resulting curves (Figure 10.2) are on the surfaces of section are nearly horizontal, the points drifting slowly from the right to the left. The curves shown in Figure 10.2 were integrated up to a final time of $t_F = 300$ canonical units. We see that the central curves (with a near perpendicular intersection of the x-axis) move much slower to the left and will thus survive longer before the central collision occurs. However, our numerical results confirm the statement by C. Marchal that a body with a large J_3 cannot have any satellites in a polar orbit.

It is possible to perform these Poincaré sections in many different ways. In particular we may choose different planes of section. Another indicated choice (besides the equator, z = 0) would be the vertical z-axis which is a symmetry axis of the problem. This section is appropriate for the study of the some of the stable periodic orbits which have been mentioned above.

Table 2

y_o	$\dot{x}_o$	$T/2$	k	J_3
0.2240139	4.879756	0.2268	410.779	-0.02
0.2821265	3.52175	6.3448	- 6.833	-0.02
1.7173395	0.411328	12.526	67.057	-0.02
1.780220	0.356986	12.440	- 0.43255	-0.02
1.887872	0.250086	12.527	-324.262	-0.02
1.862070	0.273688	12.5158	- 0.081066	-0.005
1.933694	0.187095	12.558	-402.170	-0.005
0.1381071	6.400702	0.10635	426.779	-0.005
0.1732627	4.651826	6.3052	7.52995	-0.005

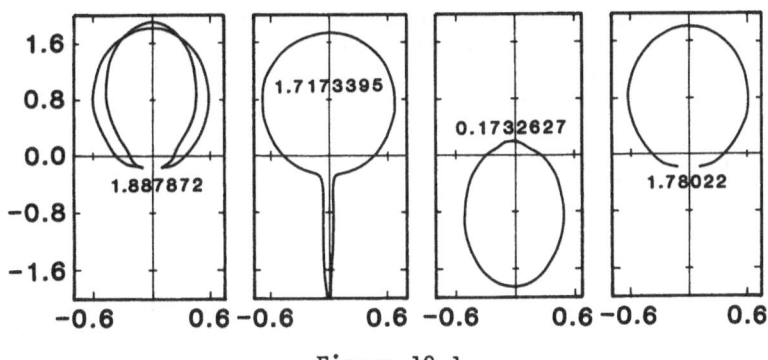

Figure 10.1

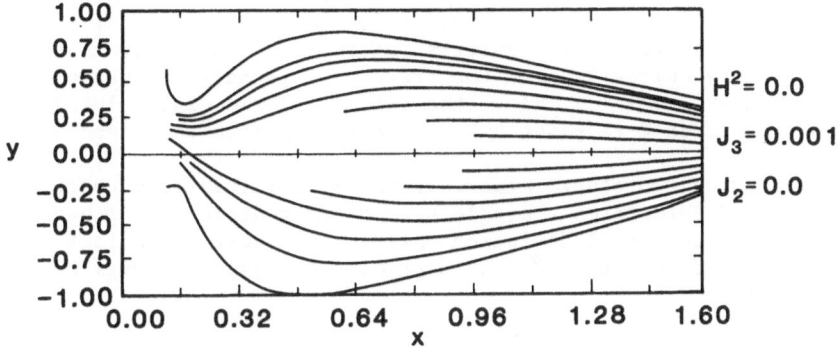

Figure 10.2

THE VARIATIONAL EQUATIONS

In order to determine the stability of the different periodic orbits, we used the standard approach consisting in integrating the variational equations:

$$\delta\ddot{\rho} = U_{\rho\rho}\ \delta\rho + U_{\rho z}\ \delta z\ ,$$

$$\delta\ddot{z} = U_{z\rho}\ \delta\rho + U_{zz}\ \delta z\ .$$

The second partial derivatives of the potential function U are found to be

$$U_{\rho\rho} = \frac{5J_3 z}{2r^7}\left(-3 + 21\frac{\rho^2}{r^2} + 7\frac{z^2}{r^2} - 63\frac{z^2\rho^2}{r^4}\right)\ ,$$

$$U_{\rho z} = U_{z\rho} = \frac{15J_3\rho}{2r^7}\left(-1 + 14\frac{z^2}{r^2} - 21\frac{z^4}{r^4}\right)\ ,$$

$$U_{zz} = \frac{5J_3 z}{2r^7}\left(-25 + 70\frac{z^2}{r^2} - 63\frac{z^4}{r^4}\right)\ ,$$

at least for the terms which are proportional to J_3. We still need to add the two-body terms which do not depend on the perturbation J_3. These terms are principally the second derivatives of $1/r(\rho,z)$ but the first equation also contains an extra term arising from the reduction from three to two degrees of freedom:

$$U_{\rho\rho} = \frac{1}{r^3} [3 \left(\frac{\rho}{r}\right)^2 - 1] - 3 \frac{H^2}{4} ,$$

$$U_{\rho z} = \frac{1}{r^3}[3 \frac{\rho z}{r^2}],$$

$$U_{zz} = \frac{1}{r^3}[3 \left(\frac{z}{r}\right)^2 - 1].$$

The numerical solution of the variational equations, with a (4-by-4) - Identity matrix as initial conditions, allows us to determine the state transition matrix for a complete revolution of each periodic orbit, also called the monodromy matrix R. The eigenvalues of this matrix give us the required stability information. Actually this information is deduced directly from the trace of this matrix because we know that Trace (R) = $1 + 1 + \lambda + \lambda^{-1}$, (the sum of all the eigenvalues). So, the stability index k is given by subtracting 2 from the trace.

We also used two other methods to verify our stability calculations. The Hill method integrates a single second-order differential equation $\delta\ddot{n} = - \theta\delta n$ for the normal deviation n from the orbit. The last method we used is the Hénon method which compares two adjacent orbits over a complete revolution.

ACKNOWLEDGEMENTS

We want to thank especially Professor J. M. Ferrandiz who invited us to the University of Valladolid, Spain, where our work on the J_3-Problem was started. We also thank the Aerospace Corporation of California, especially Dr. C. C. Chao, with whom we had many useful interactions on the application of the present problem.

REFERENCES

Aksnes, K., 1970, "A Second-Order Artificial Satellite Theory Based on an Intermediate Orbit", Astronomical Journal, Vol.75, No.9, pp. 1066-1076.

Battin, R.H., 1987, "An Introduction to the Mathematics and Methods of Astrodynamics," AIAA Education Series, New York.

Broucke, R. and Kim, M.C., 1990, "The Orbit in the J_2-Problem" Celestial Mechanics, in Press.

Brouwer, D., "The Solution of the Problem of Artificial Satellite Theory without Drag," Astronomical Journal, Vol.64, 1959, pp.378-397.

Coffey, S.. Deprit, A., Deprit E., Healy, L., 1990, "Painting the Phase Portrait of an Integrable Dynamical System," in Science, Vol. 247, 16 February 1990, pp. 769-892.

Deprit, A., 1981, "The Elimination of the Parallax in Satellite Theory," Celestial Mechanics, Vol.24 pp. 111-153.

Garfinkel, B., 1959, "The Orbit of a Satellite of an Oblate Planet," Astronomical Journal, Vol. 64, pp. 353-367.

Hall, N.S., "A Class of Orbits", ARS-Journal, Jan. 1962, pp. 96-97.

Hill, G.W., 1913, Astronomical Journal 27, p. 171.

Jupp, A., 1987, "The Critical Inclination, 30 years of progress," Celestial Mechanics, Vol. 43, No. 3-4, pp. 127-138.

Konopliv, A., "Theory of Co-orbital Motion," Ph. Dissertation, University of Texas, May 1986, Supervisor R. Broucke.

Kozai, Y., 1959, "The Motion of a Close Earth Satellite," Astronomical Journal, Vol. 64, pp. 367-377.

Zare, K., 1983, "The Possible Motions of a Satellite about an Oblate Planet", Celestial Mechanics, Vol. 30, pp. 49-58.

STABILITY OF SATELLITES IN SPIN-ORBIT RESONANCES
AND CAPTURE PROBABILITIES

Alessandra Celletti

Dipt. di Matematica Pura e Applicata
Universitá dell'Aquila
67100 Coppito (L'Aquila)-Italy
e-mail: celletti@40282.decnet.cern.ch

Abstract: The stability of satellites in spin-orbit resonances is investigated in the light of perturbation theory. By means of KAM theory we construct invariant surfaces trapping the periodic orbit associated to the resonance in a finite region of the phase space. In the last part of the work we study the probability of capture in a resonance, providing an explicit application to Mercury.

Keywords: Spin-orbit resonance, KAM theory, Capture probability.

INTRODUCTION

Consider a triaxial ellipsoidal satellite S orbiting around a central body P and at the same time rotating about its spin axis. A *spin-orbit resonance*[8],[10],[14] is an exact commensurability between the periods of revolution T_{rev} and rotation T_{rot} of the satellite. More specifically one has a $p : q$ resonance (with $p, q \in \mathbf{Z}$) when $T_{rev}/T_{rot} = p/q$, namely in q orbital revolutions around the primary body, the satellite makes p rotations about its spin-axis.

The classical example of a spin-orbit resonance is provided by the Moon-Earth system: as is well known the Moon always points the same face toward the Earth, i.e. $T_{rot} = T_{rev}$. Astronomical observations show that all the *tidally evolved* satellites of the solar system (i.e. those satellites which are close enough to the central body so that their spin rate was affected by tidal torques) are trapped in a 1:1 resonance like the Moon. Typical examples are the satellites of Mars, the galileian satellites of Jupiter and most of the satellites of Saturn. The only exception is provided by the Mercury-Sun system, since Mercury is trapped in a 3:2 resonance with the Sun: radar observations indicate that the period of rotation is $T_{rot} \simeq 58.6461 \pm 0.005$ days, while $T_{rev} \simeq 87.9693$ days, so that $T_{rev}/T_{rot} = \frac{3}{2} \left(1 + 10^{-4}\right)$.

There are several questions related to the problem of spin–orbit resonances. For example,

a) Are the satellites trapped in a resonance stable?

b) What is the mechanism of capture in a resonance by tidal evolution?

c) Why the final end-states of the satellites are the 1:1 or 3:2, while the other resonances (like the 2:1, 4:3, 6:5, etc.) are depleted?

Let us remark incidentally that the same questions can be raised in the case of orbit-orbit resonances (namely commensurabilities between orbital mean motions); for further investigations on this topic see the lecture by J. Henrard.

In this paper we try to answer the above questions applying perturbation techniques. In particular under simplifying assumptions on the model we prove the stability of a satellite in a spin-orbit resonance, confining its motion in a finite region of the phase space[2], [3]. In the last section we investigate the probability of capture in a resonance[5]. Our results support the idea that the tidal evolution drives more likely the satellite to the 1:1 or 3:2 states and that the probability of capture in the 3:2 resonance increases as the orbital eccentricity is higher.

To carry out this program we will introduce a mathematical model describing an approximation of the physical situation. Our assumptions are:

i) assume that the satellite S moves on a fixed keplerian ellipse around P (i.e. neglect secular perturbations of the orbital parameters);

ii) assume that the spin-axis of S coincides with its shortest physical axis (i.e. the axis with largest principal moment of inertia) and that

iii) the spin-axis is perpendicular to the orbit plane (i.e. we do not consider the spin-axis *obliquity*);

iv) we neglect external dissipative torques as well as forces exerted by other planets or satellites.

Under these assumptions the equation of motion can be derived from standard Euler's equations for a rigid body[6]; normalizing the mean motion to one, this equation takes the form

$$(1) \qquad \ddot{x} + \frac{3}{2}\frac{B-A}{C}\left(\frac{a}{r}\right)^3 \sin(2x - 2f) = 0 \ ,$$

where $A < B < C$ are the principal moments of inertia, a is the semimajor axis, r the instantaneous orbital radius, f the true anomaly and x the angle between the direction of the longest axis of the ellipsoid and the periapsis line (see fig.1).

Because of assumption i) the orbital radius r and the true anomaly f are both 2π-periodic functions of the time. Therefore the term $\left(\frac{a}{r}\right)^3 \sin(2x - 2f)$ can be expanded in Fourier series, so that (1) becomes

$$(2) \qquad \ddot{x} + \varepsilon \sum_{m \neq 0, m=-\infty}^{\infty} W\left(\frac{m}{2}, e\right) \sin(2x - mt) = 0 \ ,$$

where $\varepsilon = \frac{3}{2}\frac{B-A}{C}$ and the coefficients $W\left(\frac{m}{2}, e\right)$ decay as powers of the orbital eccentricity e as $W\left(\frac{m}{2}, e\right) \propto e^{|m-2|}$ (for an explicit expression of these coefficients see, e.g., [2]).

Let us simplify equation (2) retaining only a finite number of terms in the sum. To decide which terms we need to keep, we adopt the following criterion.

According to *iv)* we ignore the effect of the dissipative forces. The most important contribution would come from the force due to the internal non-rigidity of the satellite. Such a dissipative tidal torque (which should be added at the r.h.s. of (2)) has the form

$$(3) \qquad T \equiv -\frac{3}{2} k_2 \frac{GM^2 R^5}{a^6} \sin(2\delta) ,$$

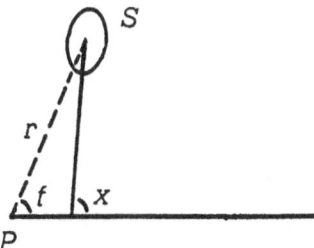

Figure 1. The notation introduced in equation (1) is displayed in the figure.

where G is the gravitational constant, M is the mass of the central body, R the mean radius of S, while k_2 ("Love number") and δ ("lag angle") are two quantities depending on the physical properties of the satellite. Now, since we ignore T (whose size is small w.r.t. the gravitational term), we will as well neglect those terms in the series expansion of (2) which are of the same order of magnitude as (the average effect of) T. Therefore we reduce our task to the study of an equation of the form

$$(4) \qquad \ddot{x} + \varepsilon \sum_{m \neq 0, m=N_1}^{N_2} W\left(\frac{m}{2}, e\right) \sin(2x - mt) = 0 ,$$

for suitable integers N_1 and N_2 depending on the physical parameters of the satellite.

TRAPPING OF PERIODIC ORBITS

A $p : q$ resonance for (4) is a periodic orbit $x = x(t)$ such that

$$x(t + 2\pi q) = x(t) + 2\pi p .$$

For $\varepsilon \neq 0$, the stable periodic orbit is surrounded by *librational* tori. The *chaotic separatrix* divides the librational region from the region in which *rotational* invariant tori can be found (see the Poincaré sectio in figure 2).

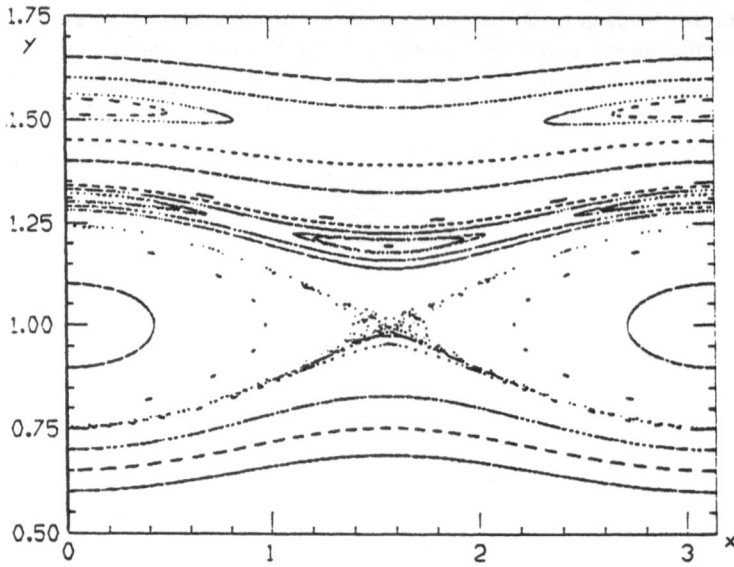

Figure 2. Poincaré sectio of (4) (with $N_1 = 1$ and $N_2 = 5$) for several initial data; here $e = 0.01$ and $\varepsilon = 0.03$.

Since the phase space $S \equiv \{(y, x, t)/y = \dot{x} \in \mathbf{R}, (x, t) \in (\mathbf{R}/2\pi\mathbf{Z})^2\}$ associated to (2) is three dimensional, any invariant KAM torus divides S into separate regions with the property that any motion starting in one of these regions would stay forever in it.

Let $P(\frac{p}{q})$ be a periodic orbit associated to the $p : q$ resonance; we try to trap $P(\frac{p}{q})$ between invariant surfaces $\mathcal{T}_1 \equiv \mathcal{T}(\omega_1)$ and $\mathcal{T}_2 \equiv \mathcal{T}(\omega_2)$ with $\omega_1 < \frac{p}{q} < \omega_2$. If $\mathcal{T}_1$ and $\mathcal{T}_2$ exist, the periodic orbit $P(\frac{p}{q})$ is trapped in the region enclosed by $\mathcal{T}_1$ and $\mathcal{T}_2$. Therefore our main task will be to construct the trapping KAM surfaces for the realistic value of the equatorial oblateness parameter ε and for the true value of the eccentricity.

In order to explain the technique used to prove the existence of the invariant tori, let us briefly illustrate the dynamics as the parameter ε is varied. Using Hamiltonian formalism equation (4) can be derived from Hamilton's equations associated to the nearly-integrable Hamiltonian

$$(5) \quad H(y, x, t) \equiv \frac{y^2}{2} - \frac{\varepsilon}{2} \sum_{m \neq 0, m = N_1}^{N_2} W(\frac{m}{2}, e) \cos(2x - mt), \quad y \in \mathbf{R}, (x, t) \in (\mathbf{R}/2\pi\mathbf{Z})^2.$$

For $\varepsilon = 0$ Hamilton's equations $\dot{y} = 0$, $\dot{x} = y$ are trivially integrated as $y = y_0$, $x = x_0 + y_0 t$ and the KAM surfaces reduce to the planes $\{y = y_0\}$ on which periodic or quasiperiodic motions take place.

For $\varepsilon \neq 0$ but small enough, there still exists an invariant surface $\mathcal{T}_\varepsilon(\omega)$ for the perturbed system with rotation number ω. However when ε reaches a critical value, say $\varepsilon_c = \varepsilon_c(\omega)$, the invariant torus $\mathcal{T}_\varepsilon(\omega)$ is destroyed and leaves place to a so-called *Mather set*[12], which is closed, invariant and is a graph of a Cantor set. Notice that in presence of Mather sets the separation of the phase space is no more valid, since the orbits can diffuse through the gaps of the Cantor set.

To determine the critical break-down threshold $\varepsilon_c(\omega)$ of the torus $\mathcal{T}_\varepsilon(\omega)$ one can apply numerical or rigorous techniques. The most reliable numerical method was developed by Greene[9] and is based on the idea that the disappearance of an invariant torus $\mathcal{T}_\varepsilon(\omega)$ is related to a sudden change, from stability to instability, of the periodic orbits $P(p_j/q_j)$ with frequencies p_j/q_j equal to the rational approximants to ω.

The Kolmogorov-Arnold-Moser theory[11],[1],[13] provides a rigorous algorithm to give an estimate $\varepsilon_r = \varepsilon_r(\omega)$ on the perturbing parameter ε, ensuring the existence of $\mathcal{T}_\varepsilon(\omega)$ for every $\varepsilon < \varepsilon_r(\omega)$. The theorem can be applied under the following conditions:

a) *non-degeneracy:* writing the Hamiltonian as $H(y,x,t) = h(y) + \varepsilon V(x,t)$ for given analytic functions h and V, one requires that

$$\frac{\partial^2 h(y)}{\partial y^2} \neq 0 \qquad \forall y \in \mathbf{R} \, ;$$

b) *non-resonance:* the rotation number ω must satisfy the diophantine inequality

$$(6) \qquad |\omega - \frac{p}{q}|^{-1} \leq C q^2 \qquad \forall p,q \in \mathbf{Z}, \; q \neq 0 \, ,$$

for some positive constant C.

The first condition is trivially satisfied by (5); as for b), we select two sequences $\{\Gamma_k\}$, $\{\Delta_k\}$ of rotation numbers of the trapping tori as

$$\Gamma_k \equiv \frac{p}{q} - \frac{1}{k+\gamma} \, , \qquad \Delta_k \equiv \frac{p}{q} + \frac{1}{k+\gamma} \, , \qquad k \in \mathbf{Z}, \; k \geq 2 \, ,$$

where γ is the golden ratio: $\gamma = \frac{\sqrt{5}-1}{2}$. Both Γ_k and Δ_k are irrational numbers satisfying (6) with a constant[2] $C = q^2(k+\gamma)$ and with the property that $\Gamma_k \leq \frac{p}{q} \leq \Delta_k$ for every k.

However, applications of standard versions of KAM theorem provide estimates which are usually far from the realistic physical value of the perturbing parameter ε. Remarkable improvements of KAM estimates can be found in [7], [4]. These works are computer-assisted, i.e. since the proof involves a huge amount of computations the use of a computer is made necessary. The rounding-off and propagation errors introduced by the machine are controlled implementing the so-called *interval arithmetic* technique. In this work we shall apply the algorithm developed in [4] to state the existence of the invariant tori.

RESULTS

Let us collect here the results of the application of the technique presented in the previous section to the Moon-Earth and Mercury-Sun systems. The mathematical model of the introduction provides a fairly good description of the realistic situation. As for the Moon, the hypothesis *iii)*, namely that the spin-axis is perpendicular to the orbit plane, is the most questionable, since the spin-axis and the orbit-normal make an angle of about $6^o41'$. Regarding Mercury the first assumption should be eliminated in a more realistic model, since due to perturbations Mercury's eccentricity undergoes strong secular variations: the current value is $e = 0.2056$, but it oscillates between 0.1 and 0.3.

Now we apply KAM theorem to state the existence of invariant surfaces trapping the 1:1 and 3:2 resonances, respectively for the Moon and Mercury. In both cases the invariant tori are constructed for the true values of the eccentricity ($e = 0.0549$ for the Moon and $e = 0.2056$ for Mercury) and of the perturbing parameter $\varepsilon = \frac{3}{2}\frac{B-A}{C}$ ($\varepsilon = 3.45 \cdot 10^{-4}$ and $\varepsilon = 1.5 \cdot 10^{-4}$, respectively). Therefore the existence of trapping surfaces guarantees the stability (in the sense of confinement of the motion) of the resonances (cfr [2], [3]).

However the question of the capture of the satellites in resonance remains still open. In [8] (see also [10]) a probability of capture was derived as follows. Let $\gamma = x - \frac{m}{2}t$ with $m \in \mathbf{Z}$; close to the $m : 2$ resonance, the angle γ changes slightly during one orbital period. Therefore averaging (1) (or (2)) holding γ fixed one obtains

$$\ddot{\gamma} + \frac{3}{2}\frac{B-A}{C} \, W(\frac{m}{2},e) \, \sin(2\gamma) \; = \frac{\langle T \rangle}{C} \, ,$$

where the effect of the dissipation has been considered by adding at the r.h.s. the average $\langle T \rangle$ of the tidal torque (3). Using for $\langle T \rangle$ the expression: $\langle T \rangle = -K(\dot{\gamma} + Z)$ (valid near $\dot{\gamma} = 0$) for some positive constants K and Z, the probability of capture in the $m : 2$ resonance is given by[8], [10]

$$(7) \quad P(\frac{m}{2}) \; = \; \frac{2}{1 + \frac{\pi Z}{2[3\frac{B-A}{C}W(\frac{m}{2},e)]^{1/2}}} \, , \quad \text{where} \quad Z = \frac{m}{2} - \frac{1 + \frac{15}{2}e^2 + \frac{45}{8}e^4 + \frac{3}{16}e^6}{(1 + 3e^2 + \frac{3}{8}e^4)(1 - e^2)^{\frac{3}{2}}}$$

(with $P = 0$ if $P \le 0$ and $P = 1$ if $P \ge 1$). Notice that such probability depends on e and $\varepsilon (= \frac{3}{2}\frac{B-A}{C})$ and in particular on the amplitude $[3\frac{B-A}{C}W(\frac{m}{2},e)]^{\frac{1}{2}}$ of the librational region surrounding the $m : 2$ resonance. In the case of Mercury, for $e = 0.2056$ and $\varepsilon = 1.5 \cdot 10^{-4}$ according to (7) the probability of capture in the $1 : 1$ resonance is 0%, in the $3 : 2$ resonance is 7%, while in the $2 : 1$ resonance is 1.6%.

An idea of the amount of the probability of capture can be obtained looking at the graph of figure 3b, where the rotation number ω is plotted versus the estimate $\varepsilon_r(\omega)$ (obtained applying the KAM theorem of [4]) ensuring the existence of the torus $T_\varepsilon(\omega)$ for every $\varepsilon < \varepsilon_r(\omega)$ (see [5] for the details). The region below the curve corresponds to the region of existence of (rotational) KAM tori.

Supposing that Mercury rotated faster in the past and that it slowed down by the effect of tidal torques, one proceeds (for a fixed $\bar{\varepsilon}$) on the line $\varepsilon = \bar{\varepsilon}$ coming from the right of the picture. Obviously, in the conservative framework one could never cross the curve due to the property of separation of the phase space by invariant surfaces. However the dissipation breaks, sooner or later, any KAM torus, driving the system to slower rotations.

For any resonance $p : q$ and for any $\bar{\varepsilon}$, let $d_{\bar{\varepsilon}}(\frac{p}{q})$ be the distance between the intersections of the line $\varepsilon = \bar{\varepsilon}$ with the two branches of the curve originating from $\frac{p}{q}$. Since $d_{\bar{\varepsilon}}(\frac{p}{q})$ measures the amplitude of the librational region around the resonance, according to (7) a capture in the $p : q$ resonance is more likely as $d_{\bar{\varepsilon}}(\frac{p}{q})$ is larger. Moreover the graph suggests that the probability of capture in minor resonances is negligible in comparison to those of the 1:1, 3:2 and 2:1.

Finally, we reproduced for $e = 0.1$ and $e = 0.3$ (the two extrema of the secular variations of the eccentricity of Mercury) the same graph of fig. 3b (see fig.s 3a and 3c). A comparison among the three pictures shows that $d_{\bar{\varepsilon}}(\frac{3}{2})$ is greater as e is larger and therefore the probability of capture in the 3:2 resonance is bigger for satellites with larger eccentricities.

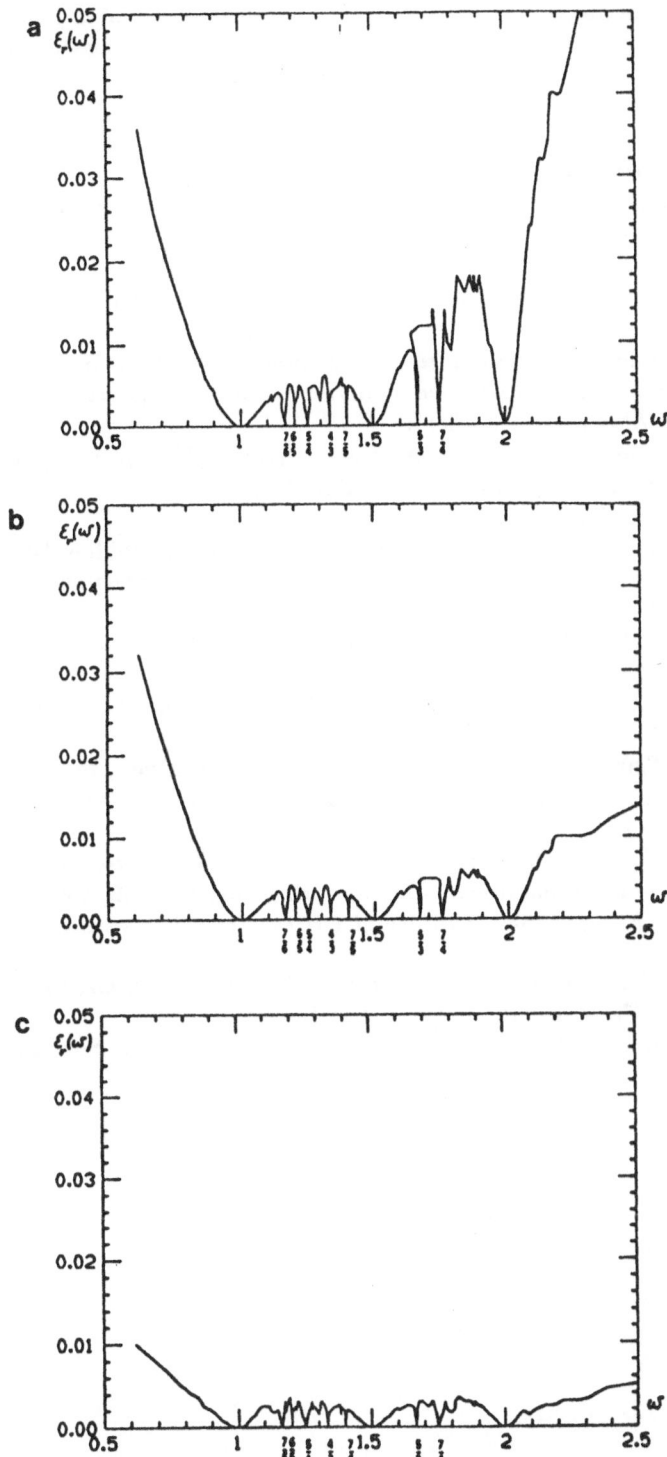

Figure 3. The rotation number ω is plotted versus the critical threshold $\varepsilon_r(\omega)$ (obtained applying KAM theorem) ensuring the existence of the torus $T_\varepsilon(\omega)$ for every $\varepsilon < \varepsilon_r(\omega)$. In fig. 3a the eccentricity is $e = 0.1$, in fig. 3b $e = 0.2056$ and in fig. 3c $e = 0.3$.

Acknowledgments: Part of this paper contains results of a joint work with C. Falcolini. I would like to thank C. Falcolini and A. Morbidelli for useful suggestions.

REFERENCES

[1] Arnold V.I., *Proof of a Theorem by A.N. Kolmogorov on the invariance of quasi-periodic motions under small perturbations of the Hamiltonian*, Russ. Math. Surveys **18**, 9 (1963)

[2] Celletti A., *Analysis of resonances in the spin-orbit problem in Celestial Mechanics: The synchronous resonance (Part I)*, J. of Appl. Math. and Phys. (ZAMP) **41**, 174 (1990)

[3] Celletti A., *Analysis of resonances in the spin-orbit problem in Celestial Mechanics: Higher order resonances and some numerical experiments (Part II)*, J. of Appl. Math. and Phys. (ZAMP) **41**, 453 (1990)

[4] Celletti A., Chierchia L., *Construction of analytic KAM surfaces and effective stability bounds*, Commun. Math. Phys. **118**, 119 (1988)

[5] Celletti A., Falcolini C., work in progress

[6] Danby J.M.A., *Fundamentals of Celestial Mechanics*, Macmillan, New York (1962)

[7] De La Llave R., Rana D., *Accurate strategies for KAM bounds and their implementation*, preprint

[8] Goldreich P., Peale S., *Spin-orbit coupling in the solar system*, Astron. J. **71**, 425 (1966)

[9] Greene J.M., *A method for determining a stochastic transition*, J. of Math. Phys. **20**, 1183 (1979)

[10] Henrard J., *Spin-orbit resonance and the adiabatic invariant*, in: "Resonances in the Motion of Planets, Satellites and Asteroids", S. Ferraz-Mello and W. Sessin eds., Sao Paulo, 19 (1985)

[11] Kolmogorov A.N., *On the conservation of conditionally periodic motions under small perturbation of the Hamiltonian*, Dokl. Akad. Nauk. SSR **98**, 469 (1954)

[12] Mather J.N., *Nonexistence of invariant circles*, Erg. theory and dynam. systems **4**, 301 (1984)

[13] Moser J., *On invariant curves of area-preserving mappings of an annulus*, Nach. Akad. Wiss. Göttingen, Math. Phys. Kl. II **1**, 1 (1962)

[14] Wisdom J., *Chaotic behaviour in the solar system*, Proc. R. Soc. Lond. **A413**, 109 (1987)

STATISTICAL ANALYSIS OF THE EFFECTS OF CLOSE ENCOUNTERS
OF PARTICLES IN PLANETARY RINGS

Filomena Pereira Gama and Jean-Marc Petit

Observatoire de la Côte d'Azur
B.P. 139
F-06003 Nice cedex (France)

INTRODUCTION

The study of the structure and dynamical evolution of planetary rings is considered nowadays as an essential part of the research in the evolution of the solar system.

As we are not able to know the exact evolution of the rings, we have to do simulations in order to try to understand better this dynamical system.

Until now, the simulations that have been done are either deterministic simulations – when gravitation between particles alone or collisions alone are taken into account – or Monte-Carlo simulations – when both effects are considered. However in these last ones there is only a variable that is kept for each particle orbit: its semi major axis. Theoretical studies show us that we must not disregard the effects on the eccentricity. So we have to do simulations which are able to follow at least these two parameters and that include both inelastic collisions and gravitation between particles.

Our purpose is thus to do a mixed simulation, that is, a deterministic integration (N-body model) for large distances and a Monte-Carlo treatment of close encounters that are far too long to follow in a deterministic simulation. So we have to start with a statistical study of encounters and collisions in order to get a preliminary set of results and conclusions that we will use in the Monte-Carlo part of the mixed simulation. A short presentation of this analysis is the aim of this paper.

PHYSICAL PROBLEM

We take two particles that describe coplanar and almost circular orbits, with slightly different radii around a heavy central body.

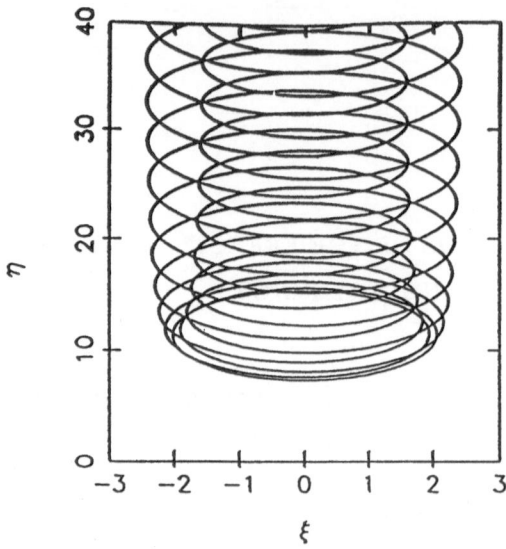

Fig.1. Epicyclic motion obtained with a non-zero initial reduced eccentricity.

Modeling the Encounters

We introduce a reference system, centered on the planet, and rotating with the velocity of the center of mass of the two particles. As we are specially interested on the relative motion of the two particles, we make a zoom which will throw the planet to the infinity and transform the circular orbits into straight orbits[1,2].

As we have considered the initial motion almost circular, we will have the orbit represented in fig.1. The orbit is defined by the impact parameter H (difference of semi-major axis), the reduced eccentricity K (amplitude of the epicyclic motion) and the phase φ which is not important in the planetary rings because there is a mixture of phases between the different encounters. If there is no interaction between particles (i.e. at a great distance) both of the parameters H and K define perfectly the orbit. Otherwise, these variables are going to evolve and our purpose is try to know how they are after the collision.

Modelling the Collision

We suppose that particles cannot break and that permanent accretion is not possible, because we are placed beyond the Roche's limit.

The particles are assumed to be perfectly smooth and rigid spheres.

If we consider that the two particles have the same density, the effect of the collision will depend on the relative velocity of the surfaces of the particles at the impact point. In order to avoid many complications we will assume that there is no surface friction between the particles. The spins are then unaffected by the collisions. So there will be no change in the tangential component of the relative velocity but, as we consider partially inelastic collisions, there will be a dissipation of energy which will be expressed by a change in the radial component of the relative velocity. So, if the relative velocity is large enough, the two particles will separate for good. There will not be any more collisions. But, if the relative velocity is small, there will be a multitude of rebounds, followed by a sliding phase and finally a separation[3].

STATISTICAL ANALYSIS (short presentation)

We take couples of initial values of the eccentricity and of the impact parameter and we make the phase vary uniformly.

Our purpose is to obtain (if possible) a theoretical model of the joint distribution of the final parameters.

We begin by testing the validity of the adjustment of a bidimensional gaussian law to the observed distribution, since it is concerning to this law we know the greatest amount of techniques and results which could be applied later.

If the data does not pass this test, the program done will be able to give us a detailed statistical description of the set of information contained in the joint distribution. In particular, it puts in a conspicuous position the global correlation and the coefficient of linear correlation.

Very briefly, and representing by HH and KK the final values of the impact parameter and the eccentricity respectively, we can say that there is a part of the (total) variance of HH, V(HH), (resp. V(KK)), which is explained by the variance of the conditional means – the explained variance – and another part which is due to the heterogeneousness of each one of the conditional distributions and that appears as a residual variance [4,5,6].

So we can write

$$V(HH) = \frac{1}{n} \sum_{l=1}^{M} n_l . \overline{HH_l^*}^2 - \overline{HH}^2 + \frac{1}{n} \sum_{l=1}^{M} n_l . V(HH_l^*)$$

(Similar for V(KK))

where

HH_l^* : the conditional variable $HH/_{KK=kk_l}$,
n : number of observed values of HH,
M : number of classes of values of the final eccentricity KK.

The square of the coefficient of global correlation is defined in the following way :

$$\eta^2 = \frac{Explained\ variance}{Total\ variance}$$

So we have two coefficients of global correlation that are different in general.

When there is a functional relation between the variables, η^2 will be 1. In the absence of correlation, the value of η^2 will be 0. Otherwise $0 < \eta^2 < 1$.

The coefficient of linear correlation is defined by

$$r = \frac{cov(HH, KK)}{s_{HH}\ s_{KK}}$$

where

s_{HH}, s_{KK} : standard deviations of HH and KK
$cov(HH, KK)$: covariance between HH and KK

and it characterizes the intensity of the correlation only when it is almost linear.

Each regression curve – that is the curve of the variation of the conditional means of one variable as a function of the values of the another variable – synthetizes the statistical relation between the variables and in a certain sense it is the curve which approaches the most the points representing the distribution.

Test of normality

γ_1 : assymmetry coefficient,
γ_2 : excess coefficient,
$\hat{\gamma}_i$: estimator of the coefficient γ_i of the population,
σ_{γ_i} : standard deviation of $\hat{\gamma}_i$,

where $i = 1, 2$.

$$\sigma_{\gamma_1} = \sqrt{\frac{6(n-2)}{(n+1)(n+3)}}$$

$$\sigma_{\gamma_2} = \sqrt{\frac{24n(n-2)(n-3)}{(n+1)^2(n+3)(n+5)}}$$

where n is the number of orbits.

If

$$\begin{cases} |\hat{\gamma}_1(n)| < 1.5\sigma_{\gamma_1} \\ |\hat{\gamma}_2(n) + \dfrac{6}{n+1}| < 1.5\sigma_{\gamma_2} \end{cases}$$

we accept the normality of the distribution with a degree of confidence of 95%.

If

$$|\hat{\gamma}_1(n)| > 2\sigma_{\gamma_1}$$

or

$$|\hat{\gamma}_2(n) + \frac{6}{n+1}| > 2\sigma_{\gamma_2}$$

we reject the normality with the same degree of confidence.

Otherwise we need other tests.

As our distribution is a bidimensional one, this test must be done to one of the marginal distributions and to every conditional distribution which is "perpendicular" to that one (test in "comb").

RESULTS

In this first phase of the work, we have selected couples of initial values of the eccentricity and the impact parameter that we have considered meaningful enough to allow the detection of a first range of regions. Later on and based on the results obtained we will refine that choice in a suitable way.

(H, K)
$H = 0.8, 1, 1.5, 2, 2.5, 3.5$
$K = 0.2, 1, 2, 3, 4$

We got then the following conclusions:

1. The joint distribution of the final parameters is never a normal one. The assumption of normality can always be rejected with a degree of confidence of 95%.

2. There is always a none vanishing correlation between the final values of the eccentricity and the impact parameter. The greater or smaller intensity of this correlation will lead us to the definition of the different regions in which we are interested.

We will only refer some of these regions (see also the graphics of the next pages) :

$H \geq 5$

There is a very strong correlation no matter the value of the initial eccentricity. We can say that there exists a real functional relation between the final impact parameter for this value of H, which tells us that we have a region where, in average, there is no collisions between the particles.

This agrees with the fact that for large values of the impact parameter and small values of the eccentricity, the minimal distance of approach of the particles is large and they move fast. So they cannot approach each other so that a collision could take place.

Jean-Marc Petit and Michel Hénon had already found an assymptotic expansion relating these two final parameters[3]. For $K = 4$, we can already refer a smaller correlation which indicates the begining of collisions.

$H = 1.5$

It is very obviously a critical region : that of the most frequent collisions. The global correlation is much smaller even for the smallest values of the eccentricity.

$H \leq 0.8$

This region is in a certain way a transition region.

The same authors have shown[2] that when the particles are taken as points without dimensions, the orbits corresponding to H = 0.8 take the "horseshoe" shape. There exists a minimal distance of approach of the particles and so there are no collisions. When we take the dimensions of the particles into account it is natural to assume that the minimal distance of approach of the particles becomes smaller and that collisions can take place, at least for the greatest values of the eccentricity. This may explain the obtained results.

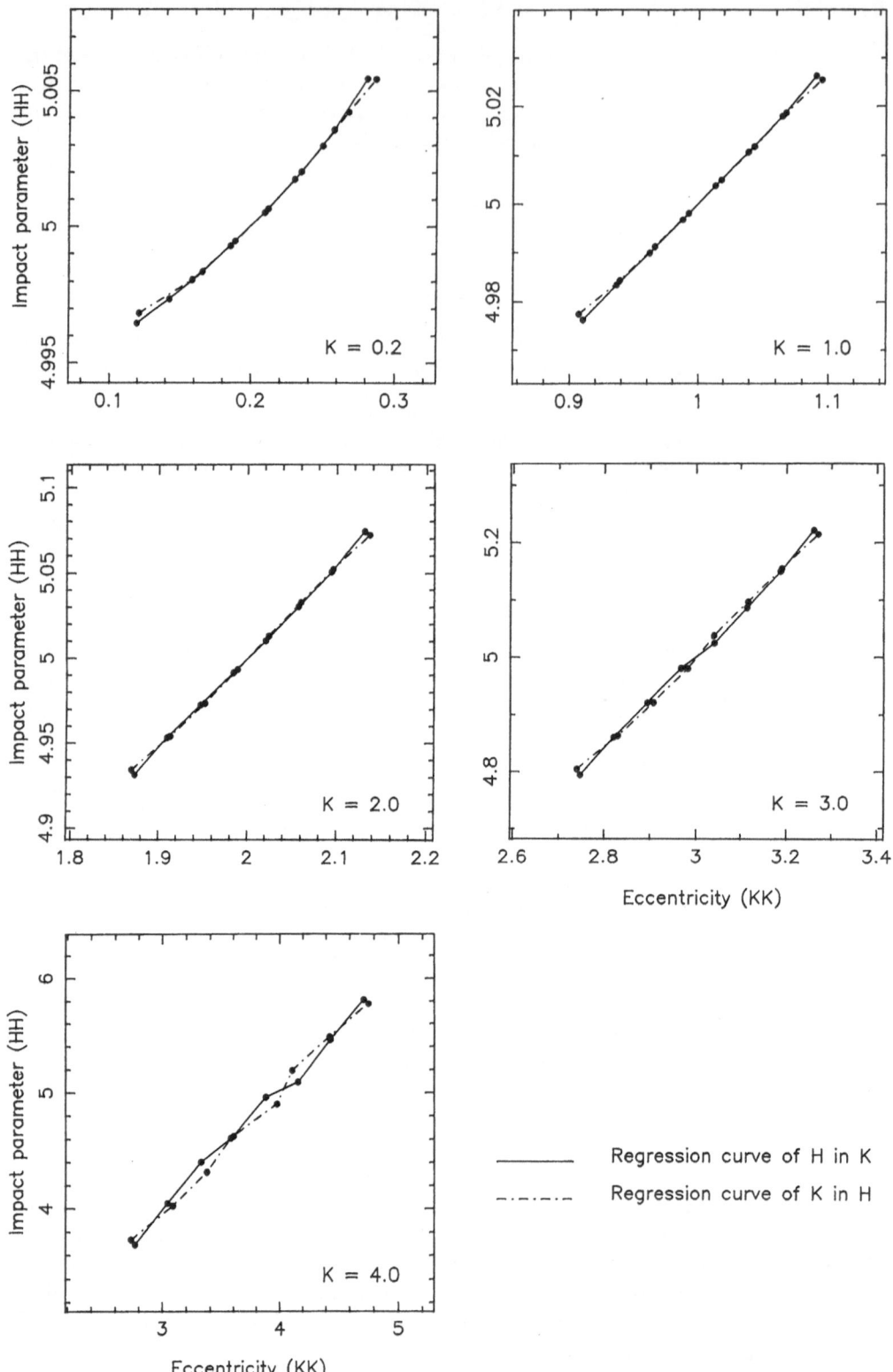

Fig.2. Regression curves corresponding to the initial impact parameter $H = 5$

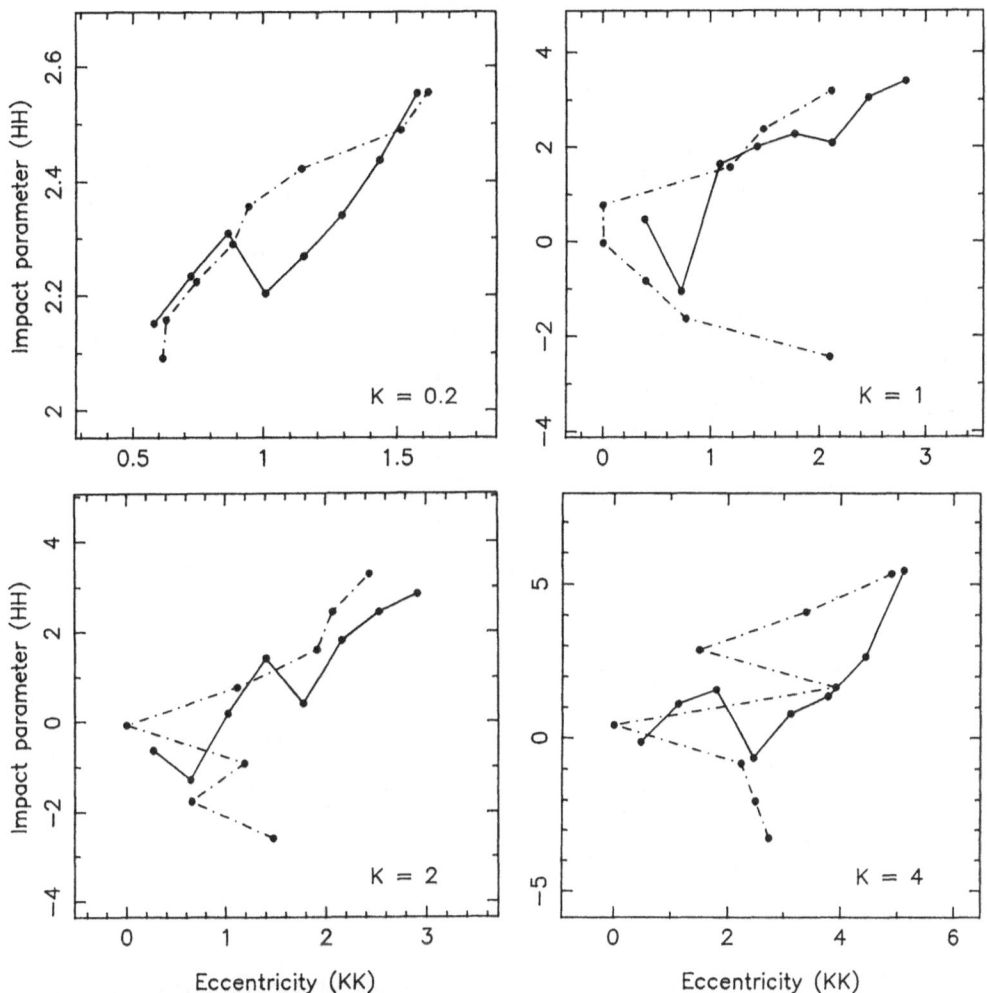

Fig.3. Regression curves corresponding to the initial impact parameter $H = 1.5$

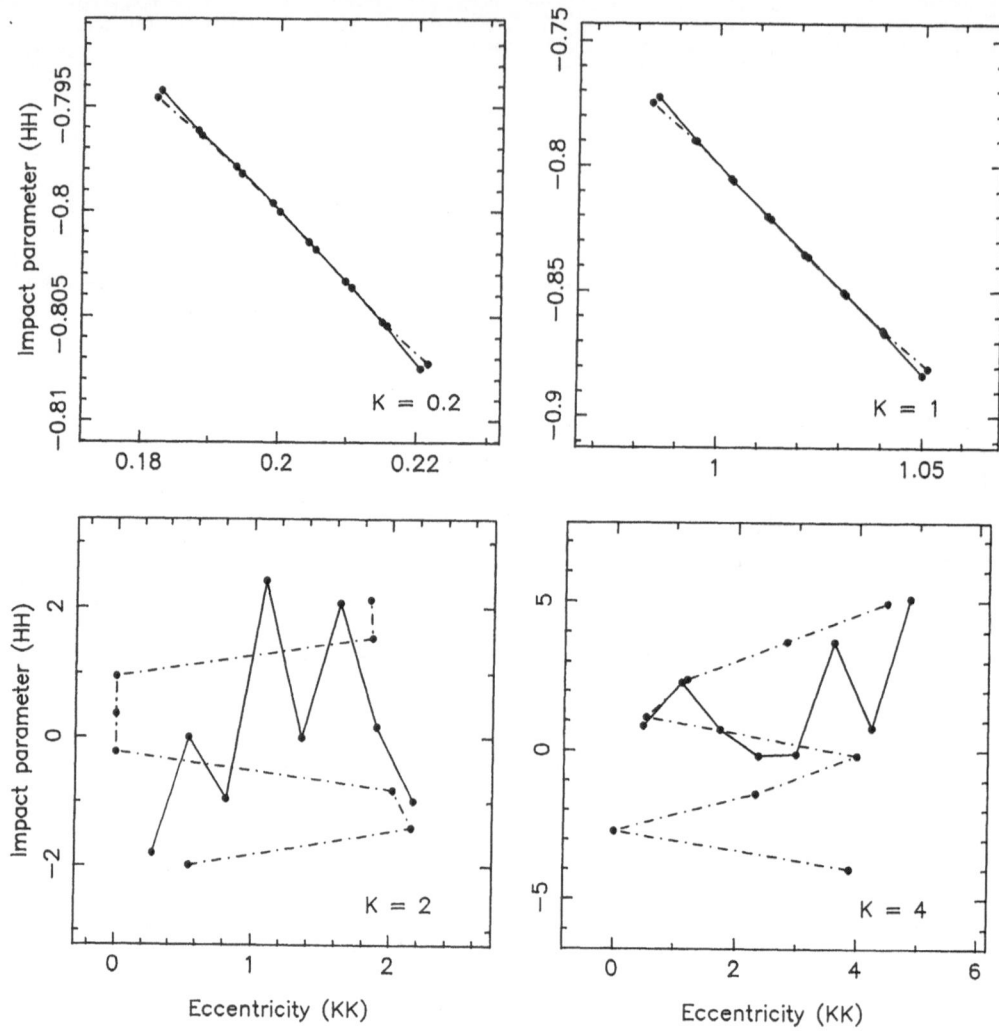

Fig.4. Regression curves corresponding to the initial impact parameter $H = 0.8$

CONCLUSION

Unhappily, we did not find gaussian distributions, at least for instance. A more refined analysis based on the present values will perhaps change a bit the results of the test of normality.

However, we have been able to distinguish quite well some regions in the distribution of the final parameters, which include an important amount of informations, that we will use in the Monte-Carlo component of the mixed simulation we are going to do later.

REFERENCES

1. M. Henon and J.- M. Petit, " Series expansions for encouter-type solutions of Hill's problem ", Cel. Mechanics 38, 67:100 (1986).

2. J.- M. Petit and M. Hénon, " Satellite encounters ", Icarus 66, 536:555 (1986).

3. J.- M. Petit and M. Hénon, " A numerical simulation of planetary rings – 1. Binary encounters ", Astron. Astrophys. 173, 389:404 (1987).

4. W. H. Press, B. P. Flannery, S. A. Teukolsky and W. T. Vetterling, " Numerical Recipes – The art of scientific computing ", Cambridge University Press (1986).

5. G. Saporta, "Probabilités, Analyse des données et Statistique ", Éditions Technip, Paris (1990).

6. B. Grais, " Méthodes Statistiques ", Dunod, Paris (1977).

THE THREE-DIPOLE PROBLEM

C.L. Goudas and E.G. Petsagourakis

University of Patras, Greece

Abstract. After investigating the two-dipole problem (first generalization of Stormer's problem of one magnetic dipole), we proceed one step further along the same lines by setting and investigating the three-dipole problem. Each of the three magnetic dipoles is assumed to be located on one member of a three-star system that performs newtonian motions. Charged particles, positive or negative, moving in the vicinity of the three moving dipoles perform motions which are the object of this study. In this paper we show that if the three stars perform the Lagrangean circular solution of the three-body problem and if the magnetic moments of their dipoles are perpendicular to their plane of motion, then three, or two, or one, closed space inside which charged particles of appropriate energy are permanently trapped . These spaces of trappings can be considered as generalized Van Allen zones.

INTRODUCTION

In 1907 Stormer investigated the motions of charged particles in the vicinity of one fixed magnetic dipole. He thus discovered the thalweg , i.e the space of trapping of radiation created by the dipole. Later, Van Allen (1958) discovered the @zones@ that are created by the magnetic dipole of the Earth and thus verified the theoretical and experimental work of Stormer. In the meantime, the problem, as set and investigated by Stormer, found essential application in the mechanism of formation of the polar aurora (Stormer, 1955). Also a number of investigators studied various aspects of the problem such as the equatorial motions (Graef et al,1938), the periodic orbits (De Vogelaere, 1949,1950,1958, Mavraganis and Goudas, 1975), etc.

The first obvious extension of this problem, namely the two-dipole problem, was set and investigated (Mavraganis,1979, Goudas et al 1985) assuming that the two magnetic dipoles are not fixed, but participating in the motion of a double-star system performing Newtonian (in fact, circular) motion. The purpose was to investigate the motions of charged particles in the field of two magnetic dipoles created within and moving with their carrier stars. The effect of the gravity field was also considered (Mavraganis, 1983). Later, Goudas et al (1985a, 1986, 1990) showed that Stormer's basic finding, namely, the formation of spaces of trapping, holds also in the case of two moving magnetic dipoles. Indeed, it was found that behind each dipole a closed

space is formed, inside which charged particles of the appropriate energy are trapped .The two spaces of trapping corrotate with
the stars.

In this paper we go one step further and consider three
magnetic dipoles each associated with one of member of a three-
star system that moves under the mutual gravitational actions of
its members. The directions and magnitudes of the magnetic
moments of the three dipoles are taken arbitrary but not time-
dependent. After setting the equations of motion of charged
mass-particles under the electromagnetic forces of the three
moving dipoles, we consider the case where the three stars move
in circles according to the well-known solution of Lagrange. We
have also considered, but do not give here any results, the case
where the three stars move according to the straight-line solution of Lagrange. These two particular cases and ·for the special
case of dipoles with magnetic moments perpendicular to the plane
of motion of the stars, are studied in some detail and two
important conclusions are drawn: First, that in the case of circular motion of dipoles, one space of trapping is formed behind
each dipole. Second, that the spaces of trapping are
time-independent and corrotate with the stars. And third, that
for critical ratios of the magnetic moments the two of the
three, or all three spaces of trapping can merge into one large
space of trapping.

FORMULATION - EQUATIONS OF MOTION

Let S_i, m_i, r_i, M_i, i=1,2,3, be the three stars, their masses
and baricentric position vectors, and their magnetic moments,
respectively. Also, let P,m,r,q, be a particle, its mass and position vector, and its electric charge. The equations governing
the newtonian motion of the three bodies are:

$$m_1 \frac{d^2 r_1}{dt^2} = - \frac{Gm_1 m_2 (r_1 - r_2)}{\|r_1 - r_2\|^3} - \frac{Gm_1 m_3 (r_1 - r_3)}{\|r_1 - r_3\|^3}, \tag{1}$$

$$m_2 \frac{d^2 r_2}{dt^2} = - \frac{Gm_2 m_1 (r_2 - r_1)}{\|r_2 - r_1\|^3} - \frac{Gm_2 m_3 (r_2 - r_3)}{\|r_2 - r_3\|^3}, \tag{2}$$

$$m_3 \frac{d^2 r_3}{dt^2} = - \frac{Gm_3 m_1 (r_3 - r_1)}{\|r_3 - r_1\|^3} - \frac{Gm_3 m_2 (r_3 - r_2)}{\|r_3 - r_2\|^3}, \tag{3}$$

where G is the gravity constant.

The equation governing the motion of the charged particle is

$$m \frac{d^2 r}{dt^2} = - \frac{q}{c} \left[\frac{\partial A}{\partial t} \right] - \frac{q}{c} \left[\frac{dr}{dt} \times \left\{ \nabla \times A \right\} \right], \tag{4}$$

where A is the vector potential of the electromagnetic field of the moving dipoles, given by the expression

$$A = \sum_{i=1}^{3} \frac{M \times (r - r_i)}{\| r - r_i \|^3} \ .$$ (5)

THE LAGRANGEAN CASE

If the three stars perform the equilibrium, equilateral or collinear, motion of Lagrange, the equations of motion (4) of the charged particle P, expressed in rotating coordinates, receive the form:

$$m \ \frac{d^2 r}{dt^2} = - 2 \left\{ \omega \times \frac{dr}{dt} \right\} + \omega \times (\omega \times r) - \frac{q}{c} \left[\frac{\partial A}{\partial t} \right] - \frac{q}{c} \left[\frac{dr}{dt} \times \left\{ \nabla \times A \right\} \right],$$ (6)

where A is given by expression (5) and ω the angular velocity of the rotating frame, taken perpendicular to the plane of m motion of the three stars, so that $\omega r_i = 0$, i=1,2,3. Let us now assume that $m_3 = 0$, $m_1 = 1-\mu$, $m_2 = \mu$, in other words adopt the conditions of the restricted 3-body problem, then Eq. (6) projected on the customary synodical frame of reference Oxyz, gives

$$m \frac{d^2 x}{dt^2} = 2m\omega \frac{dy}{dt} + m\omega^2 x - \frac{q}{c}\omega \left[y \frac{\partial A_x}{\partial x} - x \frac{\partial A_x}{\partial y} - A_y \right]$$

$$+ \frac{q}{c} \left[\left\{ \frac{dy}{dt} + \omega x \right\} \left\{ \frac{\partial A_y}{\partial x} - \frac{\partial A_x}{\partial y} \right\} - \frac{dz}{dt} \left\{ \frac{\partial A_x}{\partial z} - x \frac{\partial A_z}{\partial x} \right\} \right],$$ (7)

$$m \frac{d^2 y}{dt^2} = -2m\omega \frac{dx}{dt} + m\omega^2 y - \frac{q}{c}\omega \left[y \frac{\partial A_y}{\partial x} - x \frac{\partial A_y}{\partial y} + A_x \right]$$

$$- \frac{q}{c} \left[\left\{ \frac{dx}{dt} - \omega y \right\} \left\{ \frac{\partial A_y}{\partial x} - \frac{\partial A_x}{\partial y} \right\} - \frac{dz}{dt} \left\{ \frac{\partial A_z}{\partial y} - x \frac{\partial A_y}{\partial z} \right\} \right],$$ (8)

$$m \frac{d^2 z}{dt^2} = \frac{q}{c}\omega \left[y \frac{\partial A_z}{\partial x} - x \frac{\partial A_z}{\partial y} \right] +$$

$$\frac{q}{c} \left[\left\{ \frac{dx}{dt} - \omega y \right\} \left\{ \frac{\partial A_x}{\partial z} - \frac{\partial A_z}{\partial x} \right\} - \left\{ \frac{dy}{dt} + \omega x \right\} \left\{ \frac{\partial A_z}{\partial y} - \frac{\partial A_y}{\partial z} \right\} \right],$$ (9)

We have assumed here that $M_i = M_{iz} k$, i=1,2,3, which means that the three dipoles have moments perpendicular to the plane of motion of the stars. From the definition of the vector potential A we find the following expressions for its components A_x, A_y, A_z:

$$A_x = - \frac{y M_{1z}}{\| r - r_1 \|^3} + \frac{y M_{2z}}{\| r - r_2 \|^3} + \frac{y M_{3z}}{\| r - r_3 \|^3},$$

$$A_y = \frac{(x-\mu) M_{1z}}{\| r - r_1 \|^3} + \frac{(x+1-\mu) M_{2z}}{\| r - r_2 \|^3} + \frac{(x-.5+\mu) M_{3z}}{\| r - r_3 \|^3},$$

$$A_z = 0,$$

where
$$r = (x,y,z), r_1 = (\mu,0,0), \quad r_2 = (-1+\mu,0,0), \quad r_3 = (-.5+\mu, y_0, 0), \quad y_0 = \sqrt{3}/2,$$
and
$$\|r - r_1\| = [(x-\mu)^2 + y^2 + z^2]^{1/2}, \quad \|r - r_2\| = [(x+1-\mu)^2 + y^2 + z^2]^{1/2},$$

$$\|r - r_1\| = [(x+0.5-\mu)^2 + (y-y_0)^2 + z^2]^{1/2}.$$

INTEGRAL OF ENERGY

Equations (7),(8) and (9) admit to the energy integral

$$\frac{1}{2}\left[\left(\frac{dx}{dt}\right)^2 + \left(\frac{dy}{dt}\right)^2 + \left(\frac{dz}{dt}\right)^2\right] - \frac{1}{2}\left(x^2 + y^2\right) - \left(xA_y - yA_x\right) = C \qquad (10)$$

where ω and $q/(mc)$ are taken equal to unity and C is the energy constant. This assumption is limiting the results in regard to the dimensions of the computed quantities.

EQUILIBRIUM POINTS

We shall begin by investigating the existence and calculating the locations and energies of the equilibrium points of the field created by the three rotating dipoles. Under the assumptions made the equilibrium points must satisfy the conditions

$$F(x,y,z) = x - y\frac{\partial A_x}{\partial x} + x\frac{\partial A_y}{\partial x} + A_y = 0, \qquad (11)$$

$$G(x,y,z) = y - y\frac{\partial A_x}{\partial y} + x\frac{\partial A_y}{\partial y} - A_x = 0, \qquad (12)$$

$$H(x,y,z) = \quad y\frac{\partial A_y}{\partial z} - x\frac{\partial A_x}{\partial z} \quad = 0. \qquad (13)$$

For $M_{3z} = 0$ we have the two-dipole case, the equilibrium points of which have been fully investigated (Goudas et al, 1985). We can proceed to the investigation of the equilibrium points of this case ($M_{3z} \neq 0$) by applying the continuation method, since equilibrium points for the cases $M_{1z} \neq 0$, $M_{2z} \neq 0$, $M_{3z} = 0$ and $M_{1z} \neq 0$, $M_{2z} = 0$, $M_{3z} = 0$ are known, whereas, in general, the necessary condition for their existence

$$\partial(F,G,H)/\partial(M_{1z}, M_{2z}, M_{3z}) \neq 0$$

is true. We found it more practical and entirely satisfactory to compute directly the roots of the algebraic system (11), (12), limiting thus the calculations to planar equilibrium points. Since there are three parameters to scan, we use M_{1z} as the unit magnetic moment and identify the said parameters with 1, λ, ν, where

$$λ = M_{2z}, \qquad ν = M_{3z}.$$

We remind that the equilibrium points are, either

(a) points of self-intersection of 0-velocity curves, or
(b) 0-velocity curves of zero length.

The equilibrium points of type (a) are the interesting ones in this investigation because they belong to the category that is associated, yet not in all cases, with spaces of trapping , i.e. spaces defined by closed 0-velocity surfaces limiting the motion of the charged particles in their interior. Therefore, our main concern is to locate, calculate and follow the evolution of the equilibrium points of type (a) for a sufficiently dense set of values of the basic parameters λ,ν. And, finally, to identify those equilibrium points that indeed rest on closed 0-velocity surfaces forming spaces of trapping .

RESULTS ON EQUILIBRIUM POINTS

In Figures 1 and 2 we plot against μ , the x, in the first, and the y, in the second, of the equilibrium points computed for

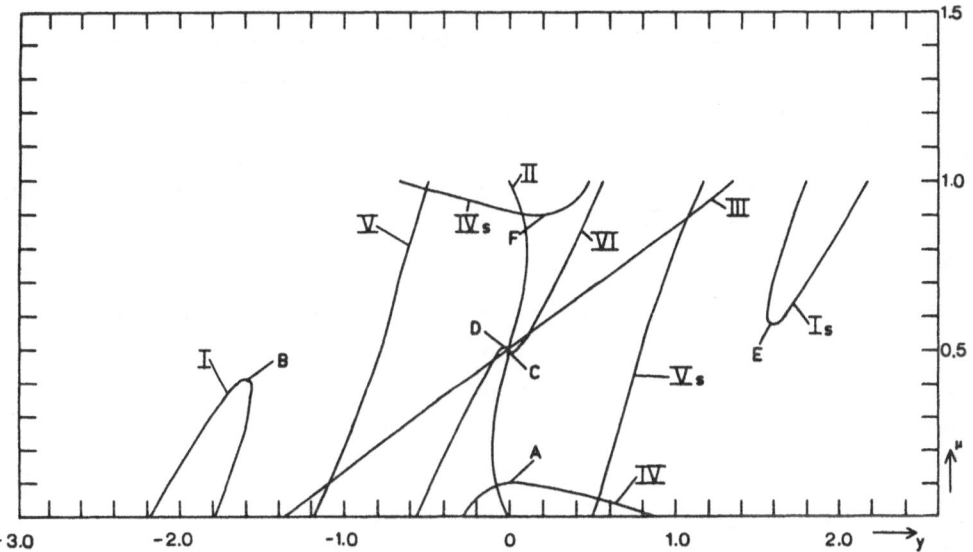

Figure 1.   The (x,μ) diagram of the equilibrium points for the case $M_1 = M_2 = M_3 = k$.

λ=ν=1. The corresponding numerical values are listed in Tables I through VI.For the particular case of three dipoles of equal magnetic moments, if (x,y,0) is an equilibrium point cor-responding to a value of the mass-parameter μ , then the point (-x ,y,0) is also an equilibrium point corresponding to 1-μ . This symmetry property is obviously fulfilled in Figure 1 and Tables I through VI.In the same Figure we observe that some fa-

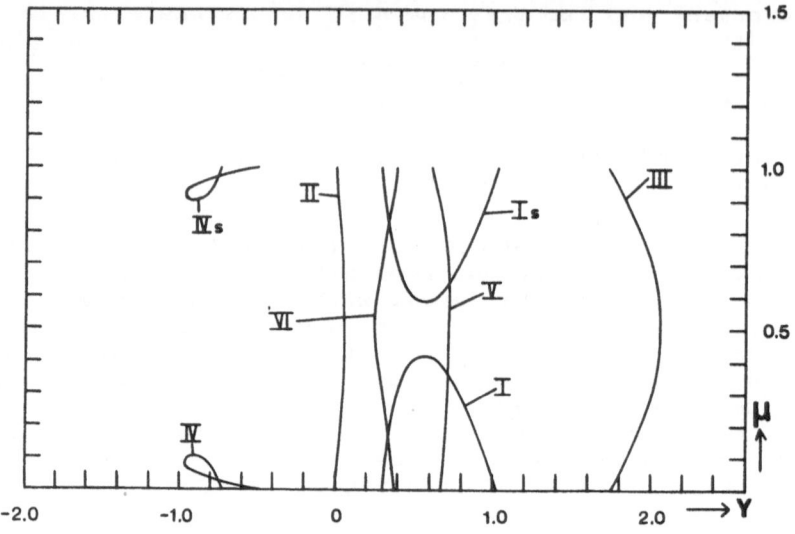

Figure 2.   The (y,μ) diagram of the equilibrium points for the case $M_1 = M_2 = M_3 = k$.

milies of equilibrium points exist only for values of the mass-parameter in an interval smaller than the interval [0,1] inside which μ receives values. For example, family I exists for $\mu \in [0, \mu_2]$, where the critical value μ_2 is roughly equal to 0.418. In fact, for each $\mu \in [0, \mu_2)$ we have two equilibrium points of this type, while for $\mu = \mu_2$ we have only one double. If the terminal points A, B, C, D and F of the families of equilibrium points presented in Figure 1, correspond to the critical values μ_1, μ_2, μ_3, μ_4 and μ_5, respectively, then taking also into account that these families are the only existing ones, we conclude that:

For $\mu \in [0, \mu_1]$ we have   nine   equilibrium points.
For $\mu = \mu_1$ "   "   eight   "   "
For $\mu \in [\mu_1, \mu_2]$ "   "   seven   "   "
For $\mu = \mu_2$ "   "   six   "   "
For $\mu \in [\mu_2, \mu_3]$ "   "   five   "   "
For $\mu = \mu_3$ "   "   six   "   "
For $\mu \in [\mu_3, \mu_4]$ "   "   seven   "   "
For $\mu = \mu_4$ "   "   six   "   "
For $\mu \in [\mu_4, \mu_5]$ "   "   five   "   "
For $\mu = \mu_5$ "   "   six   "   "
For $\mu \in [\mu_5, \mu_6]$ "   "   seven   "   "
For $\mu = \mu_6$ "   "   eight   "   "
For $\mu \in [\mu_6, 1]$ "   "   nine   "   "

As expected, the property $\mu_i = 1 - \mu_{7-i}$, $i = 1, 2, \ldots, 6$, is holds. Because the focal point of this study is to show the existence

TABLE I

Equilibrium Points for $M_1 = M_2 = M_3 = k$ *

μ	x	y	C
0.00	-2.1799	0.2988	-4.97762
0.05	-2.1218	0.3086	-4.81820
0.10	-2.0629	0.3200	-4.64846
0.15	-2.0029	0.3333	-4.48755
0.20	-1.9413	0.3493	-4.32909
0.25	-1.8778	0.3691	-4.17312
0.30	-1.8111	0.3947	-4.01969
0.35	-1.7388	0.4307	-3.86892
0.40	-1.6511	0.4943	-3.72107
0.41	-1.5677	0.6224	-3.69170
0.40	-1.5619	0.6505	-3.72024
0.35	-1.5684	0.7310	-3.86840
0.30	-1.5912	0.7851	-4.00526
0.25	-1 6201	0.8304	-4.14845
0.20	-1.6523	0.8716	-4.29234
0.15	-1.6862	0.9110	-4.43700
0.10	-1.7210	0.9501	-4.58241
0.05	-1.7554	0.9903	-4.72851
0.00	-1.7900	1.0334	-4.87518

*In the case of three dipoles of equal magnetic moments (1=n=1)
if (x,y,0) is an equilibrium point corresponding to a value m
of the mass-parameter, then so is the point (-x,y,0) for the
value 1-m of the same parameter. This see clearly in Figure 1.

and compute the forms of spaces of trapping , it is suitable to
point out here that out of the nine (maximum) or five (minimum)
equilibrium points existing in this problem for μ∈[0,1], only
three, and in some cases, only two, of these are of interest.
These are the points of self-intersection of 0-velocity surfaces
(and, of course, curves when we consider their intersection with
the plane Oxy) parts of which define closed spaces confining the
motion of charged particles of suitable energies within their
interior. The last column of Tables I to IV gives the energy
corresponding to each equilibrium point.

The symmetry property of the equilibrium points about the
value μ = 0.5 of the mass-parameter when the three dipoles are of
equal magnetic moments we checked for one equilibrium point the
locations and energy of which we give in Table VI for μ∈[0,1]
in steps of 0.05.

TABLE II*

Equilibrium Points for $M_1 = M_2 = M_3 = k$

μ	x	y	C
0.00	0.	0.	-∞
0.05	-0.0483	0.0019	-4.91695
0.10	-0.0853	0.0125	-2.31750
0.15	-0.1037	0.0279	-1.37763
0.20	-0.1070	0.0404	-0.86255
0.25	-0.1002	0.0481	-0.53146
0.30	-0.0868	0.0520	-0.30466
0.35	-0.0687	0.0534	-0.14787
0.40	-0.0474	0.0538	-0.04417
0.45	-0.0242	0.0538	0.01522
0.50	-0.	0.0538	0.03459

*In Figure 1 this family is plotted for the entire interval [0,1] of μ utilizing the symmetry property mentioned earlier for the non-listed values of μ.

A complete study of the equilibrium points of this problem necessitates follow-up of their locations found for the case $M_1 = M_2 = M_3 = k$ as the values of these parameters vary in all possible combinations. In Tables given below we present the results of the follow-up of the evolution of some equilibrium points corresponding to the cases $M_1 = k$, $M_2 = M_3 = 1.5k$, 2k, 10k, 0.8k, 0.75k, 0.5k, 0.1k.

SPACES OF TRAPPING

Throughout their motion in the field of the three revolving dipoles the charged particles satisfy the energy integral

$$\frac{1}{2}\left[\left(\frac{dx}{dt}\right)^2 + \left(\frac{dx}{dt}\right)^2 + \left(\frac{dx}{dt}\right)^2\right] - \frac{1}{2}\left(x^2 + y^2\right) - \left(xA_y - yA_x\right) = C \qquad (14)$$

and hence they satisfy the broader condition

$$\Phi(x,y,z) = \frac{1}{2}\left(x^2 + y^2\right) + \left(xA_y - yA_x\right) + C \geq 0 \qquad (15)$$

The equation

$$\Phi(x,y,z) = 0 \qquad (16)$$

defines a surface in E^3 that the particles will not cross but remain on that part of space for which

TABLE III

Equilibrium Points for $M_1 = M_2 = M_3 = k$

μ	x	y	C
0.00	−1.3488	1.7383	−4.97761
0.05	−1.2271	1.7977	−4.88171
0.10	−1.1026	1.8510	−4.79495
0.15	−0.9750	1.8986	−4.71754
0.20	−0.8441	1.9407	−4.64971
0.25	−0.7097	1.9777	−4.59174
0.30	−0.5721	2.0078	−4.54389
0.35	−0.4318	2.0320	−4.50638
0.40	−0.2892	2.0496	−4.47944
0.45	−0.1450	2.0603	−4.46321
0.50	−0.	2.0639	−4.45778

$$\Phi(x,y,z) \geq 0. \tag{17}$$

Since condition (16) – 0-velocity condition – is derived from the energy integral (14) for $\frac{1}{2}\left[\left(\frac{dx}{dt}\right)^2 + \left(\frac{dx}{dt}\right)^2 + \left(\frac{dx}{dt}\right)^2\right] = 0$, it is evident that all equilibrium points satisfy condition (16) and hence rest on 0-velocity surfaces. More specifically, each equilibrium point rests on the 0-velocity surface that corresponds to the energy constant associated with the equilibrium point.in question. The 0-velocity surfaces containing equilibrium points can be of the following types:

(a) 0-area surface

(b) Self-intersecting surface.

From relations (11)-(13) and (16) we conclude that the spaces of trapping of type (a) can exist only around equilibrium points corresponding to local minima of energy. However, spaces of trapping of the type (b) can exist for both maxima and minima of the energy constant at the equilibrium points.

From the experience of the one- and two-dipole problems, we can expect that the best procedure for identifying the spaces of trapping is to draw first the 0-velocity curves on the plane Oxy,for values of C corresponding to the equilibrium points of each case. These 0-velocity curves we call critical . Precluding the cases where the spaces of trapping, being in all cases symmetric with respect to the plane z=0, are not intersected by this plane (no such cases have been found yet), the existence of

TABLE IV

Equilibrium Points for $M_1 = M_2 = M_3 = k$

μ	x	y	C
0.00	0.8600	−0.4966	−2.01757
0.02	0.7801	−0.6508	−2.05937
0.04	0.6204	−0.8158	−2.09542
0.06	0.4169	−0.9259	−2.12354
0.08	0.2261	−0.9623	−2.14359
0.10	0.0178	−0.9223	−2.15615
0.10	−0.0339	−0.8954	−2.15619
0.08	−0.1360	−0.8252	−2.15273
0.06	−0.1826	−0.7915	−2.15153
0.04	−0.2201	−0.7663	−2.15180
0.02	−0.2536	−0.7458	−2.15239
0.00	−0.2847	−0.7285	−2.15588

spaces of trapping will be revealed by the existence of areas of trapping on the said plane. Hence, we should first identify the closed O-velocity curves and check whether they constitute "limits" for the particle motion. As soon as such an area of trapping is identified we proceed to define the entire O-velocity surface and investigate their open or closed character.

We found that the straight-forward computation of the critical O-velocity curves, the identification of those that form areas of trapping and the computation of the closed surfaces of trapping corresponding to the latter, is a successful procedure for this problem.

We shall illustrate this procedure by referring to the specific cases shown in Figures 3 and 4. In Figure 3 we give the critical O-velocity curves for $M_1 = 10M_2 = 10M_3 = k$ and μ = 0.1, while in Figure 4 we give the same curves for μ = .5. The difference between the two sets of curves is large but unessential as far as their basic features are concerned. Indeed, the important common feature in these two figures is the presence of three heart-shaped areas of trapping associated with the three dipoles and the three equilibrium points L_1, L_2 and L_3. We denote these areas of trapping with the symbols TR followed by a number, e.g. TR1, TR2 and TR3 for the cases presented in Figures 3 and 4. The motion of charged particles of energies equal or smaller than the ones corresponding to the energy at the equilibrium points L_1, L_2 and L_3, is confined in the

TABLE V

Equilibrium Points for $M_1 = M_2 = M_3 = k$

μ	x	y	C
0.00	−1.1721	0.6767	−5.53008
0.05	−1.1273	0.6812	−5.32544
0.10	−1.0832	0.6860	−5.12403
0.15	−1.0399	0.6910	−4.92600
0.20	−0.9976	0.6963	−4.73147
0.25	−0.9564	0.7017	−4.54061
0.30	−0.9164	0.7071	−4.35357
0.35	−0.8771	0.7126	−4.17054
0.40	−0.8410	0.7178	−3.99167
0.45	−0.8059	0.7226	−3.81713
0.50	−0.7727	0.7267	−3.64705
0.55	−0.7417	0.7295	−3.48149
0.60	−0.7128	0.7307	−3.32047
0.65	−0.6859	0.7297	−3.16388
0.70	−0.6607	0.7256	−3.01150
0.75	−0.6365	0.7178	−2.86298
0.80	−0.6124	0.7057	−2.71781
0.85	−0.5869	0.6890	−2.57534
0.90	−0.5587	0.6675	−2.43480
0.95	−0.5264	0.6414	−2.29532
1.00	−0.4886	0.6107	−2.15588

interior of these areas. On the other hand the closed areas present in these figures are non-permissible for the motion of charged particles and hence they are not of the TR type.

In order to find if a TR area correspond to a TRS (TRapping Space) we proceed as follows: We first calculate the maxima and minima of the 0-velocity surface (16) for the values of C corresponding to each TR area. If such max.-min. points do not exist so do the corresponding TRS's. If, on the other hand, max.-min. points exist, so do the corresponding TRS's. The verification is then ascertained by straightforward calculation of the sign of the l.h.s. of relation (15) at points with z smaller or larger than the maxima or minima, respectively. By doing so for the cases presented in Figures 3 and 4 we found that the TRS's of these cases indeed exist. In particular, for the case of Figure 3 we found that the maxima-minima (for z>0) are as follows:

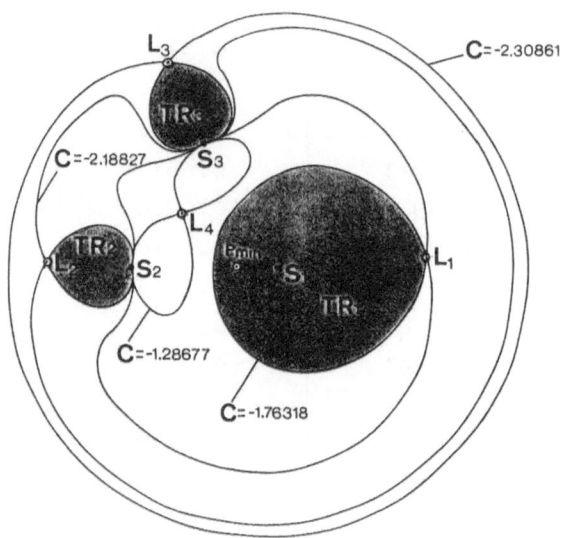

Figure 3.   Critical O-velocity curves for $\mu = 0.1$ and $M_1 = 10M_2$ $= 10M_3 = k$.

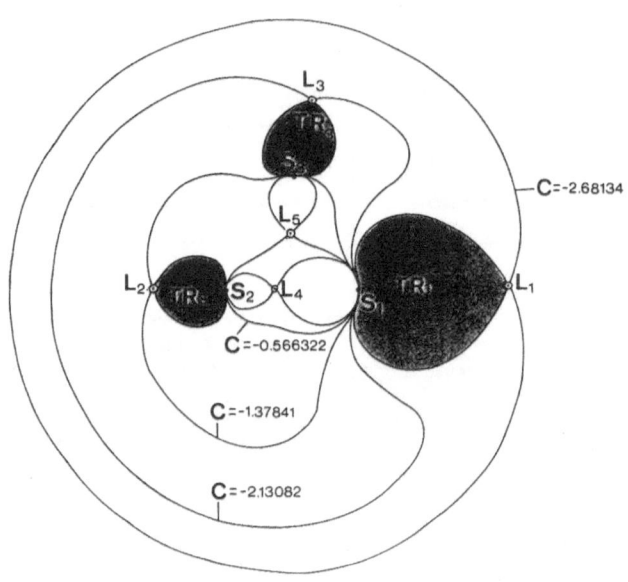

Figure 4.   Critical O-velocity curves for $\mu = 0.5$ and $M_1 = 10M_2 = 10M_3 = k$.

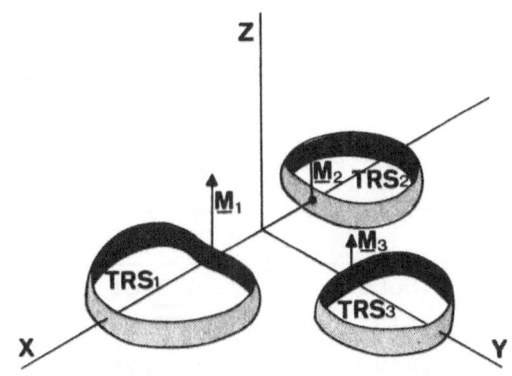

Figure 5.   For $M_1 = 2M_2 = 2M_3 = k$ the spaces of trapping are
                separate. The case shown in this figure corresponds
                to μ = 0.5. Particles with energy less than
                -3.00276, -2.46151 and -3.30680 are trapped inside
                TRS1, TRS2 and TR3, respectively.

1. For TRS1
    x= 0.5090,   y= 0.0310, z_{max} = 0.3013

    x= 0.1799,   y= 0.0096 z_{min} = 0.1195   C=-1.76318.

2. For TRS2

    x=-1.1038   y= 0.0 026 z_{max} = 0.2252   C=-2.18827

3. For TRS3
    x=-0.4844   y= 1.0529 z_{max}=0.2283.   C=-2.18827

    The corresponding quantities for Figure 4 are:

1. For TRS1

    x= 0.9031   y= 0.0019 z_{maz} = 0.3733   C=-2.68134

2. For TRS2

    x=-0.6948   y= 0.0006 z_{max} = 0.1971   C=-1.37841

3. For TRS3
    x= 0.0188   y= 1.0669 z_{max} = 0.2168   C=-2.13082.

DISCUSSION

        A first important observation of general character is that
all spaces of trapping discovered are considerably compressed in
the k-direction. For example, TRS1 of Figure 3 has a ratio of
14/3 between its longest and shortest diameters. A second
observation is that some spaces of trapping have two pairs of
extreme points along the Oz direction, as for example TRS1 of
Figure 3, while the simplest and most usual single type of TRS is
that with one pair of extreme points. A third observation based on
all examples computed is that the spaces of trapping, for
critical values of the parameters μ, M_1, M_2, M_3 can merge,   or

TABLE VI

Equilibrium Points for $M_1 = M_2 = M_3 = k$

μ	x	y	C
0.00	−0.5650	0.3261	−0.01157
0.05	−0.5171	0.3229	0.00137
0.10	−0.4692	0.3191	0.01025
0.15	−0.4212	0.3148	0.01482
0.20	−0.3730	0.3097	0.01479
0.25	−0.3243	0.3038	0.00987
0.30	−0.2751	0.2968	−0.00027
0.35	−0.2248	0.2884	−0.01593
0.40	−0.1725	0.2779	−0.03726
0.45	−0.1158	0.2639	−0.06382
0.50*	−0.0361	0.2391	−0.09186
0.50	0.0000	0.2317	−0.09258
0.59	0.0361	0.2391	−0.09186

*Three different equilibrium points of this family exist for a small interval of values of μ centered around μ=0.5.

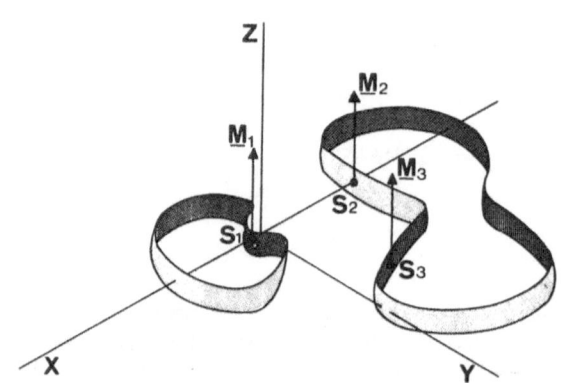

Figure 6.   Three-dimensional presentation of the spaces of trapping for $M_1 = M_2 = M_3 = k$. The spaces of trapping are not separate in this case. Indeed, TRS2 and TRS3 are in coalition.

TABLE VII

Three-Dimensional Equilibrium Points for $M_1 = M_2 = M_3 = k$

μ	x	y	z	C
0.00	−0.3351	0.3812	±0.5139	−0.0851126
0.05	−0.3086	0.3704	0.5027	−0.0865623
0.10	−0.2818	0.3587	0.4889	−0.0855814
0.15	−0.2550	0.3460	0.4720	−0.0822892
0.20	−0.2283	0.3323	0.4511	−0.0758756
0.25	−0.2019	0.3177	0.4247	−0.0696613
0.30	−0.1764	0.3024	0.3700	−0.0612464
0.35	−0.1501	0.2886	0.3470	−0.0529166
0.40	−0.1001	0.2886	0.3470	−0.0466670
0.45	−0.0500	0.2886	0.3470	−0.0429165
0.50	−0.0000	0.2886	0.3470	−0.0416666
0.55	0.0500	0.2886	0.3470	−0.0429165
0.60	0.1001	0.2886	0.3470	−0.0466667
0.65	0.1501	0.2886	0.3470	−0.0529166
0.70	0.1764	0.3024	0.3900	−0.0612464
0.75	0.2019	0.3177	0.4247	−0.0696613
0.80	0.2283	0.3323	0.4511	−0.0768756
0.85	0.2550	0.3460	0.4720	− 0.0822892
0.90	0.2818	0.3587	0.4889	−0.0855814
0.95	0.3086	0.3704	0.5027	−0.0865623
1.00	0.3351	0.3812	±0.5139	−0.0851126

get split, in all possible combinations. For example, we can have three separate TRS's become two, or even merge into one. A typical example of this case is given in Figures 5, 6 and 7. In Figure 5 the trapping spaces are three, one behind each dipole. In Figure 6 the TRS2 and TRS3 have merged into a larger one, while TRS1 is isolated. Finally, in Figure 7 all three TRS's have merged into one large space of trapping. The calculation of the combinations of the critical values of the parameters for which any two of the three TRS's can merge, or all three can become one large TRS was based on the continuation method. A first approximation of these critical values is obtained by assuming linearity between the parameters and the dimensions of the TRS's that we want to merge. For example, for μ = 0.5 and $M_1 = k$ the critical values of M_2 and M_3 (assuming them equal) is about

0.7k. For $M_2 = M_3 = 0.75k$ the trapping spaces TRS1, TRS2 and TRS3 are separate, while for $M_2 = M_3 = 3k$ all three TRS's have merged into one. Also for μ = 0.1 and $M_1 = M_2 = M_3 = k$ the spaces TRS2 and TRS3 are merged, while TRS1 is isolated.

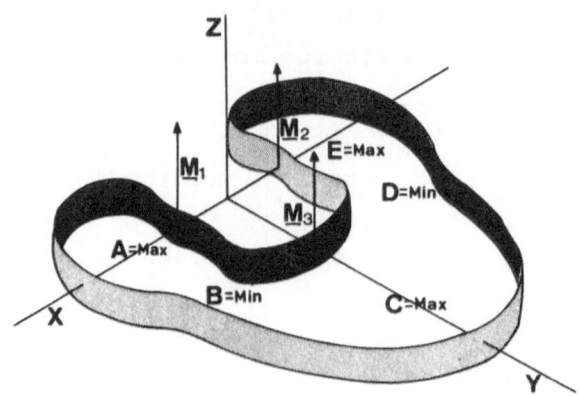

Figure 7.   For $M_1 = M_2 = M_3 = 3k$ and $\mu = 0.5$ the   spaces   of
               trapping are unified. Particles with energy less
               than −8.15763 are trapped in this extended space.
               The point (0,2.63414,0) is an equilibrium one for
               this energy.

REFERENCES

De Vogelaere, R.: 1949, Proc. Sec. Math. Congress, Vancouver,
               170—171.
De Vogelaere, R.: 1950, Can. J. Math., 2, 440.
De Vogelaere, R.: 1958, @Contribution to the Theory of Non-
               Linear Oscillations@, Vol. IV, 53—84.
Goudas, C.L. and Petsagourakis, E.G.: 1985, in @Stability of the
               Solar System and Minor Natural and Artificial
               Bodies, V. Szebehely (ed.), D. Reidel Publ.
               Co., Dordrecht, Holland, pp. 349—364.
Goudas, C.L., Leftaki, M. and Petsagourakis, E.G.: 1985a,
               Celestial Mechanics, 37, 127—148.
Goudas, C.L., Leftaki, M. and Petsagourakis, E.G.: 1986,
               Celestial Mechanics, 39, 57—65.
Goudas, C.L., Leftaki, M. and Petsagourakis, E.G.: 1990,
               Celestial Mechanics, 47, 1—14.
Graef, C. and Kusaka, S.: 1938, J. Math. Phys., 17, 43.
Mavraganis, A.G. and Goudas, C.L.: 1975, Astrophys. Sp. Sci., 32,
               115.
Mavraganis, A.G.: 1979, Astrophys. Sp. Sci., 80, 130—133.
Mavraganis, A.G. and Pangalos, C.A.: 1983, Indian J. pure appl. Math.,
               14(3), 297—306.
Stormer, C.F.: 1907, Arch. Sci. Phys. et Nat. Geneve, 24, 350.
Stormer, C.F.: 1955, @Polar Aurora@, Oxford University Press,
               Oxford, England.
Van Allen, J.:1958, J. Geophys. Res.,64, 11, 1683.

THE N-DIPOLE PROBLEM AND THE RINGS OF SATURN

C.L. Goudas

University of Patras, Greece

Abstract: N-magnetic dipoles each located on a star-member of an n-body star system, are assumed to move with their carrier stars and control the motions of charged grains in their vicinity. The case N=5, in a special configuration, where four dipoles perform rigid rotation about the fifth, while all have magnetic moments parallel to the angular velocity vector, is used as a test case to show that the "spaces of trapping" found to exist in the two and three dipole problems, receive a form similar to the rings of Saturn and that pairs of "spaces of trapping" are separated by gaps similar to the Cassini division. The effect of gravity of a rotating planet within which the five dipoles, of internal "dynamo" origin exist and corrotate are taken into account. In the Appendix a model called N-D (N-Dipole) equivalent to the Z3 model for Saturn is given.

INTRODUCTION

It is already shown that two, as well as three, magnetic dipoles in solid rotation about a fixed axis, create an equal number of closed "spaces of trapping", inside which charged particles of suitable energies are confined, theoretically, for all times (Goudas, 1990, 1990a). This result is more or less an expected extension of the result ("thalweg") of the one-dipole problem set and studied by Störmer (1907) and verified by the discovery of the "Van Allen zones". In the case of three magnetic dipoles, the corresponding three heart-shaped "spaces of trapping", can merge into two, or, most interestingly, into one single space, depending on the relative magnitude of the three dipoles.

The interest to formulate and investigate the general N-dipole problem, was encouraged by the logical expectation that for N=4 and for rigid rotation, we may have four heart-shaped "spaces of trapping". Furthermore, that these spaces, for particular magnitudes of the magnetic moments, can merge into one "torus of trapping". And, finally, that an additional, concentric to the first, "torus of trapping" may result to the formation of a brief gap, void of matter, between the two toruses, resembling the Cassini and the other divisions observed in the rings of Saturn.

While this is, indeed, so, as will be presented in this paper, the N-dipole problem stands on its own ground of interest as a real problem of Astronomy and Engineering, since it refers to the dynamics of a single, or of many charged particles moving in the electromagnetic field of N magnetic dipoles in motion.

Each of the N-dipoles is taken in this paper to participate in the motion of one member of an N-star self-gravitating system. Thus, a charged grain will move under the joint action of the electromagnetic and newtonian fields of N dipoles and N stars.

This general formulation finds practical applications in engineering when N magnetic dipoles perform solid rotation about a fixed axis, as in the case of a power generator.

THE N-DIPOLE DYNAMO FOR LARGE PLANETS

The N-dipole problem could also find practical application if the magnetic field of a star or a planet can be expressed as the resultant of N isolated dipoles located within the star, or the planet, and thus participate in its rotation. Since at the moment we know to some detail the magnetic fields of a few planets e.g. that of the Earth and Saturn, we shall consider the case of the second and attempt to express the data available in terms of the locations and magnitudes on N dipoles fixed to it. A least-squares fitting, for three values of N, is presented in the Appendix. Before going into this task a theoretical consideration is necessary.

The dynamo theory for the origin of planetary magnetic fields, attributes these fields to the convection currents of the highly ionized hot matter of their core and the adjacent parts of the inner mantle (see, e.g. Kaula, 1968). Various velocity patterns in such flows with radial and transverse components of the order of 10^{-2} cm sec^{-1}, are shown by numerical experiments (Bullard and Gellman, 1954) to be capable to maintain magnetic fields of the strength observed around planets.

Although there are doubts about the actual flow patterns as well as the conductivity of the matter in motion, it has been shown that at least a homogeneous dynamo can exist for certain artificial models (Herzenberg, 1958).

The planetary dynamo can be expressed by the magnetic moment magnitude M as follows:

$$M = kR_c^3 (\rho\lambda\omega)^{1/2} R_o^a R_M^b P_e^c R_e^d$$

where k is a dimensionless constant of unit order, R_c and ρ the core radius and density, λ the magnetic diffusivity, ω the rotation rate of the planet, R_o, R_M, P_e, R_e, the Rossby, the magnetic Reynolds, the Peclet and the ordinary Reynolds numbers, all based on R_c as a length scale and on $(F/\rho)^{1/3}$ as a velocity scale. Here F is the heat or energy flux responsible for the origin of the dynamo (Stevenson, 1979).

In the restricted formulation to be presented in this paper we assume that a pattern of N equatorial, circular, convection cells, one central and the rest peripheral, are responsible for the Kronian field. This hypothetical pattern for the case N=5 is shown in Figure 1. The centers of the N convective cells have the vector positions defined below:

$$\underline{r}_i = \underline{i}R_o \cos\left[2\pi\,\frac{2\pi(i-2)}{N}\right] + \underline{j}R_o \sin\left[\frac{2\pi(i-2)}{N}\right] + Z_o\underline{k},$$

$$i = 2, 3, \ldots, N \qquad (1)$$

where Z_o and R_o are constants. The first dipole of the model is the central one, of magnetic moment $\underline{M}_1$, located at $\underline{r}_1 = Z_o\underline{k}$, i.e. displaced by the same amount Z_o as the peripheral dipoles. This assumption reflects the finding by Acuña et al (1980) of a northward @offset@ of the magnetic field of Saturn, later verified by Connerney et al (1982, 1984), and Acuña et al (1983).

On the basis of the in situ observations of the magnetic field of Saturn by the Pioneer 11 spacecraft, in September 1979, and by the Voyager 1 and 2 space crafts, in November 1980 and August 1981, respectively, and the analysis of these observations, a more of less accurate mathematical model (the Z3 model) of Saturn's magnetic field data, is presented in the last two refs. The axisymmetry, not imposed but revealed, and the unexpected @offset@ of the dipole moment are basic characteristics of the Z3 model.

Yet the observed strong periodic modulation of the kilometer radiation from Saturn, of period commensurable to the rotation period of the planet, indicates the existence of longitude dependent terms in the magnetic field.

This expected, but yet unobserved, departure from spin symmetry of the magnetic field of Saturn was one, but not the main motive for the work on alternatives to the Z3 model to be reported here. A basic reason for alternative expressions of the Kronian magnetic field is the intrinsic contradiction between the generally accepted dynamo mechanism that postulates toroidal convection flows of ionized hot mass and the use of spherical harmonics, a tool best suited to express global properties such as gravity, for the representation of the dynamo consequences.

Imagine the core flow pattern of Saturn to resemble the one shown in Figure 1. In such a case the external magnetic field is

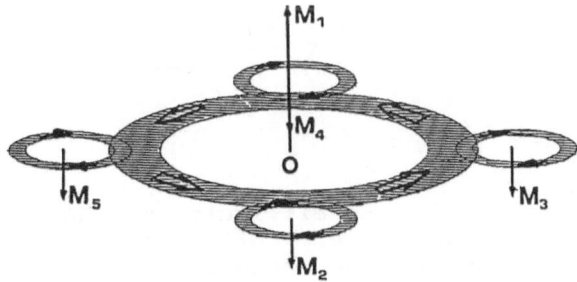

Figure 1. Equatorial section of the core of Saturn showing a hypothetical regime of convective cells, adequate to produce external magnetic field identical to the one observed in situ by Pioneer 11 and Voyagers 1 and 2.

best represented as the resultant of five dipole fields, one central of magnetic moment $\underline{M}_1$ collinear to the angular velocity $\underline{\omega}$ of the planet and four peripherals, of magnetic moments $\underline{M}_i, i=2,3,4,5$ of direction anti parallel to $\underline{\omega}$.

The above model of the magnetic field of Saturn we call N-D model (N-Dipole model), and we call 5-D the model shown in Figure 1. The number of peripheral dipoles can be any small integer, but for the case of Saturn it cannot be more than 10, or 15. In the Appendix we give the 5-D, 7-D and 9-D Models for Saturn, that fit the in situ observations to an RMS less than 2 nTesla and, therefore, are representing the Kronian magnetic field to about the same accuracy as the Z3 model.

FORMULATION OF THE GENERAL GRAVITATIONAL N-DIPOLE PROBLEM

We start with the usual formulation of the N-body problem, where N mass-points of mass m_i, at positions $\underline{R}_i$, (the origin is fixed at their center of mass), are considered and assume that each mass-point is electrically neutral but carries a magnetic dipole of magnetic moment $\underline{M}_i$, where $i=1,2,\ldots,N$.

A single mass-point P, of mass m, position vector $\underline{r}$ and electric charge q, moves in the newtonian and electromagnetic environment of the above N bodies, which it does not modify on account of its negligible mass, velocity and charge.

In order to follow the evolution of this system (N bodies with magnetic dipoles and a mass-point with charge) we have to integrate the system of equations

$$m_i \frac{d^2\underline{R}_i}{dt^2} = -G \sum_{i=1,j\neq i}^{N} \frac{m_i - m_j (\underline{R}_i - \underline{R}_j)}{\|\underline{R}_i - \underline{R}_j\|^3} , \qquad (2)$$

$$m \frac{d^2\underline{r}}{dt^2} = -\frac{q}{c} \frac{\partial \underline{A}}{\partial t} + \frac{q}{c} \left(\frac{d\underline{r}}{dt} \wedge \underline{\nabla} \times \underline{A} \right) - G \sum_{i=1}^{N} \frac{m - m_j (\underline{r} - \underline{R}_j)}{\|\underline{r} - \underline{R}_j\|^3} , \qquad (3)$$

where for the vector potential $\underline{A}$ we have

$$\underline{A} = \sum_{i=1}^{N} \frac{\underline{M}_i \wedge \underline{r}_i}{r_i^3} \qquad (4)$$

for $\underline{r}_i = \underline{r} - \underline{R}_i$; $i = 1,2,\ldots,N$.

Out of this general, and difficult to handle formulation, two practical simplified forms, of astronomical interest the first, and of engineering, the second, can be developed.

In Figure 1, we depict the first case, in which the N-Dipole problem concerns the motion of a mass-point P of mass m, charge q, (we shall also call it "grain" or "charged grain") and position vector $\underline{r}$, under the gravitational attraction of a planet, of mass M (or M_s for the mass of Saturn) and under the electromagnetic force of its N magnetic dipoles in rigid rotation with the angular velocity $\underline{\omega}$ of the planet.

The second case (engineering problem) is when there is no gravity, but only electromagnetic forces applied upon a charged grain by N dipoles in solid rotation about an axis. The first case will be treated here, since the second follows by elimination of terms.

THE 5-DIPOLE CASE

The equation of motion of a single charged grain under the combined effects of the gravity and internal magnetic field of Saturn adopting for the latter the 5-D model (see Appendix) and expressed in a frame Oxyz in solid rotation with the planet ($\underline{\omega}$ = $\omega\underline{k}$ is the common angular velocity) are

$$m\,\frac{d^2\underline{r}}{dt^2} = -2m\left(\underline{\omega}\times\frac{d\underline{r}}{dt}\right) -m(\underline{\omega}\wedge\underline{\omega}\wedge\underline{r})$$

$$-\frac{q}{c}\frac{\partial \underline{A}}{\partial t} + \frac{q}{c}\left(\frac{d\underline{r}}{dt}\wedge\underline{\nabla}\wedge\underline{A}\right) - \frac{GmM\underline{r}}{r^3}, \tag{5}$$

where

$$\underline{A} = \sum_{n=1}^{5} \frac{\underline{M}_i\wedge\underline{r}_i}{r_i^3}, \tag{6}$$

and M is the mass of Saturn.

The energy integral is

$$\frac{1}{2}m\left(\frac{d\underline{r}}{dt}\right)^2 - \frac{1}{2}m\omega^2(x^2+y^2) - \frac{q\omega}{c}(xA_y - yA_x) + \frac{GmM}{r} = C, \tag{7}$$

while the equations of the equilibrium points, essential in this problem as they were found to be in the two and three-dipole cases, are

$$x + \frac{q}{mc\omega}(xA_{yx} - yA_{xx} + A_y) - \frac{GMx}{r^3\omega^2} = 0, \tag{8}$$

$$y + \frac{q}{mc\omega}(xA_{yy} - yA_{xy} - A_x) - \frac{GMy}{r^3\omega^2} = 0, \tag{9}$$

$$\frac{q}{mc\omega}\left(xA_{yz} - yA_{xz}\right) - \frac{GMz}{r^3\omega^2} = 0. \tag{10}$$

RESULTS FOR N=5 and M=0

By eliminating gravity (M=0) and assuming $R_o=1$, $Z_o=0$ we actually extend the results reported for the two and three-dipole problems. In Figures 2 and 3 we give the section of the "spaces of trapping" by the equatorial plane. The shaded parts are "areas of trapping", isolated in Figure 2 and merged in Figure 3. The blank areas are non-permissible for particles of energies $C \leq -8.27745$ (Figure 2) or $C \leq -9.95507$ (Figure 3), while $q/(mc\omega)=1$ and

$$M_1=1, \quad M_2=2.2 \quad \text{(Figure 2)},$$
$$M_1=1, \quad M_2=3 \quad \text{(Figure 3)}.$$

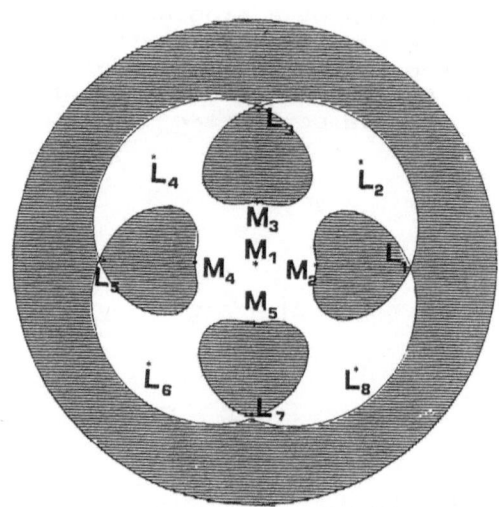

Figure 2. Equatorial section of the heart-shaped spaces of trapping in the case of five dipoles. The four closed spaces have point contacts with the outside permissible to motion space at the equilibrium points L_1, L_3, L_5, L_7. Charged particles can move only inside the shaded areas.

In the above two case the pertinent equilibrium points are eight and, in particular, they are located as follows:
(For Figure 2)

$x = +2.61352$	$y = 0,$	$z = 0,$
$x = 0$	$y = +2.61352,$	$z = 0,$
$x = -2.61352,$	$y = 0,$	$z = 0,$
$x = 0$	$y = -2.61352,$	$z = 0.$

(For Figure 3)

$x = 2.81412$	$y = 0$	$z = 0,$
$x = 0$	$y = 2.81412$	$z = 0,$
$x = -2.81412$	$y = 0$	$z = 0,$
$x = 0$	$y = -2.81412$	$z = 0.$

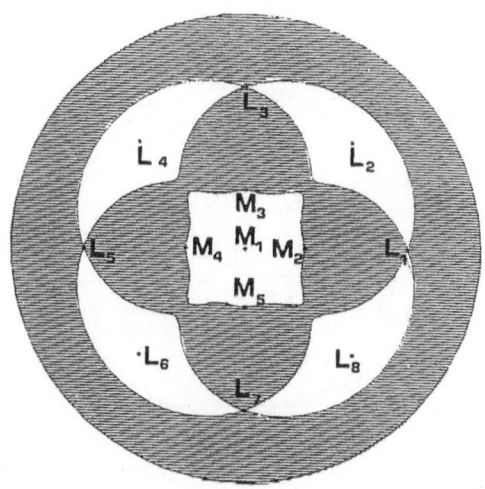

Figure 3. The case presented in Fig. 2 but for a different
ratio of the magnitudes of the central to the peripheral magnetic dipole moments. The four heart-shaped spaces of
trapping are now merged into one, thus forming an irregular
ring around the five dipoles.

RESULTS FOR N=9, M=0

In proportion to what happens for N=2,3 and 5, in the case
of N=9, with out gravity (M=0), we found that the "space of
trapping" receives the form of a torus (in Figure 4 we show its
equatorial section shaded) that has 8 point-contacts with an open
space where motion is also permissible. The thin blank ring of
Figure 4, resembles the Cassini division of Saturn's ring.

The eight point-contacts and the eight isolated points
inside the blank rings are the 16 points of equilibrium of this
problem. Their coordinates are:

x	y	x	y
2.83068	0.	2.00159	2.00159
-2.83068	0.	2.00159	-2.00159
0.	2.83068	-2.00159	-2.00159
0.	-2.83068	-2.00159	-2.00159
2.60636	1.07960	2.60636	-1.07960
-2.60636	1.07960	-2.60636	-1.07960
1.07960	2.60636	1.07960	-2.60636
-1.07960	2.60636	-1.07960	-2.60636

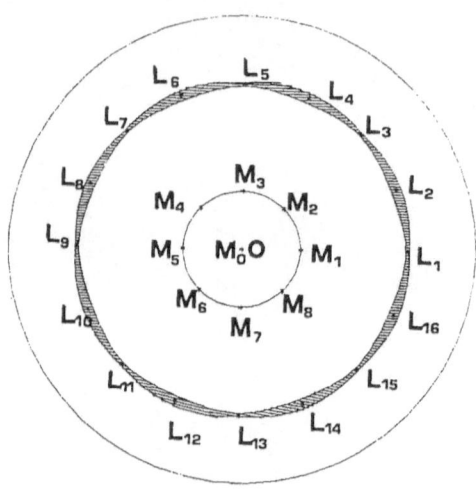

Figure 4. Equatorial section of the 0-velocity surfaces limiting the motion of charged particles in the field of 9 dipoles. The shaded strip, inside which motion is not possible, is a division between spaces where motion is possible.

FIVE DIPOLES AND GRAVITY

The inclusion of the gravity terms in the N-dipole problem does not change the basic results given earlier. Figure 5 shows the Cassini like gap corresponding to the 5-D model of the magnetic field of Saturn. The following values of the parameters are employed:

$GM_s/\omega^2 = 6.43895$ R_s^3 , $\omega = 1.63778 \cdot 10^{-4} s^{-1}$ $M_1 = 26000nT$, $M_2 = M_3 = M_4 = M_5 = -2000nT$ $R_o =$the distance of the peripheral dipoles from the center of the planet. Thus, the four dipoles are located at the points $(0.4R_s, 0, 0.04R_s)$, $(0, 0.4R_s, 0.04R_s)$, $(-0.4R_s, 0, 0.04R_s)$, $(0, -0.4R_s, 0.04R_s)$. The northward displacement of the five dipoles by $0.04R_s$, is a feature of the Z3 that we included in the 5-D model.

For the above parameters and for q/m=100 Coulomb/kg the following equilibrium points were found:

x	y	z	x	y	z
2.07733	0.	0.021380	1.47110	1.47110	0.021520
-2.07733	0.	0.021380	1.47110	-1.47110	0.021520
0.	2.07733	0.021380	-1.47110	-1.47110	0.021520
0.	-2.07733	0.021380	-1.47110	1.47110	0.021520

These eight equilibrium points correspond to only two values
of the energy constant, namely, to C = -6.51511 and C = -6.51511.
They are also displaced to the northern hemisphere by 0.02138R_s,

i.e. by less than the displacement of the dipoles. This is obviously due to the fact that the center of newtonian attraction is
taken to rest at the origin, i.e.below the dipoles. It should be
remarked that the gravity effect is indeed small since for M=0
(no gravity the z coordinate of the location of equilibrium
points would be 0.04R_s, whereas for M=mass of Saturn this component decreases by 0.01848R_s, i.e. about 46%.

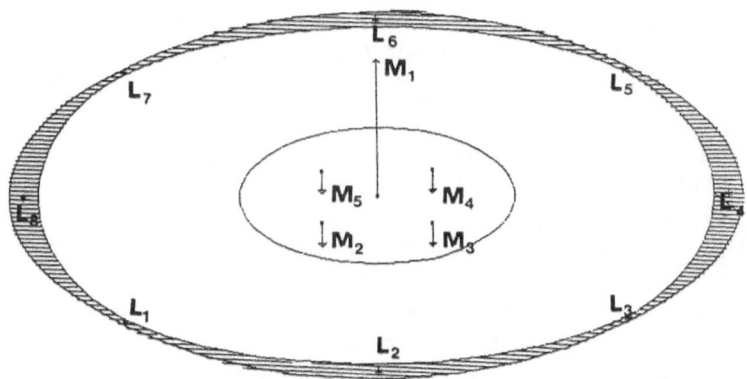

Figure 5.   A Cassini-like   division   (shaded   strip)   for   the
case of Saturn expressed by the 5-D model ($\underline{M}_1$= 26.000nT, $\underline{M}_2$=
-2000nT) and the gravity field.

In Figure 5 the equilibrium points L_1 to L_8 rest between
the two adjacent 0-velocity curves that correspond to the energy
value of the point L_1. This is so, because the curves given were
computed for z=0 and hence do not go through the equilibrium
points L_1, L_3,..,L_7. The space between these two curves is void of
charged grains of energy C < -10.03571. Grains of such energy
are either confined to move inside a torus with the equatorial
boundaries defined by the curves K_1 and K_2, or inside an open
torus with equatorial boundary defined by the curve K_3.

The position of the Cassini-like division, that   the N-D
model shows to exist, with or without gravity, depends critically on the value of q/m of the charged grains.   For the 5-D
model as given in the case 2 of the Appendix and for the mass and
rotation rate of Saturn the following numerical results for the
distance of the division from the center of the planet.

TABLE I

q/m in Coul/kg	Distance d and width D of the Cassini-like division (R_s units)		q/m	d	D
1000	3.19	0.0023	0	1.86	0.0000
800	3.00	0.0027	−50	1.73	0.0044
600	2.80	0.0031	−100	1.59	0.0084
400	2.56	0.0038	−150	1.43	0.0464
200	2.26	0.0037	−200	1.26	0.1525
100	2.08	0.0031			
50	1.97	0.0021			

This table shows the expected continuation of the distance of the Cassini-like divisions as the q/m ratio changes from positive to negative. The range of treated values of the q/m ratio is indeed very wide. The distance d of the Cassini-like division and the width D of the same division is given in columns 2 and 3 and. 5 and 6. The numerical values given are approximate since they represent the distance of the equilibrium points L_{2i-1}, i=1,2,3,4 from the center of the planet, the first, and the quantity $L_{2i-1}-L_{2i}$, which is constant for i=1,2,3,4, the second.

It also shows that divisions will appear at distances bigger bigger than $1.86R_s$ (corrotation distance) for q/m > 0, while they will be located at distances less than $1.86R_s$, for q/m < 0. This means that the magnetic field imposes a differentiation process leading to concentration of negatively charged grains at distances less than $1.86R_s$. This prediction can be used as a test for verification of the theory presented here.

SPACES OF TRAPPING AND CASSINI-LIKE DIVISIONS

This investigation concludes the effort to look into the dynamics of moving magnetic dipoles as a potential primary cause for the accumulation of charged matter. As in the cases of one, two and three dipoles, presented in previous papers, we show in this paper that four, five and many dipoles (we have treated the cases for up to N = 17) located on a circle and performing solid rotation about the axis of this circle, create a pattern of 0-velocity surfaces, some of which are closed and hence @trap@ charged grains. For N>4 the closed space receives the form of a torus, compressed along the z-axis, that has the circle of the dipoles as its inner rim. This torus is symmetrical with respect to the plane z = 0 if the dipoles rest on this plane, whence the external rim of the torus rests also upon it. The 5-D model, on account of the north-south asymmetry of the Z3 model to which is equivalent, leads to an asymmetric @torus of trapping@, the outer rim of which is lifted to the northern hemisphere of Saturn. Beyond this closed torus and concentric to it rests a second open torus that has N point-contacts with the first (internal) torus. This means that the points of contact of the two surfaces are as many as the dipoles and obviously they are points of equilibrium. Between the two toruses and near their plane of symmetry a narrow annulus is formed, giving the outlook and the dynamic function of a Cassini-like division, i.e. a region void of charged grains with the q/m ratio and the energy corresponding to the 0-velocity toruses defined above.

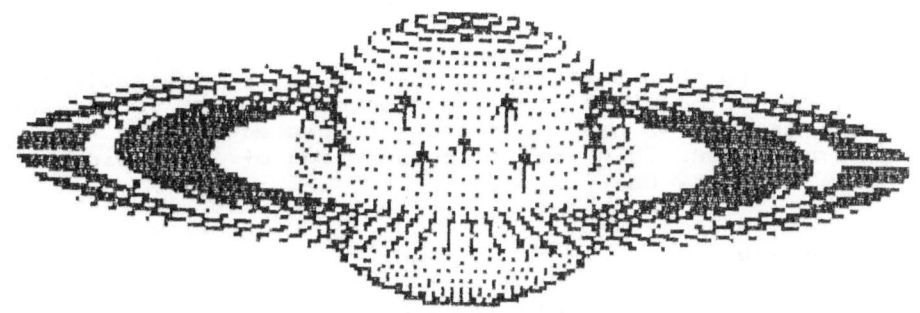

Figure 6. A computer simulation of Saturn and the rings. The shaded area on the equatorial plane is assumed to be charged grains moving in the field of 9 magnetic dipoles. The division between the rings is computed using the 0-velocity condition. This may not be the way Saturn's rings are formed but it can surely be used to form rings.

In view of the fact that the N-D model forecasts the existence of as many divisions as the groups of charged grains classified according to their q/m ratio, there is no distinction between divisions in regard to the mechanism of their production. For this reason and under this context we call all divisions as Cassini-like. The width of the divisions, on the other hand, is readily computable for each category of charged grains and since it is variable, their maxima are required, the minima being equal to zero.

THE THICKNESS OF THE RINGS

The cross-section of the toroidal "space of trapping" is oblate but not so much as to compare with the practically flat shape of the rings of Saturn. The few tens or hundreds of meters depth of the thin layer of the rings (see e.g. Bridges, 1984) is not, however, to be compared with the corresponding dimension of the space of trapping which represents a global kinetic boundary. Indeed, the motions of charged grains of energies equal or less than the unique value of the energy constant (there is one and only one such value for each configuration of the N-dipole problem) corresponding to the closed surface of trapping, are simply confined inside this space but by no means they have to occupy all the interior space.

The scenario of the stricter confinement of the charged particles inside a thin layer at the equatorial plane, is expected, but not proven yet, to be the same with the analogous gravitational cases in which a plane of preference, such as the ecliptic for the solar system, or the plane of motion of the primaries of the restricted problem where the confinement of motions close to this plane is caused by the global instability of 3-dimensional motions with high inclinations (Goudas 1963, 1967 and Jefferys et al, 1966).

In order to show that the same scenario is applicable to the case of combined gravitational and electromagnetic force acting on charged grains, we have to repeat the procedure used for the gravitational case. This implies computation of periodic orbits and their stability coefficients.

PERIODIC ORBITS

With the above motive a search program for computing periodic orbit was prepared and at least four complete families. two planar and two 3-dimensional, were discovered in the case of the 5-D problem without gravitation or relevance to the Saturn case. The problem of their stability is not examined yet.

We give below the initial and the final conditions of one member of each family as well as the fraction of the period T that corresponds to the final position given. These orbits are symmetrical with respect to two, three, or even five planes. The 5-D configuration used is defined by the following parameters: The five dipoles are of magnetic moments

$$\underline{M}_1 = \underline{k}, \ \underline{M}_2 = 0.3333 \ \underline{k} = \underline{M}_3 = \underline{M}_4 = \underline{M}_5,$$

and are located at the points
$(0,0,0), \ (1,0,0), \ (0,1,0), \ (0,-1,0), \ (-1,0,0).$
The constants involved in the equations of motion were given the values $\underline{\omega} = 0.05\underline{k}$ and $(q/mc)=1$. The initial and the final conditions of the orbits found are:

INITIAL CONDITIONS
$x_o = 1.59241$ $\qquad y_o = 0$
FAMILY I $\quad (dx/dt)_o = 0$ $\qquad (dy/dt)_o = 2.05866$

FINAL CONDITIONS $\quad (T/4 = 4.76533)$
$x = 0$ $\qquad y = 1.06286$
$(dx/dt) = -1.43882 \quad (dy/dt) = 0.$

INITIAL CONDITIONS
$x_o = 2.10128$ $\qquad y_o = 0$
FAMILY II $\quad (dx/dt)_o = 0$ $\qquad (dy/dt)_o = 0.50468$

FINAL CONDITIONS $\quad (T/4 = 39.8691)$
$x = 0$ $\qquad y = 2.10132$
$(dx/dt) = -0.50465 \quad (dy/dt) = 0.$

INITIAL   CONDITIONS

$$x_0 = 1.59241 \qquad y_0 = 0 \qquad\qquad z_0 = 0$$

FAMILY III $\quad (dx/dt)_0 = 0 \quad (dy/dt)_0 = 2.05866 \quad (dz/dt)_0 = 4.02968$

FINAL CONDITIONS $\quad (T/8 = 22.0144)$

$$x = 0.77159 \qquad y = 0.77159, \quad z = 0.18986$$
$$(dx/dt) = -0.70036 \quad (dy/dt) = 0 \quad (dz/dt) = 0$$

INITIAL   CONDITIONS

$$x_0 = 1.44182 \qquad y_0 = 0 \qquad\qquad z_0 = 0$$

FAMILY IV $\quad (dx/dt)_0 = 0 \quad (dy/dt)_0 = -0.42180 \quad (dz/dt)_0 = 0.14282$

FINAL CONDITIONS $\quad (T/4 = 5.30011)$

$$x = 1.67078 \qquad y = 0 \qquad\qquad z = 0.63272$$
$$(dx/dt) = 0 \quad (dy/dt) = 0.15873 \quad (dz/dt) = 0$$

REFERENCES

Acuña, M.H. et al, (1980). Science, 207, 444.
Acuña, M.H. et al, (1983). J. Geophys. Res., 88, No. A11, 8771-8778.
Bridges, F.G., (1984). Nature, 309, 333-335.
Bullard, E.C. et al, (1954). Phil. Trans. Roy. Soc. A., 247, 213-278.
Connerney, J.E.P. et al, (1982). Nature, 298, No. 5869, 44-46.
Connerney, J.E.P. et al, (1984). J. Geophys. Res., 89, No. A9, 7541-7544.
Goudas, C.L., (1963). Icarus, 2, 1.
Goudas, C.L. et al, (1967). The Astron. J.. 72, 202-213.
Goudas, C.L. et al, (1990). Cel. Mech. and Dynam. Astron.. 47, 1-14.
Goudas, C.L. et al, (1991). First paper in present Proceedings.
Herzenberg, A., (1958). Phil. Trans. Roy. Soc. A., 250, 534-583.
Jefferys, W.J. et al, (1966). Astron. J., 71, 568.
Kaula, W.M., (1968). @A Introduction to Planetary Physics: The Terrestrial Planets@, John Wiley & Sons, N.Y.
Stevenson, D.J., (1982). Science, 208, 746-748.
Stormer, F.C., (1907). Arch. Sci. Phys. et Natur. Geneve, 24, 350.

APPENDIX

THE N-D MODEL OF SATURN'S MAGNETIC FIELD

The Z3 model, to which the N-D model will be compared is expressed by the scalar potential

$$V = \sum_{n=1}^{3} \alpha \left(\frac{\alpha}{r} \right)^{n+1} g_n^o P_n^o (\theta) \tag{11}$$

where

α = Saturn's radius= R_s = 60.330 km

r = radial distance to point

g_n^o= Schmidt's normalized coefficients

$P_n^o(\theta)$ = Associated Legendre Functions.

θ = Kronian collatitude

Connerney et al (1984) did not assume but were led to formula (1) and in particular to the values

g_1^o= 21,535nT, g_2^o= 1,642nT, g_3^o= 2,743nT,

which when set in the expression of V reproduce the in situ measurements to an of RMS about 0.2% or 2nT. The term $\alpha(a/r)^3 g_2^o P_2^o(\theta)$ for the above value of g_2^o shows a northward displacement of the dipole field by about $\Delta z = 0.04 R_s$.

Let us now define the scalar potential, equivalent to (11), produced by N dipoles of magnetic moments $\underline{M}_1$ (central), $\underline{M}_2$, $\underline{M}_3$, ...,$\underline{M}_N$, all collinear to $\underline{\omega}$. It is

$$V = \frac{\alpha^3 M_1 (z-\Delta z)}{r^3} + \alpha^3 \sum_{n=2}^{N} \frac{M_i (z-\Delta z)}{r_n^3}, \tag{12}$$

where α, z, Δz and r are already defined and

$$\underline{r}_n^2 = \left[x-R_o \cos \left[\frac{2\pi(n-2)}{N-1} \right] \right]^2 + \left[y-R_o \sin \left[\frac{2\pi(n-2)}{N-1} \right] \right]^2 + (z-\Delta z)^2 \tag{13}$$

Expression (13) shows that the peripheral dipoles are uniformly distributed on the circumference parallel to the equatorial plane of the planet at distance Δz and of radius R_o. The functions given by the expressions (6) and (12) satisfy the relation

$$\underline{\nabla} \wedge \underline{A} = -\underline{\nabla} V$$

if we set a = 1, i.e if the distances r_i are measured in units equal to the radius of Saturn.

If we assume that $\underline{M}_i=\underline{M}_2$, i=3,4,...,N and that N=5, and compute M_1 and M_2 in order to have expressions (11) and (12) give identical results, within 2nT in the region of measurement recordings, we can obtain N-D models equivalent to the Z3 model. This was done by least-squares fit and the results obtained for three values of R_o (distance of peripheral dipoles from the center of the planet) are listed bellow:

TABLE OF N-D MODELS EQUIVALENT TO THE Z3 MODEL

R_o in R_s	N	M_1 in NanoTesla	M_i in NanoTesla	RMS in NanoTesla
0.2	5	61,802.4	−10,067.2	2.2
0.2	7	63,252.6	− 6,950.8	1.8
0.2	9	63,694.7	− 5,260.9	5.2
0.3	5	38,188.1	− 4,165.2	1.8
0.3	7	39,652.6	− 3,020.9	1.8
0.3	9	39,606.8	− 2,259.3	1.5
0.4	5	29,966.8	− 2,118.0	1.7
0.4	7	31,169.7	− 1,609.3	1.5
0.4	9	31,327.2	− 1,226.6	1.5
0.5	5	26,241.0	− 1,195.9	2.0
0.5	7	27,268.7	− 964.8	2.1
0.5	9	27,408.0	− 740.5	2.0

The Table shows that the RMS (column 5) in all cases are indeed small and it can be readily verified the N-D models fit the data of Pioneer 11 and of Voyagers 1 and 2 equally well as the Z3 model. It also shows that the magnitude of the central dipole (column 3) changes only with the distance of the peripheral dipoles (column 1) and not with the number of the peripherals (column 2). As regards the magnitude of the peripheral dipoles (column 4) it appears that it depends on N and R_o, but follows the rule $NM_i \cong$ constant for R_o=constant. Finally, the magnitude of the central and the peripheral dipoles decrease with the distance of the peripherals from the center of the planet.

LONG–TIME PREDICTIONS OF SATELLITE ORBITS BY NUMERICAL INTEGRATION

José M. Ferrándiz, M. Eugenia Sansaturio and Jesús Vigo

Departamento de Matemática Aplicada a la Ingeniería
E.T.S. de Ingenieros Industriales
Universidad de Valladolid, 47011 Valladolid, Spain

ABSTRACT

In this paper we aim at establishing limits of predictability to find the extension of time for which meaningful analytical and/or numerical predictions can be made in orbital behaviour of artificial satellites. These limits depend, of course, on the accuracy required, on the specific dynamical models formulated, on the sets of variables chosen to describe them, on the numerical or analytical techniques used and, especially, on the specific trajectories to be established. In order to check the reliability of the predictions, first integrals, constraints among redundant variables and backwards integrations from the ending point to the initial conditions have been used.

INTRODUCTION

The difficulty in providing good predictions in satellite dynamics is well known, specially when we intend to obtain them for long periods of time. Although a hope could be expected from the fact that the problem of motion of a satellite slightly deviates from a quasi–periodic one (Simó, this issue), some difficulties still remain. First, we have the Lyapunov–instability of the equations of motion, and second we have numerical instability, i.e. for given initial conditions the exact solution and the numerical one usually deviate too much.

We cannot eliminate the first drawback, although something can be done to make it more tolerable. As for the second one, a way to manage with it is by transforming the equations of motion into ones better conditioned for the numerical integration by means of changes of independent and dependent variables. Probably the simplest case reduces to carrying out a change of time into a new independent variable often with geometrical meaning, which involves an analytical step–size regulation in numerical integration.

In this line we have the so–called *linearization methods*, which provide harmonic oscillator form to the equations of motion with the advantages it involves when we integrate them numerically, such as the favorable propagation of the truncation error, and the *methods of elements*, whose aim is that the transformed dependent variables

Predictability, Stability, and Chaos in N-Body Dynamical Systems
Edited by A.E. Roy, Plenum Press, New York, 1991

are elements, that is to say, they are constant or linear functions of the independent variable along a Keplerian motion. Among them, let us quote the oldest ones, the classical elements and the linear elements associated with the variables introduced in the linearization methods and which always involve the change of time carried out to get the linearization.

Apart from this, we can also get a stabilization of the equations by means of other devices such as the introduction of integrals into the equations of motion and/or the use of additional equations (e.g. the equation of the energy integral or that of the angular momentum) and finally, by introducing a time element.

In what follows we will concentrate on highly eccentric motions, where the difficulties are more pronounced and the final behaviour strongly depends on the perturbations.

To deal with this problem we consider the following variables:

1. Cartesians, which provide the commonly called Cowell method and are well known.

2. Kustaanheimo–Stiefel (K–S) variables, including in this notation either K–S coordinates or K–S elements and the corresponding time element.

3. Burdet–Ferrándiz (B–F) variables, which also mean either B–F coordinates or B–F elements and the corresponding time element.

K–S transformation is widely described in the well known book by Stiefel and Scheifele (1971) and we will only remind the reader that by introducing the energy integral into the equations of motion and using the eccentric anomaly as independent variable, it reduces the Kepler motion to four harmonic oscillators with all the frequencies being equal to one.

As for B–F variables, they are a set of 8 redundant canonical variables, the coordinates being the direction cosines of the particle and the inverse of its distance. They have been named in this way since Burdet first used them although he did not do it in a Hamiltonian framework, which was done by Ferrándiz (1988) in recent years obtaining equations of motion similar to ones derived by Burdet. These variables also allow us to reduce the Kepler problem to four harmonic oscillators with unit frequencies by introducing the integral of the angular momentum into the equations of motion and using the true anomaly as independent variable.

The homogeneous Hamiltonian of the problem when expressed in B–F variables has the following expression

$$cK = \frac{1}{2}\mid \mathbf{p}\mid^2\mid \mathbf{x}\mid^2 + \frac{1}{2}p_z^2 z^2 - \frac{\mu}{z} + \frac{p_0}{z^2} + W = 0,$$

where $\mathbf{x}$ is the direction vector of the particle, $\mathbf{p}$ stands for the conjugate momenta of $\mathbf{x}$, $z = 1/r$, $W = z^{-2}V$ (V perturbing potential), c is the magnitude of the angular momentum and $p_0 = -h$ (h energy).

The time element associated with B–F variables was introduced by the first two authors in a Spanish–Portuguese meeting held in June 89, by defining two functions e, E of the canonical B–F variables through the expressions

$$e\sin E = -\frac{\sqrt{2p_0}}{\mu}z\,p_z = -\frac{\sqrt{2p_0}}{\mu}\frac{c\,z'}{z}, \qquad e\cos E = 1 - \frac{2p_0}{\mu z}.$$

The time element is thus defined by the generalized Kepler equation

$$l - l_0 = t - t_0 - \frac{\mu}{(\sqrt{2\,p_0})^3}\,(E - s - e\sin E),$$

s being the true anomaly.

Finally, the equation for the time element in conservative cases is given by

$$l' = \frac{\mu}{(\sqrt{2\,p_0})^3} + \frac{1}{c}\,\frac{\mu - (c^2 + 2\,W)\,z}{\mu^2 - 2\,p_0(c^2 + 2\,W)}\,\frac{\partial W}{\partial z},$$

where $()'$ stands for derivative with respect to s.

The problem we will pay attention to now is that of an artificial satellite with J_2–perturbation whose orbit is equatorial, with an eccenticity $e = 0.95$ and a perigee of approximately 0.05 E.R. (300 Km.). However, we should point out that the results obtained with these data do not differ much from the ones obtained when we vary the inclination of the orbit (up to $25°$ at least) and the location of the perigee.

Such results are sumarized in the graphics below. In these plots we have used

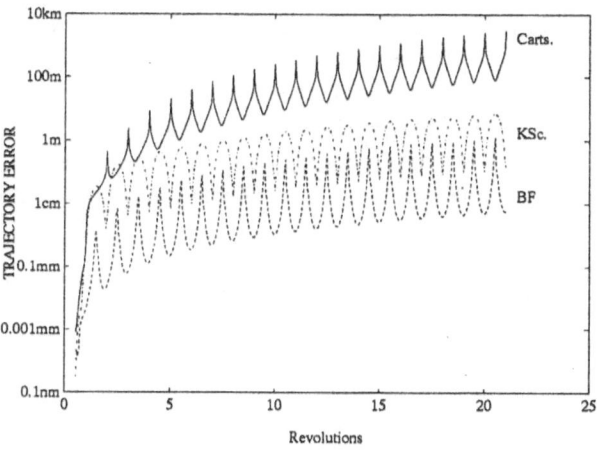

Figure 1

logarithmic scale in the vertical axis, representing errors versus number of revolutions. There will appear two kinds of errors which will be referred to as

1. Trajectory error (ε_x), which means the difference between the position given by a reference orbit, which we obtained by numerically integrating and requiring high enough accuracy, and the one obtained with the current numerical integration. Positions are compared for equal values of the independent variable, that is to say

$$\varepsilon_x = |\,\mathbf{x}_{ref}(s) - \mathbf{x}_{cal}(s)\,|.$$

2. Position error (ε), which apart from the previous error also includes an estimation of the error due to the integration of the equation of time (ε_t). If we define (ε_t) to be

$$\varepsilon_t = |\,t_{ref}(s) - t_{cal}(s)\,| \cdot |\,\text{velocity}\,|,$$

which means the linear approximation of that error, the position error is then $\varepsilon = \varepsilon_x + \varepsilon_t$.

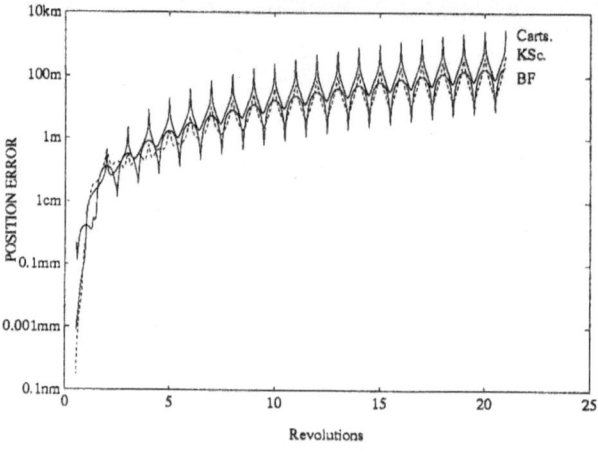

Figure 2

Figures 1 and 2 have been obtained by using a Runge–Kutta–Fehlberg (R–K–F) (7)8 code, with variable stepsize, to integrate the equations when expressed in Cartesian coordinates (Carts.), K–S coordinates (KSc.) and B–F coordinates (BF) together with the equation of time. Although B–F coordinates provide a more accurate solution for the spatial coordinates (Fig. 1), the errors in the integration of the equation of time lead to similar results for the three sets of variables (Fig. 2). However, the computation of time from the corresponding elements gives rise to a considerably different behaviour, which turns out to be favorable for BF + t.e. (time element) (Fig. 3).

The use of regular elements associated with the K–S transformation instead of

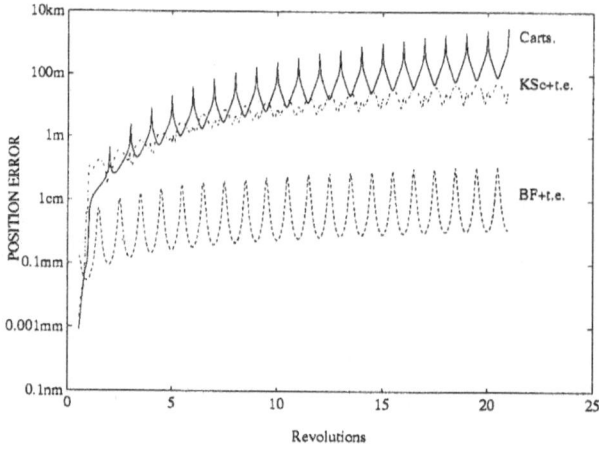

Figure 3

the coordinates does not change their behaviour significantly when integrating very eccentric orbits (in accordance with Stiefel and Scheifele, p. 95). Figure 4 shows the errors obtained by using: a) K–S coordinates together with the equation of time (KSc. + t), b) K–S coordinates together with the equation of the time element (KSc. + t.e.), c) K–S elements together with t.e. equation (KSel. + t.e.) and choosing an

Adams–Bashforth–Moulton (A–B–M) code of order 8, with 144 steps per revolution.

This is not the case for B–F coordinates, so figure 5 clearly shows the improvement got when, in B–F variables, we substitute the equation of time by the corresponding time element equation.

As a general comment about the former results let us say that, for the same

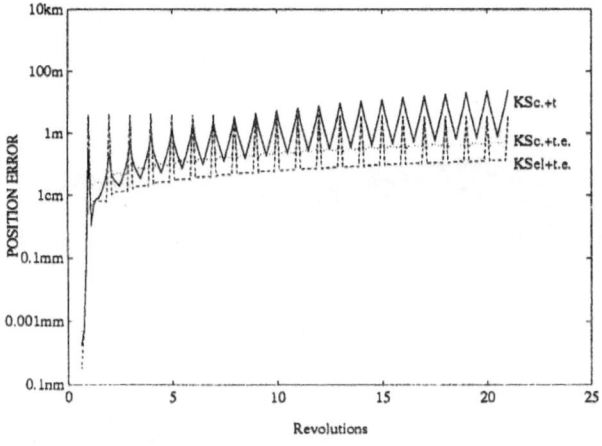

Figure 4

number of evaluations of derivatives, the use of B–F variables together with their time element clearly improves the accuracy provided by K–S methods: e.g. for A–B–M code the accuracy is improved in a factor greater or equal to 10^2, reaching in some cases 10^3.

As Cartesian coordinates behave worse than the other two sets in the above

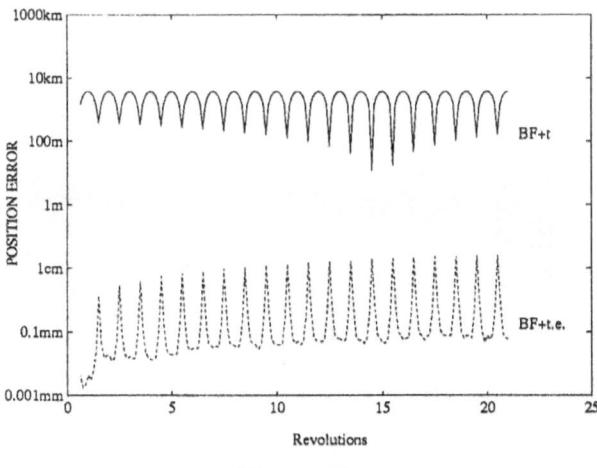

Figure 5

mentioned experiments, we will concentrate on K–S elements and B–F coordinates to study the behaviour for long periods of time.

In this way, figures 6, 7 and 8 show the position error after 1000 revolutions when we integrate the equations expressed in B–F coordinates together with the equation of their time element by using an R–K–F code (Fig. 6) and an A–B–M code with 80 p.p.r. (Fig. 7) and 300 p.p.r. (Fig. 8). Finally, figures 9 and 10 represent the position error after 1000 revolutions when an A–B–M code with 80 p.p.r. (Fig. 9) and 300 p.p.r. (Fig. 10) has been used to integrate the equations expressed in K–S elements together with the corresponding time element equation.

From the plots shown in these five figures we can conclude that the position error

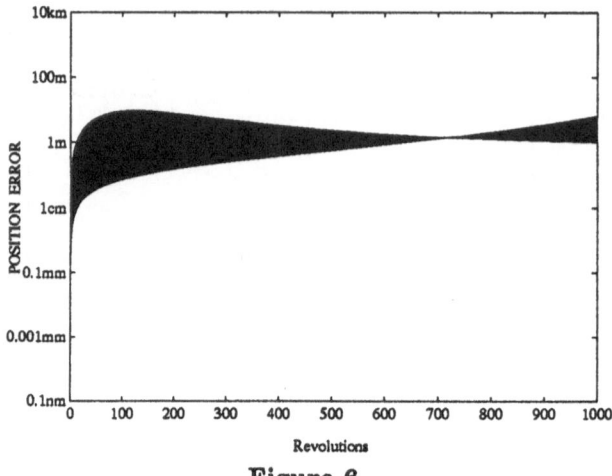

Figure 6

for long periods of time practically rules the same scheme as the one obtained for few revolutions and B–F variables still behave much better than K–S elements. We should also point out that no relevant variations have been observed when we change the values of other orbital elements.

The reliability of the reference orbits has been checked by means of the stabiliza-

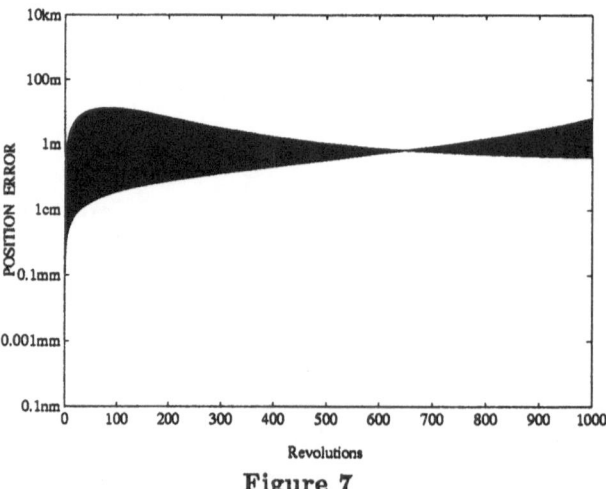

Figure 7

tion of a number of significant digits of several points of the solution distributed along the orbit and backwards integration, apart from the preservation of integrals.

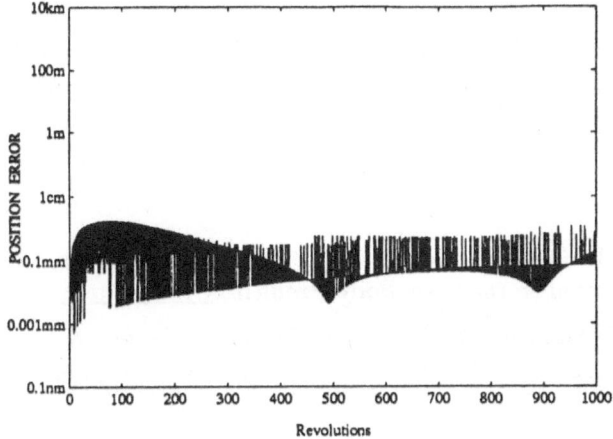

Figure 8

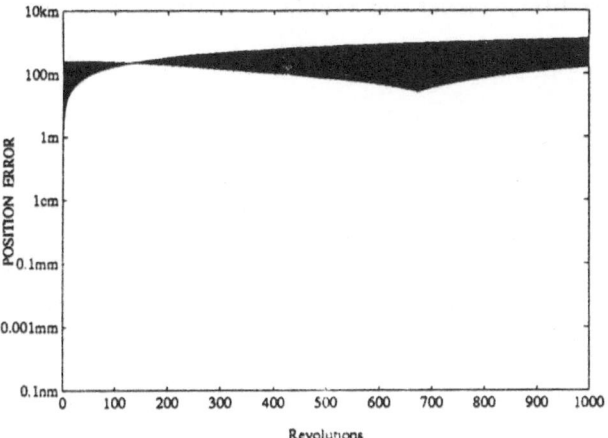

Figure 9

Figure 10

ACKNOWLEDGEMENTS

The authors would like to thank Professor V. Szebehely for his invaluable advise. Also, the authors gratefully acknowledge the financial support of the NATO cooperative research grant 0506/88 and the Spanish CICYT grant ESP88–541.

REFERENCES

Ferrándiz, J. M., 1988, *A General Canonical Transformation Increasing the Number of Variables with Application to the Two–Body Problem*, Celest. Mech., 41:343.

Ferrándiz, J. M. and Sansaturio, M. E., 1990, *Elemento de tiempo en variables de Ferrándiz*, Actas XIV Jornadas Hispano– Lusas de Matemáticas, Junio 1989, Vol. III:1231.

Ferrándiz, J. M., Sansaturio, M. E. and Pojman, J., 1990, *Increased Accuracy of Computations in the Main Satellite Problem through Linearization Methods*, Celest. Mech., (In press).

Stiefel, E. and Scheifele, G., 1971, "Linear and Regular Celestial Mechanics", Springer–Verlag, Berlin.

CHAOS IN COORBITAL MOTION

Franz Spirig and Jörg Waldvogel

Swiss Federal Institute of Technology (ETH) Zurich, Switzerland

Abstract

The motion of two coorbital satellites - at their close encounters - is adequately approximated by Hill's lunar problem. Therefore chaotic behaviour in Hill's problem implies chaos in coorbital motion. Although the nonintegrability of Hill's problem has not yet been proven, numerical evidence clearly shows complicated behaviour typical of chaotic systems. In this paper the family of solutions relevant for circular coorbital motion is explored in details, and an example of a homoclinic orbit is given.

1. Introduction

The classical problem of celestial mechanics, the two-body problem (Kepler problem), is completely integrable and shows no sign of irregular or chaotic behaviour. In contrast, many of the more complicated gravitational problems, like the N-body problem with $N \geq 3$, are non-integrable, and numerical studies clearly show stochastic behaviour. Rigorous mathematical proofs of chaotic behaviour, however, are often difficult, particularly in nearly integrable problems due to the exponential smallness of stochastic zones in the presence of a small perturbation.

In this paper we consider the motion of two small coorbital satellites m_1, m_2 of a common central body m_0 (primary). In earlier work by Colombo [1], Yoder et al. [8], Hénon and Petit [2,5] and the present authors [6,7] this motion was investigated in the framework of the planar problem of three bodies, based on near-circular satellite orbits. Hereby the relationship with Hill's lunar problem is of a particular interest.

Hill's lunar problem, as introduced in 1886 [3], approximately describes the motion of the moon under the influence of the earth (on a circular obit) and the sun. Using a geocentric rotating coordinate system and a homogeneous solar gravitational field in x-direction results in Hill's famous 4th-order system of differential equations (2).

It is no surprise that coorbital motion, too, is approximated by Hill's lunar problem as long as m_1, m_2 stay close together. The use of Jacobian coordinates and a rotating frame of reference [7] precisely yields Hill's lunar equations after a proper scaling transformation in the limit $\varepsilon \to 0$, where

$$(1) \qquad \varepsilon = (m_1 + m_2) / m_0$$

is the small parameter of the singular perturbation problem.

Predictability, Stability, and Chaos in N-Body Dynamical Systems
Edited by A.E. Roy, Plenum Press, New York, 1991

Chaos in coorbital motion can be searched for in two ways: (i) It is likely that the long-term behaviour over many close encounters is chaotic. This aspect of the full planar three-body problem will not be discussed here. (ii) Chaotic behaviour could even occur during one single close encounter of the satellites. This is the topic of the present paper. We will restrict ourselves to discussing Hill's lunar problem and its relations to chaos. Although the non-integrability of Hill's problem has not been proven, we will present strong numerical evidence for chaotic phenomena.

First we will briefly review the coordinate systems that will be used for describing coorbital motion, following [7]. Then, basic properties of Hill's lunar equations as well as asymptotic expansions will be discussed. Furthermore, the regularized form of Hill's lunar equations will be introduced as an important tool. Then we will discuss the complex evolution of the family of orbits relevant for coorbital motion. Finally, as a clear indication of chaos, numerical evidence for the existence of homoclinic orbits in Hill's problem will be presented.

2. Equations of Motion

We consider the planar motion of three bodies with respective masses m_0, m_1, m_2. If the mass m_0 of the primary is large compared to m_1+m_2, then ε of Equ. (1) is a small parameter. The bodies m_1, m_2 are referred to as satellites. We choose the unit of mass such that the gravitational parameter of the primary becomes $Gm_0=1$ (with gravitational constant G) and put $Gm_k=\varepsilon\mu_k$ (k=1,2). Thus μ_k is the relative mass of the k-th satellite.

The relative position vectors from m_0 to m_k are denoted by r_k (k=1,2) and from m_1 to m_2 by d, respectively. Then the equation of motion reads as

$$\ddot{r_1} = -(1 + \varepsilon\mu_1)|r_1|^{-3}r_1 - \varepsilon\mu_2|r_2|^{-3}r_2 + \varepsilon\mu_2|d|^{-3}d$$

$$\ddot{r_2} = -(1 + \varepsilon\mu_2)|r_2|^{-3}r_2 - \varepsilon\mu_1|r_1|^{-3}r_1 + \varepsilon\mu_1|d|^{-3}d \ .$$

Following singular perturbation techniques we obtain the outer system by letting $\varepsilon\to0$. The solutions of the outer system are unperturbed Keplerian orbits.

Next we turn to the inner system describing close encounters of the satellites. For this purpose Jacobi coordinates d, $R = \sum\mu_k r_k$ are most suitable, where R gives the position of the center of mass of the two satellites relative to the primary. In addition, we scale the distance between the two satellites by $d = \varepsilon^\alpha D$.

The exponent $\alpha>0$ will be determined below. The equations of motion in these new coordinates are

$$\ddot{R} = -(1 + \varepsilon)|R|^{-3}(R + O(\varepsilon^{2\alpha}))$$

$$\ddot{D} = -\left(|R|^{-3}+ \varepsilon^{1-3\alpha}|D|^{-3}\right)D + 3|R|^{-5}(R\cdot D)R + O(\varepsilon^\alpha) \ .$$

We choose α such that the gravitational interaction between the satellites will be of the same order of magnitude as the interaction between the pair of satellites and the primary, i.e. $1-3\alpha=0$ (distinguished limit) or $\alpha = 1/3$. Now the inner system is obtained by again letting $\varepsilon\to0$. Then the center of mass of the pair of satellites moves on a Keplerian orbit. We restrict ourselves to the circular case $R = |R|e^{it}$ (using complex notation), where t is the true anomaly. In order to describe the relative motion of the two satellites we introduce a rotating and pulsating (x,y)-coordinate system (see Fig. 1):

$$D = R z, \quad z = x + iy$$

and use t as new independent variable. Thus the inner system is described by Hill's lunar equations [3]

(2)
$$\ddot{x} - 2\dot{y} - 3x + xr^{-3} = 0, \quad r = \sqrt{x^2+y^2}$$
$$\ddot{y} + 2\dot{x} + yr^{-3} = 0$$

which admit the Jacobi integral

(3)
$$\frac{1}{2}(\dot{x}^2 + \dot{y}^2) - \frac{3}{2}x^2 - \frac{1}{r} = h$$

where h is the Jacobi constant fixed on an orbit.

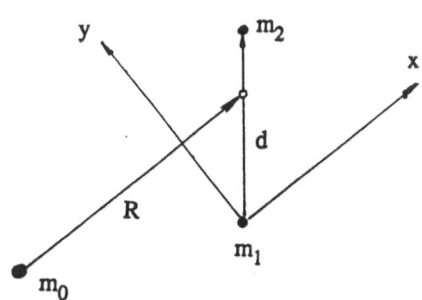

Figure 1. Coordinate systems

3. Hill's Lunar Equation

In this section we give a survey of some basic solutions of Hill's lunar equation. We restrict ourselves to solutions corresponding to circular orbits in the outer motion. These solutions have the property that the limit $c := \lim x(t)$ as $t \to \infty$ exists, i.e. the corresponding orbits, referred to as *nonoscillating orbits*, asymptotically are straight lines parallel to the y-axis. The quantity c is called the impact parameter of the orbit. More precisely, we consider solutions with the tentative asymptotic expansion

$$x = c + c_1 t^{-1} + c_2 t^{-2} + \dots$$
$$y = bt + b_1 b_2 t^{-1} + \dots$$

for $t \to -\infty$. Unfortunately, substituting this expansion into Hill's lunar equation shows that no such solution exists. This difficulty is overcome by introducing logarithmic terms

$$x = \sum_{j \geq 0} t^{-j} \sum_{k=0}^{j} c_{jk} l^k = c + \frac{8}{9} c^{-2} t^{-2} + \dots$$

$$y = \sum_{j \geq 0} t^{-j+1} \sum_{k=0}^{j} b_{jk} l^k = -\frac{3}{2} ct - \frac{4}{3} c^{-2} l + \dots, \quad l = \log |t| \;.$$

The Jacobian integral yields the relation $h = (-3/8) c^2$ between the Jacobian constant h and the impact parameter c; thus nonoscillating orbits are only possible if $h < 0$.

One way to avoid logarithmic terms in the asymptotic expansion for $t \to -\infty$ is to introduce y as the independent variable instead of t, according to Hénon and Petit [2]. After stating Hill's lunar equation in the form of a first order system,

$$\dot{x} = u$$
$$\dot{y} = v$$
$$\dot{u} = 2v + 3x - x r^{-3}$$
$$\dot{v} = -2u - y r^{-3}, \quad r = \sqrt{x^2 + y^2},$$

397

y is chosen as new independent variable instead of t ($d/dt = v \, d/dy$). Now the above expansion corresponds to

$$x = \sum_{j\geq 0} c_j y^{-j}, \quad u = \sum_{j\geq 1} a_j y^{-j}, \quad v = \sum_{j\geq 0} d_j y^{-j}.$$

An even more elegant procedure is to introduce the new independent variable τ by

$$\dot{\tau} = 1 + q/\tau \qquad \text{or} \qquad t = \tau - q \log |\tau + q|$$

where q is a parameter to be chosen appropriately. Then

$$x = \sum_{j\geq 0} c_j \tau^{-j} = c + \frac{8}{9} c^{-2} \tau^{-1} + O(\tau^{-2}), \quad y = \sum_{j\geq 0} c_j \tau^{-j+1} = -\frac{3}{2} c\tau + O(\tau^{-1})$$

and $\qquad q = \dfrac{8}{9} c^{-3}$.

It should be mentioned that asymptotic expansions for $t \to -\infty$ are important for numerical studies in order to obtain approximate initial conditions for nonoscillating orbits.

Some insight in the qualitative behaviour of the solution of Hill's lunar equation may be gained by investigating the zero velocity curves (see Fig.2). These are simply the contour lines of the function $W = (3/2) x^2 + 1/\sqrt{x^2 + y^2}$. The Jacobian integral implies that an orbit with Jacobian constant h cannot leave the region in the (x,y)-plane with $W \geq h = (3/8) c^2$, the boundary of which is the zero velocity curve $W = h$. Each unbounded zero velocity curve has the asymptotes $|x| = c/2$. For a large value of c the zero velocity curve has three separate branches approximately described as a small circle $x^2 + y^2 = (64/9)c^{-4}$ around the origin and two straight lines $x = \pm c/2$. For the critical value $c^* = 2 \cdot 3^{1/6} = 2.402$ the three curves unite. This means that for $x = c > c^*$ a nonoscillating orbit never approaches the origin but stays away from it at least at a distance $3^{-1/3} = 0.693$.

For $c < c^*$ an orbit can cross the y-axis at most at a distance $(8/3)c^{-2}$ from the origin. It is worth mentioning that the extrema of the zero velocity curves lie on the circle $x^2 + y^2 = 3^{-2/3}$.

We continue with asymptotic expansions of the nonoscillating orbits for large and small values of the impact parameter c. For *large* values of c, $c \to \infty$, we use the scaling transformation $x = c\xi$, $y = c\eta$. Hill's lunar equation and the Jacobian integral are transformed into

$$\ddot{\xi} - 2\dot{\eta} = 3\eta - c^{-3}\xi g^{-3}$$

$$\ddot{\eta} + 2\dot{\xi} = \quad -c^{-3}\eta g^{-3}, \quad g = \sqrt{\xi^2 + \eta^2}$$

$$\dot{\xi}^2 + \dot{\eta}^2 - 3\xi^2 - 2c^{-3}g^{-1} = -\frac{3}{4}.$$

The asymptotic expansion is found to be of the form

$$\xi = \sum \mu^n x_n, \quad \eta = \sum \mu^n y_n \qquad \text{with} \quad \mu = c^{-3}.$$

There is a particular solution with leading terms

$$x_0 = 1, \quad y_0 = -\frac{3}{2} t,$$

hence $x = c + O(c^{-2})$, $y = (-3/4)ct + O(c^{-2})$. This orbit is almost a straight line parallel to the y-axis corresponding to a nearly unperturbed Keplerian motion of the two satellites. Numerical experiments show that for $c > 2$ the nonoscillating orbits are almost straight lines.

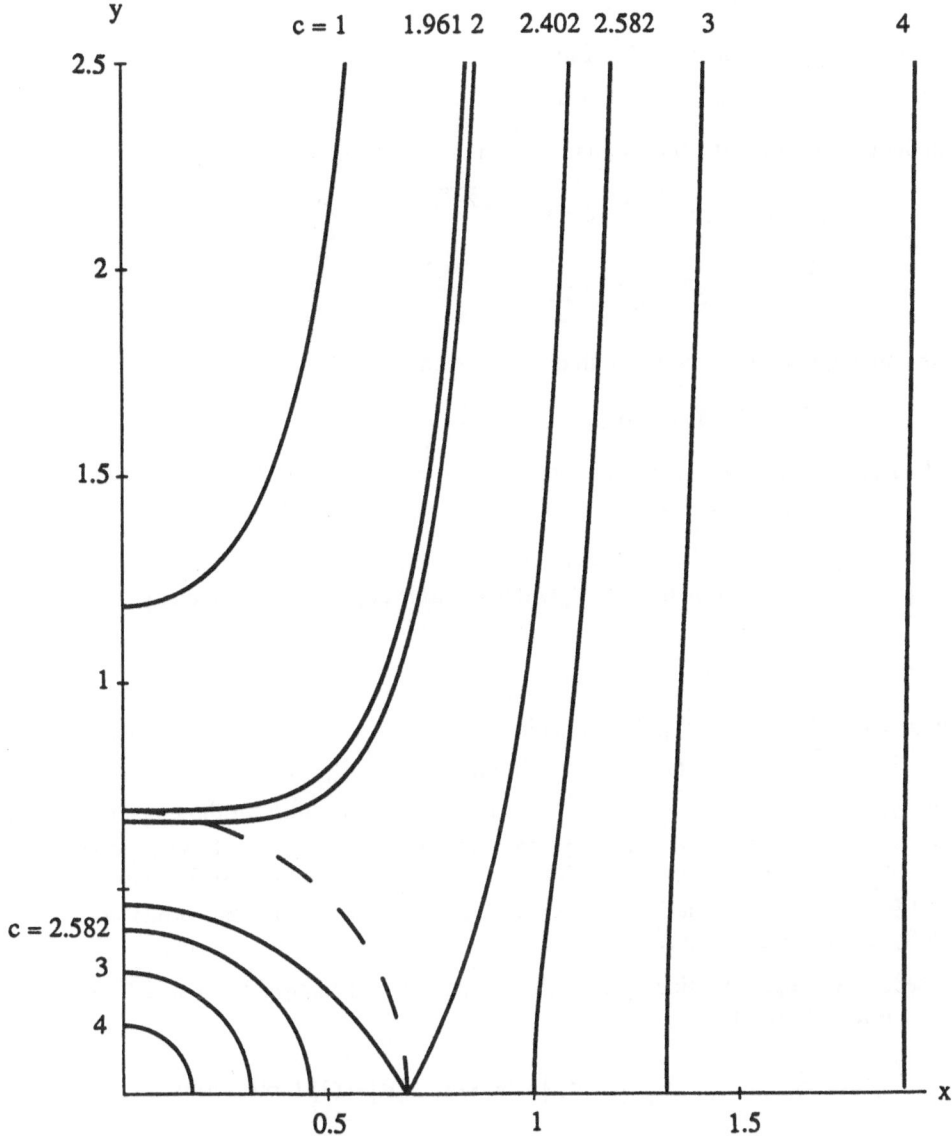

Figure 2. Zero velocity curves in Hill's lunar problem

As an example, we consider the interaction between one of the F ring shepherds of Saturn, Prometheus ($m_1=10^{-9}$, $|r_1|=139353$ km) or Pandora ($m_1=6.4\cdot10^{-10}$, $|r_1|=141700$ km), and an F ring particle ($m_2=0$, $|r_2|=140600$ km). The impact parameter c corresponds to the difference of the orbital radii of the two bodies m_1, m_2 : $||r_2|-|r_1||=\varepsilon^{1/3}|R|c$ (remember that the length unit in the (x,y)-system is stretched by a factor $\varepsilon^{1/3}|R|$). Thus we obtain in either case $c\approx9$. Therefore there is almost no interaction between the shepherds and the ring particles.

For $c\to0$ the variables are scaled as follows:

$$x = c\xi , \qquad y = c^{-2}\eta , \qquad t = c^{-3}\tau$$

Then the first one of Hill's lunar equations and the Jacobian integral become

$$c^6\ddot{\xi} - 2\dot{\eta} = 3\xi - c^6\xi g^{-3} , \quad g = \sqrt{c^6\xi^2 + \eta^2} , \quad \dot{} = \frac{d}{d\tau}$$

$$c^6\dot{\xi}^2 + \dot{\eta}^2 - 3\xi^2 - 2g^{-1} = -\frac{3}{4} .$$

These equations admit a solution with the expansion

$$\xi = \sum \mu^n x_n, \quad \eta = \sum \mu^n y_n , \quad \mu = c^6 .$$

For the leading term one immediately obtains

$$y_0 = \frac{8}{3}(1 - x_0^2) .$$

x_n, y_n are considered as functions of x_0 where the time dependence is given by

$$\dot{x}_0 = -\frac{9}{32}(1-x_0^2)^2 .$$

Thus $x = cx_0+O(c^7)$, $y=c^{-2}y_0+O(c^4)$. This represents a U-shaped orbit symmetric to the y-axis which crosses the y-axis at its minimum distance $y_m=(8/3)\,c^{-2}$ to the origin. More precisely, $y_m=(8/3)c^{-2} - (9/32)c^4+O(c^{10})$. Combining this result with the consideration on the zero velocity curves for $c<c^*$ one concludes that these orbits are only possible for values of c for which the corresponding zero velocity curves are also U-shaped, i.e. for $c<(512/9)^{1/6} = 1.961$. Numerical studies reveal that the nonosciallting orbits are almost perfectly U-shaped for $c < 0.7$.

As a second example, consider the coorbiting satellites of Saturn, Janus and Epimetheus, with the following data:

$$\varepsilon=(3.9\pm1.2)\cdot10^{-9} , \quad ||r_1| - |r_2|| = 50 \text{ km} , \quad |R| = (151460\pm2700) \text{ km} .$$

We obtain $c=0.21\pm0.03$ for the impact parameter and $|d_m|=\varepsilon^{1/3}|R|y_m=(14500\pm5500)$ km for the minimum distance .

These results may be summarized as follows: The nonosciallting orbits are U-shaped for small values of c and nearly straight lines for large c. We expect that in the transition zone in between most orbits engage in close encounters with the origin. From the discussion of the zero velocity curves we know that U-shaped orbits exist at most up to $c=1.961$ and that the transition zone has the upper bound $c=2.402$. By numerical investigation we found the transition zone to be $1.3361171883 < c < 1.7187799380$ (see Section 5).

4. Levi-Civita Regularization

For the numerical exploration of Hill's problem (Sections 5, 6) regularization of the collision at the origin is essential. It is a remarkable fact that Levi-Civita's transformation [4] not only removes the collision singularity but also transforms Hill's equations (2) into a Hamiltonian system deriving from a 6th degree polynomial Hamiltonian.

To show this we first write Equ. (2) in Hamiltonian form by using the coordinates $q_1=x$, $q_2=y$ and the canonically conjugated momenta $p_1=\dot{q}_1-q_2$, $p_2=\dot{q}_2+q_1$.

The Hamiltonian then becomes

(4) $\qquad H = \frac{1}{2}(p_1^2 + p_2^2) + p_1 q_2 - p_2 q_1 - q_1^2 + \frac{1}{2}q_2^2 - \frac{1}{r}$, $r = \sqrt{q_1^2 + q_2^2}$.

The first step of Levi-Cività's regulatization procedure is introducing a new independent variable τ and a new Hamiltionian K according to

(5) $\qquad dt = r\, dt$, $\qquad K = r\,(H - h)$,

where only orbits of the fixed energy $H(q_1,q_2,p_1,p_2)=h$ are considered (this implies $K=0$ on the orbit under consideration). Furthermore new coordinates x_1, x_2 and new momenta y_1, y_2 are introduced according to the canonical transformation

(6) $\qquad q_1 + iq_2 = (x_1 + ix_2)^2$, $\qquad p_1 + ip_2 = \dfrac{y_1 + iy_2}{2(x_1 - ix_2)}$

where complex notation is used. Finally, the new Hamiltonian K reads as

(7) $K(x_1, x_2, y_1, y_2) = \frac{1}{8}(y_1^2 + y_2^2) - (x_1^2 + x_2^2)\left(h + \frac{1}{2}(x_1 y_2 - x_2 y_1) + x_1^4 - 4x_1^2 x_2^2 + x_2^4 \right) - 1$,

and the regularized equations of motion are

(8) $\qquad \dfrac{dx_j}{d\tau} = \dfrac{\partial K}{\partial y_j}$, $\qquad \dfrac{dy_j}{d\tau} = -\dfrac{\partial K}{\partial x_j}$, $\quad (j = 1,2)$, $\qquad \dfrac{dt}{d\tau} = x_1^2 + x_2^2$

5. Orbits in Hill's Problem

For discussing orbits in Hamiltonian systems the concept of the Poincaré map is very useful. For simplicity we will again use $(x, y, \dot{x}, \dot{y})$ as coordinates in the 4-dimensional phase space. In order to define the Poincaré map we introduce a 3-dimensional surface of section S, here appropriately chosen as the non-negative x-axis $x\geq 0$, $y=0$ (in the regularized coordinates this is the hyperplane $x_2=0$). Consider now an orbit C_h of given energy h, and let $x_0\geq 0$, $y_0=0$, $\dot{x}_0$, $\dot{y}_0$ be the coordinates of one of its points of intersection with S. From the Jacobi integral we see that $\dot{y}_0$ is determined uniquely (up to the sign) from x_0 and $\dot{x}_0$:

$$\dot{y}_0 = \sqrt{2h - \dot{x}_0^2 + 3\,x_0^2 + \frac{2}{x_0}} \ .$$

Therefore the coordinates x_0, $\dot{x}_0$ determine the entire orbit C_h uniquely (up to reflexions with respect to S). In particular, the next intersection of C with S in forward time direction, denoted by x_1, $\dot{x}_1$ is determined uniquely. The Poincaré map P_h is now defined as

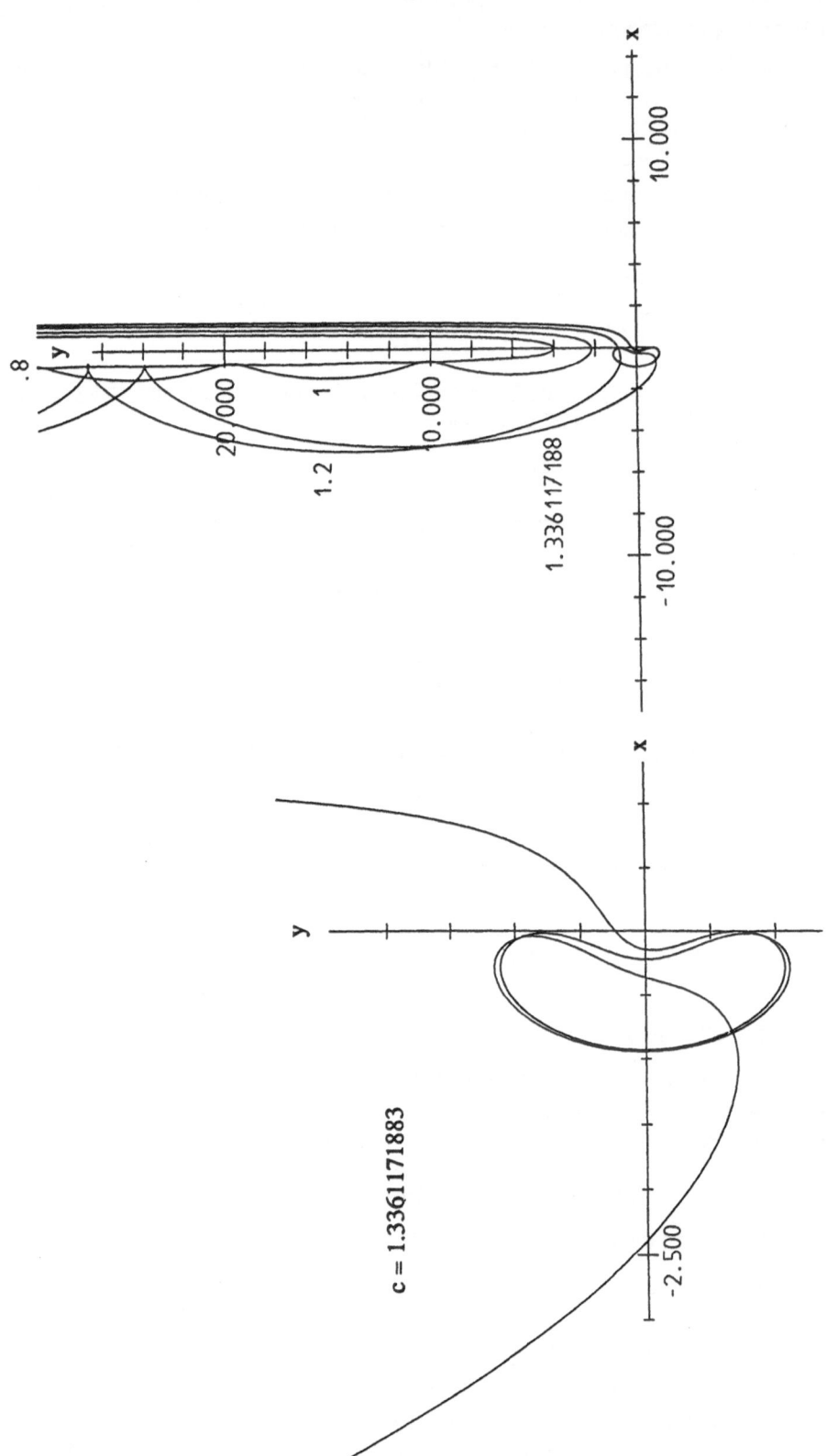

Figure 4. U-shaped orbits for small values of the impact parameter, c < 1.33611 71883

c = 1.33611171883

1.336117188

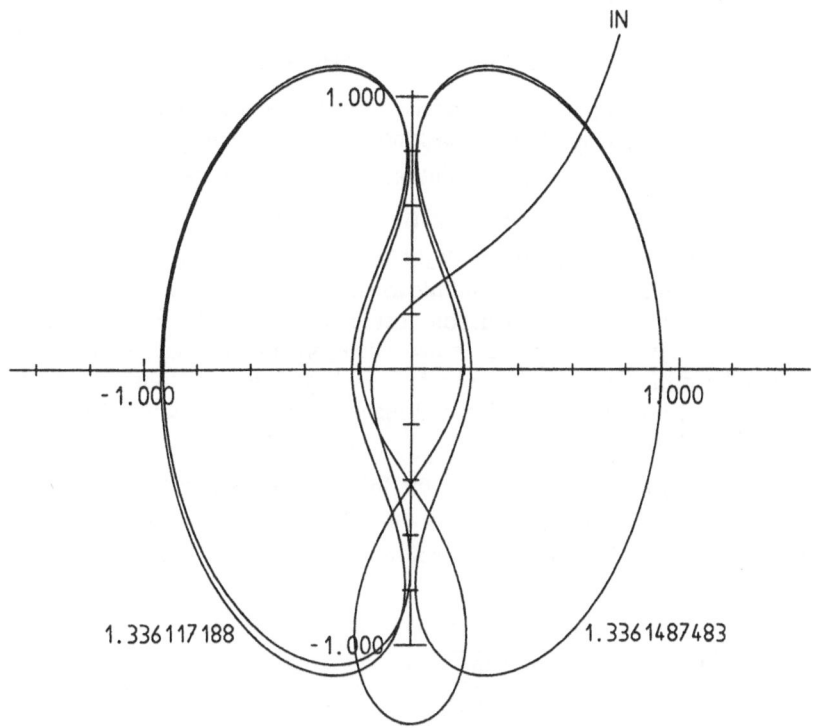

Figure 5. Limiting orbits of the first transition zone of Table 1,
c = 1.33611 71883 and c = 1.33614 87483

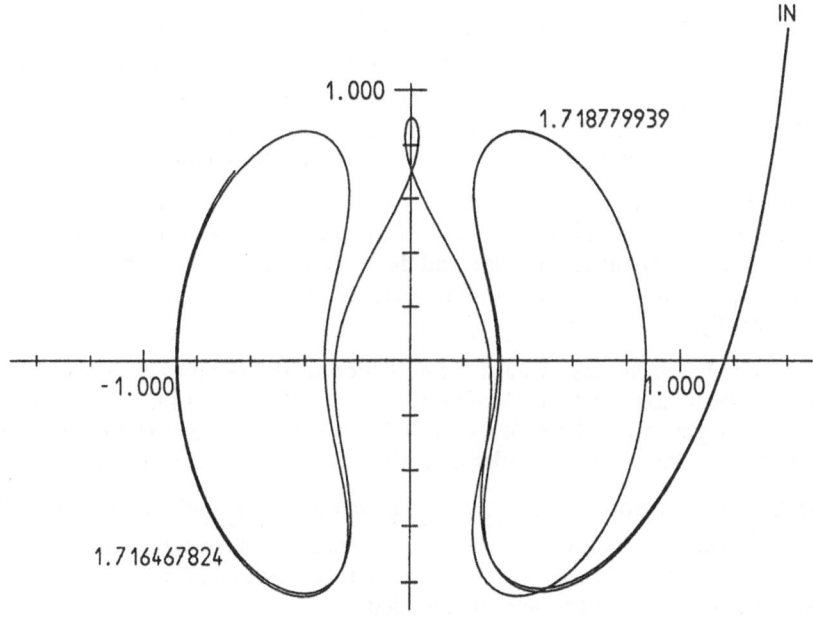

Figure 6. Limiting orbits of a transition zone,
c = 1.71646 78242 and c = 1.71877 99380

$$P_h: (x_0, \dot{x}_0) \longmapsto (x_1, \dot{x}_1) \; ;$$

it can be shown to be area preserving if it is based on a Hamiltonian system.

The periodic solutions of Hill's problem are decisive for the behaviour of the family of all solutions. Obviously, a periodic solution corresponds to a fixed point of the Poincaré map P_h in the $(x,\dot{x})$-plane if it closes after one revolution. A more complicated periodic solution (closing after k revolutions) corresponds to a fixed point of the k-th iterate P_h^k of the Poincaré map. The points on the stable manifold W^s or on the unstable manifold W^u of such a fixed point correspond to orbits asymptotic to periodic orbits in forward or backward time direction, respectively. All the periodic orbits found in the present numerical exploration correspond to hyperbolic fixed points. Therefore a neighbouring orbit typically approaches a periodic orbit during a few revolutions (e.g. 2 to 4 revolutions if 13-digit accuracy is used). Then, however, the orbit will leave the neighbourhood of the periodic solution again (see Fig. 4-9).

The numerical study described below explores the following situation: two small coorbital satellites of masses m_1, m_2 are initially moving on nearly identical circular orbits of radii r_1, r_2 where $\Delta := r_2 - r_1 \ll r_1$. Define the impact parameter c as

$$c := \frac{\Delta}{R.\varepsilon^{1/3}} \; , \quad R = \frac{m_1 r_1 + m_2 r_2}{m_1 + m_2} \; ;$$

then in the limit $\varepsilon \to 0$ the value of c determines the behaviour of the satellites at their first close encounter. The relative motion of m_2 with respect to m_1 in the rotating frame of reference (see Section 1) is given by the nonoscillating solution of Hill's problem with the impact parameter c.

A large number of nonoscillating orbits were computed by means of high precision numerical integration. Sufficiently accurate initial values on the nonoscillating orbits were obtained by means of the asymptotic expansion [7] of Section 3. By using series terms up to the order τ^{-5} the desired accuracy of tol=10^{-13} was achieved by choosing $\tau \approx -200$ as the initial value of the parameter τ. Numerical integration with a local error < tol was carried out by means of an 8th-order Runge-Kutta-Fehlberg code with automatic step size control.

The evolution of the family of nonoscillating orbits as a function of the impact parameter c is extremely complex. As mentioned at the end of Section 3, the evolution is smooth for c<1.3361171883 (Fig. 4, U-shaped orbits) and for c>1.7187799380 (Fig. 9, "straight" orbits). The corresponding coorbital motion could be termed *exchange of orbits* or *overtaking*, respectively.

In the transition zone between these values the evolution of the family alternates between regions of smooth behaviour (e.g. 1.3485868066 < c < 1.5918721433) and transition regions of violent changes in the shape of the orbit. The overall structure is a fractal; every transition region may be split up into smaller transition zones separated by smooth regions.

In Table 1 we list 17 transition zones (c_{low}, c_{up}) of length $c_{up}-c_{low}<0.000375$. All other transition zones are smaller. In Figures 5 to 8 a tiny part of the inexhaustibly complex family is illustrated. Obviously, a typical orbit may approach several periodic solutions and leave them again, before it finally becomes unbounded.

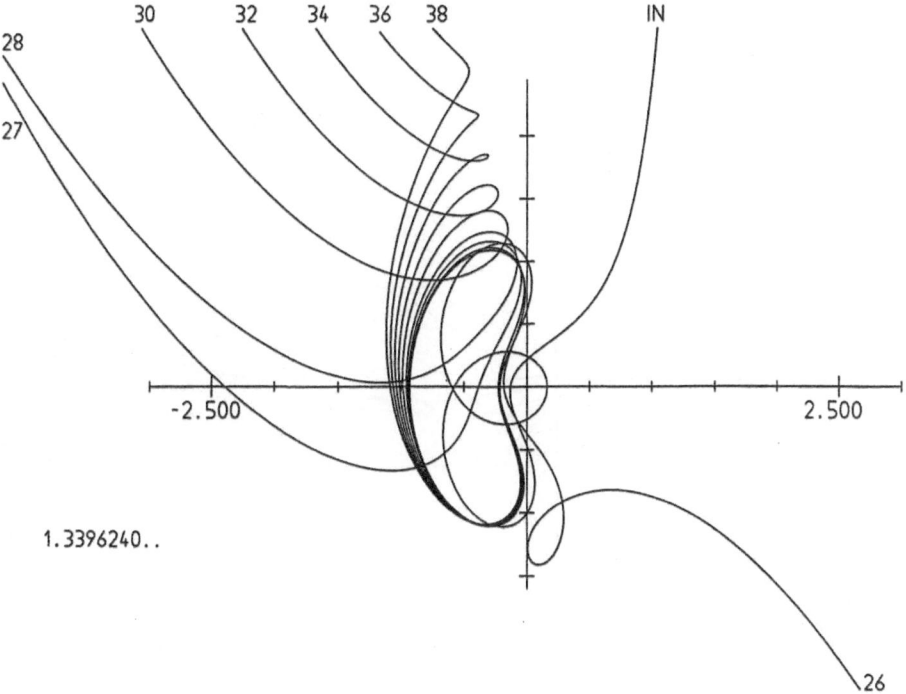

Figure 7. Smooth evolution near the limit c = 1.33962 40269

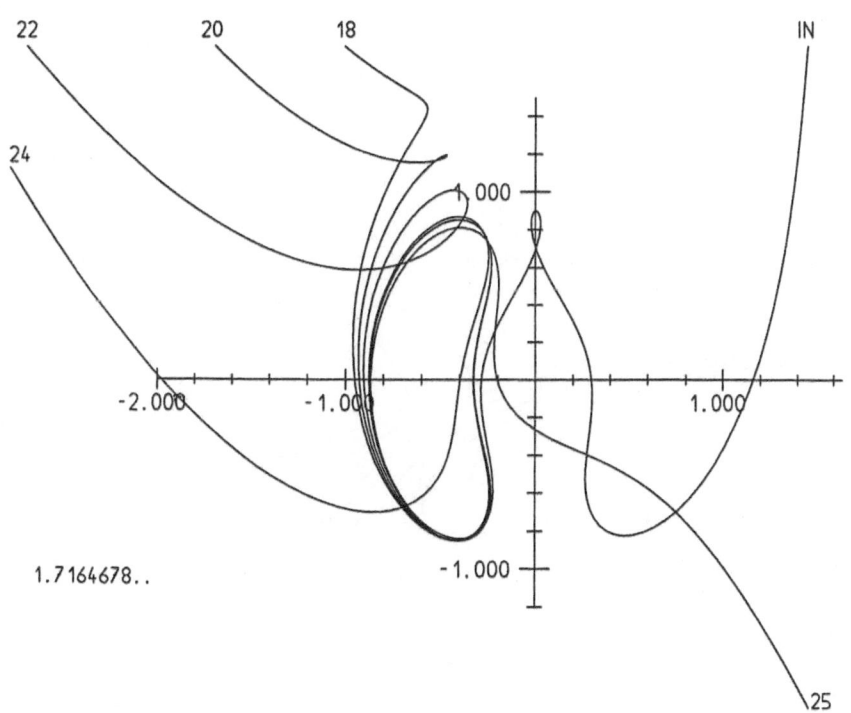

Figure 8. Smooth evolution near the limit c = 1.71646 78242

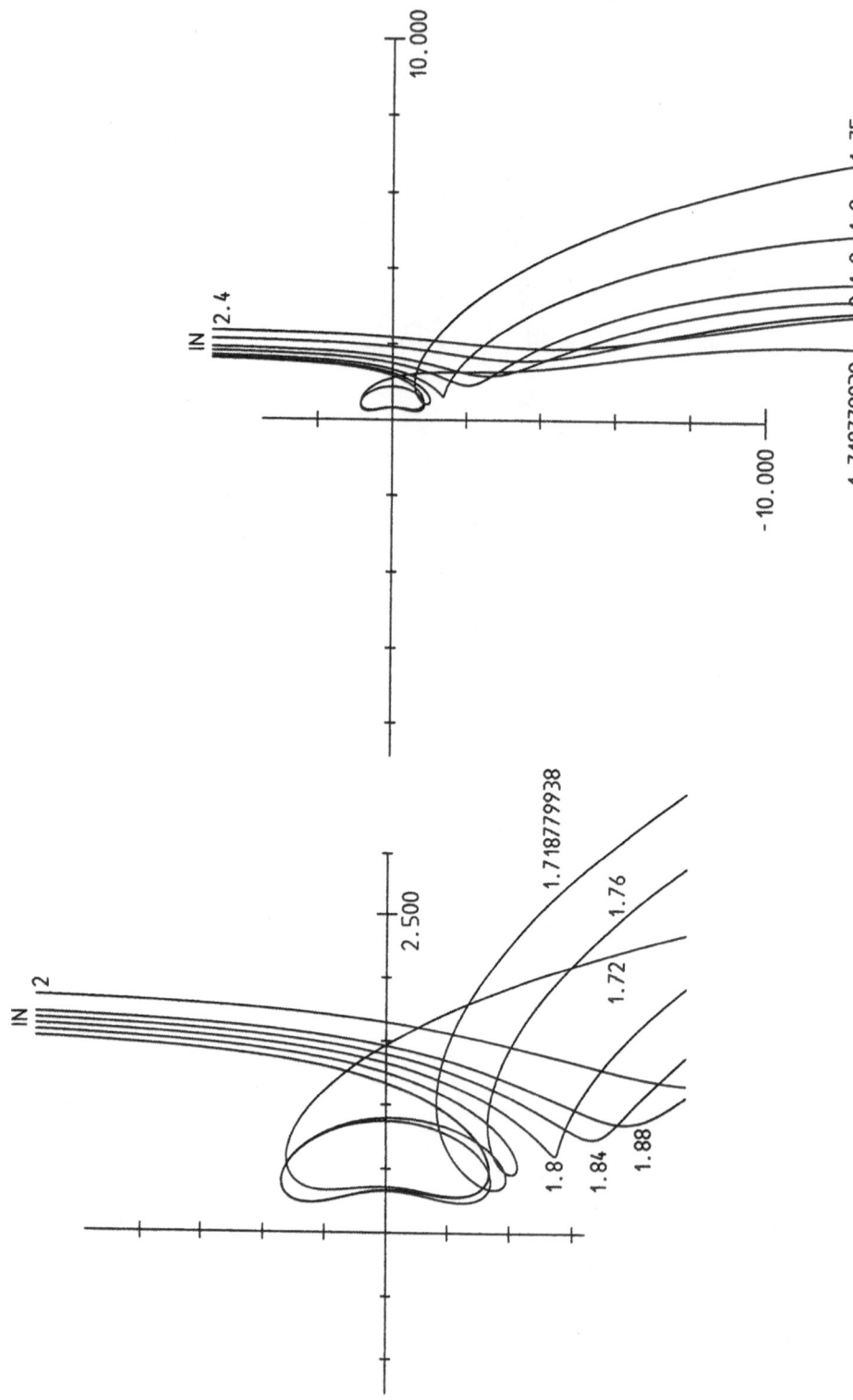

Figure 9. "Straight" orbits for large values of the impact parameter,
c > 1.71877 99380

Table 1. Transition zones in Hill's lunar problem

Initial value with $\tau = -250$, tolerance in Runge-Kutta-Fehlberg: $tol = 10^{-13}$. The values of $c := \lim x(t)$ for the boundaries of the 17 transition intervals of length < 0.000375 are listed below (the given values lie outside the transition intervals), $\Delta = c_{up} - c_{low}$:

c_{low}	c_{up}	$10^5 \Delta$
1.33611 71883	1.33614 87484	3.15601
1.33723 11033	1.33760 45911	37.34878
1.33807 85099	1.33808 66099	.81000
1.33959 648083	1.33962 40269	2.75461
1.34829 92716	1.34858 68066	28.75350
1.59187 21433	1.59199 17129	11.95696
1.59213 72069	1.59214 00053	.27984
1.59302 90180	1.59306 45977	3.55797
1.64978 77956	1.65003 38837	24.60881
1.65994 71369	1.65999 36861	4.65492
1.66410 74304	1.66468 98678	58.24374
1.66631 782141	1.66637 72547	5.94333
1.71646 78242	1.71655 02658	8.24416
1.71832 06108	1.71832 62829	.56721
1.71860 29561	1.71877 99380	17.69819

The longest transition zone ($\Delta = 58.24374 \cdot 10^{-5}$) splits into

1.66410 74304	1.66411 19891	.45587
1.66425 19122	1.66461 30872	36.11750
1.66468 72234	1.66468 98678	.26444

The orbits terminating the smooth regions on the c-axis are all asymptotic to simple (kidney-shaped) periodic solutions of Hill's problem, i.e. the iterates of the Poincaré map are on the stable manifold W^s of the corresponding hyperbolic fixed point in the $(x,\dot{x})$-plane. In the accuracy of the present computations at most 4 revolutions on the unstable periodic orbit could be carried out. The family of kidney-shaped periodic orbits emanates from the equilibrium solutions $x = \pm(1/3)^{1/3}$, $y = \dot{x} = \dot{y} = 0$ with $h = -3^{4/3}/2$ and seems to terminate in a collision solution.

We conclude this discussion with a possibly homoclinic, i.e. a truly chaotic orbit. Fig. 10 shows the initial phase of the nonoscillating solution with c=1.337339007. Surprisingly, the orbit can be continued in the same numerical approximation for more than 700 revolutions about the origin before it becomes unbounded (Fig.11). The nature of this peculiar orbit is revealed when its Poincaré map is iterated (Fig. 12). The essential feature is a "higher" periodic orbit (visible in Fig. 10) represented by the 3 "corners" in Fig. 12; they are fixed points of the 3rd iterate P_h^3. Clearly, the orbit of Fig. 11 follows the stable and unstable manifolds W^s, W^u of these fixed points for a very long time. To establish *chaoticity* of this orbit (of which there seems to be no doubt) one has to show the existence of a *homoclinic* point, i.e. a *transversal* intersection of the invariant manifolds W^s, W^u. This, in turn, would prove the non-integrability of Hill's lunar problem.

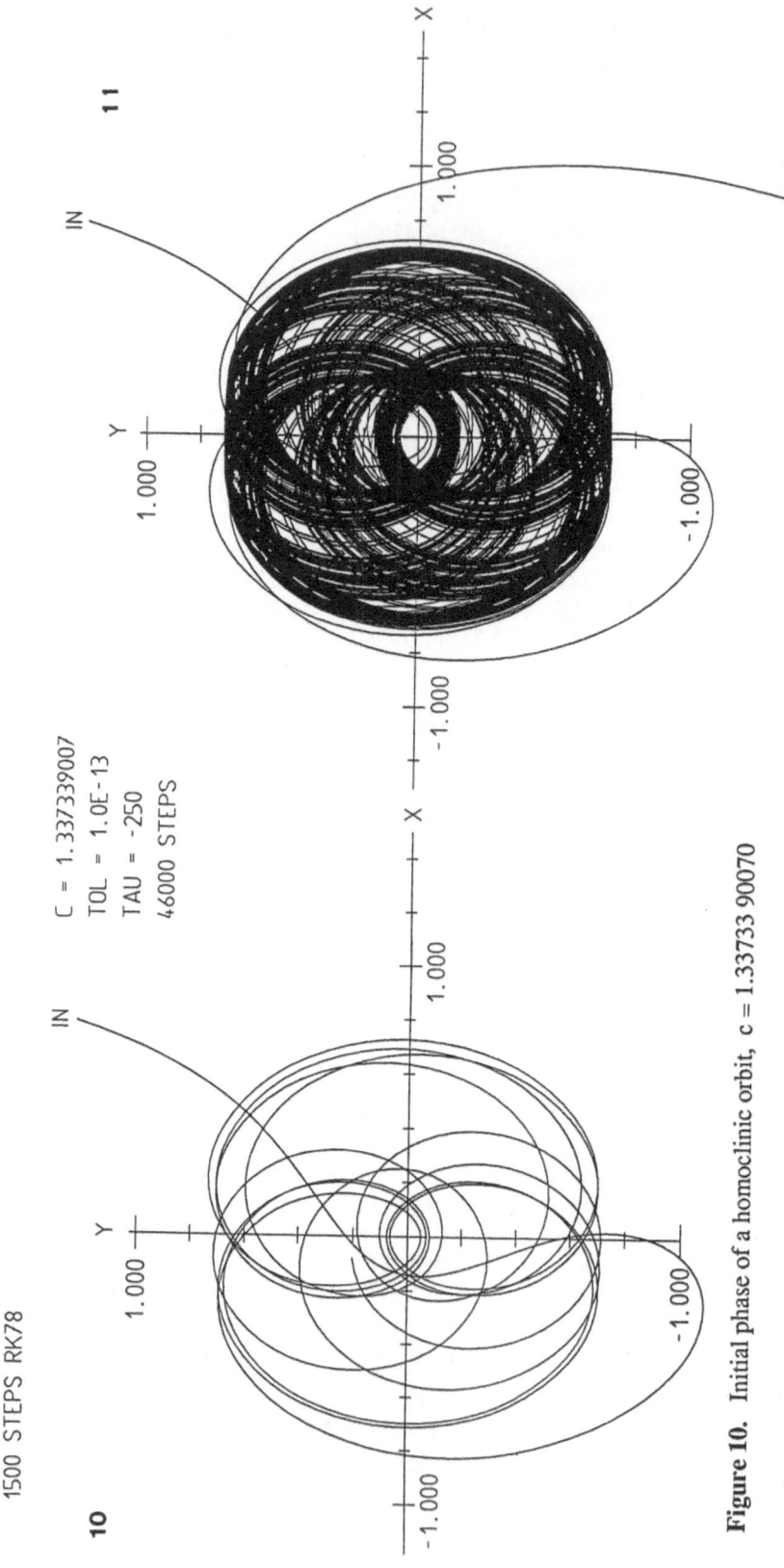

1500 STEPS RK78

C = 1.337339007
TOL = 1.0E-13
TAU = -250
46000 STEPS

Figure 10. Initial phase of a homoclinic orbit, c = 1.33733 90070

Figure 11. Homoclinic orbit until escape occurs, c = 1.33733 90070

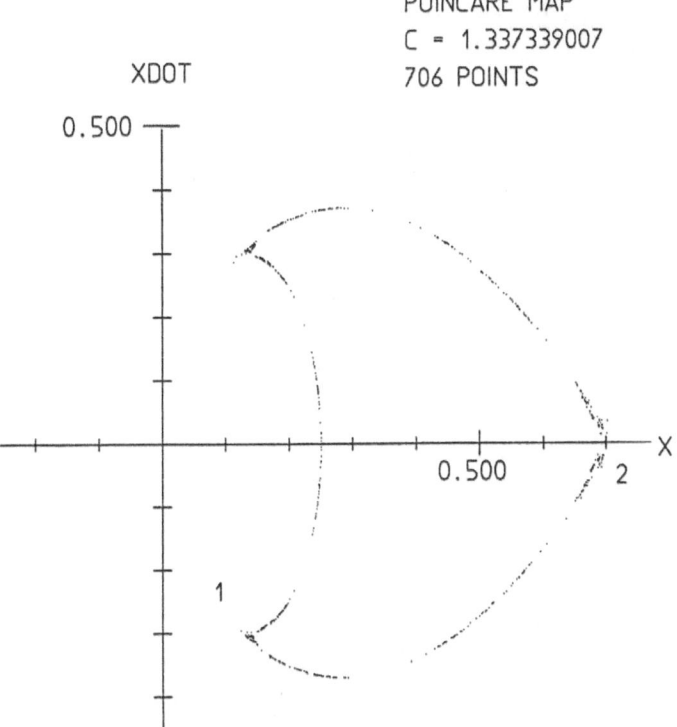

Figure 12. Poincaré map corresponding to the homoclinic orbit, c = 1.33733 90070
The first two iterates are labelled by the digits 1, 2

References

1. Colombo , G., 1982: The Motion of Saturn's Coorbiting Satellites 1980S1 and 1980S2. In: V. Szebehely (ed.), Applications of Modern Dynamics to Celestial Mechanics and Astrodynamics. Reidel Publ. Co., 21-23.

2. Hénon, M. and Petit, J.M., 1986: Series Expansions for Encounter-Type Solutions of Hill's Problem. Celest. Mech **38**, 67-100.

3. Hill, G.W., 1886: On the Part of the Motion of the Lunar Perigee which is a Fraction of the Mean Motions of the Sun and Moon. Acta Math. 8(1886), 1; also: Collected Mathematical Works of G.W. Hill. Vol 1. 1, p. 243. Cargenie Inst. of Wash., Washington, D.C., 1905.

4. Levi-Civita, T., 1920: Sur la régularisation du problème des trois corps. Acta Math. 42, 99-144.

5. Petit, J.M. and Hénon, M., 1986: Satellite Encounters. Icarus **66**, 536-555.

6. Spirig, F. and and Waldvogel, J., 1985: The Three-Body Problem with Two Small Masses: A Singular-Perturbation Approach to the Problem of Saturn's Coorbiting Satellites. In: V. Szebehely (ed), Stability of the Solar System and its Minor Natural and Artificial Bodies. Reidel, 53-63.

7. Waldvogel, J. and Spirig, F., 1988: Coorbital Satellites and Hill's Lunar Problem. In: A.E. Roy (ed.), Long-Term Dynamical Behaviour of Nature and Artificial N-Body Systems. Kluwer, 223-234.

8. Yoder, C.F., Colombo, G., Synnott, S.P., Yoder, K.A., 1983: Theory of Motion of Saturn's Coorbiting Satellites. Icarus **53**, 431-443.

PART IV

THE THREE-BODY PROBLEM

REMARKABLE TERMINATION ORBITS

OF THE RESTRICTED PROBLEM

V.V. Markellos

Dept. of Engineering Science
University of Patras, Greece

ABSTRACT

Homoclinic orbits at $L_{4,5}$ are termination orbits of families of periodic orbits of the restricted three-body problem. Very few such orbits are known in the literature. We present here a large number of new and remarkable looking non-symmetric homoclinic orbits for $\mu = 0.45$.

INTRODUCTION

Stromgren's classical results (1935) on asymptotic-periodic orbits at $L_{4,5}$ concerned few simple looking orbits symmetric w.r.t. the x-axis (the hetero-clinic orbits) or w.r.t. the y-axis (the homoclinic orbits), the latter symmetry being characteristic of the case of equal primaries (see also Szebehely, 1967). His conjecture concerning the asymptotic-periodic orbits as representing the terminations of entire families of periodic orbits was proved by Henrard (1973), while strong numerical evidence on this has been provided by Danby (1967), Szebehely and Nacozy (1967), Szebehely and Van Flandern (1967), Markellos and Taylor (1978), and Taylor (1983), among others. Danby (1984) also published a few remarkable looking (poodle) periodic orbits close to asymptotic orbits.

Gomez et al. (1988) computed symmetric heteroclinic orbits for many values of the mass parameter μ and proved the existence of an infinity of asymptotic orbits both heteroclinic and homoclinic. They also showed a 0-homoclinic orbit existing at about $\mu = 0.1$ and smaller values of μ.

In this note we present many remarkable looking asymptotic orbits homoclinic at L_4. These orbits have been computed for unequal primaries and they possess no symmetry whatsoever.

Predictability, Stability, and Chaos in N-Body Dynamical Systems
Edited by A.E. Roy, Plenum Press, New York, 1991

DETERMINATION OF THE HOMOCLINIC ORBITS

To compute homoclinic orbits we first proceed to calculate a linear approximation of the solution of the equations of the problem in the neighbourhood of the triangular point L. This is done in a coordinate system centred at L and rotated by an angle φ such that $\tan \varphi = -3^{1/2}(1-2\mu)$. The linear analysis leads to the following initial state for the outgoing (j=1) and incoming (j=2) flow at the equilibrium point:

$$x(0) = c_1$$

$$y(0) = (-1)^{j+1} c_1 A + c_2 B$$

$$\dot{x}(0) = (-1)^{j+1} c_1 a + c_2 b$$

$$\dot{y}(0) = c_1 (A a - b B) + (-1)^{j+1} c_2 a b,$$

where $\pm(a \pm i b)$ are the complex eigenvalues, and

$$a = (3 \sigma^{1/2} - 1)^{1/2} / 2, \qquad b = (3 \sigma^{1/2} + 1)^{1/2} / 2,$$

$$A = a(1-s)/2, \qquad B = b(1+s)/2,$$

where we have abbreviated,

$$\sigma = 3\mu(1-\mu),$$

$$s = [1 - (1-\sigma)^{1/2}]/\sigma^{1/2}.$$

We then put:

$$x(0) = r \cos \theta = c_1$$

$$y(0) = r \sin \theta = c_1 A + c_2 B,$$

determining c_1, c_2, and the full initial state, from r and θ. We set r to a small value, say r = 0.0001 (to ensure validity of linear approximation), leaving only θ arbitrary, and then scan the values of θ in the interval $[0, 2\pi]$ to determine the outgoing flow (similarly the incoming flow) at $L_{4,5}$.

Finally, we establish the required transversality of the two flows by a standard differential corrections procedure correcting the values of θ_{out} and θ_{in}.

Initial conditions are provided by the respective phase-plane portraits where we make sure to select points corresponding to non-perpendicular intersections of the x-axis.

The orbits are shown in the following orbit plots, in the Appendix. The scale of each plot is determined by the equilateral triangle of unit side defined by the primaries (dark dots) and the equilibrium point L_4 at which the orbits are seen to spiral in and out.

Remarkable is the apparent repetition of the simpler patterns of orbits as parts of the more complicated orbits. We note that as has been described e.g. in Henrard (1973) close to the homoclinic orbits, along the families terminating at these orbits, there exist stable periodic orbits as well as unstable ones.

REFERENCES

Danby, J.M.A.: 1967, *Astron. J. 72*, 198

Danby, J.M.A.: 1984, *Celes. Mech. 33*, 251

Gomez, G., Llibre, J., Masdemont, J.: 1988, *Celes. Mech. 44*, 239

Henrard, J.: 1973, *Celes. Mech. 7*, 449

Markellos, V.V. and Taylor, D.B.: 1978, *Astron. Astrophys. 70*, 617

Stromgren, E.: 1935, *Copenhagen Obs. Publ. 100*

Szebehely, V.: 1967, *Theory of Orbits*, Academic Press

Szebehely, V. and Nacozy, P.: 1967, *Astron. J. 72*, 184

Szebehely, V. and Van Flandern T.: 1967, *Astron. J. 72*, 373

Taylor, D.B.: 1983, *Celes. Mech. 29*, 75

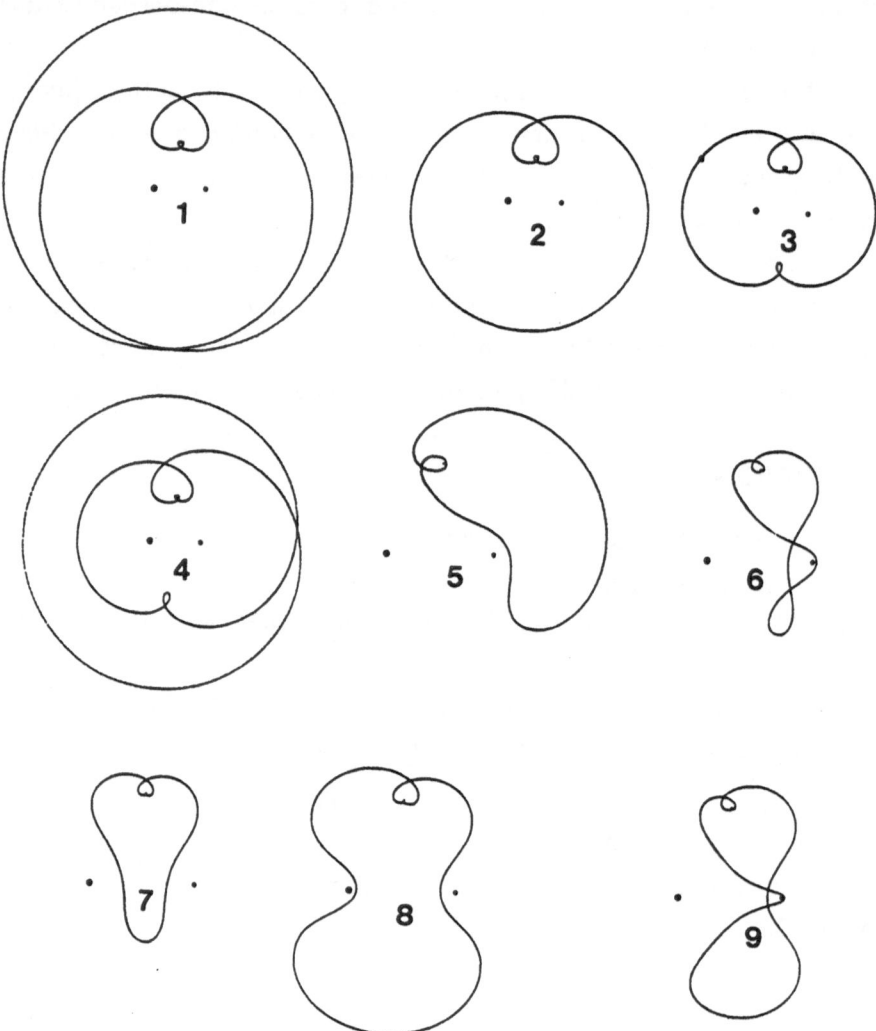

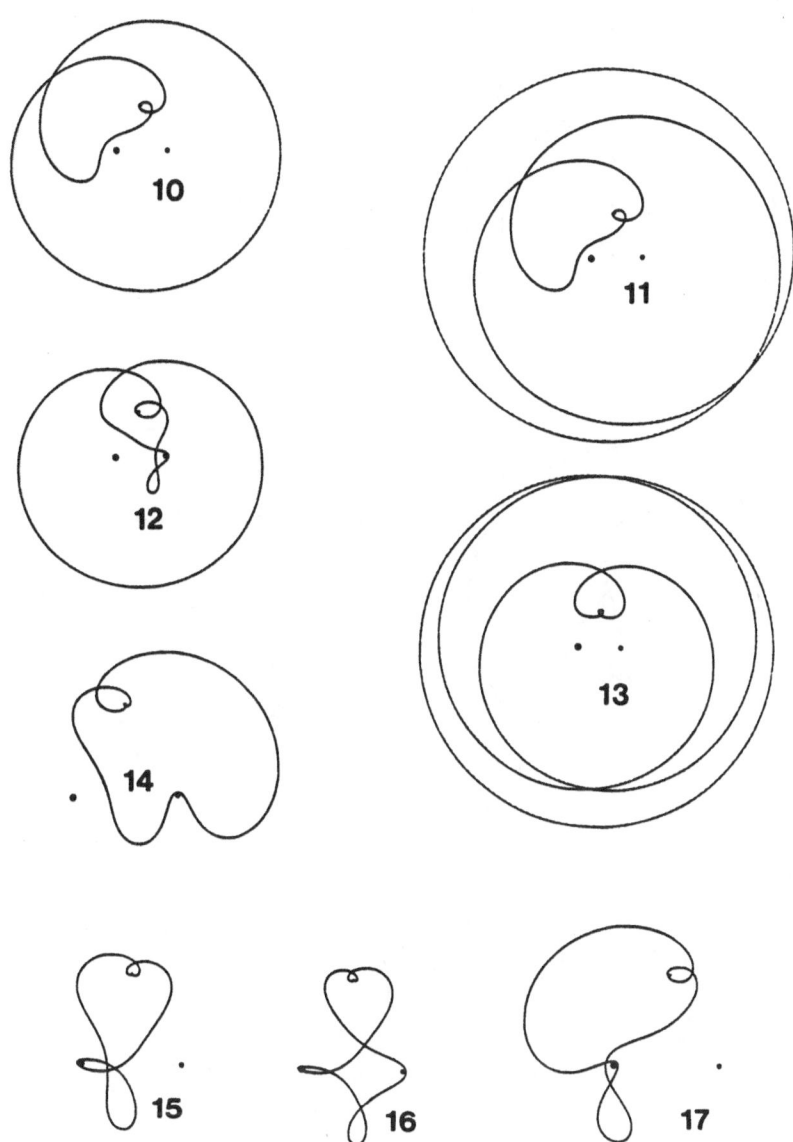

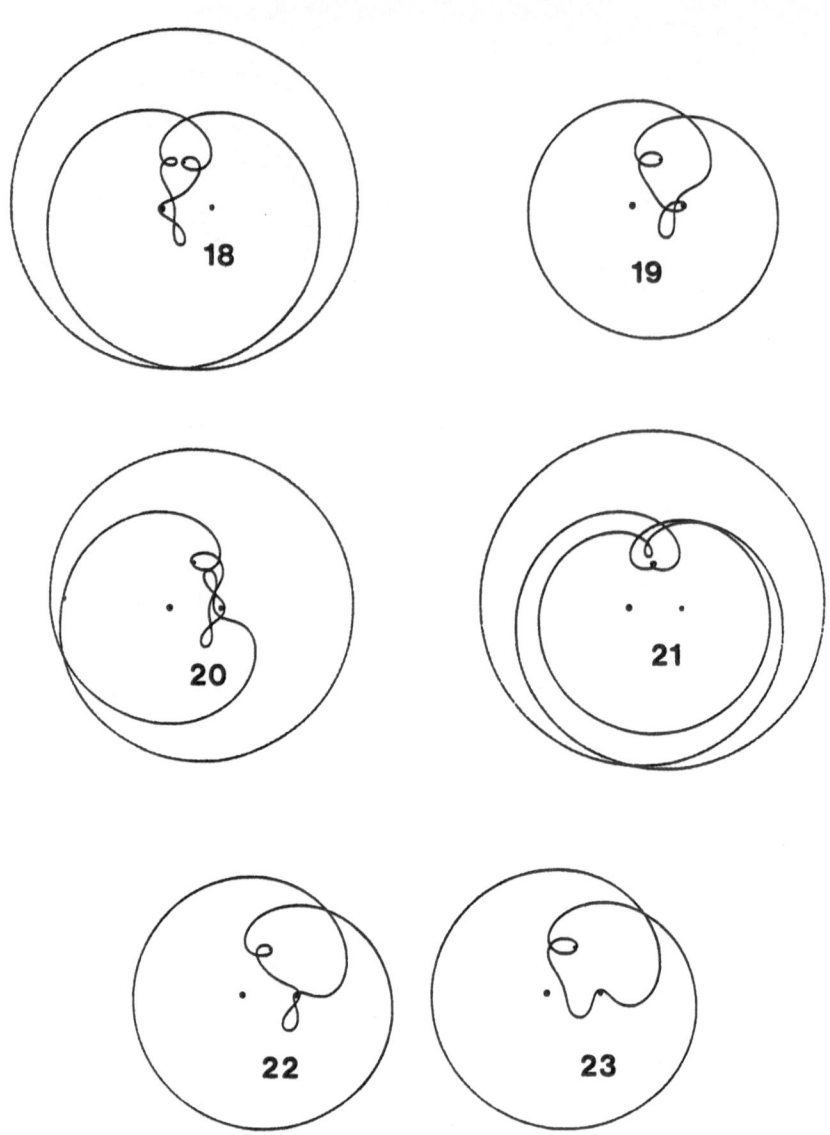

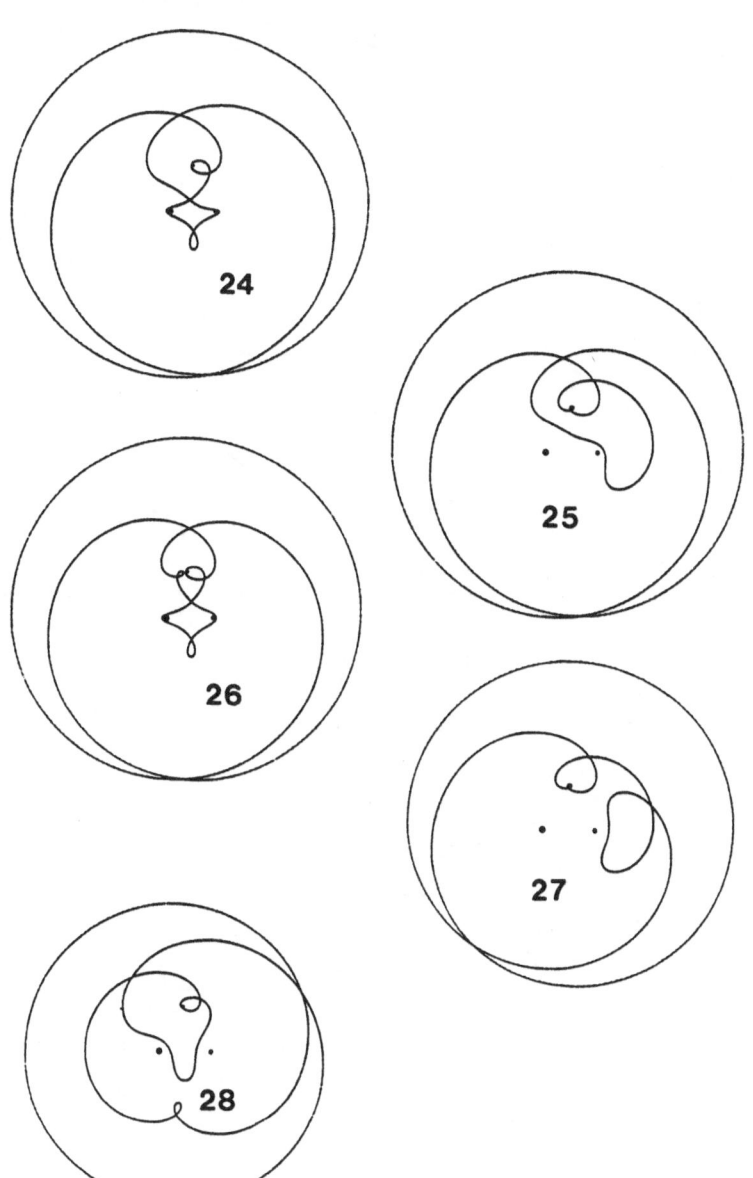

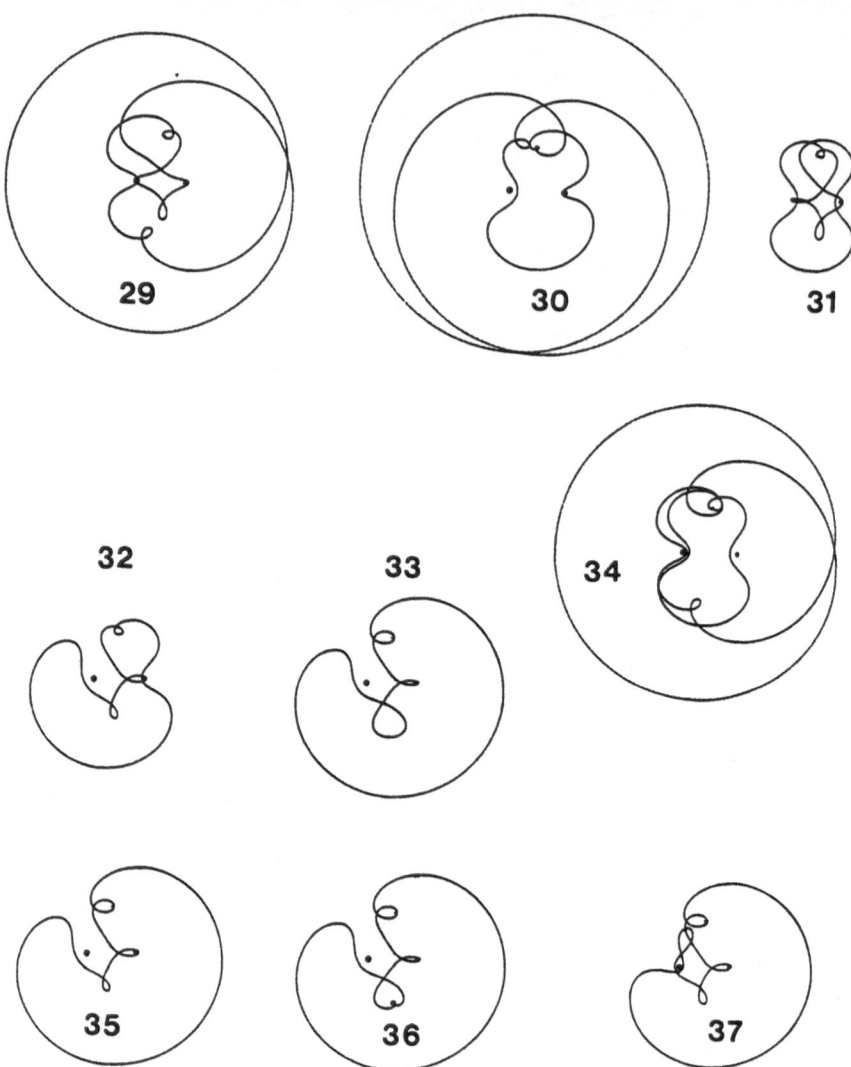

29 30 31

32 33 34

35 36 37

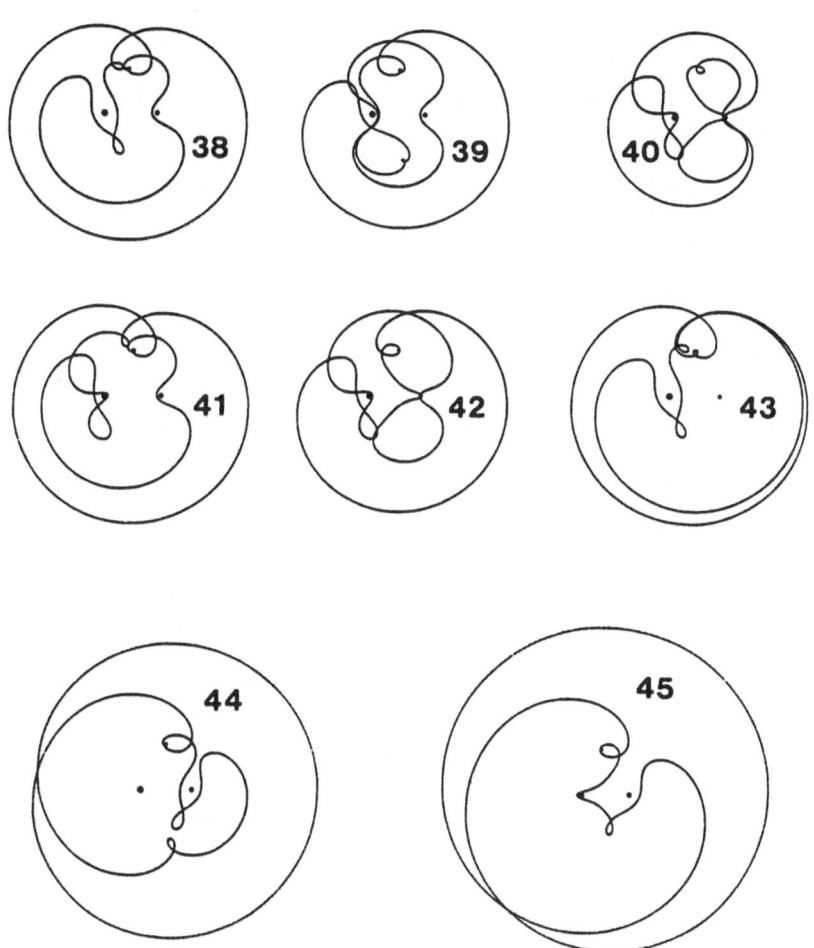

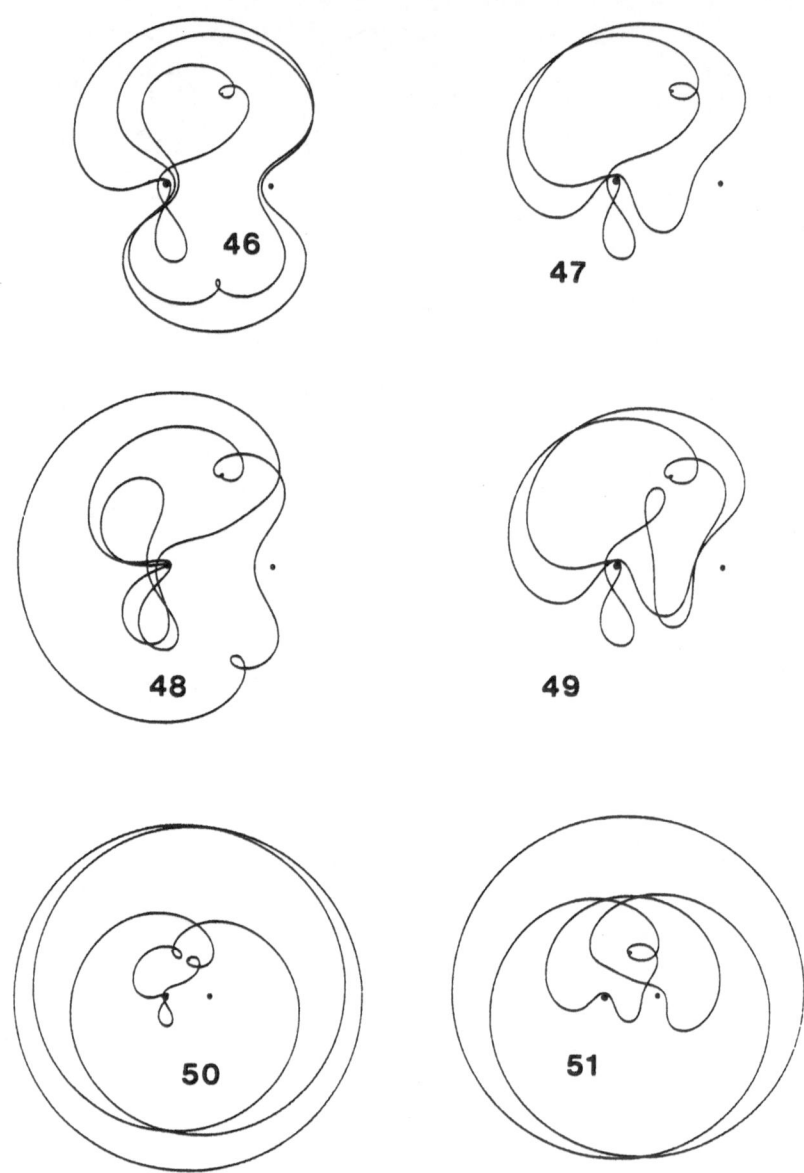

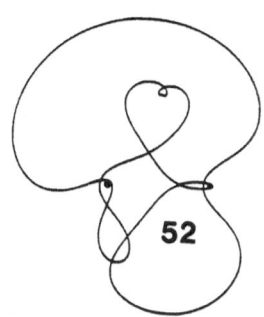

52

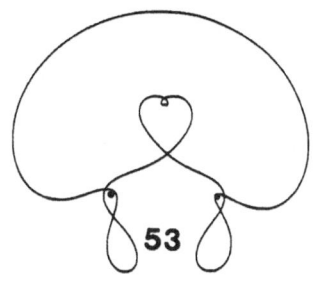

53

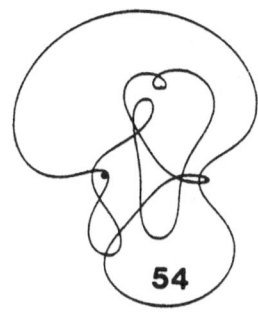

54

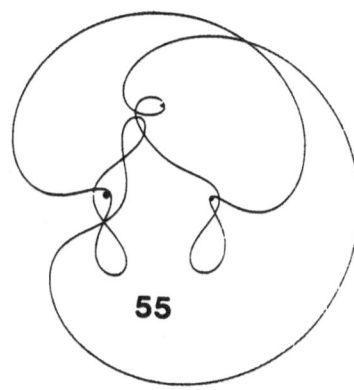

55

PERIODIC ORBITS IN THE ISOSCELES THREE-BODY PROBLEM

Cristina Chiralt Monleón and
José Martínez Alfaro

Departament de Matemàtica Aplicada i Astronomia
Facultat de Matemàtiques, Universitat de València, Spain

INTRODUCTION

The Saturn's satellites Janus and Epimetheus are the first known bodies in the Solar System that has horseshoe orbits in a frame that rotates with uniform angular velocity. Both satellites have similar masses and orbital elements when they are far from one another. Moreover, their orbits are nearly symmetric. In fact, in the past, they have been identify as a unique satellite and afterwards, some mathematical theories about their orbits has been necessaries to understand why they do not collide. In particular, the interest in planar three-body problem with two small masses has increased[6]. We assume that the two small masses have similar symmetric initial conditions. The aim of this paper is to find what kind of motion, can present these bodies. We find an infinity of different types . To prove that we analyze the limit case, when both small masses are equal and in symmetric opposite positions, i.e, the isosceles three- body problem. Afterwards, by continuity we derive properties for the initial three-body problem. The equations of motion of the isosceles three-body problem are[3]:

$$\frac{d^2x}{dt^2} = - \frac{Gm_0 x}{(x^2+y^2)^{3/2}} - \frac{Gm_1}{4x^2}$$

$$\frac{d^2y}{dt^2} = - GM \frac{y}{\left(x^2+y^2\right)^{3/2}}$$

where M is the total mass $m_0 + m_1 + m_2$.

We choose units so that $G = 1$, $m_0 = 1/3$ and we call ε the small masses $m_1 = m_2$. As our scope is the three-body problem with two small masses, we consider ε smaller than the critical value at which the Euler points of the triple collision manifold have not complex eigenvalues[3], although our results can be extended for larger values of ε. With our units $0<\varepsilon<0.02424...$. This value is complementary of the usual and, moreover, we look for orbits far from triple collision. Then, the differential system can be expressed as follows:

Predictability, Stability, and Chaos in N-Body Dynamical Systems
Edited by A.E. Roy, Plenum Press, New York, 1991

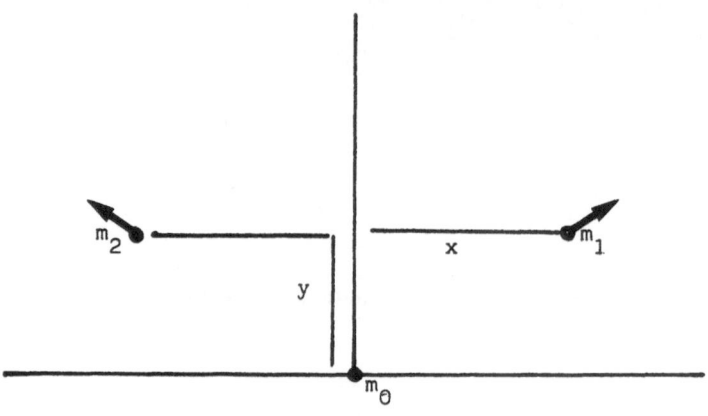

Figure 1

$$\frac{dx}{dt} = \frac{p_x}{2\varepsilon}$$

$$\frac{dy}{dt} = \frac{1 + 6\varepsilon}{2\varepsilon} \, p_y$$

$$\frac{dp_x}{dt} = - \frac{2\varepsilon x}{3 \left(x^2+y^2\right)^{3/2}} - \frac{\varepsilon^2}{2x^2}$$

$$\frac{dp_y}{dt} = - \frac{2\varepsilon y}{3 \left(x^2+y^2\right)^{3/2}}$$

corresponding to the Hamiltonian:

$$H(x, \, y, \, p_x, \, p_y) = \frac{1}{4\varepsilon} \, [p_x{}^2 + (1 + 6\varepsilon) p_y{}^2] - \frac{2\varepsilon}{3 \left(x^2+y^2\right)^{3/2}} - \frac{\varepsilon^2}{2|x|}$$

As we are interested in periodic collision orbits, we regularized double collisions by means of the Levi-Civita canonical transformation whose generating function is $p_R x^{1/2}$, i.e:

$$x = R^2 \quad , \quad p_x = \frac{p_R}{2R}$$

Then, the Hamiltonian goes over to:

$$H(R, \, y, \, p_R, \, p_y) = \frac{1}{4\varepsilon} \, [\frac{p_R{}^2}{4R^2} + (1 + 6\varepsilon) p_y{}^2] - \frac{2\varepsilon}{3r} - \frac{\varepsilon^2}{2R^2}$$

where $r = (R^4 + y^2)^{1/2}$

We make also a change of temporal variable defined implicitly by:

$$\frac{dt}{ds} = 2R^2.$$

Finally, the differential system becomes:

$$R' = \frac{p_R}{4\varepsilon}$$

$$y' = \frac{1 + 6\varepsilon}{\varepsilon} R^2 p_y$$

$$p_R' = 4R\left(H - \frac{1+6\varepsilon}{4\varepsilon} p_y^2 + \frac{2\varepsilon y^2}{3r^3} \right)$$

$$p_y' = - \frac{4\varepsilon y R^2}{3r^3}$$

By a scaling property in the study of the isosceles three-body for negative energy, we can restrict at the level H = -1. A periodic orbit will mean a periodic orbit (P.O.) of the regularized system. From each periodic orbit we derive a collision periodic orbit (C.P.O.) : if R is positive both trajectories coincide, if R is negative the C.P.O has R and p_R with opposite sign that the P.O.. See Figure 1.2,1.3.

SOME SYMMETRIC PERIODIC ORBITS

This section is a catalog of the simplest symmetric periodic orbits (S.P.O) that we have found numerically. We do not try to give a complete list; our scope is to use these simple periodic orbits to derive the existence of other type of motions. In all symmetric periodic orbit there is a cut with the surface y=0, where p_R=0, after the initial one. By simplest we mean that this cut happens after a few cuts with the same surface. They correspond to families of P.O. described in[3,5].
Quasi-circular P.O.:This is a hyperbolic periodic orbit which does not pass near the origin. In Figure 1.1 , there is a qualitative picture of this orbit . It belongs to a family such that the corresponding values of the initial conditions R and p_y when y = p_R = 0 tends to zero with ε.

From now on we will note it γ. All the other P.O. that we study are near triple collision orbits or near parabolic orbits.
Here is a table of the initial conditions of the quasi-circular P.O. depending on the parameter:

ε	R	p_y
.02	.0832643	.26077258
.015	.07175	.2303963
.01	.0584	.1912593
.005	.041118	.1379155

Moreover, for all orbits y = p_R = 0.
"Drop" periodic orbit: This is also a hyperbolic periodic orbit. The maximum absolute value of y is very small for this orbit and all the next one; therefore they are near the triple collision orbits on y=0. Figure 1.2 and 1.3 show it as a P.O. and as a C.P.O.

If ε =0.02, y = p_R = 0 and R=.116249, p_y = .012281. There is also a symmetric one with R and p_y with opposite sign. For this one R(0) = 0.002801. It is an hyperbolic orbit.

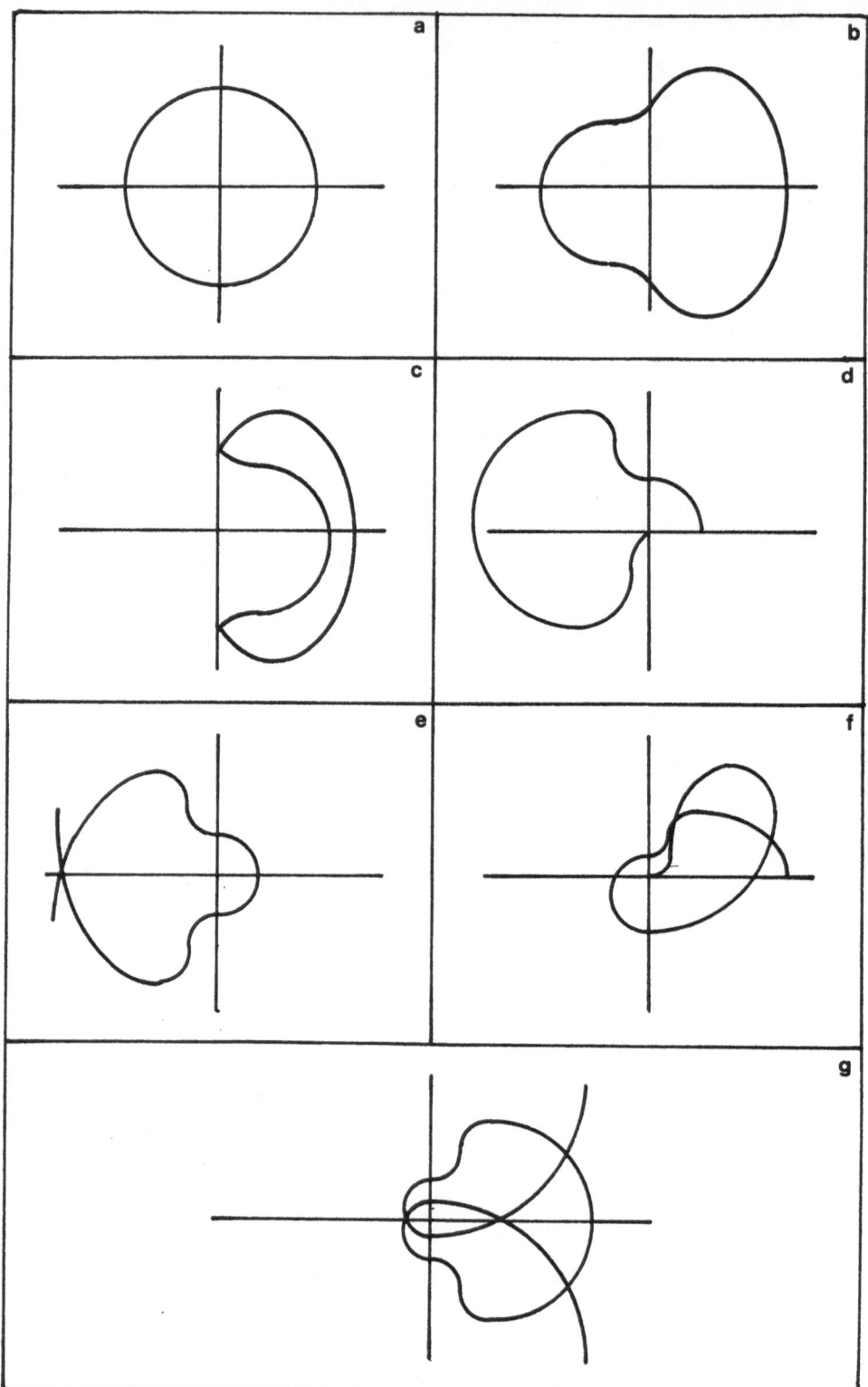

Figure 2

Triple collision periodic orbit (one cut):Initial conditions : ε = 0.02, R = .000599, p_y = 51.818. See Figure 1.4.

Trefoil reversible PO: Initial conditions: ε = 0.02, R = .0005103, p_y = 60.96. Figure 1.5.

Triple collision orbit (two cuts): Initial conditions: ε = 0.02, R = .116324, p_y = .003271. Figure 1.6.

Reversible PO (two cuts):Initial conditions: e = 0.02, R = .1163306, p_y = .0016552. Figure 1.7.

THE MANIFOLD OF CONSTANT ENERGY

The manifold of constant energy H = -1 is defined by the relation:

$$-1 = \frac{1}{4\varepsilon} \left[\frac{p_R^2}{4R^2} + (1 + 6\varepsilon) p_y^2 \right] - \frac{2\varepsilon}{3r} - \frac{\varepsilon^2}{2R^2}$$

for r > 0.

In order to determine the states corresponding to R = 0, we introduce new coordinates:

$$v = p_R \left(\frac{3r}{8\varepsilon} \right)^{1/2} \qquad u = p_y |R| \left(\frac{1 + 6\varepsilon}{2\varepsilon} 3r \right)^{1/2}$$

Then, we get the new relation:

$$v^2 + u^2 = 4\varepsilon R^2 + 3\varepsilon^2 r - 6rR^2 = k(R, y)$$

Where k is strictly positive as R belongs to the region of motion except for the value k(0,0) =0. Therefore, for each y $\neq$ 0, we have a S^2,. and for y = 0, an S^2 minus a circumference. Let paste together all the two-spheres with a radius s(y) increasing with y, lim s(y) = 0 as y tends to -∞. The manifold of constant energy V is diffeomorphic to R^3 - S^1 - one point. Introduce a frame x,y,z such that z = r, where r is R scaled in such a way that the maximum admissible value or R in the region of motion equals the radius s(y) of the sphere corresponding to y.

KNOTS AND DYNAMICAL SYSTEMS

After the description of the simplest S.P.O. we try to classify them topologically. Each P.O. can be seen as an image of the unit circumference in the manifold of constant energy. The classification of these images is the objective of the Knot Theory.

We call a knot a topological embedding of S^1 into S^3 or R^3 ; if we work in R^3 instead of S^3, we must simply compactified it. Two knots K_1, K_2 are equivalent if there is an orientation preserving homeomorphism h: $S^3 \rightarrow$ S^3, such that h(K_1) = K_2. A trivial knot is the equivalence class of the circumference $x^2 + y^2 = 1$, z=0. We represent a knot by its projection on a plane taking into account the over or undercrossings. The simplest not trivial knot is the trefoil.

A torus knot $K_{p,q}$ of type p,q (where p and q are relatively prime integers) is a curve on the surface of an unknotted torus that cuts a meridian in p points and a longitude in q points. The trefoil is the torus knot $K_{2,3}$, the figure-eight knot is not a torus knot.

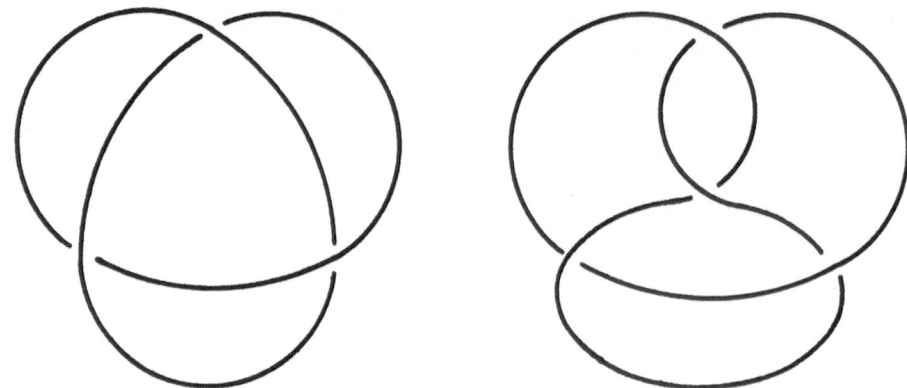

Figure 3

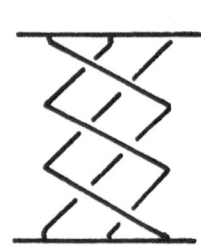 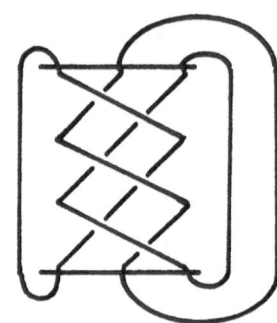

Figure 4

A n-braid B is a collection of unknotted strands, proceeding
downward from n points (top row) to n points (bottom row). The strands wind
around one another throughout the descent. Given a braid B its closure is
the knot or link obtained by attaching the n points in the top row to the
counterparts on the bottom row.

All the invariants associated to knots and braids, such as the
topologycal types, the homotopy group of the complementary space, the
associated polynomials are topological invariants.

Knots arise in differential equations through periodic orbits[1,2] J.
Franks and R.F. Williams[4] prove that in a C^1 flow on S^3 or R^3 such that
there exists a hyperbolic periodic orbit with a transverse homoclinic
point, we have closed orbits of infinitely many distinct knot types.

P.O. AS KNOTS

Among the P.O. of the isosceles problem we have only two trivial
knots: γ and the "drop" P.O.

The trefoil reversible P.O. is as its name says a trefoil of $K_{2,3}$
torus-knot. And in Figure 5, we have the reversible periodic orbit as a
knot in the constant energy manifold.

By [4], we know that near an hyperbolic P.O. such as 1.1, we can find
infinitely distints types of knots. These, and the consequences for the
full three-body problem with two small masses, will be studied in another
paper.

430

Figure 5

In the isosceles problem, the binary collision orbits (BCO) present a discontinuity at R=0, and therefore they are not real periodic orbits. At the binary collision $p_y=0$ and p_R changes it sign. This reflection makes the angle a introduced in the energy manifold to jump from 0 to π or vice versa. Therefore in the plane x-z of the manifold of constant energy V the real BCO jumps from a point (x,0) to (-x,0). That makes this orbit a braid being the top row the positive semiaxis z=0, and the bottom row the negative semiaxis. The origin is excluded because it represents a triple collision orbit.

REFERENCES

1. Birman, J. and Williams, R.F. Knotted Periodic Orbits in Dynamical Systems I: Lorenz's Equations. _Topology_ 22:47 (1983)

2. Birman, J. and Williams, R.F. Knotted Periodic Orbits in Dynamical Systems II: Knot Holders for Fibered Knots. _Contemporary Mathematics_ 20:1 (1983)

3. Broucke, R. On the Isosceles Triangle Configuration in the Planar General Three -Body Problem. _Astronomy and Astrophysics_. 73:303 (1979)

4. Franks, J. and Williams R.F. Entropy and Knots. _Transactions of the American Mathematical Society_. 291, No. 1: 241 (1985)

5. Simó, C. and Martínez, R. Qualitative Study of the Planar Isosceles Three- Body Problem. _Celestial Mechanics_. 41:179 (1988)

6. Spirig, F. and Waldvogel, J.. The Three - Body Problem with Two Small Masses: A Singular- Perturbation Approach to the Problem of Saturn's Coorbiting Satellites, in "Stability of the Solar System and Its Minor Natural and Artificial Bodies". Nato Asi Series. (1984)

QUASIPERIODIC ORBITS AS A SUBSTITUTE OF LIBRATION POINTS IN THE SOLAR SYSTEM

Gerard Gómez[1], Angel Jorba[2],
Josep Masdemont[2] and Carles Simó[1]

(1) Departament de Matemàtica Aplicada i Anàlisi, Universitat de Barcelona
Gran Via 585, 08007 Barcelona, Spain

(2) Departament de Matemàtica Aplicada I, ETSEIB, Universitat Politècnica
de Catalunya, Diagonal 647, 08028 Barcelona, Spain

Abstract

Consider the Earth-Moon-particle system as a RTBP. It is well known that there are two equilateral libration points. In the real life system, these points do not exist, due to the effect of the perturbations caused by the part of the solar system which is not taken into account. In this work the full problem is presented as a perturbation of the RTBP and we look for a dynamical equivalent of $L_{4,5}$, which seems to be a quasiperiodic orbit. In this paper we present a way to obtain these orbits, as well as their stability.

1 Introduction

Consider two bodies moving under their gravitational force, in circular orbits around their center of masses. The units of length, time and mass are chosen so that the angular velocity of rotation, the sum of masses of the bodies and the gravitational constant are all equal to one. With these normalized units, the distance between the bodies (also called primaries) is equal to one. Consider an infinitesimal particle moving under the interaction of the primaries. The study of the motion of this third particle is the so called Restricted Three Body Problem (RTBP) (for more details see [7]).

It is well known that, in a rotating frame, the RTBP has two equilateral libration points (called L_4 and L_5). These points can be defined as the ones forming an equilateral triangle with the primaries (see [7]). It is also known that, for the mass parameter μ (the mass of the small primary in the normalized units) less than the Routh critical value ($\mu_R = 0.0385\ldots$) these points are linearly stable. Applying the KAM theorem to this case we can obtain that there exist invariant tori around these points. Now, if we restrict the motion of the particle to the plane of motion of the primaries we have that these tori split the phase space, and then the equilateral points are stable (except for two values, $\mu = \mu_2$ and $\mu = \mu_3$ with low order ressonances). In the spatial case the invariant tori do not split the phase space and, due to the possible Arnold difusion, these points can be unstable. But Arnold difusion is a very slow phenomenon and we can have small neighbourhoods of "practical stability" (see [2]). The conclusion is that, if our real Earth-Moon system could be modelled as a RTBP, the problem of keeping a spacecraft near the libration points would be easy to solve.

Unfortunately, the real Earth-Moon system is rather complex. In this case, due to the fact that that the motion of the Earth and the Moon is non circular (even non elliptical!) we need to define "instantaneous" libration points as the ones forming an equilateral triangle with the Earth and the Moon at each instant. If we perform some numerical integrations starting at (or near) these points we can see that the solutions go away after a short period of time (see [5] and [3]). From this fact we can extract two conclusions. First: if we are interested in keeping a spacecraft there, we will need to use some control. Second: the RTBP is not a good model for this problem, because the behaviour displayed by it is different from the one of the real system.

For these reasons, in the subsequent sections we present a way to obtain a good model for the problem, and also a way to obtain a quasiperiodic solution near the triangular points. Then we will discuss the stability and the relation between this orbit and the real system.

A work of this kind can be found in [1], where only the planar case is considered.

2   Equations of motion

Let us consider a system of reference with the origin in one of the instantaneous equilateral points and the axes defined by the unitary vectors $\vec{e}_1$, $\vec{e}_2$, $\vec{e}_3$ given as follows:

$$\vec{e}_1 = \frac{\vec{r}_{EM}}{|\vec{r}_{EM}|}, \quad \vec{e}_3 = \frac{\vec{r}_{EM} \wedge \dot{\vec{r}}_{EM}}{|\vec{r}_{EM} \wedge \dot{\vec{r}}_{EM}|}, \quad \vec{e}_2 = \vec{e}_3 \wedge \vec{e}_1,$$

where $\vec{r}_{EM}(t)$ is the position vector of the Moon with respect to the Earth.

In order to satisfy Kepler's third law we modify the mass of the Earth, considering the remaining mass as a perturbation.

The units of length, time and mass are normalized in the usual way: the angular velocity of rotation, the sum of the masses of the Earth and the Moon, and the gravitational constant are all equal to one. With this system of reference and units the Lagrangian of the full solar system can be written (see [3] for a detailed description).

In this Lagrangian, the terms which contain Legendre polynomials (except those coming from the RTBP) are expanded as power series in x, y, z. Its coefficients are known functions of the positions of the bodies of the solar system. For a short time interval it can be assumed that these positions, and therefore the coefficients, are quasiperiodic functions of time.

A computation of these coefficients using a Fourier analysis shows that the relevant frequencies are the ones related to the following four angles:

1. The mean longitude of the Moon (equal to 1, because of the choice of the units).

2. The mean longitude of the lunar perigee.

3. The mean longitude of the ascending node of the Moon.

4. The mean elongation of the Sun.

All the contributions with amplitude less than 5×10^{-5} are dropped in order to keep a manageable number of terms. This leads to the fact that the perturbation coming from the planets, the radiation pressure and the aspherical terms coming from the Earth and Moon can be neglected (for more details about all the process of computation see [3]). The equations of the motion for this simplified model are of the form

$$
\begin{aligned}
\ddot{x} &= P(7)\left[-\tfrac{x-x_E}{r_{PE}^3}(1-\mu_M) - \tfrac{x+x_E}{r_{PM}^3}\mu_M - x_E(1-2\mu_M)\right] + \\
&\quad + P(1) + P(2)x + P(3)y + P(4)z + P(5)\dot{x} + P(6)\dot{y}, \\
\ddot{y} &= P(7)\left[-\tfrac{y-y_E}{r_{PE}^3}(1-\mu_M) - \tfrac{y-y_E}{r_{PM}^3}\mu_M - y_E\right] + P(8) + P(9)x + \\
&\quad + P(10)y + P(11)z + P(12)\dot{x} + P(13)\dot{y} + P(14)\dot{z}, \\
\ddot{z} &= P(7)\left[-\tfrac{z}{r_{PE}^3}(1-\mu_M) - \tfrac{z}{r_{PM}^3}\mu_M\right] + P(15) + P(16)x + \\
&\quad + P(17)y + P(18)z + P(19)\dot{y} + P(20)\dot{z},
\end{aligned} \tag{1}
$$

where r_{PE}, r_{PM} denote the distances from the particle to the Earth and Moon, respectively, given by $r_{PE}^2 = (x - x_E)^2 + (y - y_E)^2 + z^2$, $r_{PM}^2 = (x + x_E)^2 + (y - y_E)^2 + z^2$. We recall $x_E = -1/2$, $y_E = -\sqrt{3}/2$ for L_4 and $x_E = -1/2$, $y_E = \sqrt{3}/2$ for L_5. The functions $P(i)$ are defined as

$$
P(i) = A_{i,0} + \sum_{j=1}^{m} A_{i,j} \cos\theta_j + \sum_{j=1}^{m} B_{i,j} \sin\theta_j,
$$

with $\theta_j = \nu_j t_n + \varphi_j$ and the value t_n denotes the normalized time.

3 Method of resolution

First of all, in order to work with the equations (1) we need to expand the nonlinear functions of x, y and z in power series. This can be achieved easily by means of an algebraic manipulator and using Legendre polinomials.

Then, we are going to look for a quasiperiodic solution whose basic frequencies are the same ones of the perturbation, that is, a solution of the following form:

$$
\begin{aligned}
x &= \sum_{k \in \mathbf{Z}^4} x_1(k) \cos((k,w)t) + x_2(k) \sin((k,w)t), \\
y &= \sum_{k \in \mathbf{Z}^4} y_1(k) \cos((k,w)t) + y_2(k) \sin((k,w)t), \\
z &= \sum_{k \in \mathbf{Z}^4} z_1(k) \cos((k,w)t) + z_2(k) \sin((k,w)t).
\end{aligned} \tag{2}
$$

The problem is now to find the coefficients x_1, x_2, y_1, y_2, z_1 and z_2 of these series in order to satisfy the equations of motion. Our approach will be semianalytical, in the sense that we will not find exactly these coefficients but some numerical approximation to them. The method used is essentially to substitute the expressions (2) in the equations (1) and to make the computations analytically (using an algebraic manipulator), in order to obtain a system of equations for these coefficients that can be solved numerically by means of Newton's method (the jacobian matrix is also computed analytically using the manipulator). Note that this system has the degree of the expansion taken for the nonlinear terms of (1).

Due to the fact that the dimension of these systems is very large, we do not want to take into the Newton process all the terms appearing in this procedure but only the meaningful terms. The way this is done will be explained later.

To describe the algorithm in an easier way, we will use the following notation: let $\mathbf{x}$ denote an approximate solution of (1) ($\mathbf{x} = (x(t),\ y(t),\ z(t))$, $\mathbf{G(x)}$ denotes the residual acceleration corresponding to $\mathbf{x}$: $\mathbf{G(x)} = (G_1,\ G_2,\ G_3)$, where

$$
\left.
\begin{aligned}
G_1(x,y,z) &= f_1(x,y,z) - \ddot{x} = 0, \\
G_2(x,y,z) &= f_2(x,y,z) - \ddot{y} = 0, \\
G_3(x,y,z) &= f_3(x,y,z) - \ddot{y} = 0,
\end{aligned}
\right\}
$$

and the f_i are the right-hand sides of (1). With this, the equation we want to solve is $\mathbf{G(x)}=0$.

The algorithm is the following:

1. Take an initial condition $\mathbf{x}_0$ and a threshold value `tol`.

2. Let `M` be the initial set of terms for the Newton process.

3. $\mathbf{n} \leftarrow -1$.

4. $\mathbf{n} \leftarrow \mathbf{n}+1$.

5. compute $\mathbf{G(x_n)}$.

6. Let `N` be the set of terms of `G` whose amplitude is bigger (in absolute value) than `tol`.

7. if `N` is empty, $\mathbf{x_n}$ is the approximate solution we were looking for. Stop.

8. $\mathbf{M} \leftarrow \mathbf{M} \cup \mathbf{N}$.

9. Perform the Newton method, until the amplitude of the terms of `M` is less (in absolute value) than `tol`. The solution is called $\mathbf{x_{n+1}}$.

10. go to step 4.

The problem now is how to find a good initial condition for this process. In order to avoid this difficulty we have used a continuation method in the following way: first of all, we are going to solve the problem taking only linear terms in the expansion of the model equations. We select as `M` the empty set, and $\mathbf{x_0}$ equal to zero. With this, our algorithm performs some iterations adding terms inside the solutions (note that to perform step 9, the program needs only one iteration) until a solution of the linearized equations is found. Then, we take into account second degree terms in (1), and we start to solve this taking as initial condition the solution of the linearized problem. Due to the fact that second degree terms are big, we need to multiply them by a parameter, and we solve the problem for some intermediate values before taking the parameter equal to 1. Of course, the initial condition for each value of the parameter is the solution of the problem for the last value of the parameter. Once the second degree equations are solved, we can add the third degree terms, and so on.

The way to check this computations is to perform numerical integrations starting at a point given by the semianalytical solution and comparing the values of this integration and the ones of the semianalytical solution. This also give us a criterion to stop the later algorithm: when the agreement between the numerical integration of the model equations (1) and the semianalytical solution is good enough.

4   Stability

If we are interested in the local behaviour around this orbit, we can use a sort of Floquet theory for this case. Using an algebraic manipulator, we can substitute the expression of the solution found above inside the variational equations and to obtain

$$\dot{x} = (A + \varepsilon Q(t))x,$$

where A is a constant matrix, $Q(t)$ is a quasiperiodic matrix and ε is small (note that, in this case ε has a fixed value). Then it is possible (see [4]) to perform n quasiperiodic changes of variable and to get

$$\dot{y} = (\overline{A} + \varepsilon^{2^n}\overline{Q}(t))y. \tag{3}$$

This can be done except for a set of zero measure of values of ε. The convergence of this procedure to a constant coefficient system is proved in [4], but only for a cantorian set (with positive measure) of values of ε. For the rest of the values the method is asymptotic, but the first changes of variables still reduce the size of the quasiperiodic part. With this we can obtain the system (3), where the quasiperiodic part is small enough to apply Gronwall's lemma and to obtain that its behaviour is very similar to the one of $\dot{y} = \overline{A}y$ for very long time intervals. This is enough for practical purposes.

This procedure has the difficulty that is very hard to use and, it does not work properly if the quasiperiodic part is too large. Nevertheless, if we have in mind short time intervals (a few years) we can use easier methods. For instance, if we are interested in the behaviour during five years, we can compute the variational matrix for this time interval and its eigenvalues and eigenvectors give us the behaviour for this time interval. This is good enough if we are interested in keeping a spacecraft in this orbit (see [6]).

5   Real life system

With the later procedure we have obtained a quasiperiodic solution of the model equations. If we perform numerical integration of the real solar system (using, for instance, the JPL tapes to obtain the position of the bodies) we can see that the real orbit is close to the one found with our method, but not close enough to have a cheap (as cheap as possible) station keeping. If we want to improve this orbit we can use a parallel shooting method (for an example, see [6]) in order to find a real orbit with a similar behaviour that the one of the model equations. With this we get a solution of the real life system that looks like a quasiperiodic solution and it can be used to put the spacecraft there. Due to the fact that the nominal orbit found with this method is very good, the station keeping related to it will be cheap.

Finally, in order to perform the station keeping we can compute the unstable and the center-stable directions at the points of this orbit (for a given time span) using the numerical procedure given in the last section. Then, the control of the spacecraft can be done by killing the unstable component by a manoeuvre (essentially a jump in velocities) which lands on the center-stable manifold, when needed.

6   Conclusions

In this work we have considered equilibrium points under quasiperiodic perturbations and we have shown that there seems to exist a quasiperiodic orbit replacing the equilibrium point. This orbit has been computed for the L_4 point of the Earth-Moon system, but the method used is general enough to be applied to other libration points in the solar system.

ACKNOWLEDGEMENTS: The second and third authors have been partially supported by a GPC fund from MEC (Spain). The first and the fourth ones have been supported by CICYT Grant PB86-0527. Computer facilities were provided by CIRIT Grants. The methods presented here are being used in ESA project 8625/89/D/MD(SC).

References

[1] Díez C., Jorba A., Simó C.: A Dynamical Equivalent to the Equilateral Libration Points of the Real Earth-Moon System, preprint.

[2] Giorgilli A., Delshams A., Fontich E., Galgani L., Simó C.: Effective Stability for a Hamiltonian System near an Elliptic Equilibrium Point, with an Application to the Restricted Three Body Problem, *Journal of Differential Equations*, **77** (1989), 167-198.

[3] Gómez G., Llibre J., Martínez R. and Simó C.: Study on Orbits near the Triangular Libration Points in the perturbed Restricted Three-Body Problem, ESOC Contract 6139/84/D/JS(SC), Final Report, (1987).

[4] Jorba A., Simó C.: On the Reducibility of Linear Differential Equations with Quasiperiodic Coefficients, to appear in *Journal of Differential Equations*.

[5] Schultz, B. E. and Tapley, B. D.: Numerical studies of solar influenced particle motion near triangular Earth-Moon libration points, in Periodic Orbits, Stability and Resonances, Ed G. E. O. Giacaglia, 82-90, Reidel (1970).

[6] Simó C., Gómez G., Jorba A., Masdemont J.: Invariant Unstable Tori Computed by Lindstedt-Poincaré Method. Reduction to the Central Manifold and Applications to Space Flight Dynamics, Proceedings of the NATO/ASI held at Cortina d'Ampezzo, 1990.

[7] Szebehely, V. : Theory of Orbits, Academic Press (1967).

STABILITY ZONES AROUND THE TRIANGULAR

LAGRANGIAN POINTS

Rudolf Dvorak and Elke Lohinger

Institute of Astronomy, Vienna

1. INTRODUCTION

In the following we deal with the stability around the libration point L_4 (respectively L_5) in the circular restricted three body problem. The dependence of the largeness of the stability zones around these equilibrium points is established as a function of the mass parameter $\mu = m_2/(m_1 + m_2)$ using extensive numerical experiments. The results are compared in a first step to existing ones of the Earth-Moon case (McKenzie and V.Szebehely, 1981) and then extended to different values of μ. The shrinking and even disappearance of such zones is well explained by the existence of 2 additional critical mass parameters well below the well known value of $\mu_{crit} = 0.03852\ldots$ (A.Deprit and A.Deprit-Bartholomé, 1967).

2. THE RESTRICTED PROBLEM AND THE TRIANGULAR LAGRANGIAN POINTS

The model, which is studied here is the well known circular restricted problem of three bodies, with the two primary masses μ and 1-μ and a third body of infinitesimal mass. In the plane of the restricted problem there exist five stationary solutions known as the equilibrium points. Three of them are called collinear equilibrium points (L_1, L_2 and L_3), they are located along the axis of syzygies. The others form together with the primaries two equilateral triangles; they are called triangular equilibrium points L_4 and L_5. Whereas the collinear stationary points are unstable, the triangular points allow stable motion under certain conditions.

Much work has been done since the profound study by C.V. Charlier (1899) where he solved the linearized problem of motion of the third body close to L_4 and L_5. Both the stability of periodic orbits and the stability of the Lagrangian points have been studied up to the 3rd order by several people. Most of the important work is summarized in K.Stumpff (1965) or more detailed in V. Szebehely (1967) and recently by A.E. Roy (1988). One of the essential points is the existence of an upper limit of the mass parameter ($\mu_{crit} = 0.03852\ldots$) for the stability of the triangular libration points. This corresponds to the existence of real eigenvalues of the linearized problem.

Another interesting feature, as already pointed out by Charlier, is the existence of explicit solutions of the linearized problem. One can find two different periodic solutions around

Predictability, Stability, and Chaos in N-Body Dynamical Systems
Edited by A.E. Roy, Plenum Press, New York, 1991

L_4 – a long periodic orbit and a short periodic one. The ratio of axes (γ) of these resulting ellipses can be given as functions of the mass parameter μ: One ratio is $\gamma = \frac{1}{2}$ and therefore constant, and the other depends explicitly on μ with $\gamma \sim \frac{1}{2}\sqrt{3\mu}$. This relation shows, that the 2nd type of the periodic solutions become more and more elongated as μ decreases (e.g. for the Earth-Moon system $\gamma = 1/5$ and for the Sun-Jupiter system $\gamma = 1/19$). These results will be used in the last section.

3.   THE EARTH-MOON CASE AND THE NUMERICAL INTEGRATION METHOD

In a paper by McKenzie and V.Szebehely (= MS) published 1981 the regions of stability around the triangular equilibrium points of the Earth-Moon case were established numerically.The mass ratio in this case is equal 0.0121286 and therefore below the critical value. This means, that particles placed without velocity with respect to the equilibrium point will librate around them. Two qualitatively different motions were distinguished – Librators, which stay in the "vicinity" of the stationary points and orbits which do intersect the axis of syzygies, which are classified as unstable ones. Although it is known that special periodic orbits can cross this axis and stay in a libration around both equilibrium points (Rabe, 1961) MS kept their stability definition. This is due to the fact that in their numerical experiments after such an intersection with the x-axes the particles will suffer from a close approach to one of the primaries.

MS tested carefully the time interval of integration to be able to determine such stability zones around the libration points and they found out that about 80 periods would be the right integration time to choose. As the grid size they fixed 0.005 non-dimensional units in the x- and y- coordinate around the stability points; the distance between Earth and Moon was taken as unity. Their results will be shown in fig. 1 and will be discussed together with ours.

For our numerical investigation we had to choose an efficient integration method and to define the stability of librational motion. The stability conditions were chosen in such a way to be "precise" enough to give a qualitatively correct picture of motion. We had the same problem already in other numerical experiments concerning stability of planet-type orbits in binary stars, where this is discussed much more in detail (Dvorak 1986). For the study here we adapted the following stability definition:

An orbit is stable when it stays within a sphere of 0.86 units of the distance of primaries around L_4 during the whole integration time of 200 periods of the primaries.

Parallel we used the Lie-Integrator (Hanslmeier and Dvorak, 1984, Delva 1986 and Lichtenegger, 1987) and the Bulirsch-Stoer adapted by J.Kribbel (1987). As a grid size we took 0.005 units of the distance of primaries in the rotating x-y frame and could therefore directly compare our results with those of MS. Evidently it is a very good test of a numerical method, when results can be compared to ones derived by different authors using also different numerical integrators. As the stability region around L_4, which we have found by own calculations is quite similar to the one of MS, we could be sure about the correctness of our stability definition Fig.(1) and (2) show the region of motion around L_4. Fig.(1) represents the results derived by MS; fig.(2) resulted from own calculations.

It is evident that the features of the stability regions are very similar. Not only the area as a whole is of the same size, even the shape is surprisingly equivalent. The finger like feature

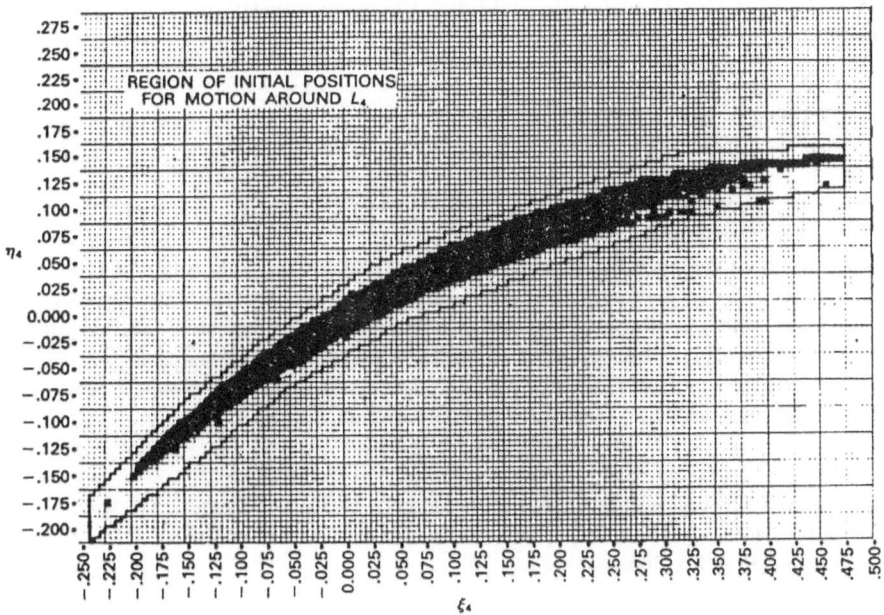

Figure 1. Stability zone around L_4 derived by MS. The black region marks all initial positions of stable orbits. The grid size is 0.005 units in both axes, which are centered at the equilibrium point.

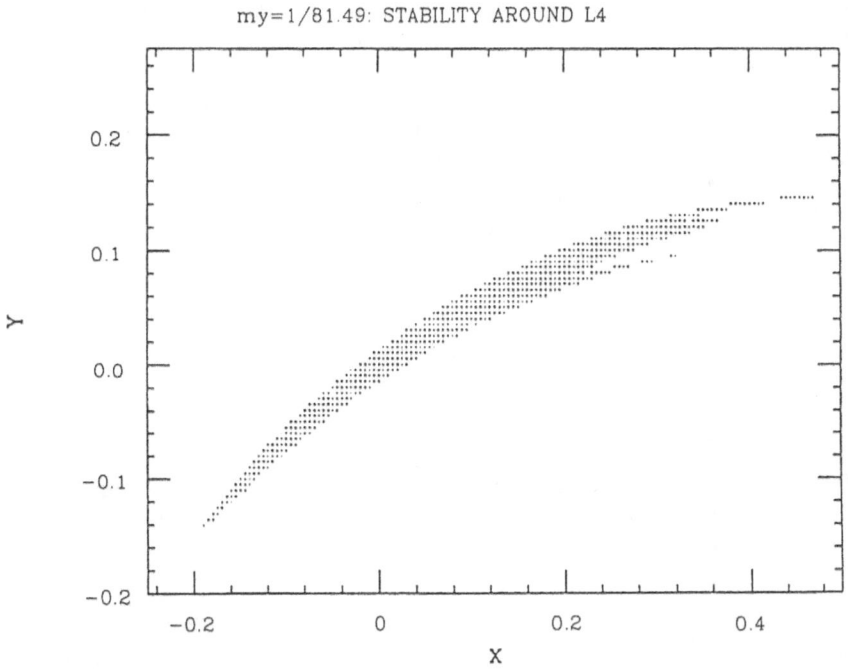

my=1/81.49: STABILITY AROUND L4

Figure 2. Stability zones around L_4 derived by own calcultions. The initial position of all stable orbits are marked by dots. The grid size is also 0.005 units in both axes, which are centered at the equilibrium point.

of the stability border (upper right side) and the spike directly below are well visible in both figures.

4.   RESULTS AND DISCUSSION

In the sequence of the fig.(3) and (4) we see the development of the size and shape of the stable zones around L_4 for decreasing values of the mass parameter μ. Every initial position for which the particle stays in libration near L_4 is marked by a point, an unstable one is not marked at all. But it should be emphasized that a large number of orbits well outside of the stable zone was integrated. In general it was expected that the size A (= number of stable orbits) will increase when the mass ratio is decreasing form μ_{crit}, (as we approach more and more the 2-body problem, where all orbits are stable). More detailed one can deduce the following important properties from fig.(3) and (4):

1st the shape of the region is more and more elongated. This corresponds to the fact, which was mentioned in the introduction: the ratio of the axes of the libration ellipses becomes smaller with decreasing μ.

2nd we have found two minima of the size A between $\mu = \frac{1}{26}$ and $\mu = \frac{1}{81}$. This feature will be discussed more detailed in connection with fig.(6).

3rd there is a slight decrease concerning the largeness of the stability zone between $\mu = 0.01$ and 0.001. This fact may be explained by the stability definition chosen in this work. When we start in the 2-body problem ($\mu = 0$) with a frequency slightly different from those of the triangular point the resulting orbit will be unstable in our definition. The fact, that we found such a good agreement with the work of MS showed the unsensibility of the qualitative picture on the chosen stability definition in the interesting range of $0.0125 \leq \mu \leq 0.01$. We therefore did not change our stability definition, although we checked some others.

In fig.5 the number of stable orbits that we found for the calculated mass ratios are plotted. The line represents the least square fit neglecting the two minima mentioned above. Fig.6 shows the most interesting part of fig.5 without smoothing. For the mass ratio μ (which was the parameter in our computations) equal 1/41 only L_4 was found to be stable. For the value of 1/76 the region is smaller than for 1/71 and for 1/66 and the size A starts to increase again from this value on.

A.Deprit and A.Deprit-Bartholomé (1967) checked in a theoretical work the stability of the Lagrangian triangular points and found three values of the mass parameter μ where they could not establish stability. Two of them are $\mu = 0.024293$ and $\mu = 0.013516$ which correspond well to the minima of fig.6 at 1/26 and close to 1/71.

Finally we can conclude that using the results of the numerical experiments we suceeded quite well in finding stability regions around the equilateral points in the restricted three body problem as functions of the mass parameter μ of the primaries. An analytical explanation of the largeness depending on it is still in progress and will make use of an additional quasi integral of motion in the sense derived by Hagel (1990).

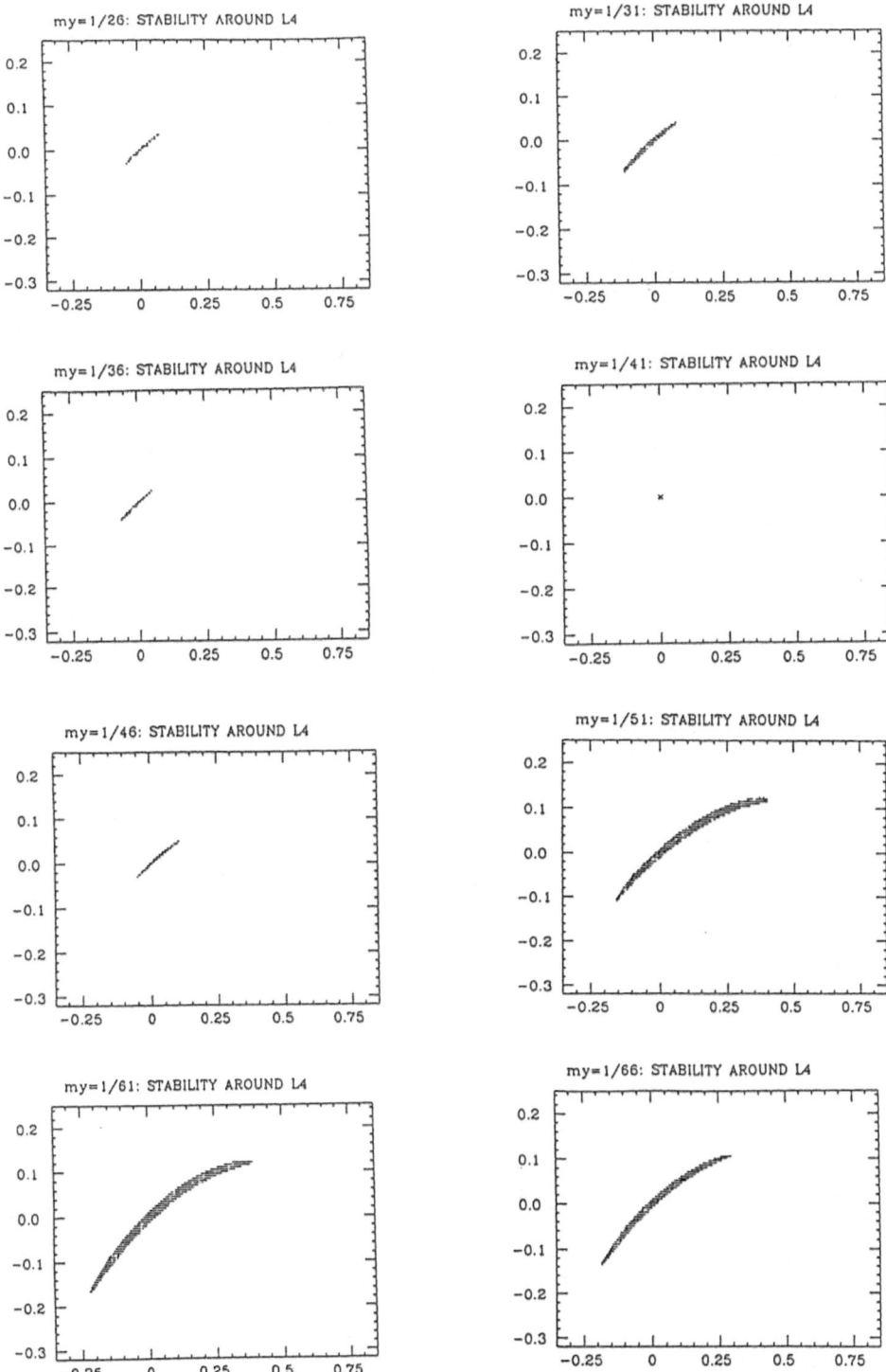

Figure 3. The development of the stability zone around L_4 for several mass ratios

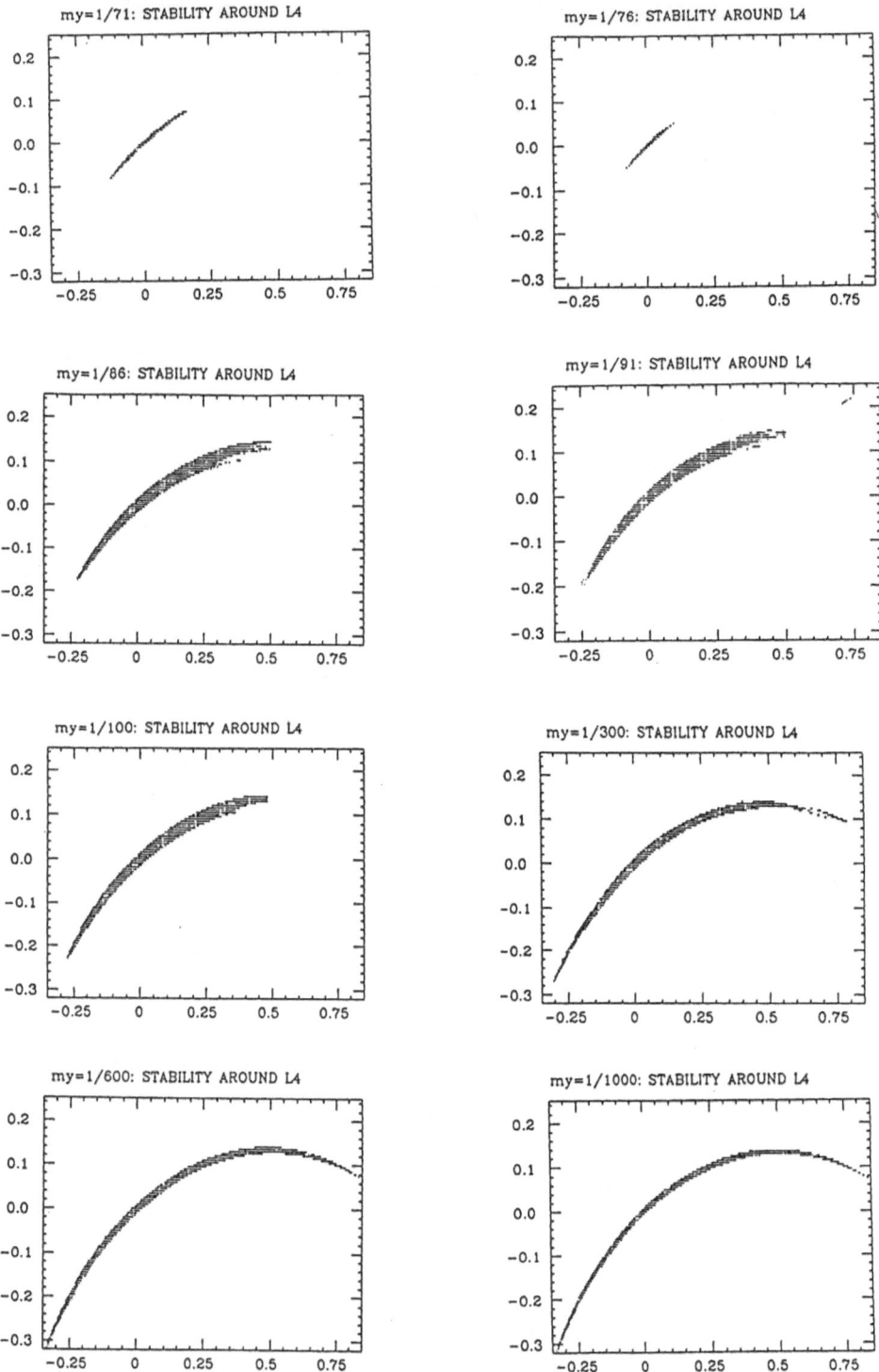

Figure 4. The continuation of the development of stable zones around L_4.

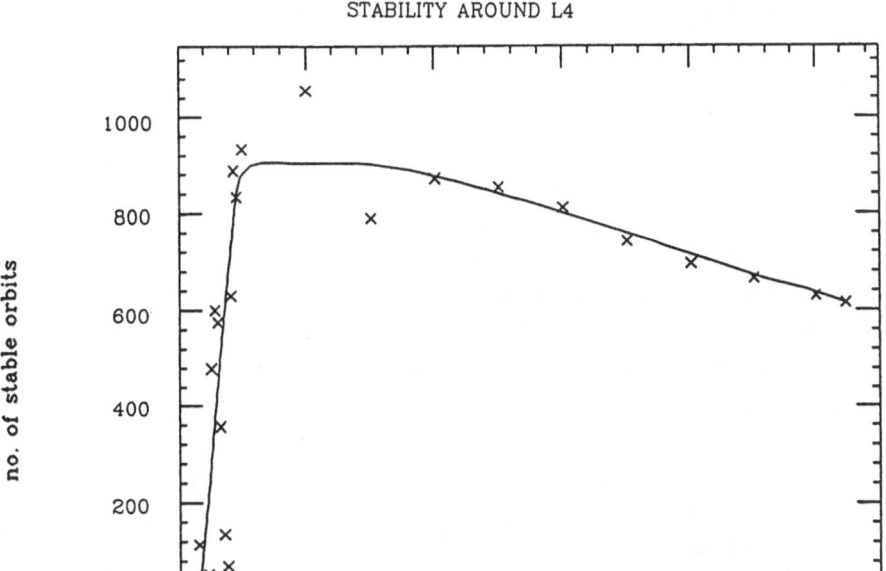

Figure 5. Size of the stability region around L_4 for all mass ratios calculated. The line was found by a least square fit.

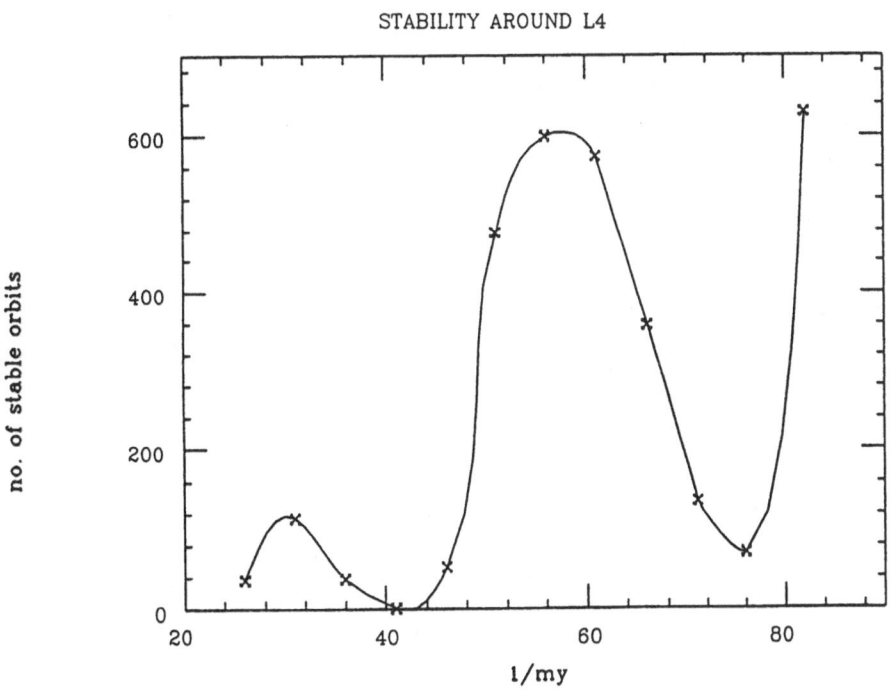

Figure 6. The enlargement of fig.5 for $\mu \geq \frac{1}{81}$

5. BIBLIOGRAPHY

Charlier, C.L.: 1907, Die Mechanik des Himmels, Bd.II, Verlag Veit & Comp., Leipzig, p.117 ff

Delva, M.: 1985, Astron. Astrophys. Suppl.60, 277

Deprit, A., Deprit-Bartholomé, A.: 1967, Astron. J. 72, 173

Dvorak, R.: 1986, Astron.Astrophys. 167, 379

Hagel, J.: 1990 (private communication)

Hanslmeier, A.,Dvorak, R.: 1984, Astron.Astrophys.132, 203

Kribbel, J.: 1987, Diplomarbeit, Universität Wien

Lichtenegger, H.: 1984, Dissertation, Universität Graz

McKenzie, R., Szebehely, V.: 1981, Cel. Mech. 23, 223

Rabe, E.: 1961, Astron. J. 66, 500

Roy, A. E.: 1988, Orbital Motion, Adam Hilger Press, p.125 ff

Stumpff, K.: 1965, Himmelsmechanik, Bd.II, VEB Deutscher Verlag der Wiss., Berlin, p.113 ff

Szebehely, V.: 1967, Theory of Orbits, Academic Press, N.Y., p.231 ff

CHAOTIC TRAJECTORIES IN THE RESTRICTED

PROBLEM OF THREE BODIES

R.H. Smith and V. Szebehely
University of Texas
Austin, TX 78712

Abstract

A complete qualitative understanding of the solutions of a set of nonlinear differential equations requires an investigation of the chaotic properties of the system. The circular restricted problem of three bodies has a long and rich history of qualitative analysis yet few studies have examined the possible existence of chaos in this problem. This study concentrates on the advent of chaos for widely different points in the phase space which correspond to closed Jacobian curves. In addition to observing the non - periodicity of the trajectory, two methods, Liapounov Characteristic Numbers and Poincaré Surface of Sections, are used to classify an orbit as chaotic. Regularized and unregularized equations of motion were used in conjunction with a variable - stepsize integrator to produce the orbits.

1 INTRODUCTION

Due to the non - integrable nature of the restricted problem of three bodies, a large portion of its history has been devoted to the study of special cases. The existence of the zero velocity curves, their restriction on the motion of the third body, and the trajectories about the libration points have been throughly investigated. Volumes have been written on the periodic orbits by members of the scientific community such as Strömgren, Darwin and Moulton. Yet all the aims of these studies have been to find periodic and quasi - periodic orbits, i. e. order in the system. Recently, nonlinear dynamics has undergone a transition which has shifted the emphasis of its study from strictly looking for order to a process that also investigates stochastic behavior. KAM theory, Hénon - Heiles conservative systems and Arnold diffusion have become standard terms in celestial mechanics. Also, several physical examples of chaotic trajectories, such as the motion of Hyperion and the 3/1 Kirkwood gap found by Wisdom [1], have been discovered. Yet in the hundreds of papers on chaos in the literature, only a handful deal with the circular restricted problem of three bodies. Hénon's extensive study [2] of the problem dealt with a complete description of the phase plane when studying the Copenhagen category of orbits. Jeffery's work on the problem has also been extensive as demonstrated in his atlas of Poincaré Surface of Sections [3] and his other study of Liapunov Characteristic Numbers [4]. As can be expected from the degree of complexity of the phase plane, these studies did not contain all the possible permutations of initial conditions. It is the aim of this investigation to expand on these works and demonstrate the existence of other chaotic trajectories.

"Chaos" is a widely and loosely used term that can indicate many things. In this study the chaotic properties of an orbit will be defined by its Liapunov Characteristic Number, its Poincaré Surface of Section and its apparent non - periodicity. The specifics of those evaluationg techniques are explained in the following sections.

2 LIAPUNOV CHARACTERISTIC NUMBERS

The Liapunov Characteristic Numbers (LCN) give an indication of the behavior of a trajectory by applying disturbances to the trajectory and integrating the disturbed trajectory forward in time. A discrete version for calculating the largest Liapunov Characteristic Number L_1, otherwise known as the Kolmogorov entropy, is given here.

The process begins by disturbing the initial conditions of the orbit that is being tested as seen in figure 1. Following that perturbation d_o, the trajectory is integrated

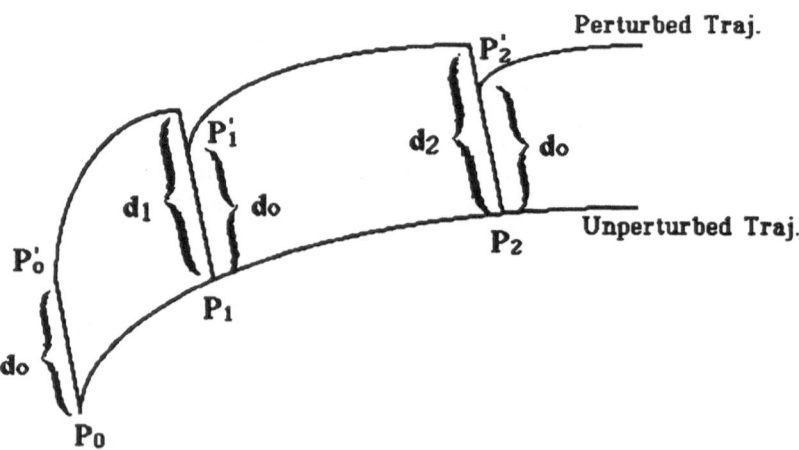

Figure 1. Computation of Liapunov Characteristic Numbers

forward in time until it reaches the next sampling time that corresponds to P_1 on the unperturbed trajectory.(Note that when the term "perturbed" is used, it refers to the numerical process and not the more common usage in celestial mechanics). The distance between the perturbed and unperturbed trajectories is determined and the process starts over again by perturbing the trajectory by d_o and integrating from $P_o{}'$. If this process is calculated for a very large number of times, the Kolmogorov entropy is expressed by,

$$L_1 = \lim_{n \to \infty} \frac{1}{n\tau} \sum_{i=1}^{\infty} ln\frac{d_i}{d_o}$$

where τ is the time interval between sampling points. The properties of Liapunov Characteristic Numbers are well known for a Hamiltonian system. Hamiltonian dynamical systems have LCNs that form opposite pairs such that

$$L_1 \geq L_2 \geq \ldots \geq L_n \geq -L_n \geq \ldots \geq -L_2 \geq -L_1$$

Furthermore, each converging integral or global integral of a Hamiltonian system corresponds to a pair of zero LCNs. By knowing this property, the integrability of a system is seen by the number of first integrals it has. For the purposes of this study, Kolmogorov entropy properties are the most important. A positive value of L_1 indicates that the region of the initial condition space is ergodic while a zero value (for Hamiltonian systems) indicates a neutral stability and regular quasi - periodic motions.

3 POINCARÉ SURFACE OF SECTIONS

In recent years, a technique developed by Poincaré has been used in conjunction with the newly acquired high - speed computational capabilities of the computer. This technique, while not limited to two degree of freedom systems, is most useful for lower dimensional systems. To plot the Poincaré section, the numerical integration of the trajectory must first be performed. This information, a four - dimensional flow in phase space $x, y, \dot{x}, \dot{y}$, can be plotted in two - dimensional space if two of the variables are constrained in some way. A simple way of doing this is to plot the values of x and $\dot{x}$ for each time $y = 0$ and $\dot{y} > 0$. This surface of section contains information about the motion of the body. If the transformation results in an invariant point, then a periodic motion has developed. If the points form a smooth curve, quasi - periodic orbits have been found. Finally, if the plane is covered with points, the motion is said to be random or chaotic.

4 THE PLANAR CIRCULAR RESTRICTED THREE-BODY PROBLEM

4.1 Equations of Motion

Using Szebehely's notation [6], consider a rotating coordinate system (x, y) where the primaries, m_1 and m_2, are located along the x-axis. The circular restricted problem of three bodies have the equations of motion,

$$\ddot{x} - 2\dot{y} = \Omega_x$$
$$\ddot{y} + 2\dot{x} = \Omega_y$$

where dot ($\dot{}$) denotes derivative with respect to time, $()_{x_i}$ denotes partial derivative with respect to that variable and,

$$\Omega = \frac{1}{2}[(1-\mu){r_1}^2 + \mu {r_2}^2] + \frac{(1-\mu)}{r_1} + \frac{\mu}{r_1}$$

The expressions for r_1 and r_2 in barycentric coordinates are

$$r_1{}^2 = (x - \mu)^2 + y^2$$

$$r_2{}^2 = (x + 1 - \mu)^2 + y^2$$

and in coordinates whose origin is located at m_1, they are

$$r_1{}^2 = x^2 + y^2$$

$$r_2{}^2 = (x + 1)^2 + y^2.$$

Both these equations contain the mass parameter μ which is the ratio of m_2 to the total mass. This system also has an integral, the Jacobian integral, which is given by,

$$\dot{x}^2 + \dot{y}^2 = 2\Omega - C.$$

The use of two coordinate systems is strictly for ease in graphical interpretation. For initial conditions where $C > C_2$ (and C_2 is value of the Jacobian constant corresponding to the colinear equilibrium point located between the two primaries), the primary - centered coordinates will be used. For the cases of $C_2 > C > C_1$, barycentric coordinates are employed.

Table 1. Typical Liapunov Characteristic Number Results

$\mu = 0, C=4, y = \dot{x} = 0$		
x position	vel. angle	L_1
0.1	90 deg.	$\rightarrow 0$
0.1	90 deg.	$\rightarrow 0$
0.1	90 deg.	$\rightarrow 0$
0.1	90 deg.	$\rightarrow 0$
0.3	45 deg.	$\rightarrow 0$
0.3	135 deg.	$\rightarrow 0$
0.3	225 deg.	$\rightarrow 0$
0.3	315 deg.	$\rightarrow 0$

5 RESULTS

As stated earlier, the aim of this study was to expand on some of the earlier work on this subject and to look at the possibility of generating families of chaotic orbits. This investigation was accomplished by studying regions of the phase plane confined to 1) $C > C_2$ and 2) $C_2 > C > C_1$. This division allows the systematic exploration of the effect of constraining the motion about a single primary. The motivation for this decision was that in previous studies chaotic regions of significant size have not been found for the case when motion is constrained to one primary. Additionally, it is generally believed that sizable regions of chaos are found in non - integrable systems as the energy of the system increases.

5.1 Motion confined to one primary

The exploration of this region of the phase plane was performed to answer the question: Do significant areas of chaos when motion is restricted to the bounded region around one primary? The search for these regions of chaos consisted of varying a multitude of parameters. Following the work done by Jeffery's in [4], five values of μ were used, $0.00095388, 0.1, 0.2, 0.3, 0.4$ for each primary and also 0.5. Additionally, the ten different positions were taken about each primary with velocities that corresponded to five different values of C above C_2. The direction of the velocity was also varied for each set in five different ways. In addition to these numerical experiments, several different versions of the case where the third body hits the zero velocity curve were performed.

Due to the large number of calculations performed in this study, a complete display of them would be overly cumbersome. Therefore, only representative results are displayed. Poincaré Surface of Sections and the corresponding Liapounov Characteristic Numbers for the set of results corresponding to $\mu = 0.1$ are found in figures 2-3 and table 1. These results are for the case where $C = 5$ and the third body's initial conditions are varied. Figure 2 presents a set of initial conditions where the position along the axis of symmetry is varied while $\dot{x} = 0$. Figure 3 exhibits the situation where the position is fixed at $x = 0.3$ and the velocity vector angle is varied. Recall that the x coordinate is measured in a primary centered coordinate system where x positive is to the right of the larger primary.

5.2 Motion about both primaries

When the zero velocity curves open at L_2 to allow motion between the two primaries, considerable complexity is added to the problem. In this study, an attempt to create a family of chaotic trajectories in a fashion similar to that which created the quasi - periodic orbits discussed above was made. Families of orbits are typically generated with initial conditions of the third body on the x - axis and a velocity perpendicular to that axis. This investigation started the chaotic family with the initial conditions of the third body on the x - axis with the velocity directed along the negative x- axis. All of the trajectories were created with $\mu = 0.5$ to observe the behavior when the system has the largest perturbations.

One of the chaotic trajectories in this family (surrounded by dots which represent the zero velocity curve) is located at the top of figure 5. This trajectory is clearly complex. It is complex both in space (as seen in the figure) and in time as a time series analysis can show. An interesting feature of this trajectory is shown in the remaining portions of this figure. The trajectory is displayed at various intervals of canonical time units. During the first interval (0 - 40), the trajectory is characterized by "spike - like" movement around one primary and smooth ovals around the other primary. In the next interval, that pattern is not found. Yet in the third interval, the pattern returns, but it is in the mirror image of the first time interval. If further time intervals were displayed, one would observe this "spike-and-oval" pattern many more times, but the recurrence of this pattern does not happen at a recognizable rate.

From this one chaotic trajectory, many others were derived. Once again, the third body's parameters of position, velocity vector angle, and Jacobian constant were varied. The Poincaré Surface of Sections for some of these trajectories are displayed in figure 6. In the bottom two plots (those which varied the velocity vector angle and position initial conditions), the surface of sections consist of widely dispersed points. This situation, as stated earlier, is an indicator of chaos. The top plot of figure 6 is of particular interest because it shows the transition to chaos. The only parameter varied in this plot is the

Jacobian constant. For $C = 4.5$, the zero velocity curves are closed and do not allow motion between the two primaries. This value of Jacobian constant is represent by the "+" symbol and its quasi - periodic motion can be observed by the smooth curve that could be drawn through its points. By $C = 4.15$, motion is allowed and the dispersed points of the surface of section demonstrate that the trajectory has become chaotic. The advent of chaos for this family does not follow the more common smooth transitions to chaotic motion found in other physical systems, but rather abruptly switches from quasi - periodic motion to a very complex one.

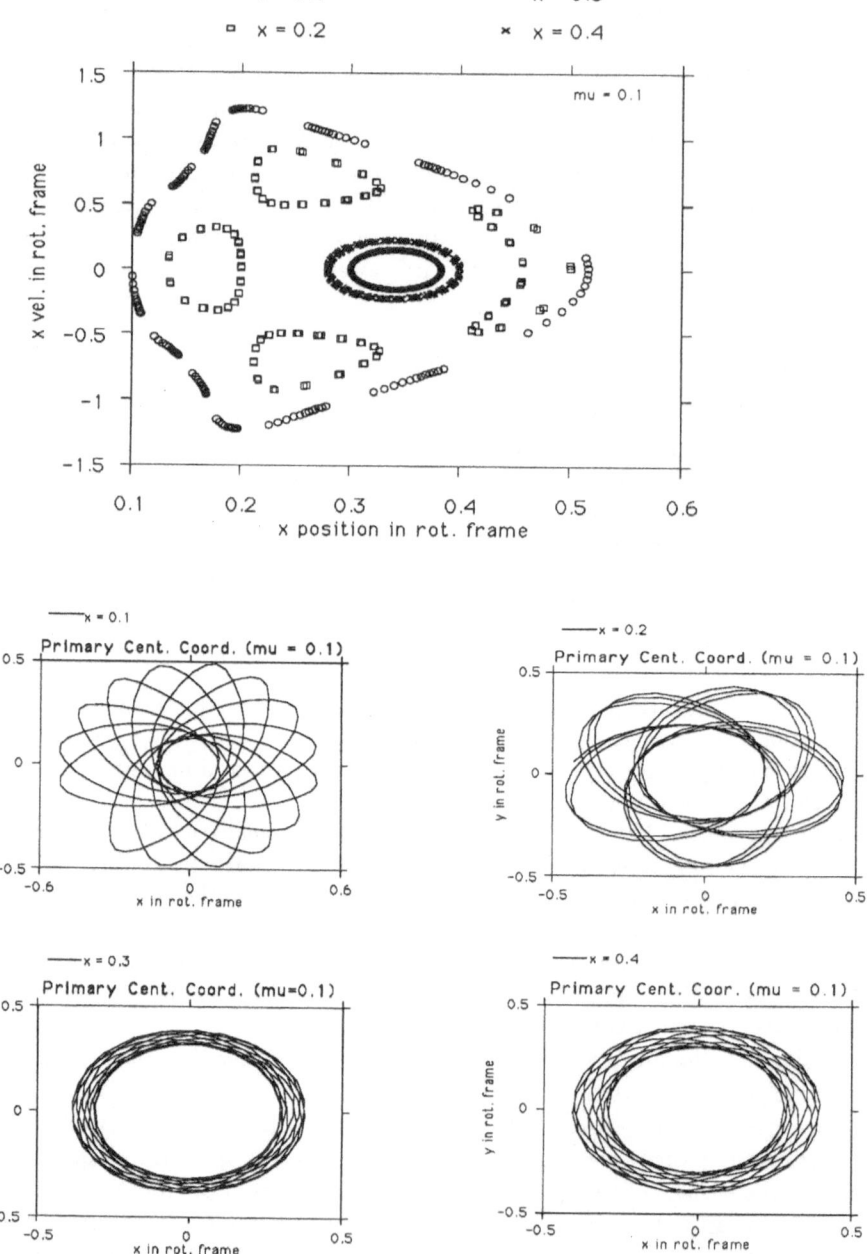

Figure 2. Varying position study Poincaré Sections and Trajectories

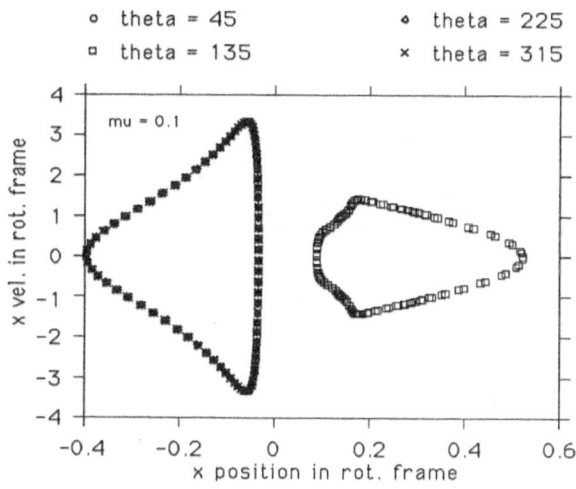

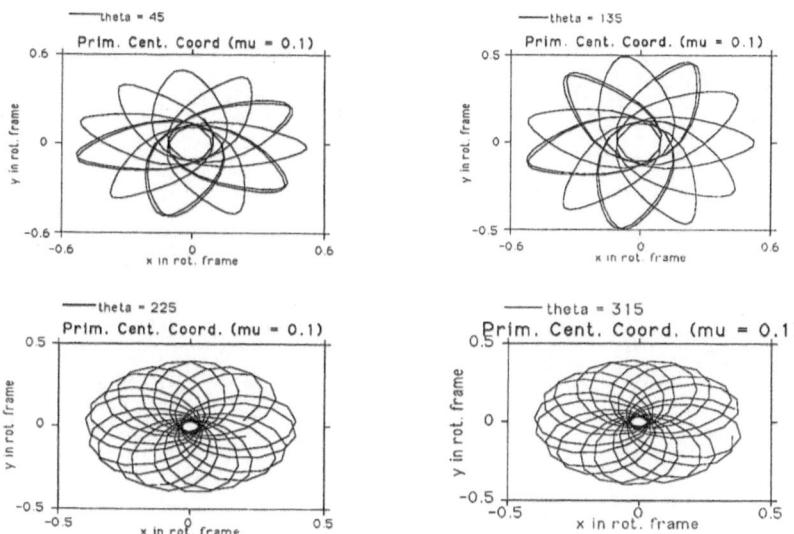

Figure 3. Varying velocity vector angle study

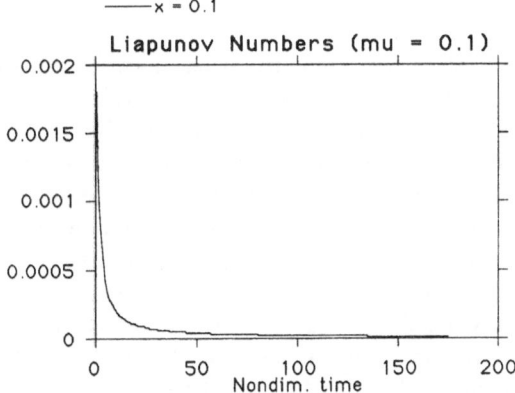

Figure 4. Liapunov Characteristic Number's decrease with time.

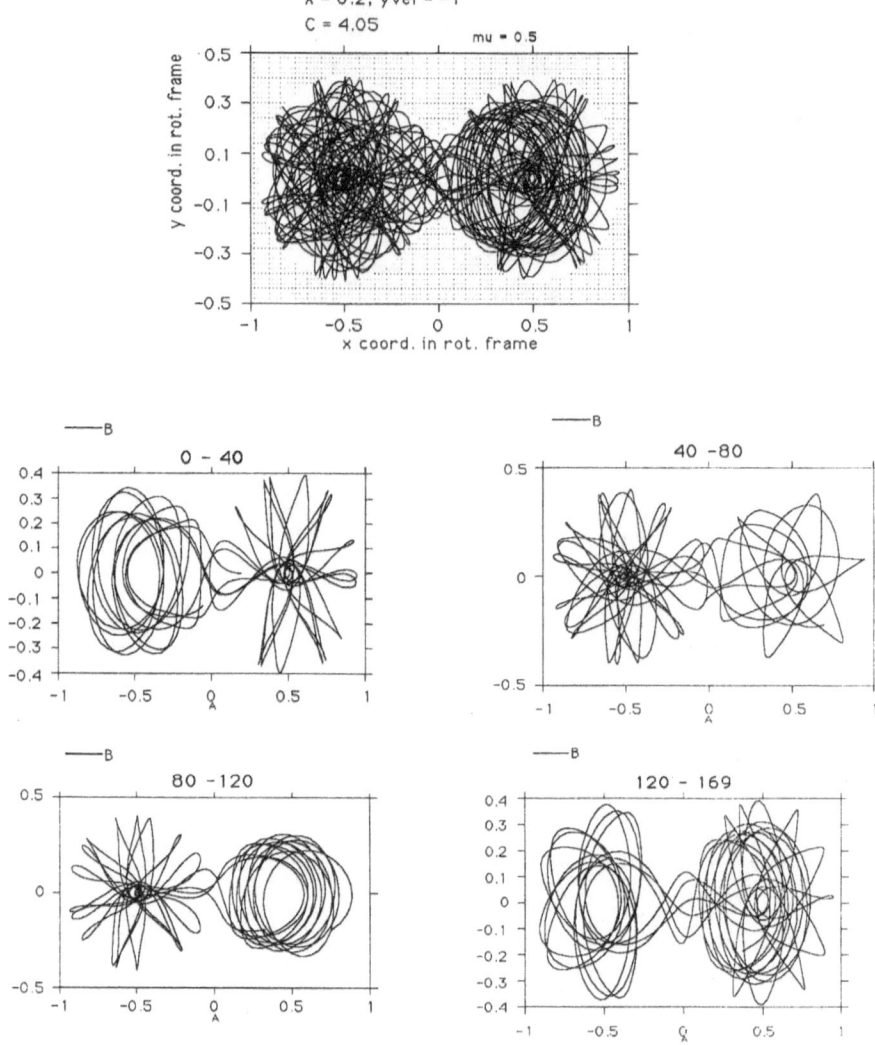

Figure 5. A chaotic trajectory

6 CONCLUSIONS

A study of regions of chaos in the phase plane has been made for the lower energy values of the restricted problem of three bodies. The totality of these calculations led to the expected result that substantial regions of chaos could not be found for $C > C_2$. For values of the Jacobian constant that permitted motion between the two primaries, a new family of chaotic trajectories was found. Additionally, some order was found in these chaotic trajectories which manifested itself in the recurrence of a pattern about the two primaries. It is conjectured that this motion is a demonstration of Poincaré's statement [5] that quasi - periodic motion is dense in the phase plane. As the chaotic trajectory evolves in time, the body is carried near quasi - periodic points in the phase plane and thereby mimics the motion of that point as close as it can. This statement is only a conjecture and should be confirmed by further investigation.

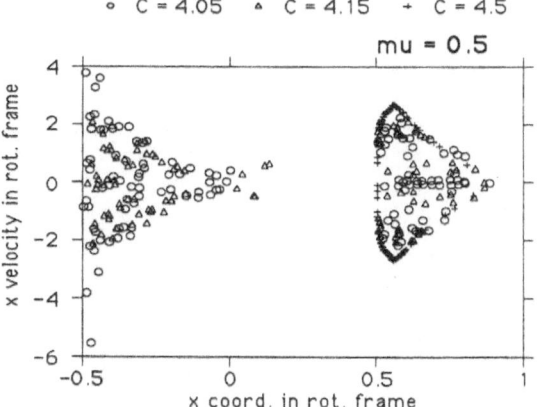

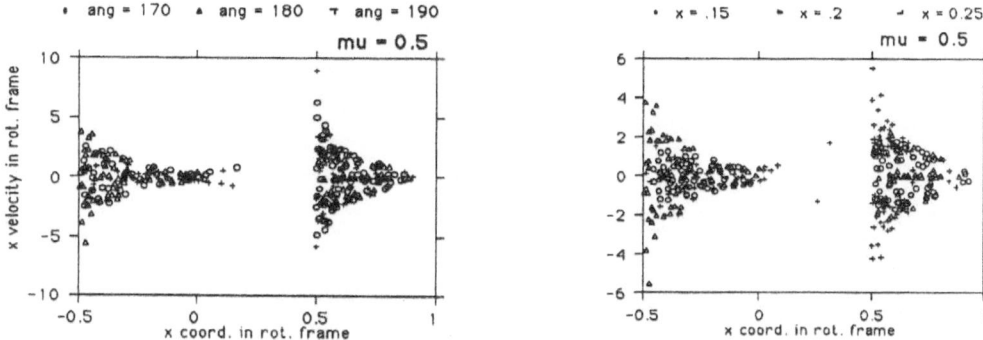

Figure 6. Poincaré Sections for different parameter variations

REFERENCES

[1] Wisdom, J.: 1987, *Icarus* **72**,241 - 275.

[2] Hénon, M.: 1966a, *Bull. Astron.* (3) 1, Fasc. 1, p. 57.

[3] Jefferys, W. H.: 1971, *'An Atlas of Surfaces of Section for the Restricted Problem of Three Bodies'*, Publications of the Department of Astronomy of the University of Texas as Austin, Ser. II., 3, 6.

[4] Jefferys, W. H. and Yi, Z.: 1983, *Celestial Mechanics* **30**, 85 - 95.

[5] Poincaré, H.: 1892-1899, *Méthods nouvelles de la mécanique célese*, 3 Vols. Gauthier-Villars, Paris. Reprinted by Dover, New York, 1957.

[6] Szebehely, V. G.: 1967, *Theory of Orbits*, Academic Press, New York.

NEW FORMULATIONS OF THE SITNIKOV PROBLEM

Karl Wodnar

Institute of Astronomy
University of Vienna, Austria

1. CONFIGURATION AND EQUATION OF MOTION

We consider the following configuration: Two primaries m_1, m_2 of non zero equal masses move around each other on congruent complanar ellipses of eccentricity ε, while the massless planet m_3 performs motion along an axis perpendicular to the primary orbit plane through the common barycenter of the primaries (see fig. 1). The system thus represents a special case of the three dimensional elliptic restricted problem of three bodies.

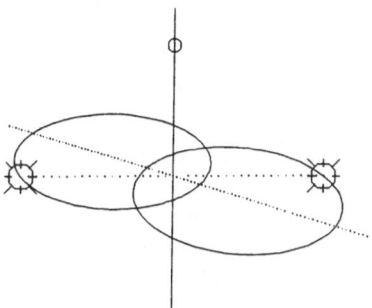

Figure 1. The Sitnikov configuration

From geometric considerations we derive the following equation of motion

$$\ddot{z} + GM \; \frac{1}{\sqrt{r(\varphi)^2 + z^2}^{\,3}} \cdot z = 0 \tag{1}$$

where z is the distance of m_3 from barycenter measured along the system axis, $M = m_1 + m_2 = 2m_1$ denotes the primary mass sum, G is the gravitational constant and

$$r(\varphi) = \frac{p}{2(1 + \varepsilon \cos \varphi)}$$

is the distance of primary m_1 to the barycenter which is of course the same for m_2. In this expression $p = a\left(1 - \varepsilon^2\right)$ is the primary ellipse parameter and a denotes the corresponding semi major axis. For this notation the ellipse of *relative* motion of m_1 w.r.t. m_2 is relevant.

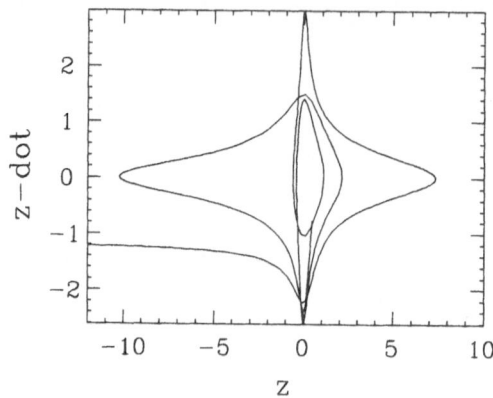

Figure 2. $(z, \dot{z})$ phase space projections

Dots have the usual meaning of derivation with respect to the time variable t. Figure 2 shows typical $(z, \dot{z})$-projections.

It is remarkable that in the case $\varepsilon = 0$, where the primaries move with constant angular velocity on opposite sides of one and the same circle around the axis of motion of m_3 we have two oscillations of different frequency – the primary orbital frequency and the small body's one – which are dynamically – i.e. w.r.t. the force of attraction acting on m_3 – completely decoupled. Yet via Kepler's third law there is a mathematical relation between these two frequencies. In the case of small and positive ε we are at the onset of chaos due to interference – and thereby shifting etc. – of this pair of frequencies.

2. HISTORICAL REMARKS

In this chapter we rely primarily on the exposition of Stumpff (1965).

The first treatment of the system was confined to the case of $\varepsilon = 0$, where the primaries move uniformly on opposite sides of a circle. In this case adopting appropriate units we derive as equation of motion due to constant $r(\varphi) \equiv \frac{1}{2}$:

$$\ddot{z} + \frac{z}{\sqrt{1/4 + z^2}^3} = 0 \tag{2}$$

which is deducable from the time independent Hamiltonian (Goldstein 1966)

$$H^{(0)}(z, \dot{z}) = \frac{\dot{z}^2}{2} - \frac{1}{\sqrt{1/4 + z^2}} \equiv H_0^{(0)} := H^{(0)}(z_0, \dot{z}_0) = const \tag{3}$$

confined to values $H^{(0)} \geq -2$ as is easily seen from the latter equation. While superscript (0) indicates $\varepsilon = 0$ subscript 0 denotes initial conditions. Resolving (3) with respect to dt leads to the following representation of the inverse of the solution $z(t)$ by means of an elliptic integral

$$t = t_0 + \int_{z_0}^{z} \frac{d\zeta}{\sqrt{2(H_0^{(0)} + 1/\sqrt{1/4 + \zeta^2})}} \tag{4}$$

This is also derivable for a special choice of parameters from the two fixed center problem solution as the primary rotation is mathematically neglectible regarding the movement of m_3.

An approach using elliptic integrals was done first by Pavanini (1907) [1] who also considered non equal masses of the primary bodies. For values of $H^{(0)} > 0$ we have hyperbolic, for $H^{(0)} = 0$ parabolic and for $H^{(0)} < 0$ elliptic orbits of m_3 which of course degenerate to part of a straight line. The elliptic case $-2 \leq H^{(0)} < 0$ leads to periodic solutions with maximum elongation $z_{max} = \sqrt{1/H_0^{(0)\,2} - 1/4}$ obtained from (2) setting $\dot{z} = 0$ and period

$$P = \frac{\pi}{\sqrt{2}}\left(1 + \frac{9}{16}\mu + \frac{489}{1024}\mu^2 + \frac{7097}{16384}\mu^3 + \quad \ldots \quad \right) \quad \text{with} \quad \mu = 1 - H_0^{(0)\,2}/4 \ . \tag{5}$$

W.D. Mac Millan (1913) gave the following Fourier representation of the solution for $H_0^{(0)} < 0$:

$$\begin{aligned}
z(t) \ = \ &z_{max}\,\{\sin\nu t + \frac{3}{64}\mu(\sin\nu t + \sin 3\nu t)+ \\
&\frac{1}{4096}\mu^2(79\sin\nu t + 108\sin 3\nu t + 29\sin 5\nu t) + \quad \ldots \quad \} \\
&\text{with} \quad \nu = 2\pi/P \quad \text{and} \quad z(t{=}0) = 0
\end{aligned} \tag{6}$$

so that the case $\varepsilon = 0$ is usually referred to as the Mac Millan case of the restricted three body problem. Equations (2)-(6) are taken from Stumpff (1965) and have been rescaled to fit our notation.

For the case of non zero ε classes of orbits were investigated first by Sitnikov (1960), who associated a sequence of integers with a solution of (1) in the following way: counting the number of full primary revolutions between two successive zeroes of $z(t)$ – i.e. passages of m_3 through the plane of primary motion – and dropping the noninteger part of this revolution number gives a non negative integer. Ordering these numbers according to their respective time succession establishes the socalled Sitnikov sequence. For non escape orbits we obtain a doubly infinite sequence $(s_k)_{k \in \mathbb{Z}}$ (type I). If we deal with an escape but non capture orbit the half infinite sequence (type II) terminates on the right side with $s_{k_{end}} = \infty$. Accordingly we find also capture and non escape orbits with half infinite (s_k) starting with $s_{k_{begin}} = \infty$ on the left side (type III). Finally capture and escape orbits are represented by a finite sequence (s_k) starting and ending with infinity: $s_{k_{begin}} = s_{k_{end}} = \infty$ (type IV). Sitnikov proved the existence of orbits where the corresponding sequence (s_k) is infinite – i.e. there are infinitely many zero passages of $z(t)$ – but unbounded, that means that the time span between successive zeroes of $z(t)$ reaches arbitrarily large values for sufficiently long time of observation.

Even the following result on the existence of special orbits reported in Moser (1973) has been established:

Given $\varepsilon \in (0,1)$ exept on a finite subset of $(0,1)$ (which is not a strong constraint !) there exists a positive M_ε such that to every sequence (s_k) of one of the above mentioned types I, II, III, IV, obeying $\quad \forall k \in Domain(s_k) : s_k \geq M_\varepsilon \quad$, there is a corresponding orbit as a solution of (1). $\quad$ (Theorem T)

This result given by Sitnikov and later refined by Alekseev, who was able to extend the theorem to positive masses of $m_3 < m_1/12$ (Marchal 1990), was further developed by Mc Gehee, who found a hyperbolic periodic orbit at infinity applying regularisation transformations. The proofs of Sitnikov, Alekseev and Mc Gehee were modified and presented newly by Conley and Moser. By defining a mapping for the time $t_{(0)}$ and velocity $\dot{z}_{(0)}$ of z for successive zeroes $z(t_{(0)}) = 0$ it is possible to prove Bernoulli shift character (Lichtenberg 1983) of a subset of the set of orbits of types I, II, III, and IV near the escape velocity (Moser 1973).

---

[1]The author thanks Prof. V. Banfi, Milano who supplied this information.

3. TRANSFORMATION FROM TIME TO TRUE ANOMALY

In order to overcome the difficulty of the occurence of primary true anomaly $\varphi(t)$ *and* time variable t in (1) because of $r(t) = p/[2(1 + \varepsilon \cos \varphi(t))]$, we perform the well known transformation to φ as the unique independent variable. In the case of arbitrary ε, where the question of scaling is a little more delicate, we do not adjust units to eliminate constants as we did for obtaining equations(2) and (3), but prefer to achieve dimensionless form step by step using Kepler's laws:

$$(I) \qquad r(\varphi) \; = \; \frac{p}{2(1 + \varepsilon \cos(\varphi))}, \qquad p = a\,(1 - \varepsilon^2)$$

$$(II) \qquad r^2 \dot\varphi \; \equiv \; c, \qquad\qquad\qquad c \text{ the area constant}$$

$$(III) \qquad \wp^2 \; = \; \frac{4\pi^2 a^3}{GM},$$

$M = m_1 + m_2,$

$\wp$ primary revolution period,

G gravitational constant.

These imply immediately:

$$(I),(II) \quad \Longrightarrow \quad (\alpha) \quad \dot r = \frac{2c\varepsilon}{p}\sin\varphi$$

$$(II) \qquad \Longrightarrow \quad (\beta) \quad c = \frac{\pi ab}{2\wp},$$

$$b = a\sqrt{1 - \varepsilon^2} \quad \text{semi minor axis of primary } \textit{relative} \text{ ellipse}$$

$$(III),(\beta) \quad \Longrightarrow \quad (\gamma) \quad c^2 = GMp/16.$$

The intended replacement of the time derivative in $\ddot z$ by one w.r.t. φ necessitates the consideration of the differential operator $(d/dt)^2$ expressed in terms of φ:

$$d/dt = \dot\varphi\, d/d\varphi \quad \Longrightarrow \quad (d/dt)^2 = \ddot\varphi\, d/d\varphi + \dot\varphi^2 (d/d\varphi)^2$$

Using primes for denoting derivation w.r.t. true anomaly φ we obtain:

$$(II),(\alpha) \quad \Longrightarrow \quad \ddot z = \frac{c^2}{r^4}\left(z'' - \frac{2\varepsilon y}{1 + \varepsilon x}z'\right), \tag{7}$$

$$x := \cos(\varphi), \quad y := \sin(\varphi)$$

$$(1),(7) \quad \Longrightarrow \quad z'' - \frac{2\varepsilon y}{1 + \varepsilon x}z' = -\frac{r^4 MG}{c^2}\frac{z}{\sqrt{r^2 + z^2}^{\,3}}$$

$$[\text{using}(\gamma)] \quad = -\frac{16rz}{p}\frac{1}{\sqrt{1 + (z/r)^2}^{\,3}}$$

$$[\text{using}(I)] \quad = -\frac{8z}{1 + \varepsilon x}\frac{1}{\sqrt{1 + 4(z/p)^2(1 + \varepsilon x)^2}^{\,3}}.$$

With $S := z/p$ we arrive at an equation containing only ε as a parameter but with a coriolis term due to the interpretation of the t to φ transformation as an oscillating ($-2\varepsilon \sin\varphi$!) rotation of the coordinate system around the z axis if we regard t as an angle (mean anomaly) in the old system:

$$(1 + \varepsilon x)\cdot S'' - 2\varepsilon y \cdot S' + \frac{1}{\sqrt{1/4 + S^2(1 + \varepsilon x)^2}^{\,3}}\cdot S = 0 \tag{8}$$

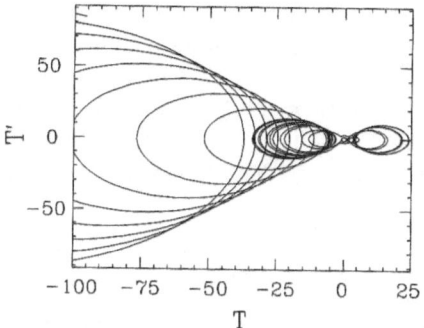

Figure 3. (T, T') phase portrait

4. COMPENSATION OF THE CORIOLIS TERM

Introducing $T := z/(2r)$ we compensate the coriolis term by straight forward substitution into (8):

$$T'' + \frac{1/\sqrt{1/4 + T^2}^3 + \varepsilon \cos \varphi}{1 + \varepsilon \cos \varphi} \cdot T = 0 \tag{9}$$

The quantity T has the geometrical meaning of half of the tangent of the angle of view of the small body m_3 seen from one of the primaries w.r.t. the plane of primary motion. An impression of how (T, T') projections look like is given in fig. 3.

This new representation allows to construct a relationship to the Hamiltonian (3) of the unperturbed $(\varepsilon = 0)$ system: Integrating $(1 + \varepsilon x) T'T'' + TT'/\sqrt{1/4 + T^2}^3 + \varepsilon x \, TT' = 0$ – obtained from (9) – w.r.t. φ, we derive

$$\frac{T'^2}{2} - 1/\sqrt{1/4 + T^2} = H^{(0)}(z{=}T_0, \dot{z}{=}T_0') - \frac{\varepsilon}{2} \int_{\varphi_0}^{\varphi} (T^2 + T'^2)' \cos \bar\varphi d\bar\varphi$$

This comparison is justified as for $\varepsilon = 0$ it follows that

$$\begin{array}{rclcrcl} r & \equiv & 1/2 & \text{i.e.} & T & \equiv & z \\ \varphi & \equiv & t & \text{i.e.} & d/dt & = & d/d\varphi \end{array}$$

It seems remarkable to the author that for positive ε, T and z, φ and t, respectively are not at all the same. With (9) we have arrived at a closed form of representation of the equation of motion – i.e. a differential equation with *one* definite independent variable φ describing the problem – with the case $\varepsilon = 0$ smoothly embedded.

Inserting the unperturbed periodic solution $z(t)$ (eq.6) obtained by Mac Millan into $z = 2Tr$ (see eq. 9) for T gives an approximate expression for $z(t)$ for the case of small amplitudes and eccentricities $(z(t), \varepsilon < .1)$. The interference phenomen between primary and fundamental frequency of m_3 respectively is well reflected by this simple transformation in spite of only zero order in ε of the perturbation expansion for T. Thus comparing systems of increasing eccentricities starting with zero, the choice of variable T implies that trajectories remain closer to the unperturbed solution than in the case of z. This feature allows easy application of perturbation methods as has been done by Juranek (1990).

The uniqueness of the independent variable φ simplifies numerical integration techniques. Usually one has to integrate the Keplerian primary ellipses additionally, or to solve Kepler's transcendental equation. In contrast to this Kepler's equation delivers t as an explicit function of true anomaly via the eccentric anomaly. We made use of this convenience in our numerical verification of the formulae presented here. Clearly the set of variables (T, T') represents an area preserving flow (Goldstein 1966). Investigating equation (9) Hagel found that it is deriveable from the Hamiltonian:

$$H(\varphi, T, T') = \frac{1}{2}(T^2 + T'^2) - \frac{1}{1 + \varepsilon \cos \varphi}(T^2/2 + 1/\sqrt{1/4 + T^2}) \qquad \text{(Hagel 1990)}.$$

While φ is a kind of pulsating time measure with pulsation strength ε, contains T mixed information from z and φ (mod 2π). This gives rise to a remarkable feature of the phase space trajectories (T, T'): Informally spoken the scale of measurement of distance z in variable T is changed every primary revolution periodically with φ and period 2π due to $T = z/(2r)$. Also the speed measure in T' with respect to new time φ shows this periodicity. Thus for roughly every primary revolution in phase space projection (T, T') a loop – whose size is changing with amplitude z – appears on the respective negative $(T < 0)$ or positive $(T > 0)$ side of phase sub space (T, T') depending on the small body being below or upside the primary plane during the current primary body revolution (see fig. 3). This means that the mentioned Sitnikov sequence appears geometrically as the number of meshes ± 1 to occur alternately on the right $(T > 0)$ and on the left $(T < 0)$ half of phase space. We will return to this correspondence later.

5. TRANSFORMATION TO POLAR COORDINATES AND COMPLEX REPRESENTATION

Polar Coordinates (R, Φ) are introduced in the following way:

$$
\begin{array}{llllll}
T & =: & R\cos\Phi & (i) & \Longrightarrow & T' & = & R'\cos\Phi - R\Phi'\sin\Phi & (iii) \\
T' & =: & R\sin\Phi & (ii) & \Longrightarrow & T'' & = & R'\sin\Phi + R\Phi'\cos\Phi & (iv)
\end{array}
$$

From (ii) and (iii) we find $R'/R = (1 + \Phi')\tan\Phi$ or – setting $A := \ln R$ –

$$A'\cos\Phi - (1 + \Phi')\sin\Phi = 0 \qquad (v).$$

In equation (9) we substitute (i) and (iv) using the abbreviation

$$\Omega^2 := \frac{1/\sqrt{1/4 + T^2}^{\,3} + \varepsilon\cos\varphi}{1 + \varepsilon\cos\varphi}$$

so that the harmonic oscillator like form of (9) is getting more visible $(T'' + \Omega^2\, T = 0$ but $\Omega^2 = \Omega^2(\varphi, T)\,!)$ which results in

$$A'\sin\Phi + (\Omega^2 + \Phi')\cos\Phi = 0 \qquad (vi).$$

Resolving (v) and (vi) w.r.t. A' and Φ' we find

$$A' = \frac{1 - \Omega^2}{2}\sin(2\Phi) \qquad \Phi' = \frac{1 - \Omega^2}{2}(1 + \cos(2\Phi)) - 1$$

which suggests the substitution $\quad B := 2A$ and $\Psi := 2\Phi \quad (vii) \quad$ leading to

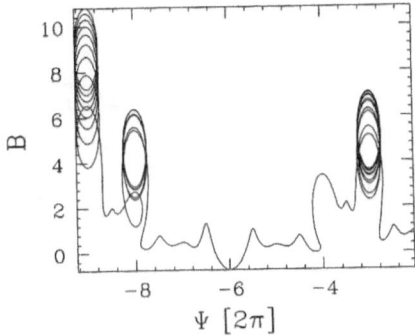

Figure 4. (B, Ψ) trajectory exhibiting loops which represent the Sitnikov sequence

$$
\begin{aligned}
B' &= f(\varphi, B, \Psi) &:= (1 - \Omega^2) \sin \Psi \\
\Psi' &= g(\varphi, B, \Psi) &:= (1 - \Omega^2)(1 + \cos \Psi) - 2
\end{aligned}
\tag{10}
$$

$$
1 - \Omega^2 = \frac{1 - 1/\sqrt{1/4 + \exp(B)(1 + \cos \Psi)/2}^{\,3}}{1 + \varepsilon \cos \varphi}.
$$

Fig. 4 shows a typical (B, Ψ)-diagram which reveals also the mentioned mesh structure proceeding from the right to the left as φ increases.

If we interpret φ and T as respective time and space variables – which is physically fulfilled only for zero ε – the above procedure is essentially a transformation to action - angle variables with logarithmic measure of the action.

A complex representation of this transformation looks like this:
With $T^* := T + iT'$, $L := B + i\Psi$ and $\bar{L} := B - i\Psi$ $(i := \sqrt{-1})$ from $(i), (ii)$ and (vii) it follows that we actually have applied the complex logarithm function to T^* deriving B and Ψ (i.e. $L = 2 \ln T^*$) and furthermore

$$
L' = F(\varphi, L) := \{(1 - \Omega^2)(\exp((\bar{L} - L)/2) + 1) - 2\} i
$$

with

$$
1 - \Omega^2 = \frac{1 - 8/\sqrt{1 + \exp(L/2) - \exp(\bar{L}/2)}^{\,3}}{1 + \varepsilon \cos \varphi}
\tag{11}
$$

where $F(L)$ is a non holomorphic complex function (Cartan 1966) in L as it involves also the complex conjugate $\bar{L}$ of L. The complex form allows to find an interesting estimation of the Sitnikov sequence by complex contour integrals:

$$
s_k = \frac{1}{2\pi i} \oint_\gamma d\zeta / \zeta \pm 1
$$

for γ appropriately chosen, namely such that the index of γ w.r.t. $(0,0)$ counts the number of the above mentioned meshes in (T, T') space provided we dispose of solution curves of (11), as γ has to consist partially of L' curves. See fig. 5 for (B', Ψ') curves, where each loop around the origin corresponds to a primary revolution without a change of sign of $z(t)$.

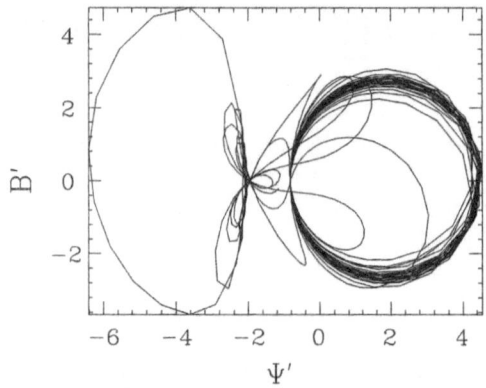

Figure 5. (B', Ψ') curve revoluting around the origin for each full
primary period without barycenter passage of m_3

6. CONNECTION WITH SITNIKOV SEQUENCES

We consider the difference quotient $\quad \Delta\Psi(\varphi) := \dfrac{\Psi(\varphi + \pi) - \Psi(\varphi - \pi)}{2\pi}$

(where Ψ is a solution of (10)) which is the average of $\Psi'(\varphi)$ taken over $[\varphi - \pi, \varphi + \pi]$ i.e.
the zero order component of the Fourier expansion of Ψ' in the same intervall and thereby an
expression for Ψ' with all the higher order contributions (i.e. overtones) of primary frequency
filtered away. Comparing fig. 6a with fig. 6d, we recognize the appearance of square like
pulses (realistic digital signals!) in the $(\varphi, -\Delta\Psi)$ diagram of approximate unity height and
length as well as area 2π corresponding to every zero of $z(t)$. The shape of these *passage
quantums* can be biased considerably by a concentration of $z(t)$ zeroes on time axis as can be
seen from figures 6a and 6d, due to superposition of single pulses thus maintaining the latter
area condition (this bias does not happen to orbits fulfilling the preliminaries of theorem (T)
above with $M_\epsilon > 1$). Remembering $\Psi = 2\Phi$ it is evident that every passage of m_3 is associated
with a mean change of Ψ by 2π (see fig. 6b); since $\int \Delta\Psi d\varphi$ is an average expression for Ψ
the square pulse area of approximately 2π is explained. The value 2π is not taken exactly as
small peaks, announcing passages to come, and also echo peaks, during primary revolutions
without passages occur; but even over long periods of time this does *not* affect the exact
balance of $\int \Delta\Psi d\varphi$ versus Ψ. This remarkable occurance is suggestive for a replacement of
$\Delta\Psi$ by approximations containing Dirac Delta (Walter 1974) or exact square or trapezoid pulse
functions and subsequent substitution back into (10) or an averaged form of (10). The exact
position on time (respectively true anomaly) axis would be determined by the corresponding
member s_k of a given Sitnikov sequence plus a presumably unknown member α_k of a sequence
of numbers between zero and one, which should be deducable from consistency with (10). It is
our aim to find solution expressions by means of approximating the mentioned *passage quants*,
which could bear even the information of how to construct a time step mapping from their
digital nature.

This is the recent state of some of our investigations leaving many ideas to be worked out in
practice. Concluding we have to state that during all the work on the Sitnikov configuration we
felt deep admiration for the variety and beauty of phenomena we encountered in the landscape
of this simple looking dynamical system. We believe that although the latter represents a very

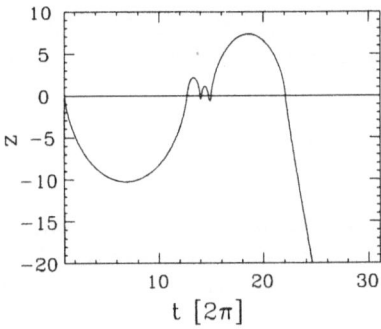

Figure 6.a. $z(t)$ pulsations showing the fluctuative character of the solution

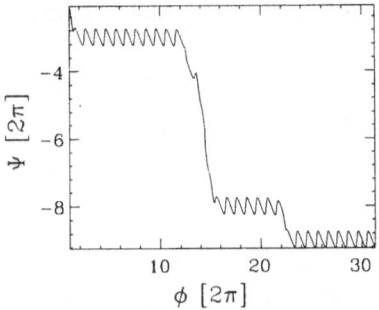

Figure 6.b. $\Psi(\varphi)$ curves, whose sawtooth pattern reveals the Sitnikov sequence

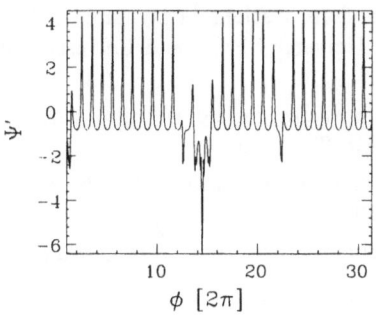

Figure 6.c. $d\Psi/d\varphi$ showing significant negative peaks corresponding to every zero of $z(t)$

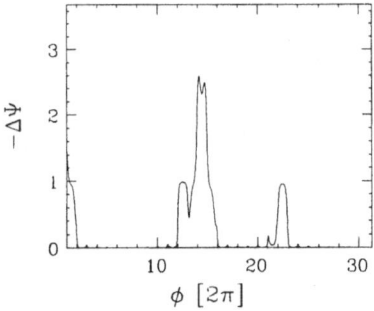

Figure 6.d. Smoothing Ψ' curves gives the *passage quantums* indicating zeroes of $z(t)$

exceptional case [2] of the elliptic restricted three body problem, the study of this system is crucial in understanding the range of dynamical properties covered by the general three body problem.

Acknowledgement

The author thanks Prof R. Dvorak, Institute of Astronomy, Vienna University, Austria for his profound support and encouragement during this work and Dr J. Hagel, CERN, Switzerland for numerous discussions in which his ideas gave me a deep insight into the nature of chaos.

References

Cartan, H., 1966, Elementare Theorien der Analytischen Funktionen einer oder mehrerer Komplexen Veränderlichen, Bibliographisches Institut, Mannheim

Goldstein, H., 1950, Classical Mechanics, Addison-Wesley, Reading Massachusetts

Hagel, J., 1990, private communication

Juranek, H., 1990, Zum Sitnikov Problem: Störungsrechnung erster Ordnung, Diploma Thesis, University of Vienna

Lichtenberg, A.J. and Liebermann, M.A., 1983, Regular and Stochastic Motion, Applied Mathematical Sciences, Volume 38, Springer, New York

Mac Millan, W. D., 1913, An integrable case in the restricted problem of three bodies, A.J. 27, 11

Marchal, C., 1990, The Three Body Problem, Studies in Astronautics, 4, Elsevier, Amsterdam

Moser, J., 1973, Stable and Random Motions in Dynamical Systems, Annals of Mathematics Studies Number 77, Princeton University Press and University of Tokio Press, Princeton, New Jersey

Pavanini, G., 1907, Sopra una nuova categoria di soluzioni periodiche nel problema dei tre corpi, Annali di Mathematica, Serie III, Tomo XIII

Sitnikov, K., 1960, Existence of oscillating motions for the three-body problem, Dokl. Akad. Nauk, USSR, 133, no. 2, 303-306

Stumpff, K., 1965, Himmelsmechanik, Band II, VEB Deutscher Verlag der Wissenschaften, Berlin

Walter, W., 1974, Einführung in die Theorie der Distributionen, Bibliographisches Institut, Mannheim

---

[2]The system is – by the way – extremly unstable in three dimensional space, as small deviation from the axis of motion of m_3 causes immediate attraction by one of the heavy masses.

PERIODIC SOLUTIONS FOR THE ELLIPTIC PLANAR RESTRICTED

THREE-BODY PROBLEM: A VARIATIONAL APPROACH

Maria Letizia Bertotti *

Department of Mathematics
University of Trento, Italy

ABSTRACT. We outline a variational approach, developed to find periodic orbits of the satellite in the elliptic restricted problem with any value of the masses of the primaries. This approach leads to a multiplicity of generalized periodic solutions (namely solutions x which satisfy a boundary condition: x(T) = x(0) and which possibly experience collisions). Those solutions are uniform limits of classical periodic solutions, having a prescribed rotational behaviour with respect to the primaries, of some suitable approximating problems.

1. INTRODUCTION AND RESULTS

Consider the planar elliptic restricted three-body problem. If μ and ν denote the masses of the two primaries and $P_\mu(t)$ and $P_\nu(t)$ are the T-periodic functions representing their (prescribed) motions, the potential $U(t,x)$ is defined as

$$U(t,x) = \frac{\mu}{|x - P_\mu(t)|} + \frac{\nu}{|x - P_\nu(t)|}$$

and the equations of motion for the satellite in an inertial coordinate system are

$$(1) \qquad \ddot{x} = \frac{\partial}{\partial x} U(t,x)$$

(here $x=(x_1,x_2)\in \mathbb{R}^2$, $\ddot{x}$ denotes $\frac{d^2x}{dt^2}$ and $\frac{\partial U}{\partial x} = (\frac{\partial U}{\partial x_1}, \frac{\partial U}{\partial x_2})$; the constant of gravitation as well as the sum of the masses μ and ν is normalized to one).

In this note we will refer about an investigation directed to find analytically T-periodic solutions for system (1).

The question of existence of periodic solutions for the restricted problem (more properly for the <u>circular</u> restricted problem) has a long history and is at the origin of a large amount of research in analytical as well as numerical setting. For an extensive

---

* Partially supported by G.N.F.M. - C.N.R.

bibliography we refer to the book by Szebehely (1967) (see also Marcolongo (1919), Siegel - Moser (1971) and Bertotti (1989)).

As far as our work is concerned, we point out that the main motivation was the purpose of considering the problem without any restriction on the value of the masses of the primaries. In other words our aim was to establish existence of periodic motions of the satellite (having the same period of the primaries) admitting any value of the primaries masses. As far as we know, the <u>analytical</u> results available on periodic orbits for the r.t.b.p. require a smallness assumption for the mass of one of the primaries and rely to some extent on some perturbation argument. (See paragraph 1.c in Bertotti (1989)). The only exception we know is represented by a work by Conley (1963), where the circular problem is studied with arbitrary positive masses of the primaries. In that work values of the Jacobian integral are considered s.t. two compact Hills regions surrounding the primaries and separating them exist and infinitely many long period closed orbits of the satellite close to one of the primaries are found.

In order to formulate our result the notion of "generalized" solution has to be introduced. The origin of this notion will be illustrated in the following paragraph, where we will be in the position to give also a precise definition. Roughly speaking a generalized T-periodic solution of (1) coincides with a classical solution $x(t)$ for each t, but a set of t's of measure zero, on which collision with one of the P_j's takes place. In other words a "generalized" solution (which in particular may be classical) is obtained by glueing together different solutions x_i, $i \in J \subset N$: if the cardinality of J is strictly greater than 1, a collision of x_i is followed by an ejection of some x_j for $i,j \in J$.

We prove (Bertotti (1989)):

THEOREM 1: *Let T > 0 be the first time after which in the elliptic restricted three-body problem the two primaries occupy the same position. Then there exist infinitely many distinct generalized T-periodic solutions.*

The solutions of Theorem 1 are actually obtained as uniform limits of classical (regular) periodic solutions of some approximating problems. We illustrate next this point by presenting a Theorem which is in this context an auxiliary result, but which may be of some interest in itself at least from the mathematical point of view.

Some notations are in order.
Let us recall that

$$P_j \in C^2(\mathbb{R}, \mathbb{R}^2), \qquad P_j(t+T) = P_j(t) \qquad\qquad t \in \mathbb{R}, \quad T>0, \quad j = \mu, \nu$$

and

$$P_\mu(t) \neq P_\nu(t) \qquad \text{for all } t \in \mathbb{R}$$

(in our case the sets $\{P_\mu(t): t \in [0,T]\} \subset \mathbb{R}^2$ and $\{P_\nu(t): t \in [0,T]\} \subset \mathbb{R}^2$ are ellipses). We are dealing with a "singular" system. Indeed $U(t,x)$ is not defined for $x = P_\mu(t)$ and $x = P_\nu(t)$ and it tends to infinity when $x - P_\mu(t)$ or $x - P_\nu(t)$ tends to zero.
The set S defined as

$$S \quad := \{(t,P_\mu(t))\}_{t \in \mathbb{R}} \ \cup \ \{(t,P_\nu(t))\}_{t \in \mathbb{R}} \ \subset \ \mathbb{R}^3$$

will be called the "singularity set".
We will refer to homotopy classes of continuous curves $\{(t,x(t))\}_{t \in \mathbb{R}}$ with values in $\mathbb{R}^3 \backslash S$ and s.t. $x(T) = x(0)$.

Let us also introduce the following degree which expresses the rotational behaviour of the loop x: to every continuous curve $x \in C(\mathbb{R}, \mathbb{R}^2)$, s.t.

$$x(t + T) = x(t) \qquad \text{all } t,$$
$$x(t) \neq P_j(t) \qquad \text{all } t, j = \mu, \nu$$

we associate a couple of maps (here and henceforth we will identify S^1 with $\mathbb{R}/[0,T]$):

$$\xi_j(x) : S^1 \rightarrow S^1 \qquad\qquad j = \mu, \nu.$$

They are defined as follows:

$$\xi_j(x) (t) := \frac{x(t) - P_j(t)}{|x(t) - P_j(t)|}.$$

The $\xi_j(x)$ are well defined (the denominators never vanish) and continuous functions. Denoting the mapping degree of these circle maps by $\deg(\xi_j(x))$, we associate to the curve x the vector:

$$\deg(x) := (\deg(\xi_\mu(x)), \deg(\xi_\nu(x))) \in \mathbb{Z}^2.$$

By nontrivial homotopy class we will mean a class of curves $\{(t, x(t))\}_{t \in \mathbb{R}}$ s.t. for the loop x $\deg(x) \neq (0,0) \in \mathbb{Z}^2$.

In order to state the announced auxiliary Theorem, we postulate the following hypotheses:
Consider a potential U such that

(H1)

    (i) $U \in C^2(\mathbb{R}^3 \backslash S, \mathbb{R})$,

    (ii) $U(t+T, x) = U(t, x)$ for all $(t, x) \in \mathbb{R}^3 \backslash S$,

    (iii) $U(t, x) > 0$ for all $(t, x) \in \mathbb{R}^3 \backslash S$,

    (iv) $U(t, x), \dfrac{\partial}{\partial x} U(t, x) \rightarrow 0$ uniformly in t as $|x| \rightarrow \infty$,

    (v) $U(t, x) \rightarrow +\infty$ as $(t, x) \rightarrow S$.

and suppose

there exists a function $W \in C^1(\mathbb{R}^3 \backslash S, \mathbb{R})$, satisfying

(H2)

    (i) $W(t+T, x) = W(t, x)$ for all $(t, x) \in \mathbb{R}^3 \backslash S$,

    (ii) $W(t, x) \rightarrow -\infty$ as $(t, x) \rightarrow S$,

    (iii) $|\nabla W(t, x)|^2 \leq aU(t, x) + b$ a, b positive constants.

$(\nabla W$ denotes $(\dfrac{\partial W}{\partial t}, \dfrac{\partial W}{\partial x_1}, \dfrac{\partial W}{\partial x_2})).$

Remark:
(H1) is a very natural hypothesis in the sense that it is satisfied by the gravitational potential we are dealing with.
In contrast (H2) is not satisfied by gravitational potentials. It concerns the local behaviour of U near the singularity and it is a generalization of a condition (the "Strong Force" condition) which was introduced in Gordon (1975). At a first sight (H2) may look mysterious. To give a feeling of its meaning, we point out that it is satisfied for example if U(t,x) behaves near the singularity as

$$U^*(t,x) = \frac{c}{|x - P_\mu(t)|^\alpha} + \frac{c}{|x - P_\nu(t)|^\alpha} \quad \text{with } \alpha \geq 2, c > 0.$$

For example if $\alpha = 2$ a function W suitable to show that assumption (H2) is satisfied if $U(t,x) = U^*(t,x)$ is

$W : \mathbb{R}^3 \backslash \{(t,P_\mu(t)) \cup (t,P_\nu(t)) : t \in \mathbb{R}\} \rightarrow \mathbb{R}$, defined by

$W := \log (|x - P_\mu(t)|) + \log (|x - P_\nu(t)|)$.

The announced auxiliary result is:

THEOREM 2: *Assume that* U *satisfies* (H1) *and* (H2). *Then in every nontrivial homotopy class of continuous curves* $\{(t,x(t))\}_{t \in \mathbb{R}}$ *s.t.*

$$x(T) = x(0) \quad and \quad (t,x(t)) \in \mathbb{R}^3 \backslash S \quad for\ all\ t$$

there is a curve $\{(t,x(t))\}_{t \in \mathbb{R}}$ *s.t.* $x(t)$ *is a classical T-periodic solution of the system*

(1) $$\ddot{x} = \frac{\partial}{\partial x} U(t,x).$$

A weaker, but more expressive statement is: consider a restricted three-body problem with potential U satisfying (H1) and (H2). Then, for every prescribed $k = (k_1, k_2) \in \mathbb{Z}^2$, $k \neq (0,0)$, there is a solution which is periodic of period T and according to which, in an interval of time of length T, the satellite winds $|k_1|$ times around one and $|k_2|$ times around the other of the primaries, counterclockwise or clockwise, depending on the sign + or - of k_1 and k_2.

The relation between Theorem 1 and Theorem 2 is illustrated by the following work-strategy: we first study systems for which assumption (H2) holds true and we obtain for them Theorem 2. Specifically the solutions guaranteed by Theorem 2 are obtained variationally as local minima of a suitable functional on disjoint open subsets of a loop space. In a second time we face systems of the form (1) (for which (H2) is not satisfied): we approximate them by means of suitable systems satisfying (H2). This is possible, because (H2) involves only the local rate of growth of the potential near the singularity. We establish a limiting process, and we obtain convergence of the regular solutions for the approximating problems to solutions of system (1). What get lost in passing to the limit is the certainty that the limiting solutions are bounded away from the singularity. Consequently we end up with generalized solutions. We are able anyway to prove that there are infinitely many of them.

2. SOME COMMENTS ON THE VARIATIONAL APPROACH

The approach we choose to attack the problem is via the calculus of variations: the goal is finding the solutions as critical points of a suitable functional defined on a Hilbert space of periodic functions.

The study of periodic solutions for dynamical systems of the form (1) carried out with the use of variational methods has been largely developed in the last, say, ten years (see e.g. the monograph by Rabinowitz (1986) and the book by Mahwin and Willem (1989)). More recently also dynamical systems with singular potentials have been considered (for a discussion we refer to paragraph 1.d in Bertotti (1989)).

Here we point out that the approach we will outline below was actually developed for more general singular dynamical systems than the restricted problem.

For a more complete presentation, for proofs, details and further comments we refer to Bertotti (1989). Our purpose in this paragraph is simply to give an idea of the work setting, motivating at the same time the appearance of the notion of generalized solutions.

The basic ingredients are as follows: consider

$$E := \{ x \in H^{1,2}(\mathbb{R}, \mathbb{R}^2) : x(0) = x(T) \}$$

equipped with the norm

$$\| x \|^2 := \int_0^T | \dot{x}(t) |^2 \, dt + (|[x]|)^2 \qquad\qquad x \in E,$$

where

$$[x] := \frac{1}{T} \int_0^T x(t) \, dt \ \in \mathbb{R}^2.$$

E is an Hilbert space compactly embedded in $C([0,T], \mathbb{R}^2)$.

Set now

$$\Lambda := \{ x \in E : \text{for all } t, 0 \le t \le T : (t, x(t)) \notin S \}.$$

(Λ is the set of loops belonging to E s.t. $x(t) \ne P_\mu(t)$ and $x(t) \ne P_\nu(t)$ for all $t \in \mathbb{R}$).

Then $\Lambda \subset E$ is open and we define on Λ the Lagrangian functional:

$$(2.8) \qquad I(x) := \int_0^T \{ \frac{1}{2} |\dot{x}(t)|^2 + U(t, x(t)) \} \, dt$$

One shows that $I \in C^1(\Lambda, \mathbb{R})$.

I can be extended to all of E, by setting $U(t,x) = + \infty$ if $(t,x) \in S$. Then:

$I : E \to \mathbb{R} \cup \{ + \infty \}$ is (sequentially) weakly lower semicontinuous.

Denoting by $I'(x)$ the Frechet derivative of I, it is easily seen that $I'(x) = 0$ (equivalently x is a critical point of I) if and only if x is a weak solution of

$$\ddot{x} = \frac{\partial}{\partial x} U(t,x).$$

That's the reason, why we look for critical points of I.

A main difficulty in the proposed variational approach is due to the fact that ejection-collision orbits (namely orbits originating and dying at a singular point) are read just as periodic ones. However solutions which experience collisions do not have the repetitive character of the classical periodic solutions; therefore, one would like to distinguish those solutions by the classical ones. In other words, having found the existence of some critical points and equivalently of some solutions, one needs an argument in order to recognize whether or not they are bounded away from the singularity. The problem can be successfully faced for certain singular potentials. But it turns out to be a particularly hard task when gravitational potentials are involved. Indeed the following fact can be checked: for potentials which behave near the singularity as $\frac{1}{r^\alpha}$ with $\alpha \ge 2$ (if r denotes the distance from the singularity set) the Lagrangian functional becomes infinite in correspondence to loops which touch the singularity. In contrast to that the Lagrangian functional corresponding to gravitational potentials may attain finite values on collision orbits as well as on regular ones. Think for example to the Kepler problem:

$$(2) \qquad\qquad\qquad \ddot{x} = \frac{\partial}{\partial x} \frac{1}{|x|} \qquad\qquad x \in \mathbb{R}^2 \backslash \{0\}.$$

We recall that for negative values of the energy the solutions of (2) (ellipses in the configuration space) occur in families, where all orbits have the same minimal period and the same energy. To these families belong also ejection-collision homotetic orbits. Well: the value of the Lagrangian functional is the same on all orbits (classical and collision ones) belonging to one of the above mentioned families.

This example gives an insight of the difficulty of the problem of distinguishing collision orbits from regular ones in the case of gravitational potentials.

Also according to further investigations this problem appears as a deep one, at least when there are not symmetries allowing the use of suitable devices (as is the case for the Kepler problem, where the potential is radial).

All that should explain the genealogy of the notion of generalized solution (which has been introduced by Bahri and Rabinowitz (1989)) and which we can finally define (in a version adapted to the case of interest for us):

<u>Definition</u>: we call generalized T-periodic solution of (1) a continuous loop
$$x \in C(\mathbb{R}, \mathbb{R}^2), \text{ s.t. } x(T) = x(0) \text{ and}$$

(i) $x \in H^{1,2}(\mathbb{R}, \mathbb{R}^2)$ and $I(x) < +\infty$;

(ii) $x(t) - P_\mu(t)$ as well as $x(t) - P_\nu(t)$ vanishes on a set $\mathcal{V}$ of measure zero and whose complement is open;

(iii) x is of class C^2 on $\mathbb{R} \setminus \mathcal{V}$ and satisfies (1) on $\mathbb{R} \setminus \mathcal{V}$.

Summarizing: once again we emphasize that the main advantage provided by this variational approach is that no smallness assumptions are required for the masses μ and ν of the primaries. Moreover, it allows to study at a time also the elliptic, not only the circular, problem. (Typically the search for periodic orbits is carried out for the circular problem, and this is studied in a rotating coordinate system, where the primaries are fixed and an integral of motion (the Jacobi integral) exists. For the elliptic problem it is no possible to pass to an uniformly rotating system where an integral of motion exists). In contrast, a disadvantage is due to the weaker kind of solutions one is able to get.

REFERENCES

Bahri A. - Rabinowitz P.H., 1989, A minimax method for a class of Hamiltonian Systems with singular potential, Journal of Functional Analysis, Vol. 82, pp. 412-428

Bertotti M.L., 1990, Forced Oscillations of Singular Dynamical Systems with an Application to the Restricted Three Body Problem, Journal of Differential Equations, in print

Conley C.C., 1963, On some new long periodic solutions of the plane Restricted Three Body Problem, Comm. P.A.M., vol. 16, pp. 449-467

Gordon W.B., 1975, Conservative dynamical systems involving strong forces, Transac. A.M.S., Vol. 204, pp. 113-135

Mahwin J. - Willem M., 1989, Critical Point Theory and Hamiltonian Systems, Springer Verlag, New York

Marcolongo R., 1919, Il problema dei tre corpi, Hoepli, Milano

Rabinowitz P., 1986, Minimax methods in Critical Point Theory with applications to Differential Equations, CBMS Reg. Conf. Ser. in Math. #65, A.M.S., Providence

Siegel C.L. - Moser J.K., 1971, Lectures on Celestial Mechanics, Springer Verlag, Berlin

Szebehely V., 1967, Theory of orbits, Academic Press, Orlando

HILL-TYPE STABILITY AND HIERARCHICAL STABILITY

OF THE GENERAL THREE-BODY PROBLEM

Yan-Chao Ge

Department of Physics and Astronomy
The University of Glasgow
Glasgow G12 8QQ, Scotland, U.K.

1. INTRODUCTION

Hierarchical stability (HS hereafter) was defined by Walker and Roy (1983) in connection with the Jacobian coordinate system. A dynamical N-body system is held to be HS if, during an interval of time substantially longer than the periods of revolution of the bodies in the system, the following conditions hold:

HS-(A). none of the bodies escapes to infinity from the system;

HS-(B). no dramatic changes occur in any orbit's size, shape or orientation to the invariable plane of the system;

HS-(C). $\rho_i < \rho_j$ for any $i < j$; where $\rho_i = |\rho_i|$ (i=2, ..., N), ρ_i being the Jacobian vectors which connect the barycentre of the first (i-1) masses and the ith mass.

These conditions will be referred to as stability conditions HS-(A), HS-(B) and HS-(C) respectively. When any one of them is contradicted, we shall call it instability condition (A), (B) and (C) respectively.

Because of the non-integrability nature of the N-body problem, there is no general analytical criterion to guarantee such stability conditions. Nevertheless, in the context of the general 3-body problem, there are analytical criteria to guarantee the stability condition HS-(C), no matter what the sign of the total energy integral is (Marchal and Bozis, 1982). In the case of negative total energy, this is essentially based on Sundman's inequality and an inequality directly deduced from it, viz.

$$IU^2 \geq -2C^2H \tag{1}$$

where C is the value of the total angular momentum integral, H being the total energy integral, U the potential energy, and I the system's moment of inertia. Being held in both the barycentre frame and an inertial frame for any N-body system, these inequalities give useful results if they are applied to the barycentre frame of the 3-body problem.

The parameter C^2H (a combination of integrals) on the right side of equation (1) imposes restrictions on the function of positions, IU^2, on the left side. Such restrictions lead to possible and forbidden regions of motion analogous to the Hill regions in the circular restricted 3-body problem (see Poincare, 1892; Golubev, 1967, 1968; Smale, 1970; Easton, 1971; Marchal and Saari, 1975; Bozis, 1976; Zare, 1976, 1977; Saari, 1976, 1984, 1987; Ge, 1990). This is called **Hill-type stability,** and the condition for this stability may be written in two equivalent forms, viz.

$$C^2H \leq (C^2H)_c \quad \Leftrightarrow \quad \alpha \leq \alpha_c \quad (\alpha \equiv a_2/a_3) \tag{2}$$

where (C^2H) and α_c are the critical values. The second one is more convenient in application, for it is is expressed as the ratio of the semi-major axes (eg. Szebehely and Zare, 1976; Walker et al, 1980). In this connection it is also conventional to introduce the normalised masses, μ and μ_3, and Walker and Roy's (1983) empirical stability parameters, ε_{23} and ε_{32}, namely,

$$\begin{cases} \mu = m_2/(m_1 + m_2) \in [0, 0.5] \\ \mu_3 = m_3/(m_1 + m_2) \end{cases} \qquad \begin{cases} \varepsilon_{23} = \mu(1-\mu)(\rho_2/\rho_3)^2 \\ \varepsilon_{32} = \mu_3(\rho_2/\rho_3)^3 \end{cases} .$$

The value of α_c depends mainly on these two pairs of parameters and the eccentricities. The choice of these two pairs of parameters does not change the basic feature of the problem; however, in order to compare with the work of Walker and Roy (1983), their ε parameters are used in the present paper.

The importance of Hill-type stability in relation to hierarchical stability is based on the following points. It is generally held that a Hill-type stability guarantees hierarchical stability, in spite of the fact that this analytical stability criterion precludes neither collision nor escape instability. Supports on this view may be found not only from comparison with real systems (Szebehely and Zare, 1976; Markellos and Roy, 1981; Walker, 1983), but also from the systematic numerical experiments on initially circular 3-body systems (Walker and Roy, 1983). In fact the above intuitive view can only be analytically established for the condition HS-(C) (see Ge, 1990), and cases against the other two conditions will be shown in the next section.

Since the Hill-type stability is only a sufficient condition for Hierarchical stability, the existence of empirical stability regions outside the Hill-type stability regions observed by Walker and Roy (1983) and the drastic collapse in the value of α_c shown by Valsecchi et al (1984) due to the introduction of orbital eccentricities have been interpreted as a confirmation of the above generally adopted view. Although it has also been stressed that collision of the inner binary and escape of the outer mass may happen even when a system is stable in the generalised sense of Hill, the lack of such examples leads to the intuition that they may be very unlikely to happen. In the next section, systematic numerical experiments show many such examples.

2. NUMERICAL EXPERIMENTS ON INITIALLY ELLIPTIC, PROGRADE, COPLANAR 3-BODY SYSTEMS

In this section we present the result of several hundred numerical integration experiments on initially elliptic, prograde, coplanar 3-body systems. All the experiments were carried out on the ICL 3980 mainframe computer at Glasgow University, using the same numerical routine that Walker and Roy (1983) used. In this routine the mutual radius vectors are calculated by a tenth order Taylor series, where the derivatives are evaluated by recurrence relations. The programme incorporates an automatic step-length regulator which shortens or lengthens the integration length of the computer in order that the error caused by truncating the Taylor series after the tenth order is less then a given tolerance (10^{-12} in this approach).

The accuracy of the integration routine is affected by both truncation error and round-off error. The accumulated error can be estimated by running the programme for fictitious 3-body systems with μ's -> 0 or for the linearly stable equilateral triangle motion, whose orbital elements should remain constant. Such an estimation gives the result of about 6000 synodic periods for an 0.1% relative error in the position. Programmes have been run up to 1000 synodic periods if no instability sets in before this time limit. The energy and angular

Table 1.   Schematic Plot of the Numerical Experiment Result

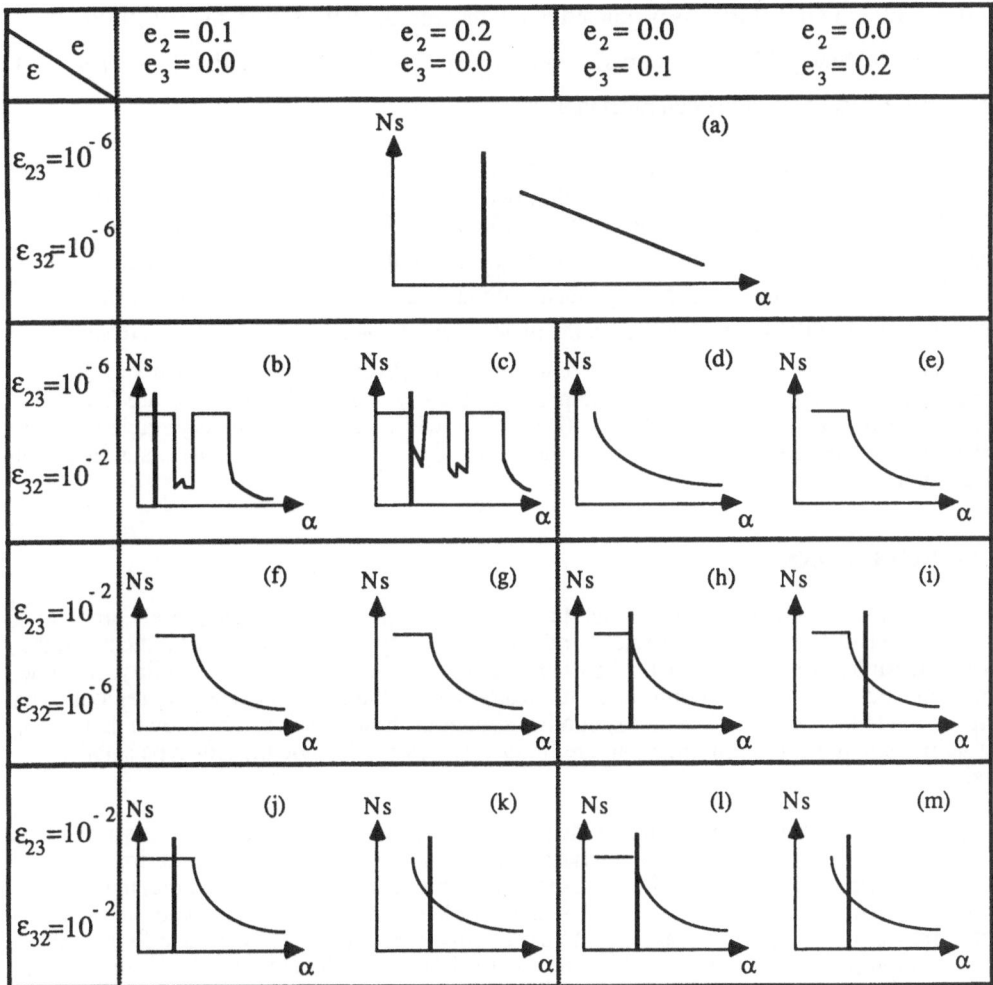

momentum integrals were used to check the integration error, though they are not very adequate for this role (an integral of motion is not sensitive to integration error even if the motion is irregular). The relative error of them on commencing and at the end of the integration is found always below 10^{-7}.

The initial conditions are chosen such that the masses form a mirror configuration on commencing the integrations (Roy and Ovenden, 1955). The advantage of this choice is that by studying one direction of time we also gain knowledge of the other one, so that the time-scale is cut down. To further this point the initial conditions are actually chosen at what are believed to be the worst configurations, ie. at the maximum perturbation: the body in the inner orbit at apocentre, with the outer mass at its pericentre, and all masses collinear. We are confident that no generality is lost because of this specific choice of initial conditions.

The results are schematically collected and shown in Table 1, where the lifetime, Ns, of the systems (in synodic periods) is plotted against the initial α, and ε parameters are used. Each diagram in the Table contains about thirty systems with the same initial parameters [e_2 e_3 ε_{23} ε_{32}] but different initial α's. As is pointed out in the introduction section, the conclusion should not be changed if the set of parameters [e_2 e_3 μ μ_3] is used in stead of [e_2 e_3 ε_{23} ε_{32}]. The thick vertical bars in the diagrams indicate the values of α_c.

It is seen from this Table that the initially elliptical 3-body systems display more complex behaviour than the systems begun from initially circular orbits. Firstly, in contrast to the initially circular systems studied by Walker and Roy (1983), linear rather than exponential curves may be fitted to data of the families of systems with $\varepsilon_{23}=\varepsilon_{32}=10^{-6}$. Secondly, new phenomena in the form of groups of plateau and valley are observed (eg. (b) and (c)). In fact, up to the time limit of the present study, the plateau with bigger initial α can be more stable than that with smaller initial α. We believe that this is a reflection of the complexity of the phase space structure (eg. islands) of the problem. Finally, instabilities A and B are found inside the Hill-type stability regions (eg. (i) and (k)), while outside which very stable systems exist (eg.(j)). Therefore, we conclude that the elliptical C^2H stability criterion fails in indicating practical hierarchical stability.

3. CONCLUSIONS

Systematic numerical experiments show the complexity of the general 3-body problem and that the Hill-type stability criterion (or the C^2H stability criterion) does not indicate, in the case of **initially elliptic**, coplanar 3-body problem, a practical hierarchical stability. In some situations, this analytical criterion seems too restrictive, so that empirical stability regions exist outside the Hill-type stability regions; in other situations, however, it can be shown that instabilities A and B (see introduction) are frequently observed inside the Hill-type stability regions.

However, this does not mean that the analytical Hill-type stability criterion is without value in other situations. Firstly, all analysis has shown that it is a good indicator for hierarchical stability of initially circular 3-body systems. Secondly, it is found that the conservation of C^2H can be used to explain the correlated variation of semi-major axes and eccentricities with respect to time (Ge, 1990). Thirdly, inequalities stronger than Sundman's and equation (1) have been obtained for the spatial 3-body problem (Saari, 1987; Ge, 1990). Fourthly, the inequalities can be used to facilitate the study of escape or collision conditions (Marchal et al, 1984).

ACKNOWLEDGEMENT

My thanks are due to Professor Archie E. Roy for invaluable advice and numerous discussions. I also acknowledge the support of a Technical Co-operation Programme award.

REFERENCES

Easton, R. (1971): J. Diff. Eqs., 10:371.
Ge, Y.C. (1990): "Ph.D. Thesis", The University of Glasgow.
Golubev, V.G. (1967): Soviet Phys. Doklady, 12 (No. 6):529.
Golubev, V.G. (1968): Soviet Phys. Doklady, 13 (No. 5):373.
Marchal, C. and Bozis, G. (1982): Celest. Mech., 26:311.
Marchal, C. and Sarri, D.G. (1975): Celest. Mech., 12:115.
Marchal, C., Yoshida, J. and Sun, Y.S. (1984): Celest. Mech., 33:193.
Poincare, H. (1892): "Les Methods Nouvelles de la Mechanique Celeste", Gauthier Villars.
Roy, A.E. and Ovenden, M.W. (1955): M.N.R.A.S., 115:296.
Saari, D.G. (1976): Celest. Mech., 14:11.

Saari, D.G. (1984): Celest. Mech., 33:299.
Saari, D.G. (1987): Celest. Mech., 40:197.
Smale, S. (1970): Invent. Math., 10:305.
Smale, S. (1970): Invent. Math., 11:45.
Szebehely, V.G. and Zare, K. (1976): Astro. Astrophys., 58:145.
Valsecchi, G.B., Carusi, A. and Roy, A.E. (1984): Celest. Mech., 32:217.
Walker, I.W., Emslie, A.G. and Roy, A.E. (1980): Celest. Mech., 22:371.
Walker, I.W. and Roy, A.E. (1983): Celest. Mech., 29:117.
Walker, I.W. (1983): Celest. Mech., 29:215.
Zare, K. (1976): Celest. Mech., 14:73.
Zare, K. (1977): Celest. Mech., 16:35.

EQUILIBRIUM CONNECTIONS ON THE TRIPLE
COLLISION MANIFOLD

A. SUSÍN [1] AND C. SIMÓ [2]

[1]Dept. de Matemàtica Aplicada I , Universitat Politècnica de Catalunya
 Diagonal 647, 08028 Barcelona, Spain
[2] Dept. de Matemàtica Aplicada i Anàlisi , Universitat de Barcelona
 Gran Via 585, 08007 Barcelona, Spain

Abstract. In the three body problem the triple collision manifold plays a fundamental
role to describe passages near triple collision. To study the possible transitions from the
approach to collision to the escape from it, the invariant submanifolds on that manifold
are essential. In this paper we study mainly the connections between the equilateral
approaches and escapes.

1. Introduction

We consider the planar 3-body problem with masses m_1, m_2, m_3. If $q_i, p_i \in \mathbf{R}^2$,
$i = 1, 2, 3$ are the related positions and momenta, $q = (q_1^T, q_2^T, q_3^T)^T$, $p = (p_1^T, p_2^T, p_3^T)^T$
and $M = \text{diag}(m_1, m_1, m_2, m_2, m_3, m_3)$ denotes the mass matrix, the Hamiltonian
of the problem is

$$(1) \qquad\qquad H(q,p) = \frac{1}{2}p^T M p - U(q),$$

where $U(q) = \sum_{1 \leq i \leq j \leq 3} \dfrac{m_i m_j}{|q_i - q_j|}$ is the potential energy.

The system has singularities due to double and triple collision. We are interested in
the behaviour of the system close to a triple collision where the angular momentum
is zero as it is required to have an exact triple collision.

To put the equations of motion in a suitable form, one needs to do several changes
(blow up, regularization, etc) which have been described in [6], [4] and [5]. In this
way we introduce variables, $\alpha_1, \alpha_2, \alpha_3 \in \mathbf{R}$ essentially related to the normalized form
of the configuration triangle which satisfy the relation

$$(2) \qquad\qquad \sum_{1 \leq j \leq k \leq 3} m_j m_k (\alpha_j^2 + \alpha_k^2)^2 = m_1 + m_2 + m_3$$

and variables $\pi_1, \pi_2, \pi_3 \in \mathbf{R}$ which play the role of momenta and satisfy the energy relation

$$(3) \quad K(\alpha, \pi) = \frac{1}{8} \pi^T B(\alpha) \pi - \sum_{1 \le j \le k \le 3} m_j m_k (\alpha_{j+1}^2 + \alpha_{j+2}^2)(\alpha_{k+1}^2 + \alpha_{k+2}^2) = 0,$$

where $\alpha_s = \alpha_{s-3}$ if $s > 3$, $\alpha = (\alpha_1, \alpha_2, \alpha_3)^T$ and $\pi = (\pi_1, \pi_2, \pi_3)^T$. In (3) $B(\alpha)$ is a symmetric 3×3 matrix with components

$$(4) \quad \begin{aligned} b_{jj} &= \frac{\alpha_k^2 + \alpha_\ell^2}{m_j}(\alpha_1^2 + \alpha_2^2 + \alpha_3^2) + \frac{\alpha_j^2 + \alpha_\ell^2}{m_k}\alpha_\ell^2 + \frac{\alpha_j^2 + \alpha_k^2}{m_\ell}\alpha_k^2, \\ b_{jk} &= -\frac{\alpha_j^2 + \alpha_k^2}{m_\ell}\alpha_j \alpha_k, \end{aligned}$$

where the indices (j, k, ℓ) are $(1, 2, 3)$ or a cyclic permutation of it.

If we introduce also $v = \sum_{i=1}^{3} \alpha_i \pi_i$ then the equations of motion are

$$\frac{d\alpha}{d\tau} = \frac{\partial}{\partial \pi} K(\alpha, \pi) - \frac{1}{4} z \cdot \alpha, \quad \frac{d\pi}{d\tau} = -\frac{\partial}{\partial \alpha} K(\alpha, \pi).$$

where $z = (\alpha_1^2 + \alpha_2^2)(\alpha_2^2 + \alpha_3^2)(\alpha_3^2 + \alpha_1^2) \cdot v$ and τ is a suitable scaled time (we refer to [6], [4] again for details).

The set of variables $(\alpha, \pi) \in \mathbf{R}^6$ with the constrains (2), (3) is called the non-rotating triple collision manifold and will be denoted by N.

On N the flow is gradient like with respect to v. The topological type of N may be derived from (2) and (3). The integral (2) can be considered as S^2 in the configuration space and, given α, (3) can be considered as an ellipsoid except when $\alpha_j = \alpha_k = 0$ (binary configurations) where it degenerates into a cylinder. Therefore, N is topologically $S^2 \times S^2$ with 6 holes corresponding to the binary configurations. Adding these 6 points one obtains a compactified manifold that we will denote by $\tilde{N}$. On the compactified manifold we found the critical points in pairs. We denote them by P^i (respectively P^s) when the corresponding level of v is negative (respect. positive). The critical points of $\tilde{N}$ are related to the classical central configurations. Then we have 6 Euler's configurations (collinear) denoted by $E_j^{i,s}$, $j = 1, 2, 3$ and 4 Lagrange's configurations (equilateral) denoted by $L_{+,-}^{i,s}$. We can add also the 6 binary configurations as critical points on $\tilde{N}$ denoted by $B_j^{i,s}$, $j = 1, 2, 3$. The corresponding dimensions of the stable and unstable manifolds associated to the critical points are shown on table I.

Table I

Table I

	B_j^i	E_j^i	$L_{+,-}^i$	$L_{+,-}^s$	E_j^s	B_j^s
dim W^n	4	3	2	2	1	0
dim W^s	0	1	2	2	3	4

We show on figure 1 the position of the critical points in the configuration space. The equator of this sphere corresponds to the collinear configurations.

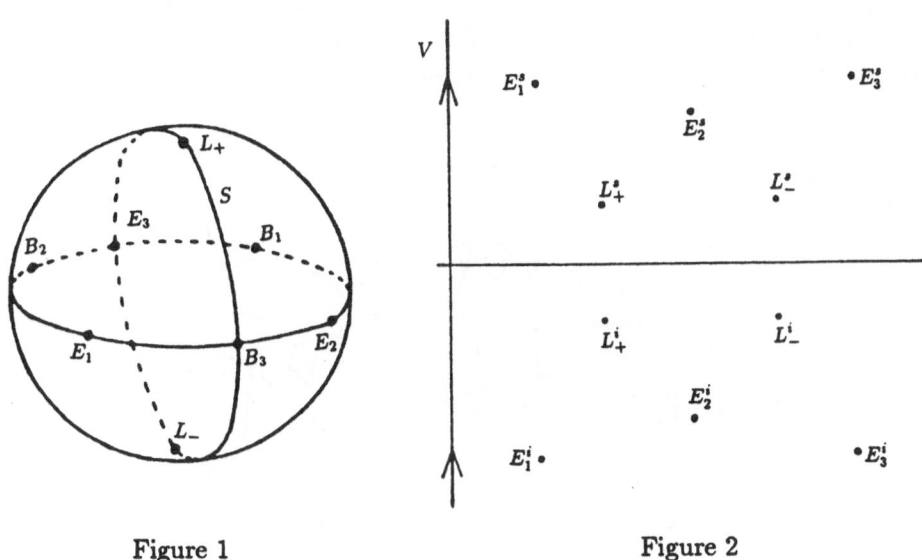

Figure 1   Figure 2

On figure 2 we show the relative positions of the equilibrium points on $\widetilde{N}$ with respect to the increasing variable v.

Our goal is to describe all the possible connections between equilibrium points in the planar triple collision manifold and how do they depend on the masses of the bodies.

2. Qualitative description of the $v = 0$ section

Now we want to give a qualitative description of the equator of $\widetilde{N}$ that is $\widetilde{N} \cap \{v = 0\}$. On table II we give the topological structure of $W_{P^s}^s \cap \{v = d\}$ for different values of $d = v_i$, $i = 0, 1, 2$ where P^s stands for one of the critical points $L_{+,-}^s$, E_j^s or B_j^s. We consider $v_2 > \max_j v(E_j^s)$, $0 < v_1 < \min_j v(E_j^s)$, $v_0 = 0$ (see [4]).

Table II

	B_j^s	E_j^s	$L_{+,-}^s$
v_2	S^3	–	–
v_1	$S^2 \times (0,1)$	S^2	–
v_0	$E^2 \times S^1$	$S^1 \times (0,1)$	S^1

On figure 3 we show the relative positions between the stable manifolds $W_{P_\bullet}^s \cap \{v = 0\}$. One can see the circles corresponding to the Lagrange's points and the open cylinders associated to the Euler's points which are the boundaries of the open solid tori corresponding to the binaries.

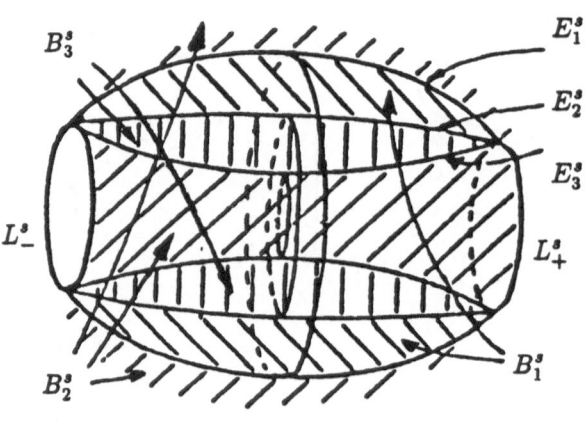

Figure 3

The full description of the possible transitions between collision and ejection close to triple collision would be given by the intersections of the type $W_{P_i}^u \cap \{v = 0\} \cap W_{Q^\bullet}^s$, P^i (resp. Q^s) being one of the critical points with $v < 0$ ($v > 0$). The main objective will be to know where to put the circles corresponding to the Lagrange's points and in particular when we will obtain connections between these critical points. These connections will classify all the other ones.

3. On the numerical computation of the unstable manifolds

We consider the simplex of masses T, defined by $m_1 + m_2 + m_3 = 1$. If we denote by λ_1, λ_2 the two positive eigenvalues associated to $W_{L_+^i}^u$ they have the expression

$$\lambda_{1,2} = 8^{-1/2} \cdot \mu^{-3/4} \cdot \left[\sqrt{13 \pm 12\sqrt{1-3\mu}} - 1\right]$$

where $\mu = m_1 m_2 + m_2 m_3 + m_3 m_1$. We obtain $\lambda_1 = \lambda_2$ when we consider equal masses $m_i = 1/3$, $i = 1, 2, 3$.

A linear aproximation of $W_{L_+^i}^u$ does not work because the quotient $\frac{\lambda_1}{\lambda_2}$ can be far away from 1. On figure 4 we display level curves of the quotient $\frac{\lambda_1}{\lambda_2}$ for values of this quocient equal to $(1.2)^k$, $k = 0, \ldots, 17$.

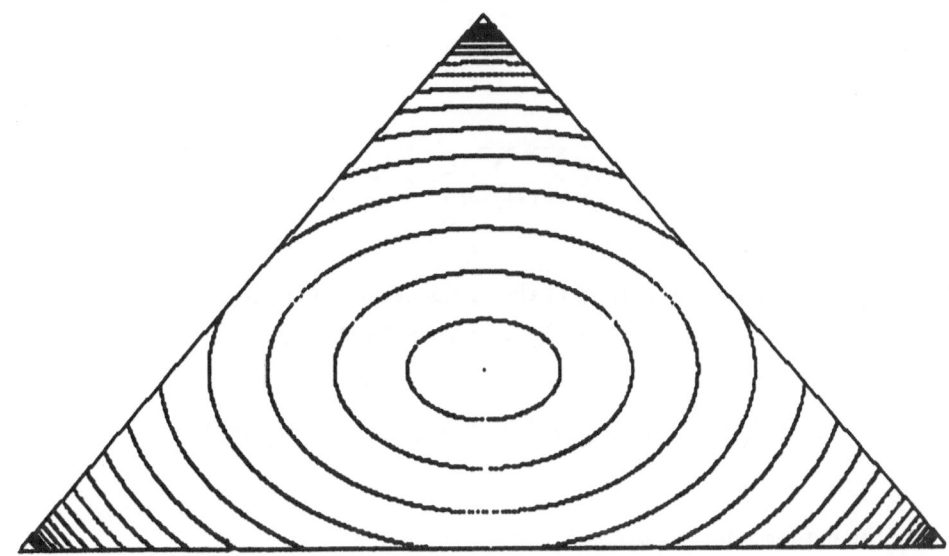

Figure 4

To decrease the difficulties we have used a higher order approximation to the invariant manifold. The algorithm to compute this approximation can be described as follows. We consider a new reference taking the Lagrange critical point L_+^i as origin and a new base made of the eigenvectors e_i, $i = 1, ..., 6$, associated to this critical point. Considering that the system of differential equations is polynomial we can write the equations in the new reference as

(5) $$\dot{X} = B^{(1)}X + B^{(2)}(X)^2 + B^{(3)}(X)^3 + \cdots + B^{(9)}(X)^9,$$

with $X \in \mathbf{R}^6$. We want to remark that there is not independent term in the equations (of course) and that they are polynomial in 6 variables of degree 9.

We can express $W^u_{L^i_+}$ up to order n as

$$(6) \qquad \bar{X}^{(n)} = C_1^{(1)} s_1 e_1 + C_2^{(1)} s_2 e_2 + \sum_{p=2}^{n} \sum_{k=1}^{6} \sum_{i+j=p} C_{i,j,k}^{(p)} s_1^i s_2^j e_k,$$

where $s_1 = e^{\lambda_1 \tau}$ and $s_2 = e^{\lambda_2 \tau}$, being τ the time variable. We can compute recursively the coefficients of $W^u_{L^i_+}$ of order n as a function of their coefficients up to order $n-1$. The idea is to impose the invariance of the manifold (6) when we substitute in the differential system (5). If we do that, we obtain the following expression for the coefficients of order n:

$$\sum_{k=1}^{6} \sum_{i+j=n} C_{i,j,k}^{(n)} (i\lambda_1 + j\lambda_2) s_1^i s_2^j e_k =$$

$$= \sum_{k=1}^{6} \sum_{i+j=n} C_{i,j,k}^{(n)} \lambda_k s_1^i s_2^j e_k + \sum_{k=1}^{6} \sum_{i+j=n} d_{i,j,k}^{(n)} s_1^i s_2^j e_k,$$

where $d_{i,j,k}^{(n)}$ denote the coefficients of the terms of order n in s_1, s_2 obtained from the substitution $X = \bar{X}^{(n-1)}$ in the nonlinear terms of (5), that is

$$B^{(2)}(\bar{X}^{(n-1)})^2 + B^{(3)}(\bar{X}^{(n-1)})^3 + \cdots + B^{(n)}(\bar{X}^{(n-1)})^9.$$

Finally the expression for the coefficients of order n is

$$(7) \qquad C_{i,j,k}^{(n)} = \frac{d_{i,j,k}^{(n)}}{i\lambda_1 + j\lambda_2 - \lambda_k}.$$

This has been implemented up to order 20 by means of and ad hoc algebraic manipulator. Let us discuss the presence of resonances in (7). As $\lambda_1, \lambda_2, \lambda_5, \lambda_6 > 0, \lambda_3, \lambda_4 < 0$, $i+j \geq 2$, $\lambda_1 > \lambda_2$, $\lambda_5 = \lambda_6$, $\lambda_1 > \lambda_5$, the only possible resonances are of the form $j\lambda_2 = \lambda_5 = \lambda_6$ with $j \geq 2$. There is a decreasing sequence of values of μ, $\mu_j > \mu_{j+1}$, $j \geq 2$ for which a resonance appears. One has $\mu_2 \approx 0.001984$. Hence the problem of resonance occurs only for small values of μ

When we have this local approximation we can start the numerical computations far away enough from the critical point to avoid the troubles we mention before. We can choose the optimal distance to start the computations comparing the numerical results obtained, when we cut the unstable manifold for differents values of v equal constant, whith the analytical results predicted for these levels by the local approximation. In other words, we found a practical radius of convergence of our expansion. Once we have decided this radius r we start the computations from a circle of radius r until we get the points on the equator of the manifold, $v = 0$. These points on the equator are obtained controlling the maximum distance between two consecutive points and controlling also the maximum angle between three of them. That way, we obtain a sufficiently regular curve on $v = 0$ which is compared to symmetrical copies of it (because of the symmetry of the problem) to found the possible connections between Lagrange's points.

4. Some numerical results

We summarize the Lagrange's connections we have found at the moment in figure 5, where we show only one sixth of T.

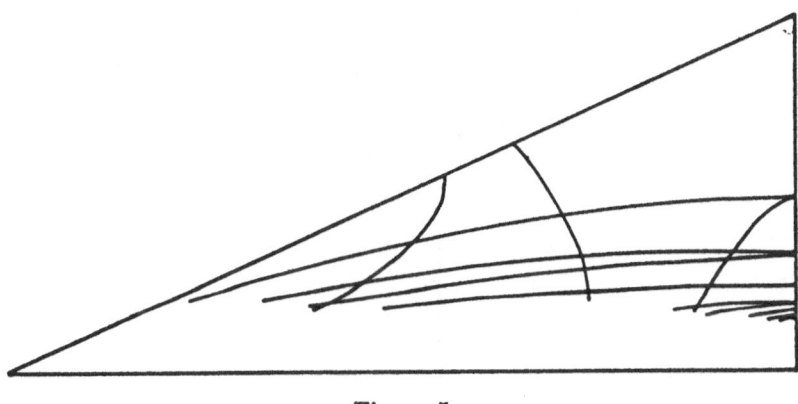

Figure 5

In figure 5 we can see a family of curves beginning in an isosceles configuration of masses (in this case $m_1 = m_2$) and also three other curves not belonging to this family.

The difference between them is that the curves corresponding to the family do not give an isosceles connection when we consider the points of these curves corresponding to an isosceles configuration of masses (see [3]).

On the curves of this family we can distinguish two differents subfamilies when we look to the birth of the curves in the points with isosceles configuration of masses. We found alternatively connection points which give rise to one or two different curves and they correspond also to different type of connections.

On the table III we give the values of the masses of these points. We remark that they satisfy $m_1 = m_2$, $m_1 + m_2 + m_3 = 1$. We use $+ \longrightarrow +$ (resp. $-$) to denote Lagrange's connections between $L_+^i \longrightarrow L_+^a$ (resp. L_-^a).

Table III

m_3	num. curves	Type of connection	
0.632912	1	$+ \longrightarrow +$	non isosceles
0.5710061	1	$+ \longrightarrow -$	isosceles
0.1597719	1	$+ \longrightarrow -$	non isosceles
0.1591555	1	$+ \longrightarrow +$	isosceles
0.1081336	2	$+ \longrightarrow +$	non isosceles
0.0808633	1	$+ \longrightarrow -$	non isosceles
0.0638546	2	$+ \longrightarrow +$	non isosceles
0.0522381	1	$+ \longrightarrow -$	non isosceles

The shape of the curves corresponding to the intersection $W_{L_+^i}^u \cap \{v = 0\}$ is shown in figures 6,7,8. In the last section we will give an explanation to the increase of the intersections in the curves observed when the mass parameter m_3 decreases.

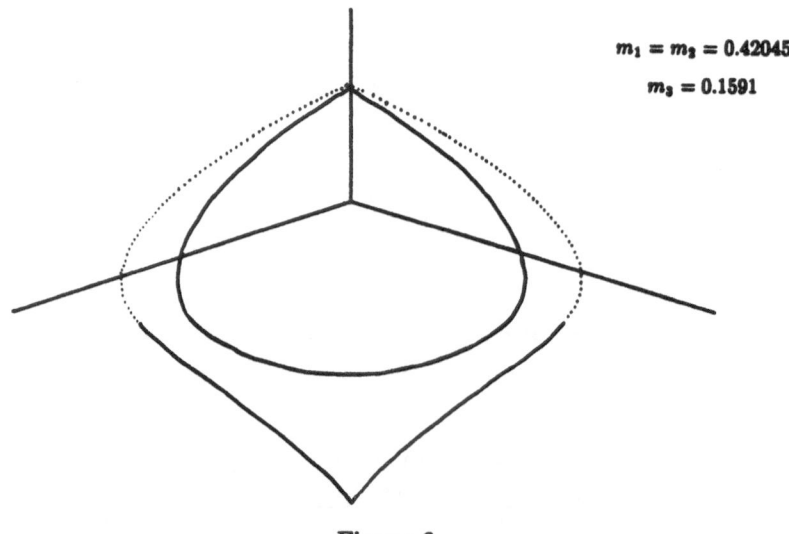

$m_1 = m_2 = 0.42045$

$m_3 = 0.1591$

Figure 6

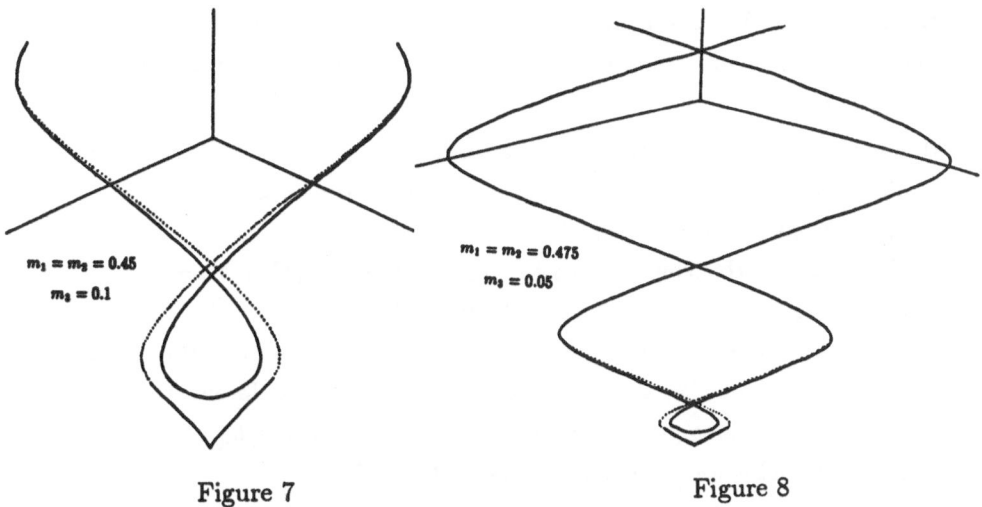

$m_1 = m_2 = 0.45$
$m_3 = 0.1$

$m_1 = m_2 = 0.475$
$m_3 = 0.05$

Figure 7   Figure 8

As it can be seen we do not display the termination of the curves on figure 5. This is due to the numerical troubles we mentioned before. Moreover we can not reach the level $m_3 = 0$ because the equations become unbounded when one of the masses vanishes. As a conjecture we think that these curves will finish at the corner point corresponding to the masses configuration $m_2 = m_3 = 0$, $m_1 = 1$. We base our conjecture on the relation, that we will comment later, between the planar and the collinear problem. For this last problem we have the family of curves given in figure 9 (see [2], [4]).

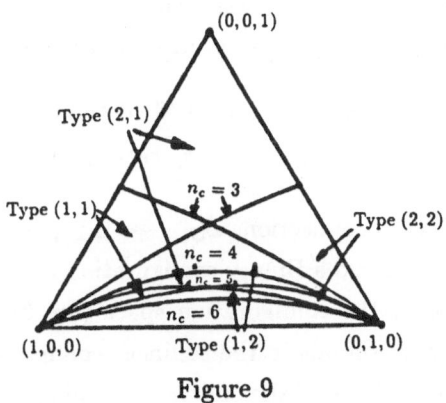

Figure 9

5. Theoretical considerations

As a final point, we want to give an explanation of why we conjecture the existence of an infinity of curves giving Lagrange's connections. In [4] we gave a complete description of the different connections between critical points known up to that moment. We can give here a summary of the connections known for the planar problem (see [1])

(i) For all masses there are orbits connecting $L^a_{+,-}$ with E^a_i and B^a_i, $i = 1, 2, 3$. There are also the symmetric connections in the $v < 0$ region. All these connections are transversal.

(ii) If there is an orbit connecting E^i_j to L^a_+ transversally, then there are also orbits connecting E^i_j to L^a_- and to E^a_k, B^a_k, $k = 1, 2, 3$.

(iii) If the index of the projection of the curve $W^u_{L^i_+} \cap \{v = c+\varepsilon\}$ on the configuration space with respect to the projections of L^a_+ or L^a_- is different from zero, then all the connections given in table IV occur. Here $c = v(L^a_+)$ and $\varepsilon > 0$. This condition holds for the equal masses configuration.

Table IV

	B^i_k	E^i_k	$L^i_{+,-}$	$L^a_{+,-}$	E^a_k	B^a_k
B^i_j		$\times(*)$	$\times$	$\times$	$\times$	$\times$
E^i_j			$\times$	$\times$	$\times$	$\times$
$L^i_{+,-}$					$\times$	$\times$

$$j, k = 1, 2, 3$$
$$(*)k \neq j$$

The connections we have shown in this paper are connections between Lagrange's points. When we have an isosceles masses configuration, $m_1 = m_2$, two sufficient conditions to assure that one of these connections exist are

(8) $$(\alpha, \alpha, 0, -\pi, \pi, \pi') \in W^u_{L^i_+} \cap \{v = 0\},$$

(9) $$(0, \alpha, 0, \pi, 0, \pi') \in W^u_{L^i_+} \cap \{v = 0\}.$$

Condition (8) corresponds to connections $L^i_{+,-} \longrightarrow L^a_{-,+}$ and (9) to $L^i_{+,-} \longrightarrow L^i_{+,-}$. They are related to an Euler and a Binary configuration respectively, both associated to points in the equator of the configuration space (see fig. 1). We may have in mind that this equator corresponds to the collinear configurations and also that the collinear problem is a subproblem of the one we are studing. As one can see there is a similitude between figures 5 and 9, which correspond to the planar and the collinear problem respectively. Considering the collinear problem included in the planar one, the invariant manifold $W^u_{E^i_3}$ (3–dimensional) contains the submanifold corresponding to the collinear problem (1–dimensional) and it lies on the equator of the configuration space (see fig. 1) between B_1 and B_2. The collinear submanifold has two branches (see [2]) which get successive binary configurations. That is, on the configuration space they go from B_1 to B_2. The curves corresponding to $W^u_{L^i_+} \cap \{v = \text{ctant.}\}$ have points in the equator of the configuration space and they lie between the two branches mentioned before. The behaviour of these branches (see fig.7,8,9) forces the mouvement of the points of the curve. In particular for the isosceles configurations it means that we will find points with configuration corresponding to (8) and (9).Therefore, we can conjecture that we will find an infinity of curves giving Lagrange connections.

REFERENCES

1. Moeckel, R., *Chaotic Dynamics Near Triple Collision*, Arch. Rat. Mechanics and Analysis **107** (1989), 37–70.
2. Simó, C., *Masses for which Triple Collision is Regularizable*, Celestial Mech. 21 (1980), 25–36.
3. _____, *Analysis of triple collision in the isosceles problem*, in "Classical Mechanics and Dynamical Systems," Ed.R.L. Devaney and Z. Nitecki, Marcel Dekker, 1981, pp. 203–224.
4. Simó, C.; Susín, A., *Connections between critical points in the collision manifold of the planar 3-body problem*, To appear Proceed. Workshop on the Geometry of Hamiltonian Systems, Berkeley (1989).
5. Susín, A., *Passages Near Triple Collision*, in "Long-Term Dynamical Behaviour of Natural and Artificial N-Body Systems," Ed. A.E.Roy. Reidel, 1988, pp. 505–513.
6. Waldvogel, J., *Symmetric and regularized coordinates on the plane triple collision manifold*, Celestial Mech. **28** (1982), 69–82.

ORBITS ASYMPTOTIC TO THE OUTERMOST KAM IN THE

RESTRICTED THREE-BODY PROBLEM

Masayoshi Sekiguchi
Astrometry & Celest. Mech. Div. National Astron. Obs.
Mitaka, Tokyo 181, Japan

Kiyotaka Tanikawa
Theor. Astrophys. Div. National Astron. Obs. Mitaka, Tokyo 181, Japan

Abstract

We checked next **Conjecture** by numerical integration for $C = 2.98$ and $\mu = 0.001$. We will briefly describe our procedure and will show the numerical results in this paper(Sekiguchi and Tanikawa,1990).

> **Conjecture:** There exist collision orbits with the planet in any neighborhood of the outermost KAM around the retrograde satellite orbit in the restricted three-body problem for certain values of the Jacobi constant C and the mass parameter μ.

1. Introduction

In the circular planar restricted three-body problem (hereafter referred to as RTBP), there are periodic solutions crossing the x-axis at two right angles. The symmetric and simply-periodic one is *retrograde satellite orbit* which is classified as *Class f* in Copenhagen category. It revolves around the planet clockwisely. The retrograde satellite orbit is stable for certain range of parameters(Fig.1). It becomes smaller and shrinks toward the planet as C increases. It becomes bigger as C decreases. It includes the collision with the sun as $C = 1.000798$ and $\mu = 0.001$ (Fig.2). It becomes a cometary orbit revolving around the sun as $C < 1.000798$ and $\mu = 0.001$, therefore the retrograde satellite orbit ends then. It is notable that it never collide with the planet.

Our assertion is that there exist orbits coming from somewhere, colliding with the planet and winding around the outermost KAM torus of the retrograde satellite orbit in the RTBP. We consider that it is quite rare to wind around prograde

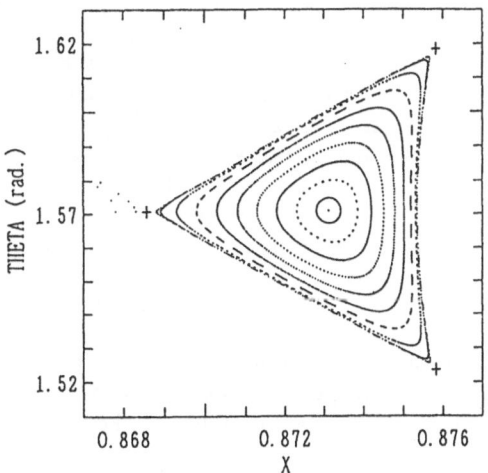

Fig.1 . The invariant region of the retrograde satellite orbit for $C = 2.98, \mu = 0.001$. Concentric curves are the KAM curves. The unstable 3-periodic points are denoted by '+'.

Predictability, Stability, and Chaos in N-Body Dynamical Systems
Edited by A.E. Roy, Plenum Press, New York, 1991

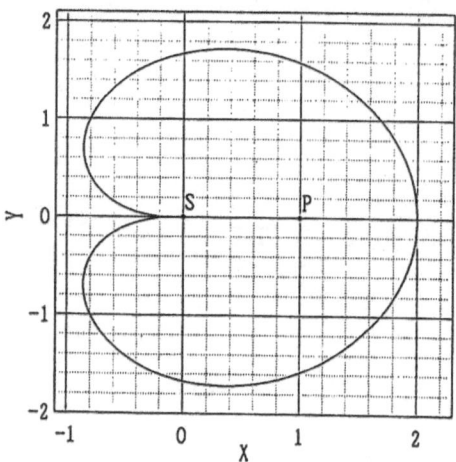

Fig.2 . The retrograde satellite orbit for $C = 1.000798$.

satellite orbits because projected particles from collision move clockwisely near collision due to the Coriolis force. Therefore we will not be able to apply our assertion to prograde satellites, but we can apply it to other retrograde satellite orbits, for instance, two or three periodic ones. We can also discuss the origin of retrograde satellites in the solar system by taking some dissipative processes into account, but we limited ourselves to only mathematical problem in this paper.

We performed numerical integrations only for short time span and we used a method of surface transformation instead of some long-time-span numerical integrations in order to check Conjecture. It is a merit of our procedure that it is sufficient to carry out only short-time-span numerical integrations.

2. The equations of motion

The equations of motion of the zero-mass particle in the RTBP are given as following(Fig.3).

$$\ddot{x} = 2\dot{y} + \Omega_x, \quad \ddot{y} = -2\dot{x} + \Omega_y, \qquad (1)$$

$$2\Omega = x^2 + y^2 + \frac{2(1-\mu)}{r_S} + \frac{2\mu}{r_P},$$

$$r_S^2 = (x+\mu)^2 + y^2, \quad r_P^2 = (x+\mu-1)^2 + y^2,$$

where a dot above letters denotes the derivative with respect to time t, Ω_x and Ω_y are partial derivatives of Ω with respect to x and y, respectively. The solution of Equations(1) will be denoted by $z(t) = (x(t), y(t))$. We can obtain the Jacobi constant C from (1).

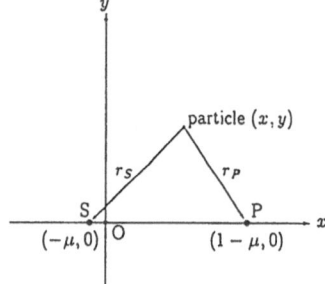

Fig.3. The coordinate system of the R.T.B.P.

$$C = 2\Omega - \dot{x}^2 - \dot{y}^2.$$

When the so-called *zero velocity curve* passes through the Lagrange's equilibrium points L_j $(j = 1, 2, \cdots, 5)$, C is denoted by C_j. When $C \geq C_4 (= C_5)$, the zero-velocity curve appears. In this paper we limit ourselves to the case $C < C_4 = 2.999001$ and $\mu = 0.001$.

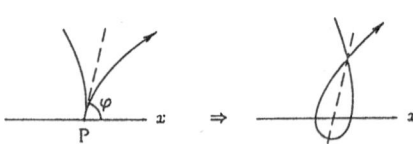

Fig.4 . The schematic explanation of our treatment of the collision orbit with the planet.

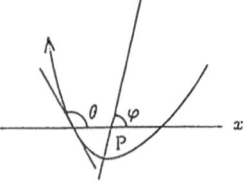

Fig.5 . Definition of the angles θ and φ at collision.

3. Collision

We treat collision especially in order to construct a surface of section. We regard the collision with the planet as a limit of a sequence of retrograde close approaches with peri-planet distance decreasing to zero(Fig.4).

Collision orbit becomes parabolic in the neighborhood of collision. The angles θ and φ are illustrated in Fig.5. The focus of the parabola is at rest on the planet. The particle moves on this parabola to the direction of the arrow, i.e., retrogradely. Then we easily obtain the relation: $\theta = (\varphi + \pi)/2$ for $\varphi \in (-\pi, \pi)$. We define $\theta = 0$ (resp. $\theta = \pi$) as $\varphi = -\pi$ (resp. $\varphi = \pi$).

4. Definitions

Let us define surfaces of section Σ, Σ_1, Σ_2 and Σ_3. Take any orbit $z(t) = (x(t), y(t))$ intersecting the x-axis.

Definition 1: $\Sigma \equiv \{(x, \theta) \mid x \neq -\mu, \; y = 0, \; \theta \equiv \arctan(\dot{y}/\dot{x}) \in [-\pi, \pi)\}$.

Aux.Def. 2: $\Sigma_1, \Sigma_2, \Sigma_3$ are subsets of Σ, and
$$\Sigma_1 \equiv \{(x, \theta) \mid x \in (-\mu, 1 - \mu], \theta \in (0, \pi)\},$$
$$\Sigma_2 \equiv \{(x, \theta) \mid x \in [1 - \mu, \infty), \theta \in (-\pi, 0)\},$$
$$\Sigma_3 \equiv \{(x, \theta) \mid x \in [1 - \mu, \infty), \theta \in [-\pi, 0]\}.$$

We shall explain the reason why we select the variable θ which is not a canonical variable. The velocity becomes infinity at collision, but the intersection angle θ is bounded still at collision. Therefore we can make the surface of section a compact space by using θ.

Next, we shall define *point sequence of m+n+1 elements* $\{p_{-m}, \cdots, p_0, \cdots, p_n\}$ and *orbit revolving around the planet retrogradely once at least*.

Definition 3: Suppose that there exists a sequence of times
$$t_{-m} < \cdots < t_0 < \cdots < t_n,$$
such that an orbit $z(t)$ intersects the x-axis at $t_{-m}, \cdots, t_0, \cdots, t_n$, and that it never intersects the x-axis at any time in (t_j, t_{j+1}) for any integer $j \in [-m, n-1]$. Then $\{p_{-m}, \cdots, p_0, \cdots, p_n\} \equiv \{p(t_{-m}), \cdots, p(t_0), \cdots, p(t_n)\}$ is called the point sequence of $m + n + 1$ elements of the orbit $z(t)$. We can put $t_0 = 0$ without losing generality.

Definition 4: Take a point sequence of 5 elements $\{p_{-1}, p_0, \cdots, p_3\}$. If the point sequence satisfies next conditions, the corresponding orbit $z(t)$ is called the orbit revolving around the planet retrogradely once at least.

- $p_{-1} \in \Sigma_3$ and
- $p_0 \in \Sigma_1$ at time $t = 0$ and
- $p_1 \in \Sigma_2$ and
- $p_2 \in \Sigma_1$ and
- $p_3 \in \Sigma_3$.

The set of these orbits is termed $W_1(C, \mu)$. We can also write $\{p_{-1}, p_0, \cdots, p_3\} \in W_1$. Then the sets of p_0 and p_2 are termed $\omega_1(C, \mu)$ and $\omega_1^{(1)}(C, \mu)$, respectively. These can be written to W_1, ω_1 and $\omega_1^{(1)}$ in abbreviation since we fix C and μ.

Definition 5: Take a point sequence of 5 elements $\{p_{-1}, p_0, \cdots, p_3\} \in W_1$. The mapping $T : \omega_1 \to \omega_1^{(1)}$ is defined as below.

$$T\, p_0 = p_2.$$

We use $T\omega_1$ instead of $\omega_1^{(1)}$.

Aux.Def. 6: W_1^c is a subset of W_1. and W_1^c orbits include *one* collision with the planet at least.

Aux.Def. 7: ω_1^c is a subset of ω_1, and the set of representative point p_0 of W_1^c.

Definition 8: W_n, ω_n, W_n^c and ω_n^c are defined similarly to W_1, ω_1, W_1^c and ω_1^c, respectively. W_n and W_n^c orbits revolve around the planet n times at least successively.

Definition 9: W is a set of orbits revolving around the planet retrogradely an infinite times.

Definition 10: ω is a set of representative points of W. ω is strictly defined as followings.

$$\omega \equiv \lim_{n \to \infty} T^n \omega_\infty,$$

$$\omega_\infty \equiv \lim_{n \to \infty} \omega_n.$$

$\omega_1, \omega_1^c, \omega_n, \omega_n^c$ and ω all appear in the surface of section Σ_1. Fig.1 is exactly a numerical observation of ω.

5. Three theorems and three properties

We obtained next three theorems from the definitions in the previous section.

Theorem 1: The surface transformation T is a homeomorphism, i.e., there exists the inverse transformation T^{-1}, and both T and T^{-1} are continuous on their domains, respectively.

Theorem 2: $T^{n-1}\omega_n = \omega_1 \cap (T^{n-1}\omega_{n-1})$ for any integer $n \in [2, \infty)$.

Theorem 3: $\omega_n \supset \omega_{n+1} \supset T^{m-1}\omega_\infty \supset T^m\omega_\infty \supset \omega,$ for any integer $n, m \in [1, \infty)$.

Let us define some terms in order to state properties and propositions which will appear in the next section. For any set A, $\overline{A}$ is the closure of A, $\text{int}A$ is the interior of A and $\partial A = \overline{A} - \text{int}A$ is the boundary of A. Let $U_\epsilon(p)$ denote an ϵ-neighborhood of $p \in \Sigma$ with the Euclidean metric.

Property 1: $T\partial\omega_n = \partial T\omega_n$.

Property 2: $T^{n-1}\partial\omega_n = \{\partial\omega_1 \cap (T^{n-1}\omega_{n-1})\} \cup \{\omega_1 \cap (T^{n-1}\partial\omega_{n-1})\}$.

Property 3: $T^{n-1}\omega_n^c = \{\omega_1^c \cap (T^{n-1}\omega_{n-1})\} \cup \{\omega_1 \cap (T^{n-1}\omega_{n-1}^c)\}$.

Property 1 is derived from **Theorem 1**. We obtained **Property 2** from **Theorem 2** and **Property 1**. **Property 3** is obtained from the definition of W_n^c. The proof of **Property 2** will be shown as following.

The proof of **Property 2**

$$
\begin{aligned}
T^{n-1}\partial\omega_n &= \partial T^{n-1}\omega_n \\
&= \partial\{\omega_1 \cap (T^{n-1}\omega_{n-1})\} \\
&= \{\partial\omega_1 \cap (T^{n-1}\omega_{n-1})\} \cup \{\omega_1 \cap \partial(T^{n-1}\omega_{n-1})\} \\
&= \{\partial\omega_1 \cap (T^{n-1}\omega_{n-1})\} \cup \{\omega_1 \cap (T^{n-1}\partial\omega_{n-1})\}.
\end{aligned}
$$

Q.E.D.

6. Three propositions

Let us divide **Conjecture** into next three propositions. We can complete the proof of Conjecture by proving these propositions.

Proposition 1: For a certain value of (C, μ), $\partial \omega_1 = \omega_1^c$.

Proposition 2: If $\partial \omega_1 = \omega_1^c$, then $\partial \omega_n = \omega_n^c$.

Proposition 3: Suppose that $\partial \omega_n = \omega_n^c$. For any $\varepsilon > 0$ and any point $p \in \partial \omega$, there exist a sufficiently large integer N and a point $q \in \omega_N^c$ such that

$$q \in U_\varepsilon(p).$$

We checked **Proposition 1** by numerical integrations. The results are shown in Fig.6 and 7. We can see that the boundary of ω_1 consists of collision orbits. We proved **Propositions 2 and 3** by using **Theorem 3, Properties 2 and 3**. The proof of **Propositions 2** will be shown as following.

The proof of **Proposition 2**

We use **Properties 2 and 3** in the case of $n = 2$. $\partial \omega_1 = \omega_1^c$ holds true from assumption.
$$
\begin{aligned}
T\partial \omega_2 &= \{\partial \omega_1 \cap (T\omega_1)\} \cup \{\omega_1 \cap (T\partial \omega_1)\} \\
&= \{\omega_1^c \cap (T\omega_1)\} \cup \{\omega_1 \cap (T\omega_1^c)\} \\
&= T\omega_2^c.
\end{aligned}
$$
By repeating this operation, this proposition can be proved.

$$Q.E.D.$$

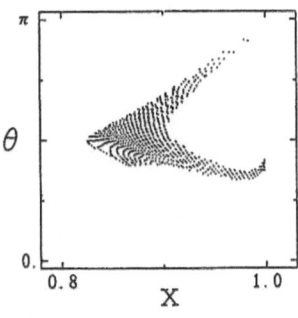

Fig.6 . The set ω_1 obtained numerically.

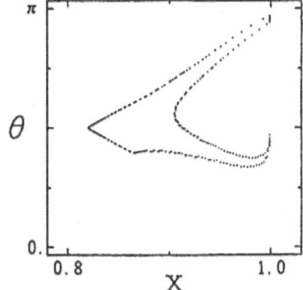

Fig.7 . The set ω_1^c obtained numerically.

Acknowledgments

Computations were carried out on the FACOM M780/10S computer at the Astronomical Data Analysis Center of the National Astronomical Observatory. One of the authors M.Sekiguchi participated this NATO ASI under the financial aid of Inoue Foundation for Science.

References

Arnold,V.I.:1963, Russ. Math. Surv. **18**, 85.

Hagihara,Y.:1975, *Celestial Mechanics*, (Japan Society for the Promotion of Science), Vol.4, Part 1, p.36.

Moser,J.:1962, Nachr. Akad. Wiss. Goettingen Math. Phys. K1, 1.

Sekiguchi,M. and Tanikawa,K.:1989, *Proc. Symp. 23 "Celestial Mechanics"*, H.Kinoshita and H.Yoshida (eds.), (*in Japanese*), p.96.

Sekiguchi,M. and Tanikawa,K.:1990, submitted to J. of Math. Phys.

Stiefel,E.L. and Scheifele,G.:1971, *Linear and Regular Celestial Mechanics*, (Springer-Verlag Berlin Heidelberg New York), p.20.

Szebehely,V.:1967, *Theory of Orbits*, (Academic Press, New York), pp. 455, 463.

Tanikawa,K.:1979, *'Dynamics of the Solar System'*, (R.L.Duncombe(ed.), D.Reidel), p.181.

Tanikawa,K.:1983, Celest. Mech. **29**, 367.

Tanikawa,K. and Yamaguchi,Y.:1990, submitted to J. of Math. Analy. Appl.

PART V

SELECTED TOPICS IN DYNAMICS

A NEW INTERPRETATION OF COLLISIONS IN THE N-BODY PROBLEM

John G. Bryant

47, avenue Felix Faure
75015 Paris, France

INTRODUCTION

At the previous meeting in Cortina [1], I showed how it was possible to reformulate the N-Body Problem in such a way that the velocities remain bounded, even at collision, and the motion of particles with zero mass can be described; at the same time the classical properties of planetarty motion are preserved. In this paper we will see how the modified formulation can shed new light on an important aspect of the N-Body Problem, i.e. the behaviour at collision, which can now be interpreted as a process for the *emission* or *absorption* of particles, in an essentially *classical* context.

1. THE REFORMULATED N-BODY PROBLEM

1.1 We first of all give a brief reminder of how to construct the new Hamiltonian for the 3-Body problem. We start by writing the classical Hamiltonian in a system of Jacobi coordinates

$$H(p_{Gi}, p'_i, p''_i, q_{Gi}, q'_i, q''_i) = \frac{p_G^2}{2m} + \frac{p'^2}{2m'} + \frac{p''^2}{2m''} - f\frac{\bar{m}m'}{r'} - f\frac{mm''}{r''} \quad (1.1.1)$$

where

$$p_G = \sqrt{\sum_{i=1}^{3} p_{Gi}^2} \quad ; \quad p' = \sqrt{\sum_{i=1}^{3} p'^2_i} \quad ; \quad p'' = \sqrt{\sum_{i=1}^{3} p''^2_i}$$

$$r' = r_{12} = \sqrt{\sum_{i=1}^{3} q'^2_i} \quad ; \quad \frac{m_1 + m_2}{r''} = \frac{m_1}{r_{13}} + \frac{m_2}{r_{23}}$$

$$r_{13} = \sqrt{\sum_{i=1}^{3}(q''_i - \frac{m_2}{\bar{m}}q'_i)^2} \quad ; \quad r_{23} = \sqrt{\sum_{i=1}^{3}(q''_i + \frac{m_1}{\bar{m}}q'_i)^2}$$

$$m = m_1 + m_2 + m_3 \; ; \; \bar{m} = m_1 + m_2 \; ; \; m' = m_1 m_2 / \bar{m} \; ; \; m'' = \bar{m}m_3 / m$$

Predictability, Stability, and Chaos in N-Body Dynamical Systems
Edited by A.E. Roy, Plenum Press, New York, 1991

N.B. To obtain the Hamiltonian for 4 and more bodies, we must add the term $p'''^2/2m'''$, as well as the term

$$f\frac{m'''(m_1 + m_2 + m_3 + m_4)}{r'''} = fm_4\left(\frac{m_1}{r_{14}} + \frac{m_2}{r_{24}} + \frac{m_3}{r_{34}}\right)$$

where $m''' = \dfrac{m_4(m_1 + m_2 + m_3)}{m_1 + m_2 + m_3 + m_4}$, and so on for each successive body.

The Hamiltonian system associated with (1.1.1) is written

$$\frac{dq_{Gi}}{dt} = \frac{\partial H}{\partial p_{Gi}} = \frac{p_{Gi}}{m} \quad ; \quad \frac{dp_{Gi}}{dt} = -\frac{\partial H}{\partial q_{Gi}} = 0$$

$$\frac{dq'_i}{dt} = \frac{\partial H}{\partial p'_i} = \frac{p'_i}{m'} \quad ; \quad \frac{dp'_i}{dt} = -\frac{\partial H}{\partial q'_i} = m'\alpha'\frac{\partial}{\partial q'_i}\left(\frac{1}{r'}\right) + m''\alpha''\frac{\partial}{\partial q'_i}\left(\frac{1}{r''}\right)$$

$$\frac{dq''_i}{dt} = \frac{\partial H}{\partial p''_i} = \frac{p''_i}{m''} \quad ; \quad \frac{dp''_i}{dt} = -\frac{\partial H}{\partial q''_i} = m''\alpha''\frac{\partial}{\partial q''_i}\left(\frac{1}{r''}\right) \qquad (1.1.2)$$

where $\alpha' = f\bar{m}$; $\alpha'' = fm$

1.2 In place of the classical Hamiltonian (1.1.1), we now consider the following Hamiltonian

$$E = c\sqrt{m_o^2 c^2 + p_G^2} + c\sqrt{\frac{m_o'^2 c^2 + p'^2}{1 + \dfrac{2\alpha'}{c^2 r'}}} + c\sqrt{\frac{m_o''^2 c^2 + p''^2}{1 + \dfrac{2\alpha''}{c^2 r''}}} \qquad (1.2.1)$$

$$= \qquad M_G c^2 \qquad + \qquad M' c^2 \qquad + \qquad M'' c^2$$

where m_o, m'_o, and m''_o are *modified* reduced masses replacing the classical reduced masses m, m', and m", and c is the velocity of light. (E therefore has the dimension of energy.) The associated system of Hamiltonian equations is written

$$\frac{dq_{Gi}}{d\tau} = \frac{p_{Gi}}{M_G} \quad ; \quad \frac{dp_{Gi}}{d\tau} = 0$$

$$\frac{dq'_i}{d\tau} = \frac{p'_i}{M'\left(1 + \dfrac{2\alpha'}{c^2 r'}\right)} \quad ; \quad \frac{dp'_i}{d\tau} = \frac{M'\alpha'}{\left(1 + \dfrac{2\alpha'}{c^2 r'}\right)}\frac{\partial}{\partial q'_i}\left(\frac{1}{r'}\right) + \frac{M''\alpha''}{\left(1 + \dfrac{2\alpha''}{c^2 r''}\right)}\frac{\partial}{\partial q'_i}\left(\frac{1}{r''}\right)$$

$$\frac{dq''_i}{d\tau} = \frac{p''_i}{M''\left(1 + \dfrac{2\alpha''}{c^2 r''}\right)} \quad ; \quad \frac{dp''_i}{d\tau} = \frac{M''\alpha''}{\left(1 + \dfrac{2\alpha''}{c^2 r''}\right)}\frac{\partial}{\partial q''_i}\left(\frac{1}{r''}\right) \qquad (1.2.2)$$

Note that contrary to the classical system (1.1.2), the above system is well defined *even when the reduced masses m'_o and m"_o are equal to zero.* (We always assume that $m_o \neq 0$.) We have the following relation between M_G, M', M", r', r", and the Jacobi velocities

$$w_G = \sqrt{\sum_{i=1}^{3}\left(\frac{dq_{Gi}}{d\tau}\right)^2} \quad ; \quad w' = \sqrt{\sum_{i=1}^{3}\left(\frac{dq'_i}{d\tau}\right)^2} \quad ; \quad w'' = \sqrt{\sum_{i=1}^{3}\left(\frac{dq''_i}{d\tau}\right)^2}$$

$$\left(\frac{w_G}{c}\right)^2 = 1 - \frac{m_o^2}{M_G^2} \quad ; \quad \left(\frac{w'}{c}\right)^2 = \frac{1}{1 + \frac{2\alpha'}{c^2 r'}}\left(1 - \frac{m_o'^2/M'^2}{1 + \frac{2\alpha'}{c^2 r'}}\right)$$

$$\left(\frac{w''}{c}\right)^2 = \frac{1}{1 + \frac{2\alpha''}{c^2 r''}}\left(1 - \frac{m_o''^2/M''^2}{1 + \frac{2\alpha''}{c^2 r''}}\right) \tag{1.2.3}$$

which show that we always have

$$\frac{w_G}{c} \leq 1 \quad ; \quad \frac{w'}{c} \leq 1 \quad ; \quad \frac{w''}{c} \leq 1 \tag{1.2.4}$$

and that

$$r' \longrightarrow 0 \implies w' \longrightarrow 0 \; ;$$

$$r'' \longrightarrow 0 \text{ (i.e. } r_{13} \longrightarrow 0 \text{ or } r_{23} \longrightarrow 0) \implies w'' \longrightarrow 0 \tag{1.2.5}$$

When $m_o' \neq 0$ and $m_o'' \neq 0$, relations (1.2.3) can be solved for M_G, M' and M''

$$M_G = \frac{m_o}{\sqrt{1 - \left(\frac{w_G}{c}\right)^2}} \quad ; \quad M' = \frac{m_o'}{\sqrt{1 + \frac{2\alpha'}{c^2 r'}}\sqrt{1 - \left(1 + \frac{2\alpha'}{c^2 r'}\right)\left(\frac{w'}{c}\right)^2}} \quad ;$$

$$M'' = \frac{m_o''}{\sqrt{1 + \frac{2\alpha''}{c^2 r''}}\sqrt{1 - \left(1 + \frac{2\alpha''}{c^2 r''}\right)\left(\frac{w''}{c}\right)^2}} \tag{1.2.6}$$

It follows that if

$$\frac{w_G}{c} \ll 1 \quad ; \quad \frac{w'}{c} \ll 1 \quad ; \quad \frac{w''}{c} \ll 1 \quad ;$$

$$\frac{2\alpha'}{c^2 r'} \ll 1 \quad ; \quad \frac{2\alpha''}{c^2 r''} \ll 1 \quad ; \tag{1.2.7}$$

which is the case for planetary motion, then we have

$$M_G \simeq m_o \quad ; \quad M' \simeq m_o' \quad ; \quad M'' \simeq m_o''$$

and system (1.2.2) is practically *identical* to the classical system (1.1.2).

N.B. To obtain the modified Hamiltonian for 4 and more bodies, we must add the term

$$M'''c^2 = c\sqrt{\frac{m_o'''^2 c^2 + p'''^2}{1 + \frac{2\alpha'''}{c^2 r'''}}}$$

and so on for each successive body. Relation (1.2.6) remains valid for M''', so that in the case of planetary motion we also have $M''' \simeq m_o'''$

The differences between the two formulations will appear *outside* the domain of planetary motion, when conditions (1.2.7) no longer hold. This is the case for motions near collision, as we shall see in the following section.

2. A NEW INTERPRETATION FOR BINARY COLLISIONS

2.1. In this section. we first of all consider the Three-Body Problem , and assume that bodies 2 and 3 collide at a finite instant τ_0 . The modified Hamiltonian E gives rise to a new "energy" integral

$$E = M_G c^2 + M' c^2 + M'' c^2 = \text{constant} = mc^2 \qquad (2.1.1)$$

(Note that m no longer designates the total Newtonian mass.) Since, according to the first set of equations (1.2.2), M_G is also constant. it follows that

$$M' + M'' = \text{constant} \qquad (2.1.2)$$

During the motion, M' and M" are therefore bounded *above and below* since, by definition,

$$M' \geq 0 \; ; \; M'' \geq 0$$

and this is true *even at collision*. We now propose to show that, in the case of collision between bodies 2 and 3 (as well as between 1 and 3), *M" goes to zero*, i.e.

$$r'' \longrightarrow 0 \quad \Rightarrow \quad M'' \longrightarrow 0 \qquad (2.1.3)$$

To see this. we first of all write again the expression for M"

$$M'' c^2 = c \sqrt{\frac{m_0''^2 c^2 + p''^2}{1 + \frac{2\alpha''}{c^2 r''}}}$$

When r" goes to zero, we see that $M'' \simeq p'' \sqrt{\frac{r''}{2\alpha''}}$, so it is enough to prove that p" is bounded. Consider the canonical change of variables

$$(p'_1 , p''_1 , q'_1 , q''_1) \longrightarrow (\bar{p}_1 , p_{231} , \bar{q}_1 , q_{231})$$

defined by

$$\bar{p}_1 = p'_1 + \frac{m_{10}}{m_{10} + m_{20}} p''_1 \; ; \; p_{231} = p''_1$$

$$\bar{q}_1 = q'_1 \; ; \; q_{231} = q''_1 - \frac{m_{10}}{m_{10} + m_{20}} q'_1 \qquad (2.1.4)$$

(It is easily checked that $\sum(p'_1 dq' + p''_1 dq''_1) = \sum (\bar{p}_1 d\bar{q}_1 + p_{231} dq_{231})$.)

The Hamiltonian E (less the constant term $M_G c^2$) is written in the new variables

$$E = c \sqrt{\frac{m_0'^2 c^2 + \sum(\bar{p}_1 - \frac{m_{10}}{m_{10} + m_{20}} p_{231})^2}{1 + \frac{2\alpha'}{c^2 r'}}} + c \sqrt{\frac{m_0''^2 c^2 + p_{23}^2}{1 + \frac{2\alpha''}{c^2 r''}}}$$

$$(2.1.5)$$

with $r' = \sqrt{\sum \bar{q}_1^2}$

and
$$\frac{m_{1o} + m_{2o}}{r''} = \frac{m_{1o}}{r_{13}} + \frac{m_{2o}}{r_{23}} = \frac{m_{1o}}{\sqrt{\sum(\bar{q}_i + q_{23i})^2}} + \frac{m_{2o}}{\sqrt{\sum q_{23i}^2}}$$

We can write

$$\frac{d\bar{p}_i}{d\tau} = \frac{\partial E}{\partial \bar{q}_i} = \frac{M'\alpha'}{1 + \frac{2\alpha'}{c^2 r'}} \frac{\partial}{\partial \bar{q}_i}\left(\frac{1}{r'}\right) + \frac{M''\alpha''}{1 + \frac{2\alpha''}{c^2 r''}} \frac{\partial}{\partial \bar{q}_i}\left(\frac{m_{1o}}{m_{1o}+m_{2o}} \frac{1}{r_{13}}\right)$$

and, since $r' = r_{12}$ and r_{13} are bounded away from zero, we see that $\dfrac{d\bar{p}_i}{d\tau}$ remains finite when $r'' \longrightarrow 0$. It follows that $\bar{p}_i$ is also bounded. At the same time p'_i is necessarily finite because M' is bounded as we know. Therefore

$$p''_i = \frac{m_{1o} + m_{2o}}{m_{1o}}(\bar{p}_i - p'_i)$$

as well as $\quad p'' = \sqrt{\sum_{i=1}^{3} p''^2_i}\quad$ are bounded, and M'' goes to zero with r''.

N.B. A closer examination of the equations of motion shows in fact that M'' behaves like $|\tau - \tau_o|^{1/2}$. and r_{23} like $|\tau - \tau_o|$.

2.2 The term $M'c^2$ in E can be interpreted as the relative "energy" of body 2 with respect to body 1. and $M''c^2$ as the relative "energy" of body 3 with respect to the system formed by bodies 1 and 2. In the event of a collision of body 3 with body 2 (or body 1), the above result shows that body 3 loses all of its energy to the system formed by bodies 1 and 2, since M'' goes to zero, while at the same time, the sum $M' + M''$ remains constant.

We are therefore led to the remarkable interpretation that the collision of body 3 with body 2 or body 1 results in the *absorption* of body 3. And by reversing the process, we can state that the ejection of body 3 from body 1 or 2, leading to the acquisition of a non-zero value for M'' at the expense of M', results from the *emission* of body 3.

We note that. although the classical Hamiltonian (1 .1 .1) can also be written in the form $H_G + H' + H''$. neither term corresponding to the relative energies is necessarilly bounded at collision. and in fact the relative velocity becomes infinite as is well known. Using standard techniques, the classical equations can be regularized, and the collision motion can be interpreted as an "elastic bounce" of body 3 against body 1 or 2, i.e. *the total number of bodies is preserved*

With the new interpretation, *there is no need to suppose that this is in fact the case*: we could just as well assume that before ejection, or after collision, we are in the presence of an ordinary 2-Body Problem where the relative energy of the two bodies is constant and equal to the partial energy $M'c^2 + M''c^2$ of the subsequent or previous 3-Body Problem. More generally, we could even assume that we are in the presence of entirely different 3-Body Problems,where not even the values of the masses are necessarily preserved, so long as we have the conservation of the total energy.

2.3 The above interpretation can be easily extended.to binary collisions in the N-Body Problem. For example, in the case of a collision between bodies 2 and

4 in the 4-Body Problem, one can show, after a suitable canonical change of variables, that p''' remains finite and M''' goes to zero with r'''. Body 4 loses all of its energy with respect to the system formed by the other three bodies, and is in fact *absorbed* by body 2. As before we do not have to assume that the total number of bodies is constant before and after collision, only the total energy.

A process of the above type corresponds exactly to what occurs in particle physics when we consider the collision of two elementary particles, *if we make the assumption that the masses of all the other particles are negligible with respect to that of Body 1, which plays the role of the observer*. It follows from this assumption that

$$m'_o \simeq m_{2o} \; ; \; m''_o \simeq m_{3o} \; ; \; m'''_o \simeq m_{4o} \quad \text{etc.}$$

and it is easily verified that the standard laws of the conservation of energy and momentum for the system formed by the particles, less the observer, remain valid, as long as the attraction of the observer is neglected with respect to the mutual attraction of the particles (as is done classically). Now however, and contrary to what happens classically, *it is possible to follow the motion right up to the instant of collision*, i.e. we no longer have to assume (as in [2]) that the collision interaction takes place in some unspecified "black box".

Finally, we note that the interaction does *not* have to be solely *gravitational* in nature. More general types (depending on the *charges* of the particles for example) can be obtained by simply modifying the nature of the constants α', α', α''', etc. that appear in the new N-Body Hamiltonian.

2.4 In the case of celestial mechanics, where the dimensions of the bodies must be taken account of in the vicinity of collision, neither the classical point of view nor the one given above are valid, since they assume that each body is in fact *point-like*. Their validity is limited to near-collision orbits, and they give essentially the *same* description, since conditions (1.2.7) still apply.

Contrary to the classical theory however, the new one can be applied when the bodies are for all practical purposes point-like, such as naked *black holes* and *photons*. For instance, we could consider a system of two self-gravitating black holes, A and B, and assume that at an instant τ_o, A emits a photon C. The Hamiltonian for the motion is the one given by (1.2.1), where we must set $m'' = 0$. Among the possible motions, there could be the temporary capture of C by B, or even its reabsorption. In the case where the energy of C is small with respect to that of the binary, the general motion resembles that of the classical Restricted Three-Body Problem, except for the (essential) fact that *the motion of the massless particle C affects that of the massive ones A and B* (even if only very slightly). And nothing prevents us from assuming that the energy of C is *not* negligible with respect to that of A and B, in which case the motion of the binary is *strongly* affected by C (as in the regular Three-Body Problem).

2.5 As is well known, higher order collisions play a central role in the classical N-Body Problem, owing to the non-regularizable nature of the corresponding singularity. The fact that such collisions actually occur is guaranteed by the existence of a special class of motions known as *central configurations*, i.e. motions where the q'_1, q''_1 have the form

$$q'(t) = q'_{1o} \lambda(t) \; ; \; q''(t) = q''_{1o} \lambda(t)$$

with q'_{1o} = constant ; q''_{1o} = constant. However, a close examination of the modified Hamiltonian system (1.2.2) shows that the existence of central configurations appears highly unlikely in the new formulation of the N-Body Problem ; we therefore do not have at our disposal a simple method for determining multiple-collision orbits as we do classically. In fact, the only "obvious" example of a triple-collision motion that comes to mind occurs when two equal masses are symmetrically placed about a third one, with opposite velocities towards each other, the third mass being fixed. (This is really a double double-collision motion.) Note that in the new formulation, it is possible to obtain motions that remain very close to the classical central configurations, as long as conditions (1 .2.7) hold.

In a forthcoming article, we will give a much more detailed description of the above and other properties of the reformulated N-Body Problem. Although this is done in a basically classical framework, the relativistic and quantum mechanical aspects will also be examined.

REFERENCES

[1] Bryant, J.G. : 1988, 'A Formulaton of the N-Body Problem where the velocities are bounded', in Long Term Dynamical Behaviour of Natural and Artificial N-Body Systems, ed. A.E.Roy, NATO ASI Series, Kluwer.

[2] Goldstein. H. : 1980, 'Classical Mechanics' (2nd edition), Addison-Wesley

AN IMPULSIONAL METHOD TO ESTIMATE THE LONG–TERM

BEHAVIOUR OF A PERTURBED SYSTEM: APPLICATION TO

A CASE OF PLANETARY DYNAMICS

Bertrand Chauvineau

Observatoire de la Côte d'Azur, Avenue Copernic, 06130 Grasse, France

Abstract. This paper aims to present a method to investigate the long term evolution of a perturbed system. The unperturbed system is supposed to possess an integral of the form:

$$\dot{r}^2 + h(r)$$

and the perturbation to be time-dependant. The results are compared to direct analytical and numerical computations in the case of a perturbed harmonic oscillator. Then, it is shown how this method applies to the lifetime of a binary asteroid perturbed by Jupiter.

I) Introduction

The long term evolution of planetary systems is of great interest in astronomy to study their stability. Lagrange and Laplace have initiated the study of the long term evolution of the solar system at the first order with respect to the masses. This problem is studied in great details with help of numerical methods (Milani et al., 1987; Carpino et al., 1987; Laskar, 1985, 1986, 1988, 1989). However, the long time required to obtain precise numerical integrations limits the possibilities of such explorations. This problem appears as soon as the precise evolution of the orbits or of the osculating elements corresponding to a given system (defined by its elements and its initial conditions) is explored.

This paper presents an analytical method to evaluate the statistical properties of the long term evolution of a dynamical system. The particularity of this method is to replace the continuous perturbations acting on the system by impulsional perturbations even in the case where the characteristic time of the continuous perturbation is not small, compared to the other characteristic times. In such an approach, the precise evolution of the orbits is rapidly lost. The precise evolution of the secular elements is lost too. However, in some cases, the application of the method to a simple physical system shows that it gives the correct mean secular evolution of the system in a sense more precisely defined latter. As a planetary application, the method is applied to the long term evolution of a binary asteroid orbiting the Sun and perturbed by Jupiter.

II) The perturbed harmonic oscillator

1) Description of the system

Let us consider a one degree of freedom perturbed harmonic oscillator, the evolution of which is described by the differential equation:

$$\ddot{x} = -x + \epsilon(t)x$$

where $\epsilon(t)$ is a step function of the time taking alternatively the values 0 and μ such that $0 \leq \mu << 1$. The so called unperturbed case is obtained for $\mu = 0$. In this case, the energy:

$$E = \dot{x}^2 + x^2$$

is preserved through the evolution of the system. If $\mu > 0$, E is not preserved. Let us note N the number of the perturbation (the Nth step with $\mu > 0$). Let be τ the duration of the perturbation, not small compared to the evolution time θ of the system, and $T(N)$ the elapsed time between the beginnings of the Nth and the $(N + 1)^{th}$ perturbation. Let us take $T(N)$ under the form:

$$T(N) = T_0 \left(1 + \lambda \sin \frac{2\pi N}{N_0} \right) \tag{1}$$

where T_0 is the mean value of $T(N)$, λ a constant such that $0 \leq \lambda < 1$ and N_0 any *real* number. We would like to derive the mean form of the value of $E(N) = \dot{x}^2 + x^2$ after the Nth perturbation.

During one perturbation, the variation of E is given by:

$$\frac{dE}{dt} = \frac{d}{dt}(\dot{x}^2 + x^2) = 2\mu x \dot{x}$$

and the value of the energy is shifted by the value:

$$\Delta E = \mu \Delta(x^2)$$

For the unperturbed harmonic oscillator:

$$x = \sqrt{E} \sin(t - t_0)$$

τ being roughly greater or equal than the evolution time θ of the system, it is easy to derive the mean value and the dispersion of ΔE for a lot of initial conditions, and one finds:

$$< \Delta E >= 0$$

$$\sigma_{\Delta E} = \frac{\mu E}{2}$$

assuming that the initial and final values of the position are uncorrelated.

2) Impulsional modelisation

We would like to replace the previous step perturbations by Dirac impulsions, the amplitudes of which being such that the statistical global evolution of $E(N)$ is preserved. Let us then consider the system:

$$\ddot{x} + x = Ax\delta^{\tilde{}}(t)$$

where $\delta^{\tilde{}}$ is a succession of Dirac's distributions, the Nth and the $(N + 1)^{th}$ being separated by $T(N)$. A is an impulsional coefficient to be adjusted. For a single perturbation, the energy variation is:

$$\Delta E = \Delta E_1 + \Delta E_2$$

with:

$$\Delta E_1 = 2Ax\dot{x} = AE \sin(2t - 2t_0)$$

$$\Delta E_2 = A^2 x^2 = A^2 E \sin^2(t - t_0)$$

At the first order in A (neglecting in a first time ΔE_2), the values of $< \Delta E >$ and $\sigma_{\Delta E}$ are recovered if:

$$A = \mu/\sqrt{2}$$

This gives for the mean value of ΔE, for one perturbation:

$$< \Delta E > = < \Delta E_2 > = \frac{\mu^2 E}{4}$$

Assuming that the system has an ergodic behaviour, its mean long time evolution is governed by:

$$\frac{dE}{dN} = \frac{\mu^2 E}{4} \qquad [2]$$

which implies an exponential long term evolution of the energy:

$$\mathrm{Ln}(E(N)/E(0)) = \mu^2 N/4 \qquad [3]$$

The total energy of the system presents then a secular effect.

3) Comparison with numerical experiments

Because $\tau \sim$ or $> \theta$, the solutions for the impulsional and the step perturbed systems are quit different. But with $A = \mu/\sqrt{2}$, the statistical properties of the two types of perturbations are sensitively similar, and it is hoped that the long time mean evolutions are comparable in these two cases. It is hoped too that the previous analytical treatment gives the correct mean value for the secular effect of the long term energy variation. Besides, the differential equation [2] has been obtained assuming that the system presents an ergodic behaviour, which is probably not the case for simple functions $T(N)$ like [1].

The numerical simulations show that, for given values of μ, T_0, τ and λ, the energy of the system presents a secular effect for some values of N_0 and no secular effect for the other values. The figure [1] shows the typical aspect of the numerically computed function

$$p = < \frac{d \, \mathrm{Ln} E(N)}{dN} >$$

on an interval of values N_0. p is sometimes positive, but never negative. Then, it seems very unprobable that the cases $p \neq 0$ do not correspond to real secular comportments.

Several such numerical experiments have been made for step perturbed harmonic oscillators, for different values of μ, T_0, τ and λ. For each such experiment, an interval of values of N_0 is chosen, in which the values of p are computed as in the figure [1]. The mean value $< p >$ of the founded values p is computed in each such interval, and one finds that the values $< p >$ are generally close to the theoretical value $\mu^2/4$. In fact, it is clearly the case for $\lambda \neq 0$. For $\lambda = 0$, the perturbation is periodic and we are in a case of parametric resonance. In this case, an analytical analysis of the comportment of the system can be made, and it shows that the size of the interval of N_0 for which there is secular effect is very weak. However, it is possible to show that the mean value of p over a large interval in N_0 remains of the order of μ^2.

Then, the founded law [3] does not correspond to the evolution of each system, but to the mean evolution of a great number of such systems.

III) Jovian perturbations on a binary asteroid

1) Equations of the relative motion

We aim in this chapter to show how the previous method gives an estimate of the lifetime of a binary asteroid. Let us consider a binary asteroid (two close asteroids) orbiting the Sun. The motions of the two objects are supposed to be perturbed by the Sun and Jupiter only. The jovian's perturbation is supposed to be weak enough so that the orbital motion of the binary system about the Sun is circular or elliptic. Taking only the jovian's tidal force into account, the equations of the relative motion write (Chauvineau, 1990; Chauvineau and Mignard, in press):

$$\ddot{x} - 2\dot{y} = 3x - \frac{3x}{r^3} + \mu f(t)[(3\cos^2 S' - 1)x + 3\cos S' \sin S'.y]$$

$$\ddot{y} + 2\dot{x} = -\frac{3y}{r^3} + \mu f(t)[(3\sin^2 S' - 1)y + 3\cos S' \sin S'.x]$$

for a circular motion about the Sun. (x, y) are the coordinates of the vector separating the two asteroids in a rotating frame. μ is the ratio M'/M of the jovian and Sun's masses. $f(t) = (R/D)^3$, where R is the radius of the orbit about the Sun and D the distance Jupiter-Asteroid's center of mass, which is a function of time. S' is the angle between the directions of the vectors Sun-Asteroids and asteroids-Jupiter. For $\mu = 0$ (the unperturbed Hill's problem), the value of:

$$J = -\dot{x}^2 - \dot{y}^2 + 3x^2 + 6/r \qquad [4]$$

remains constant along each orbit. It is the Jacobi's integral. For the perturbed problem, one has:

$$\frac{dJ}{dt} = -2\mu f(t)[(3\cos^2 S' - 1)x\dot{x} + (3\sin^2 S' - 1)y\dot{y} + 3\cos S' \sin S'(x\dot{y} + \dot{x}y)]$$

Then, J is no longer an invariant, and its numerical value is changing with time. But because of the small value of μ, a significant change in J requires a long time. J will be called the "Jacobi's function" in the following.

For $\mu = 0$, it is possible to show that all the close relative orbits for the binary system are bounded if $J \geq 9$ (Chauvineau and Mignard, 1990). In the case of the elliptical Hill's problem, this is approximately the case if $J > J_{stab} = 12$ (Chauvineau, 1990). The problem is then to give an estimation of the typical time involved such that the Jacobi's function [4] passes from an initial value $J > J_{stab}$ to the value $J = J_{stab}$.

Monte-carlo simulations and simplified analytical models show that the dispersion of the variation of J during one conjonction is given by:

$$\sigma(J) = \frac{\mu}{(p-1)^3} \frac{200}{3J^2}$$

where $p = R'/R$ is the ratio of the jovian and asteroidal orbits radius about the Sun.

2) Impulsional model

Let us replace the continous force by an impulsional one. The equations of the motion take the simplified form:

$$\ddot{x} = 2\dot{y} + 3x - \frac{3x}{r^3} + 2Ax\delta^{\tilde{}}(t)$$

$$\ddot{y} = -2\dot{x} - \frac{3y}{r^3} - Ay\delta^{\tilde{}}(t)$$

where $\delta^{\tilde{}}$ is a succession of Dirac's impulsions. During one conjonction, the variation of the Jacobi's function is:

$$\Delta J = 2A(-2x\dot{x} + y\dot{y}) - A^2(4x^2 + y^2)$$

Comparing the dispersion $\sigma_{\Delta J}$ in the impulsional case (neglecting the term proportional to A^2) to the value obtained by the analytico-numerical method, one finds (Chauvineau and Mignard, in press):

$$A = \frac{200\mu}{3\sqrt{30}} \frac{1}{(p-1)^3} \frac{1}{J^{3/2}}$$

Taking into account the term proportional to A^2, one obtains a differential equation analogous to [2]:

$$\frac{dJ}{dN} = -\frac{\psi(A,E)}{(P_0-1)^6} \frac{K}{J^5} \tag{5}$$

where N is the number of the perturbation, $\psi(A,E)$ a function of the orbital elements of the motion about the sun, $p_0 = R'/A$ where A is the semi-major axis of the binary asteroid about the sun in the case of an elliptic orbit. K is a constant of the order of $5000\mu^2$.

3) Stability of binary asteroids

The constant sign of the derivative in [5] shows that the Jacobian function presents a secular effect which is a long term decreasing comportment. The resolution of the equation [5] shows that the number of conjonctions for decreasing J from an initial value J_{in} to a final value $J_f < J_{in}$ is:

$$N = \frac{1}{6K} \frac{(p_0-1)^6}{\psi(A,E)} (J_{in}^6 - J_f^6)$$

For $J < J_{stab}$, the Hill's stability is no longer preserved. This gives a typical lifetime for a binary asteroid whose the initial value of the Jacobi's function is J (Chauvineau and Mignard, in press):

$$\tau \sim (450 \text{ years}) \frac{(p_0-1)^6}{(p_0^{3/2}-1)\psi(A,E)} (J^6 - J_{stab}^6)$$

It appears that this lifetime is not very dependent of the precise value of J_{stab} because of the sixth power in J. It is important to point out that all the supposed binary asteroids are such that their lifetime τ is largely greater than the age of the solar system.

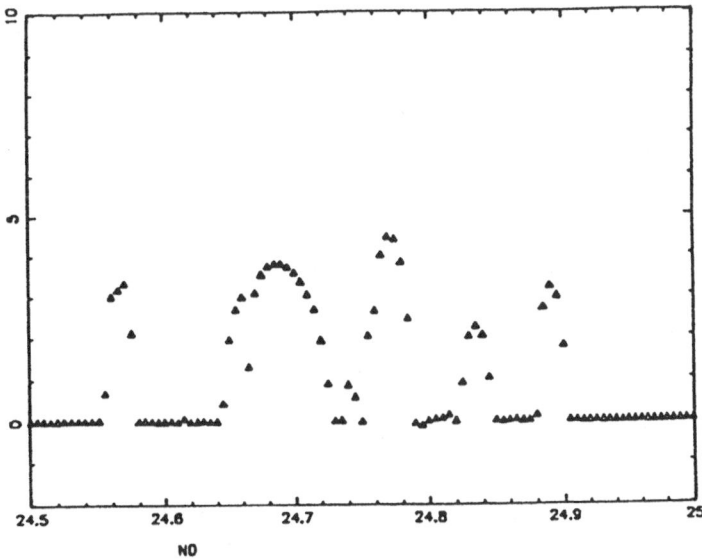

Fig. 1 . The values of the slope p of the logarithm of the energy normalized by the analytically computed value, as a function of the parameter N_0. One has taken $\mu = 0.1$, $\lambda = 0.5$, $\tau = 8.1$ and $T_0 = 22.7$. The number of perturbations is 4000.

REFERENCES

Carpino M., A. Milani, A.M. Nobili 1987;
Long-term numerical integrations and synthetic theories for the motion of the outer planets. Astron. Astrophys. **181**, 182-194.

Chauvineau B. 1990;
Dynamique à long terme des astéroides binaires. Bilan des observations et approche théorique. Thèse de Doctorat. Université de Nice.

Chauvineau B. and F. Mignard 1990;
Dynamics of binary asteroids. I - Hill's case. Icarus **83**, 360-381.

Chauvineau B. and F. Mignard in press;
Dynamics of binary asteroids. II - Jovian perturbations. To appear in Icarus.

Laskar J. 1985;
Accurate methods in general planetary theory. Astron. Astrophys. **144**, 133-146.

Laskar J. 1986;
Secular terms of classical planetary theories using the results of general theory. Astron. Astrophys. **157**, 59-70.

Laskar J. 1988;
Secular evolution of the solar system over 10 million years. Astron. Astrophys. **198**, 341-362.

Laskar J. 1989;
Numerical experiment on the chaotic behaviour of the solar system. Nature **338**, 237-238.

Milani A., A.M. Nobili, M. Carpino 1987;
Secular variations of the semimajor axes: theory and experiments. Astron. Astrophys. **172**, 265-279.

IMPROVED BETTIS METHODS FOR LONG-TERM PREDICTION

José M. Ferrándiz and Sylvia Novo

Departamento de Matemática Aplicada a la Ingeniería
E.T.S. de Ingenieros Industriales
Universidad de Valladolid, 47011 Valladolid, Spain

ABSTRACT

In this paper we introduce a modification of Bettis' method in order to improve the long-term numerical integration of perturbed oscillators. We give several examples, involving both single and coupled oscillators, to illustrate the efficiency of this approach with respect to the methods of Bettis and Adams-Bashforth.

INTRODUCTION

The integration of perturbed harmonic oscillators is a very common problem in many fields, and of course in Celestial Mechanics. Sometimes the original equations are already in this form and more often, after adequate changes of variables (Stiefel and Scheifele, Szebehely, Bond, Ferrándiz , etc) the problem reduces into an oscillator one.

Due to Lyapunov or orbital instability, numerical solutions can move or spiral away from the exact solution, which means that the integration remains valid only for a span of time that could be too short if long-term accurate solutions are needed. So far two kinds of techniques are commonly used to deal with this difficulty. One of them is to make transformations of the original equations of motion, which modify the unpleasant features mentioned above, prior to numerical integration; the other one consists in the creation of a special method to reduce errors and remove numerical instability. Usually, both are combined in order to first, transform the equations to different ones possessing stable solutions, and then integrate these equations by a method of numerical integration known to be stable.

A number of new techniques have been developed for the numerical solutions of differential equations in Celestial Mechanics. Bettis (1969) modified the classical difference methods of numerical integration to integrate certain products of an ordinary polynomial and a Fourier polynomial without truncation error. Scheifele (1971) obtained a refinement of Taylor's methods based on his G-functions. Recently Kirchgraber (1988) described an ODE-solver based on the method of averaging. Ferrándiz and Pérez (1990) succeeded in constructing spherically exact algorithms. Canonical or symplectic integrators (i.e. numerical integration schemes for Hamiltonian systems, which conserve the symplectic 2-form exactly) are proposed by several authors. Kinoshita, Yoshida and Nakai (1990) tested, in a satellite problem, one intended only for particular Hamiltonians.

DESCRIPTION OF THE METHOD AND EXAMPLES

Consider the Cauchy problem

$$\begin{cases} \mathbf{x}' = \mathbf{f}(t, \mathbf{x}) \\ \mathbf{x}(t_0) = \mathbf{x_0} \end{cases}$$

and let us denote

$$\mathbf{f}_m = \mathbf{f}(t_m, \mathbf{x}_m) \ , \quad t_m = t_0 + mh \ ,$$

where m represents an integer and h a positive real number. The method of Adams-Bashforth of n+1 steps with fixed stepsize h is given by the formula

$$\mathbf{x}_i = \mathbf{x}_{i-1} + h \sum_{j=0}^{n} \alpha_j \nabla^j \mathbf{f}_{i-1} \ ,$$

where ∇^j are the backward differences, that is,

$$\nabla^{q+1} \mathbf{f}_m = \nabla^q \mathbf{f}_m - \nabla^q \mathbf{f}_{m-1} \ ,$$
$$\nabla^0 \mathbf{f}_m = \mathbf{f}_m \ ,$$

and

$$\alpha_j = (-1)^j \int_0^1 \begin{pmatrix} -s \\ j \end{pmatrix} ds \ ,$$

hence

$$\alpha_0 = 1, \ \alpha_1 = \frac{1}{2}, \ \alpha_2 = \frac{5}{12}, \ \alpha_3 = \frac{3}{8}, \ \alpha_4 = \frac{251}{720}, \ \alpha_5 = \frac{95}{288}, \cdots$$

It is known that, if the Adams-Bashforth method of fourth or higher order is used for the integration of circular Keplerian motion, the resulting points spiral away from the exact circular orbit. To avoid this, Bettis devised in 1969 the following alternate scheme

$$\mathbf{x}_i = \mathbf{x}_{i-1} + h \sum_{j=0}^{n} \alpha_j^* \nabla^j \mathbf{f}_{i-1} \ ,$$

in terms of the modified coefficients given by the equalities

$$\alpha_j^* = \alpha_j \ , \quad j = 0, 1, \cdots, n-2 \ ,$$

$$\alpha_{n-1}^* = -\mu u R_{-1-n} + \sum_{k=0}^{n-2} \alpha_k S_{k-n} \ ,$$

$$\alpha_n^* = \mu u R_{-n} - \frac{1}{u} \sum_{k=0}^{n-2} \alpha_k S_{-k-n-1} \ ,$$

in which h is the step length,

$$\sigma = \omega \frac{h}{2} \ , \quad u = \frac{4}{\sin^2 \sigma} \ , \quad \mu = \frac{\sin \sigma}{2\sigma \cos \sigma} \ ,$$

and $R_m(u), S_m(u)$ are defined (for positive and negative integers m) by the recurrences

$$R_{m+1} = u(R_m - R_{m-1}) \ , \quad R_0 = 2 \ , \quad R_1 = u \ ,$$
$$S_{m+1} = u(S_m - S_{m-1}) \ , \quad S_0 = 0 \ , \quad S_1 = u \ .$$

This method of n+1 steps integrates exactly the function

$$P_{n-2}(t) + a_0 \cos \omega t + b_0 \sin \omega t \ ,$$

where $P_{n-2}(t)$ is an ordinary polynomial of degree $n - 2$ and ω a fixed frequency. Therefore we are able to obtain by this procedure the exact solution of the equation $x'' + \omega^2 x = 0$ (written as a first order system). It is thus natural to try it also for the perturbed oscillator

$$x'' + \omega^2 x = \epsilon g(x, x')$$

for small ϵ.

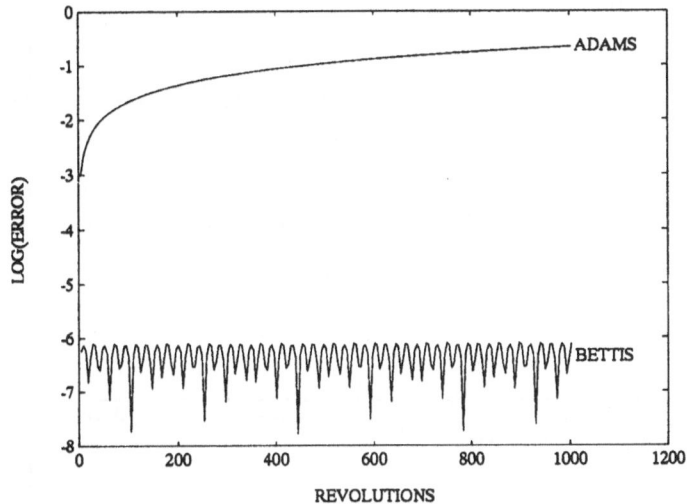

Figure 1. EXAMPLE 1 : $x'' + x = \epsilon \cos 2t$, $x(0) = 1$, $x'(0) = 0$. The exact solution is $x(t) = (1 + \epsilon/3) \cos t - \epsilon/3 \cos 2t$. Step h=0.1, order 4 and $\epsilon = 10^{-3}$.

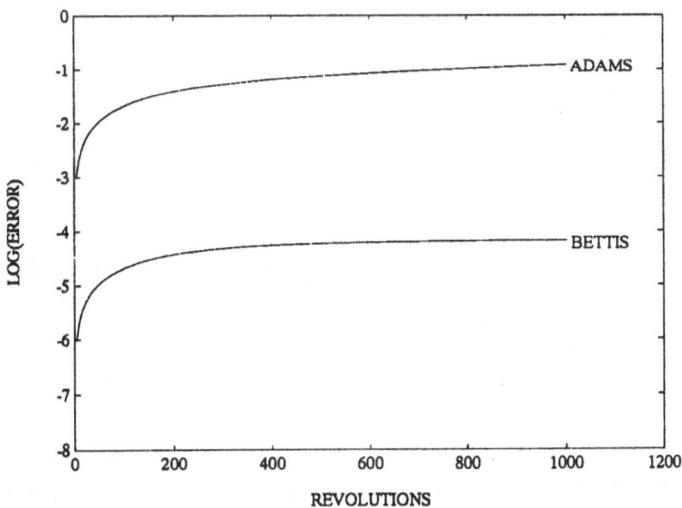

Figure 2. EXAMPLE 2 : $x'' + x = \epsilon \cos t$, $x(0) = 1$, $x'(0) = 0$. The error is computed with respect to the exact solution $x(t) = \cos t + \epsilon/2 \ t \sin t$. Step h=0.1, order 4 and $\epsilon = 10^{-3}$.

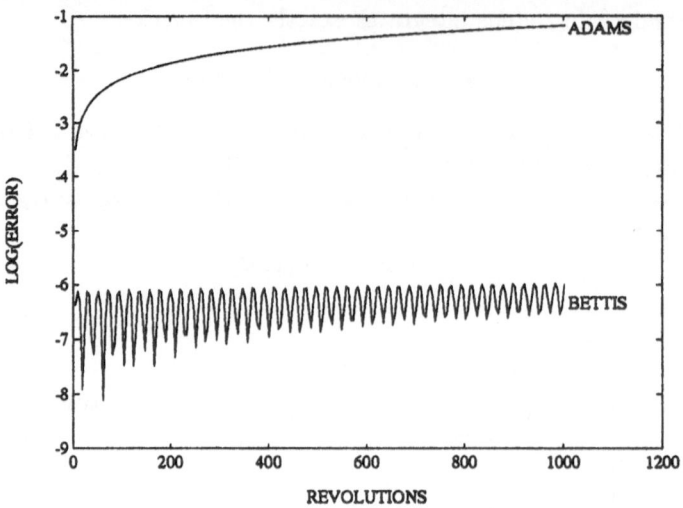

Figure 3. EXAMPLE 3 : $x'' + x = \epsilon x^2$, $x(0) = 1$, $x'(0) = 0$. The error is computed with respect to the first integral $H(x, x') = 1/2 \, (x^2 + x'^2) - \epsilon/3 \, x^3$. Step h=0.1, order 4 and $\epsilon = 10^{-3}$.

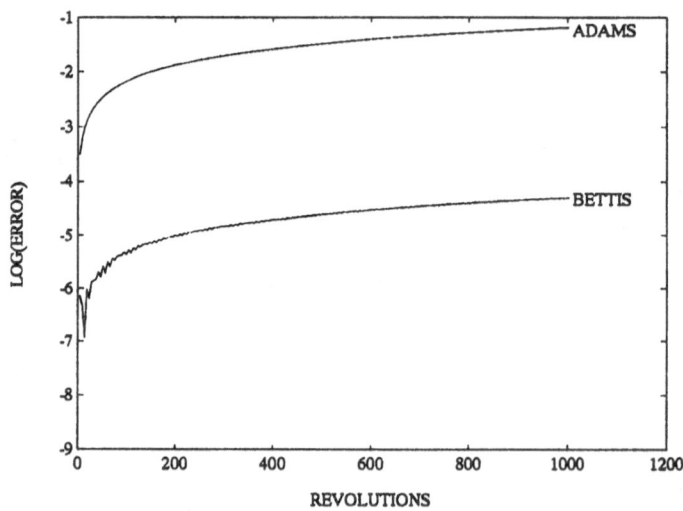

Figure 4. EXAMPLE 4 : $x'' + x = \epsilon x^3$, $x(0) = 1$, $x'(0) = 0$. The error is computed with respect to the first integral $H(x, x') = 1/2 \, (x^2 + x'^2) - \epsilon/4 \, x^4$. Step h=0.1, order 4 and $\epsilon = 10^{-3}$.

Unfortunately, when this disturbing term gives rise to a real or fictitious resonance the error grows with time and the algorithm breaks down for long-term prediction. To illustrate this behaviour we have chosen Examples 1-4 (see Fig. 1-4). In the first one, with an easy time-dependent perturbed term, the error is bounded, almost as in the unperturbed case. However, in the second example a time-dependent resonant perturbation produces an error growth similar to the Adams one (of course 3 orders of magnitude lower because $\epsilon = 10^{-3}$ and the error in Bettis'method is proportional to ϵ). In examples 3 and 4 we show that even without real resonance different perturbing terms like ϵx^2, ϵx^3 induce parallel behaviours, namely, boundedness in one case and a linearly growing error in the other. To solve this problem we may apply much more complicated methods derived by Bettis modifying more than two coefficients to integrate exactly products of ordinary polynomials and Fourier polynomials. The consequence of this is that higher order methods with a more involved implementation must be used. For instance in order to integrate exactly

$$a_0 + \sum_{k=1}^{n}(a_k \cos \omega_k t + b_k \sin \omega_k t) \, ,$$

we need a method of order $2n$ at least. The same happens with the function

$$(c_{n-1}t^{n-1} + c_{n-2}t^{n-2} + \cdots + c_0)(a_0 + a \cos \omega t + b \sin \omega t) \, .$$

In this paper we give an alternative for fictitious resonance. It is possible to control the error growth just by properly changing the frequency in the initial Bettis modification of the Adams method. The way of choosing the right frequency is found by perturbation techniques, as shown in the following examples. It is worth remarking that the right choice is the usual one in analytical integrations, i.e., resulting from the Lindstedt-Poincaré device.

Let us consider again the Cauchy problem shown in example 3

$$\begin{cases} x'' + x = \epsilon x^2 \\ x(0) = 1 \\ x'(0) = 0 \, . \end{cases}$$

Assuming a perturbation expansion for $x(t)$ of the form

$$x(t) = \sum_{n=0}^{\infty} \epsilon^n x_n(t) \, ,$$

where

$$x_0(0) = 1, \quad x_0'(0) = 0 \, ,$$

and

$$x_n(0) = 0, \quad x_n'(0) = 0 \quad (n \geq 1),$$

we find that

$$x(t) = \cos(t) + \epsilon(\frac{1}{3}(2 - \cos^2 t - \cos t)) + \cdots \, ,$$

without secular term in the powers 0 and 1 of ϵ. Thus we conclude that the frequency we need is $\omega = 1$. (See fig. 3).

Let us try the perturbing term used in example 4. Consider

$$\begin{cases} x'' + x = \epsilon x^3 \\ x(0) = 1 \\ x'(0) = 0 \, . \end{cases}$$

By applying a rough perturbation technique we have

$$x(t) = \cos(t) + \epsilon\left(\frac{1}{8}(\cos t - \cos^3 t + 3t \sin t)\right) + \cdots$$

and there is a secular-mixed term. To avoid this, (Lindstedt) we may write the problem in the form

$$\begin{cases} x'' + \omega^2 x = \epsilon(x^3 - ax) \\ x(0) = 1 \\ x'(0) = 0 \end{cases}$$

where a is a parameter and the new frequency depends on a through

$$\omega = \sqrt{1 - a\epsilon}.$$

Hence, we can take as unperturbed solution $x_0(t) = \cos \omega t$ and $x_1(t)$ is the solution of

$$\begin{cases} x_1'' + \omega^2 x_1 = \left(\frac{3}{4} - a\right)\cos \omega t + \frac{1}{4}\cos 3\omega t. \\ x_1(0) = 0 \\ x_1'(0) = 0. \end{cases}$$

Thus, the only way to avoid resonance is by taking $a = 3/4$, therefore, the right frequency we propose to use with the Bettis method in this case is

$$\omega = \sqrt{1 - \frac{3}{4}\epsilon}.$$

In fig. 5 we show the behaviour of the error function with respect to the first integral for several frequencies. The right choice leads to a remarkable stabilization of the error.

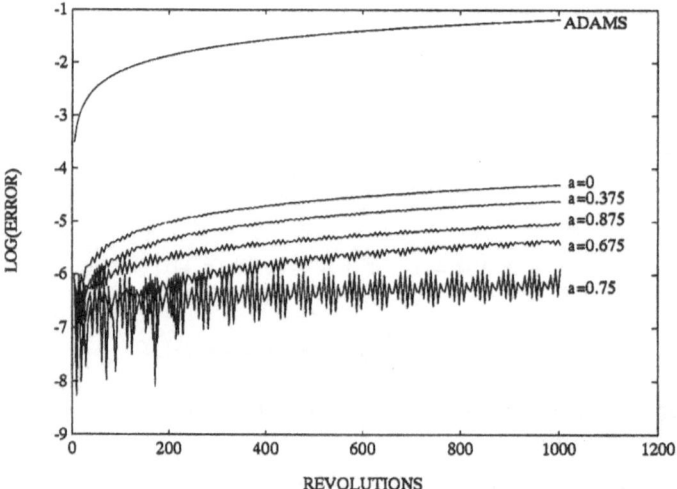

Figure 5. EXAMPLE 4 : $x'' + x = \epsilon x^3$, $x(0) = 1$, $x'(0) = 0$. The error is computed with respect to the first integral $H(x, x') = 1/2 \ (x^2 + x'^2) - \epsilon/4 \ x^4$. Step h=0.1, order 4 and $\epsilon = 10^{-3}$. Frequency $\omega = \sqrt{1 - a\epsilon}$.

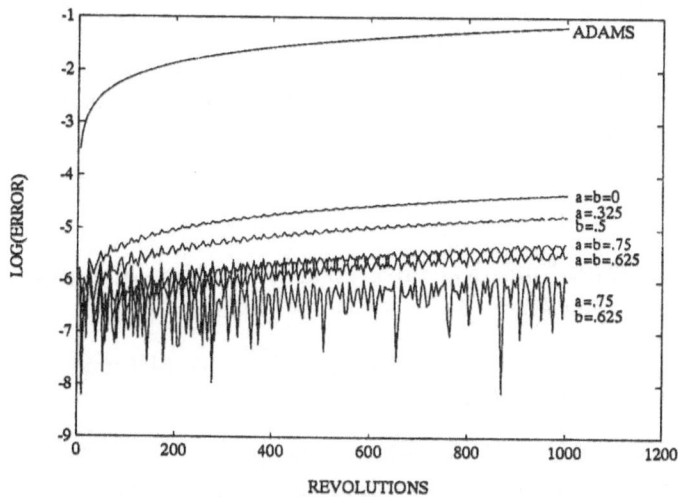

Figure 6. EXAMPLE 5. The error is computed with respect to the first integral $H(x, x', y, y') = 1/2(x^2 + x'^2 + y^2 + y'^2) - \epsilon/4\ x^4 + \epsilon xy^2 - \epsilon/6\ y^6$. Frequencies $\sqrt{1 - a\epsilon}$ and $\sqrt{1 - b\epsilon}$. Step h=0.1, order 4 and $\epsilon = 10^{-3}$.

When we integrate coupled oscillators the modification of the frequencies is done separately in the associated components. We illustrate that fact in fig. 6 with several pairs of frequencies, where the Cauchy problem (EXAMPLE 5)

$$\begin{cases} x'' + x = \epsilon(x^3 - y^2) \\ y'' + y = \epsilon(y^5 - 2xy) \\ x(0) = 0\,, x'(0) = 1 \\ y(0) = 0\,, y'(0) = 1 \end{cases}$$

is integrated. Although this is not the most general case, similar problems often appear in applications.

FINAL REMARKS

Notice that the calculation of the right frequency, which has been done in the examples by perturbation techniques, may be done symbolically on the computer and of course also numerically. To do the latter assume the perturbed term of the problem is $\epsilon g(x, x')$; once the solution of the unperturbed problem $x_0(t)$ is known we may calculate a such that the first Fourier coefficient of the function $g(x_0(t), x_0'(t)) - ax_0(t)$ vanishes and then we take the frequency $\omega = \sqrt{1 - a\epsilon}$.

In some examples, when an exact or approximate solution is available, the error has been computed with respect to this solution and the behaviour of the error is the same as the one with respect to the first integral. We remark that fact because it is known that some methods integrate well or even conserve the energy but not the position.

A thorough study including applications to satellite motion problems is now being done by the authors.

ACKNOWLEDGEMENTS

The authors thanks Professors Szebehely, Simó and Milani for their comments after the exposition of this paper. They also acknowledge the financial support received from C.I.C.Y.T. of Spain under Project ESP.88-0541.

References

[1] Bettis, D.G., 1970, Numerical Integration of Products of Fourier and Ordinary Polynomials. *Numer.Math.*, **14**, 421:434.

[2] Bettis, D.G., 1970, Stabilization of Finite Difference Methods of Numerical Integration. *Celest.Mech.*, **2**, 282:295.

[3] Ferrándiz, J.M. and Pérez M.T.,1990, Application of Spherically Exact Algorithms to Numerical Predictability in Two-Body Problems. In this issue.

[4] Kinoshita H., Yoshida H. and Nakai H., 1990, Symplectic Integrators and Their Merits in Application to Dynamical Astronomy, *in*: Proceedings of the twenty-third Symposium on "Celestial Mechanics", Kinoshita, H. and Yoshida, H. eds., Kyoto, Japan.

[5] Kinoshita H. and Nakai H., 1989, Numerical Integration Methods in Dynamical Astronomy. *Celest.Mech.*, **45**, 231:244.

[6] Kirchgraber U., 1989, An ODE-Solver Based on the Method of Averaging. *Numer.Math.*, **53**, 621:652.

[7] Stiefel, E. and Scheifele, G., 1971, "Linear and Regular Celestial Mechanics". Springer. Berlin, Heidelberg, New York.

APPLICATION OF SPHERICALLY EXACT ALGORITHMS TO NUMERICAL PREDICTABILITY IN TWO–BODY PROBLEMS

José M. Ferrándiz and M. Teresa Pérez

Departamento de Matemática Aplicada a la Ingeniería
E.T.S. de Ingenieros Industriales
Universidad de Valladolid, 47011 Valladolid

ABSTRACT

Numerical predictions are strongly dependent on the algorithms used in the integration, even in cases as simple as the two–body problem, perturbed or not. In this contribution we show some numerical experiments comparing the results obtained by applying different codes. Among them we include some with special preservation properties, such as being spherically exact.

INTRODUCTION

The design of special numerical methods, taking advantage of the regularity of the problem to be solved, is a common task in numerical analysis. The integration by means of such codes allows us, in many cases, to obtain accurate numerical solutions useful for long–term predictions. Those methods turn out to be much more efficient than the classical ones when applied to the problem for which they have been created although, as a matter of fact, they do not improve the results when an arbitrary problem is considered.

We are specially interested in those concerning Celestial Mechanics where special algorithms designed to integrate orbital problems are not exceptional. Requirements of high accuracy or long–term validity of the solution have forced some incursions in the subject.

Consider the two–body problem. It is well known that there are some changes of coordinates that reduce it to harmonic oscilators and then solutions of the form $a \cos \omega t + b \sin \omega t$ appear. In this direction, algorithms exact for problems whose solution is:

$$P_{n-2}(t) + a \cos \omega t + b \sin \omega t$$

were introduced by Bettis (1969). They are basically linear multistep schemes for which the set of integration coefficients are obtained from those of the classical finite difference methods. Some extensions of these methods have the property of integrating without truncation error products of Fourier polynomials and ordinary polynomials ([1]). A more recent improvement has been introduced by Ferrándiz and Novo in [3].

Another property of the two–body problem is the preservation of the area in the phase–space. In this sense the so–called symplectic methods are area–preserving ones and have some other caractheristics—such as not having a secular term in the discretization error in the energy integral—that make them of great interest. On the other hand simple symplectic methods presently known ([8]) exhibit some disadvantages such as being of low order or requiring a special form of the Hamiltonian which means that new, more efficient sets of coordinates cannot be used in the integration.

Moreover, in the Kepler motion some magnitudes lie on circumferences or spheres of certain spaces. Then what we call geometrically–exact methods can be used in the integration. Geometrically designed algorithms were introduced by Lambert and McLeod ([9]) without connexion with Celestial Mechanics to integrate what they called trajectory problems, providing an efficient procedure to integrate circular orbits. Later Ferrándiz and Pérez have developed several schemes that have the property of being spherically exact, that is to say when the solution of the problem is on a sphere and the initial values are on it, the points generated by the method remain on that surface provided that no round–off errors are present.

In this work we make a preliminary study of the benefits of these new methods. The results of the numerical experimentation show that it is worth going ahead with the investigation in this direction.

DESCRIPTION OF THE METHODS

In ([4]) and ([11]) particular methods of the same kind have been introduced and studied and a wider generalization of them will appear in ([5]). They belong to this sort of algorithms derived geometrically to integrate trajectory problems; we shall understand for trajectory problems those where the interest lies in obtaining the curve traced by the solution rather than this as a correspondence between the values of the parameter and the point of the trajectory. One of their advantages is that their study can be made by means of the geometrical elements of the solution. They have some conservation properties, the aforementioned spherical exactness being the most important. Therefore, they are suitable to integrate unitary problems.

We shall now make a short description of the algorithms. Assume that the initial value problem to be solved is given by an autonomous differential equation of the form:

$$\begin{cases} \dfrac{d\vec{y}}{ds} = \vec{\Phi}(\vec{y}), & 0 \le s \le L, \\ \vec{y}(0) = \vec{y}_0, \end{cases} \tag{1}$$

with $\|\vec{\Phi}(\vec{y})\| = 1$ when $\vec{y}$ lies in a certain domain Ω verifying some properties and that $\vec{y}(s_0)$ is known. If $\vec{u}$ is an approximation to $\vec{y}(s) - \vec{y}(s_0)$, $\vec{y}(s)$ being a point of the solution at a distance h from $\vec{y}(s_0)$, then a new approximation to $\vec{y}(s)$, being on the sphere S^n and with exact normal component, can be generated from $\vec{u}$ by means of the following formula:

$$(1 - \frac{h^2}{2})\vec{y}(s_0) + h\sqrt{1 - \frac{h^2}{4}}\,\frac{\vec{c}}{\|c\|}, \tag{2}$$

where $\vec{c} = \vec{u} - (\vec{u}, \vec{y}(s_0))\vec{y}(s_0)$ and $(.,.)$ stands for the dot product. From that expression, predictor–corrector codes can be derived, varying only the way of obtaining the previous approximation $\vec{u}$.

As can be seen, the methods are obtained projecting on the sphere previous approximations to the solution. The way the projection is done can be changed to obtain different algorithms. In this sense, we introduce here a new method given by the following expression:

$$\vec{y}_{n+1} = \vec{y}_n + \vec{u}_n(1 + \sin^2(\beta_n)) - <\vec{u}_n, \vec{y}_n> \frac{\vec{y}_n}{\|\vec{y}_n\|^2}, \tag{3}$$

where $\beta_n = \frac{1}{2}\text{arc}\sin(\|\vec{y}_n^*\|)$, with $\vec{y}_n^* = \vec{u}_n - <\vec{u}_n, \vec{y}_n> \frac{\vec{y}_n}{\|\vec{y}_n\|^2}$ and $\vec{y}_n + \vec{u}_n$ a previous approximation to $\vec{y}_{n+1}$.

We have built predictor–corrector schemes using explicit Adams–Bashforth and implicit Adams–Moulton methods to obtain the required former approximation in formulae (2) and (3). In the next section, the results before and after the projection are compared.

NUMERICAL EXPERIMENTATION

In the figures that appear in this section, we have drawn the integration errors versus the number of revolutions. The errors show the deviation of the computed solution with a desired number of steps per revolution from a reference orbit. This

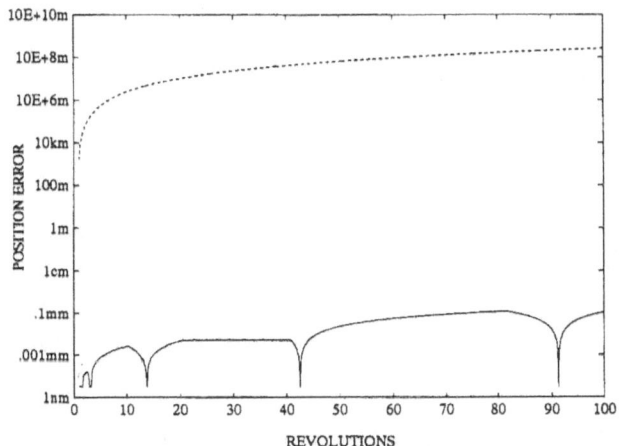

Figure 1. Kepler problem in KS variables, e=0, I=0.

reference is a numerical solution of the problem obtained with the same integrator. It has been selected satisfying strong requirements on conservation of three independent first integrals and stabilization of a number of digits in the computed solutions while the step–size is being reduced. As the methods are convergent when the step–size tends to zero, this procedure gives a trustworthy estimation. Nevertheless, its validity has been assured using other classical integrators.

After a suitable scaling the Kepler problem in canonical KS variables is given by the homogeneous Hamiltonian:

$$H = \frac{1}{8}\|\vec{p}_u\|^2 + p_0\|\vec{u}\|^2 - 1 = 0.$$

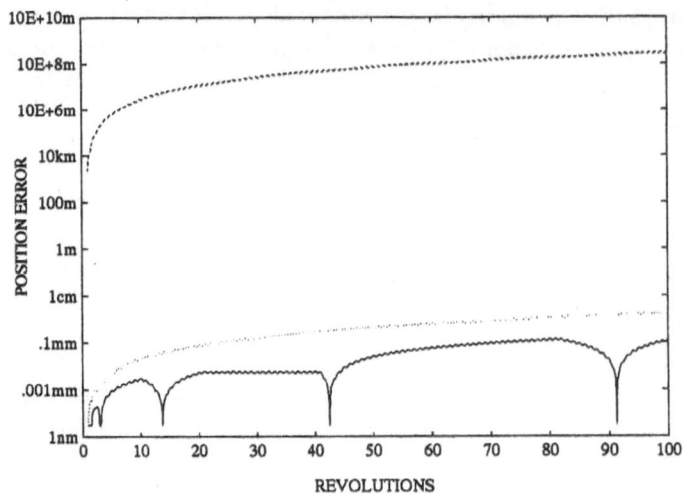

Figure 2. Kepler problem in KS variables, e=0.1, I=0.

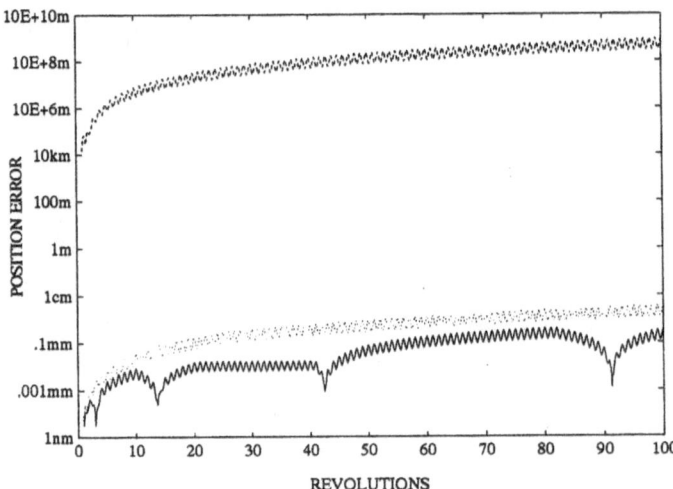

Figure 3. Kepler problem in KS variables, e=0.5, I=0.

This can be transformed with the non–canonical change $\vec{y} = \dfrac{\vec{p_u}}{\sqrt{8}}$, $\vec{v} = \vec{u}\sqrt{p_0}$, into the unity sphere in the 8–dimensional Euclidean space.

So, in those modified KS variables the Kepler problem reduces to the form (1) and methods (2) are suitable to integrate it.

We have integrated the Kepler problem in KS variables by means of a predictor–corrector pair in PECE mode where the predictor is an explicit Adams–Bashforth and the corrector, an implicit Adams–Moulton, both of order seven. To integrate the Kepler problem in modified KS variable we have projected the former code with help of the expression(2). The resulting integrator is, as well, a predictor–corrector scheme in PECE mode. In the first four figures we show the results obtained. The significant parameter in them is the eccentricity, that varies assuming the values 0.0, 0.1, 0.5 and 0.99. The inclination is always zero. The dashed and dotted upper curves display the

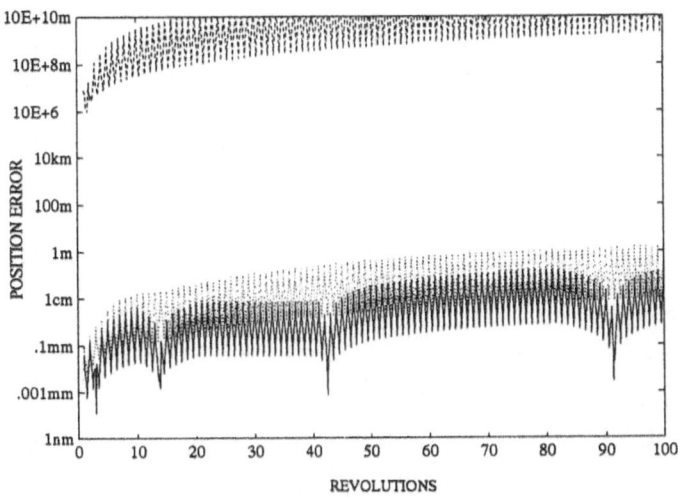

Figure 4. Kepler problem in KS variables. e=0.99, I=0.

error performed by the Adams integrator with fixed step–size corresponding to 16 and 200 steps per revolution, respectively. The lower solid graph shows the error of the spheric method with only 16 steps per revolution. The last one is about round–off error while the classic Adams algorithm turned out to be highly inefficient for long step–size and although much better when this is reduced, it still behaves worse than the spherical one.

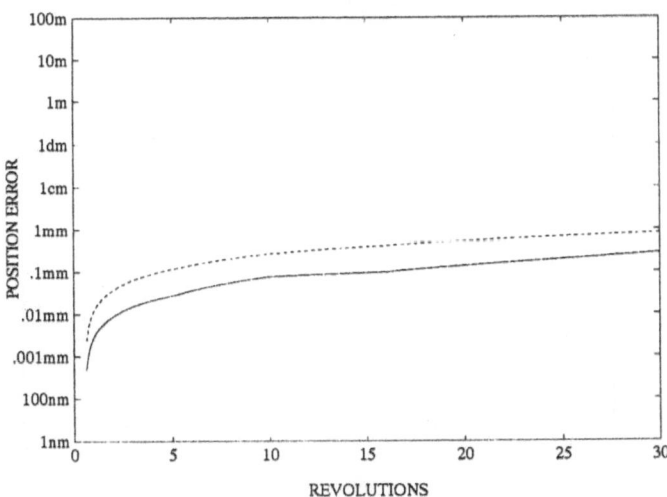

Figure 5. Kepler problem in BF variables, e=0, I=0.

The eccentricity has small effect in the integration except when it is very big, f.i. 0.99. In this case, the methods integrate worse but the difference between the errors is still the same. For the spherical method the results are remarkable although its extension to perturbed problems is notimmediate. The results of the experiments we have just commented on, make it worthwhile to continue the investigation in this direction. More numerical results concerning these methods are included in [11] and [6].

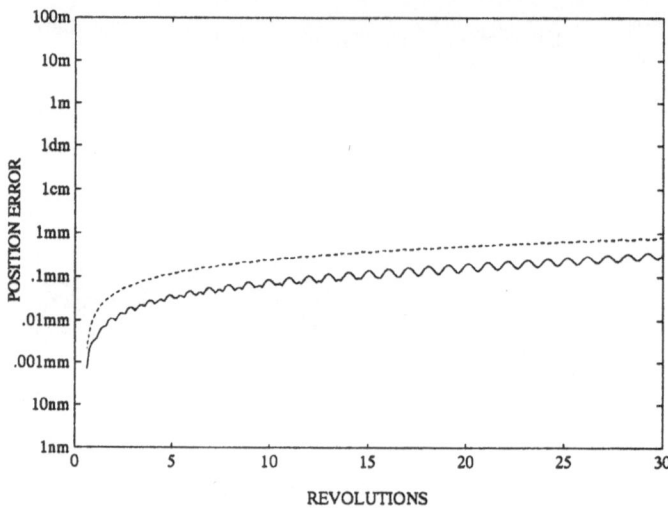

Figure 6. Main problem in BF variables, e=0, I=0, J_2=0.001.

The two–body problem in Bf variables (see [7]):

$$\begin{cases} x''_i + x_i = P_i, \\ z'' + z = \dfrac{\mu}{c^2} + Q, \\ t' = \dfrac{1}{cz^2}, \end{cases} \qquad (4)$$

where P_i, Q are disturbing terms, has been integrated by means of Adams methods and of a mixed code.

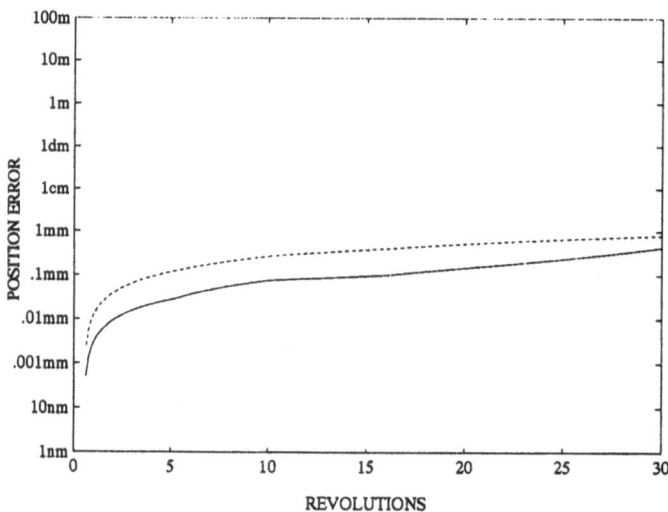

Figure 7. Kepler problem in BF variables, e=0, I=10.

This mixed code consists in a spherically exact algorithm of the kind (3) to integrate the direction vector and its derivative and an Adams method for the remaining equations. In the following figures we show some typical results. The solid graph corresponds to the mixed code and the dashed one to the Adams, both with fixed step corresponding to 144 steps per revolution. Therefore, equal number of derivative evaluations are made.

In figures 5 and 6 there are represented the errors for the two–body problem with zero eccentricity and inclination, the first without perturbation. Figures 7 and 8 show what happens when the inclination varies. Inclination 90 has practically no effect meanwhile inclination 10 deteriorates the solid graph slightly.

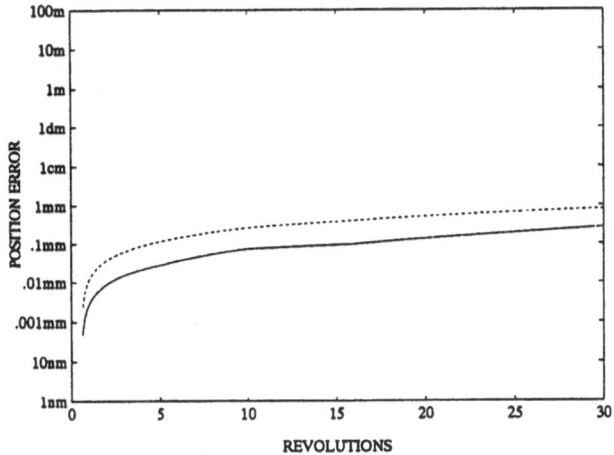

Figure 8. Kepler problem in BF variables, e=0, I=90.

In all the experiments performed, the new method happens to be more accurate although the mixed code is not refined; it can be improved including other special algorithms, such as circularly exact ones, for the remaining equations that now have been integrated by Adams. The benefits of this new method are similar to those of other special algorithms and sometimes even better–see for instance [10].

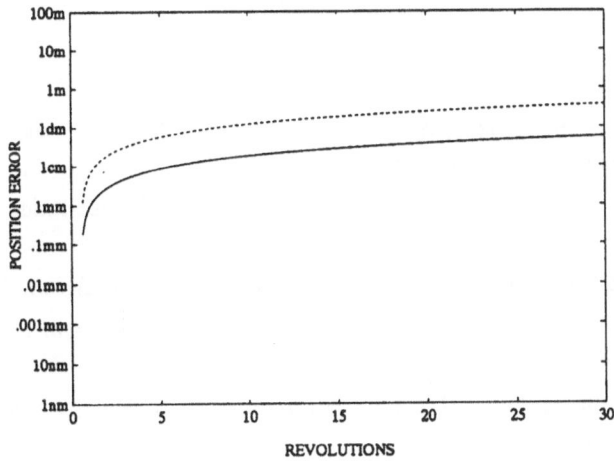

Figure 9. Kepler problem in BF variables, e=0, I=0, Order 5.

To finish we change the order of the Adams integrator chosen. In figure 9 it is of order five and of order six in figure 10. The behavior observed here is the same we have noticed for other orders: when the order of the method is odd the projected scheme integrates better than the linear multistep one but when the order is an even number it performs worse. This can be overcome changing the way the projection is done depending on the order.

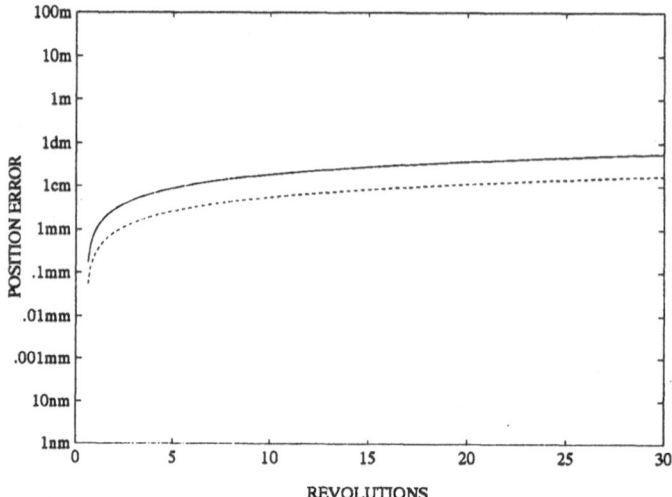

Figure 10: Kepler problem in BF variables, e=0, I=0, Order 6.

ACKNOWLEDGEMENT

The authors acknowledge the financial support received from CICYT of Spain under Project ESP.88–0541 and from the NATO cooperative research grant 0506/88.

REFERENCES

[1] Bettis D.G.: 1970, *Numer. Math.* **14**, 421-434.

[2] Ferrándiz, J.M.: 1988, *Celest. Mech.* **41**, 343-357.

[3] Ferrándiz, J.M. and Novo, S.: 1990, *in this issue*, Cortina.

[4] Ferrándiz, J.M. and Pérez, M.T.: 1987, *Actas del X C.E.D.Y.A.*, Valencia, Spain.

[5] Ferrándiz, J.M. and Pérez, M.T., "A family of multi–step methods to integrate trajectories on spheres" (in progress).

[6] Ferrándiz, J.M. and Pérez, M.T.: 1989, *Actas de las XIV Jornadas Hispano–Lusas*, Tenerife, pp. 1225-29.

[7] Ferrándiz, J.M., Sansaturio, M.E. and Pojman, J., *Celest. Mech.* (submitted).

[8] Kinoshita, H., Yoshida, H. and Nakai, H.: 1990, *Proceedings of the XXIII Symposium on "Celestial Mechanics"*, Japan, pp.1-6.

[9] Lambert, J.D. and McLeod, R.J.Y.: 1979, *Numerical Analysis Proceedings* (Dundee). Springer Verlag 1980.

[10] Moore, P.: 1978, *Celest,. Mech.* **17**, 281-297.

[11] Pérez, M.T.: 1989, Doctoral Dissertation, Univ. Valladolid, Spain.

[12] Stiefel, E., and Scheifele, G.: 1980, *Linear and Regular Celestial Mechanics*. Berlin - Heildelberg - New York, Springer.

ARE THERE IRREGULAR FAMILIES OF CHARACTERISTIC CURVES?

Joaquim Font and Carles Simó

Departament de Matemàtica Aplicada i Anàlisi
Universitat de Barcelona
Gran Via 585, Barcelona 08071 (Spain)

Abstract

For Hamiltonian systems of two degrees of freedom the symmetric periodic orbits of a given type appear in continuous families. Every periodic orbit can be represented by one point in some suitable plane of parameters, and the full family is represented by the so called characteristic curve. Some of these curves have components which are isolated, and they are called irregular characteristic curves. In this work we consider one of the examples of this kind of behaviour and we show that if we embed the given Hamiltonian in a one parameter family the components are no longer isolated. Furthermore we give a full explanation of the structure and evolution of those characteristic curves, by using several invariant manifolds.

1 INTRODUCTION

The one parameter family of Hamiltonians that we consider is given by

$$H(x,y,\dot{x},\dot{y}) = \frac{1}{2}(\dot{x}^2 + \dot{y}^2) + \frac{1}{2}(x^2 + y^2) - xy^2 + ay^4 \ , \tag{1}$$

a being a real parameter. The system (1) is a perturbation of the Contopoulos potential [4] obtained by adding a quartic term. Recent studies on (1) can be found in [1],[2],[3],[7],[6],[11] and [5]. We summarize some properties of (1) :

1. The equations of motion are invariant under the involution

$$(x, y, \dot{x}, \dot{y}, t) \rightarrow (x, -y, -\dot{x}, \dot{y}, -t) \ .$$

2. For $a \geq \frac{1}{2}$ the origin is the only fixed point. For $a < \frac{1}{2}$ there are two additional fixed points, $P_{1,2}$, at $x = (2 - 4a)^{-1}$, $y = \pm(2 - 4a)^{-1/2}$ which are of saddle–center type, with related eigenvalues $\pm\lambda$, $\pm\nu i$, $\lambda, \nu \in \mathbb{R}^+$, sitting on $H = h_c = (8 - 16a)^{-1}$.

3. The Hill's region, given by $\frac{1}{2}(x^2 + y^2) - xy^2 + ay^4 \leq h$ has only one connected component (which is bounded) for all $a \geq \frac{1}{2}$ and all $h \geq 0$. For $a < \frac{1}{2}$ it has three connected components (two of them unbounded) if $h < h_c$ and only one component (unbounded) for $h > h_c$.

4. For $a < \frac{1}{2}$ and $h > h_c$ there are Lyapunov periodic orbits emanating from $P_{1,2}$ which have two points in the zero velocity curve (z.v.c., the boundary of the Hill's region). They are hyperbolic (see [10]).

5. For some values of a there exists a double heteroclinic connection between P_1 and P_2, that is, $\mathcal{W}^u(P_1) \equiv \mathcal{W}^s(P_2)$, $\mathcal{W}^s(P_1) \equiv \mathcal{W}^u(P_2)$ if we only consider the left hand branches. Both connections have the same projection on the (x, y) plane. We mention the value $a = a_c \simeq 0.49188633722$ for which this connections occur.

The system (1) has two important families of periodic orbits, the so called "principal families" emanating from the normal modes at the origin. They exist for all $h \in (0, \infty)$. These families are (see figure 1):

1. The symmetry axis (family B_1) with constant period 2π, given by $x = \sqrt{2h}\cos t$.

2. The "central" family (family B_2) made of orbits as follows: take an initial point $(x_0, 0, 0, \dot{y}_0)$ with $\dot{y}_0 > 0$. Then they reach the z.v.c. and return in a symmetric way.

The family B_1 bifurcates to symmetric 3–periodic orbits (3PO) (see [2]), that is, they are periodic orbits symmetrical w.r.t. the x axis such that they cut the Poincaré section $y = 0$ with $\dot{y} > 0$ exactly three times, and they reach the z.v.c. at two points (see figure 1). The bifurcation is produced by duplication of the period, that is, when the monodromy matrix of the normal variational equations

$$\ddot{\eta} + (1 - 2\sqrt{2h}\cos t)\eta = 0 \ ,$$

has an eigenvalue equal to -1. This happens for $h \simeq 8.5399674963$, and also for $h \simeq 43.65153458$. We do not consider here additional bifurcations to the families of periodic orbits with limit period $2m\pi$, $m > 2$, nor other bifurcations as the 6–periodic orbits obtained from the 3PO ones (see figure 1).

Each 3PO has a point, A, in $y = 0$ with $\dot{x} = 0$. This point characterizes the 3PO and it is determined by the values (h, x). Given a the locus of (h, x) is called the characteristic curve (c.c.) of the family of 3PO for this value of the parameter. For $a = \frac{1}{2}$ and $a = 1$ the c.c. are given in [2],[3] (see also figure 2). Other values of a have been studied in [6] and [5]. We summarize some numerical results. For $a = a_c$ inwards revolutions of the spiral c.c. were computed without going again outwards. In fact, there are two of these spirals and both of them appear to be infinite. For $a = 0.494, 0.498, 0.54$ and 0.6 the c.c. has a behaviour similar to the case $a = \frac{1}{2}$ but now with n inwards revolutions ($n = 13, 9, 4$ and 2, respectively) before going outwards. For $a = 0.6$ there is just 1 bubble inside. For $a = 0.54$ there are no bubbles. For 0.498 there is a finite number of bubbles. However in the case $a = 0.494$ it seems that the number of bubbles is not bounded. When $a > a_t \simeq 0.4946439$ the c.c. is made of a finite double spiral bounding some nested bubbles (the so called c.c. of irregular orbits in the sense that they are not connected, on that level of a, to any one of the principal families).

From the numerical results it seems interesting to characterize the c.c. of the 3PO of (1) when a ranges near $\frac{1}{2}$, that is, to study the evolution of the spirals and the bubbles. This is a really involved problem because the phenomenon is far from being local or semiglobal.

The object of this work is to summarize the evolution of the c.c. when a changes, including the life cycle (birth, growth and death) of the bubbles. The tools used are numerical computations and local analysis of bifurcations.

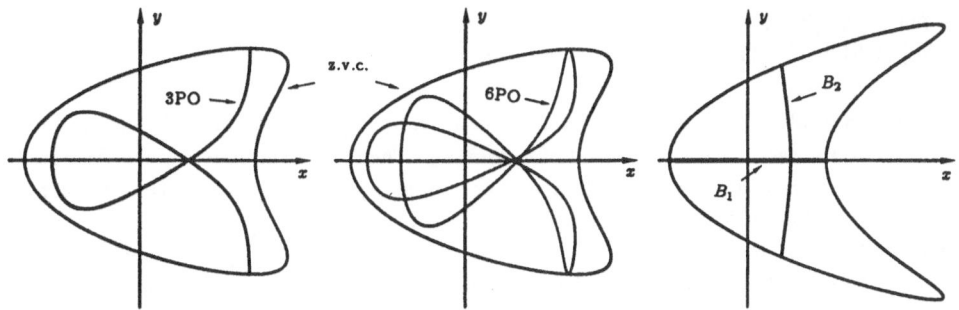

Figure 1: Zero velocity curve with a 3PO, a 6PO and the "principals families" B_1 and B_2.

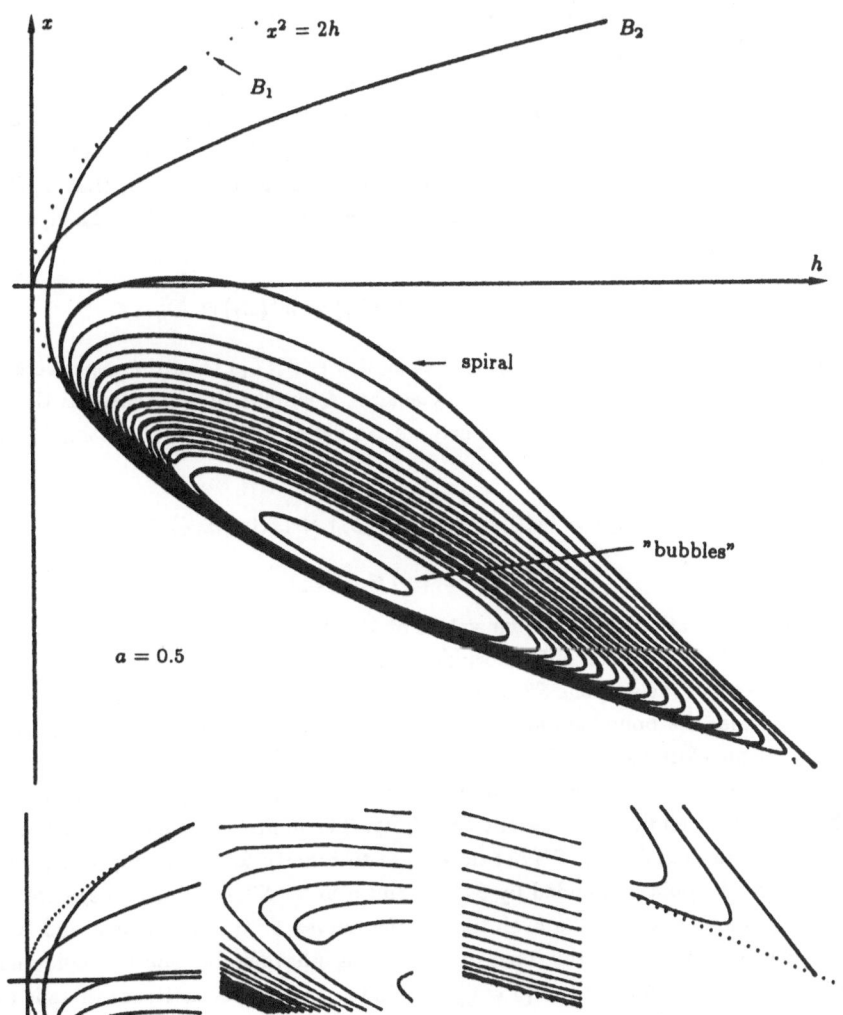

Figure 2. C.C. of 3PO for $a = \frac{1}{2}$, $x \in [-9.5, 5]$ and $h \in [-1, 44]$. The spiral and two bubbles of the eight that exist. At the bottom, four enlargements of different regions.

2 THE LYAPUNOV ORBITS AND THEIR INVARIANT MANIFOLDS: THE ESCAPE REGION

Consider an analytical Hamiltonian system with two degrees of freedom with an equilibrium point of saddle–center type on a level of energy h_0. Then the Lyapunov–Moser theorem [10] assures that there is an analytical change of coordinates such that the new Hamiltonian is of the form $K = \lambda \xi \eta + \frac{\kappa}{2}(u^2 + v^2) + O_2(\xi \eta, \frac{1}{2}(u^2 + v^2))$. The level $H = h_0$ is now $K = 0$. For values of $K > 0$ there is an hyperbolic periodic orbit, named the Lyapunov orbit, at least for K small. This applies to (1) for $a < \frac{1}{2}$ around P_1, P_2 for $h \gtrsim h_c$. We obtain two families (with h as natural parameter), L_1 and L_2, emanating form P_1 and P_2 respectively. Their invariant manifolds, $\mathcal{W}^{s,u}(L_i)$, $i = 1,2$, are locally diffeomorphic to cylinders. Due to the symmetry of the equations of motion the projections of $\mathcal{W}^s(L_i)$ and $\mathcal{W}^u(L_i)$ on the (x, y) plane coincide, both for $i = 1$ and $i = 2$. Furthermore the projection on that plane of $\mathcal{W}^s(L_2)$ is symmetrical of the projection of $\mathcal{W}^s(L_1)$ with respect to the x axis. In fact, for (1) those orbits and the related manifolds exist for all $h > h_c(a)$.

For a given value of $a < \frac{1}{2}$ and $h > h_c$ the left branches of $\mathcal{W}^{u,s}(L_i)$, $i = 1, 2$ can be computed numerically until they reach $y = 0$ with $\dot{y} > 0$ for $\mathcal{W}^u(L_1)$ and $\mathcal{W}^s(L_2)$ and with $\dot{y} < 0$ for $\mathcal{W}^s(L_1)$ and $\mathcal{W}^u(L_2)$ (see figure 3). Let $\Sigma_h^\sigma = \{(x, y, \dot{x}, \dot{y}) \in \mathbb{R}^4 | y = 0, \sigma \dot{y} > 0, H(x, y, \dot{x}, \dot{y}) = h\}$ for $\sigma = \pm 1$. Σ_h^σ is a surface parametrized by $(x, \dot{x})$, that is, it can be considered as the open disc in the $(x, \dot{x})$-plane defined by $x^2 + \dot{x}^2 < 2h$. Then we defined $S^{u,s}(L_i)$ as follows:

$$S^u(L_1) = \mathcal{W}^u(L_1) \cap_1 \Sigma_h^1 \ , \quad S^s(L_2) = \mathcal{W}^s(L_2) \cap_1 \Sigma_h^1 \ ,$$

where $\cap_1$ denotes the "first" intersection, that is, if $p \in S^u(L_1)$ then the negative semiorbit through p (which goes to L_1) never cuts Σ_h^1 again. A similar thing occurs for the positive semiorbit of any point of $S^s(L_2)$. Analogously we can define $S^u(L_2)$, $S^s(L_1)$ using Σ_h^{-1}. Again due the symmetry we have:

$$(x, 0, \dot{x}, \dot{y}) \in S^u(L_1) \Leftrightarrow (x, 0, \dot{x}, -\dot{y}) \in S^u(L_2) \Leftrightarrow (x, 0, -\dot{x}, \dot{y}) \in S^s(L_2) \Leftrightarrow$$

$$\Leftrightarrow (x, 0, -\dot{x}, -\dot{y}) \in S^s(L_1) \ .$$

From several computations for different values of (h, a) one has the numerical evidence that all the sections $S^{u,s}(L_{1,2})$, are diffeomorphic to circles (at least for a not too far from $\frac{1}{2}$). Therefore each one of the sections $S^{u,s}(L_{1,2})$ divides the $(x, \dot{x})$-plane in two connected components. We consider the bounded one and, for definiteness, we study the case $S^s(L_1)$. Points in this component close to $S^s(L_1)$ go, under the positive flow, close to L_1 following closely the left branch of $\mathcal{W}^s(L_1)$. Then they follow the right branch of $\mathcal{W}^u(L_1)$ and, as we shall see later, they are good candidates to escape to infinity. The numerical simulations show, in fact, that all the points bounded by $S^s(L_1)$ do escape to infinity for increasing time. A similar thing occurs for the points bounded by $S^s(L_2)$ and also for $S^u(L_{1,2})$ using negative time.

Now we want to describe the evolution of the curves $S^{u,s}(L_{1,2})$ when h and a change. Before we note that for $a < \frac{1}{2}$ the fixed points $P_{1,2}$ also have stable and unstable manifolds. For $a = a_c$ one has $\mathcal{W}^u(P_1) \cap_1 \Sigma_h^1 \in \{\dot{x} = 0\}$ as said in §1, property 5. Hence there is a double heteroclinic connection between P_1 and P_2. For (h, a) close to $(h_c(a_c), a_c)$ the curves $S^u(L_1)$ have points close to the x axis. The variations of the size and the position of $S^u(L_1)$ are related, for a fixed value of a, to the variations of h. It has been seen that, for some range of a, there is an interval of energies $[h_0, h_f]$, depending on a for which $S^u(L_1) \cap \{\dot{x} = 0\} \neq \emptyset$.

For $h = h_0$ and $h = h_f$ the curve $S^u(L_1)$ is tangent to the x axis. For $a = a_t$ one has $h_0 = h_f$, and for $a > a_t$ we have found $S^u(L_1) \cap \{\dot{x} = 0\} = \emptyset$ (see figure 4).

Let us consider $a < a_t$ and $h \in (h_0(a), h_f(a))$. Then $S^u(L_1)$ cuts $\{\dot{x} = 0\}$ in at least two points and the numerical explorations show that the intersection is reduced exactly to two points. We call them $n(h,a)$ and $m(h,a)$, $n < m$. As the points in (n,m) are contained in the bounded component of the complementary of $S^u(L_1)$ they escape to infinity in backwards time. For a fixed value of a, when h ranges in $[h_0, h_f]$ the points $n(h,a)$, $m(h,a)$ give rise to a closed curve, $\Gamma(a)$, in the (h, x) plane where the c.c. are represented. For all points in the region bounded by $\Gamma(a)$ the orbits with initial conditions $(x, 0, 0, \sqrt{2h})$ escape to infinity for negative time without additional crossing of Σ_h^1. For $S^s(L_1)$, $S^{u,s}(L_2)$ symmetrical results hold but $\Gamma(a)$ is common. The curve $\Gamma(a)$ can be drawn for $a < a_t$, reduces to one point if $a = a_t$ and becomes empty if $a > a_t$.

Next we give a proposition (see [5]) which gives a criterion ensuring that orbits starting to the right of the Lyapunov orbit and with a suitable initial velocity escape to infinity.

Proposition 2.1 Given $a \in (0, \frac{1}{2})$ and a value of $h > h_c(a)$ then there is a curve of the form $q(x,y) = ye^{dx} = k$, d being equal to $\frac{6a-1+\sqrt{36a^2-28a+9}}{2}$ and k a function of h and a, such that, if the initial point is to the right of $q(x,y) = k$ (i.e.. $q(x_i, y_i) > k$) and the initial velocity satisfies $y_i d\dot{x}_i + \dot{y}_i \geq 0$, then the point escapes to infinity. For a close to $\frac{1}{2}$ and h large enough the value of k is close to

$$\frac{2}{a}\left(\frac{1}{\sqrt{1-2a}} + \sqrt{1+a}\right)\sqrt{h}\exp\left(d(a)\frac{2\sqrt{h}}{\sqrt{1-2a}}\right) .$$

3 BIFURCATIONS GIVING RISE TO THE STRUCTURE OF THE CHARACTERISTIC CURVE

As a first step we would like to know in which cases, given values of a and h, we have a finite or infinite number of 3PO.

1. Keeping a fixed, $a \leq a_t$ we split the (h, x) plane in three regions:

   (a) $\{(h, x) | h < h_0\}$, (b) $\{(h, x) | h_0 \leq h \leq h_f\}$, (c) $\{(h, x) | h_f < h\}$,

   where $h_0 = h_0(a)$ and $h_f = h_f(a)$ have been defined in § 2. Taking a value of h in (a) or (c) we have only a finite number of 3PO, while in (b) the number is infinity. To show this we consider the set $I = \{(x, \dot{x}) \in \Sigma_h^1 | \dot{x} = 0, |x| < \sqrt{2h}\}$ and the map $\pi_h : \Sigma_h^1 \to \Sigma_h^1$ obtained by following the flow backwards. If $I \cap S^u(L_1)) = \{n(h,a), m(h,a)\} \neq \emptyset$ then $\pi_h(I)$ contains two infinite spirals tending towards $S^u(L_1)$. By symmetry the infinite points fixed by the action of π_h give rise to 3PO (other points in $I \cap \pi_h(I)$ appear in pairs and are related to 6PO). Otherwise, if $I \cap S^u(L_1) = \emptyset$, the curve $\pi_h(I)$ is finite and cuts a bounded number of times the line I. We note also that in all the cases $h_0 \geq h_c(a)$. If $h < h_c(a)$ then there are not Lyapunov orbits and $S^u(L_1)$ is empty. In this case the reasoning to show the finiteness of the number of 3PO uses the variational equations associated to $\mathcal{W}^{u,s}(P_1)$ if h is close to $h_c(a)$.

2. For $a \in (a_t, \frac{1}{2})$ we are in the same situation as in cases 1. (a) or (c) and the finiteness of the number of 3PO follows.

3. When $a \geq \frac{1}{2}$ the topological structure of the problem is different because there are not equilibrium points $(P_{1,2})$ and the Hill's region is bounded for any value of the energy.

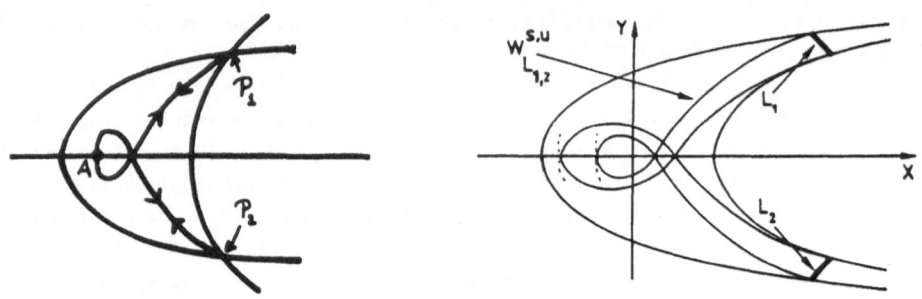

Figure 3. The z.v.c. with two equilibrium points, $P_{1,2}$, and $\mathcal{W}^{u,s}(P_{1,2})$ and two Lyapunov orbits, $L_{1,2}$, and $\mathcal{W}^{u,s}(L_{1,2})$ projected on the (x,y)-plane.

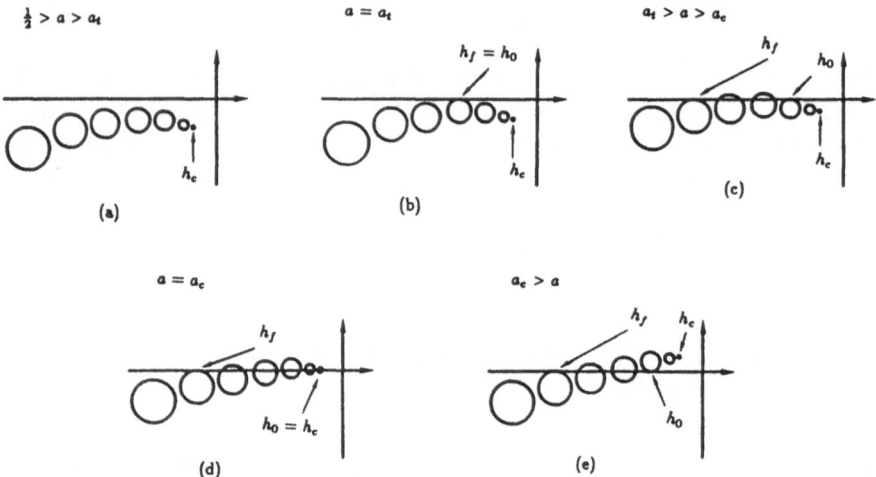

Figure 4. $\mathcal{S}^{u,s}(L_{1,2})$ for different values of a and h.

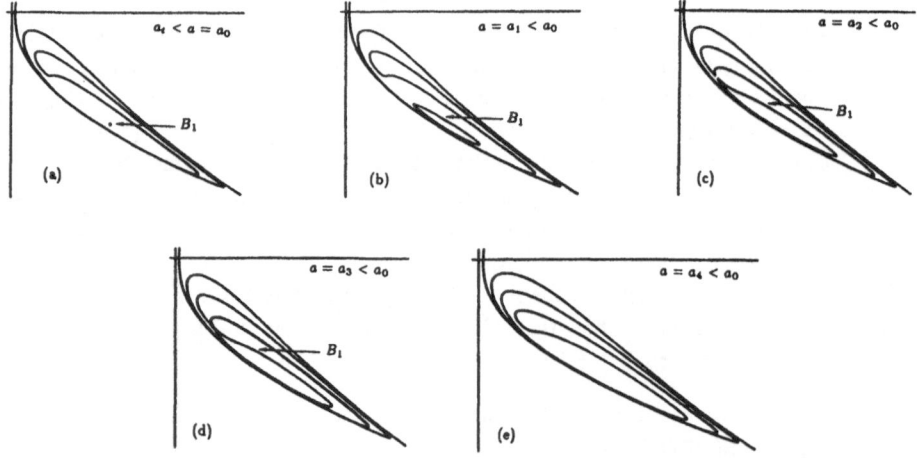

Figure 5. In (a), the bubble appears. In (b) and (c), the bubble increases. In (d), the spiral and the bubble coincide in one point. In (e), the spiral wins a revolution.

Lots of numerical simulations have been done for different values of a and h. The result has been always a finite number of 3PO and 6PO. Indeed, in any one of those cases the curve $\pi_h(I)$ is a double finite spiral, as it should be by continuity with respect to the parameters a and h. Furthermore, if a decreases the number of "revolutions" of the spiral $\pi_h(I)$ decreases.

If we keep a fixed the c.c. has different kinds of bifurcations. Near the turning points (points where the c.c. has a local extremum of the variable h) there are two bifurcation points: one of them is a saddle–node (or fold) where one elliptic and one hyperbolic 3PO are born, and the other is period doubling (or flip) where one elliptic 6PO is born and the 3PO goes from elliptic to hyperbolic with reflection.

We recall that an orbit is said "irregular" if it belongs to a connected component of the c.c. which is not connected to any one of the "principal" families (eventually by means of a chain of bifurcations). This is the case of the "bubbles" which appear in (1) for the 3PO. An important question is to locate the birth of a bubble. Figure 5 (a) displays the birth of one of them.

To study this phenomenon with more detail we use again π_h. We consider a subinterval $I' \subset I$ such that $\pi_h(I')$ is the part of spiral $\pi_h(I)$ where the sense of rotation changes (the "end" of the spiral). In [5] one can see the evolution of I' and $\pi_h(I')$ leading to the appearition of a bubble of 3PO and its satellite bubble of 6PO. The table 1 shows some values of a for which a bubble is born.

In section 1 we have described the spiral c.c. of the 3PO for different values of a. We have noticed that the (finite) number of inwards revolutions of the spiral before going outwards can be different for different values of a. Now we shall display the mechanism allowing to the spiral to increase the number of revolutions when a decreases (see [8] and [5]).

We have seen that for some values, say $a = a_0$, a bubble is born. Let us denote it by B_1 (see figure 5). When a decreases, B_1 evolves by increasing its size. However the bubble is bounded by finite spiral and therefore B_1 can not increase indefinitely. It has been observed, numerically, that B_1 increases until, by means of a suitable bifurcation, it becomes part of the spiral. In this way the spiral wins, essentially, one revolution (the figure 5 displays this process). This phenomenon can be studied using the map π_h, considering the interval I' and its image $\pi_h(I')$ (see [5]).

The birth of a bubble and its death (the connection with the spiral) are related to saddle–node bifurcations. They happen when two points of the c.c. coincide. In the first case both points belong to the bubble. In the second, one is from the spiral and the other from the bubble. In [5], curves of fold bifurcation have been computed, and the points where the birth and the death of a bubble are produced, are located on those curves. The set of birth points accumulates to a limit (h_t, a_t) (roughly $h_t \simeq 15.787$). The set of death points accumulates also to a limit $(h_c(a_c), a_c)$.

Using suitable models for classical Hamiltonians ($H = T + U$) with the same symmetries that we have in our problem one can prove birth and death Theorems as 3.1 below (see [5] and [9]).

We consider a two degrees of freedom Hamiltonian system with Hamiltonian function $H = \frac{1}{2}(p_x^2 + p_y^2) + V(x, y)$ satisfying the following assumptions. For simplicity we suppose it is analytical, but this is not strictly necessary.

1. It is invariant under the involution $(x, y, \dot{x}, \dot{y}, t) \rightarrow (x, -y, -\dot{x}, \dot{y}, -t)$ and depends on some parameter a.

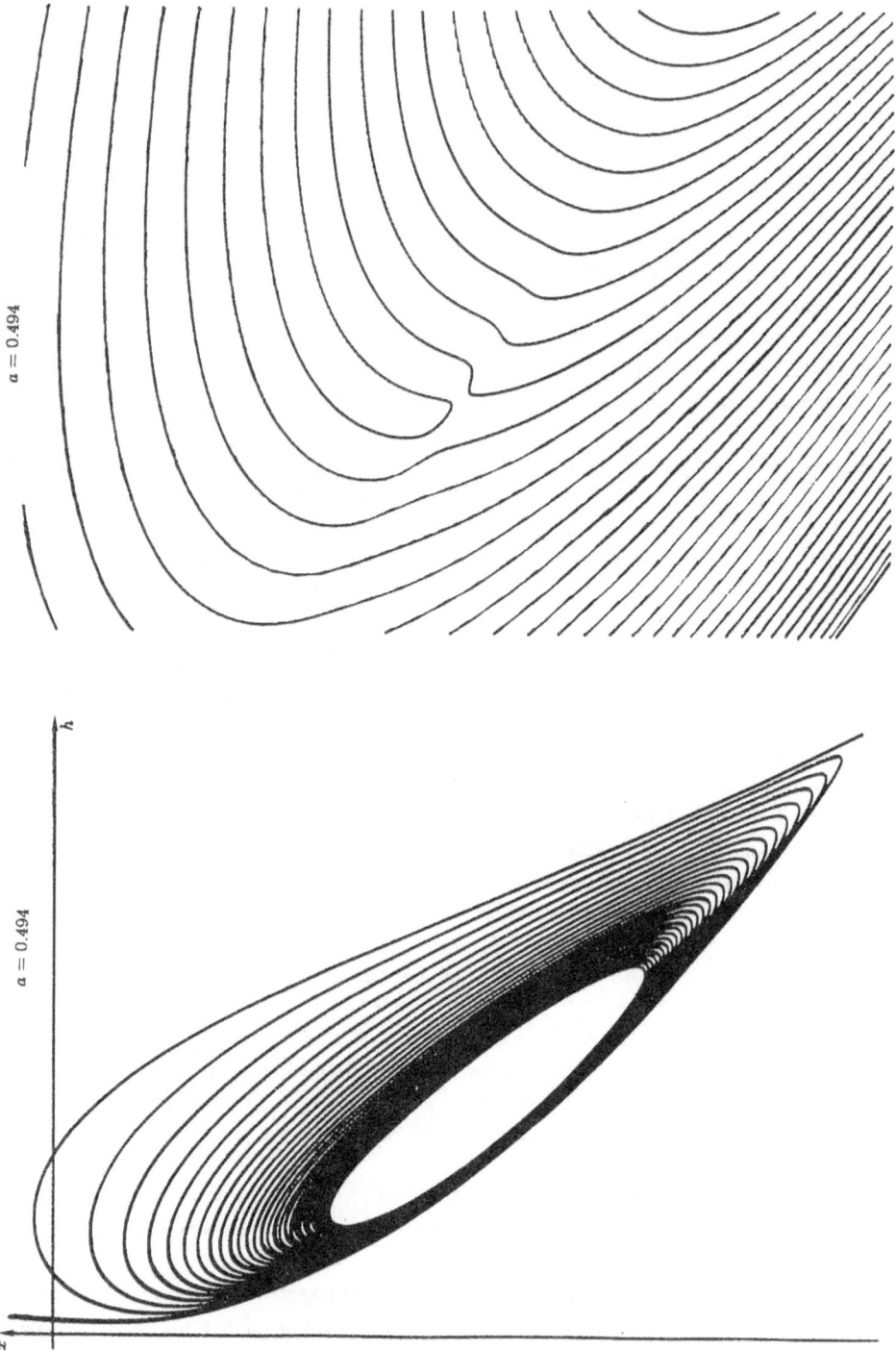

Figure 6. C.C. for $a = 0.494$, $x \in [-9.5, 0.5]$ and $h \in [-1, 44]$. Enlargement of the C.C., $x \in [-3.8, -2.8]$ and $h \in [7, 8.8]$.

2. For some range of values of a it has a fixed point of saddle–center type, P_1 located at a point of coordinates (x_p, y_p) with $y_p \neq 0$. Then also P_2, located at $(x_p, -y_p)$ is of center–saddle type.

3. For a value, a_t, of a, and a suitable value, h_t, of the energy, the Lyapunov periodic orbits, L_1, L_2 born at P_1 and P_2 have double heteroclinic connections. That is, there is a couple of orbits of the system, O_1 and O_2 such that O_1 is contained in $\mathcal{W}^u(L_1) \cap \mathcal{W}^s(L_2)$ and then, by the symmetry, O_1 is contained in $\mathcal{W}^u(L_2) \cap \mathcal{W}^s(L_1)$. Furthermore the intersections of the invariant manifolds are tangential.

4. Keeping $a = a_t$ fixed, if the energy is changed, either by increasing or by decreasing h with respect to h_t, there are not heteroclinic orbits.

5. When a is increased from a_t for all values of the energy (at least locally near h_t) there are not heteroclinic connections. If a is decreased from a_t, locally near h_t, there is a range of values of the energy $[h_0(a), h_f(a)]$ such that for $h = h_0$ and $h = h_f$ there is exactly one tangential heteroclinic connection from L_1 to L_2 (and the symmetrical one from L_2 to L_1), and for $h \in [h_0(a), h_f(a)]$ there are two transversal heteroclinic connections from L_1 to L_2 (and the symmetrical one from L_2 to L_1).

Theorem 3.1 A Hamiltonian system under the assumtions 1) to 5) has characteristic curves of periodic orbits which are closed curves. These curves increase in size when the parameter a of the Hamiltonian decreases. New curves are born for values of the parameter, a_m, which behave in a geometrical way. The limit value corresponds to the double heteroclinic connection occurring for $a = a_t$. Then

$$\frac{a_m - a_t}{a_{m-1} - a_t} \longrightarrow \lambda^2 \text{ when } m \to \infty ,$$

where λ is the unstable eigenvalue of the fixed point of the Poincaré map associated to the Lyapunov periodic orbit.

The proof of 3.1 uses the tangential heteroclinic connections between the periodic Lyapunov orbits for $a = a_t$ on the level of energy h_t, as reference orbit. The described behaviour is obtained by an analysis of the motion in the vicinity of this orbit including the passage near the Lyapunov orbits in an hyperbolic way.

Using similar reasonings but slightly more delicate, requiring the inclusion of non-linear terms, which should satisfy some conditions, one can prove a theorem concerning the accumulation of the connections between the bubbles and the spiral.

The table 1 also gives some numerical values for the death of the bubbles. Each bubble exists in the interval of a between the birth and the death. So, from table 1 one can obtain easily the number of bubbles to be found for every value of a. For $a_c < a \leq a_t$ there are infinity many bubbles. For $a = a_c$ all of them have been eaten by the spiral that becomes infinite.

4 CONCLUSION

We have presented an example of a one parameter family of Hamiltonian systems with two degrees of freedom, having families of periodic orbits of irregular type. The evolution of these families has been studied as a function of the parameter. It becomes clear that the irregular character is lost if we allow for the variation of one paramenter. That is, irregular orbits can exist for a given Hamiltonian but can not when we embed the Hamiltonian in a suitable family. This behaviour is persistent for a wide class of one parameter families of Hamiltonian of the type described.

Table 1. Some values of a for which birth and death of bubbles are found

birth	death	birth	death	birth	death
1.079975	0.916503	0.514814	0.505296	0.504024	
0.729705	0.670396	0.512069	0.503280	0.503064	
0.632072	0.595751	0.509832	0.501655	0.502236	
0.562393	0.541663	0.507985	0.500326	0.501514	
0.535937	0.521235	0.506440		0.500882	
0.522601	0.511106	0.505134		0.500325	

REFERENCES

[1] Barbanis, B.: Celestial Mechanics **33** (1984) 385.

[2] Barbanis, B.: Celestial Mechanics **36** (1985) 257.

[3] Barbanis, B.: Celestial Mechanics **39** (1986) 345.

[4] Contopoulos, G.: Astrophysical Journal **138** (1963) 1297.

[5] Font, J.: *The role of homoclinic and heteroclinic orbits in two–degrees of freedom Hamiltonian systems* (1990), PhD Thesis. University of Barcelona.

[6] Font, J. and Grau, M.: Proc. NATO-ASI Series C vol. **246** (1988) 385.

[7] Font, J. and Simó, C.: CEDYA Universidad de Santander (1985) (to appear).

[8] Font, J. and Simó, C.: European Conference on Iteration Theory (ECIT 87) World Scientific (1989) 421.

[9] Font, J. and Simó, C.: *Spirals and bubbles in characteristic curves of periodic orbits*, in preparation .

[10] Moser, J.K.: Comm. Pure and Appl. Math. **XI** (1958) 257.

[11] Simó, C.: *Homoclinic and heteroclinic phenomena in some Hamiltonian systems* in Hamiltonian Dynamical systems, ed. K.R. Meyer and D.G. Saari, Contemporary Mathematics **81** (1988) 193.

Non-Linearity in the Angles–Only Initial Orbit Determination Problem

Denise Kaya, Mathematician and Daniel Snow, Aerospace Engineer

Air Force Space Command, Colorado Springs, CO, USA

Abstract

An investigation has been initiated to examine the concept of determining an initial orbit using angles–only observations over a short arc from a single space based sensor. The methods of Laplace and Gauss for angles–only initial orbit determination were the methods of choice for this study. To apply these methods, one has to solve a system of non–linear equations which have multiple solutions. Typical solution schemes recommend using an iterative solution technique with a first guess to start. However, since only one solution will be found using such a technique, the success of this approach will depend entirely upon how good (or bad) the first guess is. We have found that this technique is not satisfactory for performing initial orbit determination over short arcs of data. In this paper we will discuss past methods used in solving these non–linear equations and then we will present our approach which involves finding all the necessary solutions and determining which solution is correct.

1 Introduction

The methods of Laplace and Gauss [Ref [2]] and [Ref [3]] to determine initial orbits using three angles–only observations have been used for almost two centuries. These methods were used mainly by astronomers in observing planets, moons and asteroids and recently by astrodynamicists to observe artifical satellites from the earth. A newer application includes determining initial orbits of artificial satellites from observations taken by orbiting sensors.

Both the Laplace and Gauss methods are very straight forward and use simple techniques with two–body dynamics to produce an orbit. One of the major drawbacks of these methods is that in order to get the final solution, one has to solve a system of non–linear equations which have multiple solutions. For the Laplace method, this system consists of two equations which can be rewritten as an eighth–order polynomial in r (the magnitude of the satellite's position vector). Similarly, an approximation of the Gauss method, that uses a second order approximation in time (hereafter referred to as the Truncated Gauss method), produces the same eighth–order polynomial in r.

Most applications in Laplace and Gauss methods that we have investigated have used a Newton–Raphson method (or other similar methods) to obtain the solution of the system of non–linear equations. This requires that one makes an initial guess of the solution to start the iterative process. However, to obtain the "correct" solution using these methods, the initial guess must be within the region of convergence to the correct solution. When multiple solutions exist, how does one guarantee this?

How has this problem of finding the correct solution been addressed in the past? The purpose of this paper is to discuss the problem of "root solving" in angles–only initial orbit determination. Specifically, we are interested in determining initial orbits using angles–only observations over a short arc from a single space based sensor. Initially some time will be spent reviewing past work in root solving, but the bulk of this paper will present our contributions to the solution of this problem.

Predictability, Stability, and Chaos in N-Body Dynamical Systems
Edited by A.E. Roy, Plenum Press, New York, 1991

2 Watson's Approach to Root Solving

Unfortunately, due to size limitations, we must request that the reader study many of the background details. The past work that we will be referencing most is that of Watson [Ref [4]]. Due to the age of this document it may be easier for the reader to reference Moulton [Ref [3]], Escobal [Ref [2]], or Danby [Ref [1]]. These documents do not give the detail of Watson but are much easier to obtain.

Watson took the eighth–order polynomial in r that is produced by both the Laplace method and the Truncated Gauss method and redefined it in terms of an angular variable ϕ so that the solutions occur when

$$y_1 - y_2 = 0$$

where

$$y_1 = \sin^4 \phi$$

and

$$y_2 = M \sin (\phi + m)$$

M and m are defined by the geometry of the problem (as well as two–body dynamics) and, by using Watson's variable transformation, ϕ can be related to a value of range. Using this formulation, Watson determined that for a space based sensor there must exist three real roots for ϕ between the values of 0 and π. One of these roots must be the location of the sensor itself or the trivial solution. (This is not true for a ground based sensor. The solutions to either the Laplace or Gauss methods must satisfy the equations of two–body motion. Since the motion of a ground based sensor does not satisfy these equations, this problem does not have a trivial solution.) Of the remaining two roots, one may be excluded if it has a ϕ greater than the ϕ belonging to the trivial solution. However, if this is not the case, both roots will satisfy the physical conditions of the problem and will produce entirely different orbital elements. This is true to the extent that if a differential correction is performed over the interval of the initial orbit determination, both solutions will produce equally good results. Watson then pointed out that it will be necessary to compare the elements computed from each of the two values of ϕ with other observations in order to determine the correct root.

3 Problems in Applying Initial Orbit Determination Methods

While Watson's work was an extremely thorough examination of the initial orbit determination problem there are certain facets which were not covered. These are problem areas which usually are not discussed much in the literature. The selection of the correct solution is a problem which has received minimal discussion in the literature whereas finding "non–existent" roots has not been discussed at all. This section will focus on past work in root selection and its shortcomings.

As mentioned earlier, cases exist where there are two equally satisfactory solutions to the initial orbit determination problem. This is particularly true for a space based sensor. For the space based scenario there will exist three roots for ϕ between 0 and π (as we discussed previously, this is not always true for ground based sensors). One would therefore expect to see more cases where two roots are equally satisfactory solutions for space based initial orbit determination than for ground based. Choosing the correct roots is extremely important if one wants to reacquire and/or correctly identify the objects at a later time.

Past suggestions for root selection (e.g. Watson, Moulton, Escobal, etc) have centered on either using the calculated ranges or using the orbital elements which correspond to these ranges. By combining these quantities with either additional observations or "partial information" (e.g., using a semi–major axis value of 6.6 earth radii for a suspected geosynchronous object), it was felt that the correct solution could be chosen. This is not always true in practice.

The first suggested approach is to solve for a new value of range, at the same time as was used previously, using two of the same observations as used previously. By using a new third observation, two new range values can be calculated. One of these new values will, in theory, agree with one of the old values while the other will not agree. This agreed upon value is the correct value, therefore a root can be chosen based upon the value of range. Another approach involving range is based upon Laplace's method. Using a fourth observation, a value for the third derivative of the angular measurements with respect to time can be obtained. With this additional information and by reformulating certain equations, a linear equation for range can be developed. Since this equation will have only one solution, the dilemma of choosing a root is avoided.

The other route for root selection is to use the orbital elements which correspond to each solution. Some feel (and Watson implies) that comparing both solutions to one or a few observations (other than those used in the initial solution) will render root selection a moot point. One school of thought is that the differences between the new observations and the orbit for the wrong root will be far greater than for the orbit for the correct root. Others feel that the differential correction will either converge to the correct solution or that the correction will yield a solution with a vastly smaller root mean square (RMS). Finally, there is the "common sense" method which says that by inspecting the range or the orbital elements it will be obvious which root is correct.

All of these methods, though sound on paper, can run into problems in practice. The first approach, in which the initial orbit determination problem is resolved with a slightly different set of observations, usually does not eliminate one of the values of range for short arcs of data. Both values of range tend to change approximately the same order of magnitude. This is probably due to numerical error accumulating during the formulation of the equations. As mentioned earlier, the approach using the linear equation in range needs the triple derivatives of the angular measurements with respect to time. These are determined with numerical derivatives which are extremely unstable for short arcs of data because the observations usually do not contain enough accuracy. The presence of noise or systematic error can make these numbers worthless. Occasionally this approach will yield a value of range which is not close to either of the two values produced by the initial orbit determination process. Rather than simplifying things, this new value for range now gives a third value to choose from.

The methods using orbital elements also have problems. For short arcs of angles–only data the differential correction has at least two solutions, one near each of the element sets produced by the initial orbit determination process. The RMS for both solutions is approximately the same order of magnitude and the solution with the smaller RMS is not always the correct solution. Due to the similarity in character of the two solutions, there is not really anything that one can look at to aid in the selection of the proper root. This inability to use "common sense" with the differential correction results would tend to point toward an inability to use this method with the initial orbit determination results, and this is true. In the realm of astrodynamics there exist many different types of orbits of concern. These orbits are as different as suborbital objects to interplanetary probes leaving the Earth. Even the resident population (those objects which orbit the Earth) vary from near–Earth circular orbits to orbits with four–day periods and eccentricities near 0.9. There is also no plane containing most of the resident objects (as opposed to the Solar System with the ecliptic plane). This lack of "rhyme or reason" to the orbits of these objects eliminates inspection of the semi–major axis, eccentricity and inclination as parameters to use in eliminating roots.

4 New Approaches to Root Solving and Selection

In many cases the past approach to initial orbit determination was to use one standard guess (for example $\rho = 3$ earth radii) to start a Newton iteration with the chosen initial orbit determination method. This approach yields only one solution and does not always work when two non–trivial roots exist. The current approach used in this study is similar to one Escobal implies should be used. We start with Watson's formulation of the problem and, by defining

$$\phi_n = \frac{n\pi}{1000}$$

calculate $y_1 - y_2$ for all values of ϕ_n from 0 to π, where $y_1 - y_2 = 0$ is a solution. The "one thousand" in the denominator is simply an arbitrarily large number. If the sign of $y_1 - y_2$ differs for ϕ_n and ϕ_{n+1} we know that a root exists between ϕ_n and ϕ_{n+1}. With this search scheme we find approximate values to all the necessary roots. These values are then used to start the iteration scheme for the initial orbit determination method of choice. This approach differs from Escobal's in its use of Watson's formulation to find the initial guesses for the roots. Watson's formulation was chosen because of the geometric interpretation and the smaller search interval.

As mentioned earlier, for space based sensors there should exist three roots between 0 and π. What if only one root is found? This can (and does) happen when errors cause the y_1 and y_2 curves to shift relative to each other. The two "missing" roots are located where y_1 and y_2 approach closely but do not intersect. This can mathematically be described as the point where

$$\frac{d}{d\phi}(y_1 - y_2) = 0$$

Since other points satisfy this relationship, care must be taken to initially guess a ϕ near the desired point. A Newton's iteration will then quickly converge to the desired point. Since ground based sensors can have only one root between 0 and π, it is impossible to tell when roots can not be found due to error. Therefore, this method is not applied to ground based sensors.

Another problem is that of root selection. How do you select the correct root from two equally viable candidates? As covered previously, many methods (each with their flaws) have been proposed. Our concept is based on the fact that the real solution is constrained to a real orbit. Real orbits have certain parameters which do not vary greatly over time. The "phantom" (or fictitious) orbit associated with the correct solution is not a real orbit, so it may be nearly circular for one data span but hyperbolic for the next. Therefore, the correct solution will have "stable" (i.e. slowly changing) orbital characteristics when compared to the fictitious solution. We chose to compare perigee heights for this study although other orbital parameters may be as good. This approach is very similar to the first suggested range approach with the exception that perigee is a more uniform parameter in time than range.

5 Numerical Results

To test the suggested solutions to the discussed problems, observations from an orbiting sensor were simulated for seven satellites; two circular near–earth orbits, one eccentric near–earth orbit, a molynia type orbit (12 hr., 0.7 eccentricity), a GPS orbit, a sun synchronous orbit, and a geosynchronous orbit. Four initial orbit determination methods were used; Laplace (correct through t^2), Truncated Gauss (t^2), Gauss-Gibbs (t^4) and Full Gauss (no approximations made). These methods were chosen to test various levels of accuracy and efficiency. Note that no conclusions will be made as to which of these methods is the "best", since the definition of best varies from user to user. Three types of data were generated in order to gain an understanding of the limitations of each method. The data types are referred to as two–body double precision or TBD (generated with two–body dynamics and using observations with eleven significant digits after the decimal point), two–body or TB (two– body dynamics using observations with four significant digits after the decimal point) and Perturbed (dynamics include perturbations and the observations have four significant digits after the decimal point).

At this point, some details on the implementation of the various methods should be mentioned. First, the Laplace method needs derivatives of the line–of–sight unit vectors. To generate these derivatives, a quartic fit of the unit vectors was used. Second, higher order methods (Gauss–Gibbs and Full Gauss) iterate on three values of range, one at each time point. In order to start these methods, the solution from a lower order method (range at the middle time point from Laplace, for example) is used as the first guess for all three values of range in the higher order methods. Also, in order to perform a Newton iteration with the Full Gauss method, the derivatives from the Gauss–Gibbs method were suitably modified and used.

In order to test the concept of the stability of the correct perigee, plots of perigee height (height above the earth's surface) were generated. These were produced by processing the data in a "moving window" fashion, choosing the correct perigees from the two possible solutions, and then plotting both the correct and incorrect perigees. If only one non–trivial solution existed, the incorrect perigee height was set to zero. The correct root was chosen by an engineering approach and not by the approach described within. This was done in order to quickly produce results which could initially test the concept. While this engineering approach did produce results useful in this vein, it was observed that the engineering approach did not always choose the truly correct root. One final note about the perigee plots is that in order to keep the scale of the plots such that the critical structure could be observed, upper bounds were placed on the values of the perigee plotted.

Due to space limitations, only a few of the many test cases that were performed will be included here. Figures 1 and 2 are perigee plots for the Molynia satellite using the TBD and Perturbed data types, respectively. Both sets of data were processed using the Full Gauss Method. Five minute data spans were used to represent a short arc of data for this satellite. For the double precision data, the correct perigee curve is reasonably flat, as it is for all four initial orbit determination methods, and is much flatter than the incorrect perigee curve. The slight bumps (which occur to a varying degree in all the methods) are due to the three line–of–sight vectors being nearly planar and/or the method used causing a loss of precision. The correct perigee curve in Figure 2, while still flatter than the incorrect perigee curve, is much rougher than in Figure 1. Much of the degradation in the smoothness of the correct perigee curve is caused by going from TBD data to single precision data. Also, the fact that the Perturbed data does not satisfy the two body dynamics implicit in all initial

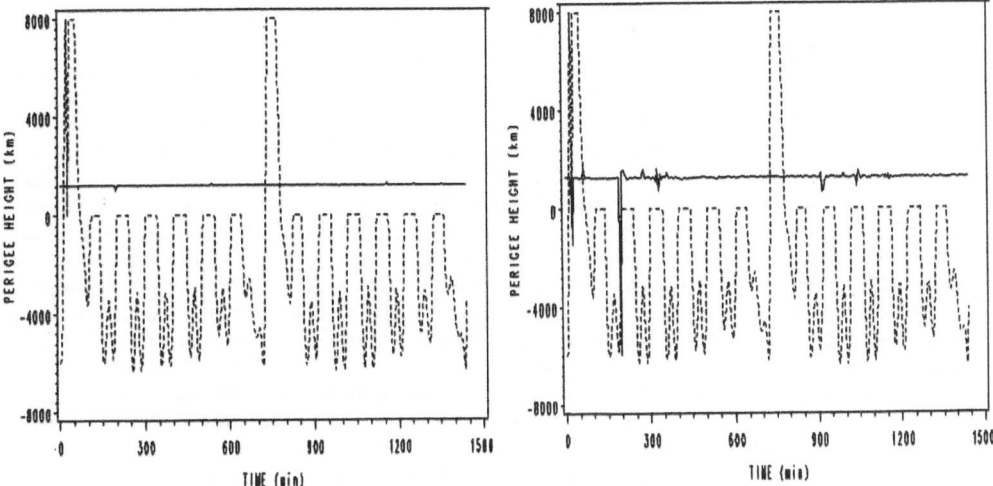

Figure 1. Full Gauss-Molynia/TBD (5min).

Figure 2. Full Gauss-Molynia/Perturbed (5 min).

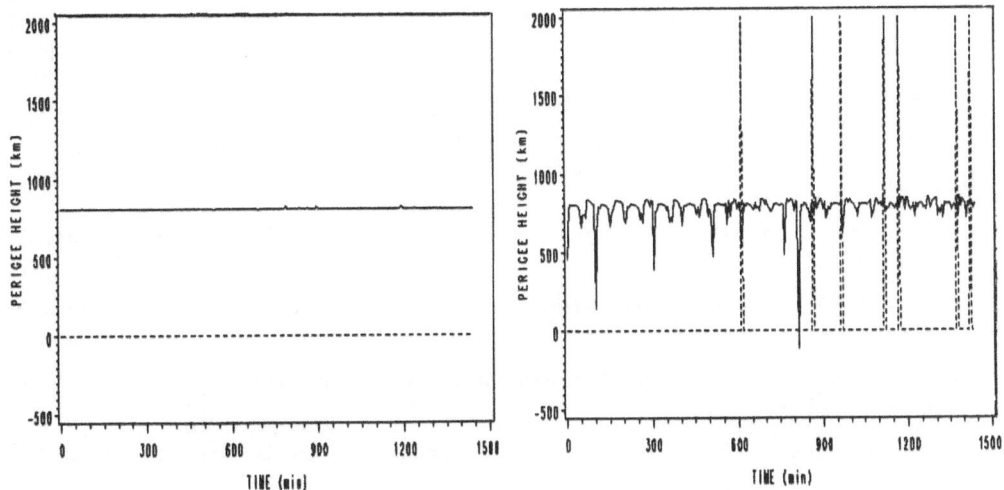

Figure 3. Full Gauss-Sun Synch/TBD (1 min).

Figure 4. Full Gauss-Sun Synch/ Perturbed (1 min).

orbit methods contributes some to the increased roughness seen. This increased roughness is seen to a similar degree when any of the four methods are used.

Since the correct perigee curve was always smoother than the incorrect perigee curve, one is tempted to conclude that the stability of perigee criterion for root selection is a success for these test cases. However, this is not entirely true. In the true application of this method, five minutes of data may be all that is available. If this data is taken near an inflection point in the incorrect perigee curve, the value of the incorrect perigee can appear to be more stable for that short time span. However, in most cases the method appears to have some merit for these test cases.

Figures 3 and 4 are for the sun–synchronous satellites, once again using TBD and Perturbed data, respectively, and processing the data using the Full Gauss method. A one minute data span was considered to be representative of a short arc of data for this satellite. For the TBD data in Figure 3, we see that the correct perigee curve for the Full Gauss method appears perfectly flat (the actual variations are two small to be seen on this scale). This is also true for the Gauss-Gibbs method, but not for the Laplace or Truncated Gauss. These two lower order methods show periodic variations of approximately 3000 kilometers. A most surprising result is the total lack of fictitious roots, as shown by the incorrect perigee value always being zero. It is not known if this result is true for all sun–synchronous orbits viewed from this sensor. For Perturbed data, it is seen that the stability of the perigee is vastly degraded, but due to the relative lack of non–trivial second solutions, the method still works. However, if more non–trivial second solutions existed, the method may have trouble.

The algorithm developed to find missing roots was used approximately 2000 times (in 35,000 test cases) in processing the data. The range from this algorithm was used approximately 900 times, with a failure rate (where failure is defined as producing a hyperbolic element set or the trivial solution) of 11.4%. Over 99% of the usage of this algorithm occured when using the single precision quality data (either Perturbed or un–Perturbed), illustrating that the algorithm may prove useful when noise is added to the data.

6   Conclusions

For a spaced–based sensor,it is necessary to solve for all viable roots when using a short arc of data. Only then does the possibility of getting the correct solution exist. The use of the stability of perigee for root selection was somewhat successful, but since the data examined did not include noise, further work is needed. The algorithm for finding "non–existent" roots has shown to be valuable in this study, and may be even useful in future studies with noisy data. Finally, the perigee plots developed appear to be a very useful tool to gain an overall view of the success or failure of initial orbit determination for a certain sensor–satellite set.

References

[1] Danby, J.M.A., *Fundamentals of Celestial Mechanics*, Willmann– Bell, Inc., Richmond, Virginia, 1988, pp. 213-251.

[2] Escobal, P.R., *Methods of Orbit Determination*, Robert E. Krieger Publishing Co., Malabar, FL, 1985, pp. 239-292.

[3] Moulton, F.R., *An Introduction to Celestial Mechanics*, Dover Publications, Inc., New York, 1970, pp. 191-260.

[4] Watson, James C., *Theoretical Astronomy*, J.B. Lippincott and Co., London, 1881, pp. 235-243.

* Other references are available upon request.

A PERTURBATION OF THE RELATIVISTIC KEPLER PROBLEM

Ana Nunes[1]*, Josefina Casasayas[2]**, and Jaume Llibre[3]+

(1) Departamento de Física, Faculdade de Ciencias, Universidade de Lisboa
Lisboa, Portugal

(2) Departament de Matemàtica Aplicada i Anàlisi, Facultat de Matemàtiques
Universitat de Barcelona, 08071 Barcelona, Spain

(3) Departament de Matemàtiques, Facultat de Ciències, Universitat Autònoma
de Barcelona Bellaterra, 08193 Barcelona, Spain

Abstract. We consider the Kepler Problem with the first order relativistic correction and show that, for a suitable class of perturbations, "almost all" the invariant tori and cylinders of the unperturbed system persist and that the perturbed system has strong evidences of non–integrability.

1. Introduction

Let us consider the first order relativistic correction to the Kepker problem, given by the Hamiltonian

$$\bar{H}_\varepsilon : \Re^+ \times S^1 \times \Re^2 \to \Re$$
$$(r, \theta, p_r, p_\theta) \to \frac{p_r^2}{2} + \frac{p_\theta^2 - 2\varepsilon}{2r^2} - \frac{1}{r},$$

where (r, θ) are polar coordinates in the plane, p_r and p_θ are the conjugate momenta and ε is a positive real parametre (when $\varepsilon = 0$ this is just the Kepler Hamiltonian). The associated equations of motion are

(1)
$$\dot{r} = p_r,$$
$$\dot{\theta} = p_\theta r^{-2},$$
$$\dot{p}_r = (p_\theta^2 - 2\varepsilon)r^{-3} - r^{-2},$$
$$\dot{p}_\theta = 0.$$

---

* Partially supported by Instituto Nacional de Investigaçao Cientifica.

** Partially supported by a grant of the DGICYT no.BE90-135.

\+ Partially supported by a grant of the DGICYT no.PB86-0351.

We shall study a perturbation of these equations of the form

$$\dot{r} = p_r + \mu \frac{\partial \bar{H}_1}{\partial p_r},$$

(2)
$$\dot{\theta} = p_\theta r^{-2} + \frac{\mu}{1+r^2} \frac{\partial \bar{H}_1}{\partial p_\theta},$$

$$\dot{p}_r = (p_\theta^2 - 2\varepsilon) r^{-3} - r^{-2} - \mu \frac{\partial \bar{H}_1}{\partial r},$$

$$\dot{p}_\theta = - \frac{\mu}{1+r^2} \frac{\partial \bar{H}_1}{\partial \theta},$$

where μ is a small real parameter and $\bar{H}_1 = \bar{H}_1(r, \theta, p_r, p_\theta, \mu)$.

System (2) has the first integral

$$\bar{H}(r, \theta, p_r, p_\theta, \varepsilon, \mu) = \bar{H}_\varepsilon(r, \theta, p_r, p_\theta) + \mu \bar{H}_1(r, \theta, p_r, p_\theta, \mu) + \frac{p_\theta^2 - 2\varepsilon}{2}$$

and, in general, is not Hamiltonian. In fact, it is easy to prove that if equations (2) have Hamiltonian form then the system is integrable.

For $\mu = 0$, system (2) reduces to the relativistic Kepler problem $\bar{H}_\varepsilon$ which is integrable. For every value of the energy parameter, the energy levels of $\bar{H}_\varepsilon$ are foliated by invariant tori and cylinders. In [1] it has been proved that, in a certain region of every energy level, most of these tori and cylinders persist for the perturbed system (2). In this paper, our main purpose is to discuss the integrability of (2).

2. The extended phase space

Consider the change of variables given by

$$F : \Re^+ \times S^1 \times \Re^2 \to (0, \pi/2) \times S^1 \times \Re^2$$
$$(r, \theta, p_r, p_\theta) \to (x = \arctan r, \theta, p_x = p_r, p_\theta)$$

which amounts just to taking $r = \tan(x)$. In these variables, and choosing an appropiate time scale given by $dt/d\tau = \cos^{-2} x$, system (2) becomes

$$x' = p_x + \mu \frac{\partial H_1}{\partial p_x},$$

(3)
$$\theta' = p_\theta \sin^{-2} x + \mu \frac{\partial H_1}{\partial p_\theta},$$

$$p_x' = (p_\theta^2 - 2\varepsilon) \cos x \sin^{-3} x - \sin^{-2} x - \mu \frac{\partial H_1}{\partial x},$$

$$p_\theta' = -\mu \frac{\partial H_1}{\partial \theta},$$

where $H_1 = \bar{H}_1 \circ F^{-1}$, which is Hamiltonian with Hamiltonian function

(4)
$$H(x, \theta, p_x, p_\theta, \mu) = \frac{p_x^2}{2} + \frac{p_\theta^2 - 2\varepsilon}{2 \sin^2 x} - \frac{\cos x}{\sin x} + \mu H_1(x, \theta, p_x, p_\theta, \mu),$$

and the prime denotes the derivative with respect to the new time variable τ.

The image by F of the phase space of (2) is $(0, \pi/2) \times S^1 \times \Re^2$, but (3) may be analytically extended to $(0, \pi) \times S^1 \times \Re^2$. In particular, the infinity singularity of the original system (2) is regularized. From now on, we shall study system (3) in the extended phase-space.

3. Persistence of the invariant cylinders

In this section, we shall briefly review the results obtained in [1] concerning the persistence for $\mu \neq 0$ of the invariant tori and cylinders of the unperturbed system. Consider then the unperturbed extended system, that is, we shall take $\mu = 0$ and $x \in (0, \pi)$.

When $\mu = 0$ system (3) has two independent first integrals, the energy $H = h$ and the angular momentum $p_\theta = c$. So, the study of a given energy level $H = h$ of (3) may be reduced to the study of the family of one degree of freedom Hamiltonians

$$H_c(x, p_x) = \frac{p_x^2}{2} + V_c(x), \quad V_c(x) = \frac{c^2 - 2\varepsilon}{2 \sin^2 x} - \frac{\cos x}{\sin x}, \quad c^2 \in [0, h + 2\varepsilon + \sqrt{h^2 + 1}],$$

together with the complementary equations

$$\dot{\theta} = c \sin^{-2} x,$$

$$\dot{p}_\theta = c.$$

In Figure 1 we have represented the curves $V_c(x)$. When $h < -\varepsilon$, the projection of the energy level $H = h$ on the x-axis is contained in $(0, \pi/2)$. This corresponds to the case when the original problem has no unbounded orbits. But, for $h > -\varepsilon$, the projection of $H = h$ intersects the interval $(\pi/2, \pi)$ and some of the invariant surfaces of constant energy and angular momentum of (3) will strictly contain invariant surfaces of the original problem.

The foliation of the energy levels of the extended unperturbed system by the surfaces of constant angular momentun is represented in Figure 2. For $h < (4\varepsilon)^{-1} - \varepsilon$, motions for the extended system are bounded away from the $x = \pi$ singularity, and the surfaces of constant angular momentum are tori whenever $c^2 > 2\varepsilon$. In particular, when $h > 0$, the tori which correspond to $2\varepsilon < c^2 < 2\varepsilon + h$ contain invariant cylinders of the original problem. Taking action–angle variables on these families of tori, it is straightforward to prove that the extended unperturbed system is isoenergetically non–degenerate whenever $\varepsilon > 0$ (the relativistic correction removes the degeneracy that we would get for the Kepler problem). Applying KAM theorem to this system and going back through the change of variables we have the following result, (see [1]):

Theorem. **Consider system (2) and denote by I_h the invariant region $\{(r, \theta, p_r, p_\theta) : \bar{H} = h\}$. Assume that the extended perturbing term $H_1 = \bar{H}_1 \circ F^{-1}$ is analytical in $(0, \pi) \times S^1 \times \Re^2 \times [-\mu^*, \mu^*]$. Then, for $|\mu|$ small enough, system (2) has invariant tori and cylinders close to the tori and cylinders that foliate the region $I_h \cap \{p_\theta^2 > 2\varepsilon\}$ for $\mu = 0$. Moreover, the measure of the complement in $I_h \cap \{p_\theta^2 > 2\varepsilon\}$ of the union of these invariant tori and cylinders tends to zero when μ tends to zero.**

The question now arises of knowing whether this persistence property is trivial or not, i.e. if the perturbed system is integrable or not.

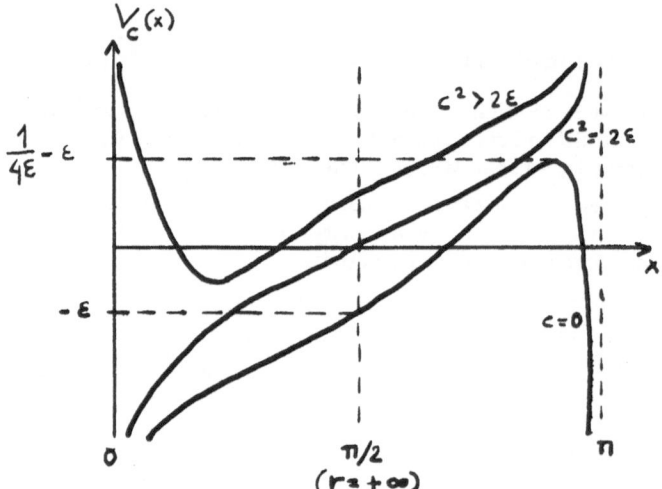

Figure 1

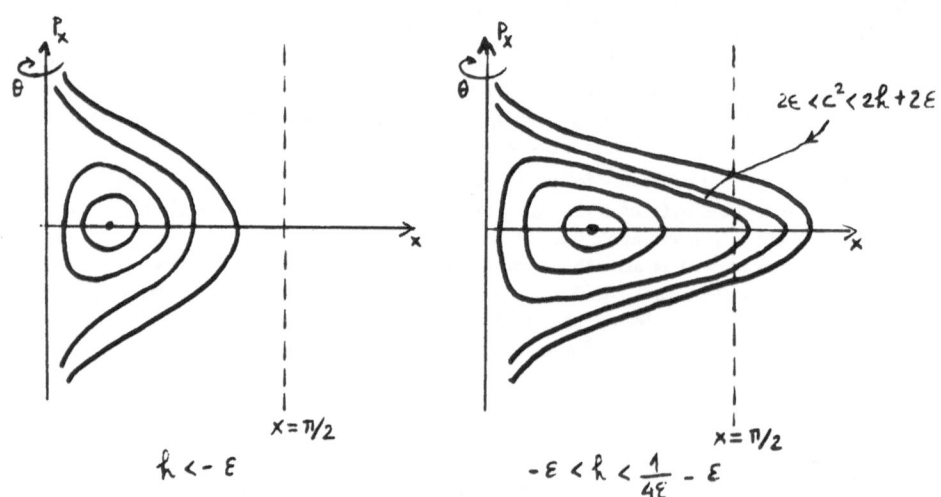

Figure 2

4. Non-integrability

Our goal is to prove that, for certain families of perturbing functions H_1, system (2) is non-integrable (in the sense that there are no analytical integrals independent of the energy integral) by showing that there exist transversal homoclinic orbits, (see [2]).

First, we regularize the collision singularity ($x = 0$) of the extended system (3) by using an appropiate version of McGehee's variables ([3]). Changing from $(x, \theta, p_x, p_\theta, t)$ to (x, θ, v, u, τ) by

$$v = p_x \sin x,$$
$$u = p_\theta,$$
$$\frac{dt}{d\tau} = \sin^2 x,$$

the equations of motion (3) become

$$\begin{aligned}
\dot{x} &= v \sin x + \mu \sin^2 x \frac{\partial H_1}{\partial p_x}, \\
(5) \qquad \dot{\theta} &= u + \mu \sin^2 x \frac{\partial H_1}{\partial p_\theta}, \\
\dot{v} &= 2h \cos x \sin^2 x + 2 \sin x \cos^2 x - \sin x - \mu \sin^2 x \cos x H_1 - \\
&\quad - \mu \sin^3 x \frac{\partial H_1}{\partial x} - \mu v \cos x \sin x \frac{\partial H_1}{\partial p_x}, \\
\dot{u} &= - \mu \sin^2 x \frac{\partial H_1}{\partial \theta},
\end{aligned}$$

and the energy relation (4) becomes

$$(6) \qquad v^2 + u^2 = 2\varepsilon + 2 \sin x \left(h \sin x + \cos x \right) + 2\mu \sin^2 x H_1.$$

We note that this set of variables, (x, θ, v, u), regularizes simultaneously the collision ($x = 0$) and the infinity ($x = \pi/2$) singularities of the original system (2) when $\mu = 0$.

The collision manifold $\Lambda = \{(\theta, v, u) \in S^1 \times \Re^2 : v^2 + u^2 = 2\varepsilon\}$ defined by setting $x = 0$ in (6) is a torus and the flow on Λ, obtained by setting $x = 0$ in (5), is formed by circular orbits for $u \neq 0$ and circles of equilibrium points for $u = 0$, see Figure 3.

For $h < (4\varepsilon)^{-1} - \varepsilon$, the energy levels $\bar{I}_h = I_h \cup \Lambda$ are solid tori which, for $\mu = 0$, are foliated by tori and cylinders as shown in Figure 4 (topological representation of a $\theta = constant$ section of the solid torus $\bar{I}_h$ foliated by the invariant sets I_{hc}; the shadowed region corresponds to points whose projection on the x-axis is contained in $[\pi/2, \pi)$; moreover, $c_{max} = \sqrt{h + \sqrt{h^2 + 1} + 2\varepsilon}$). We see that there are two cylinders, which correspond to $p_\theta = c = \pm\sqrt{2\varepsilon}$, of homoclinic orbits to the circular orbits $S^\pm = (x = 0, \theta, u = \pm\sqrt{2\varepsilon}, v = 0)$ on the collision manifold.

We shall prove the following result:

Proposition. Suppose that the stable and unstable manifolds of the (non-hyperbolic) circular orbits $S^\pm$ on the collision manifold persist for the perturbed system (5) with $H_1(x, \theta, p_x, p_\theta, \mu) = \cos \theta$. Then, for $|\mu|$ and ε sufficiently small, system (5) has at least two transversal homoclinic orbits associated to each $S^\pm$.

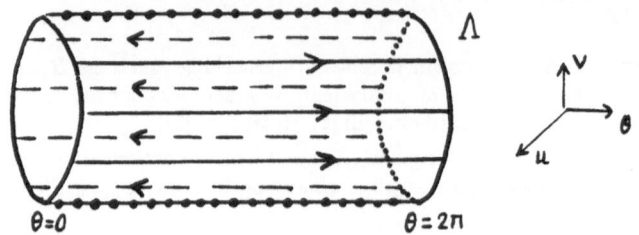

Figure 3.

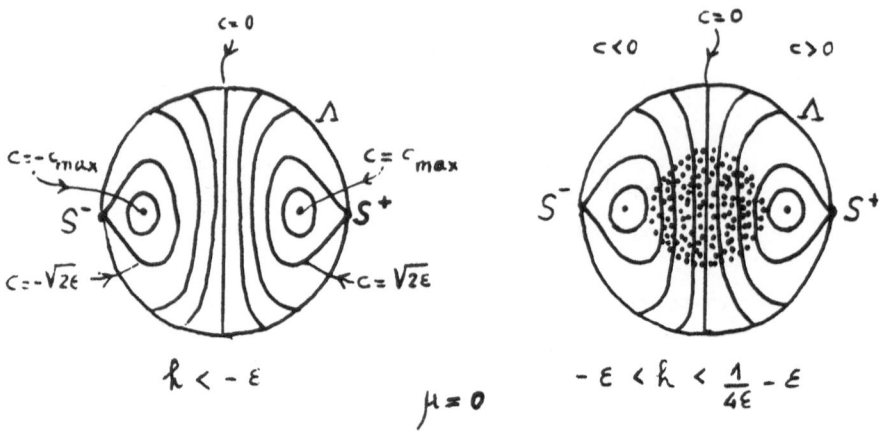

Figure 4

Proof. For $H_1(x, \theta, p_x, p_\theta, \mu) = \cos \theta$, the perturbed system (5) becomes

$$
\begin{aligned}
\dot{x} &= v \sin x, \\
\dot{\theta} &= u, \\
\dot{v} &= 2h \sin^2 x \cos x + 2 \sin x \cos^2 x - \mu \sin^2 x \cos x \cos \theta, \\
\dot{u} &= \mu \sin^2 x \sin \theta,
\end{aligned}
\tag{7}
$$

which is invariant under the symmetry

$$
\Sigma : (x, \theta, v, u, \tau) \rightarrow (x, -\theta, -v, u, -\tau).
$$

Let $\mathbf{r}_\mu(\tau) = (x_\mu(\tau), \theta_\mu(\tau), v_\mu(\tau), u_\mu(\tau))$ with $\mu \in [0, \delta)$, denote the unique orbit in the unstable manifold $W^u(S^+)$ of S^+ such that $v_\mu(0) = 0$, $\theta_\mu(0) = \theta_0$. For δ small enough $\mathbf{r}_\mu(\tau)$ is well defined and, moreover, it is approximated by $\mathbf{r}_0(\tau) = (x_0(\tau), \theta_0 + \sqrt{2\varepsilon}\tau, v_0(\tau), \sqrt{2\varepsilon})$ uniformly in $\tau \in (-\infty, 0]$ (see [4]). Then, we may write:

$$
\begin{aligned}
x_\mu(\tau) &= x_0(\tau) + \mu x_1(\tau) + O(\mu^2), \\
\theta_\mu(\tau) &= \theta_0 + \sqrt{2\varepsilon}\tau + \mu\theta_1(\tau) + O(\mu^2), \\
v_\mu(\tau) &= v_0(\tau) + \mu v_1(\tau) + O(\mu^2), \\
u_\mu(\tau) &= \sqrt{2\varepsilon} + \mu u_1(\tau) + O(\mu^2),
\end{aligned}
\tag{8}
$$

552

for $\tau \in (-\infty, 0]$. Substituting (8) in (7) we obtain for the first variational equations in the u variable

$$(9) \qquad \frac{du_1}{d\tau} = \sin^2 x_0(\tau) \sin(\sqrt{2\varepsilon}\tau + \theta_0).$$

Since $\mathbf{r}_\mu(\tau)$ belongs to the unstable manifold, the solution we seek must verify the asymptotic condition

$$\lim_{\tau \to -\infty} u_1(\tau) = 0,$$

and so, from (9),

$$(10) \qquad u_1(0, \theta_0, \varepsilon) = \int_{-\infty}^0 \sin^2 x_0(\tau) \sin(\sqrt{2\varepsilon}\tau + \theta_0) d\tau \ .$$

The integral in (10), regarded as a function of θ_0, gives the intersection of $W^u(S^+)$ with the annulus $A = \{v = 0, u > 0, \theta_0 \in S^1\}$ to first order in μ and apart from the constant term $u_0(\tau) = \sqrt{2\varepsilon}$. In fact, it can be proved that the integral converges independently from the persistence hypothesis.

Now, since equations (7) have the symmetry Σ, the intersection of $W^s(S^+)$ with the annulus A will be given by

$$\sqrt{2\varepsilon} + u_1(0, -\theta_0, \varepsilon) + 0(\mu^2).$$

The equation in θ_0, $u_1(0, \theta_0, \varepsilon) - u_1(0, -\theta_0, \varepsilon) = 0$ has always at least the solutions $\theta_0 = 0$, $\theta_0 = \pi$. If we prove that the curves $W^u(S^+) \cap \Lambda$ and $W^s(S^+) \cap \Lambda$ at these points are transversal, we may assure that $W^u(S^+)$ and $W^s(S^+)$ have at least two transversal homoclinic orbits for $|\mu|$ small enough.

From (10), we may write

$$u_1(0, \theta_0, \varepsilon) = A(\varepsilon) \cos \theta_0 + B(\varepsilon) \sin \theta_0,$$

where

$$A(\varepsilon) = \int_{-\infty}^0 \sin^2 x_0(\tau) \sin \sqrt{2\varepsilon}\tau d\tau, \ \text{and} \ B(\varepsilon) = \int_{-\infty}^0 \sin^2 x_0(\tau) \cos \sqrt{2\varepsilon}\tau d\tau.$$

Hence,

$$u_1(0, \theta_0, \varepsilon) - u_1(0, -\theta_0, \varepsilon) = 2B(\varepsilon) \sin \theta_0,$$

and transversality is guaranteed provided that $B(\varepsilon) \neq 0$. But

$$2B(\varepsilon) = \int_{-\infty}^{+\infty} \sin^2 x_0(\tau) \cos \sqrt{2\varepsilon}\tau d\tau$$

is the Fourier transform of the $L^1(\Re)$ function $\sin^2 x_0(\tau)$, and so it is a continuous function of ε. Since $B(0) > 0$, we must have $B(\varepsilon) > 0$ for ε small enough, and the transversality follows.

As to the hypothesis that $W^u(S^+)$, $W^s(S^+)$ persist for the perturbed system, we have no proof so far. However, we have calculated the Poincaré map of the perturbed system in a neighbourhood of S^+ up to third order terms and found candidates to invariant manifolds up to this order.

Acknowledgement

We would like to thank C.Simó for many helpful comments and suggestions.

References

1. E. Lacomba, J. Llibre, A. Nunes: 'Invariant tori and cylinders for a class of perturbed Hamiltonian systems', to appear in the Proceedings of "Geometry of Hamiltonian Systems", Berkeley.

2. Moser, J.: 1973, "Stable and random motion in dynamical systems", Princeton Univ. Press, Princeton.

3. McGehee, R.: 1974, 'Triple collision in the collinear three–body problem', Invent. Math. 27, 191-227.

4. Sanders, J.A.: 1982, 'Melnikov's method and averaging', Celestial Mechanics 28, 171-181.

INTEGRABLE 3-DIMENSIONAL DYNAMICAL SYSTEMS AND THE PAINLEVÉ PROPERTY

C. Polymilis

Section of Astrophysics, Astronomy, and Mechanics
Department of Physics, University of Athens
Panepistimiopolis, 157 83 Zagrafos
Athens, Greece

ABSTRACT

We investigate some classes of 3-dimensional potentials and find all the cases that have a second integral of motion quadratic in the velocities. Whenever there is no third integral of motion the Painlevé or weak Painlevé property is not satisfied, because two of the resonances are complex.

1. INTRODUCTION

We consider 3-D Hamiltonian systems of the form

$$H_1 = (p_x^2 + p_y^2 + p_x^2)/2 + (ax^2 + by^2 + cz^2)/2 + dxz^2 + eyz^2 , \qquad (1.1)$$

$$H_2 = (p_x^2 + p_y^2 + p_z^2)/2 + (ax^2 + by^2 + cz^2)/2 + dxz^2 + fxy^2 , \qquad (1.2)$$

$$H_3 = H_2 + gx^4 + hy^4 + kz^4 + lx^2y^2 + mx^2z^2 . \qquad (1.3)$$

The Hamiltonian (1.1) has been introduced by Contopoulos and Magnenat (1985) and has been explored in detail for a large set of parameters by Zachilas (1989). This Hamiltonian has several interesting properties such as complex instability and collisions of bifurcations, that do not appear in 2-D systems. The Hamiltonians (1.2) and (1.3) are of the form of the truncations at 3rd and 4th order, respectively, of the reduced 4-particle Toda lattice, but with undefined coefficients (see Appendix).

We find which Hamiltonians of the form (1.1,2,3) possess other integrals of motion quadratic in the velocities, besides the energy. Also, we check if the respective equations of motion possess the Painlevé or the weak Painlevé property.

2. QUADRATIC INTEGRALS

We want to find quadratic integrals for the Hamiltonians under consideration, of the form

Predictability, Stability, and Chaos in N-Body Dynamical Systems
Edited by A.E. Roy, Plenum Press, New York, 1991

$$\phi = A p_x^2 + B p_y^2 + C p_z^2 + D p_x p_y + E p_x p_z + F p_y p_z + K \quad , \tag{2.1}$$

where A, B, C, D, E, F, K are functions of x, y and z only.
The Poisson bracket of H and ϕ is zero

$$[H, \phi] = 0. \tag{2.2}$$

Setting the coefficients of x^ℓ, y^m and z^n of (2.2) equal to zero we obtain the following system of equations:

$$A_x = 0 \quad , \quad A_y + D_x = 0 \quad , \quad A_z + E_x = 0 \quad ,$$

$$B_y = 0 \quad , \quad B_x + D_y = 0 \quad , \quad B_z + F_y = 0 \quad , \tag{2.3}$$

$$C_z = 0 \quad , \quad C_x + E_z = 0 \quad , \quad C_y + F_z = 0 \quad ,$$

$$F_x + E_y + D_z = 0 \quad ,$$

$$K_x = 2 A V_x + D V_y + E V_z = 0 \quad ,$$

$$K_y = D V_x + 2 B V_y + F V_z = 0 \quad , \tag{2.4}$$

$$K_z = E V_x + F V_y + 2 C V_z = 0 \quad ,$$

where the subscripts x, y, z denote derivatives and V is the potential of the Hamiltonian.

From Equations (2.3) we find the solutions A, B, C, D, E, F which are of the form

$$A = \alpha_0 + \alpha_1 y + \alpha_2 z + \alpha_3 y^2 + \alpha_4 yz + \alpha_5 z^2 \quad ,$$

$$B = \beta_0 + \beta_1 x + \beta_2 z + \alpha_3 x^2 + \beta_4 xz + \beta_5 z^2 \quad ,$$

$$C = \gamma_0 + \gamma_1 x + \gamma_2 y + \alpha_5 x^2 + \gamma_4 xy + \beta_5 y^2 \quad , \tag{2.5}$$

$$D = \delta_0 - \alpha_1 x - \beta_1 y + \delta_1 z - 2\alpha_3 xy - \alpha_4 xz - \beta_4 yz + \gamma_4 z^2 \quad ,$$

$$E = \epsilon_0 - \alpha_2 x + \epsilon_1 y - \gamma_1 z - \alpha_4 xy - 2\alpha_5 xz - \gamma_4 yz + \beta_4 y^2 \quad ,$$

$$F = \theta_0 + \theta_1 x - \beta_2 y - \gamma_2 z - \beta_4 xy - \gamma_4 xz - 2\beta_5 yz + \alpha_4 x^2 \quad ,$$

where $\alpha, \beta, \gamma, \delta, \epsilon, \theta$ are constant coefficients, and we have also the relation $\delta_1 + \epsilon_1 + \theta_1 = 0$.

From Equations (2.5) we derive the following compatibility conditions

$$(2A_y - D_x) V_x + (D_y + 2B_x) V_y + (E_y - F_x) V_z +$$

$$+ D(V_{yy} - V_{xx}) + 2AV_{xy} - 2BV_{yx} - FV_{zx} + EV_{zy} = 0 \quad .$$

$$(2A_z - E_x)V_x + (D_z - F_x)V_y + (E_z - 2C_x)V_z +$$

$$+E(V_{zz} - V_{xx}) + 2AV_{xz} - 2CV_{zx} - FV_{yx} + DV_{yz} = 0 \quad, \tag{2.6}$$

$$(D_z - E_y)V_x + (2B_z - F_y)V_y + (F_z - 2C_y)V_z +$$

$$+F(V_{zz} - V_{yy}) + 2BV_{yz} - 2CV_{zy} + DV_{xz} - EV_{xy} = 0 \quad.$$

Substituting equations (2.5) in these compatibility relations, we find certain relations among $\alpha, \beta, \gamma, \delta, \varepsilon, \theta$ and the parameters of the Hamiltonian by equating the coefficients of $x^\ell y^m z^n$ of equations (2.6) to zero for each type of Hamiltonians (1.1,2,3).

Hamiltonian (1.1)

Case 1.1 : $e.d \neq 0$ and $a = b$, c = arbitrary. $\tag{2.7}$
Then there is a second integral of the form

$$\phi = (p_x^2 + ax^2)e/2d + (p_y^2 + ay^2)d/2e - (p_x p_y + axy). \tag{2.8}$$

The special case $a=b=c$ was found to be integrable by Contopoulos and Barbanis (1985).

Case 1.2 : $ed = 0$. Then d and/or e is zero and the problem is reduced to a 2-D system plus an uncoupled oscillation in the third dimension.

Hamiltonian (1.2)

Case 2.1 : $d = f$, $b = c$. $\tag{2.9}$
This Hamiltonian has axial symmetry around the x-axis. Thus there is a second integral

$$\phi = zp_y - yp_z \tag{2.10}$$

that is the angular momentum along the x-axis.

Case 2.2 : $d f = 0$. This case is reduced to a 2-D system, plus an independent oscillation, as in case 1.2 above.

Hamiltonian (1.3)

Case 3.1 : $d = m = 0$, a,b,c,f,g,h,k,ℓ = arbitrary. $\tag{2.11}$
Then there is a second integral of the form

$$\phi = p_z^2 + cz^2 + 2kz^4 \quad. \tag{2.12}$$

(We find, also, the case $f = \ell = 0$, a,b,c,d,g,h,k,m = arbitrary, but this case is the same with case 3.1 if we interchange the variables y and z. Thus we consider the case 3.1 only).
The integral (2.12) expresses the fact that the z-motion is uncoupled from the x-y motion.
The Hamiltonian (1.3) in this case is partially integrable and the x-y

motion is a 2-D system. But the system becomes completely integrable in the following subcases:

Subcase 3.1.1 : $a = b$, $f = 0$, $h = g$, $\ell = 6g$

c,k = arbitrary. (2.13)

The third integral ϕ is of the form

$$\phi' = p_x p_y + 4gx^3 y + xy(4gy^2 + a) \ .$$ (2.14)

Subcase 3.1.2 : $b = 4a$, $f = 0$, $h = 16g$, $\ell = 12g$

k = arbitrary. (2.15)

and

$$\phi' = yp_x^2 - xp_x p_y - x^2(8gy^3 + ay) - 4gx^4 y.$$ (2.16)

Subcase 3.1.3 : $h = g$, $\ell = 2g$, $f = 0$

a,b,c,k = arbitrary. (2.17)

The third integral is of the form

$$\phi' = (yp_x - xp_y)^2 + (\alpha - \beta)(p_x^2 + ax^2 + 2gx^4 + 2gx^2 y^2) \ ,$$ (2.18)

and $2g(\alpha - \beta) + a - b = 0$, where α, β are arbitrary.

Subcase 3.1.4 : $g = 16h$, $\ell = 12h$

a,b,c,f,k = arbitrary. (2.19)

The third integral is of the form

$$\phi' = (\alpha - \beta)p_x^2 + xp_y^2 - yp_x p_y + 2y^2 x + 32h(\alpha - \beta)x^4 -$$
$$8hx^3 y^2 + x^2[2fy^2 + a(\alpha - \beta)] - x[-4hy^4 + y^2(b - a/2)]$$ (2.20)

and $8h(\alpha - \beta) - f = 0$, $4f(\alpha - \beta) + 4a - b = 0$, (2.21)

where α, β are arbitrary.

Subcase 3.1.5 : $f = h = k = \ell = 0$, a,b,c,g = arbitrary (2.22)
This case is a trivial one, because the system is reduced to a 1-D Duffing oscillator without damping.

Case 3.2 : $b = c$, $f = d$, $\ell = m$, $h = k = 0$

a,g = arbitrary. (2.23)

The integral ϕ in this case is the angular momentum $(yp_z - zp_y)$ and the system is partially integrable.

Case 3.3 : $c = b$, $a = 4b$, $g = 8k$, $h = k$, $\ell = m = 6k, f = d = 0$.   (2.24)

This case is a known integrable case (Dorizzi et al, 1986).

3. PAINLEVÉ ANALYSIS

    In order to find if the Hamiltonians (1.1,2,3) possess the Painlevé or the weak Painlevé property we use the standard method (Dorizzi et al., 1983, Grammaticos et al. 1983).

We express the solutions of the differential equations in the form

$$x = \alpha_0 \tau^p + \sum_{i=1}^{r_{max}} \alpha_i \tau^{p+i}$$

$$y = \beta_0 \tau^q + \sum_{i=1}^{r_{max}} \beta_i \tau^{q+i} \qquad (3.1)$$

$$z = \gamma_0 \tau^s + \sum_{i=1}^{r_{max}} \gamma_i \tau^{s+i}$$

where p,q,s are integers of rational and r_{max} is the maximum of the "resonances". Since we have 3-D systems we need 6 arbitrary coefficients in the series solutions (3.1) at the resonances. For each of the Hamiltonians)1,1,2,3) we find, from the leading order behaviour various possible values of p,q,s . Next we find the resonances and we check what coefficients $\alpha_i, \beta_i, \gamma_i$ are arbitrary at the resonances.

Hamiltonian (1.1)

We find that the only possibility for p,q,s is

$$p = q = s = -2 \quad , \qquad (3.2)$$

and we must also have $d = e$.
In this case the resonances are

$$r = -1,\ 2,\ 3,\ 6 \text{ and } \frac{5 \pm i(23)^{\frac{1}{2}}}{2} \qquad (3.3)$$

that is 4 are integers and 2 complex. Thus in this case we have not the necessary 6 integer or rational values of r and the system does not possess the Painlevé property.

Hamiltonian (1.2)

For this system we find the following 3 possibilities for p,q,s

(1) $p = q = s = -2$

then $f = d$.

(2) $p = q = -2$, $s > -2$,

then $2f\alpha_0 = s(s-1)$, $2\alpha_0^2 = \beta_0^2$, $- d\beta_0^2 = 6\alpha_0$. $\qquad (3.4)$

(3) $p = s = -2$, $q \gtrless -2$, symmetrical to (2).

For the possibility (1) the resonances r are

$$r = -1, 0, 5, 6 \quad \text{and} \quad \frac{5 \pm i(23)^{\frac{1}{2}}}{2} \qquad (3.5)$$

and for the possibility (2)

$$r = -1 (\text{double}) \ , \ 6, \ -2s \ (s < 1/2) \text{ and } \frac{5 \pm i(23)^{\frac{1}{2}}}{2} \qquad (3.6)$$

The possibility (3) is symmetric to 2.
For all the possibilities we have not 6 integers or rational values of r. Thus the Hamiltonian (1.2) does not possess the Painlevé property.

We find 6 possibilities for p,q,s:

 (1) $p = q = s = -1$

 (2) $p = q = -1$, $s > -1$

 (3) $p = s = -1$, $q > -1$ (3.7)

 (4) $p = -1$, $q > -1$, $s > -1$

 (5) $q = s = -1$, $p = 0$

 (6) $q = s = -1$, $-1 < p < 0$

We exclude from our analysis the case 3.3 (relations 2.24) because it is known to possess the Painlevé property (Dorizzi et al. 1986).

From the possibility (1) and for the case 3.1 (relation (2.11)) we find the values of $r = -1$ (double) , 4 (double) and two more resonances whose type (complex, integers, etc.) depends on the coefficients of the Hamiltonian. Since -1 is a double root always, independently of the values of the coefficients we cannot check the Painlevé property.

For the case 3.2 (relations (2.23)) and for the same possibility (1) we find the values of $r = -1,0,3,4$, and 2 more values from the relation

$$r = (\pm(32g/\ell - 7)^{\frac{1}{2}} + 3)/2$$

It is evident from this last relation that these resonances are integers or rationals for certain relations between g and ℓ . Two such relations for integer resonances are $32g = 8\ell$ and $2g = \ell$. In these cases the values of r are 1 and 2, and 0 and 3 respectively but the coefficients of the solutions (3.1) are not arbitrary at the resonances. The same is true and for all possible relations between g and ℓ for which there are integer or rational values of r.

From the possibility 2, and for the case 3.1 (relation (2.11)), we find 4 values of $r = -1,0,1,4$ and 2 values determined by the relation

$$r = [\pm \left(\frac{4g(25h - 8\ell) - \ell(32h - 7\ell)}{(4gh - \ell^2)} \right)^{\frac{1}{2}} + 3]/2 \qquad (3.8)$$

Substituting in (3.8) the relations (2.13), we find for the subcase 3.1.1, the values $r = 1, 2$, and $p = q = -1$, $s = 0$. The coefficients of the solutions at the values of r are arbitrary and thus this subcase possesses the Painlevé property. The same is true for the subcase 3.1.3 for which from (3.8) and the relations (2.17) we find $r = 0, 3$, and the same values of p,q,s as above. Substituting in (3.8) the relations (2.15) we find $r = -2, 5$. Thus for the subcase 3.1.2 we cannot check for the Painlevé property. The same is true for the subcase 3.1.4 for which we find the same values of r as above.

For the case 3.2 we find from the possibility (2) that one of the 6 values of r is -3. We find the same value for r in this case from the possibility (3) and thus we cannot check for the case 3.2 if it possesses the Painlevé property.

In case 3.1 from the possibility (3) we find that $r = -1$ is double. Thus we cannot check this case and its subcases if they possess the Painlevé property.

From the possibility (4), we find that for the subcase 3.1.4, relations (2.19), the values of r are: -1, 0 (double), 1,2,4 and $p = -1$, $q = -\frac{1}{2}$, $s = 0$. The coefficients of the solutions (3.1) are arbitrary at the resonances and this subcase possesses the Painlevé property. The values of r in this possibility for the case 3.2 are $r = -1$, 0 (double), 4 and the other 2 depend on relations between ℓ and g. That is, we have the relations (2.23) and if $4\ell = 3g$ then we find the double value $r = 2$, and $p = -1$, $q = s = -\frac{1}{2}$. But, the coefficients α_2, β_2, γ_2 of the solutions (3.1) are not arbitrary. Also, for $= 0$ we find the double value $r = 1$, and $p = -1$, $q = s = 0$. But the coefficients α_0, β_0, γ_0 in the solutions (3.1) are not arbitrary. From the possibilities 5 and 6 we find that in all cases the double value $r = -1$, and thus we cannot check for the Painlevé property.

4. DISCUSSION

The Hamiltonian (1.1) possesses a second integral of motion if e.d.=0, and a=b but the Painlevé analysis gives 2 complex values of the resonances r and e=d only. In the particular case e=d=1 Steeb et al. (1986a,b) have found the same values as ours, and they conclude by using the theorem of Yoshida (1983) that the Hamiltonian system (1.1) cannot be algebraically integrable. However this system is partially integrable.

The Hamiltonian (1.2) possesses a second integral of motion if $f = d$, $b = c$; thus the system is partially integrable. With the Painlevé analysis we find $f = d$ or $f/d = s(1-s)/6$, $(s < 1/2)$ and 2 complex values of the resonances always.

The Hamiltonian (1.3) when $d = m = 0$ is reduced to an integrable system in which the motion is on the x,y plane, and the z-motion is an independent Duffing oscillator without damping. Thus the system is partially integrable. But when

(1) $h = g$, $\ell = 6g$, $f = 0$, $a = b$
(2) $h = 16g$, $\ell = 12g$, $f = 0$, $b = 4a$
(3) $h = g$, $\ell = 2g$, $f = 0$
(4) $g = 16h$, $\ell = 12h$

the system is completely integrable. The integrals (2.16,18) of the above cases (2) and (3) respectively are known, but the integrals (2.14,20) (above cases (1) and (2) respectively) do not seem to be included in the known cases of Hietarinta (1987), Grammaticos et al (1985) or Ramani et al (1985). From the Painlevé analysis we find that in the partially integrable cases 2 values of the resonances may be integers, rational, irrational or complex depending on the values of the coefficients of the Hamiltonian. But when the coefficients satisfy certain relations the system is completely integrable and possesses the Painlevé property, except in the above case (2) in which we cannot check for the Painlevé property, because one of the resonances values is -2. Also, the Hamiltonian system with

$m = \ell$, $h = k = 0$, $f = d$, $b = c$

is partially integrable. From the Painlevé analysis we find that 2 values

of the resonances depend on the values of the coefficients. When the
values of the coefficients are such that the values of r are 6 integers
or rationals the corresponding coefficients in the series solutions are
not arbitrary. That is, the system does not possess the Painlevé property.

5. ACKNOWLEDGEMENTS

I would like to thank Professor Contopoulos for his interest and for
many helpful comments that led in the final presentation of this work.

6. REFERENCES

Contopoulos, G., Magnenat, P.: 1985, Celest. Mech. _37_, 387.

Contopoulos, G., Barbanis, B.: 1985, Astron. Astroph. _153_, 44.

Dorizzi, B., Grammaticos, B., Hietarinta, J., Ramani, A. Schwarz, F.:
   1986, Phys. Lett. A 16(9), 432.

Dorizzi, B., Grammaticos, B., Ramani, A.: 1983, J. Math. Phys. 24(9), 2282.

Grammaticos, B., Dorizzi, B., Ramani, A.: 1983, J. Math. Phys. 24(9), 2289.

Grammaticos, B., Dorizzi, B., Ramani, A., Hietarinta, J.: 1985, Phys. Lett.
   _109_, A(3), 81.

Hietarinta, J.,: 1987, Phys. Rep. 147(2).

Ramani, A., Hietarinta, J., Dorizzi, B., Grammaticos, B.: 1985, Phys.
   Lett. 108 A(2), 55.

Steeb, W.H., Louw, J.A., Leach, P.G.L., Mahomed, F.M.: 1986a, Phys.
   Review A 33(4), 2131.

Steeb, W.H., Louw, J.A., Vilet, C.M.: 1986b, Phys. Review A 34(3), 3489.

Yoshida, H.: 1983, Celest. Mechanics _31_, 363, 381.

Zachilas, L.: 1989, Ph.D. Thesis, University of Athens.

7. APPENDIX

Reduction of the 4-particle Toda lattice to a 3-D system

The Hamiltonian of the 4-particle periodic Toda lattice is:

$$H = (1/2) \sum_{i=1}^{4} p_i^2 + \sum_{i=1}^{4} (\exp(q_n - q_{n+1})) - 4, \tag{1}$$

where q_i, p_i are positions and momenta respectively and $q_4 = q_1$.
We apply the transformation q_i, $p_i \longrightarrow Q_i, P_i$ through the equations

$$q_i = (1/2) \sum_{i=1}^{4} A_{i\,j} Q_j \quad ,$$

$$j = 1,...4 \tag{2}$$

$$p_i = (1/2) \sum_{i=1}^{4} A_{i\,j} P_j$$

where A is the matrix

$$A = \begin{pmatrix} -1 & -1 & 1 & 1 \\ -1 & 1 & -1 & 1 \\ 1 & 1 & 1 & 1 \\ 1 & -1 & -1 & 1 \end{pmatrix} \tag{3}$$

The Hamiltonian (1) in the new coordinates Q_i, P_i

$$H = (P_1^2 + P_2^2 + P_3^2 + P_4^2)/2 + \exp(-Q_2 + Q_3) +$$

$$+\exp(-Q_1 - Q_3) + \exp(Q_2 + Q_3) + \exp(-Q_3 + Q_1) - 4. \tag{4}$$

Since Q_4 is ignorable the momentum P_4 is a constant of motion, and we can omit it. Then the reduced 4-particle Hamiltonian is

$$H = (P_1^2 + P_2^2 + P_3^2)/2 + \exp(-Q_2 + Q_3) + \exp(-Q_1 - Q_3) +$$

$$+ \exp(Q_2 + Q_3) + \exp(-Q_3 + Q_1) - 4 \tag{5}$$

If we set $Q_1 = z$, $Q_2 = y$, $Q_3 = x$ and expand the exponentials we find the 3rd and 4th order approximations

$$H_3 = (x^2 + y^2 + z^2 + 4x^2 + 2y^2 + 2z^2)/2 + xy^2 - xz^2 \tag{6}$$

$$H_4 = H_3 + (2x^4 + y^4 + z^4)/12 + (x^2y^2 + x^2z^2)/2$$

These Hamiltonians are of the form (1.2), (1.3), with

$a = 4$, $b = c = 2$, $d = -1$, $f = 1$, $g = 1/6$, $h = k = 1/12$

$\ell = m = 1/2$.

GENERIC AND NONGENERIC HOPF BIFURCATION

Franz Spirig

CH-9400 Rorschacherberg
Switzerland

ABSTRACT: A system of ordinary differential equations depending on
a parameter ε is considered. The origin is assumed to be an equilibrium
point for all ε. Furthermore the following hypotheses are made: The
linearized system admits a pair of complex conjugate eigenvalues
$\alpha(\varepsilon) \pm i\beta(\varepsilon)$, $\beta(\varepsilon) > 0$; for $\varepsilon = 0$ these eigenvalues are purely imaginary
whereas the other eigenvalues are not integer multiples of $i\beta(0)$. Generi-
cally $\alpha'(0) \neq 0$. In this case the well known Hopf bifurcation theorem
states that there exists one family of periodic solutions. If however the
genericity condition is violated then more than one family of periodic
solutions may branch off from the origin. In this paper a procedure is
proposed allowing to determine all small periodic solutions of the system
considered and their stability.

1. INTRODUCTION

Consider a system of d autonomous ordinary differential equations
depending on a parameter ε. Assume that the origin is an equilibrium
point for all ε and that the linearized system has a pair of complex con-
jugate eigenvalues $\alpha(\varepsilon) + i\beta(\varepsilon)$, $\beta(\varepsilon) > 0$, which lie on the imaginary
axis for $\varepsilon = 0$, i.e. $\alpha(0) = 0$. Using an appropriate scaling of the independent variable $\beta(\varepsilon) \equiv 1$ is achieved. The other eigenvalues are
supposed not to be integer multiples of the imaginary unit i. Thus
without loss of generality the system can be written as

$$
\begin{aligned}
\dot{x} &= \alpha(\varepsilon)x - y && + f(x,y,z,\varepsilon) \\
\dot{y} &= x + \alpha(\varepsilon)y && + g(x,y,z,\varepsilon) \\
\dot{z} &= \Gamma(\varepsilon)z && + h(x,y,z,\varepsilon),
\end{aligned} \qquad (1)
$$

with $z = (z_3,\ldots,z_d)^T$. $'\cdot'$ denotes the derivative with respect to the independent variable t. The functions f,g,h represent terms of order $O(x^2+y^2+|z|^2)$. Generically one has $\alpha'(0) \neq 0$. Then it is well known (see e.g. [5]) that there exists one family of periodic solutions. More generally we will assume $\alpha'(0) = \ldots = \alpha^{(m-1)}(0) = 0$, $\alpha^{(m)}(0) \neq 0$, i.e.

$$\alpha(\varepsilon) = \varepsilon^m \sigma + O(\varepsilon^{m+1}), \sigma \neq 0, \text{ for some } m \in \mathbb{N}. \qquad (2)$$

The second order equation

$$\ddot{y} - 2\sigma\varepsilon^m \dot{y}+y = f(\dot{y},y,\varepsilon), \quad f = O(\dot{y}^2+y^2) \qquad (3)$$

may be considered as a special case of Equation (1) with d=2 (see [6]). In this paper a procedure will be proposed allowing to determine all small periodic solutions of Equation (1) and Equation (3), respectively, as well as the stability of these solutions. Such a procedure was already developed by Flockerzi [1], [2], but the method of averaging used there is rather demanding. We think that the approach presented here is instead more elementary and more tractable. Equation (3) representing essentially the planar case (d=2) will be treated in Section 2. In Section 3 it will be indicated how the higher dimensional case (d≥3) may be reduced to the planar one. In this paper we just give an outline of the procedure. More details concerning the planar case may be found in [6]. A rigorous proof of the validity of the reduction in Section 3 will be given in a forthcoming paper.

2. THE PLANAR EQUATION

In this section Equation (3) is investigated. This problem is best solved in polar coordinates φ,r in the phase-plane, i.e. in the $\dot{y}$-y-plane:

$$\dot{y} = rc, \quad y = rs, \quad c = \cos\varphi, \quad s = \sin\varphi .$$

If φ is used as new independent variable instead of t Equation (3) is transformed to

$$r' = r(\varepsilon^m 2\sigma c^2 + K(\varphi,r,\varepsilon) + O(\varepsilon^m r + \varepsilon^{2m})), \qquad ' = \frac{d}{d\varphi}, \quad (4)$$

where $\quad K = \frac{cF}{1-sF}$, $\quad rF(\varphi,r,\varepsilon) = f(rc,rs,\varepsilon)$.

Of course, in the function K only terms up to the order m-1 with respect to ε have to be considered; thus the expansion of K with respect to ε reads $\quad K = \sum_{i=0}^{m-1} \varepsilon^i k_i(\varphi,r)$, where $k_i = O(r)$. Either $k_i \equiv 0$ or

$$k_i = k_{in_i}(\varphi)r^{n_i} + O(r^{n_i+1}).$$

Let $r(\varphi, r_0, \varepsilon)$ denote the solution of Equation (4) with initial condition r_0 at $\varphi = 0$: $r(0, r_0, \varepsilon) = r_0$. Due to the fact that $K = 0(r)$ the solution of Equation (4) exists for small r_0 at least for $\varphi \in [0, 2\pi]$ and the following asymptotic representation holds

$$r(\varphi, r_0, \varepsilon) = r_0(1 + 0(r_0 + \varepsilon^m)). \tag{5}$$

A periodic solution $y(t)$ of Equation (3) corresponds to a 2π-periodic solution of Equation (4). In order to establish the 2π-periodic solutions Equation (4) we introduce the averaging operator $M[.] = \frac{1}{2\pi} \int_0^{2\pi} . d\varphi$. A positive solution r of Equation (4) is 2π-periodic if and only if $M\left[\frac{r'}{r}\right] = 0$. If the asymptotic representation (5) is used this leads to the bifurcation equation

$$M[K(., r(., r_0, \varepsilon), \varepsilon)] + \varepsilon^m \sigma + 0(\varepsilon^m r_0 + \varepsilon^{2m}) = 0 . \tag{6}$$

Applying Equation (5) once more we get

$$K(\varphi, r(\varphi, r_0, \varepsilon), \varepsilon) = \sum_{i \in I} \varepsilon^i (k_{in_i}(\varphi) r_0^{n_i} + 0(r_0^{n_i + 1})),$$

where I denotes the set $\{i \mid k_i \neq 0, i = 0, \ldots, m-1\}$. Thus the bifurcation Equation (6) reads

$$\sum_{i \in I} \varepsilon^i (M[k_{in_i}] r_0^{n_i} + 0(r_0^{n_i + 1})) + \varepsilon^m \sigma + 0(\varepsilon^m r_0 + \varepsilon^{2m}) = 0. \tag{7}$$

Equation (7) may be written in the form in the form

$$\sum_{i \in J} \varepsilon^i (u_{in_i} r_0^{n_i} + 0(r_0^{n_i + 1})) + 0(\varepsilon^m r_0 + \varepsilon^{2m}) = 0, \tag{8}$$

with $J = I \cup \{m\}$, $u_{in_i} = M[k_{in_i}]$ for $i \in I$, $u_{mo} = \sigma$ and $n_m = 0$.

For $\varepsilon = 0$, each solution r of Equation (4) is periodic if the left-hand side of Equation (8) for $\varepsilon = 0$ vanishes for all r_0. This situation is called a vertical bifurcation. For $\varepsilon \neq 0$, all small solutions r_0 of Equation (8), i.e. solutions with

$$r_0 = |\varepsilon|^q a(\varepsilon), \quad q > 0, \quad a(0) > 0,$$

may be found by Newton's diagram.

Newton's diagram is obtained as follows. Plot all points (i, n_i), $i \in J$, in an i-j-coordinate system. Let s be the smallest $i \in J$. (If there is no vertical bifurcation then $s = 0$.) Draw a line from the point (s, n_s) to the point for which the slope of the line is negative and as small as possible. From this point continue in the same way. As an example Figure 1 shows Newton's diagram in a case where $I = \{0, 1, 2, 4\}$, $n_0 = 6$, $n_1 = 4$, $n_2 = 2$, $n_4 = 1$, $m = 5$.

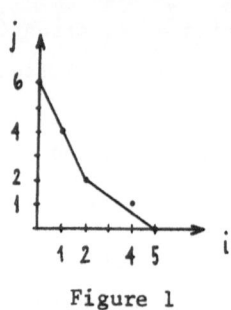

Figure 1

Points corresponding to an admissible q are all those which lie on a straight line in Newton's diagram. J_q denotes the set of all $i \in J$ such that the point (i, n_i) belongs to this straight line. In the above example two values q_1, q_2 are possible for q : $J_{q_1} = \{0, 1, 2\}$ and $J_{q_2} = \{2, 5\}$. Now, for each set J_q found by Newton's diagram instead of Equation (8) the equation

$$\sum_{i \in J_q} \varepsilon^i u_{in_i} b^{n_i} = 0 \tag{9}$$

is considered. Equation (9) gives approximate solutions to Equation (8). For the time being it is assumed that none of the u_{in_i} vanishes in Equation (9), i.e. $M[k_{in_i}] \neq 0$ for all $i \in J_q$. It is easy to prove that if b is a positive root of Equation (9) with odd multiplicity k then the bifurcation Equation (8) has at least one and at most k solutions $r_o = b + O(1)$ as $\varepsilon \to 0$. On that way each solution b of Equation (9) produces one or more 2π-periodic solution r of Equation (4) and the following asymptotic representation holds: $r = b(1 + O(1))$ (see Equation (5)).

The stability of the derived 2π-periodic solution is decided as follows. If Equation (4) is written in the form

$$r' = R(\varphi, r, \varepsilon) , \quad R(\varphi + 2\pi, r, \varepsilon) = R(\varphi, r, \varepsilon)$$

then a 2π-periodic solution $r(\varphi, \varepsilon)$ is (asymptotically) stable if $M \left[\frac{\partial R}{\partial r} (., r(., \varepsilon), \varepsilon) \right] < 0$. This means that the derivative of the left-hand side of Equation (9) with respect to b is negative:

$$\frac{\partial}{\partial b} \sum_{i \in J_q} \varepsilon^i u_{in_i} b^{n_i} < 0. \tag{10}$$

Example: $\ddot{y} + \varepsilon^m \dot{y} + y = \varepsilon \dot{y}^3 + \dot{y}^5$, $m = 2, 3, 5$

Equation (7) reads $\frac{5}{16} r_o^4 + \frac{3}{8} \varepsilon r_o^2 - \frac{1}{2} \varepsilon^m + \ldots = 0$.

568

Newton's diagram as sketched in Figure 2 provides the following equations of type (9).

(a) $\frac{5}{16} b^4 + \frac{3}{8} \epsilon b^2 - \frac{1}{2} \epsilon^2$ for m = 2.

(b) $\frac{5}{16} b^4 + \frac{3}{8} \epsilon b^2 = 0$ and $\frac{3}{8} \epsilon b^2 - \frac{1}{2} \epsilon^m$ for m = 3,5.

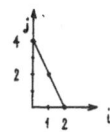   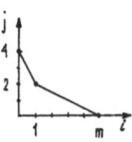

Figure 2a   Figure 2b

The bifurcating solutions are

(a) $r = \sqrt{\frac{4}{3}} \epsilon + O(\sqrt{\epsilon})$ and $r = \sqrt{-2\epsilon} + O(\sqrt{-\epsilon})$

(b) $r = \sqrt{-\frac{6}{5}} \epsilon + O(\sqrt{-\epsilon})$ and $r = \sqrt{\frac{4}{3}} \epsilon^{m-1} + O(\sqrt{\epsilon^{m-1}})$.

The corresponding bifurcation diagrams are given in Figure 3, where solid lines indicate stable solutions.

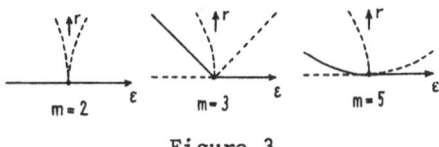

Figure 3

The case where an $i \in J_q$ exists such that $M[k_{in_i}] = 0$ can be reduced to the case just considered. This is done by means of the following simple lemma.

Lemma: If $U(\varphi)$ is a 2π-periodic function with $M[U] = 0$ and
         $r(\varphi, r_o, \epsilon)$ a solution of Equation (4) then
         $M[ur^j] = M[K \, vr^j] + O(\epsilon^m r_o^j)$ for all $j \in \mathbb{N}$,
         where $v(\varphi) = - j \int_o^\varphi u(s) ds$.

This lemma is easily proven using integration by parts. The crucial point is that Kvr^j is of order $O(r^{j+1})$ whereas ur^j is of order $O(r^j)$. The following example shows how the lemma is applied.

Example: $\ddot{y} + \epsilon^3 \dot{y} + y = y^2 + \epsilon^2 \dot{y}^3$

Equation (6) reads $M[k_o] + \frac{3}{8} \epsilon^2 r_o^2 - \frac{1}{2} \epsilon^3 + \ldots = 0$,

with $k_o = \frac{c \, s^2 \, r}{1 - s^3 r} = \sum_{j=0}^{\infty} k_{oj} r^j$, $k_{oj} = s^{3j-1} c$, $M[k_{oj}] = 0$.

Now the lemma with $K = k_o + \varepsilon^2 O(r^2)$ is applied to obtain

$$M[k_{oj}\, r^j] = -\frac{1}{3} \sum_{\ell=j+1}^{\infty} M[k_{o\ell}\, r^\ell] + O(\varepsilon^2 r_o^3 + \varepsilon^3 r_o),$$

From this one easily concludes by induction

$$M[k_{oj}\, r^j] = O(r_o^{j+n} + \varepsilon^2 r_o^3 + \varepsilon^3 r_o),$$

with n being an arbitrary positive integer. Taking the analyticity into account the term of order $O(r_o^{j+n})$ can be cancelled. Thus the bifurcation Equation (6) reads

$$\frac{3}{8}\, \varepsilon^2 r_o^2 - \frac{1}{2}\, \varepsilon^3 + \ldots = 0.$$

For $\varepsilon = 0$, all r_o are solutions, so there is a vertical bifurcation. Equation (9) becomes

$$\frac{3}{8}\, \varepsilon^2 b^2 - \frac{1}{2}\, \varepsilon^3 = 0.$$

Thus $r = \sqrt{\frac{4}{3}}\,\varepsilon + O(\sqrt{\varepsilon})$ is the only family of bifurcating periodic orbits for $\varepsilon \neq 0$ (see Figure 4).

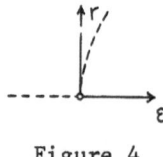

Figure 4

3. THE HIGHER DIMENSIONAL SYSTEM

Consider now Equation (1) with $d \geq 3$. By assumption, the eigenvalues of the matrix $\Gamma(\varepsilon)$ are not integer multiples of the imaginary unit i. In addition, a necessary condition for the stability of the derived periodic solutions is that the eigenvalues of $\Gamma(\varepsilon)$ have negative real parts. Again polar coordinates are introduced in the x-y-plane, i.e. $x = rc$, $y = rs$, and the vector z is scaled by $z = r\zeta$. Then Equation (1) is transformed to

$$r' = r(\alpha(\varepsilon) + P(\varphi,r,\zeta,\varepsilon) + O(\varepsilon^m r))$$

$$\zeta' = \Gamma(\varepsilon)\zeta + Z(\varphi,r,\zeta,\varepsilon) + O(\varepsilon^m), \qquad ' = \frac{d}{d\varphi}, \qquad (11)$$

where $P = \dfrac{cF + sG}{1 - sF + cG}$, $Z = \dfrac{H + (sF - cG)\Gamma\zeta - (cF + sG)\zeta}{1 - sF + cG}$.

The functions F,G,H are defined as

$$rF(\varphi,r,\zeta,\varepsilon) = f(rc,rs,r\zeta,\varepsilon), \quad rG = g, \quad rH = h.$$

Thus F ,G and H are of order $O(r)$. In P and Z only terms up to ε^{m-1} have to be considered. The expansion of P and Z with respect to r,ζ is written as $P = \Sigma P_{jk}(\varphi,\varepsilon)r^j\zeta^k$, $Z = \Sigma Z_{jk}(\varphi,\varepsilon)r^j\zeta^k$. If $d > 3$ then $\zeta = (\zeta_3,\ldots,\zeta_d)^T$ and k has to be interpreted as a multiindex:

$k: = (\ell_1,\ldots,\ell_k)$, $\zeta^k: = \zeta_{\ell_1}\ldots\zeta_{\ell_k}$, $3 \leq \ell_j \leq d$.

For ζ the following expansion is used

$$\zeta = \Sigma \eta_j(\varphi,\varepsilon)r^j , \qquad (12)$$

when η_j are 2π-periodic with respect to φ. Substituting the expansion (12) into Equation (11) one gets

$$\eta_1^1 = \Gamma \eta_1 + Z_1$$

$$\vdots \qquad\qquad (13)$$

$$\eta_j^1 = \Gamma\eta_j + Z_j(c,s,\eta_{j-1},\ldots\eta_1,\varepsilon) ,$$

where Z_j is a polynomial in $c,s,\eta_{j-1},\ldots,\eta_1$. System (13) has to be solved recursively. If the uniquely determined 2π-periodic solutions $\eta_1,\ldots,\eta_{j-1}$ have been found the solution η_j may be constructed as follows. Let $Z_j = \sum_k \sum_{\ell=0}^k c^\ell s^{k-\ell} A_\ell^k$ and make the ansatz $\eta_j = \sum_k \sum_{\ell=0}^k c^\ell s^{k-\ell} B_\ell^k$ to obtain a system of linear algebraic equations

$$(k-\ell+1)B_{\ell-1}^k - \Gamma B_\ell^k - (\ell+1)B_{\ell+1}^k = A_\ell^k . \qquad (14)$$

If one puts the expression (12) into the r-equation of Equation (11) one is led to the planar problem of Section 2 (see Equation (4)). As an easy consequence of this procedure the direction-of-bifurcation formula for the generic Hopf bifurcation $(m=1)$ as given in [3] , [4] e.g. may be obtained.

REFERENCES

[1]   D. Flockerzi, Existence of small periodic solutions of ordinary differential equations in $\mathbb{R}^2$. Arch. Math. 33, 263-278 (1979).

[2]   D. Flockerzi, Bifurcation formulas for ODE's in $\mathbb{R}^n$. Nonlinear Analysis Vol.5 , No.3, 249-263 (1981).

[3]   U. Kirchgraber, On the method of averaging. In V. Szebehely and B. D. Tapley (eds.), Long-time predictions in dynamics, 111-117. Reidel Publ. Comp., Dordrecht 1976.

[4]   U. Kirchgraber and E. Stiefel, Methoden der analytischen Störungs-
      rechnung und ihre Anwendungen, 88-101. Teubner, Stuttgart 1978.

[5]   D. S. Schmidt in J.E. Marsden and M. McCracken, The Hopf bifurcation
      and its application, 95-103. Springer, New York 1976.

[6]   F. Spirig, Bifurcation equation for planar systems of differential
      equations. ZAMP Vol.39, 504-517 (1988).

THE CHAOTIC MOTION OF A RIGID BODY

ROTATING ABOUT A FIXED POINT

F. El-Sabaa and M. El-Tarazi

Department of Mathematics
Faculty of Education
Ain Shamis University
Cairo, Egypt and
Department of Mathematics
Kuwait University, Kuwait

1. INTRODUCTION

The dynamical system with two degrees of freedom has been studied by many authors. It is well known, that if the system possesses a first integral besides the integral of energy, then the system is completely integrable. After the fundamental works of Kolmogorov [1], Arnold [2] and Moser [3] (KAM) in 1960-1968, many theoretical and numerical results were presented by authors who treated the problem, when the system contained only the integral of energy (Jacobi's integral). The results of (KAM) have clarified the picture of nonintegrable systems through the small perturbations of integrable systems; for small perturbations we get very regular orbits, lying apparently on invariant tori, while for larger perturbations a part of the tori seems to be destroyed, and erratic orbits appear instead, filling the so-called stochastic region.

On the other hand, the problem of a heavy rigid body rotating about a fixed point has not been solved except in three cases, where the mass distributions satisfy certain relations.

These are the Euler, Lagrange, and Kovaleveskaya cases. Euler's case was reduced to quadrature in elliptic functions [4]. Deprit reduced the problem to only one degree of freedom [5]. This reduction permits the representation of all possible solutions of the problem by isoenergetic curves in the phase plane. The perturbation of Euler's case in Deprit's variables has been treated by many authors to prove the existence of periodic solutions with small parameters, in a Newtonian field. (see for example [6], [7], [8]). In the Lagrange case, the problem was integrated in terms of elliptic functions [9], but no one simplified this problem as in Euler's case, while in the Kovaleveskaya case the problem was integrated in terms of the Riemann θ-functions of two variables which is a very complicated solution. After the solution of Kovaleveskaya, there are many works concerned almost entirely with the consideration of special cases of the problem, starting with the work of Applerot [10], and including the work of Kozlov and El-Sabaa [4], [12], attempting to find enough special cases to be able to know more about the general behaviour of the problem.

In the present work we transform the equations of motion of a heavy rigid body into the dynamical system of two degree of freedoms. The new system has Jacobi's integral and we investigate numerically the existence of the second integral. This integral occurs when the regular orbits lie on invariant tori in the phase space. Stochastic regions indicate that the system is not integratable.

2. THE EQUATIONS OF MOTION

Consider a set of Cartesian coordinates OXYZ, fixed to a rigid body with respect to a reference system Oxyz fixed in the inertial space. The moving system Oxyz is chosen such that the axes are directed along the principal axes of inertia for point 0. The orientation of the fixed system relative to the moving one is specified by means of the Eulerian angles θ, ϕ and ψ. The unit vector γ along the axis of symmetry of the body has the components γ_1, γ_2 and γ_3 connected with Eulerian angles by the relations:

$$\gamma_1 = \sin\theta\sin\phi \ , \ \gamma_2 = \sin\theta\cos\phi \ , \ \gamma_3 = \cos\theta \tag{1}$$

while the components of the angular velocity of the body $\underline{\omega}$ can be expressed in terms of Eulerian angles and their temporal derivatives as follows:

$$p = \dot{\psi}\sin\theta\sin\phi + \dot{\theta}\cos\phi \tag{2}$$
$$q = \dot{\psi}\sin\theta\cos\phi - \dot{\theta}\sin\phi$$
$$r = \dot{\psi}\cos\theta + \dot{\phi}$$

The Lagrangian function of the system can be written in the form
$$L(\theta,\phi,\psi,\dot{\theta},\dot{\phi},\dot{\psi}) = \frac{1}{2}[A(\dot{\psi}\sin\theta\sin\phi + \dot{\theta}\cos\phi)^2$$
$$+ B(\dot{\phi}\sin\theta\cos\phi + \dot{\theta}\sin\phi)^2 - C(\dot{\psi}\cos\theta + \dot{\phi})^2] - V(\theta,\phi)] \tag{3}$$

where A,B,C are the principal moments of inertia and V is the potential energy defined as
$$V(\theta,\phi) = mg(X_o\sin\theta\sin\phi + Y_o\sin\theta\cos\phi + Z_o\cos\theta) , \tag{4}$$

where X, Y, Z are the components of the radius vector of the center of mass in the reference system which is fixed to the body, m is the mass of the body and g is the acceleration due to gravity.

The equations of motion are

$$\frac{d}{dt}(\frac{\partial L}{\partial\dot{\theta}}) - \frac{\partial L}{\partial\theta} = 0 ,$$

$$\frac{d}{dt}\frac{\partial L}{\partial\dot{\phi}} \quad \frac{\partial L}{\partial\phi} = 0 , \tag{5}$$

$$(\frac{\partial L}{\partial\dot{\psi}}) = f$$

where ψ is a negligible coordinate. The system (5) is a nonlinear system of differential equations in the unknown functions $(\theta,\phi,\psi,\dot{\theta},\dot{\phi})$. This system can be solved if there exist four time-independent integrals. There are three known cases where the problem can be solved: Euler, Lagrange and Kovaleveskaya, where the mass distribution satisfies certain relations. In general the problem of a rigid body has not been solved and is in a sense unsolvable.

3. THE REDUCTION OF THE EQUATIONS

Aside from the three cases mentioned above, many other cases are known for solving the rigid body problem either by assuming some restrictions on the constants of integration or by transforming the equations of motion into an easier reduced system. According to [13] the system of equation (5) can be transformed into a system of two degrees of freedom using isothermal coordinates as follows:

The Routhian function of the system is written as

$$R = \frac{D}{2} \{ C(A \sin^2\phi + B \cos^2\phi) \sin^2\theta \, \dot{\phi}^2 - \frac{1}{2} C(A-B)\sin 2\theta \, \sin 2\phi \, \dot{\theta} \, \dot{\phi}$$

$$+ [\frac{1}{D} (A \cos^2\phi + B \sin^4\phi) - (A-B)^2 \sin^2\theta \sin^2\phi \cos^2\phi] \, \dot{\theta}^2 \}$$

$$+ f D \sin\phi \cos\phi \cos\theta \frac{d}{dt} [C \ln|\tan\phi| - (A-B)\ln|\cos\theta|]$$

$$+ V(\theta,\phi) - \frac{1}{2} f^2 D, \tag{6}$$

where

$$D = (A \sin^2\phi + B \cos^2\phi) \sin^2\phi + C \cos^2\theta . \tag{7}$$

If we introduce the new variables ξ, η, ζ related to the Eulerian angles by the relations:

$$\sin\theta \sin\phi = \sqrt{A} \, \xi, \quad \sin\theta \cos\phi = \sqrt{B} \, \eta, \quad \cos\theta = \sqrt{C} \, \zeta \tag{8}$$

then (ξ, η, ζ) are the coordinates of any point on the surface of ellipsoid inertia

$$A\xi^2 + B\eta^2 + C\zeta^2 = 1. \tag{9}$$

The Routhian function becomes

$$R = \frac{1}{2} ABCD (\dot{\xi}^2 + \dot{\eta}^2 + \dot{\zeta}^2) + \frac{fD \sqrt{ABC}}{1 - C\zeta^2} [C\zeta(\eta\dot{\xi} - \xi\dot{\eta})$$

$$- (A-B)\xi\eta\dot{\zeta}] + V - \frac{1}{2} f^2 D. \tag{10}$$

Let $(\xi, \eta, \zeta) = \dfrac{1}{\sqrt{(1-n\sigma^2)(1+m\rho^2)}} [\dfrac{\sqrt{B}}{A} \sigma \sqrt{1 - k'^2 \rho^2},$

$$\frac{1}{\sqrt{B}} \sqrt{(1-\sigma^2)(1-\rho^2)}, \frac{\sqrt{B}}{C} \rho \sqrt{1 - k^2\sigma^2}]$$

where $k^2 = 1-k'^2 = \dfrac{A-B}{A-C}, \quad n = \dfrac{A-B}{A}, \quad m = \dfrac{B-C}{C} . \tag{11}$

The quadratic terms of the Routhian function in the old velocities $\dot{\xi}, \dot{\eta}$ and $\dot{\zeta}$,

$$R_2 = \frac{1}{2} ABCD (\dot{\xi}^2 + \dot{\eta}^2 + \dot{\zeta}^2)$$

are transformed into quadratic terms in the new velocities $\dot{\sigma}$ and $\dot{\rho}$ as follows:

$$R_2 = \frac{B\kappa}{2} \left[\frac{C}{A} \frac{\breve{\sigma}^2}{(1-n\sigma^2)^2(1-k^2\sigma^2)(1-\sigma^2)} + \frac{A}{C} \frac{\breve{\rho}^2}{(1+m\rho^2)^2(1-k'^2\rho^2)(1-\rho^2)} \right] (12)$$

where $\kappa = 1 - k^2\sigma^2 - k'^2\rho^2$

Introducing new variables x,y such that

$$x = \sqrt{\frac{C}{A}} \int_0^\sigma \frac{d\sigma}{(1-n\sigma^2)\sqrt{(1-k^2\sigma^2)(1-\sigma^2)}} \quad , \quad (13)$$

$$y = \sqrt{\frac{A}{C}} \int_0^\rho \frac{d\rho}{(1+m\rho^2)\sqrt{(1-k'^2\rho^2)(1-\rho^2)}} \quad ,$$

with $\quad d\tau = \frac{dt}{B\kappa}$ $\quad\quad\quad\quad\quad\quad$ (14)

The Routhian function takes the form

$$R = \frac{1}{2}(x'^2 + y'^2) + \frac{f}{M}(PTx' - QSy') + U,$$

where
$$P = \sqrt{1-n\sigma^2}\,(1 - \frac{A+B-C}{C}k^2\sigma^2), \quad\quad (15)$$

$$Q = n\sqrt{1+m\rho^2}\,(1 - \frac{A+B+C}{C}\rho^2),$$

$$S = \sigma\sqrt{(1-\sigma^2)(1-k^2\sigma^2)(1-n\sigma^2)},$$

$$T = \rho\sqrt{(1-\rho^2)(1-k'^2\rho^2)(1+m\rho^2)},$$

$$M = 1 - n\sigma^2 - \rho^2 + (1 - \frac{B}{A}k'^2)\sigma^2\rho^2.$$

The equations of motion in the new variables are

$$x'' + \Omega y' = \frac{\partial U}{\partial x} \quad , \quad y'' - \Omega x' = \frac{\partial U}{\partial y} \quad . \quad (' \equiv \frac{d}{d\tau}) \quad (16)$$

The system (16) is a plane motion system of two degrees of freedom under action of potential and gyroscopic forces U and Ω defined by

$$U = Bx[E + V - \frac{f^2}{2B}(1 - n\sigma^2)(1 + m\rho^2)] \quad . \quad\quad (17)$$

$$\Omega = \frac{f\kappa}{\sqrt{AC}}\sqrt{(1-n\sigma^2)(1+m\rho^2)}[A-B+C - 2(A-B)\sigma^2 + 2(B-C)\rho^2] \quad (18)$$

and possesses the Jacobi's integral

$$x'^2 + y'^2 = 2U \quad\quad\quad\quad\quad\quad\quad (19)$$

3. THE PERIODIC ORBITS OF THE MOTION

The non-integrable system (16) is quasi-integrable in the sense that there exist invariant tori near the stable periodic solutions, in accordance with KAM theory which states that "almost all the invariant tori of the unperturbed system continue to exist in spite of the presence of a small perturbation". The system (16) was solved analytically to get the periodic solution about the equilibrium positions for small per-turbation. The results are compatible with KAM theory, since the invar-iant tori exist in the neighbourhood of the stable periodic solution [14].

Henon and Heiles [15] used computer calculations to show that the invariant tori exist for small values of the energy constant and consequently the system is close enough to its periodic solutions; they show also that for large values of the energy constant (escape energy), stochastic regions appear instead of the invariant tori.

In this paper we use the fixed point method introduced by Poincaré [16] and continued by Birkhoff [17] and Moser [3] , to obtain the periodic orbits of the system (16). The solution of (16) represents the trajectory in the phase space (x, x', y, y'), and along this trajectory the value of the Jacobi's constant is fixed. Thus for a given value of this constant, the trajectory of the problem will be treated in three dimensional space (x, y, x'). So the periodic orbits are obtained by the successive intersections of the trajectory with the (x, x') plane, in the positive direction.

4. THE RESULTS

For the initial conditions $x = x_o$, $y = 0$, $x' = 0$ and y'_o obtained from the Jacobi's integral

$$y'^2_o = 2 \, B\kappa [E + V - f^2/_{2B} \, (1 - n\sigma^2) \, (1 + m\rho^2)] \qquad (20)$$

and a fixed value for the constant E, the equation (16) can be integrated numerically. We shall take the Kovaleveskaya case of a rigid body where the problem reduced to quadrature. The general behaviour of the solution in this case is still not clear, so the numerical calculations will provide more information about it. In addition to the conditions of the above case, if we take $x=0$, then from (2), the real motion will satisfy the conditions

$$\rho^2 \; < \; \frac{2 \, B \, E \, - \, f^2}{m \, f^2} \qquad (21)$$

$$E \; > \; f^2/_{2B} \qquad (22)$$

with any value of σ.

The points of intersections are always inside the curve representing the periodic orbits $y = y' = 0$.

If the curve is closed, then in consequence all other orbits will lie inside this curve and then we have an ordered motion. If the curve is open, then the orbits will be on one side of the surve and then we have chaotic motion.

In our problem, we took $A = B = 0.5$ and $C = 0.25$. The results show that when x'_o is close to zero; each orbit is close to its periodic. These are shown in Figures 1-3, for $x' = 0.1$, 0.05 and 0.01 respectively. Figures 4-6 show the periodic orbits when $x'_o = 0$ and for $E = 1.02$, 1.1 and 1.21. When E is close to the value 3, the orbit is closed but not periodic, where its symmetry is absent. Figures 7-9 show these orbits, while Figure 10 shows that the orbit becomes open and goes to ∞. Figures 11-12 show the invariant curves for $E = 2$ and $E = 3$. The invariant curve is closed. This means that there is no stochastic region and consequently the motion is ordered.
(Note: the curve is an ellipse with a very small minor axis which causes the figures to appear like a vertical straight line).

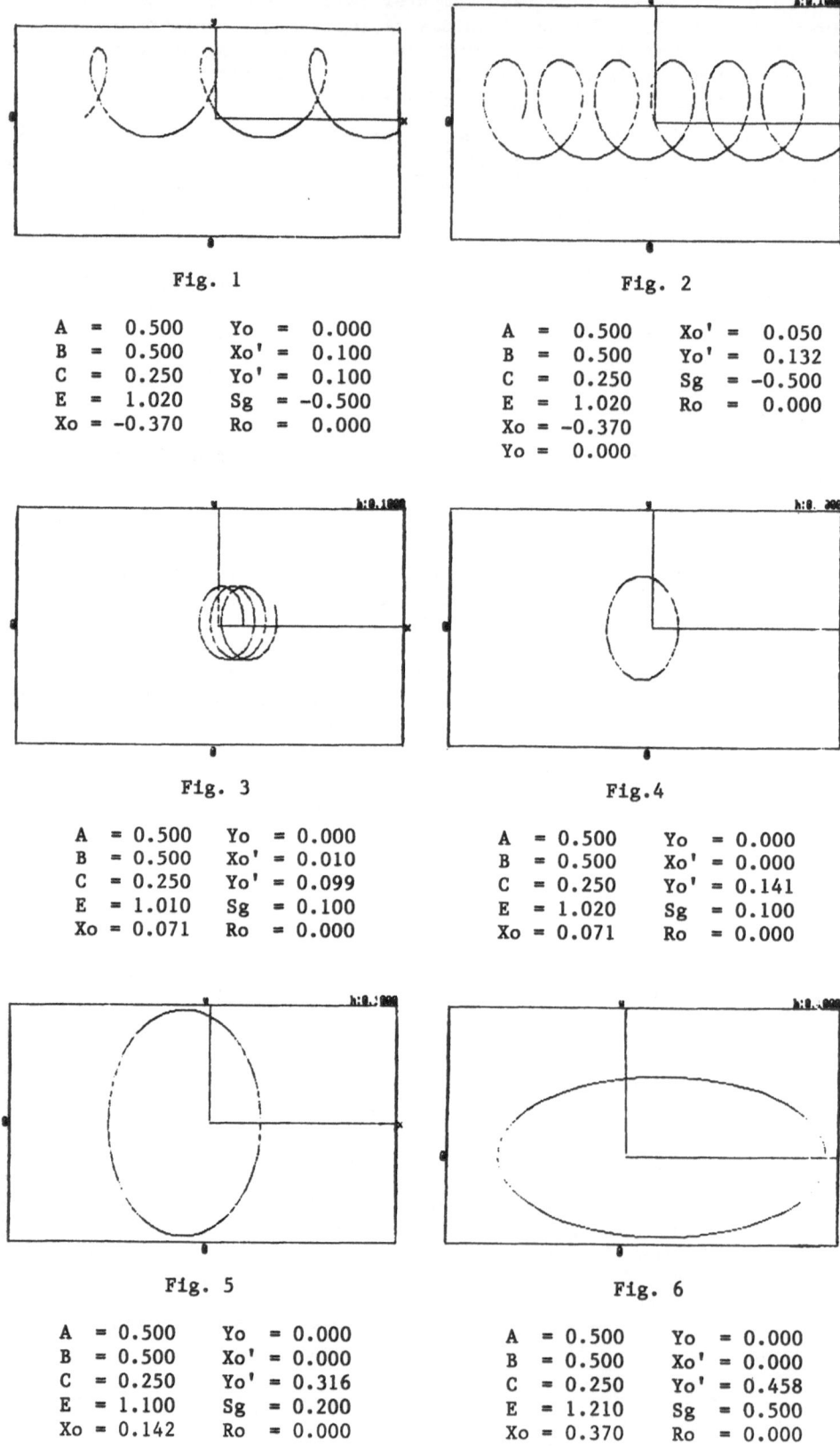

Fig. 1

A	= 0.500	Yo	= 0.000
B	= 0.500	Xo'	= 0.100
C	= 0.250	Yo'	= 0.100
E	= 1.020	Sg	= -0.500
Xo	= -0.370	Ro	= 0.000

Fig. 2

A	= 0.500	Xo'	= 0.050
B	= 0.500	Yo'	= 0.132
C	= 0.250	Sg	= -0.500
E	= 1.020	Ro	= 0.000
Xo	= -0.370		
Yo	= 0.000		

Fig. 3

A	= 0.500	Yo	= 0.000
B	= 0.500	Xo'	= 0.010
C	= 0.250	Yo'	= 0.099
E	= 1.010	Sg	= 0.100
Xo	= 0.071	Ro	= 0.000

Fig.4

A	= 0.500	Yo	= 0.000
B	= 0.500	Xo'	= 0.000
C	= 0.250	Yo'	= 0.141
E	= 1.020	Sg	= 0.100
Xo	= 0.071	Ro	= 0.000

Fig. 5

A	= 0.500	Yo	= 0.000
B	= 0.500	Xo'	= 0.000
C	= 0.250	Yo'	= 0.316
E	= 1.100	Sg	= 0.200
Xo	= 0.142	Ro	= 0.000

Fig. 6

A	= 0.500	Yo	= 0.000
B	= 0.500	Xo'	= 0.000
C	= 0.250	Yo'	= 0.458
E	= 1.210	Sg	= 0.500
Xo	= 0.370	Ro	= 0.000

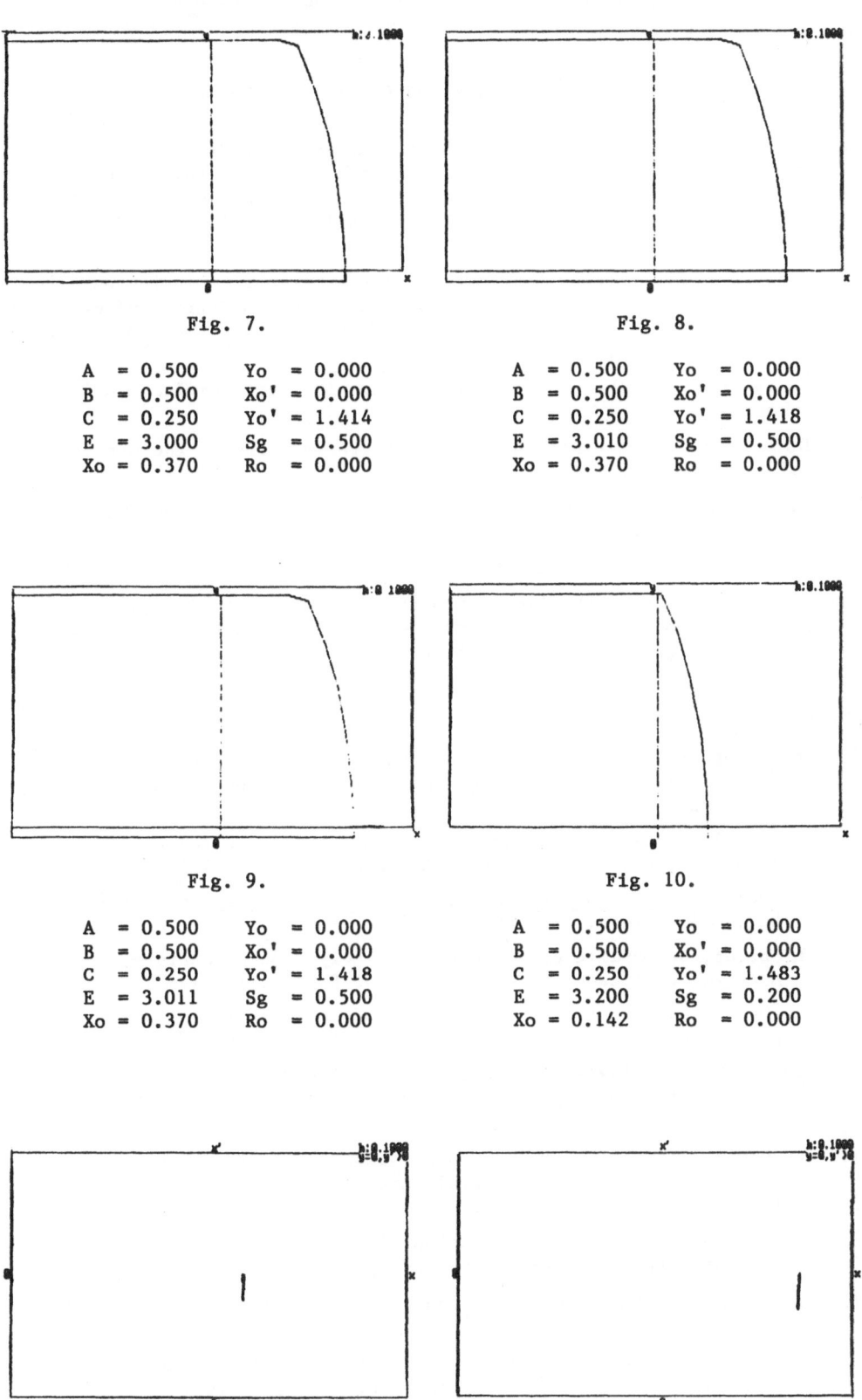

Fig. 7.

A	= 0.500	Yo	= 0.000
B	= 0.500	Xo'	= 0.000
C	= 0.250	Yo'	= 1.414
E	= 3.000	Sg	= 0.500
Xo	= 0.370	Ro	= 0.000

Fig. 8.

A	= 0.500	Yo	= 0.000
B	= 0.500	Xo'	= 0.000
C	= 0.250	Yo'	= 1.418
E	= 3.010	Sg	= 0.500
Xo	= 0.370	Ro	= 0.000

Fig. 9.

A	= 0.500	Yo	= 0.000
B	= 0.500	Xo'	= 0.000
C	= 0.250	Yo'	= 1.418
E	= 3.011	Sg	= 0.500
Xo	= 0.370	Ro	= 0.000

Fig. 10.

A	= 0.500	Yo	= 0.000
B	= 0.500	Xo'	= 0.000
C	= 0.250	Yo'	= 1.483
E	= 3.200	Sg	= 0.200
Xo	= 0.142	Ro	= 0.000

Fig. 11.

Fig. 12.

579

4. CONCLUSIONS

1 - In the case of Kovaleveskaya, we have a real motion for $1 < E \leqslant 3.011$ and $0 < \sigma < 1$.

2 - While the old system in our case is reduced to quadrature and the system can be integrated by θ functions in two variables, the new system possesses only one first integral and the question about the existence of another integral (numerically) is still open for future investigation.

3 - Theoretically, the new system possesses a periodic solution only in the vicinity of equilibrium points, where the conditions of existence of this solution is found according to Liapounov's theorem of holomorphic integral.

4 - We confirm that the other two cases have the same results mentioned above.

5 - There is no chaotic motion in the integrable cases of the problem of a rigid body.

6 - In general, the periodic solution in the old system is not equivalent to the periodic solution in the new system.

7 - Is the inverse problem of the mentioned point true?

8 - Does there exist a chaotic motion in the non-integrable cases of the rotation of a rigid body problem? The two points mentioned above are open questions for future investigation.

REFERENCES

1. A Kolmogorov, On Conservation of Conditionally periodic motions under small perturbations of the Hamiltonian, Dokl. Akad. Nauk SSSR. 98: 527 (1954)

2. V. Arnold, Small divisor problems in classical and celestial mechanics, Russian Mathematical Survey. 18: 581 (1963)

3. J. Moser, Stable and random motions in dynamical systems, Princeton University Press, Princeton (1973)

4. L. Euler, Decouverte d'une nouveau principle de mechanique, Memoires de l'Academie des Sciences de Berlin, 14: 154 (1758)

5. A. Deprit, Free rotation of a rigid body studied in the phase plane. American Journal of Physics. 35: 224 (1967)

6. Iu. Barkin, E. Ievlev, Periodic motion of a rigid body with a fixed point in the gravity field of two centers, Prikladnia Matemati e Mekhanik. 41: 574 (1977)

7. V. Demin, F. Kiselev, On periodic motions of a rigid body in a central Newtonian field, Prikladnia Matematik e Mekhanik. 38: 224 (1974)

8. F. El-Sabaa, The periodic solution of a rigid body in a central Newtonian field, Astrophysics and Space Science, 162: 235 (1989)

9. J. Lagrange, Mechanique analytique, Gauthier-Villars, Paris (1888)

10. G. Applerot, 1893. Some additions of paper N. Delone "algebraic integrals of a heavy rigid body about fixed point". Troda Otdelenii Fizicheskikh Obshchestve Liubitelei Estesvoznaiia. G: 1 (1893)

11. V. Kozlov. Qualitative analysis method in dynamics of a rigid
     body, Moscow State University Press, Moscow (1980)

12. F. El-Sabaa. About the periodic solution of Kovaleveskaya's top by
     using Liapunov's method. Journal of the University of Kuwait
     (Science) 16: 21 (1989)

13. E. Kharlamov. On the equations of motion for a heavy body with
     fixed point. Priklad Matematik e Mekhanik. 27:1070 (1963)

14. A Liapunov. Stability of motion. Academic Press, New York (1966)

15. M. Hénon, C. Heiles, The applicability of the third integral of
     motion: some numerical experiments. Astronomical Journal 69:73
     (1964)

16. H. Poincaré, Methodes Nouvelles de la Mecanique Celeste. Dover,
     New York (1957)

17. G. Birkhoff. Dynamical systems. American Mathematical Society
     New York (1927)

PARTICIPANTS AND SPEAKERS

Baille, P.	(Canada)	Mathematics Dept. R.M.C. K7K-5LO Kingston Ontario Canada
Banfi, V.	(Italy)	Corso Sempione 82 20154 Milano, Italy
Bendjoya, P.	(France)	Observatoire de Nice BP 139 06003 Nice Cedex, France
Bertotti, M.L.	(Italy)	Dep. Matematica, Universita di Trento 38050 Povo-Trento, Italy
Bois, E.	(France)	Obs. de la Cote d'Azur, Dept. CERGA, Avenue Copernic, 06130 Grasse, France
Broucke, R.	(U.S.A.)	Dept. of Aerospace Engineering University of Texas, Austin, Texas 78712, U.S.A.
Bryant, J.G.	(France)	47 Av. Felix Faure, 75015, Paris, France
Carpino, M.	(Italy)	Osservatorio Astronomico di Brera via Brera 28, 20121 Milano, Italy
Casasayas, J.	(Spain)	Dep. Matem. Aplic. v Anal... Fac. de Matematiques. Univ. Barcelona Gran Via 585, 08071 Barcelona, Spain
Celletti, A.	(Italy)	Dip. Matematica Pura e Appl.. Univ. L'Aquile, 67100 Coppito - L'Aquila Italy
Chauvineau, B.	(France)	OCA - CERGA, Avenue Copernic, 06130 Grasse, France
Chiralt, C.	(Spain)	Facultat de Matematiques. Univ. de Valencia, 46100 Buriassot, Valencia, Spain
Contopoulos, G.	(Greece)	Dept. of Astronomy, University of Athens, Athens, Greece
Conway, B.	(U.S.A.)	Dept. Aeron. Astron. Eng., 101 Transp. Bldg. 104, S. Matthews Av. Urbana, Illinois 61801, U.S.A.
Barre, De La, C.M.	(U.S.A.)	Dept. of Earth and Space Sciences UCLA Los Angeles. California 90024 U.S.A.
Dvorak, R.	(Austria)	Dept. of Astronomy, University of Vienna, Turkenschanzstr. 17, 1180 Wien, Austria

El-Sabaa, F.M.	(Kuwait)	Dept. of Mathematics, Univ. of Kuwait
Enright, P.	(U.S.A.)	Dept. Aeron. and Astronautical Eng., 101 Transportation Bldg. 104 S. Mathews Avenue, Urbana, Illionois, 61801, U.S.A.
Exertier, P.	(France)	Obs. de la Cote D'Azur - OCA/CERGA Avenue Copernic, 06130 Grasse,France
Ferrandiz, J.M.	(Spain)	Dept. Matematica Aplicada a la Tech. Univ. de Valladolid, 47011 Valladolid, Spain
Ferraz-Mello, S.	(Brazil)	Inst. Astron. E. Geofisico Univ. Sao Paulo, Caixa Postal 30627 01051 Sao Paulo, SP. Brazil
Floria, L.	(Spain)	Dept. Matematica Aplicada a la Tech. Univ. de Valladolid, 47011 Valladolid, Spain
Font, J.	(Spain)	Dept. Matematica Aplicada i Analisi, Univ. de Barcelona, Gran Via 585, 08071 Barcelona, Spain
Froeschlé, Ch.	(France)	Observatoire de Nice, BP 139, 06003 Nice Cedex, France
Froeschlé, Cl.	(France)	Observatoire de Nice, BP 139, 06003 Nice Cedex, France
Galgani, L.	(Italy)	Dipartmento di Matematica, Univ. di Milano, via Saldini 50, 20113 Milano, Italy
Gama, F.	(France)	Observatoire de Nice, BP 139, 06003 Nice Cedex, France
Ge, Yen-Chao	(U.K.)	Dept. of Physics and Astronomy, University of Glasgow, G12 8QQ. Glasgow, U.K.
Goudas, C.	(Greece)	Dept. of Mathematics, University of Patras, 26110 Patras, Greece
Hadjidemetriou, J.	(Greece)	Dept. of Theoretical Mechanics, University of Thessaloniki, 54006 Thessaloniki, Greece
Heggie, D.C.	(U.K.)	Dept. of Mathematics. Univ of Edinburgh, King's Buildings, EH9 3JZ Edinburgh, U.K.
Henrard, J.	(Belgium)	Dept. de Mathematiques - FUNDP 8 Rempart de la Vierge 5000 Namur, Belgium
Jorba, A.	(Spain)	Dept. Matematica Aplicada I. ETSEIB-UPC, Diagonal 647, 08028 Barcelona, Spain
Kaya, D.	(U.S.A.)	Headquarters Air Force Space Command, Peterson Air Force Base Colorado 80914-5001 U.S.A.
Laskar, J.	(France)	Bureau des Longitudes, 77 Av. Denfert Rochereau, 75014 Paris, France

Marchal, C.	(France)	D.E.S. - Onera 92320 Chantillon France
Markellos, V.V.	(Greece)	Dept. of Engineering Mathematics University of Patras, 26110 Rion Patras, Greece
Martin, P.	(Spain)	Dept. de Matematica Aplicada a la Tecnica, Univer. de Valladolid, Valladolid, 47011, Spain
Martines Alfaro J.	(Spain)	Facultad de Matematiques. Universidad de Valencia, 46100 Buriassot Valencia Spain
Masdemont, J.	(Spain)	Dept. de Matematica Aplicada I. ETSEIB-UPD Diagonal 647, 08028 Barcelona, Spain
McDonald, A.	(U.K.)	Logica Aerospace and Defence Ltd. 68 Newman Street, W1A 4SE London,U.K.
Message, P. J.	(U.K.)	Dept. of Applied Math. & Theoretical Physics, Univ. of Liverpool, P.O. Box 147, Liverpool L69 3BX,U.K.
Milani, A.	(Italy)	Gruppo Meccanica Spaziale. Dip. Matematica, via Buonarroti, 2. 56127 Pisa, Italy
Morbidelli, A.	(Belgium)	Dept. de Mathematiques - FUNDP 8 Rempart de la Vierge, 5000 Namur, Belgium
Nobili, A.	(Italy)	Gruppo Meccanica Spaziale. Dip. Matematica via Buonarroti, 2. 56127 Pisa, Italy
Nunes, A.	(Portugal)	Departmento de Fisica, Univ. de Lisboa, Campo Grande. ed.C1.Piso 4 1700 Lisboa, Portugal
Oberti, P.	(France)	Observ. de la Cote d'Azur. Dept. CERGA Avenue Copernica 06130 Grasse France
Onargan, G.	(Turkey)	Faculty of Engineering, Div. of Appl. Math., Dokuz Eylul Univer. Bornova-Izmir, Turkey
Perez, M.T.	(Spain)	Dept. de Matematica Aplicada a la Technica. Univ, de Valladolid 47011 Valladolid, Spain
Perozzi, E.	(Italy)	Telespazio S.P.A. via Tiburtina 965 00156 Roma, Italy
Pojman, J.	(U.S.A.)	Dept. Aerospace Engineering. Univ. of Texas at Austin, Austin. Texas 78712, U.S.A.
Puel, F.	(France)	Observatoire de Besancon, 41 Bis Av. de l'Observatoire, 25044 Besancon Cedex, France
Robutel, P.	(France)	Rue G. Wodli, 77220 Gretz, Paris France

Rossi, A.   (Italy)   Gruppo Meccanica Spaziale. Dip.
                                        Matematica via Buonarroti, 2.
                                        56127 Pisa, Italy

Roy, A.E.   (U.K.)   Dept. of Physics and Astronomy
                                        University of Glasgow G12 8QQ,
                                        Glasgow, U.K.

Saturio, M.E. San   (Spain)   Dept. de Matematica Aplicada a la
                                        Tecnica, Univ. de Valladolid,
                                        47011 Valladolid, Spain

Sekiguchi, M.   (Japan)   National Astron. Obs., Astrometry
                                        & Celestial Mechanics Div. Mitaka
                                        181 Tokyo, Japan

Simo, C.   (Spain)   Dept. Matematica Aplicada. Univ.
                                        Barcelona Gran Via 585,
                                        08007 Barcelona, Spain

Slezak, E.   (France)   Observatoire de Nice BP 139,
                                        06003 Nice Cedex, France

Smith, R.   (U.S.A.)   Dept. of Aerospace Engineering,
                                        Univ. of Texas at Austin, Austin,
                                        Texas 78712 U.S.A.

Snow, D.   (U.S.A.)   Headquarters Air Force Space Command
                                        Bld. 1, Stop 7 Peterson Air Force
                                        Base, Colorado 80914-5001, U.S.A.

Spirig, F.   (Switzerland)   Wilenstrasse 10, CH-9400
                                        Roschacherberg, Switzerland

Steves, B.   (U.K.)   Dept. of Physics and Astronomy,
                                        Glasgow University, G12 8QQ,
                                        Glasgow, U.K.

Susin, A.   (Spain)   Dept. Matematica Aplicada I.
                                        ETSEIB-UPC Diagonal 647
                                        08028 Barcelona, Spain

Szebehely, S.   (U.S.A.)   Dept. of Aerospace Engineering,
                                        Univ. of Texas at Austin, Austin,
                                        Texas, 78712, U.S.A.

Udry, S.   (Switzerland)   Observatoire de Geneve
                                        Ch.de Maillettes 51
                                        CH-1290 Sauverny, Switzerland

Valsecchi, G.B.   (Italy)   IAS - Planetologia Viale dell
                                        'Universita' 11 00185, Roma Italy

Varvoglis, H.   (Greece)   Dept. of Physics, Aristoteleion
                                        Univ. of Thessaloniki, 54006
                                        Thessaloniki, Greece

Vigo, J.   (Spain)   Dept. de Matematica Aplicada a la
                                        Tecnice. Univ. de Valladolid, 47011
                                        Valladolid, Spain

Vokrouhlicky, D.   (Czechoslovakia)   Dept. of Astron & Astrophys.
                                        Charles Univ. Svedska 8.
                                        1500 00 Prague 5 Czechoslovakia

Volpi, G.	(Italy)	via Washington, 102 20145 Milano, Italy
Waldvogel, J.	(Switzerland)	ETH - Zurich, Applied Mathematics Fliderstrasse 23, CH-8092 Zurich, Switzerland
Williams, I	(U.K.)	Queen Mary & Westfield College, Univ. of London, Mile End Road, El 4NS. London, U.K.
Wodnar, K.	(Austria)	Dept. of Astronomy, University of Vienna, Turkenschanzstr. 17 1180 Wien, Austria
Wytrzyszczak, I.	(Poland)	Astron. Observatory of A. Mickiewicz Univ. ul. Sloneczna 60, 60-286, Poznan, Poland

Aarseth, S.J.   48, 58, 62
Abergel, A.   249, 254
Abraham, R.   23, 32
Abu-El-Eta, N. 103, 104, 114
Acuna, J.C.   519, 525, 526, 527,
        528, 529
Aksnes, K.   314, 316, 334
Alekseev,   459
Alexander, W.M. 238
Alfaro, J.M.   425 - 431
Alotham-Alragheb, A. 90
Angoul, F.   206, 212
Angstrom, A.J. 247
Applerot, G.   573, 580
Arenstorf, R.F. 68, 70
Arnéodo, A.   212
Arnold, V.I.   25, 33, 52, 60, 62,
        87, 91, 141, 153, 344, 573,
        580
Asebiomo, A.S. 238
Atela, P.   9
Avez, A.   91

Babadzhanov, P.D.   231, 238
Bahri, A.   472
Bailey, M.E.   145, 146, 155
Baille, Ph.   146, 148, 153
Baines, M.J.   227, 237
Banfi, V.   459
Barbanis, B.   138, 540, 557, 562
Barkin, Iu.   580
Bashforth, F.   525, 526
Battin, R.H.   316, 334
Bendjoya, Ph. 205-213
Benest, D.   114
Bennetin, G.   35, 46, 128, 153
Bennett, A.   90
Belton, M.J.S. 249, 254
Bergé, P.   91
Bergstrahl, J.   196
Bertaux, J.L.   249, 254
Bertotti, M.L. 467-473
Bettis, D.G.   515, 516, 519, 522,
        523, 530

Bijaoui, A.   207, 210, 213
Binney, J.L.   47, 57, 58, 62, 119,
        122
Birkhoff, G.   577, 581
Birman, J.   431
Bois, E.   249-254, 257-264,
        265-271, 291-295
Bond, V.R.   301, 303, 515
Borderies, N. 259, 264
Borel, E.   86
Bozis, G.   475, 478
Brackhill, J.U. 62
Breeden, J.L. 61, 62
Bretagnon, P. 185, 186, 189, 190,
        192
Bridges, P.G. 381, 383
Broucke, R.   68, 69, 70, 146, 147,
        154, 158, 175, 311-335,431
Brouwer, D.   104, 105, 114, 189,
        190, 192, 208, 213, 230,
        232, 238, 303, 311, 334
Bryant, J.G. 501, 507
Bullard, E.C. 372, 383
Burns, J.   194
Burdet   386
Burton, W.M. 238
Bussoleti, E. 238

Calame, O.   264, 266, 267, 271
Caranicolas, D. 170, 174, 175, 179
Carnevali, P.   47, 62
Carpino, M.   145, 154, 184, 192,
        213, 240, 247, 493, 511,
        514
Cartan, H.   463, 466
Casasayas, J. 3-9, 547-554
Cassini, J.D. 258
Celletti, A.   90, 337-344
Cellino, A.   213
Cercignani,   46
Chabaudie, J.E. 264
Chandrasekhar, S. 57
Channel, P. J. 61, 62
Chao, C.C.   334

Chapman, C.R.   223
Chapront, J.   103, 104, 114
Charlier, C.   94, 105, 114, 439
    446
Chauvineau, B. 509-514
Chebotarev, G.A. 238
Chenciner, A. 97, 114
Chierchia, L. 344
Chirikov, B.V. 126, 142, 146, 147,
    150, 153, 159, 246, 247
Cleary, P.W.   119, 122
Clemence, G.M. 104, 105, 114, 232,
    238, 303
Coffey, S.   309, 334
Cohen, B.I.   62
Cohen, C.J.   240, 247
Cohn, H.   62
Collet, P.   91
Colombo, G.   395, 410
Combes, F.   120, 122
Conley, C.C.   459, 468, 472
Connerney, J.E.P. 373, 383, 384
Contopoulos, G. 35-46, 65, 70, 116,
    122, 138, 175, 531, 540, 555,
    557, 562
Coullet, P.   35, 46

Danby, J.M.A. 90, 344, 397, 413, 415,
    542, 546
Dallas, S.S.   291, 295
Darwin, G.H.   447
Daubechies, I. 206, 208, 213
Davidson, M.C.   68, 70
Davie, A.M.   55
Debbash, F.   122
Deprit, A.   90, 297, 300, 302, 303,
    309, 316, 329, 334, 439, 442,
    446, 573, 580
Deprit, E.   334
Deprit-Bartholomé, A. 90, 439, 442,
    446
De La Llave, R. 344
De Luccia, M.R. 246, 247
Dejonghe, H.   49, 62
Delaunay, Ch.   282
Delshams, A.   90, 438
Delva, M.   440, 446
Demin, V.   580
Dermott, S.F.   194, 195, 196
Deslambres, J.B.J.   273, 282
Devanay, R.L.   9, 491
De Vogelaere, R.   355, 370
de Zeeuw, T.   115, 116, 118, 122
Diez, C.   438
Doggett, L.E.   246, 247
Dorizzi, B.   557, 560, 562
Dormand, J.R.   232, 234

Drossart, P.   254
Dubru, P.   192
Duncan, M.   146, 148, 150, 151,
    153
Duncombe, R.L. 155, 498
Duriez, L.   244, 247
Dvorak, R.   147, 153, 246, 247,
    439-446, 466

Easton, R.   475, 478
Eckhart, D.H. 259, 264, 265, 271
Eckmann, J.P.   91
Egorov, W.A.   68, 69, 71
Einstein, A.   63, 153
El Bakkali, L. 91
El Mikkai, M.E.A. 238
El-Sabaa, F.   573-581
El-Tarazi, M.   573-581
Elezgaray, J.   212
Emslie, A.G.   479
Encrenaz, T.   254
Erdi, B.   68, 69, 71
Escobal, P.R. 542, 543, 546, 933
Euler, L.   573, 580
Evans, G.C.   238
Evans, S.T.   238

Falcolini, C.   344
Farinella, P.   213
Farquhar, R.W. 289, 290
Faulkner, J.   185, 186, 190, 192
Feigenbaum, M.J. 35, 46
Feraudy, D.   264
Ferrandiz, J.M. 297-303, 334, 387,
    394, 515-530
Ferrari, A.J.   267, 271
Ferraz-Mello, S. 177-184, 344
Festou, M.C.   249, 254
Fierberg, M.A. 216, 223
Firth, J.G.   238
Flannery, B.P. 62, 353
Flockerzi, D.   566, 571
Floria, L.   297-303
Font, J.   531, 540
Fontich, E.   90, 309, 438
Ford, J.   70, 71, 126
Fox, K.   143, 154, 228, 231,
    232, 238, 247
Franks, J.   430, 431
Friedli, D.   120, 121, 122
Froeschlé, C. 114, 158, 175, 249-
    254, 264
Froeschlé, Ch. 192
Froeschlé, Cl. 19, 33, 85, 125-
    155, 205-213, 214-223,
    233, 238

Gabor, D.   205, 213
Galgani, L.   46, 90, 153, 438
Gama, F.P.   345-353
Garfinkel, B.   311, 335
Gauss, C.F.   148, 541
Ge, Y-C   475-479
Gehrels, T.   154, 213
Gellman   372
Giacaglia, G.E.O.   291, 295, 438
Giffen,   140
Giorgilli, A.   46, 84, 90, 153, 177
    184, 438
Giuducelli, M. 212
Gleick, J.   53, 65, 71
Glentzlin, M.   264
Goldreich, P.   193, 196, 344
Goldstein, H.   169, 175, 266, 271,
    461, 466, 507
Gollub, J.P.   71
Golubev, V.G.   478
Gomez, G.   283-290, 397, 413,
    415, 433-438
Gonczi, R.   136, 146, 147, 154
Goodman, J.   49, 50, 55, 62
Gordon, W.B.   469, 472
Goudas, C.L.   355-370, 371-385
Goupillaud, P. 205, 213
Graef, C.   353, 370
Graf, O.   303
Grais, B.   353
Grammaticos, B. 557, 561, 562
Grard, R.J.L.   238
Grassberger, P. 65, 71
Grasseau, G.   212
Grau, M.   526
Grebogi, C.   155
Green, S.F.   238
Greenberg, R.   141, 154
Greene, J.M.   341, 344
Grossman, A.   206, 213
Grun, E.   238
Guckenheimer, J. 61, 62, 127, 154
Gurzadyan, V.G. 53, 57, 62
Gustafson, B.A.S. 233, 238

Hadamard, J.   65, 71
Hadjidemetriou, J.   25, 33, 116,
    122, 149, 150, 157-175
Hagel, J.   442, 446, 461, 466
Hagihara, Y.   177, 184, 231, 235,
    309, 497
Hahn, G.   145, 154, 184
Hajduk, A.   233, 238
Hall, N.S.   328, 335
Hamid, S.E.D.   230, 238
Hanner, M.S.   238
Hanslmeier, A. 420, 446

Hartmann, P.   33
Hawkes, R.   232
Healy, L.   309, 334
Heggie, D.C.   47-62
Heiles, C.   125, 126, 154, 239,
    247, 309, 577, 581
Heissler, J.   116, 119, 122, 146
Helleman, R.H.G. 67, 71
Hénon, M.   37, 40, 46, 68, 69,
    71, 91, 116, 122, 125, 126,
    127, 131, 132, 141, 142,
    148, 154, 239, 247, 290,
    309, 349, 353, 395, 410,
    447, 455, 577, 581
Hénrard, J.   28, 33, 37, 46, 144,
    154, 158, 170, 174, 175,
    179, 186, 188, 190, 192,
    193-196, 303, 338, 344,
    413, 415
Heppenheimer, T.A. 192
Heugerger, H.S.   289
Herzenberg, A.   372, 383
Hietarinta, J.   561, 562
Hill, G.W. 314, 335, 395, 410, 476
Hipparcus,   273
Hirayama, K.   205
Hirsch, F.   15, 33
Holmes, P.   61, 62, 127, 154
Holsapple, K.A. 205, 213
Holschneider, M. 205, 213
Hooimeyer, J.R.A. 119, 122
Hori, G.   94, 114, 291
Housen, K.R.   205, 213
Hubbard, E.C. 240, 247
Hughes, D.W. 231, 235, 238
Hunt, J.   232, 238
Hut, P.   49, 50, 62
Ictiaroglou, S. 171, 175
Ievlev, E.   580
Igenbert, E.   238
Innanen, K.A. 198, 204
Ioos, R.   67, 71
Ioos, G.   67, 71
Irigoyon, M.   309
Iszak, I.G.   316

Jacobi, C.G.J. 67
Jaffe, A.   91
Jeffreys, W.J. 382, 383, 447, 455
Jones, J.   231, 232, 233, 238
Jorba, A.   283-290, 433-438
Joseph, D.D.   67, 71
Juranek, H.   461, 466
Jourdain, N.   126
Jupp, A.   312, 335

Kamel, A.A.   290

Kaula, W. M.   264, 291, 295, 372, 383
Kaya, D.   541-546
Keesey, M. S. W. 282
Kepler, J.   291
Khabaza, I. M. 232, 238
Kharlamov, E. 581
Kiang, T.   245, 247
Kim, M. L.   312, 328, 334
Kinoshita, H. 186, 188, 192, 497, 516,
        522, 540
Kirchgraber, U. 515, 522, 571, 572
Kiselev, F.   580
Kissel, J.   238
Klafke, J. C. 177-184
Knezevic, Z.   190, 191, 192, 205, 206,
        207, 213
Kohl-Moreira, J.L. 254
Kolmogorov, A.N. 87, 344, 573, 580
Konopliv, A.   316, 335
Kovaleveskaya, 573
Kovalevsky, J. 291, 292, 295
Kozai, Y.   62, 187, 192, 291, 311,
        335
Kozlov, V.   573, 581
Kresak, L.   238
Kribbel, J.   147, 153, 246, 247, 440,
        446
Krogh, F.T.   217, 223
Kronland-Martinet, R. 205, 213
Krylov, N. S.   52, 54, 62
Kuczera, H.   238
Kusaka, S.   370

Lacomba, E.   553
Lagerkvist, C.I. 145, 153, 154, 238
Lagrange, J. L. 67, 71, 148, 291, 356,
        357, 509, 573, 580
Lambert, J. D. 523, 530
Landau, L.   266, 271
Langevin, Y.   238
Laplace, P.S. 63, 70, 94, 101, 104,
        114, 509, 541
Larson, H. P. 223
Laskar, J.   89, 93-114, 190, 192,
        240, 247, 514
Lauer, T.R.   120, 122
Lazutkin, V. F. 309
Leach, P. G.L. 562
Lecacheaux, J. 254
Lecar, M.   48, 49, 62
Leftaki, M.   370
Lefschetz, S.   21, 33
Leibnitz, G.W. 63
Lemaitre, A.   144, 154, 158, 171,
        174, 175, 190, 192, 194, 196
Lemaitre, G.   204
Leontovitch, A.B. 90
Leverrier, U. J. J. 93, 104, 114

Levi-Civita, T.   410
Levin, B. Y. 231, 238
Lichtenberg, A. J. 50, 62, 127, 140,
        142, 150, 154, 168, 175,459,
        466
Lichtenegger, H.   440, 446
Liebermann, M. A. 50, 62, 127, 140,
        142, 150, 154, 168, 175,
        466
Lifchitz, E.   84, 266, 271
Lindblat, B. A.   153, 238
Lingren, M.   153, 238
Lissauer, J. J.   249, 250, 254
Liu, J.   158, 175
Llibre, J.   9, 415, 438, 547-554
Lohinger, E.   439-446
Lorenz, E. N.   65, 71, 91
Louw, J. A.   562
Lovell, A. C. B. 230, 238
Lundstedt, H.   238
Lyapunov, A. A. 65, 66, 71, 126,
        581
Lyddane, R. H. 303
Lynden-Bell, D. 120, 122

Macdonald, A.J. 247
MacMillan, W.D. 459, 461, 466
Maddox, J.   65, 71
Magnenat, P. 116, 119, 122, 135,
        154, 555, 568
Mahomed, F. M. 562
Mahwin, J.   470, 472
Malhotra, R. 194, 196
Mallat, S.   205, 207, 213
Mandeville, J.C. 238
Mangin, J. F.   266
Marchal, C. 73-91, 335, 459, 466,
        475, 478
Marchioro, C. 33
Marcolongo, R. 468, 473
Markellos, V.V. 413-423, 476
Mars, G.   212, 218
Marsden, J.E.   22, 32, 572
Martinet, L.   115-122, 135, 154
Martinez, R. 9, 431, 438
Masdemont, J. 283-290, 415,
        433-438
Mather, J. N. 340, 344
Matthews, M. 196
Mavraganis, A.G. 355, 370
McCracken, M. 23, 572
McDonnell, J.A.M. 228, 238
McGehee, R. 9, 459, 551, 554
McIntosh, B.A. 231, 233, 238
McKenzie, R. 439, 440, 446
McLeod, R.J.Y. 524, 530
McMillan, S. 62
Melendo, B.   144, 155

Merritt, D. R. 122
Message, P. J. 28, 33, 164, 175,
    239-247
Meton,   273
Meyer, K. R.   540
Meyer, Y.   206, 213
Mignard, F.   146, 154, 511, 513,
    514
Migues, A.   266, 271
Milani, A.   11-33, 127, 145, 154,
    183, 184, 190, 191, 192, 205,
    206, 207, 213, 247, 509, 513,
    514, 522,
Miller, R.H.   49, 50, 62, 66, 71
Miralda-Escudé, J. 115, 118, 122
Moeckel, R.   491
Monleon, C.C.   425-431
Moore, P.   530
Morbidelli, A.   177, 184, 185-192, 344
Morlet, J.   206, 213
Moser, J.K.   6, 9, 344, 459, 466,
    468, 473, 497, 540, 553,
    573, 577, 580
Moulton, F.R.   497, 525, 526, 542,
    546
Muhonen, D.P.   289
Mulder, W.A.   119, 122
Mulholland, J.D. 267, 271
Murenzi, R.   212
Murray, C.D.   143, 144, 154, 196,
    238, 247

Nacozy, P.   240, 247, 291, 295,
    397, 413
Nakai, H.   186, 188, 192, 516,
    522, 530
Nappo, S.   238
Newhall, X.X. 259, 264, 282
Newhouse, S.E. 32, 33
Newman, C.R.   289
Newton, I.   63, 70, 230
Nitecki, A.   491
Nobili, A.M.   145, 154, 184, 189,
    190, 192, 247, 513, 514
Nova, S.   515-520, 523, 530
Nunes, A.   3-9, 547-554

Oberti, P.   249-254, 264
Obrubov, Y.Y. 231, 238
Oesterwinter, C.   240, 247
Oliver, C.P.   238
Oseledec,   126
Ovenden, M.W. 274, 282, 477, 478

Packard, N.H.   62
Palacios, M.P. 303

Pangalos, C.A. 370
Pankiewicz, G.S.A. 238
Paolicchi, P.   213
Papayannopoulos, T. 116, 122
Patsis, P.A.   119, 122
Pavanini, G.   459, 466
Peale, S.   193, 196, 250, 254,
    344
Perez, M. T.515, 522, 523-530
Perozzi, E. 273-282
Perry, C.H. 238
Peters, S.F. 48, 49, 62
Petit, J.-M. 148, 150, 154, 158,
    175, 345-353, 395, 410
Petrosky, T.Y. 146, 147, 154,
    158, 175
Petsagourakis, E.G. 355-370
Pfenniger, D.   115, 116, 118, 119,
    120, 121, 122
Pham - Van, J. 264
Pinotis, A.   35, 46
Plavec, M.   230, 238
Poincaré, H. 11, 16, 23, 25, 31, 32,
    33, 63, 64, 65, 66, 67, 71,
    93, 94, 95, 97, 98, 99, 101,
    104, 114, 125, 126, 140, 141,
    174, 185, 186, 192, 292, 298,
    306, 307, 449, 455, 458, 475,
    478, 577, 581
Pojam, J.   197-204, 394, 530
Pollard, H. 303
Polymilis, C. 555-563
Pomeau, Y.   91
Press, W.H.   50, 63, 353
Prigogine, I. 66, 67, 70, 71
Prince, P.I.   232, 238
Puel, F.   254

Quinn, T.   148, 153

Rabe, R.   440, 446
Rabinowitz, P.H. 470, 472, 473
Ramani, A.   561, 562
Ramamani, N.   61, 62
Rana, D.   344
Rees, M.   120, 124
Remy, F.   146, 154
Reynolds, O. 67, 71
Richardson, D.L. 240, 241, 247
Rickman, H.   145, 146, 148, 153,
    154, 238, 250, 254
Rom, A.   303
Roy, A.E.   33, 46, 62, 175, 204,
    240, 241, 247, 271, 273-282,
    295, 410, 446, 475, 476, 477,
    478, 479, 491, 507

Ruelle, D.   65, 71, 91

Saari, D.G.   475, 478, 479, 540
Sagdeev, R.Z.   147, 154
Sanders, J.A.   554
Saporta, G.   353
Sargent, W.L.   120, 122
Sansaturio, M.E.   387-394, 530
Santangelo, P. 47, 62
Sato, M.   177, 178, 184, 193, 196
Savvidy, G.K.   53, 57, 62
Scheidecker, J.   126
Scheifele, G.   297, 300, 303, 388, 390
        394, 498, 515, 522, 530
Schiaparelli, G.V.   220
Schmidt, D.S.   572
Scholl, H.   140, 141, 142, 143, 144,
        154, 155, 192, 215-223, 233,
        238
Schultz, B.E.   438
Schubert, J.   140, 141, 142, 143, 155
Schwarz, F.   562
Schwarzschild, M.   115, 118, 122
Schwehm, G.H.   238
Schweizer, F.   120, 122
Scovel, C.   61, 62
Sekiguchi, M.   493-498
Seidelmann, P.K.   246, 247
Sekanina, Z.   238
Sessin, W.   344
Sherbaum, L.M. 231, 238
Siegel, C.L.   468, 473
Sidlichovsky, M.   144, 155
Silva, G.   177, 184
Simo, C.   9, 84, 90, 127, 283-290,
        305-309, 387, 431, 433-438,
        481-491, 522, 531, 540, 554
Simonenko, A.N.   238
Simons, S.   238
Simpson, I.C.   238
Simpson,   228
Sinclair, A.T.   247
Sinclair, W.S.   271
Sitnikov, K.   459, 466
Sjogren, W.L.   271
Slade, M.A.   264
Slezak, E.   205-213, 228
Smale, S.   7, 8, 15, 32, 33, 475, 479
Smith, H.Jr.   58, 62
Smith, R.   68, 69, 71, 447-455
Snow, D.   541-546
Spirig, F.   395-410, 431, 565-572
Spitzer, L.   48, 58, 59, 61, 62
Spline,   150
Stagg, C.R.   145, 146, 155
Standish, E.M.   49, 62, 264, 282

Steeb, W.H.   561, 562
Stengers, I.   71
Stevenson, D.J.   372, 383
Stevenson, T.J.   238
Steves, B.A.   273-282
Stewart, H.B.   67, 71
Stewart, I.   65, 71
Stiefel, E.   303, 388, 390, 394,
        498, 515, 522, 530, 572
Stora, R.   67, 71
Stormer, C.F.   355, 370, 371, 383
Streit, L.   213
Strelcy, J.M.   153
Strömgren, E.   413, 415, 447
Stumpff, K.   303, 439, 456, 459,
        466
Sundman, K.   177, 475, 478
Susin, A.   481-491
Swinney, H.L.   67, 71
Synnott, S.P.   410
Szebehely, V.   33, 39, 46, 48,
        49, 62, 63-71, 155, 197-
        204, 247, 370, 394, 410,
        413, 415, 438, 439, 440,
        446, 447-455, 468, 473,
        476, 479, 570

Takens, F.   91
Tanikawa, K.   493-498
Tapley, B.D.   33, 438, 571
Taylor, D.B.   397, 413, 415
Taylor,   150
Tchamitchian, Ph.   213
Teulolsky, S.A.   62, 353
Thompson, J.M.T.   61, 67
Thüring, B.   68, 71
Tisserand, M.F.   185, 192
Tittermore, N.C.   194, 196
Torre, J.M.   264
Tremaine, S.   47, 57, 58, 62, 146,
        148, 153
Tresser, C.   35, 46
Tschauner, J.   90
Turner, R.F.   238

Udry, S.   115-122

Vaghi, S.   148, 154
Valsecchi, G.   205, 213, 273-282,
        479
Valtonen, M.J.   58, 62, 213
Van Allen, J.   355, 370
Van Flandern, T.   197, 204, 397, 413
Van Woerkom, A.J.J.   190, 192
Varosi, F.   150, 155
Vecheslavov, V.V.   146,147,153,246,247

Veillet, C.   257, 264, 265, 271
Vetterling, W.T. 62, 353
Vidal, C.   91
Vigo, J.   387-394
Vilet, C.M.   562

Waldvogel, J.   395-410, 431, 491
Walker, C.F.   240, 241, 247
Walker, I.W.   247, 475, 476, 478,
      479
Wallis, M.K.   238
Walter, W.   464, 466
Watson, J.C.   542, 543, 546
Weishaupt, U.   238
Weissman, P.R.   146, 155
West, R.M.   238
Wetherill, G.W.   216, 223
Whipple, A.L.   198, 204, 227, 229,
      230, 238
Whittaker, E.T.   64, 67, 71
Wielen, R.   60, 62
Wilkening, L.L.   155
Willem, M.   470, 472
Williams, I.P.   225-238, 247
Williams, J.D.   185, 186, 189, 192
Williams, J.G.   190, 192, 213, 257,
      264, 271
Williams, R.F.   430, 431
Wisdom, J.   70, 71, 126, 142, 143,
      144, 153, 155, 157, 170, 174,
      175, 179, 184, 194, 196, 216,
      223, 239, 247, 344, 447, 455

Wnuk, E.   291, 295
Wodnar, K.   457-466
Wytrzyszcak, I.   257-264, 265-271,
      291-295

Yeomans, D.K. 230, 235, 246, 247
Yi, Z.   455
Yi-Sui Sun   158, 175
Yoder, C.F.   267, 271, 395, 410
Yoder, K.A.   410
Yorke, K.A.   155
Yoshida, H.   478, 497, 516, 522
      530, 561, 562
Yoshikawa, M. 186, 188, 192, 216
      223
Yuasa, M.   94, 114, 187, 190,
      192

Zachilas, L.   119, 122, 555, 562
Zappala, V.   206, 208, 210, 211
      213
Zare, K.   317, 335, 475, 476
      479
Zarnecki, J.C. 238
Zaslavsky, G.M. 147, 154
Zhang, S.P.   198, 204
Zidian, W.   225-238

SUBJECT INDEX

Adams-Bashford method,232, 516,
    525, 526
Adams-Bashford-Moulton method,
    391, 392, 525, 526
Adeona, 212
Adiabatic invariant theory,
    194, 195, 196
Agnia, 212
Alinda, 182
Alpha Capricornids, 234
Amalasuntha, 212
Ancient Chaldeans, 273, 274
Apollo, 183
Appenzela, 212
Arnold diffusion, 73, 75, 78,
    83, 87, 447
Arnold tori, 73, 78, 87, 88, 99
Artificial satellite, *see*
    satellite
Asteroid belt, 185, 215-223
Asteroid family, 205, 207, 208,
    211, 212
Asteroids, 125-155, 157-175,
    177-184, 186, 187,
    188, 205, 207, 208,
    211, 212
Asteroids, binary, 197-204, 493-498
Asteroids, planet-crossing, 145
Astrodynamics, 541, 542, 544
Asymptotic orbits, 127, 413-423,
    493-498
Aurora, 355
Averaging, 28-30, 140-142

Babylonians, *see* Ancient Chaldeans
Baby's rattle effect, 65
Berolina, 212
Bettis methods, 515-522
Birkoff invariant, 79
Bulirsch-Stoer method, 440
Burdet-Ferrandiz variables, 388,
    390, 391, 392

Burrau problem, 48, 49
Butterfly effect, 65, 89

Cantor set, 340
Capture, 193-196, 337, 338, 342,
    344
Cassini division, 371, 377
Cassini-like gap, 378, 379, 380,
    381
Cassini's laws ,258
Celestial mechanics, 47
Chaos, 3-9, 11-33, 35-46, 47-62,
    63-71, 73-91, 115-122, 157,
    173, 193, 195, 225, 226,
    233, 239-247, 395-410,
    447-455, 573-581
Chaotic behaviour, 93-114, 125-155
Chaotic layer, 195, 196
Chaotic motion, 73, 182
Chaotic separatrix, 339
Chaotic solution,181
Chaotic trajectories, 447-455
Chaotic zone, 143, 144, 216, 249
Characteristic curves, 531-540
Close encounters, 345
CNES, 266
Collision, 4, 5, 54, 193-196, 334,
    345, 347, 349, 471, 489,
    494, 495, 501-507
Comet Biela, 226
Comet Encke, 230
Comet Halley, 146, 147, 228, 231,
    232, 233, 239, 245, 246,
    249-252
Comet nuclei, 249-254
Comet orbit, 125-155
Comet Tempel-Tuttle, 226, 230
Commensurability, 187, 191, 193,
    194, 195, 272-282, 337,
    348
Coorbital motion, 395-410
Copenhagen problem, 39

Core collapse, 48, 49, 60
Cowell method, 388
Critical inclination, 311, 312, 323, 328, 329
Crossing time, 47

Daytime Arietids, 234
Daytime ζ Perseids, 234
Daytime β Taurids, 234
Daytime Sextantids, 234
Deformations, 265, 266, 267, 269, 271
Deformations, elastic, 265, 266, 269, 271
Deformations, anelastic, 265, 266
Delaunay variables, 23, 96, 97
Delaunay- similar variables, 297, 298, 300, 301, 302, 303
Delta Aquarids, 234
Deprit's radial intermediary, 297 302
Determinism, 84
Digital Orrery, 140
Dirac impulsions, 510
Dora, 212
Dynamical Ephemeris of Moon, 257
Dynamo theory, 372, 373
DVDQ code, 217

Earth, 64, 153, 215, 216, 217, 220, 220, 222, 223, 225, 258, 259, 264, 283, 284, 285, 288, 289, 292, 306, 311, 337, 341, 372, 439, 440
Earth-Moon figure - Figure interactions, 257, 259
Earth-Moon system, 273-282, 433, 438, 439, 440
Eclipse, 273, 274, 275, 276
Elliptic restricted three-body problem, 467-473
Encounter, 146, 250, 254, 345-353
Eos, 205, 212
Epicyclic motion, 346
Epimetheus, 400, 425
Equinoctial variables, 316
Error growth, 54-58, 476
Error position, 389
Error trajectory, 389
Eta Aquarials, 234
Euler case, 573, 574
Euler's angles, 249, 258
Euler's equations, 266, 338
Euler-Poinsot motion, 266

Evolution 48, 115, 182, 183, 205, 225, 345, 509

FAST consortium, 291
Feigenbaum sequence, 35
Floquet multipliers, 20, 21, 25
Floquet theory, 437
Flora, 223
Fokker-Planck equation, 59, 61
Fractals, 73, 88

Gauss planetary equations, 248
Gauss's equations, 154
Gauss's method, 541, 542, 544, 546
Gauss-Gibbs method, 544, 546
Gauss-Halphen-Goryachev method, 231
Gauss-Jackson method, 231
Geminid stream, 226, 231, 232, 234
Giotto, 228
Global correlation, 347
Gronwall's lemma, 437

Halley's comet, see Comet Halley
Halo orbit, 283-290
H Aquarid stream, 232
Hartmann-Grobman theorem, 17
Hénon method, 334
Hénon-Heiles problem, 308
Hierarchical stability, 475-479
Hill variables, 314, 315
Hill's lunar equation, 397-400
Hill's problem, 395-410
Hill-type stability, 475-479
Hipparcos, 291
Homoclinic orbits, 413-423
Hopf bifurcation, 565-572
Hori's perturbation method, 291
Horseshoe orbits, 425
Hubble profile, 115

Iduberga, 212
Impact parameter, 346, 347, 349
Impulsional method, 509-514
Integrality, 66, 67
Integrals, N-body problem, 64
Intermittency, 66, 67
International Sun-Earth Explorer, 3, 283
Invariant manifolds, 17, 18
Irregular families, 531-540
Irreversibility, 66, 67

Jacobi coordinates, 94, 244, 479, 501
Jacobian constant 36, 68–70, 152,
        450–452, 493, 494
Jacobi's equation, 53
Jacobi integral, 397
Janus, 400, 425
JPL DE 301 ephemeris, 257, 276
Jupiter, 140, 143, 169, 170, 174,
        177, 184–186, 193, 197, 200,
        207, 215–217, 236, 237, 240,
        241, 245, 246, 249, 250, 254,
        259, 337, 439, 509, 512

KAM theorem, 342, 433, 549
KAM theory, 126, 134, 342, 417,
        573, 576
Kepler equation, 389
Kepler problem, 4, 185, 314, 388,
        471, 525, 547–554
Keplerian map, 147
Keplerian orbit, 250
Kirkwood gaps, 125, 140, 143, 205,
        216
Knot theory, 429–431
Kolmogorov entropy, 448, 449
Koronis, 205, 212
Kovaleveskaya case, 573–574, 577, 580
Kronian field, 356–358
Kustaanheimo-Stiefel (KS) variables,
        385, 390–392, 525

Lag angle, 339
Lagrange case, 357, 573–574
Lagrange planetary equations, 148,
        242, 291
Lagrange solution, 356
Lagrange stable 75
Lagrangian equilibrium point,
        84, 439–446
Lagrangian motion, 73, 75–83
Laplace's demon, 63
Laplace's method, 541–542, 544, 546
LeJeune-Dirichlet theorem, 82
Leonce, 212
Leonid stream, 226, 230, 234
Leto, 212
Levi-Civita canonical transformation,
        426
Libration, 239–247, 250, 254,
        257–271, 342, 433, 439, 440
Libration, physical, 257–271
Librators, 439–440
Lie series, 207
Lie transform technique, 298
Liouville's theorem, 60
Lissajous orbit, 283
Local theory, 15, 16
LONGSTOP, 190, 240
Long term predictions, 387–394,
        509–522
Love number, 266, 339

Lucretia, 212
Lunar cycle, 273–282
Lunar Laser Ranging Stations, 257
Lunar theory, 276, 277
Lyapunov characteristic exponents,
        16, 17, 19, 21, 32, 50–52,
        58, 61, 65, 66, 85, 116,
        126, 136, 143, 144, 146,
        241, 311, 447–455
Lyapunov instability, 88, 387, 515
Lyapunov orbits, 283, 534–536, 539
Lyapunov stability, 75, 79, 82, 83
Lyrids, 234

Magnetic diffusivity, 370
Magnet dipole, 355
Mapping, 142–153, 157–175
Maria, 212
Mars, 215–217, 220, 223, 337
McGehee coordinates, 5, 551
Melnikov theory, 321
Mercury, 185, 337, 341, 342
Meteor streams, 225–238
Meteorites, 215–223
Meteorology, 65
Metonic cycle, 273
Miranda, 194, 195
Mirror theorem, 274
Modelling, 63, 64, 115, 116, 125–155,
        510, 511
Molniya satellite, 312, 546
Monocerotid stream, 232
Monte-Carlo treatment, 145, 146, 216,
        345, 352
Month, 273–282
Moon, 94, 257–271, 283–286, 292, 337,
        341, 342, 439, 440,
Moon's physical libration, 257–271
Morse theory, 27

NASA, 266
N-body problem, 47–71, 73, 371, 374,
        395, 501–507
N-body simulations, 58–60
N-dipole problem, 371
Neptune, 152, 185
Newtonian dynamics, 63–71, 125
Newtonian motion, 125, 355, 356
Numerical errors, 48–50

Oblateness, 193–195, 305–309
One-dipole problem, 355, 363, 371
Oort cloud, 146
Orbit determination, 541–546
Orionids, 234
Oseledec's ergodic theorem, 126

Painleve property, 555–563
Pandora, 400
Particles, 345, 347, 355–385
Peclet number, 372

Periodic orbits, 11, 19–27, 35–46, 115–116, 272, 276, 279–284, 311–335, 355, 425–431, 439–446, 467–473, 531–532
Periodic solution, 404, 407
Perseid stream, 226, 232
Perturbation theory, 239–247
Perturbed circular motion, 291–295
Phaethon, 226
Phoenicids, 234
Physical libration of Moon, 257–271
Pioneer, 13, 373, 384
Planetary effects, 257, 259, 261
Planetary rings, see Rings
Plummer model, 58
Plummer potential, 120
Plummer sphere, 121
Poincaré characteristic exponents 116
Poincaré invariants, 60, 61
Poincaré map, 6, 8, 21–25, 27, 30, 31, 116, 125, 126, 128, 151, 152, 287, 306, 307
Poincaré section, 74, 306, 307, 340, 447–455
Poincaré stability, 75, 82, 83
Poincaré surface of section, 65, 306, 307, 311, 320–322, 328, 329, 332
Poincaré variables, 23, 93, 94, 98, 99, 106
Poincaré-Birkhoff fixed point theorem, 168
Poincaré-Hadamard-Perron theorem, 16
Poincaré-similar variables, 297–303
Poisson stable, 75
Poisson's method, 243, 244
Poisson's theorem, 243
Predictability, 63–71, 73–91, 240, 241, 523
Prediction horizon, 85, 89
Prometheus, 400

Quadrantid meteor stream, 226, 231, 233–237
Quasiperiodic orbits, 433–438

Random law, 86
Relaxation time, 48, 57
Resonance, 11, 28, 77–83, 115, 118–120, 140, 142–145, 157–175, 177–196, 209, 215–223, 239–247, 337–344, 438, 486, 561, 562
Restricted four-body problem, 3–9, 197–204
Restricted many-body problem, 197–204

Restricted three-body problem, 64, 66, 77, 78, 136, 140, 146, 148, 151, 169, 193, 239, 283–285, 287–288, 357, 413–423 433–439, 447–455, 467–473, 493–498
Reynolds number, 372
Riemann curvature, 53
Rigid body rotation, 561–569
Rings, 345, 346, 371, 381
Roche's limit, 346
Rossby number, 372
Rotation, 249–254, 257–271, 573–581
Routh's value, 433
Runge-Kutta-Fehlberg, 404
Runge-Kutta-Nystrom method, 232

Saros, 273–282
Satellite, 64, 193–196, 283, 291–295, 297–303, 305–335, 337–344, 395–410, 467–473
Satellite, geosynchronous, 291, 292
Saturn, 185, 186, 193, 207, 215, 217, 240, 241, 337, 371, 373–375, 377–382, 384–385, 400, 425
Separatrix, 195, 196, 307
Shepherd satellites, 400
Simulations, 345
Simulations, deterministic, 345
Simulations, Monte-Carlo, 345
Sitnikov problem, 89, 457, 466
Sitnikov sequence, 459, 462, 465
Smale's horseshoe theorem, 7, 8, 32
Solar system, 89, 93–114, 185, 228, 345, 382, 425
Spherically exact algorithms, 529–530
Spin-orbit resonance, 337–344
Spiral characteristics, 35–46
Stability, 63–71, 73–91, 197–204, 311
Stability zones, 439–446
Statistical analysis, 345, 347
Stellar dynamics, 47–62
Stellar systems, 115
Stochastic event, 86
Stochastic zone, 306–308
Stormer's problem, 355
Strange attractor, 73, 85, 89, 90
Sun, 140, 152, 169, 174, 185, 186, 197, 200, 230, 249, 259, 260, 283, 285, 292, 341, 439, 509, 512
Sundman's inequality, 475, 478

Taurid meteors, 230, 234
Taylor and Spline interpolation, 150
Termination orbits, 413–423
Thalweg, 355, 371
Themis, 205, 212
Three-body problem, 11–33, 273–282, 460–463, 481–491, 504

Three-body problem, restricted, 11-33, 435-466
Three-dimensional potentials, 555-563
Three-dipole problem, 355-371, 375
Theory of consequents, 30-32
Tidal evolution, 194, 340
Tides, 265
Triple collision manifold, 481-491
Trapping space, 337, 342, 355-382
Triangular points, 439-446
Triaxial models, 115-122
TRIP, 97, 101, 106
Two fixed centre problem, 458
Two-dipole problem, 355, 363, 371, 375

Umbriel, 195
Uranus, 193-195
Ursids, 234

Van Allen zones, 371
Venus, 217, 220, 222, 259, 260
Veritas, 212
Vesta, 211, 212
Virial ratio, 47
Voyager, 373, 385

Watson's method, 548, 549
Wavelet transform, 209-213

Zero velocity curves, 399-400
Zero velocity surface, 68-70, 318, 361, 363-364, 366